INTERDISCIPLINARY SURFACE SCIENCE

INTERDISCIPLINARY SURFACE SCIENCE

PROCEEDINGS OF THE THIRD INTERDISCIPLINARY
SURFACE SCIENCE CONFERENCE

UNIVERSITY OF YORK, ENGLAND
27–30 MARCH 1977

Guest Editors:

D.P. WOODRUFF
University of Warwick

and

D.A. KING
University of Liverpool

1977

NORTH-HOLLAND PUBLISHING COMPANY – AMSTERDAM

© *North-Holland Publishing Company, 1977*

All rights reserved. No part of this publication may be reproduced, stored in a retrieval system, or transmitted, in any form or by any means, electronic, mechanical, photocopying, recording or otherwise, without the prior permission of the copyright owner

ISBN North-Holland 0 444 85051 1

Reprinted from:
SURFACE SCIENCE 68 (1977)

PRINTED IN THE NETHERLANDS

PREFACE

The Third Interdisciplinary Surface Science Conference organised by the Thin Films and Surfaces Group of the Institute of Physics in collaboration with the Vacuum Group of the Institute of Physics, the Surface Reactivity and Catalysis Group of the Faraday Division of the Chemical Society and the Institution of Corrosion Science and Technology was held in the Physics Department at the University of York from 27th to 30th March 1977. In common with earlier meetings the 180 delegates from 18 countries engaged in an intensive 3 day meeting; as a new innovation for this meeting a poster session was included which ran one evening and allowed the organising committee to accept a higher proportion of submitted papers; 23 of the papers in the proceedings were presented in this way. A further 14 post-deadline papers were also included in the poster session but as these were intended to allow new work to be presented without committing the authors to print, these are not included in this set of proceedings papers.

As at the 1975 Warwick Conference (Surface Science volume 53) the main theme of the conference was adsorption at surfaces. This topic seems to be one in which the interdisciplinary approach is well-established and not only is the distinction between physics and chemistry unclear, but even the distinction between physicists and chemists is becoming diffuse. Following a period of explosion in the number of surface techniques we now appear to be entering a phase where these techniques are put to work both to improve the understanding of adsorption under well-defined conditions, but also in a wider range of applications. A small number of papers in this proceedings represent this broader field.

The Guest Editors and Conference Organising Committee would like to thank all authors of both invited and contributed papers for their efforts and look forward with interest to the new developments which will be reported at the Fourth Interdisciplinary Surface Science Conference (also to be the Second European Conference on Surface Science) to be held at Cambridge in March 1979.

D.P. Woodruff
D.A. King

CONTENTS

Preface ... vii

W.J. van Ooij, The role of XPS in the study and understanding of rubber-to-metal bonding ... 1
S. Storp and R. Holm, ESCA investigation of the oxide layers on some Cr containing alloys ... 10
G.E. Rhead, Surface defects ... 20
K. Besocke, B. Krahl-Urban and H. Wagner, Dipole moments associated with edge atoms; a comparative study on stepped Pt, Au and W surfaces ... 39
H. Albers, W.J.J. van der Wal and G.A. Bootsma, Ellipsometric study of oxygen adsorption and the carbon monoxide–oxygen interaction on ordered and damaged Ag(111) ... 47
G. Martin and B. Perraillon, A model for morphological changes driven by step–step interaction on clean surfaces ... 57
M. Gettings and J.C. Rivière, Precipitation and re-solution of impurities at the surface of indium on traversing the melting-point ... 64
J. Erlewein and S. Hofmann, Segregation of tin on (111) and (100) surfaces of copper ... 71
A.R. Waugh and M.J. Southon, Surface studies with an imaging atom-probe ... 79
J. Howard and T.C. Waddington, An inelastic neutron scattering study of C_2H_2 adsorbed on type 13X zeolites ... 86
W. Heiland and E. Taglauer, The backscattering of low energy ions and surface structure ... 96
P. Bertrand, F. Delannay, C. Bulens and J.-M. Streydio, Angular dependence of the scattered ion yields in $^4He^+ \to Cu$ scattering spectrometry ... 108
K. Wittmaack, The use of secondary ion mass spectrometry for studies of oxygen adsorption and oxidation ... 118
M. Barber, J.C. Vickerman and J. Wolstenholme, The application of SIMS to the study of CO adsorption on polycrystalline metal surfaces ... 130
A.R. Williams and N.D. Lang, Atomic chemisorption on simple metals: chemical trends and core-hole relaxation effects ... 138
D.W. Bullett, Localized orbital approach to chemisorption: H and O adsorption on Ni, Pt and W(001) surfaces ... 149
H. Hjelmberg, O. Gunnarsson and B.I. Lundqvist, Theoretical studies of atomic adsorption on nearly-free-electron-metal surfaces ... 158
D.R. Hamann, Theoretical studies of the electronic structure of semiconductor surfaces ... 167
F. Humblet, H. Van Hove and A. Neyens, Photoemission studies on ZnO(0001), (000$\bar{1}$) and (10$\bar{1}$0) ... 178
M. Housley, R. Heckingbottom and C.J. Todd, The interaction of Ag with Si(111) ... 179
P. Wagner, K. Müller and K. Heinz, Adsorption studies of Cs on Si(111) ... 189
G. Laurence, B.A. Joyce, C.T. Foxon, A.P. Janssen, G.S. Samuel and J.A. Venables, Adsorption–desorption studies of Zn on GaAs ... 190
F. Jona, Past and future surface crystallography by LEED ... 204
J.A. Walker, C.G. Kinniburgh and J.A.D. Matthew, LEED calculations of exchange reflections from antiferromagnetic NiO(100) ... 221
R. Feder, Spin-polarized LEED from low-index surfaces of platinum and gold ... 229
H.P. Bonzel, The role of surface science experiments in understanding heterogeneous catalysis ... 236
R.A. Wille, F.P. Netzer and J.A.D. Matthew, Electron energy loss spectrum of cyanogen on Pt(100) ... 259
A.M. Bradshaw, P. Hofmann and W. Wyrobisch, The interaction of oxygen with aluminium (111) ... 269

M. Housley, R. Ducros, G. Piquard and A. Cassuto, The adsorption of carbon monoxide on rhenium: basal (0001) and stepped |14 (0001) × (10$\bar{1}$1)| planes 277
C.F. Battrell, C.F. Shoemaker and J.G. Dillard, A study of the interaction of sulfur-containing alkanes with clean nickel 285
F. Garbassi, G. Petrini, L. Pozzi, G. Benedek and G. Parravano, An AES study of the surface composition of cobalt ferrites 286
F.J. Kuijers and V. Ponec, The surface composition of the nickel–copper alloy system as determined by Auger electron spectroscopy 294
M. Abon, G. Bergeret and B. Tardy, Field emission study of ammonia adsorption and catalytic decomposition on individual molybdenum planes 305
S.J.T. Coles and J.P. Jones, Adsorption of gold on low index planes of rhenium 312
R. Browning, M.M. El Gomati and M. Prutton, A digital scanning Auger electron microscope 328
C. Le Gressus, D. Massignon and R. Sopizet, Low beam current density Auger spectroscopy and surface analysis 338
G. Le Lay, M. Manneville and R. Kern, Desorption kinetics of condensed two-dimensional phases on a single crystal substrate 346
M.C. Muñoz and J.L. Sacedón, AES analysis of oxygen adsorbated on Si(111) and its stimulated oxidation by electronic bombardment 347
P. Légaré, G. Maire, B. Carrière and J.P. Deville, Shapes and shifts in the oxygen Auger spectra 348
M.B. Gordon, F. Cyrot-Lackmann and M.C. Desjonquères, On the influence of size and roughness on the electronic structure of transition metal surfaces 359
G. Lindell, Exchange corrections to the density–density correlation function at a surface 368
J.L. Moran-Lopez and A. ten Bosch, Changes in work function due to charge transfer in chemisorbed layers 377
R. Dorn and H. Lüth, The adsorption of oxygen and carbon monoxide on cleaved polar and nonpolar ZnO surfaces studied by electron energy loss spectroscopy 385
E.W. Kreutz, E. Rickus and N. Sotnik, Oxidation properties of InSb(110) surfaces 392
N. Garcia, G. Armand and J. Lapujoulade, Diffraction intensities in helium scattering; topographic curves 399
M. Garcia, Threshold and Lennard-Jones resonanaces and elastic lifetimes in the scattering of atoms from a corrugated wall and an attractive well surface model 408
V.T. Binh, Y. Moulin, R. Uzan and M. Drechsler, Grain-boundary groove evolution in the presence of an evaporation 409
M.F. Felsen and P. Regnier, Influence of some additional elements on the surface tension of copper at intermediate and high temperatures 410
D.R. Lloyd, C.M. Quinn and N.V. Richardson, The oxidation of a Cu(100) single crystal studied by angle-resolved photoemission using a range of photon energies 419
J.K. Sass, S. Stucki and H.J. Lewerenz, Plasma resonance absorption in interfacial photoemission from very thin silver films on Cu(111) 429
M.R. Welton-Cook and M. Prutton, Calculations of rumpling in the (100) surfaces of divalent metal oxides 436
M.K. Debe, D.A. King and F.S. Marsh, Further dynamical and experimental LEED results for a clean W{001}-(1 × 1) surface structure determination 437
F. Soria, J.L. Sacedon, P.M. Echenique and D. Titterington, LEED study of the epitaxial growth of the thin film Au(111)/Ag(111) system 448
S.J. White, D.P. Woodruff, B.W. Holland and R.S. Zimmer, A LEED study of the Si(100) (1 × 1)H surface structure 457
J.C. Fuggle, XPS, UPS and XAES studies of oxygen adsorption on polycrystalline Mg at ~100 and ~300 K 458

C.R. Brundle, T.J. Chuang and K. Wandelt, Core and valence level photoemission studies of iron oxide surfaces and the oxidation of iron 459
G.C. Allen, P.M. Tucker and R.K. Wild, High resolution LMM Auger electron spectra of some first row transition elements 469
M. Šunjić, Ž. Crljen and D. Šokčević, Photoelectron spectroscopy of localized levels near surfaces: scattering effects and relaxation shifts 479
B.W. Holland, Theory of angle-resolved photoemission from localised orbitals at solid surfaces 490
G.L. Price and B.G. Baker, Angle-resolved UPS measurements in a modified LEED system 507
C. Backx, R.F. Willis, B. Feuerbacher and B. Fitton, Infrared vibration spectroscopy of molecular adsorbates on tungsten using reflection inelastic electron scattering 516
A. Crossley and D.A. King, Infrared spectra for CO isotopes chemisorbed on Pt{111}: evidence for strong absorbate coupling interactions 528
J.C. Bertolini, G. Dalmai-Imelik and J. Rousseau, CO stretching vibration of carbon monoxide adsorbed on nickel(111) studied by high resolution electron loss spectroscopy 539
D.R. Lloyd, C.M. Quinn and N.V. Richardson, Experimental aspects of angle-resolved photoemission from clean and adsorbate-covered metal surfaces 547
T.T.A. Nguyen and R.C. Cinti, Directional uv photoemission from clean and sulphur saturated (100), (110) and (111) nickel surfaces 566
C. Webb and P.M. Williams, Angle resolved photoemission studies of the band structures of semiconductors: PbI_2 576
J. Castle, The use of X-ray photoelectron spectroscopy in corrosion science 583

Author index 603
Subject index 609

THE ROLE OF XPS IN THE STUDY AND UNDERSTANDING OF RUBBER-TO-METAL BONDING

W.J. VAN OOIJ
Akzo Research Laboratories, Corporate Research Department, Arnhem, The Netherlands

XPS has been used to study the mechanism of the adhesion of rubber to well-characterized metal surfaces. By means of a special sample preparation technique the rubber-to-brass interface could be analyzed by X-ray photoelectron spectroscopy following vulcanization of rubber against a CuZn alloy surface. Combination of these XPS results with quantitative data on the adhesion levels of brass has led to the development of a new adhesion model. Reaction of brass with rubber results in the formation of both Cu_xS and ZnS. The amount of Cu_xS formed must be carefully controlled. On the one side, it improves the adhesion as a result of a catalytic effect on the rubber vulcanization. On the other hand, excessive Cu_xS formation leads to embrittlement of the interfacial Cu_xS/ZnS film and a loss of adhesion.

1. Introduction

The adhesion of rubber to steel is of considerable practical importance in view of the present-day application of steel cords as a reinforcement material in belted radial tires.

The rubber compounds normally used in tires cannot be bonded directly to steel. A very high bond strength can, however, be obtained between sulfur-vulcanizable rubbers and brass. Therefore, in steel tire cords manufacture a step is incorporated in which the wires are electrolytically brass-coated. Although the performance of the cords has been optimized empirically, the mechanism of the rubber-to-brass bonding has never been understood. It has, however, been recognized that a reaction between Cu and S is involved in the bonding [1–3]. Others have postulated a catalytic oxidation of the rubber surface by Cu atoms [4].

The advent of surface-sensitive techniques in recent years has prompted interest in a relation between surface composition and adhesive properties of the brass. Analyses by XPS, AES and SIMS techniques have clearly shown that the surface of the brass coating is invariably enriched in Zn. However, relations between surface composition and adhesion behaviour can not be straightforward as other effects, such as the presence of residual drawing lubricants on the cords, may interfere. We have obviated these problems by studying the bonding between rubber and small brass sheet samples. In this approach relations between adhesion level and initial brass surface composition can be investigated unambigously. Further, the chemical composition of the rubber-metal interface can more easily be determined.

2. Experimental

The alloys used were vacuum melted cold-rolled strips of 0.2 or 0.5 mm thickness made from copper and zinc of 99.999 wt% purity. They contained 71.26 at% Zn as per X-ray microanalysis. The samples used for surface analysis and for adhesion experiments were punched out of the alloy strips (fig. 1a). This sample shape was selected as it represents the most favorable sample for the XPS instrument used. Prior to surface analysis the samples were mechanically polished. They were adhered to rubber by combining two samples with a slab of vulcanized rubber of 1×1 cm^2 and 2 mm thickness (fig. 1b). Ten of these composite samples could be vulcanized simultaneously in a heated vulcanization press at 150 or 180°C under high pressure for 25 min using a specially designed vulcanization mold.

The rubber was a commercially available test compound. Wet rubber (water content 1 wt%) was prepared by storing unvulcanized rubber at 90% rh for at least 10 days. The adhesion level (in N/64 mm^2) was determined in an Instron tensile tester (fig. 1c, lap-shear test). For interface analysis the samples were ripped apart under liquid nitrogen. Under those conditions adhesion failure was observed, i.e. the samples broke at one of the two metal-rubber interfaces, which permitted both sides of the interface to be analyzed by XPS.

XPS was performed on a Vacuum Generators model III spectrometer using AlK$_\alpha$ for spectrum excitation. The instrument was equipped with a nitrogen-flushed glove box in which the samples were prepared for interface analysis without exposure to the atmosphere. Vacuum conditions were in the 10^{-9} Torr region throughout this study. Depth profiling was performed in the sample preparation chamber using an A$^+$ ion beam of 3 keV and 20 μA at 10^{-5} Torr pressure (normal incidence). The

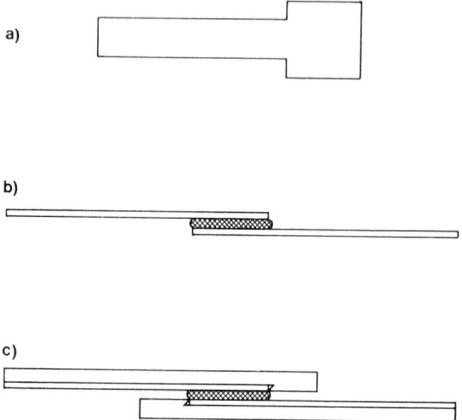

Fig. 1. Sample arrangement for determination of adhesion level of the rubber-to-brass bond. (a) brass sample; (b) vulcanized rubber-to-brass sample; (c) determination of adhesion level in lap-shear test using reinforcement templates.

absolute sputter rates of Cu and Zn were found to be 2.0 ± 0.2 and 2.2 ± 0.2 nm/min, respectively. Deconvolution was done with a DuPont Curve resolver type 310 assuming photoelectron peaks intermediate between Gaussian and Lorentzian. Quantitative analysis was performed with the use of Jørgensen's set of elemental sensitivities [5].

3. Results

3.1. Surface composition of brass

A separate study of the surface composition of 70/30 brass specimens following treatments will be published elsewhere [6].

In general, XPS analysis of 70/30 brass indicates the presence of a very thin film of Cu_2O on top of a thicker film consisting of ZnO which contains some copper. Upon heating in air the thickness of the ZnO layer increases, thereby decreasing the surface copper content markedly. Apparently, only the zinc is oxidized under those conditions. However, if zinc is preferentially dissolved out of the surface layers, subsequent oxidation at elevated temperatures will form a thick layer of CuO on top of the ZnO layer. Prolonged heating will lead to a further growth of the ZnO layer only, which will eventually overgrow and reduce the CuO layer. The surface of mechanically polished 70/30 brass samples used in this study contained approximately 50% Cu and a ZnO surface oxide thickness of less than 10 nm. Ion bombardment of homogeneous brass samples does lead to some preferential sputtering of zinc as indicated by the deviation of the observed Cu/Zn ratio from that of the bulk. This ratio remains constant, indicating that a steady state is readily obtained. Comparison of the Cu/Zn ratios derived from the $Cu2p_{3/2}$ and $Zn2p_{3/2}$ lines (electron range 0.5 nm) and from their 3p lines (range approximately 2.5 nm [7]) also shows that sputtering causes a copper gradient in the outermost 2.5 nm or more of the bombarded material. At the surface the Cu3p/Zn3p ratio is higher than the Cu2p/Zn2p ratio. In the steady state situation Cu3p/Zn3p < Cu2p/Zn2p.

3.2. Interface analysis of rubber-to-brass samples

The results of an interface analysis and depth profiling of a rubber-to-brass sample are shown in fig. 2. The bond strength would exceed the rubber strength, if tested at room temperature. Therefore the results of fig. 2 are typical of a sample with a high adhesion level. On the brass side sulfur is detected. The thickness of this film is 10 nm at most. In the rubber part of the sample considerable amounts of zinc and copper are detected, originating from the brass, as copper was originally absent and zinc is present in rubber at a much lower level. The maxima observed in the in-depth concentrations of Cu, Zn, S and O imply that the reaction of brass with rubber has led to the formation of an interfacial zone consisting of copper sul-

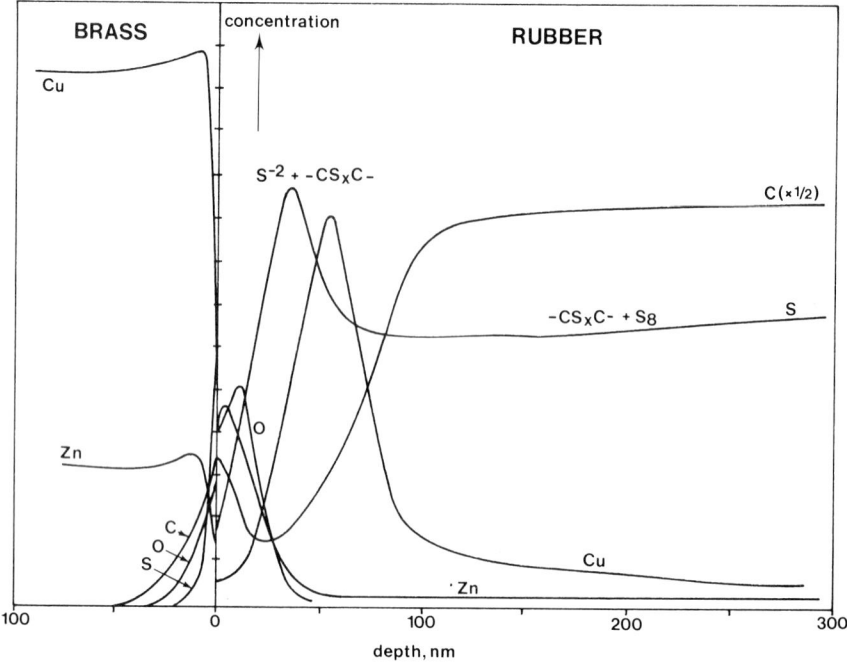

Fig. 2. In-depth concentration profiles of rubber-to-brass sample broken at liquid nitrogen temperature following vulcanization of polished 70/30 brass to dry rubber at 150°C for 25 min.

fide on top of a layer of zinc sulfide and zinc oxide. The observation that the curves for Cu- and Zn-sulfides do not coincide is probably related to the low miscibility of these sulfides [8]. Although the binding energies of Cu, Zn and S of argon ion treated surfaces are inconclusive qualitative changes can be noted in the S 2p signal as a function of depth. At depths greater than 100 nm the sulfur signal can be deconvolved into two peaks (cf. table 1). They can be attributed to uncombined sulfur (S_8) and to sulfur in the form of rubber crosslinks ($-CS_xC-$) which are mixtures of various bonds of the type:

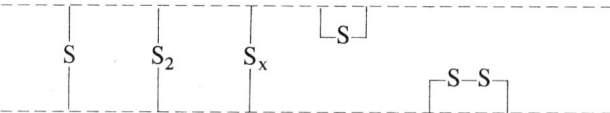

In the region where sulfides are detected no contribution from elementary sulfur is detected but an additional peak at low BE instead (table 1) due to S^{-2} ions. The Cu2p binding energy as a function of depth does not change. It is close to the value for metallic copper (table 1) but the $L_3M_{4,5}M_{4,5}$ Cu Auger line shows that the copper is not metallic, but probably in a sulfide form. More conclusive information can

Table 1
Binding energies of Cu $2p_{3/2}$, Zn $2p_{3/2}$ and S 2p in some selected compounds and at the rubber-to-brass interface

Sample	Binding energy (eV) [a]
Cu	932.2
Cu_2O	932.6
Cu_2S (powder)	932.4
Cu_2S (film grown on copper)	931.6
Interface rubber–brass [b]	931.6
Zn	1021.2
ZnO	1022.0
ZnS	1021.9
Interface rubber–brass [b]	1022.0
S_8 (unvulcanized rubber)	164.1
Vulcanized rubber	163.5, 164.1
Cu_2S	161.3, 164.0
ZnS	161.5
Interface rubber–brass [b]	161.3, 163.5, 164.0

[a] Using Au $4f_{7/2}$ = 83.6 eV as reference; estimated accuracy ±0.2 eV.
[b] Binding energies on both sides of the interface were identical.

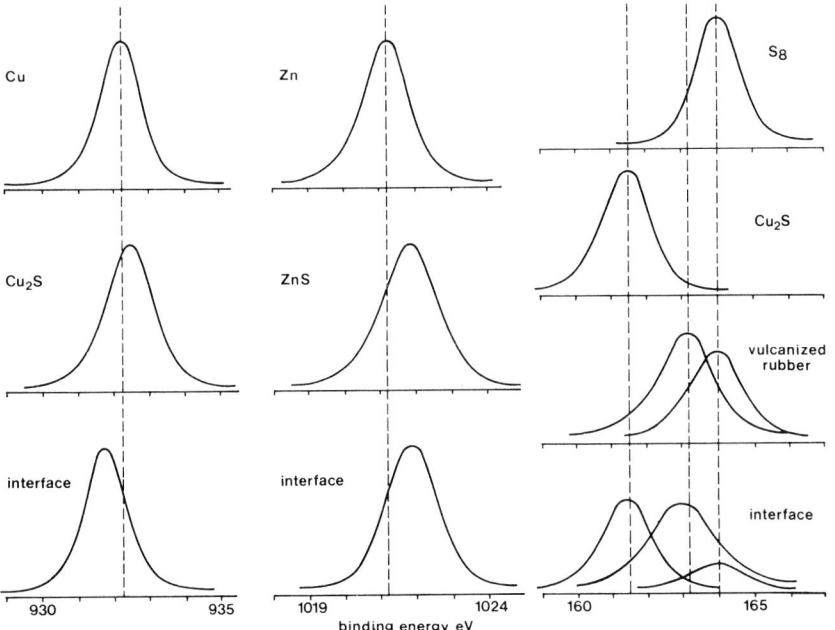

Fig. 3. 2p photo lines of Cu, Zn and S at the rubber-to-brass or rubber-to-polyester interface following vulcanization at 150°C for 25 min.

be obtained from the photoelectron lines observed at the two surfaces, i.e. prior to argon bombardment (fig. 3). Zn is identified as ZnO/ZnS. The copper is completely in the monovalent state, but its binding energy is unusually low (931.6 eV), unlike pure Cu_2S. If a brass surface is reacted for 30 min, with sulfur dissolved in paraffin at 180°C, the same $Cu2p_{3/2}$ binding energy is observed. Consequenty, a compound containing only Cu and S must be involved.

The S 2p spectrum can conclusively be deconvolved in peaks of S^{-2} and $-CS_xC-$ units. Fig. 3 also gives the observed S 2p spectrum of the rubber surface following vulcanization against a non-adhesive surface. Comparison of the two S 2p spectra shows that the ratio $-CS_xC-/S_8$ is greater in the case of vulcanization against brass. Fig. 4 depicts the results obtained after vulcanization of a similar brass sample against wet rubber. The detrimental influence of small amounts of water on the adhesion is well known but no explanation has been offered. Comparison of figs. 2 and 4 shows that the interfacial layer formed in the presence of H_2O is considerably different. The copper on the brass sample is identified as cuprous sulfide (table 1). The zinc, however, is largely in the form of ZnO as evidenced by the depth profiles of Zn and O. In the rubber side of the sample Cu and Zn are not concentrated in the first 100 nm.

Their concentrations are lower than in fig. 2 but copper and zinc signals extend

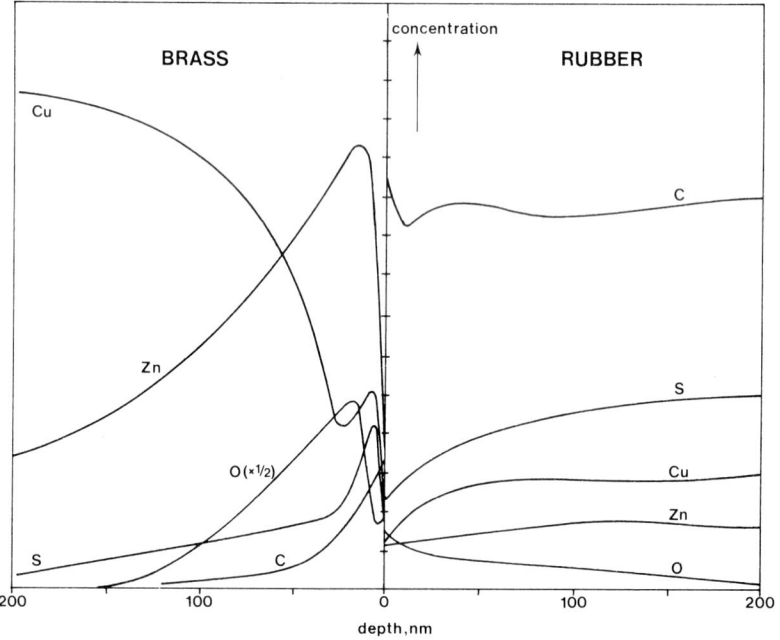

Fig. 4. In-depth concentration profiles of rubber-to-brass sample broken at liquid nitrogen temperature following vulcanization of polished 70/30 brass to wet rubber at 150°C for 25 min.

Table 2
Adhesion of some selected materials to rubber [a]

Material	Adhesion level [b]	XPS of interface	Remarks
1. iron, steel	0	–	no adhesion
2. copper sheet	0	excess Cu_2S	some adhesion if undercured
3. copper-plated steel [c]	700–900	–	good adhesion if plating thickness <50 nm
4. steel + Cu_2S coating [d]	700–800	–	good adhesion for fresh Cu_2S layer (<50 nm)
5. zinc sheet	100–200	ZnS formation	poor adhesion
6. copper-plated zinc [c]	700–800	Cu_2S formation	good adhesion if plating thickness <50 nm
7. 70/30 brass sheet	700–1000	formation of Cu_2S and ZnS	good adhesion; level depends on surface preparation

[a] Vulcanized at 150°C for 25 min.
[b] In N/64 mm^2.
[c] By electroless immersion plating.
[d] Prepared from sample 3 by reaction with sulfur in liquid paraffin at 180°C.

to a much greater depth, i.e. the total amount of copper in particular is considerably higher. This was confirmed by means of more quantitative techniques (X-ray fluorescence). It thus seems that small amounts of water increase the thickness of the interfacial film and also modify its chemical composition.

3.3. Adhesion experiments

In order to use the XPS results for the development of an adhesion model they were combined with quantitative data on the adhesion of samples with different properties or compositions. Therefore, some experiments were conducted to evaluate the adhesive properties of materials other than 70/30 brass. A summary of these results is given in table 2. The qualitative conclusion can be reached that a high adhesion level can only be obtained with brass of 60–70% Cu, pure copper of thickness less than 50 nm or with a thin Cu_2S layer. Results of other adhesion experiments will be published elsewhere.

4. Discussion

The main components of the rubber compounds used throughout this study are natural rubber (polypentadiene), zinc oxide, carbon black, elementary sulfur, vulcanization accelerators and antioxidants. During vulcanization of rubber ZnO reacts with the vulcanization accelerator and with sulfur (S_8) which, in a very complex reaction scheme, results in the formation of active S atoms which react with the rubber molecules to form S crosslinks. The reaction between rubber and elementary sulfur itself is very slow. The experiments of table 2 suggest that adhesion of rubber to brass is the result of the in situ formation of Cu_2S. Adhesion to ZnS or FeS formed during vulcanization of rubber with Zn or Fe is negligible.

Quantitative analysis of thin films of Cu_2S grown on copper or brass in a solution of sulfur in paraffin (180°C) shows that the surface composition of the film is in fact $Cu(I)_xS$ where x is approximately 1.6. The S 2p signal indicates that some sulfur at the surface is in the form of S instead of S^{-2}. We can thus describe the surface of the sulfide films as $S \rightarrow Cu_y(I)S \leftarrow S$ where y is 1.90 to 1.97. The results further show that the total amount of Cu_xS formed during vulcanization is of paramount importance. This can be explained on the basis of the cohesive strength of epitaxial Cu_xS films as a function of their thickness [9]. Thin films are strong and adherent, but they change their composition and modification as they grow thicker. Their cohesive strength decreases with thickness due to lattice contraction. The optimum Cu_xS film thickness is of the order of 30–40 nm. Therefore, pure copper sheet will form too much Cu_2S leading to a complete loss of adhesion. In 70/30 brass the reactivity of the surface towards S is lower as a result of the formation of a ZnS/Cu_xS sulfide (fig. 2). Cu_xS is a p-type semiconductor, whose growth mechanism is based upon diffusion via cation vacancies. ZnS is an n-type semiconductor, whose growth mechanism is via interstitial Zn ions. Further, the ionic radii of Cu^+ and Zn^{+2} are considerably different (0.96 and 0.74 Å respectively). Consequently, a thin continuous film of ZnS will slow down the diffusion of Cu^+ ions to the surface of the sulfide film by an order of magnitude resulting in a surface reactivity which is adequate for a good adhesion.

The differences in adhesive behaviour of Zn, brass and steel must be related to some property of the surface of the sulfide film, probably the presence of free sulfur atoms at the $Cu(I)_xS$ surface.

By analogy to Cu_2O which is an oxidation catalyst [10], it seems likely that the sulfur atoms at the $Cu(I)_xS$ surface are considerably more reactive towards rubber molecules than undissociated S_8 molecules (cf. section 3.2). This is supported by our observation that vulcanization of rubber in contact with brass leads to a higher $-CS_xC-/S_8$ ratio. Since the modulus of vulcanized rubber is dependent on the crosslink density, these observations imply that the Cu_xS film induces a higher crosslink density and hence a higher rubber modulus by catalytic action. The effect of modulus on observed bond strength is demonstrated by the observation that the bond strength of steel to rubber – which is zero at room temperature – increases considerably if measured at low temperatures (−30 to −50°C) [11]. The temperature dependence curve of the adhesion closely matches the temperature curve of the rubber modulus, which indicates that the steel-rubber bond is a physical one. Since the brass-to-rubber bond shows the same temperature dependence as the steel-to-rubber bond we may assume that the former is also of the physical type, the relatively high bond strength at room temperature being the result of the catalytic activity of the Cu_xS formed at the interface.

Acknowledgements

The author is indebted to Messrs. N. van Veenendaal and H.E.C. Schuurs for their experimental assistance.

References

[1] G. Rutz, Plaste Kautschuk 17 (1970) 909; 20 (1973) 33.
[2] W.E. Weening, Gummi, Asbest, Kunstst. 29 (1976) 749.
[3] N. Stuart, Plastics 21 (1956) 308.
[4] C.M. Blow, India Rubber J. (1947) 519.
[5] H. Berthou and C.K. Jørgensen, Anal. Chem. 47 (1975) 482.
[6] W.J. van Ooij, Surface Technol., accepted for publication.
[7] I. Lindau and W.E. Spicer, J. Electron Spectrosc. Related Phenomena 3 (1974) 409.
[8] A.A. Bundel, A.V. Vishnyakov and V.N. Zubkovskaya, Izv. Akad. Nauk SSSR, Neorg. Mater. 6 (1970) 1248.
[9] C. Labar and R. Breckpot, Bull. Soc. Chim. Belges 81 (1972) 565.
[10] B.J. Wood, H. Wise and R.S. Yolles, J. Catalysis 15 (1969) 355.
[11] S. Buchan and J.R. Shanks, Trans. Inst. Rubber Ind. 21 (1946) 266.

ESCA INVESTIGATION OF THE OXIDE LAYERS ON SOME Cr CONTAINING ALLOYS

S. STORP and R. HOLM

Bayer AG, Ing. Bereich Angew. Physik, D-5090 Leverkusen, Germany

Cr is an important constituent of many technically important alloys, such as stainless steels and those used for metal implants. The following characteristics of such alloys are revealed by ESCA measurements:

(1) Heating the sample in air below 400°C leads to oxide layers with high contents of Fe or Co oxides. For temperatures higher than 400°C the Cr becomes mobile enough to migrate to the surface and to build up Cr-enriched oxide layers.

(2) In H_2O and HNO_3 a Cr-rich oxide layer is always observed.

(3) The chemical shift for Fe oxide in stainless steels is smaller than for pure Fe or Fe–Si alloys; i.e. there is a larger amount of Fe(II) oxide.

(4) The intensity ratio $Fe_{metal} : Cr_{metal}$ or $Co_{metal} : Cr_{metal}$ after exposure of the sample to H_2O or HNO_3 is lower than for oxidation in air, i.e. Cr is enriched not only in the oxide layer but in the metal underneath.

(5) If the alloy also contains Mo, the intensity ratio $Cr_{oxide} : Mo(VI)_{oxide}$ is about 10 : 1 for exposure to H_2O, as well as to diluted and concentrated HNO_3. In H_2O and diluted HNO_3, Mo(IV) is found also.

The ESCA results were obtained without any ion bombardment because the sampling depth is large enough to permit analysis of oxide layers below contamination and even of the metal below a thin oxide layer. This non-destructive method is of great advantage because in ion bombardment of the sample Cr is preferentially sputtered.

1. Introduction

Cr is an important constituent of a number of technically important alloys, especially because it improves their corrosion behaviour. This study is devoted to steels (Fe, Cr, Ni and Mo alloys) and to alloys used for metal implants (Co, Cr and Mo alloys). The two groups are related in as far as the functions of Fe are taken over by Co in the latter group.

The surface layers, which were formed under the influence of thermal and chemical treatments, were investigated by ESCA (Electron Spectroscopy for Chemical Analysis) [1]. The information provided by the method is well known [1–3].

The thicknesses of the oxide and passive layers observed are often less than 100 Å. Advantage can therefore be taken of the fact that, as ESCA comprehends a relatively high layer thickness, it provides information, not only about oxide layers below the contamination, but also about the metal below the oxide layer, without removal of the surface layers. In addition to the rapid determination of the oxide layer, this avoids the artefacts associated with the removal of layers by ion bombardment (e.g. the selective sputtering of the Cr).

2. Experimental

Table 1 shows the compositions of the steels investigated. The wet chemical analyses differed only insignificantly from the average analyses of these types of steel. Of the alloys used from metal implants, we examined, above all, Vitallium®, which is used particularly often in orthopaedics; its average analysis is 28% Cr, 6% Mo, Mn + Ni + Fe + W < 2%, balance Co.

Test specimens ($3 \times 6 \times 13$ mm^3) were produced from all the alloys. Before each experiment the surface to be investigated was polished mechanically (initial state). The specimens were then kept in laboratory air and in H_2O at various temperatures. In addition, corrosion tests with HNO_3 were performed on the steels, while Vitallium was used for several implantation tests.

The ESCA measurements were made with an electron spectrometer, model ES 200 (AEI Scientific Apparatus, Manchester).

3. Results and discussion

With Cr the metallic state and the trivalent and hexavalent states are easily distinguished by ESCA (fig. 1). In the case of the trivalent state the 2p lines are broadened as a result of spin–spin interactions. Therefore the Cr lines do not enable one to distinguish between oxide and hydroxide [*]. In every case only the lines of the metallic and trivalent Cr were found. Even the treatment with highly concentrated HNO_3 did not result in a higher oxidation state.

On pure Cr the oxide layer stable in air at room temperature is very thin (≤15 Å, fig. 1). But at temperatures exceeding 200°C it rapidly becomes thicker than the layer thickness comprehended by ESCA (about three times the value of the mean excape depth for Cr oxide, i.e. about 60 Å). Until then the oxide and metal appear side by side in the ESCA spectrum. The oxide: metal intensity ratio is a measure of the thickness of the oxide layer. Therefore it is possible to observe the growth of the oxide layer on metals stored in air or H_2O and, quite generally, the composition and rates of formation of oxide and passive layers on metallic materials as functions of thermal and chemical treatments [4–6].

With multi-component systems (e.g. alloys) [7] the oxide and metal are compre-

---

[*] Only in idealized experiments it is normally possible to distinguish between oxides and hydroxides from the O 1s line.

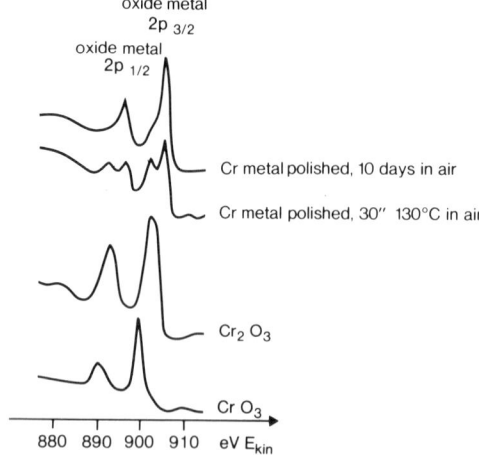

Fig. 1. Chemical shifts in Cr 2p spectra.

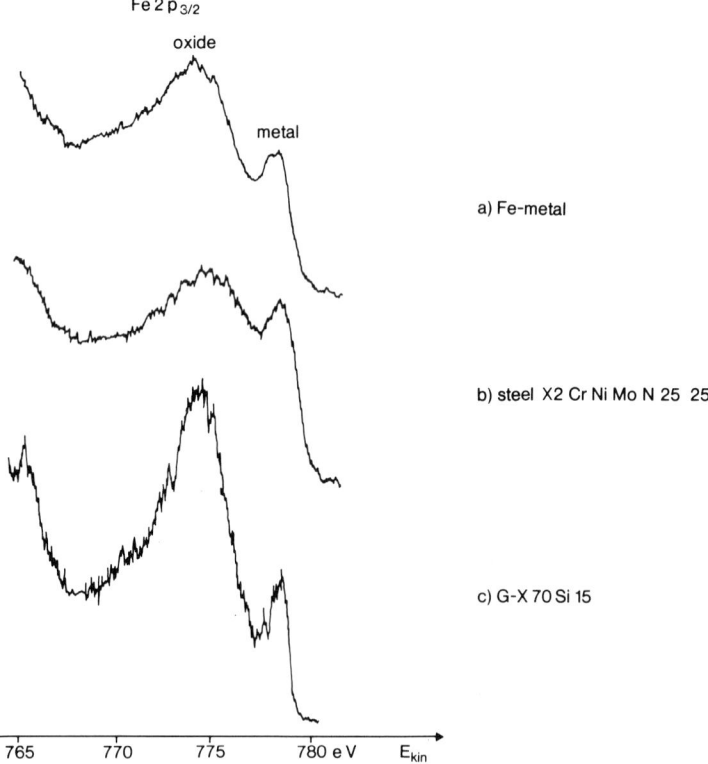

Fig. 2. Fe2p spectra of Fe metal, steel X2 CrNiMoN 25 25 and G-X 70 Si 15.

hended separately for each alloying constituent. If the intensity ratios of the individual elements are formed separately in the oxide and metal phases, one can see from an ESCA spectrum immediately what changes of concentration have occured in the oxide as compared with the metal. After storage of the specimen in air for thirty days the oxide layer is so thin that the oxide and metal lines of each alloying constituent can be detected in the ESCA spectrum [4]. The intensities of the oxide lines for Fe and Cr are greater than those of the freshly polished specimen, the increase being slightly larger in the case of Cr than in that of Fe.

The $Fe_{metal} : Cr_{metal}$ intensity ratio does not differ from the inital state. A careful examination of the chemical shift of the Fe $2p_{3/2}$ lines shows that the difference between Fe metal and Fe oxide is about 0.5 eV smaller than in the case of air oxition of pure Fe. Asami et al. [8] interpreted this as an increase in the proportion of bivalent Fe in the oxide layer. We have made the same observation with all the

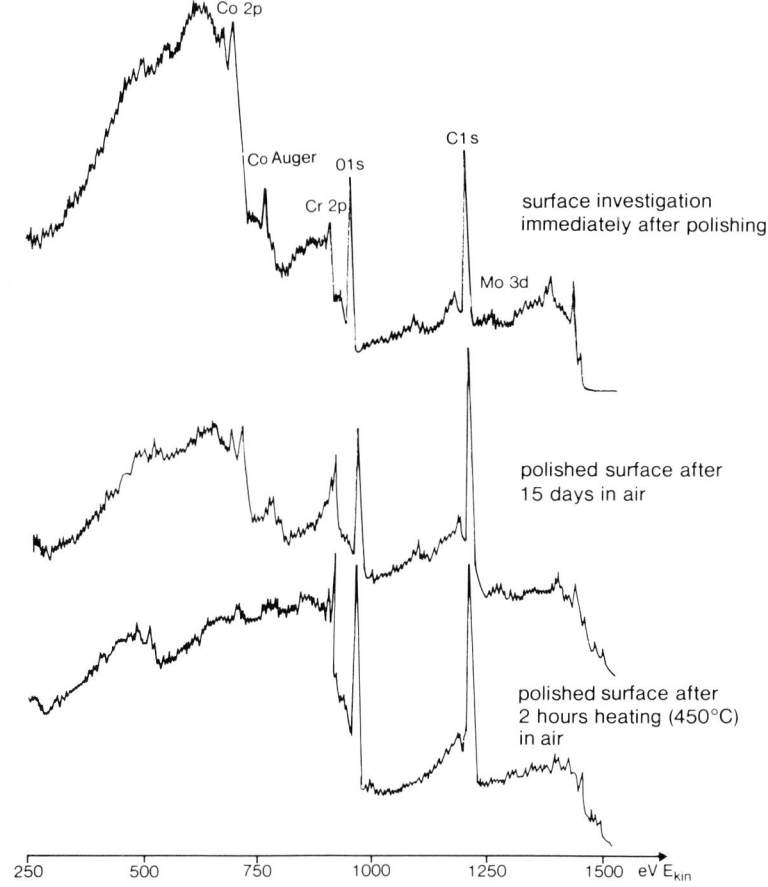

steels we have investigated, no matter whether they were exposed to air or attacked by H_2O and acids. But the effect occurs only when an oxide layer rich in Cr is built up [8]. Thus, for example, after the air oxidation of the substantially Cr-free material G-X 70 Si 15, the same Fe oxide shift was found as with an oxide layer on pure Fe (fig. 2). Hence Cr exerts a decisive influence on the oxidation of the Fe. With Vitallium (fig. 3) there is no analogous effect; here, on the whole, Co, even in the pure state, is oxidized in air only to the bivalent state. Co is indeed less represented in the oxide layer formed on a freshly polished specimen through exposure to air than is Fe in steels with a comparable amout of Fe (fig. 3b). Here again, incidentally, the oxide layer is so thin that the alloying constituents are also detected in the metallic state. The $Co_{metal} : Cr_{metal}$ intensity ratio then corresponds to the initial state.

In earlier publications [4,9–11] it has been shown that the rate of formation of the oxide layer in air increases rapidly at elevated temperatures. The information provided by ESCA concerning the relatively thick layers then formed is confined to the uppermost 50–100 Å. At temperatures up to 400°C Fe (as Fe oxide) or Co, as the main alloying constituent, is represented most of all. Only at temperatures above 400°C does Cr become fairly mobile, with the result that an oxide layer rich in Cr oxide is formed above 500°C. This can be seen from the $Fe_{oxide} : Cr_{oxide}$ intensity ratio for certain steels as a function of temperature [4] and from the spec-

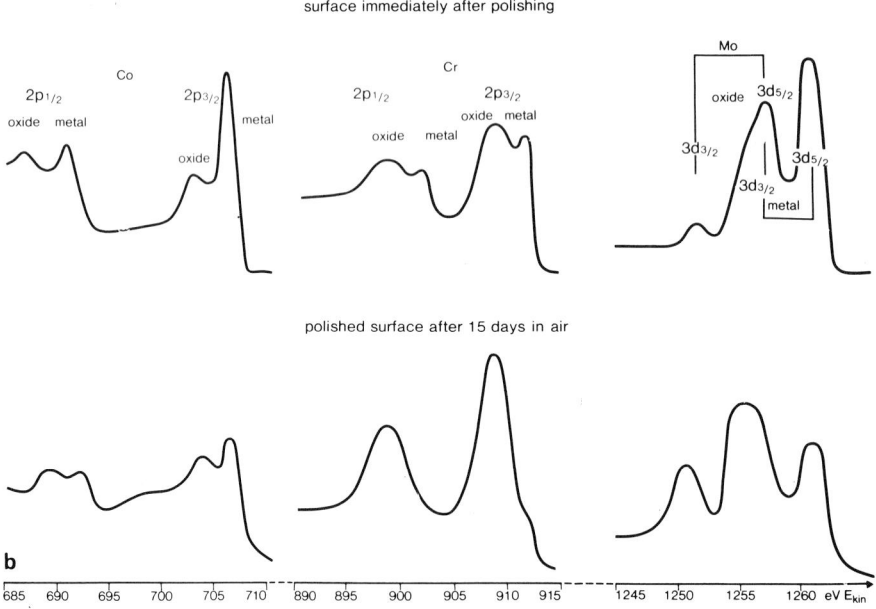

Fig. 3. Photoelectron spectra of Vitallium before and after exposure to air: (a) wide scan spectra; (b) Co2p, Cr2p and Mo3d spectra.

tra of Vitallium (fig. 3a). Ni plays practically no part in the formation of the oxide layers.

When freshly polished steel and Vitallium specimens are kept in water the effects differ as follows from those of exposure to air:

(a) A Cr-rich oxide layer in which the Cr is present in the trivalent form is always found. The Cr enrichment relative to Fe and Co may be up to twenty times greater than that of an oxide layer formed through exposure of the specimen to air. The immersion of Vitallium in boiling water leads within a short time (about 15 min) to the formation of an approximately 30 Å thick Cr oxide layer, whose thickness increases only slightly if the immersion time is extended (fig. 4).

(b) The $Fe_{metal} : Cr_{metal}$ intensity ratio is lower (table 2). It follows that in the metal layer below the oxide layer Fe is depleted in relation to Cr. This suggests that the mechanism of the oxide layer formation is as follows: the individual alloying constituents are dissolved at different rates, the rate for Fe being higher than that for Cr [12]. Oxidized Fe is likewise dissolved, whereas oxidized Cr remains on the surface. As soon as the Cr content of the oxide layer reaches a high level a state of

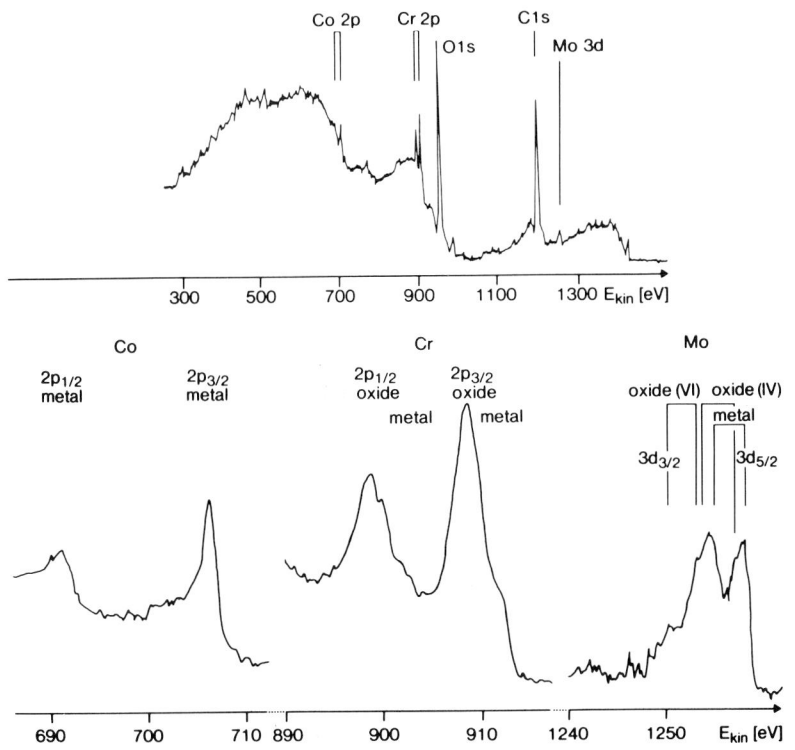

Fig. 4. Composition of the oxide layer on Vitallium after sterilization (15 min in H_2O, 100°C).

Table 1
Composition of steels investigated

Steel (DIN 17006)	Composition (%)									
	C	Cr	Ni	Mo	Ti	Mn	Si	P	S	N
X10 CrNiMoTi 18 10	0.051	17.4	12.7	2.36	0.46	1.80	0.57	0.018	0.004	–
X2 CrNiMoN 25 25	0.016	24.12	24.71	2.35	–	1.35	0.30	0.008	0.012	0.13
X2 CrNi 18 9	0.015	18.43	10.52	0.03	–	1.16	0.69	0.020	0.017	–

Table 2
ESCA intensity ratios of important alloying constituents at the surface of steel X10 CrNiMoTi 18 10 in H_2O and HNO_3.

Sample	Fe $2p_{3/2}$: Cr $2p_{3/2}$		Fe $2p_{3/2}$: Mo $3d_{5/2}$ Oxide : Oxide	Fe $2p_{3/2}$: Si 2p Oxide : Oxide
	Metal : Metal	Oxide : Oxide		
Sample polished	3.4	2.3	20	–
H_2O (20°C) mean value for 2 h–3 d exposure	3.1	0.7	12	–
HNO_3 (10%, 20°C) mean value for 20 min–1 h exposure	2.9	0.5	4.5	35
HNO_3 (98%, 20°C) 5 min	3.0	0.5	5.0	26
2 h	3.2	0.5	4.2	2.4
HNO_3 (98%, 86°C) 1 h	–	0.2	1.6	0.2
4 h	–	–	–	SiO_2 only

equilibrium is reached or the oxide layer acts as a passive layer. Analogous processes were observed in the case of Vitallium.

(c) Quadrivalent Mo is also found on alloys [5], as also when Mo surfaces react with H_2O (fig. 4).

A number of steel specimens was also exposed to dilute, concentrated and highly concentrated HNO_3. The results obtained for X 10 CrNiMoTi 18 10, which are compiled in table 2, are representative of those obtained for the other specimens, which will not be specified in detail. They make it clear that the effects of exposure to water are also caused by treatment with HNO_3: the formation of a Cr-rich oxide layer and accumulation of the Cr in the metal beneath the oxide layer also. As the alloying constituents are dissolved at different rates, other components of the alloy, too, are accumulated in the oxide layer. If the steel contains Si, an SiO_2 layer with a thickness of more than 100 Å may be formed, especially if the specimen is kept for a long time in highly concentrated HNO_3. This, again, is only understandable on the assumption that in the formation of the oxide or passive layers not merely the uppermost monolayers of the material are attacked. When concentrated HNO_3 is used, the oxidizing properties of the acid are more apparent and mainly Mo(VI) is formed. With regard to the amount of Mo it has been found that Mo accumulates in the oxide layer side by side with Cr and that this process is substantially independent of the HNO_3 concentration; the Cr_{oxide} : $Mo(VI)_{oxide}$ intensity ratio is about 10.

It should be noted that the behaviour of Cr-containing alloys in HNO_3 is not necessarily matched by their behaviour in other acids. In 10% HCl, for example, Fe and Cr are dissolved very rapidly, a passive layer is not formed and Mo and Ni accumulate.

To ascertain whether the thin Cr oxide layer on Vitallium is indeed passive towards biological tissues, a number of samples was implanted in rabbits after sterilization in boiling water. When examined four weeks later, after having been removed from the animals and cleaned in an ultrasonic bath, they showed practically no changes as compared with the initial state. Thus, as a rule, the approximately 30–50 Å thick Cr oxide layer suffices to prevent interactions between the metal implant and the tissue with which it is in contact. Nevertheless, care must be exercised, especially if the passive layer is damaged with an operating instrument during the implantation.

In the studies described we deliberately refrained from investigating depth profiles with the aid of ion bombardment and used only depth information obtained non-destructively by ESCA. For, as earlier studies [13] had shown, oxides on metals are easily reduced by the ion doses normally used to remove contamination and oxide layers. In addition, ion bombardment may result in the ion-induced formation of carbides [13], which, together with implanted Ar, greatly change the oxidation and corrosion behaviour [14]. Finally, removal of the oxide layer by ion bombardment results in substantial depletion of Cr, as shown by table 3, in which values obtained in this way are compared with ordinary chemical analysis and with values obtained non-destructively from ESCA spectra. This can be attributed to the higher sputtering rate of Cr as compared with that of Fe or Mo and to the fact that Cr reacts preferentially with oxygen (only a small proportion of which originates

Table 3

Fe : Cr and Fe : Ni ratios at the surfaces of various steels; comparison of the integral quantity ratios with the intensity ratios after mechanical polishing and ion bombardment

Sample	Fe : Cr		Fe : Ni
	Metal : Metal	Oxide : Oxide	Metal : Metal
Steel X 10 CrNiMoTi 18 10			
Chem. analysis	3.7	–	5.1
Mech. polished [a]	3.4	2.3	3.9
Ion bombardment [b]	6.0	–	4.3
Steel X 2 CrNiMoN 25 25			
Chem. analysis	1.9	–	1.9
Mech. polished [a]	1.7	1.3	1.6
Ion bombardment [b]	3.1	–	1.8

[a] Fe $2p_{3/2}$: Cr $2p_{3/2}$, Fe $2p_{3/2}$: Ni $2p_{3/2}$.
[b] Ion bombardment 6×10^{-2} A sec cm^{-2} Ar$^+$, 5 keV.

from the residual gas, most of it being formed by the selective sputtering of oxygen from the oxide layer).

In view of the difficulties caused by ion bombardment advantage should be taken wherever possible of the fact that ESCA permits non-destructive investigation, not only of the uppermost monolayer, but also of a fairly large range of depths, that is to say — in the present case — the chemically relevant layers situated beneath the contamination or oxide layer at a depth of up to 50 Å. There is then no need to remove the contamination or oxide layer just mentioned [6,15]. Herein lies one of the most significant advantages of this method in the investigation of real surfaces, surfaces, that is to say, which are not prepared under UHV [16].

References

[1] K. Siegbahn et al., Atomic, Molecular and Solid State Structure Studied by Means of Electron spectroscopy (Uppsala, 1969).
[2] R. Holm, G-I-T Fachz. Lab. 16 (1972), 122.
[3] R. Holm, G-I-T Fachz. Lab. 17 (1973) 929, 1025.
[4] R. Holm and E.M. Horn, Metalloberfläche-Angew. Electrochem. 29 (1974) 490.
[5] R. Holm and S. Storp, Vakuum-Technik 25 (1976) 175.
[6] R. Holm, Vakuum-Technik 23 (1974) 208.
[7] R. Holm and S. Storp, J. Electron Spectr. 8 (1976) 139, 459.
[8] K. Asami, K. Hashimoto and S. Shimodaira, Corrosion Sci. 16 (1976) 387.
[9] R. Holm and J. Ohnsorge, Sonderbände der Prakt. Metallographie 6 (1976) 126.
[10] H. Fischmeister and I. Olefjord, Monatsh. Chemie 102 (1971) 2693.
[11] I. Olefjord, Metal Sci. 9 (1975) 263.
[12] Y.M. Kolotyrkin, Electrochim. Acta 18 (1973) 593.
[13] R. Holm and S. Storp, Appl. Phys. 12 (1977) 101.
[14] V.T. Cherepin, A.A. Kosyachkow and M.A. Vasilyev, Surface Sci. 58 (1976) 608.
[15] R. Holm and S. Storp, Vakuum-Technik 25 (1976) 41, 73.
[16] R. Holm and S. Storp, Phys. Bl. 32 (1976) 342.

SURFACE DEFECTS

G.E. RHEAD

Université Pierre et Marie Curie, ENSCP, 11, rue Pierre et Marie Curie, 75231 Paris Cedex 05, France

Lack of information on defects may impede progress in understanding many surface properties. A survey is made of the evidence for the effects of defects in various types of experiment. Work function measurements, LEED, decoration and chemical reactivity, atomic steps and high temperature properties are among the topics briefly reviewed. An extensive bibliography is given. There is a need for an agreed experimental protocol to deal with the problem of characterizing the degree of perfection of a specimen surface. The routine measurement of the work function may be provide the best simple means for monitoring defect concentrations.

1. Introduction

This survey is mainly bibliographical. It is hoped that it will help focus attention on an important topic and encourage work in new directions.

Until about ten years ago the biggest problem in surface science — for experimentalists at least — was the control and detection of contamination. Now it seems that Auger spectroscopy has resolved that problem in many experiments, although there are still difficulties in quantitative measurements. The next most important fundamental problem — atomic scale structure — has only been partly solved. LEED experiments in particular have tended to overemphasize the apparent crystalline perfection of the surface. In many types of experiment the procedures, the kind of information recorded and the structural models used in interpretation may not adequately take into account the effects of surface defects.

The purpose of this paper is to review available information in different areas of surface science with the aim of assessing the sensitivity of various properties to surface imperfection; this should suggest possible experimental protocols for the control and detection of defects.

Each of the topics that follow is examined from this viewpoint and an attempt is made to summarize some of the most significant recent work and to bring out an overall picture of the possible role of defects in different phenomena.

The discussion is restricted mainly to the solid metal-to-vacuum (or gas) interface and to single crystal specimens. It generally excludes grain boundaries, thin films and processes in crystal growth [1].

2. Field-emitter techniques [2]

It is appropriate to mention field-ion microscopy first since this technique has been successful in imaging directly all the various defects that can be expected from simple models [3]: atomic steps, vacancies [4], adatoms, dislocations, stacking faults and clusters of adatoms. There is currently a special interest in the binding and mobility of clusters [5,6].

The following are some of the most important conclusions from FIM observations: Steps on thermally equilibrated emitter surfaces are of monoatomic height. Equilibrium vacancies and self-adsorbed adatoms are rare at the cryogenic temperatures used for observation but they form spontaneously on heating to quite low temperatures and they lead to a high degree of structural disorder [7] (see section 10). Adatom interactions can be quite short-range. Properties of clusters may show strong chemical specificity.

Developments with atom-probe field-ion microscopy and field desorption spectroscopy may also lead ultimately to more information on defects but the surface processes involved in these techniques are complex [8].

FIM provides images that are powerful in forming conceptual pictures of surface structure. However, these pictures may be much less appropriate at higher temperatures where states of disorder and high mobility may prevail.

3. Work function measurements

It is well established that the work function is sensitive to crystal structure and perfection [9]. The effect of atomic-scale surface roughness can be understood in terms of the smoothing of the electronic charge density parallel to the surface, as suggested by Smoluchowski [10]. This smoothing leads to a more positive surface potential and a lowering of the work function. For a review (up to 1966) of experimental data on work functions for different metals and different crystallographic orientations see ref. [11].

Advances in the last ten years have been concerned with measuring changes due to defects deliberately introduced into the surface. Plummer and Rhodin [12,13] showed that a single tungsten adatom deposited onto a tungsten (110) field emitter tip produces a detectable change in the emitted current: from this an estimate can be made of the adatom dipole moment. Other measurements on tungsten [14,15] have explored ranges of orientations close to low-index poles and so detected the lowering of the work function due to increasing densities of atomic steps.

Besocke and Wagner [16,17] have extended this type of work to examine, on large single crystals, the clustering of adatoms, the correlation between step structure, kinks, and work function and the migration of adatoms to step sites. Linear variations are observed between the work function decrease and the step density up to densities of the order 10^7cm^{-1} [18]. Step densities corresponding to as low as 2%

of the surface sites can be detected. In a comparative study of platinum, gold and tungsten it has been possible to distinguish different degrees of electronic rearrangements at steps on these metals [19].

The production of defects by ion bombardment and their subsequent annealing-out has also been followed by work function measurements — for example on palladium [20], nickel [21], copper [22], molybdenum [23] and recently on alluminium [24] and platinum [25]. The ion bombardment damage produced by conventional cleaning techniques can produce work function reductions of the order of several hundred meV.

A rough assessment based on the results in the above papers suggests that the changes, extrapolated to a complete structural disordering, amount to about 10% of the perfect-structure work function and that a 1% defect concentration can produce a reduction, depending on the metal, of the order 10–20 meV. This is well within the range of detectability.

From the slow changes in work function due to annealing-out of defects, observed even at ambient temperatures [25], it may be possible to obtain information on low-temperature defect diffusivities.

Because of the dependence of the work function on defect concentrations there might be a strong temperature-dependence due to thermally activated surface roughening (section (10). Although many other factors can produce a temperature-dependence of the work function [9] this possibility does not appear to have been examined: it would be of fundamental interest to search for evidence of the onset of surface roughening at a critical temperature. Some recent results for the temperature variation for different planes of copper up to near the melting point [26] show changes comparable to those produced by ion bombardment.

4. Low-energy electron diffraction

Although not strictly defects in the normal sense, microfacets were among the first surface imperfections to be examined by LEED [27–29]. If facet widths are large enough compared with the electron coherence width (several hundred Å) each set of facet planes will produce its own pattern centred on its specular reflexion. Thus a faceted surface is characterized by a mixture of patterns each with a different centre (00 spot) and distinguishable by the displacement of the spots as the wavelength is varied. Facets are not usually equilibrium structures on clean metals and generally they form only in the presence of adsorbed impurities, particularly oxygen [30].

In the late 1960s several investigators reported effects of ion bombardment damage on LEED patterns from low-index faces of various materials: germanium [31], nickel [32], molybdenum and tungsten [33] and palladium [20]. Attention was focussed on the changes in intensity-versus-electron energy, $I(E)$, plots: it is found that bombardment produces a decrease in peak intensities and a broadening over a wider energy range. Monitoring of these effects can be used to characterize the degree of surface damage and to follow its annealing-out during heat treatment. Such measurements can be useful in determining the minimum anneal required to restore the surface in cases where too long an anneal produces unwanted segregation of impurities from the bulk of the crystal [20].

The creation of vacancies and lattice strains produced by occluded ions was first suggested as the origin of the degradation of the diffraction patterns after ion bombardment. With hindsight it now appears that the formation of surface hillocks with sides of vicinal orientations, i.e. with atomic steps, is the more likely cause of these changes. The interpretation is much more evident from the angular distribution $I(\theta)$ than from $I(E)$ plots. Thus, beam broadening is also observed in the angular display on the fluorescent screen in post-acceleration equipment but only certain spots are broadened while others remain quite sharp [34]. This result is entirely compatible with the interpretation of diffraction from stepped surfaces discussed below.

The annealing-out of bombardment damage can be interpreted mainly in terms of a smoothing of surface undulations by surface diffusion. One puzzling observation is that ion bombardment of vicinal surfaces does not produce the same degradation of the LEED patterns [34]. Possibly a high density of steps results in a high self-diffusivity and consequently leads to a rapid surface smoothing even at ambient temperatures. Alternatively, it may be that the initial interaction with the ions is affected by the presence of steps.

Ion bombardment damage can produce an increase in the background intensity in the angular LEED display [35] but this effect appears to be due primarily to an increase in the inelastically scattered electrons which, because of field penetration, manage to escape filtering by the grids. With a properly biased Faraday collector such increases in background are not observed [20].

The relative insensitivity of the visually displayed LEED pattern to surface disorder was well demonstrated in the work of Jona [36,37] who showed by depositing amorphous layers of silicon on silicon that visually "good" patterns can be obtained even when only 80% of the surface is well crystallized.

The theory of diffraction from imperfect lattices has been developed in connection with X-ray and optical diffraction [38,39]. A basic result can be understood as follows in terms of the convolution theorem [40,41]. An imperfect lattice (I) can be represented by a perfect lattice (P) "multiplied" by a phase object (O) — which can be thought of as being placed in from of P so as to simulate the irregular wavefront scattered by I. The principle of correspondence between multiplication and convolution (between real and reciprocal spaces) shows that the amplitude of the diffraction pattern of I is the convolution of the Fourier transforms of O and P. The transform of O gives the perfect lattice pattern. If P is aperiodic its transform will consist of a strong "forward" peak broadening into low amplitude, irregularly shaped, wings. Thus *random* defects will only change the intensity profile of beams and since the phase information is lost it is not possible to deduce any details of the imperfect structure [39]. Even if the intensity profile could be measured accurately it would not be possible to ascribe it to a unique arrangement of imperfections. Moreover, with most LEED equipment the diffraction spots are broadened by low instrumental resolution and this can mask the effect of quite high degrees if imperfection. In this connection Heckingbottom [42] has discussed limitations due to the small electron coherence widths of LEED beams. LEED has, however, provided useful information on surface defects and especially on atomic steps in situations where the spatial distribution has a strong *periodic* component.

Since 1968 many studies have been made on surfaces deliberately cut at small angles to low-index orientations (i.e. vicinal surfaces). The first patterns obtained from uranium dioxide [43], copper [34] and germanium [44] all showed the — now familiar — spot splittings that characterize the average periodicity of step spacings on these surfaces. In these splittings (doublets) the separation is inversely proportional to the average step spacing on an idealized terrace—ledge—kink (TLK) model of the surface. Instrumental resolution usually permits detection of step spacings up to about 40 atomic diameters which corresponds to a surface with about 2% step sites. As the electron wavelength is varied there occurs an alternation between the positions of doublet and singlet spots. This is easily interpreted in terms of the phase relationships between waves scattered from adjacent terraces. In fact the geometry of the apparently complex patterns is easily explained by simple kinematic diffraction theory — via computations of structure factors [43] or Fourier transform analysis (graphical [34] or optical simulation [45]) or in terms of the Ewald construction [46]. For clean metal surfaces [34,47—49] the patterns are found to conform to the TLK model with steps one atom high. Insofar as this model would correspond to the ideal minimum surface energy for a clean metal this result is not surprising since the variation of surface energy with orientation is probably small (see section 10).

The early LEED work on vicinal surfaces was followed by papers treating the theory of diffraction from imperfect surfaces and in particular showing that very sharp, nearly ideal, patterns can be produced by surfaces having markedly irregular arrays of steps [50,51]. As mentioned above, this result is also known in the field of optical diffraction [39]. Similar conclusions have come from work on anti-phase domains in ordered alloys [52]. It is, however, remarkable to observe sharp, highly contrasted, patterns with doublet spacings corresponding in some cases to terrace widths of an irrational number of atomic spacings, appropriate to the average spacing for an orientation that contains a mixture of two atomic terrace widths. (It is worth noting that in LEED studies of adsorption on vicinal surfaces the interpretation of patterns is facilitated by using only orientations for which the ideal structures have only one terrace width).

It must be emphasized that the insensitivity of LEED to random steps and other non-periodic features is not peculiar to vicinal surfaces. LEED is relatively insensitive to departures from the ideal periodic structure in any surface.

In principle it is possible to examine certain properties of surface kinks by means of LEED. Observed patterns from kinked stepped surface (i.e. surfaces oriented along a high-index zone) agree with the pattern expected from an ideal model: doublets appear oriented in the direction appropriate to the mean step direction and with the appropriate separation [34]. However, there are no additional spots that would correspond to the average kink spacing along the step. The explanation (supported by results from optical simulation [45]) would appear to be that the kink sites are not correlated in position from one step to another and so do not form a two-dimensional periodic net. Recent observations [53] suggest that it may be possible to observe average kink spacings by RHEED, with the beam nearly parallel to the step direction (diffraction essentially by one-dimensional chains).

The degradation of LEED intensities due to ion-bombardment damage, mentioned above, may be understood in terms of the patterns for vicinal surfaces. Formation of hillocks gives a range of vicinal orientations so that the broadened spots are effectively composed of doublets smeared out over a range of intervals and directions. This interpretation can be checked by comparing the observed wavelength dependence of the diffuse-to-sharp alternation of the spots with the behaviour for a vicinal surface. Degradation of the pattern due to point defects does not give the same wavelength dependence [46].

The alternation of doublet and singlet spots as a function of wavelength can be used as a sensitive measurement of step height and the terrace widths. An accuracy of 1% can be achieved in determining the step height and 1 to 3% in the terrace width [49]. Intensity measurements may also give information on surface relaxation around the step. Data for germanium, silicon and tungsten [44,54,49] suggest that atoms at the top of a step are preferentially relaxed inward towards the nulk so that the profile of the terraces is slightly curved. It appears that an alternative explanation in terms of a special scattering phase shift by step atoms can be ruled out [46]. This type of information is important for an understanding of electronic rearrangements at step sites.

The presence of random steps on a nominally flat low-index surface can affect LEED intensity data – $I(E)$ plots – used for structure determinations. Houston et al. [55,56] and also Cowley and Shuman [51] have made model calculations using various step distribution functions. It is found that peaks (Bragg maxima) can be shifted towards higher energies (i.e. energies without the inner potential correction) and secondary structure can be enhanced. By appropriate curve fitting it might be possible to obtain information on real step distributions but instrumental limitations tend to make this impracticable. Steps can also change the absolute intensities scattered into individual beams [56].

With only a couple of exceptions the structure determinations from intensity data published so far make no quantitative assessment of the possible effect of steps. Aberdam et al. [57] have recently reported results for Al(100) and Al(110) from an analysis of LEED intensities as a function of angle of incidence, averaged over the azimuthal angle. A contraction is reported for the top layer spacing on the (110) face. Significant effects of surface roughness especially at high angles of incidence are also reported and the authors point to the "inability to compensate completely for (these effects)"; Earlier work on Al(110) encountered similar difficulties [58]. In another recent LEED study, of Ag(110) [59] only mediocre agreement was obtained between theory and experimental [60]. Corrections of the $I(E)$ plots for the effect of steps produced some small improvements but it seems that a full treatment of these corrections may have to include changes in the scattering properties and relaxation near the step edge. Systematic studies of the effect of imperfections on LEED intensities are very much needed. It would be ironic, at least, if ten years of effort towards understanding LEED intensities are thwarted by out inability to control and account for surface imperfections.

A few LEED observations have been made of the effect of steps on anomalous (reconstructed) clean surface structures. For gold (100) vicinal surfaces, for example, it appears that if the terraces are large enough — at least 5 atoms wide — anomalous patterns corresponding to hexagonal reconstructed layers are observed but with the supression of one of the two orientations that occur on the low-index face [61]. Highly stepped surfaces such as (311), (210) and (320) have "normal" unreconstructed arrangements [62].

The suppression of certain domains of an oriented adsorbed overlayer structure due to the presence of steps is frequently observed. It can be useful in interpreting complex LEED patterns.

Finally, there are some interesting observations, on uranium dioxide, of diffraction rings around integral-order spots that can be explained by randomly separated triangular clusters [63]. The unusual patterns can be simulated by means of optical transforms [45].

5. Adsorption at defects, decoration and chemical reactivity

The subject of prefered adsorption at defect sites is linked to the larger question of the influence of substrate structure and surface heterogeneity in adsorption. These problems were touched on in very early surface studies. The concept of "active centres" in catalysis was introduced by Taylor in 1925 [64] and by the late 1930s observations of the crystallographic dependence of adsorption had already been made by thermionic and field-emission experiments [9]. It is outside our scope to trace the full history of these topics but it is worth mentioning certain papers that started directions of research that are still being followed today.

Gwathmey and co-workers [65] and Rhodin [66–68] were especially active in the late 1940s in promoting the use of single crystals in catalytic and adsorption studies [69]. Both spheres and flat plate specimens were used. Thermodynamic data and relative reaction rates for individual faces were obtained for the first time. Even in these early studies the importance was noted of the possible role of surface imperfections — both those originally present and those formed during the reaction. A summary of this work will be found in ref. [70].

Farnsworth and co-workers [71] perfected the (now very familiar) method of surface cleaning by outgassing, inert gas ion-bombardment and annealing. They went on to look at the effects of ion-bombardment damage on catalytic activity. In the hydrogenation of ethylene on nickel, for example, bombardment was found to increase the activity by a hundredfold over that of a well annealed surface [72, 73].

A comprehensive survey has been made (up to 1967) of studies of effects of various ionizing radiations on solid catalysts [74]. It is of interest to note that this survey emphasizes the role of point defects and does not mention atomic steps. It is only quite recently that LEED experiments have shown the importance of steps introduced via roughening of the surface topography.

An example of a more recent study of the effect of ion-bombardment on adsorption has been reported by Chung et al. [75] who used low-energy electron loss spectroscopy to examine acetylene and oxygen adsorbed on silicon. Evidence was found for stronger binding to the disordered surface.

The role of surface imperfections in oxide and sulphide formation has been the subject of several studies [76]. One of the most interesting observations concerns selective reactivity at subgrain boundaries in single crystals [77]. Even on carefully prepared specimens a "polygonized" structure can be formed (probably due to mechanical work during polishing). The subgrains may be disoriented by less than a minute of arc but they can be revealed by the action of an adsorbate that preferentially diffuses into the sub-boundary and provokes a change in the surface topography [78]. At a later stage, or at a higher chemical potential, compound formation occurs preferentially at these sites. This is an example of significant changes in chemical reactivity brought about by surface defect concentrations as low as 10^{-5}.

Burton, Cabrera and Frank's classic work on crystal growth and the role of dislocations and surface steps [79] led to methods for observing the growth patterns produced by steps. In particular methods of "decoration" were used based on enhanced chemical reactivity at steps [80] and on the ability of steps to provide nucleation sites in thin film growth. A large number of studies have been made of step decoration by metallic particles deposited in vacuum and observed by replica electron microscopy. (For first references see refs. [81–86].)

Bénard et al., using radioactive tracer techniques, obtained the first adsorption isotherms for single crystal surfaces for sulphur on silver [87]. Sharply varying, step-like, isotherms were obtained but it was noted that long "tails" in the low-pressure range could be ascribed to adsorption at defects. (It is interesting that these tails were especially important for the (110) face [59].) For this system a study was also made of the effect of atomic steps on different vicinal orientations. The variation of the adsorbed quantity with step density (at a chemical potential too low for adsorption on atomic terraces) suggests that the number of sites was larger than that predicted by the simple TLK model, either because of high kink densities or because steps influence adsorption over part of the terraces.

In addition, a very unusual effect of adsorption at steps is observed for sulphur on silver at temperatures above about 700°C. Conditions can be found for which the contribution of steps to the surface energy goes negative (Gibbs' adsorption equation). When this happens low-index and a range of vicinal surfaces become unstable and break up into conical pits and humps. The phenomenon is reciprocally analogous to thermal faceting and can be ascribed to a spontaneous increase in the step length [88,89].

Decoration by a suitable adsorbate might provide a general approach to monitoring defect concentrations if there is available a convenient method for measuring the adsorbed quantity [79]. The use of quantitative Auger spectroscopy with metal adsorbates that are unlikely to modify the substrate structure has been suggested [90] but first results are not encouraging [91]. It might be more interesting to look at work function changes during adsorption. It has been found, for example, that the surface potential of a physisorbed layer can be a sensitive indication of surface perfection [92]. Significant effects of defects on work function changes have also been observed recently for hydrogen on stepped platinum surfaces [25] and nitrogen on roughened tungsten field-emitter planes [93].

6. Recent work on adsorption on stepped surfaces

Examining crystals cut at different vicinal orientations is an obvious, potentially useful, approach to studying adsorption at steps. It appears that the preconceptions about the effects of the randomness of steps deterred experimenters from applying diffraction techniques until the late 1960s. Some early RHEED observations were made of adsorbed oxygen structures on copper single crystal spheres [94,95] but in this work, although apparently the whole stereographic triangle was explored, high temperature annealing in the presence of oxygen caused microfaceting and in fact only certain crystal planes were observed.

Observations by LEED in 1969 [34] showed that adsorption and a whole range of step properties can be readily investigated by applying available techniques to vicinal surfaces. Subsequent work using LEED and RHEED to monitor surface structure is reported in refs. [25,48,96–113]. Parallel with this work there have been investigations of vicinal surfaces using molecular beam techniques to follow the kinetic aspects of adsorption and reactivity [114–116]. The following conclusions are based mainly on the first set of cited papers.

Topographical stability of the substrate. Stepped surfaces may facet in the presence of an adsorbed layer. The facet orientations generally lie on low-index zones containing the original orientation, i.e. the process corresponds to a regrouping of the steps at closer spacings or even complete elimination of the terraces to form two sets of low-index orientations. Faceting is more likely to be due to adsorbed oxygen or carbon than to any other element. It can happen even at ambient temperatures but more usually occurs on heating. Desorption and annealing generally restores the original clean surface structure. In some cases the adsorbate can produce a reconstruction of the substrate structure with double or other multiple-height steps and corresponding increases in terrace widths [97]. Evidence has been found from RHEED experiments on tungsten [104] of short-range displacement fields at steps, induced by CO adsorption. One case has been observed (oxygen on copper [34]) where the adsorbate caused a rearrangement from a partially kinked stepped structure to a completely kinked structure (a sort of one-dimensional faceting). There is some evidence that bombardment damage and point imperfections are annealed out more rapidly on stepped surfaces than on low-index faces [34,17].

Ordering of adsorbed layers. Stepped surfaces have a greater tendency than low-index surfaces to produce disordered adsorbate structures. Frequently streaks are observed in the LEED patterns that correspond to a well ordered periodicity along the steps but to disorder in the perpendicular direction. This effect has been ascribed to disorder on the terraces [97] but it could also arise partly from a lack of registry between ordered domains on successive terraces — this in turn could be a consequence of the roughness of the steps and a high concentration of randomly distributed kinks. Steps might also promote the mobility of an adsorbed layer and in this way induce disorder. Some preliminary results suggest that this effect may have been observed for low-melting point metal overlayers [91]. On the other hand coincidence lattices indicating well ordered structures are also frequently observed. Their formation can be facilitated by a coincidence between the overlayer periodicity and the average terrace width. For certain adsorbates, and especially where lateral interactions dominate, the structure may be remarkably insensitive to the presence of steps: this appears to be the case for carbon layers [96], surface oxides [100] and inert gas overlayers [107,110]. In the adsorption of hydrocarbons the formation of ordered or disordered structures has been found to be quite complex and influenced by a number of competing factors each dependent on the step density [103].

Saturation coverage. Because of experimental difficulties little is known of how steps affect saturation coverages. Available data [25,107,110] suggest that variations are small, of the order 5%.

Sticking probabilities. Many qualitative observations have shown that sticking probabilities can be increased by steps and that the increases correlate with step densities [34,96,98]. Detailed quantitative measurements have been made only for a few systems. For oxygen on silicon [101,117] it has been found that a fourfold increase in step concentration causes the stricking probability to change exponentially over three orders of magnitude. This variation has been attributed to a non-local effect of the steps — the ability to dissociate molecules being spread out over the atomic terraces and related to a non-localised change in work function and a surface relaxation induced by the step [117].

For hydrogen adsorption on platinum the reported increase in sticking probability is only a factor of 4 for a change in step density of 11% [25,118].

Adsorption thermodynamics. Few quantitative data have been obtained. Two effects can be attributed to steps: the activation energy for dissociative adsorption can be lowered (so that adsorption that takes place only with difficulty on low-index surfaces is promoted by the presence of steps) [97,98,109] and the binding of the adsorbate to the stepped surface may be stronger. Adsorption isotherms obtained from work function measurements on palladium with adsorbed hydrogen show a small increase in the isosteric heat of adsorption at low coverages on stepped surfaces [48]. This increase can be ascribed to preferential adsorption at step sites. Similar investigations of hydrogen on platinum [25] also show evidence of higher heats of adsorption at steps and even the existence of two distinct sites — at the top and bottom of steps. AES investigations of segregated carbon on nickel also show higher binding at steps as well as evidence for differences in binding at different parts of the atomic terraces [106].

Bond breaking and reaction kinetics. Steps can produce significant changes in chemical reactivity. For example in the platinum catalysed dehydrocyclization of n-heptane to toluene steps can increase the yield by up to ten times [98]. Steps on platinum can break H–H, C–H and C–C bonds in adsorbed hydrocarbons (in order of increasing difficulty); kink sites appear to be about five times more effective than kinkless steps [109]. However, bond breaking is only one stage and in a complete reaction the role of steps may be quite complex [103]. Steps do not always produce changes in reactivity. For example in a molecular beam study of hydrogen on copper the behaviour of (100) and stepped (310) surfaces was nearly identical [115]. The hydrogen–deuterium exchange reaction on platinum is increased by steps although the effect appears to be less pronounced than originally reported [118].

Chemical specificity. A better understanding of the role of steps might be obtained by comparing different elements. With this in mind investigations have been made of gold, platinum and iridium stepped surfaces with the same structure [105, 111,112]. Results for various adsorbates are quite complex and not easily rationalized in a simple way. Small differences in electronic structure from one element to another appear to be as, or more, important than the atomic surface geometry [11]. Information on the electronic structure at stepped surfaces both from theory [125–127] and from photoemission and work function experiments [128,19] may help to unravel this problem.

7. Electron microscopy

The resolution from replication techniques in electron microscopy is limited to about 1.5 nm. This makes then unsuitable for examining atomic scale surface imperfections — except by the decoration methods already mentioned. Recently, however, much progress has been made in resolving surface detail in both transmission and reflection high-energy electron microscopy [129–135]. Part of this progress stems from the recognition of the need for ultrahigh vacuum conditions to reduce contamination.

By transmission microscopy images have been obtained of atomic steps, kinks and even adsorbed molecules on MgO particles [129–131,133] while atomic steps have been seen on gold [132,134] and platinum [134] films. Steps, possible of multiple atomic height, have been observed on copper using reflection microscopy [135].

These important developments offer the most direct means for observing individual imperfections on extended crystal surfaces and ultimately for looking at their role in dynamic surface processes.

8. Molecular and ionic beam studies

In addition to the work already cited on chemical reactivity and catalysis using molecular beams an interesting study has been made of the scattering of helium beams from stepped platinum surfaces [136]. Steps significantly alter the angular distribution of the scattered beam. By varying the azimuthal orientation of the surfaces it is shown that when the steps are perpendicular to the beam direction the angular distribution is strongly modulated (rainbow scattering) whereas in the parallel direction the scattering is specular. There is evidence that steps, and atomic-scale roughness in general, increase the energy exchange between the gas and the surface. Little is known about the energy transfer (accommodation) between a surface and the tangental component of an impinging beam: further studies on vicinal surfaces could be useful here.

Effects of defects can be observed in ion scattering experiments and the energy and angular distributions of reflected ions can be used to distinguish different types of imperfection [137]. Results for 6 keV krypton ions on Cu(100) show pronounced effected of atomic steps [138]. The technique may be limited in its usefulness for looking at defects by the fact that the ion beam itself may induce surface damage.

9. Electrochemical studies

Electroplating has been used as an alternative to vacuum deposition as a means of decorating steps on germanium [139] and silicon [140]. The surfaces were prepared by cleavage while immersed in the electrolyte, thus avoiding oxidation.

There is a growing interest in electrochemical studies of single crystal metal surfaces and marked effects of crystallographic orientation have been observed [141]. Some work has been done on stepped surfaces [142,143]. Many experimental techniques in this field could be developed to examine surface defects. For example the zero charge potential for single crystal gold electrodes immersed in NaF solutions is found to vary with crystallographic orientation and is very sensitive to steps near low-index orientations [143].

10. High temperature properties

In a typical experiment the means of detecting and controlling surface defects at ambient temperatures are usually uncertain. We turn now to the question of defects at thermal equilibrium at "high temperatures". Topographical surface equilibrium (i.e. the annealing out of accidental defects) can be reached in times that depend on the scale (transport distances) and on the various transport mechanisms [144]. From available mass transport data for metals we may arbitrarily define "high temperature" as above half the absolute melting temperature, T_m. In that range equilibrium can be reached on at least the micron scale in reasonable experimental times [145]. From the following survey it will be seen that we do not really know from experiment what equilibrium defect concentrations are, nor how they vary with temperature. There are strong indications that surfaces may be very rough on an atomic scale and that it is important to understand the dynamic behavious of defects.

The degree of atomic-scale roughening depends on the competition between low total bond energy and high configuration entropy. The theory of this cooperative phenomenon has been examined by many authors, generally using simple models of crystal structure and binding (e.g. simple cubic lattices and nearest neighbour bonds) (see ref. [79] for early references). The emphasis has been on establishing mathematical methods and on examining whether a discontinuous "surface roughening" transition is theoretically possible. Although some early results did not show evidence for a transition temperature they did predict quite high defect concentrations — up to 15% at the melting point [146]. As well as analytical approaches [79, 146–150] there have been many computer simulations using the Monte Carlo method [151–155]. The effect of atomic steps has been of particular interest in recent work [153] and it has been shown that a step can become very rough and its energy can effectively vanish at a critical temperature (expressed in terms of the bond strength) that would correspond to roughening. Only below this temperature does the surface have a well defined finite profile width. The computer simulations provide pictures of roughness [153] that are in dramatic contrast to the more familiar, relatively smooth, model of surfaces.

It is difficult to relate quantitatively these theoretical results to the properties of real crystal surfaces with much certainty. A large number of factors are not included in the calculations, for example: surface vibrations and their dependence on structure and contribution to the entropy; translational freedom of defects (surface diffusion) and its dependence on structure and entropy contribution; cluster formation; more realistic models for binding; interactions other than pairwise; effects of relaxation and electronic rearrangements at defects. The problem is one of the most formidable in surface physics.

Experimentally some information on roughening is available from a number of sources. The most direct evidence comes from some FIM observations on tungsten. Bassett [7] has found that heating *as low as 850 K* ($\sim 0.23\ T_m$) produces considerable surface rearrangements with extensive roughening of stepped and high-index surfaces and formation of steps of multiple atomic height. (After quenching for imaging low-index surfaces were smooth but it is not known if rapid diffusion on these surfaces permitted a more rapid return to a low temperature structure during quenching.) On the basis of homologous temperatures these results suggest that the surfaces of many metals might be quite rough at only a few hundred degrees centigrade.

Some observations on the melting of single crystal copper spheres may be evidence for the pre-melting of high-index surfaces below T_m [156]. During slow melting sharp differences were observed in the optical emissivity from different crystallographic regions. Areas of high-index orientations appeared bright (molten) while darker discs appeared around low-index poles. Such differences are likely to be caused by a thick liquid layer and an alternative explanation to surface roughening could be that nucleation of the liquid phase is easier on the atomically rough surfaces. However, nucleation of bulk liquid could be via the prior formation of a "two-dimensional" liquid.

Sharp disordering or melting transitions have been recorded from LEED observations of metal monolayers deposited on metals. Monolayers can melt at temperatures significantly below the bulk melting point [90]. No systematic study using LEED intensity measurements has yet been made for clean surfaces. Some visual observations [157] on lead, bismuth and tin crystals showed that LEED patterns can be visible right up to the bulk melting point — but this does not prove that the top layer is not disordered. Some NMR studies [158] of organic monolayers and multiple layers on silver and silica showed lowered melting points for the thin layers with the bulk melting temperature being reached at four to ten layers.

The problem of surface roughness is closely linked to the question of the shape of the γ-plot (variation of surface energy (work) with crystallographic orientation). The absence of a cusp at any particular orientation is a necessary consequence of roughening. γ-plots have been obtained for several metals from studies of equilibrium topographies formed at high temperatures (see ref. [159] for references). The observed variations of γ are small — a few percent — and evidence is found for cusps only at low-index orientations. From a particularly detailed study of copper in the range $0.8\ T_m$ to $0.97\ T_m$ McLean [159] showed that the anisotropy of γ decreases with temperature. He was able to obtain both the step energies and — for the first time — step entropies. From these data it can be concluded that the steps are rough in this temperature range. Similar evidence for roughening comes from work on platinum for which it is thought that the critical temperature could also be as low as $0.8\ T_m$ [160]. The evidence from various γ-plots suggests that vicinal and high-index surfaces can roughen but that the low-index planes do not go through a transition temperature below the bulk melting point.

Direct observations of the equilibrium shape of inclusions (negative crystals) in diphenyl also show evidence of surface melting well below the bulk melting point for all but the most dense face [161].

Finally, strong evidence that surfaces are not only rough in the static structural sense but also in a dynamic sense comes from experimental results on surface self-diffusion. This evidence has been discussed elsewhere [162]. Near the melting point surface diffusivities are more characteristic of two-dimensional liquids or dense gases than solids. A more complete understanding of high temperature surface properties will require much more information not only on defect concentrations but also on the dynamics of their interactions, association and migration.

11. Conclusions

The subject of defects cuts across the whole field of surface science and a fuller understanding of defect properties will require information from many different kinds of experiment. There is a need for an agreed experimental protocol to deal with the problem of characterising the degree of perfection of specimen surfaces. Without routine methods for monitoring defects the reproducibility of many results may be doubtful and agreement between theory and experiment may be sometimes only fortuitous. (There is a tendency to overlook this problem because the question of reproducibility is often not fully examined).

Of the available methods that could be used for assessing and monitoring surface perfection the routine measurement of the work function appears to be the most promising. The application of LEED optics to such measurements offers a simple practical approach in many experiments [19]. Work function changes on clean substrates can give indications of changes in perfection but without identifying the type of defect. It is possible that by following work function changes during adsorption at defect sites (decoration) different types of site could be distinguished [25]. For studies of the solid/liquid interface and for the assessment of methods of surface preparation, prior to studies under vacuum, electrochemical methods should be worth developing.

Of the many problems to be solved some of the most interesting are concerned with the detailed structure of atomic steps — the relaxation of lattice positions, extended influence over atomic terraces, electronic rearrangements, step roughening and the properties of kink sites. Experimental investigations of high-temperature surface roughening is another challenging topic.

But above all there needs to be initiated a wide-ranging programme of *systematic* studies of defect properties based on controlled variations in surface perfection.

References

[1] For articles on the role of defects in crystal growth phenomena see for examples refs. [79,163–167].

[2] For a recent review of field electron emission, field-ionization and field-evaporation

techniques see:
M.J. Southon, E.D. Boyes, P.J. Turner and A.R. Waugh, Surface Sci. 53 (1975) 554.
[3] E.W. Müller and T.T. Tsong, Field-Ion Microscopy: Principles and Applications (Elsevier, New York, 1969).
[4] T.E. Schmid and R.W. Balluffi, Surface Sci. 28 (1971) 32.
[5] D.W. Bassett, J. Phys. C9 (1976) 2491.
[6] K. Stolt, W.R. Graham and G. Ehrlich, J. Phys. Chem. 65 (1976) 3206.
[7] D.W. Bassett, Proc. Roy. Soc. (London) A286 (1965) 191.
[8] A.R. Waugh, E.D. Boyes and M.J. Southon, Surface Sci. 61 (1976) 109.
[9] C. Herring and M.H. Nichols, Rev. Mod. Phys. 21 (1949) 185.
[10] R. Smoluchowski, Phys. Rev. 60 (1941) 661.
[11] J.C. Rivière, Solid State Surface Sci. 1 (1969) 179.
[12] E.W. Plummer and T.N. Rhodin, Appl. Phys. Letters 11 (1967) 194.
[13] E.W. Plummer and T.N. Rhodin, J. Chem. Phys. 49 (1968) 3479.
[14] G.A. Haas and R.E. Thomas, J. Appl. Phys. 40 (1969) 3919.
[15] W. Korner, Vide 163(5) (1973) 75.
[16] K. Besocke and H. Wagner, Phys. Rev. B8 (1973) 4597.
[17] K. Besocke and H. Wagner, Surface Sci. 53 (1975) 351.
[18] B. Krahl-Urban, E.A. Niekisch and H. Wagner, Surface Sci. 64 (1977) 52.
[19] K. Besocke, B. Krahl-Urban and H. Wagner, Surface Sci. 68 (1977) 39.
[20] R.L. Park and H.H. Madden, Surface Sci. 11 (1968) 188.
[21] H.E. Farnsworth and H.H. Madden, J. Appl. Phys. 32 (1961) 1933.
[22] P.O. Gartland, S. Berge and B.J. Slagsvold, Phys. Rev. Letters 30 (1973) 916.
[23] S. Berge, P.O. Gartland and B.J. Slagsvold, Surface Sci. 43 (1974) 275.
[24] J.K. Grepstad, P.O. Gartland and B.J. Slagsvold, Surface Sci. 57 (1976) 348.
[25] K. Christmann and G. Ertl, Surface Sci. 60 (1976) 365.
[26] G.A. Haas and R.E. Thomas, J. Appl. Phys. 48 (1977) 86.
[27] J.W. May, Ind. Eng. Chem. 57 (1965) 19.
[28] C.W. Tucker, J. Appl. Phys. 38 (1967) 1988.
[29] C.W. Tucker, Acta Met. 15 (1967) 1465.
[30] A.J.W. Moore, in: Metal Surfaces, ASM-AIME Symposium (Am. Soc. Metals, Cleveland, Ohio, 1963).
[31] R.L. Jacobson and G.K. Wehner, J. Appl. Phys. 36 (1965) 2674.
[32] R.L. Park, J. Appl. Phys. 37 (1966) 295.
[33] H.E. Farnsworth and K. Hayek, Surface Sci. 8 (1967) 35.
[34] G.E. Rhead and J. Perdereau, Compt. Rend. (Paris) C269 (1969) 1183, 1261, 1425.
G.E. Rhead and J. Perdereau, in: Structure et Propriétés des Surfaces des Solides, Colloque CNRS, 1969 (CNRS, Paris, 1970).
[35] A.M. Mattera, R.M. Goodman and G.A. Somorjai, Surface Sci. 7 (1967) 26.
[36] F. Jona, Surface Sci. 8 (1967) 478.
[37] F. Jona, in: Surfaces and Interfaces. I. Chemical and Physical Characteristics (Syracuse Univ. Press, 1967).
[38] A. Guinier, X-ray Diffraction (Freeman, San Francisco, 1963).
[39] G.W. Stroke, Progr. Opt. 2 (1963), 3, 68.
[40] H. Lipson and C.A. Taylor, Fourier Transforms and X-ray Diffraction (Bell, London, 1958).
[41] C.A. Taylor and H. Lipson, Optical Transforms (Bell, London, 1964).
[42] R. Heckingbottom, Surface Sci. 17 (1969) 394.
[43] W.P. Ellis and R.L. Schwoebel, Surface Sci. 11 (1968) 82.
[44] M. Henzler, Bull. Am. Phys. Soc. Ser. II, 14 (1969) 794; Surface Sci. 19 (1970) 159.
[45] W.P. Ellis, in: Optical Transforms, Ed. H. Lipson (Academic Press, 1972);
W.P. Ellis, Surface Sci. 45 (1974) 569.

[46] M. Henzler, Appl. Phys. 9 (1976) 11.
[47] B. Lang, R.W. Joyner and G.A. Somorjai, Surface Sci. 30 (1972) 440.
[48] H. Conrad, G. Ertl and E.E. Latta, Surface Sci. 41 (1974) 435.
[49] K. Besocke and H. Wagner, Surface Sci. 52 (1975) 653.
[50] J.E. Houston and R.L. Park, Surface Sci. 26 (1971) 269.
[51] J.M. Cowley and H. Shuman, Surface Sci. 38 (1973) 53.
[52] K. Fujiwara, J. Phys. Soc. Japan. 12 (1957) 7.
[53] F. Hottier, J.B. Theeten, A. Masson and J.L. Domange, Surface Sci. 65 (1977) 563.
[54] M. Henzler and J. Clabes, Japan. J. Appl. Phys. Suppl. 2 Part 2 (1974) 389.
[55] J.E. Houston, G.E. Laramore and R.L. Park, Surface Sci. 34 (1973) 477.
[56] G.E. Laramore, J.E. Houston and R.L. Park, J. Vacuum Sci. Technol 10 (1973) 196.
[57] D. Aberdam, R. Baudoing, C. Gaubert and E.G. McRae, Surface Sci. 57 (1976) 715.
[58] F. Jona, J.A. Strozier and C. Wong, Surface Sci. 30 (1972) 225.
[59] The (110) surfaces of fcc metals have given unexplained results in several instances. The anomalous structures of (110) Au and Pt have not yet been satisfactorily explained. This crystallographic orientation is probably represented by saddle points in the γ-plot and the topographical stability on an atomic scale may be anomalous.
[60] E. Zanazzi, F. Jona, D.W. Jepsen and P.M. Marcus, J. Phys. C (1977) in press.
[61] J.P. Bibérian, Thesis, Université Pierre et Marie Curie (Paris VI) (1975).
[62] M.G. Barthes, unpublished research.
[63] W.P. Ellis, J. Chem. Phys. 48 (1968) 5695.
[64] H.S. Taylor, Proc. Roy. Soc. (London) A108 (1925) 105.; J. Phys. Chem. 30 (1926) 145.
[65] H. Leidheiser and A.T. Gwathmey, J. Am. Chem. Soc. 70 (1948) 1200, 1206.
[66] T.N. Rhodin, J. Am. Chem. Soc. 72 (1950) 4343, 5102, 5691.
[67] T.N. Rhodin, J. Am. Chem. Soc. 73 (1951) 3143.
[68] T.N. Rhodin, J. Appl. Phys. 21 (1950) 971.
[69] As early as 1934–5 Gwathmey, in spite of great technical difficulties, tried to use LEED to look at surface anisotropy of catalytic activity [70].
[70] A.T. Gwathmey and R.E. Cunningham, Advan. Catalysis 10 (1958) 57.
[71] H.E. Farnsworth, R.E. Schlier, T.H. George and R.M. Burger, J. Appl. Phys. 26 (1955) 252.
[72] R.K. Sherburne and H.E. Farnsworth, J. Chem. Phys. 19 (1951) 387.
[73] H.E. Farnsworth and R.F. Woodcock, Advan. Catalysis 9 (1957) 123.
[74] E.H. Taylor Advan. Catalysis 18 (1968) 111.
[75] Y.W. Chung, W. Siekhaus and G.A. Somorjai, Surface Sci. 58 (1976) 341.
[76] J. Bénard, in: Proc. 4th Intern. Symp on the Reactivity of Solids (Elsevier. Amsterdam, 1960) p. 362,
J. Bénard, Met. Rev. 9 (1964) 473.
[77] J. Oudar, Mém. Sci. Rev. Metall. 62 (1965) 47.
[78] J. Perdereau and G.E. Rhead, Compt. Rend. (Paris) 262 (1966) 257.
[79] W.K. Burton, N. Cabrera and F.C. Frank, Phil. Trans. A243 (1951) 299.
[80] A.J. Forty and F.C. Frank, Proc. Roy. Soc. (London) A217 (1953) 262.
[81] G.A. Bassett, Phil. Mag. 3 (1958) 1042.
[82] C. Sella, P. Conjeaud and J.J. Trillat, in: 4ème Congres Interns. Microscopie Electronique, Vol. 1 (1958) p. 508.
[83] H. Bethge, Phys. Status Solidi 2 (1962) 3, 375.
[84] H. Bethge, Surface Sci. 3 (1965) 33.
[85] J.G. Allpress and J.V. Sanders, Phil. Mag. 9 (1964) 645.
[86] A. Masson and R. Kern, J. Crystal Growth 2 (1968) 227.
[87] J. Bénard, J. Oudar and F. Cabané-Brouty, Surface Sci. 3 (1965) 359.
[88] G.E. Rhead and J. Perdereau, Acta Met. 14 (1966) 448.

[89] J. Perdereau and G.E. Rhead, Acta Met. 16 (1968) 1267.
[90] G.E. Rhead, J. Vacuum Sci. Technol. 13 (1976) 603.
[91] A. Sepulveda and G.E. Rhead, Surface Sci. 66 (1977) 436.
[92] D.F. Klemperer and J.C. Snaith, Surface Sci. 28 (1971) 209.
[93] S.P. Singh-Bopoarai and D.A. King, Surface Sci. 61 (1976) 275.
[94] L. Trepte, Chr. Menzel-Koop and E. Menzel, Surface Sci. 8 (1967) 223.
[95] L. Trepte, Z. Natursorschung 23a (1968) 1241.
[96] J. Perdereau and G.E. Rhead, Surface Sci. 24 (1971) 555.
[97] B. Lang, R.W. Joyner and G.A. Somorjai, Surface Sci. 30 (1972) 454.
[98] G.A. Somorjai, R.W. Joyner and B. Lang, Proc. Roy. Soc. (London) A331 (1972) 335.
[99] R.W. Joyner, B. Lang and G.A. Somorjai, J. Catalysis 27 (1972) 405.
[100] F. Grønlund and P.E. Højlund Nielsen, Surface Sci. 33 (1972) 399.
[101] H. Ibach, K. Horn, R. Dorn and H. Lüth, Surface Sci. 38 (1973) 433.
[102] D.R. Kahn, E.E. Petersen and G.A. Somorjai, J. Catalysis 34 (1974) 291.
[103] K. Baron, D.W. Blakely and G.A. Somorjai, Surface Sci. 41 (1974) 45.
[104] K.J. Matysil, Surface Sci. 46 (1974) 457.
[105] M.A. Chesters and G.A. Somorjai, Surface Sci. 52 (1975) 21.
[106] L.C. Isett and J.M. Blakely, J. Vacuum Sci. Technol. 12 (1975) 237.
[107] H. Papp and J. Pritchard, Surface Sci. 53 (1975) 371.
[108] E. Legrand-Bonnyns and A. Ponslet, Surface Sci. 53 (1975) 675.
[109] G.A. Somorjai and D.W. Blakely, Nature 258 (1975) 580.
[110] R.H. Roberts and J. Pritchard, Surface Sci. 54 (1976) 687.
[111] D.I. Hagen, B.E. Nieuwenhuys, G. Rovida and G.A. Somorjai, Surface Sci. 57 (1976) 632.
[112] B.E. Nieuwenhuys, D.I. Hagen, G. Rovida and G.A. Somorjai, Surface Sci. 59 (1976) 155.
[113] J.E. Demuth and D.E. Eastman, J. Vacuum Sci. Technol. 13 (1976) 283.
[114] S.L. Bernasek, W.J. Siekhaus and G.A. Somorjai, Phys. Rev. Letters 30 (1973) 1202.
[115] M. Balooch, M.J. Cardillo, D.R. Miller and R.E. Stickney, Surface Sci. 46 (1974) 358.
[116] S.L. Bernasek and G.A. Somorjai, Surface Sci. 48 (1975) 204.
[117] H. Ibach, Surface Sci. 53 (1975) 444.
[118] There has been some controversy over this system. Earlier work suggested a much greater influence of steps. See ref. [25] for a discussion and refs. [97,114,116,119–124].
[119] J.A. Joebstl, J. Vacuum Sci. Technol. 12 (1975) 347.
[121] K.E. Lu and R.R. Rye, Surface Sci. 45 (1974) 677.
[122] K.E. Lu and R.R. Rye, J. Vacuum Sci. Technol. 12 (1975) 334.
[123] R.J. Madix, J. Vacuum Sci. Technol. 13 (1976) 253.
[124] I.E. Wachs and R.J. Madix, Surface Sci. 58 (1976) 590.
[125] 3L.L. Kesmodel and L.M. Falicov, Solid State Commun. 16 (1975) 1201.
[126] Y.W. Tsang and L.M. Falicov, J. Phys. C9 (1976) 51.
[127] M.C. Desjonquères and F. Cyrot-Lackman, Solid State Commun. 18 (1976) 1127.
[128] J.E. Rowe, S.B. Christman and H. Ibach, Phys. Rev. Letters 34 (1975) 874.
[129] A.F. Moodie and C.E. Warble, Phil. Mag. 16 (1967) 891.
[130] A.F. Moodie and C.E. Warble, J. Crystal Growth 10 (1971) 26.
[131] A.F. Moodie and C.E. Warble, in' Proc. 8th Intern. Congr. on Electron Microscopy, Canberra, 1974, Vol. I, p. 230.
[132] D. Cherns, Phil. Mag. 30 (1974) 549.
[133] A.F. Moodie and C.E. Warble, in: Materials Science Research, Vol. 10, Sintering and Catalysis, Ed. J.C. Kuczynski (Plenum, New York, 1975) p. 1.
[134] R.L. Hines, Thin Solid Films 35 (1975) 229.
[135] P.E. Højlund Nielsen and J.M. Cowley, Surface Sci. 54 (1976) 340.

[136] S.T. Ceyer, R.J. Gale, S.L. Bernasek and G.A. Somorjai, J. Chem. Phys. 64 (1976) 1934.
[137] E.P. Th.M. Suurmeijer and A.L. Boers, Surface Sci. 43 (1973) 309.
[138] S.H.A. Begemann and A.L. Boers, Surface Sci. 30 (1972) 134.
[139] W. Mehl and M.D. Coutts, J. Appl. Phys. 34 (1963) 2120.
[140] M.D. Coutts and A.G. Revesz, J. Appl. Phys. 37 (1966) 3280.
[141] J.W. Schultze and D. Dickertmann, Surface Sci. 54 (1976) 489.
[142]3A. Hamelin and J.P. Bellier, Electroanal. Chem. and Interfacial Electrochem. 41 (1973) 179.
[143] A. Hamelin and J. Lecoeur, Surface Sci. 57 (1976) 771.
[144] H.P. Bonzel, in: Structure and Properties of Metal Surfaces, Ed. S. Shimodaiva (Maruzen, Tokyo, 1973) p. 248.
[145] G.E. Rhead, in: Electronic Structure and Reactivity of Metal Surfaces, Eds. E.G. Derouane and A.A. Lucas (Plenum, New York, 1976) p. 229.
[146] W.W. Mullins, Acta Met. 7 (1959) 746.
[147] E.E. Gruber and W.W. Mullins, J. Phys. Chem. Solids 28 (1967) 875.
[148] J.D. Weeks, G.H. Gilmer and H.J. Leamy, Phys. Rev. Letters 31 (1973) 549.
[149] R.A. Hunt and B. Gale, J. Phys. C7 (1974) 507.
[150] G.H. Gilmer, K.A. Jackson, H.J. Leamy and J.D. Weeks, J. Phys. C7 (1974) L 123.
[151] H.J. Leamy and K.A. Jackson, J. Appl. Phys. 42 (1971) 2121.
[152] R.A. Hunt and B. Gale, J. Phys. C6 (1973) 3571.
[153] H.J. Leamy, G.H. Gilmer and K.A. Jackson in: Surface Physics of Materials, Vol. I, Ed. J.M. Blakely (Academic Press, New York, 1975).
[154] R.A. Hunt and B. Gale, Surface Sci. 61 (1976) 241.
[155] R.H. Swendsen, Phys. Rev. Letters 37 (1976) 1478.
[156] K.D. Stock and E. Menzel, Surface Sci. 61 (1976) 272.
[157] R.M. Goodman and G.A. Somorjai, J. Chem. Phys. 52 (1970) 6325, 6331.
[158] G. Karagounis, E. Papayannakis and C.I. Stassinopoulos, Nature 221 (1969) 655.
[159] M. McLean, Acta Met. 19 (1971) 387.
[160] M. McLean and H. Mykura, Surface Sci. 5 (1966) 466.
[161] A. Pavlovska and D. Nenow, Surface Sci. 27 (1971) 211.
[162] G.E. Rhead, Surface Sci. 47 (1975) 207.
[163] R.L. Schwoebel and E.J. Shipsey, J. Appl. Phys. 37 (1966) 3682.
[164] R.L. Schwoebel, J. Appl. Phys. 38 (1967) 3154.
[165] R.L. Schwoebel, J. Appl. Phys. 40 (1969) 614.
[116] R. Ghez and G.H. Gilmer, J. Crystal Growth 21 (1974) 93.
[167] C. van Leeuwen, R. van Rosmalen and P. Bennema, Surface Sci. 44 (1974) 213.

DIPOLE MOMENTS ASSOCIATED WITH EDGE ATOMS; A COMPARATIVE STUDY ON STEPPED Pt, Au AND W SURFACES

K. BESOCKE, B. KRAHL-URBAN and H. WAGNER

Institut für Grenzflächenforschung und Vakuumphysik, Kernforschungsanlage Jülich, 5170 Jülich, Germany

Work function measurements have been performed on stepped Pt and Au surfaces with (111) terraces and on W surfaces with (110) terraces. In each case the work function decreases linearly with increasing step density and depends on the step orientation. The work function changes are attributed to dipole moments associated with the step edges. The dipole moments per unit step length are larger for open edge structures than for densely packed ones. The dipole moments for Pt are about twice as large as for Au and W.

1. Introduction

The structural arrangement of atoms in the outermost planes of solid surfaces affects the surface dipole contribution to the electron work function. Work function values for various orientations of single crystal surfaces may thus differ by several tenth of an eV. Densely packed planes exhibit higher values than planes with more open structures. Smoluchowski [1] pointed out already back in 1941 that the physical reason for this crystallographic dependence consists in a "smoothing effect" of the electron charge distribution in the topmost surface layer. By this effect negative charges accumulate between surface atoms leading to a dipole contribution which tends to decrease the work function. Recent self consistent work function calculations [2] lend support to this physical understanding.

One may expect that the electron charge distribution is altered in the vicinity of atomic steps on surfaces as compared to the distribution around atoms located in terraces. The "smoothing effect" of the electrons should lead quite similarly to dipole moments associated with atoms in edge positions which causes a lowering of the work function below the value of a flat surface. In a recent publication [3] Krahl-Urban et al. showed that the work function of W single crystal planes with terraces of [110] direction decreases linearly with increasing step density. This result could be interpreted in terms of additional dipole moments connected with step atoms. Different dipole moments per unit step length have been obtained for different edge structures. i.e. for step edges along different crystallographic directions.

An increasing number of publications deal with the modified properties of single crystal surfaces due to steps especially as far as adsorption and catalytic reaction kinetics are concerned. Henzler [4] has recently reviewed this subject. Attempts have been made to correlate the altered adsorption kinetics due to steps with corresponding work function changes [5,6] and to rationalize enhanced catalytic activities associated with steps in terms of an electron charge transfer from edge atoms [7,8].

In view of these implications it seemed desirable to investigate the influence of steps on the work function of various metals and to deduce correspondingly dipole moments connected with the studied edge structures. The following paper describes work function measurements on stepped Pt, Au and W single crystal surfaces as function of step density and step structure. For all three metals a linear decrease with step density was found from which dipole moments per unit step length could be derived. Different step structures led to different dipole moments for a given metal. Considerable differences between dipole moments exist among the various metals. The results are discussed qualitatively by considering existing theoretical models.

2. Experimental

The platinum and gold surfaces were prepared from MRC single crystals of 6 mm diameter. By suitable spark erosion machining the surfaces were given cylindrical shapes in order to produce a locally varying step density in one direction. The cylinder radius was about 20 mm. The samples were oriented by Laue back reflection in such a way that the cylinder axis coincided with the $[1\bar{1}0]$ direction and the middle part of the cylinder surface consisted of the (111) plane as indicated in fig. 1. The exposed surfaces encountered by moving across the cylinder belong to the $(01\bar{1})$ zone and are inclined towards the (111) plane up to about 7° on both sides. After erosion the surfaces were gently polished by hand and mounted in an UHV chamber equipped with LEED and Auger optics, mass spectrometer, ion sputter gun and gas inlets. Turbo and ion getter pumping provided a base pressure in the 10^{-9} Pa region. The samples were annealed in an oxygen atmosphere of 10^{-5} Pa

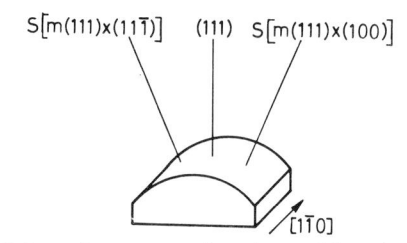

Fig. 1. Schematic representation of curved Pt and Au sample.

for about 10 h at 900°C and 700°C in the case of Pt and Au, respectively, to remove carbon the main contaminant. Subsequent argon ion sputtering at around 500°C over several hours effectively cleaned the surfaces of calcium. Auger spectroscopy revealed that C and Ca impurities were less than 0.5% of a monolayer as estimated by the procedure of Palmberg et al. [9].

Local work function changes across the cylinder surface were obtained from corresponding changes of the low energy cut off of the secondary electron spectrum. For this purpose the sample surface was moved perpendicular to the cylinder axis and to an impinging electron beam from the central LEED gun having a diameter of about 0.2 mm and an acceleration voltage of 200 V. The first derivative of the secondary electron spectrum was recorded using the retarding field analyzer of the LEED system with a modulation voltage of 0.02 V. Local changes in work function $\Delta\phi$ give rise to shifts of the derivative plots with regard to the retarding voltage. The shift equals the work function change $\Delta\phi$. The method yields a local resolution of about 0.2 mm and a sensitivity for relative work function changes of 0.02 eV.

In the case of tungsten 7 flat samples have been prepared. Four samples of (110), (10 9 0), (650) and (750) orientation, respectively belong to the (001) zone and exhibit (110) terraces but different step densities. The steps run parallel to the [001] direction. Two further samples with (110) terraces have steps parallel to the [110] and [111] directions and step densities close to the one of the (650) surface (4×10^6 cm^{-1}). Absolute work function measurements were carried out using thermionic emission. Details of the sample preparation, experimental set up and measuring procedure have already been published [3].

3. Results

3.1. LEED observations

The flat W surfaces inclined various degrees towards the (110) plane showed the characteristic LEED features of stepped surfaces, i.e. spot splitting at distinct beam voltages [10]. Fig.2 shows as an example a LEED pattern of the clean W(750) surface at 90 V with splitted (11) reflexes. The evaluation of the LEED patterns reveal that all investigated W surfaces are formed by terraces of (110) orientation separated by monoatomic steps. The ratio of the step height to the average terrace width for each sample surface as determined by LEED [10] gives the tangent of the inclination angle towards the (110) plane. These values agree within experimental errors with the iclination angles as obtained by Laue back reflection.

In the case of the curved Pt and Au surfaces the LEED patterns showed again spot splitting at distinct beam voltages. The splitting separation increased continuously by moving the incident beam from the center line of the cylinder surface across the surface. Fig. 3 shows a LEED pattern of the Au surface taken at 125 V and a position where the average surface direction forms an angle of about 7° with

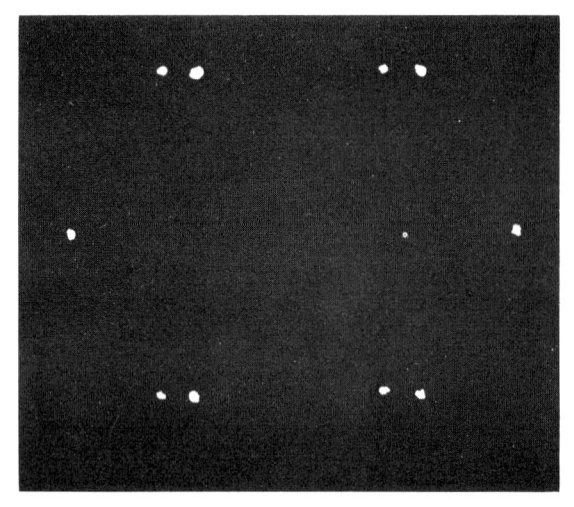

Fig. 2. LEED pattern from W(750) surface at 90 V showing spot splitting of (11) reflections.

the normal to the (111) plane. The LEED patterns and their evaluation indicate that the curved Pt and Au surfaces are formed by (111) terraces separated by monoatomic steps and that the terrace width decreases continuously (increasing step density) with increasing inclination angle towards the (111) plane. The inclination angle as function of surface position was measured by an optical microscope.

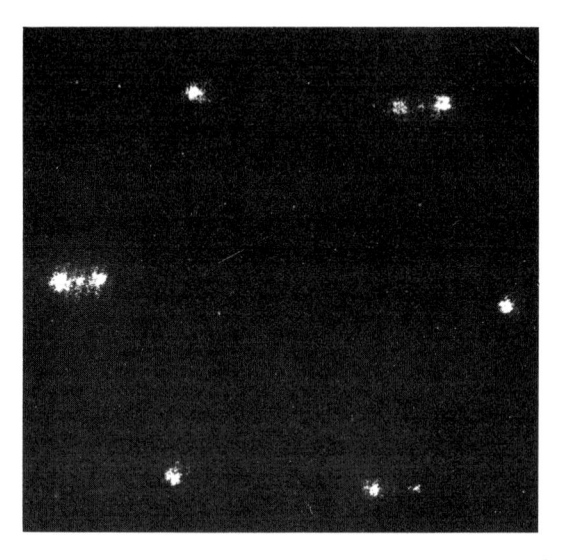

Fig. 3. LEED pattern from part of the curved Au surface inclined about 7° towards the (111) plane taken at 125 V.

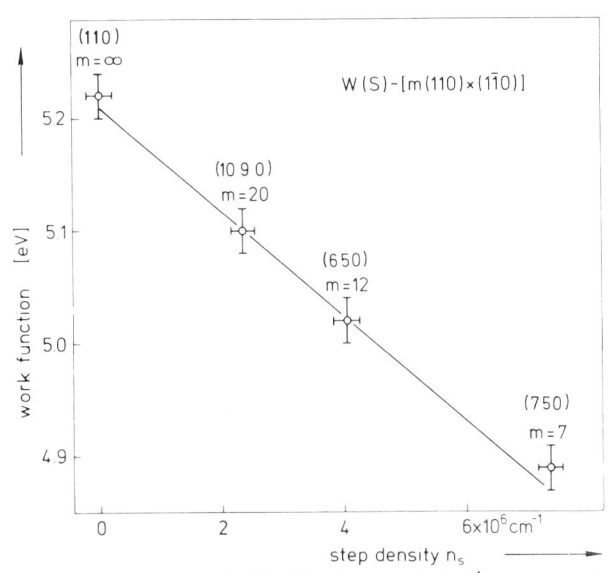

Fig. 4. Work function versus step density for W surfaces with (110) terraces and steps parallel to the [001] direction.

Inclination angle and step height yield the step density which varied across the surface between zero and about 3.5×10^6 cm^{-1}.

3.2. Work function

Fig. 4 reproduces the results [3] of the work function measurements for the W surfaces belonging to the (001) zone, i.e. the steps are parallel to the [001] direc-

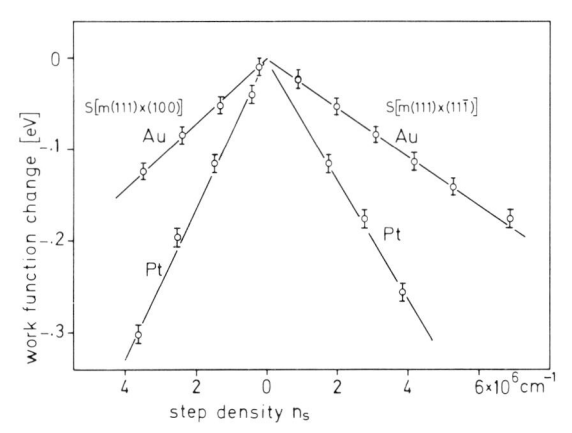

Fig. 5. Relative work function change versus step density for Pt and Au surfaces with (110) terraces and steps parallel to the [1$\bar{1}$0] direction.

[1̄1̄1] →   [1̄10] →   [001] →

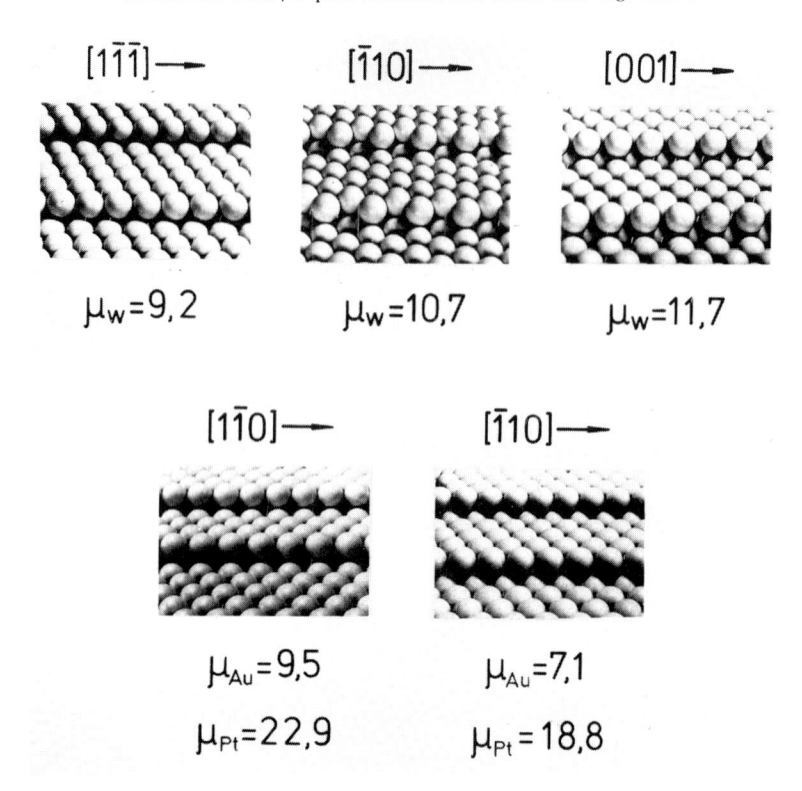

$\mu_W = 9.2$ $\mu_W = 10.7$ $\mu_W = 11.7$

[11̄0] →   [1̄10] →

$\mu_{Au} = 9.5$ $\mu_{Au} = 7.1$
$\mu_{Pt} = 22.9$ $\mu_{Pt} = 18.8$

Fig. 6. Hard sphere models of edge structures investigated on (a) W surfaces and (b) Pt and Au surfaces. The dipole moments per step length are given in 10^6 D cm^{-1}.

tion. The Miller indices of the investigated surface planes are indicated in the figure.

The work function decreases linearly with step density. For the W(10 9 1) plane with a step density of 4.05×10^6 cm^{-1} and steps parallel to the [1̄11] direction a work function of 5.08 ± 0.02 eV was obtained. The corresponding value for the (15 15 2) plane with steps parallel to the [1̄10] direction and a step density of 4.20×10^6 cm^{-1} was 5.05 ± 0.02 eV.

Fig. 5 shows the relative work function changes with step density for the curved Pt and Au surfaces, respectively. For both metals the work function decreases again linearly with increasing step density. The slopes for the two directions from the central part of the curved surface are however different. Because of the trigonal symmetry of the (111) plane the two directions are not equivalent. Although the step edges are parallel to the [11̄0] direction on both sides the step orientations are different: [100] on one side and [111̄] on the other. This distinction is indicated in fig. 5 by the respective characterization of stepped surfaces [11]. The two different step structures are shown in fig. 6.

4. Discussion

The change in work function with surface orientation can be correlated with the respective step structures of the vicinal surfaces. The contribution of random steps will be neglected by assuming that the density of random steps is small compared to the density of steps on the vicinal planes and that it might not depend strongly on surface orientation. A potential influence of random steps on the LEED patterns [12] has not been observed.

The linear decrease of the work function as function of step density for all three investigated metals Pt, Au and W can be interpreted by attributing additional dipole moments to the atomic steps of the vicinal step structures. We are therefore going to evaluate the work function results in terms of dipole moments per unit length of a step using the Helmholtz equation,

$$\Delta\phi = 300 \times 10^{-18} 4\pi n\mu ,$$

with $\Delta\phi$ in V, n the number of steps per cm and μ the dipole moment per cm step length in Debye. Fig. 6 summarizes the results obtained and illustrates by hard-sphere models the corresponding edge structures.

Let us first consider the magnitude of the dipole moments and compare it with dipole moments of adsorbed electropositive atoms on metal surfaces which decrease the work function too. For this comparison we do not relate the dipole moment to a step length of 1 cm but rather to the average interatomic distance. We obtain values of about 0.6 D for Pt and about 0.3 D for W. For W atoms adsorbed on the W(110) plane Besocke and Wagner [13] obtained a dipole moment of 1 D which is about a factor of 3 larger than the corresponding value for a W atom in an edge position. In the same paper a dipole moment for an average edge position of a W island of 0.3 D was derived in excellent agreement with the present results on stepped W surfaces. For the electropositive alkali atoms Na, K, Cs on various metal planes values of dipole moments ranging between 1.5 to 5 D [14] have been measured with Na yielding the lower and Cs the higher values. The adsorbed alkali atoms are expected to be nearly completely ionized in the low coverage limit. K has an ionic radius of 1.33 Å which is pretty close to the atomic radius of W and Pt. K atoms adsorbed on the close packed W(110) and Ta(110) planes give rise to a dipole moment of about 3.3 D caused by a charge transfer of 0.8 electron charges [14]. According to these values an adsorbed W atom on the W(110) plane should carry a positive charge of 0.24 e and an edge atom correspondingly a charge of somewhat less than 0.1 e. These values should be a factor of two larger for Pt as indicated by the obtained dipole moments.

As far as the dependence of the dipole moment on the edge structure is concerned we find, in the case of W, that the close packed step parallel to the [111] direction exhibits a smaller dipole moment than the more open steps. The step formed by the more open (100) plane in the case of Pt and Au yields a larger dipole moment than the step formed by the (111) plane. These observations are in qualitative agreement with Smoluchowski's argument [1] that the more open structures cause a larger smoothing effect of the electron charge distribution.

Let us finally compare the dipole moments among the different metals. The Au and W values are roughly the same whereas for Pt the values are about two times larger. We may compare the present experimental results with theoretical investigations by Kesmodel, Tsang and Falicov [8,15]. These authors consider an electronic charge transfer from d orbitals of atoms located in edge positions. The charge transfer is caused by the electronic potential due to the s-like electrons confined in a 90° metal wedge [15]. The states of the s-like electrons treated as free electrons act as a reservoir for the charge transfer from the localized d orbitals. The local charge transfer is considered to be proportional to the density of d states at the Fermi level. This density is much larger for Pt than for both W and Au. Therefore, the charge tranfer and hence the dipole moment should be larger for Pt than for W and Au. This finding is in qualitative agreement with our experimental results.

A quantitative comparison is certainly not very meaningful at present because of the following reasons: (i) the theoretical derivation of the charge transfer is a very crude one and does not take into account self-consistency; (ii) the proper dipole length necessary for calculating the dipole moment is not known.

References

[1] R. Smoluchowski, Phys. Rev. 60 (1941) 661.
[2] N.D. Lang and W. Kohn, Phys. Rev. B3 (1971) 1215.
[3] B. Krahl-Urban, E.A. Niekisch and H. Wagner, Surface Sci. 64 (1977) 52.
[4] M. Henzler, Appl. Phys. 9 (1976) 11.
[5] H. Ibach, Surface Sci. 53 (1975) 444.
[6] H. Hopster, H. Ibach and G. Comsa, J. Catalysis 46 (1977) 37.
[7] M.A. Chesters and G.A. Somorjai, Surface Sci. 52 (1975) 21.
[8] Y.W. Tsang and L.M. Falicov, J. Phys. C (Solid State Phys.) 9 (1976) 51.
[9] R.W. Palmberg, G.E. Riach, R.E. Weber and N.C. McDonald, Handbook of Auger Electron Spectroscopy (Physical Electronics Industries, Edina, MN, 1972).
[10] K. Besocke and H. Wagner, Surface Sci. 52 (1975) 653.
[11] B. Lang, R.W. Joyner and G.A. Somorjai, Surface Sci. 30 (1972) 440.
[12] M. Henzler, Surface Sci. 22 (1970) 12.
[13] K. Besocke and H. Wagner, Phys. Rev. B8 (1973) 4597.
[14] J.W. Gadzuk, in: The Structure and Chemistry of Solid Surfaces, Ed. G.A. Somorjai (Wiley, New York, 1976). The dipole moment values given in this paper are by a factor of two larger because of a different definition of the dipole length.
[15] L.L. Kesmodel and L.M. Falicov, Solid State Commun. 16 (1975) 1201.

ELLIPSOMETRIC STUDY OF OXYGEN ADSORPTION AND THE CARBON MONOXIDE–OXYGEN INTERACTION ON ORDERED AND DAMAGED Ag(111)

H. ALBERS, W.J.J. VAN DER WAL and G.A. BOOTSMA

Van 't Hoff Laboratory, University of Utrecht, Padualaan 8, Utrecht, The Netherlands

The adsorption of oxygen on Ag(111) has been studied by ellipsometry in conjunction with AES and LEED. The oxygen pressure varied between 10^{-5} and 10^{-3} Torr and the crystal temperature between room temperature and 250°C. Changes in the Auger spectrum and the LEED pattern upon oxygen adsorption are very small. Oxygen coverages were derived from the changes in the ellipsometric parameter Δ. At room temperature a maximum coverage is reached within a few minutes. Its value increases with the damage produced by the preceding argon ion bombardment. The sticking coefficient derived from the initial rate of Δ-change amounts to 3×10^{-5} for well-annealed surfaces and $2.5 - 5 \times 10^{-4}$ for damaged surfaces. After evacuation no desorption takes place. Other types of adsorption, associated with much larger changes in Δ, were observed upon bombardment with oxygen ions and with oxygen activated by a hot filament. The reaction of CO with adsorbed oxygen was studied ellipsometrically at room temperature in the CO pressure range $10^{-7} - 10^{-6}$ Torr. The initial reaction rate is proportional to the CO pressure. The reaction probability (number of oxygen atoms removed per incident CO molecule) is 0.36.

1. Introduction

The adsorption of oxygen on silver has been the subject of many studies because of the catalytic interest, especially in the oxidation of ethylene. The results are often contradictory, even those obtained with well-defined single crystal surfaces. On Ag(111) Degeilh [1] and Dweydari and Mee [2] found changes in the work function ϕ which indicate adsorption at room temperature at oxygen pressures below 10^{-6} Torr. In LEED and thermal desorption experiments Rovida et al. [3] observed ordered adsorption at temperatures above 100°C and pressures greater than 10^{-3} Torr, and small and insignificant variations of ϕ at lower pressures. Very recently Engelhardt and Menzel [4] also noted very small and irreproducible changes in ϕ, and ascribed these to a general inertness of the close-packed Ag(111) face and to some adsorption of oxygen at irregularities.

Surface techniques which use electrons and ions as probe particles are subject to serious drawbacks in the study of adsorption. It is possible that these particles interfere with the adsorption processes, and *in situ* measurements at pressures exceeding about 10^{-5} Torr become impossible. Aside from these common problems the study of the adsorption of oxygen on silver is complicated by the low Auger sensitivity for oxygen [3,4] and the interaction of oxygen with small amounts of carbon monoxide [5]. To overcome some of these difficulties the interaction of oxygen with silver surfaces of various orientations is being studied in our laboratory with the optical method of ellipsometry in conjunction with LEED and AES. Ellipsometry appeared to be also a suitable method for the evaluation of the extent of damage produced by ion bombardment [6].

2. Experimental

For the investigation of the interaction of O_2 and CO with Ag(111) two UHV systems were used. Most experiments were performed in system I, pumped by a turbomolecular pump and a titanium sublimator and equipped with an ellipsometer, LEED optics and a quadrupole mass-spectrometer (details in ref. [6]). The AES measurements were carried out in system II pumped by a turbomolecular pump, titanium sublimator and ion pump, with facilities for ellipsometry, LEED–AES and mass-spectrometry (details in ref. [7]). The sample was prepared as described before [6], the final cleaning being achieved by argon ion bombardment.

Preliminary experiments showed that for reproducible oxygen adsorptions the CO pressure had to be less than ~0.2% of the oxygen pressure. Several precautions were taken to fulfil this condition: (i) the oxygen was continuously renewed by pumping with the turbomolecular pump; (ii) the titanium film was renewed before each adsorption, followed by a short bombardment of the crystal surface (1 min, 500 eV Ar^+, 1.5 $\mu A/cm^2$); (iii) the ion pump was "cleaned" by operating in 10^{-4} Torr oxygen before a set of measurements and switched off at least 15 min before the start of an adsorption experiment; (iv) hot filaments were excluded, except the filament of the pressure gauge for a few seconds at the start of an adsorption; control experiments showed that such a short operation had no influence on the results.

With the ellipsometer transient phenomena were followed by off-null irradiance measurements which were checked by two-zone measurements (details in ref. [6]). The wave-length was 632.8 nm and the angle of incidence 71.7 ± 0.2°. The Auger spectra were taken with a retarding field analyzer using an electron gun at an angle of incidence of 8° with the plane of the surface. Oxygen peaks were recorded within 2 min after switching on the electron gun. The effect of the electron beam on the adsorption was checked ellipsometrically. No indication was found for electron stimulated desorption; in fact after long oxygen exposures quite often an increase of the oxygen peak due to the operation of the electron gun was observed.

The gases used, i.e. argon (purity 99.9997%), oxygen (99.998%) and carbon monoxide (99.997%) were purchased from l'Air liquide.

3. Results

The results of the ellipsometric measurements will be given as changes in one of the ellipsometric angles: $\delta\Delta = \overline{\Delta} - \Delta$, where $\overline{\Delta}$ refers to the clean surface and Δ to a surface with adsorbed species. All observed variations in the other ellipsometric parameter, $\delta\psi = \overline{\psi} - \psi$, were of the same sign as those in Δ, and generally $\delta\psi \sim \frac{1}{3}\delta\Delta$. The relation between $\delta\Delta$ and the oxygen coverage θ is discussed in section 4, to a good approximation $\delta\Delta$ is proportional to θ.

AES results are presented as the ratio of the peak heights in the second derivative spectra $R_x = h_x/h_{Ag}$, where h_{Ag} is the height of the main silver peak in the clean silver spectrum.

3.1. Oxygen adsorption on Ag(111)

Depending on the nature of the interacting species, three different types of adsorption of oxygen were distinguished.

(i) With *molecular oxygen* the adsorption was found to depend on the extent of damage introduced by the preceding ion bombardment. The most reliable results were obtained with an oxygen pressure of about 10^{-4} Torr. At 10^{-5} Torr the degree of CO contamination was too high and at 10^{-3} Torr the adsorption was too rapid to permit accurate measurements.

On a clean and "completely annealed" surface [6], $\delta\Delta$ reached a saturation value, $\delta\Delta_{sat} = 0.04 \pm 0.01°$ (fig. 1a), irrespective of the substrate temperature (up to 250°C) and oxygen pressure. Damage was introduced by argon ion bombardment

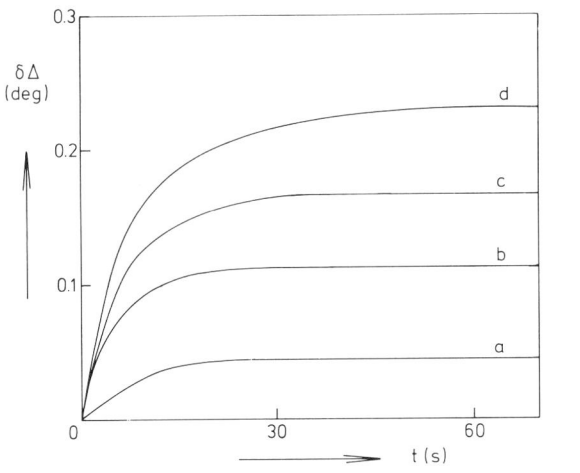

Fig. 1. $\delta\Delta$ versus time during oxygen adsorption at room temperature and 10^{-4} Torr on Ag(111) surfaces with different extent of damage (a) "completely annealed" surface; (b), (c) and (d) damaged surfaces with $\delta\Delta_{bomb} = 0.68°$, $3.94°$ and $8.14°$ respectively.

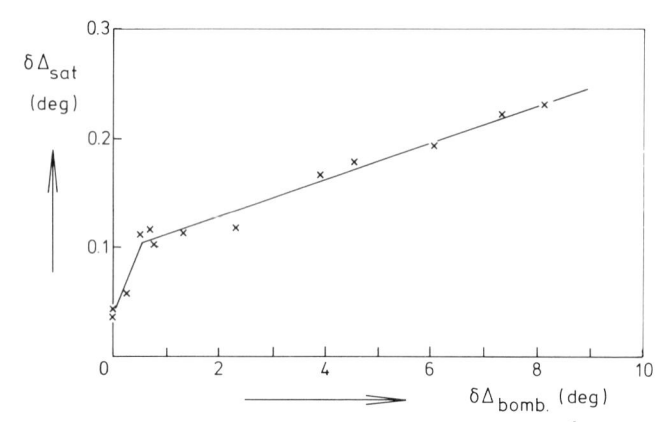

Fig. 2. $\delta\Delta_{sat}$ due to adsorption of oxygen at room temperature and 10^{-4} Torr plotted against the change in Δ caused by previous argon ion bombardment of Ag(111).

(500 eV, 1.5 µA/cm^2) for varying periods up to 24 h. Figs. 1 and 2 show the dependence of the oxygen adsorption on the extent of damage as characterized by the change in Δ ($\delta\Delta_{bomb}$) upon bombardment. Measurements at other pressures indicated that $\delta\Delta_{sat}$ is independent of the pressure and the initial adsorption rate is proportional to the pressure. At room temperature no desorption was observed upon evacuation.

The oxygen peaks observed in the Auger spectrum of damaged Ag(111) after interaction with molecular oxygen are hardly discernible. Fig. 3 gives such an exam-

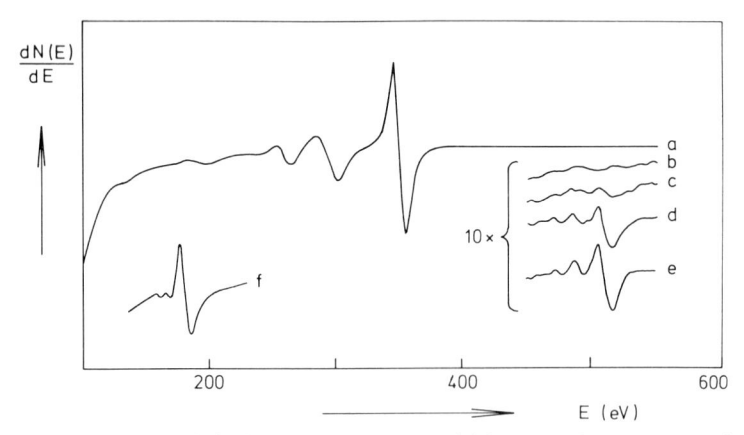

Fig. 3. Auger spectra recorded with a primary energy of 1450 eV, a beam current of 150 µA and a modulation voltage of 10 V p-p. (a), (b) clean Ag(111); (c) oxygen peak for damaged surface with $\delta\Delta_{sat} = 0.08°$; (d) oxygen peak after interaction with O_2^+ ions, $\delta\Delta_{sat} = 0.6°$; (e) and (f) oxygen and chlorine peak after interaction with oxygen atoms.

ple (c) with $R_O = 0.007$ for $\delta\Delta_{sat} = 0.08°$. The LEED pattern of clean Ag(111) showed a slight increase in background intensity upon adsorption.

(ii) Much larger effects were obtained upon bombardment with O_2^+ ions as produced in the ion gun (500 eV, 1 μA/cm^2, 10^{-4} Torr oxygen). Independent of the extent of damage of the surface and its temperature (23–250°C) saturation was reached after a flux of about 10^{15} ions/cm^2 at $\delta\Delta_{sat} = 0.6°$ and $R_O = 0.03$ (fig. 3d). The initial adsorption rate $\delta\Delta$/ion dose $\approx 1°/10^{15}$ ions/cm^2. No desorption was noted upon evacuation.

With LEED a (4 × 4) superstructure was observed both for adsorptions above 160°C and upon heating an adsorbed layer formed at lower temperatures to beyond this temperature. The superstructure remained after cooling, and disappeared after heating to temperatures above 260°C. In the latter case ellipsometric measurements indicated the removal of the adsorbed layer.

(iii) A third type of interaction was observed with oxygen activated by the hot filaments of ionization gauges or ion guns and is most probably ascribable to *atomic oxygen*. Fig. 4 gives some curves obtained at various temperatures of a tungsten filament; the temperature was estimated from the electron emission. AES shows that the $\delta\Delta_{sat}$ of 0.6° at the lowest filament temperature (1810 K) is mainly due to chlorine; after removal of the oxygen with CO, $\delta\Delta = 0.5°$ and $R_{Cl} = 0.29$. The same results were obtained with thoria-coated rhenium and iridium filaments. With tungsten filaments at higher temperatures no saturation in Δ occurred even up to $\delta\Delta = 10°$, while $\delta\psi$ levelled off at 0.3°. The observed changes in Δ are partly due to chlorine but mainly to oxygen. After an initial increase the oxygen Auger peak remained constant at $R_O = 0.04$. The LEED patterns were very diffuse with a high

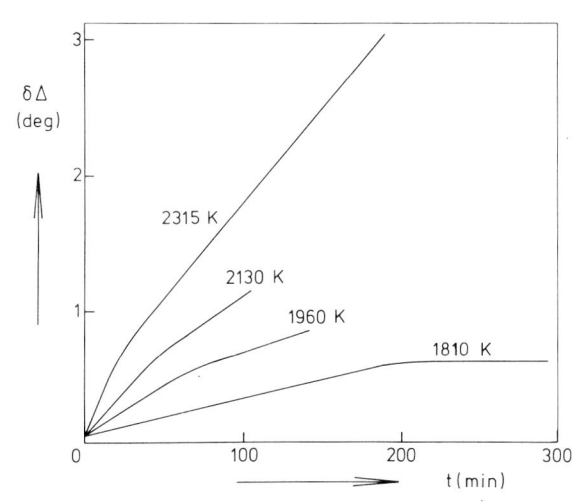

Fig. 4. $\delta\Delta$ versus time during adsorption on Ag(111) induced by a tungsten filament at various temperatures in 10^{-4} Torr oxygen, with the crystal at room temperature.

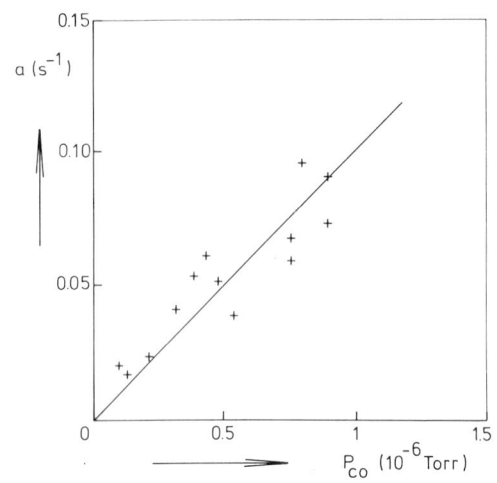

Fig. 5. Pressure dependence of the rate coefficient a (eq. (1)) of the reaction of CO at room temperature with oxygen adsorbed on damaged Ag(111).

background intensity. No changes in Δ and R_O were observed after evacuation.

During sputtering of a layer formed by interaction with atomic oxygen the oxygen Auger peak first remains constant and then starts to decrease after Δ has already attained the value of that of a clean surface, in contrast with the behaviour of the chlorine peak, which is readily removed.

3.2. Reaction of carbon monoxide with adsorbed oxygen

Ellipsometric measurements with the crystal at room temperature showed that at pressures up to 10^{-3} Torr CO and CO_2 do not adsorb on clean Ag(111), nor does CO_2 on Ag(111) with preadsorbed oxygen. A more detailed kinetic study was made of the reaction of CO with oxygen adsorbed on damaged surfaces (type (i) of section 3.1). The time dependence of Δ could be described by the relation

$$\ln(\delta\Delta/\delta\Delta_{sat}) = -at .\qquad(1)$$

In the range of CO pressures studied ($10^{-7} - 10^{-6}$ Torr) the rate coefficient a appeared to be proportional to the pressure (fig. 5).

For the other types of sorbed oxygen the rate of the clean-off reaction with CO was found to have the same order of magnitude. Remarkable was the almost complete removal of a layer, obtained by interaction with oxygen atoms (type (iii) of section 3.1) with $\delta\Delta \sim 8°$, within 1 min at a CO pressure of 10^{-4} Torr.

4. Discussion

4.1. Oxygen coverage

From changes in Δ the oxygen coverage θ (number of O atoms/surface Ag atom) can in principle be derived in two ways; namely by model calculation or by experimental calibration.

The first method requires a suitable optical model for chemisorption. Such a model, which takes into account changes in the substrate, has been developed for semiconductor surfaces but not yet for metal surfaces [8,9]. On the assumption that the effective refractive index of the adsorbed layer may be calculated from the polarizability and degree of coverage of oxygen with the Lorentz–Lorenz equation [8] and that the substrate is unperturbed by the adsorption process, we find for a monolayer of oxygen atoms on Ag(111): $\delta\Delta = 0.23°$ and $\delta\psi = -0.0007°$ [6]. This (physisorption) model does not explain the observed changes in ψ and it would be fortuitous if the relation between $\delta\Delta$ and θ were correct.

In the second method the coverages must be determined by another independent technique, e.g. AES, LEED, $\Delta\phi$ measurements. For oxygen on Ag(111) such a coverage determination is not available. Combined ellipsometric and AES measurements of oxygen adsorption on Cu(111) showed a linear relationship between $\delta\Delta$ and R_O [10]. Preliminary results obtained by ellipsometry, AES, LEED and $\Delta\phi$ measurements for O_2 on Ag(110) indicate the same relationship to exist. For the (2 × 1) structure with $\theta = 0.5$ [5], we found $\delta\Delta_{sat} = 0.45°$. Neglecting possible differences in optical response of oxygen adsorbed on Ag(110) and Ag(111), taking into account the atomic densities of the planes and assuming $\delta\Delta$ proportional to θ on both faces, we find for oxygen on Ag(111):

$$\theta = 0.7\delta\Delta \tag{2}$$

On applying (2) to the example given in fig. 3c one finds $\theta_{sat} = 0.06$. From the Auger signal, with the ionization cross sections derived from the data of Vrakking and Meyer [11], a coverage of 0.04 is calculated, assuming that at the glancing incidence only one monolayer of silver is detected. Considering the possibility that adsorption may take place in the valleys of the rough surface the agreement may be taken as reasonable.

4.2. Oxygen adsorption

The rate of adsorption of *molecular oxygen* on damaged surfaces appeared to be porportional to $(\theta_{sat} - \theta)$ and the oxygen pressure. The sticking coefficient s may thus be described by the relation

$$s \equiv \frac{N(d\theta/dt)}{n_O} = k_1(\theta_{sat} - \theta), \tag{3}$$

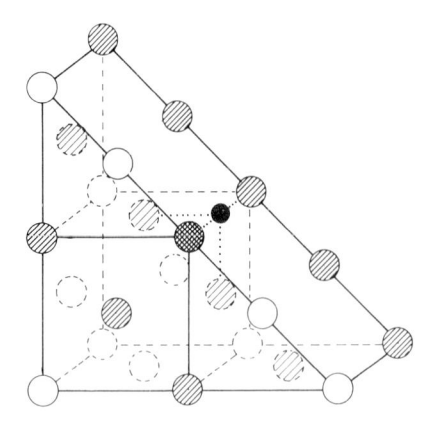

Fig. 6. Adsorption model for oxygen on a rough Ag(111) surface. Filled circle: oxygen atom, other circles: part of the silver fcc lattice, hatched circles: silver atoms in the same (111) plane, doubly-crossed circle: silver atom on top of this plane.

where N is the number of oxygen atoms in a monolayer ($\theta = 1$) and n_O is two times the collision frequency of oxygen molecules. The magnitude of the rate coefficient k_1 shows considerable scatter, $k_1 = (3.0 \pm 0.8) \times 10^{-3}$, but no correlation with θ_{sat} was found. The initial sticking coefficient derived from eq. (3) ranges from 2.5×10^{-4} to 5×10^{-4}, in accordance with the values derived directly from the initial changes in Δ. For the completely annealed surface s is not described by eq. (2); the initial value is $(3 \pm 1) \times 10^{-5}$.

The increase of θ_{sat} with increasing damage is probably due to surface roughness. On (111) surface rough on the atomic scale, the same kind of adsorption sites may be present as proposed for Ag(110) [5,12] (fig. 6). The small adsorption on the "completely annealed" surface may be accounted for by residual irregularities, such as introduced by a small deviation from the ideal (111) orientation.

The $\delta\Delta_{sat}$ observed upon bombardment with O_2^+ ions corresponds to a coverage of $\theta_{sat} = 0.4$. The initial sticking coefficient is ~ 0.5. The behaviour of the adsorbed layer (superstructure, desorption temperature) resembles that reported by Rovida et al. [3] for much higher pressures of molecular oxygen.

The interaction with oxygen in the presence of tungsten filaments at temperatures above 1960 K is most likely ascribable to *oxygen atoms* [13]. Under comparable experimental conditions Wood [14] (flash filament measurements) and McBee and Yolken [15] (ellipsometry) also noted sorption of atomic oxygen at a nearly constant rate well beyond monolayer coverage. The Auger data obtained upon sputtering suggest a homogeneous composition of the Ag_xO layer with $x = 3$. This layer appeared to be stable against decomposition at room temperature just as Ag_2O [16].

Table 1
Oxygen adsorption from CO/O_2 mixtures

p_{CO}/p_{O_2}	θ_{st}/θ_{sat}	
	Experimental	Calc. with eq. (5)
0.005	0.90	0.76
0.02	0.45	0.44
0.04	0.30	0.28

4.3. Reaction of carbon monoxide with adsorbed oxygen

According to eq. (1) the reaction probability P, defined as the number of oxygen atoms removed per incident CO molecule, may be expressed by

$$P \equiv \frac{N(-d\theta/dt)}{n_{CO}} = k_2\theta \quad (4)$$

The value of $k_2 = 0.36 \pm 0.03$.

For mixtures of oxygen and CO the stationary oxygen coverage θ_{st} may be calculated by combining eqs. (3) and (4):

$$\frac{\theta_{st}}{\theta_{sat}} = \frac{k_1}{k_1 + k_2(n_{CO}/n_O)} = \frac{k_1}{k_1 + 0.53\, k_2(p_{CO}/p_{O_2})}. \quad (5)$$

In table 1 experimental and calculated coverages are compared for some mixtures. Considering the experimental errors (CO content) the agreement is satisfactory.

References

[1] R. Dégeilh, in: Proc. 4th Intern. Congr. Catalysis, Moscow, 1968 (Akad. Kaido, Budapest, 1971) Vol. 2, p. 35; Vide 139 (1969) 29.
[2] A.W. Dweydari and C.H.B. Mee, Phys. Status Solidi (a) 17 (1973) 247.
[3] G. Rovida, F. Pratesi, M. Maglietta and E. Ferroni, Surface Sci. 43 (1974) 230.
[4] H.A. Engelhardt and D. Menzel, Surface Sci. 57 (1976) 591;
    H.A. Engelhardt, Thesis, Technische Universität, München (1975).
[5] H.A. Engelhardt, A.M. Bradshaw and D. Menzel, Surface Sci. 40 (1973) 410.
[6] H. Albers, J.M.M. Droog and G.A. Bootsma, Surface Sci. 64 (1977) 1.
[7] F.C. Schouten, E.W. Kaleveld and G.A. Bootsma, Surface Sci. 63 (1977) 460.
[8] G.A. Bootsma and F. Meyer, Surface Sci. 14 (1969) 52.
[9] F. Meyer, Surface Sci. 56 (1976) 37.
[10] F.H.P.M. Habraken, E.P. Kieffer and G.A. Bootsma, Proc. 3rd ICSS, Vienna, September 1977, to be published.

[11] J.J. Vrakking and F. Meyer, Phys. Rev. A9 (1974) 1932.
[12] W. Heiland, F. Iberl, E. Taglauer and D. Menzel, Surface Sci. 53 (1975) 383.
[13] P.O. Schissel and O.C. Trulson, J. Chem. Phys. 43 (1965) 737.
[14] B.J. Wood, J. Phys. Chem. 75 (1971) 2186.
[15] M.J. McBee and H.T. Yolken, Natl. Bur. Std. (US) Report No. 9802 (1968).
[16] J.A. Allen, Australian J. Chem. 13 (1960) 431.

A MODEL FOR MORPHOLOGICAL CHANGES DRIVEN BY STEP–STEP INTERACTION ON CLEAN SURFACES

Georges MARTIN and Bernard PERRAILLON

Section de Recherches de Métallurgie Physique, Centre d'Etudes Nucléaires de Saclay, BP 2, 91190 Gif-sur-Yvette, France

The model describes the migration of surface steps in the field of interaction with their neighbors. For well choosen mean surface orientations, this model predicts an exponential decay of the amplitude of sinusoidal undulations on the surface of a semi-infinite body. The initial rate of decay is proportionnal to the sum of the second and fourth powers of the wave vector of the ondulation. A systematic study of this rate should provide us with new data on step–step interactions. Reliable data from the literature are better fitted by this model than by Mullins' standard model [7] allthough some of the conditions of validity of the present model are not fulfilled. This calls for more theoretical and experimental work on the subject.

1. Introduction

The energy of interaction between monatomic steps on clean surfaces is still now not a well documented problem [1]. Several mechanisms for step interaction have been postulated: configurational repulsion [2], Van der Waals type interaction [3], broken bonds [4], elastic interaction [1,5,6]. Experimental studies, however, are rare and have provided only qualitative answers [6].

The purpose of this note is to show that for certain well choosen surface orientations, a study of the kinetics of the flattening of surface undulations could bring new data on the step step interaction potential.

2. The model surface

Let us consider a two dimensional semi infinite crystalline solid bounded by a macroscopically flat surface ($y = 0$ in fig. 1). Let the closest singular surface make an angle θ_0 with respect to this surface. On the atomic scale, the $y = 0$ surface is built of a uniform distribution of monatomic steps of height h, the density of steps (number of steps per unit length in the x direction) is independent of x and is given by:

$$n_0 = (1/h) \sin \theta_0 . \tag{1}$$

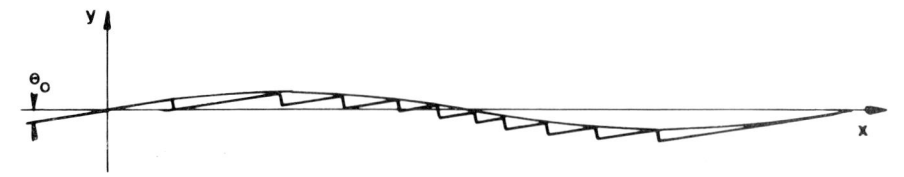

Fig. 1. Sinusoidal undulation on a vicinal surface. The step height is exaggerated.

Any small deviation of the surface orientation with respect to $y = 0$ results in a variation of the local density of steps given by:

$$n(x) - n_0 = -\frac{\cos \theta_0}{h} \frac{dy}{dx} \qquad (2)$$

As shown by fig. 1 there is no need to introduce steps of oposite sign as long as:

$$dy/dx < \text{tg}\, \theta_0 \qquad (3)$$

For a sinusoidal undulation of amplitude A and wavelength λ ($\lambda = 2\pi/k$), this condition is:

$$kA < \text{tg}\, \theta_0. \qquad (4)$$

Provided that this condition is fulfilled, the flattening of the undulation will change neither the overall number of steps, nor the overall area of terraces. It merely modifies the relative position of the steps.

The only driving force for this process, therefore, is the decrease of the free energy of interaction between the steps. An experimental study of the kinetics of this flattening process should thus shed some light on this interaction energy.

Interpretation of such a study would require a model for the time evolution of the amplitude of the undulation under consideration. Such a model is proposed below.

3. The kinetic model

A detailed model of the problem would of necessity be very complicated since three elementary processes contribute to the decay of the undulation: The emission of an adatom at a step, the propagation of this adatom on one of the adjacent terraces and its adsorption on the next step. These processes result in a cooperative motion of the steps which seems difficult to describe in detail in analytical form.

On the other hand, Mullins' classical model [7], extended to the case of anisotropic surface tension [8], relies on the assumption that the rate controlling process of the morphological change is diffusion, the source and sinks of mobile defects beeing sufficiently effective for the concentration of ad-defects to be locally in equilibrium with the surface.

It is not clear how interaction energy between steps affects these processes. We rather propose a very crude description which we believe contains the important point of the problem: i.e. the step step interaction potential as a driving force for the morphological change.

Let us assume that in the absence of the other steps, each step would perform a Brownian like migration, and that the presence of other steps merely adds a driving force to the migration of each step.

Under these assumptions, the flux J of steps is:

$$J = +\Gamma a^2 \left(-\frac{\partial n}{\partial x} + \frac{Fn}{k_B T} \right), \tag{5}$$

where Γa^2 is a step "diffusion coefficient", which we need not define in more detail now, $k_B T$ is the thermal energy and F the force which derives from the step step interaction potential $V(x' - x)$

$$F = -\frac{\partial}{\partial x} \int_{-\infty}^{+\infty} n(x') V(x' - x) \, dx' \tag{6}$$

Expression (5) for the flux of steps is built in a somewhat artificial way, but it expresses the proportionality between the flux of steps and the decrease of free energy which initiates this flux: indeed the first term in the brackets of the RHS of eq. (5) gives the contribution of the configurationnal entropy, while the second term gives the contribution of the internal energy. A similar approach has been used by Flynn [9] to present Cahn's model of the evolution of solid solutions towards equilibrium configuration [10]. It has been given a more rigorous form recently for the problem of phase separation in fluid systems [11]. A similar approach has also been used to describe the evolution of a population of dislocations [12].

Since for the evolution which we consider, the steps are conservative units (cf. section 2) the time dependence of the local concentration of the steps is given by the divergence of J, i.e.:

$$\frac{\partial n(x,t)}{\partial t} = \Gamma a^2 \left[\frac{\partial^2 n}{\partial x^2} - \frac{1}{k_B T} \frac{\partial}{\partial x} n(x,t) F(\{n(x,t)\}) \right]. \tag{7}$$

This equation may be solved for small perturbations of the uniform steady state solution n_0. To do this, we introduce $[n(x,t) - n_0]$ in (7), make a Fourier transformation of this equation, and keep only the terms linear in the perturbation. We obtain the following equation for the time evolution of the kth harmonic of the perturbation:

$$\partial \eta(k,t)/\partial t = -k^2 \Gamma a^2 [1 + n_0 V(k)/k_B T] \eta(k,t), \tag{8}$$

where

$$\eta(k,t) = \int_{-\infty}^{+\infty} e^{-ikx} [n(x,t) - n_0] \, dx, \qquad V(k) = \int_{-\infty}^{+\infty} e^{-ikx} V(x) \, dx.$$

The solution of eq. (8) is:

$$\eta(k, t) = \eta(k, 0) \, e^{-t/\tau}, \tag{9}$$

with

$$1/\tau = k^2 \, \Gamma a^2 [1 + n_0 V(k)/k_B T]. \tag{10}$$

In the next section, we discuss the experimental consequences of eqs. (9) and (10).

4. Discussion

The above model has the following characteristics:

(a) The amplitude of surface undulations decays * exponentially with the time according to eq. (9),

(b) The relaxation time (eq. (10)) is a function of the wavelength of the undulation ($2\pi/k$), of the step–step interaction potential $V(k)$, and of the misorientation θ_0 of the unperturbed surface ($y = 0$) with respect to the closest singular surface: the larger the misorientation, the more rapid the relaxation (eq. (1)).

(c) In the limit of large wavelength (i.e. wavelength large compared to the lattice spacing) of the undulation (small values of k) the relaxation time given by eq. (10), may be rewritten:

$$1/\tau = k^2(\alpha + \beta k^2), \tag{11}$$

with

$$\alpha = \Gamma a^2 (1 + n_0 V_0/k_B T) \qquad \beta = \Gamma a^2 \, n_0 V_2/k_B T \tag{12}$$

where V_0 and V_2 are the coefficients of a Taylor expansion $V(k)$ to second order in k. Such an expansion exists in the limit of small values of k as long as $V(x)$ varies more rapidly than x^{-3} (3 = dimensionality of space +2 [13]), which can be assumed without much loss of generality. That k does not appear to the first power in the RHS of (11) results from the parity of $V(x)$. This suggests a means of collecting information on the step step interaction potential. As shown in fig. 2, the measurement of $1/\tau$ at a given temperature T and surface misorientation (n_0), for several harmonics of the surface undulation allows the determination of the parameters α and β (eq. (11)). Similar measurements done on different surface orientations (i.e. various values of n_0) would provide us with V_0 and V_2 as shown in fig. 2b. It is interesting to note that this procedure eliminates the "step diffusion coefficient" which was introduced in a rather artificial way in section 3.

(d) As shown by (eq. (11)), the wavelength dependance of the relaxation time

---

* We assume $1 + n_0 V(k)/k_B T > 0$. A negative value would lead to facetting which we do not discuss here.

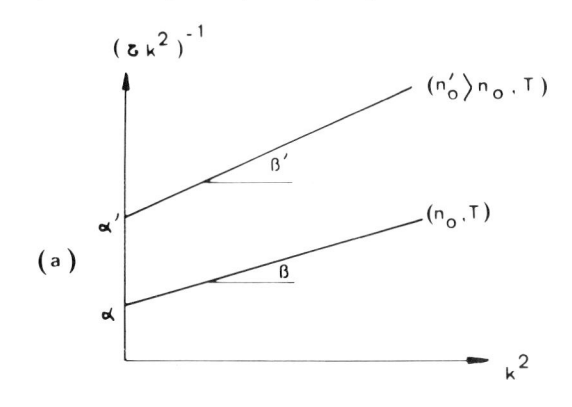

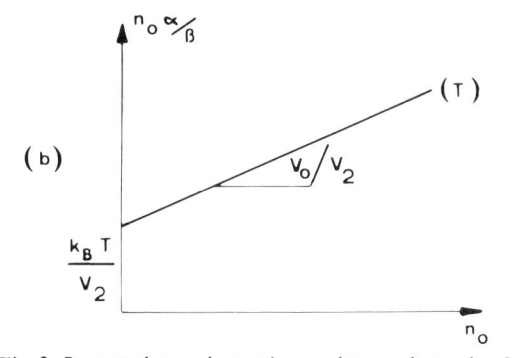

Fig. 2. Suggested experimental procedure to determine V_0 and V_2.

predicted by this model is different from that predicted by Mullins [7]:

$$1/\tau = Ak + Bk^2 + Ck^3 + Dk^4. \tag{13}$$

The k and k^3 terms do not appear in our model.

The two models however are not valid in the same range of wavelength. Mullins' model which assumes the equilibrium of mobile surface defects with the local curvature of the surface is certainly correct in the limit of small curvatures and small curvature gradients. The present model which makes no such assumption is more appropriate to shorter wavelength: indeed the k^4 term in (11) results from the variation of the density of steps inside the range of the step–step interaction potential.

Is it possible to distinguish between the two models owing to the k^3 dependence of $1/\tau$?. As a check, we have reexamined the results of two experimental studies taken from the literature [14,15]. Figs. 3a and b show a plot of these data both in their original form ($1/\tau k^3$ versus k) and according to our model ($1/\tau k^2$ versus k^2). This latter plot is slightly better than the original one, although these experiments

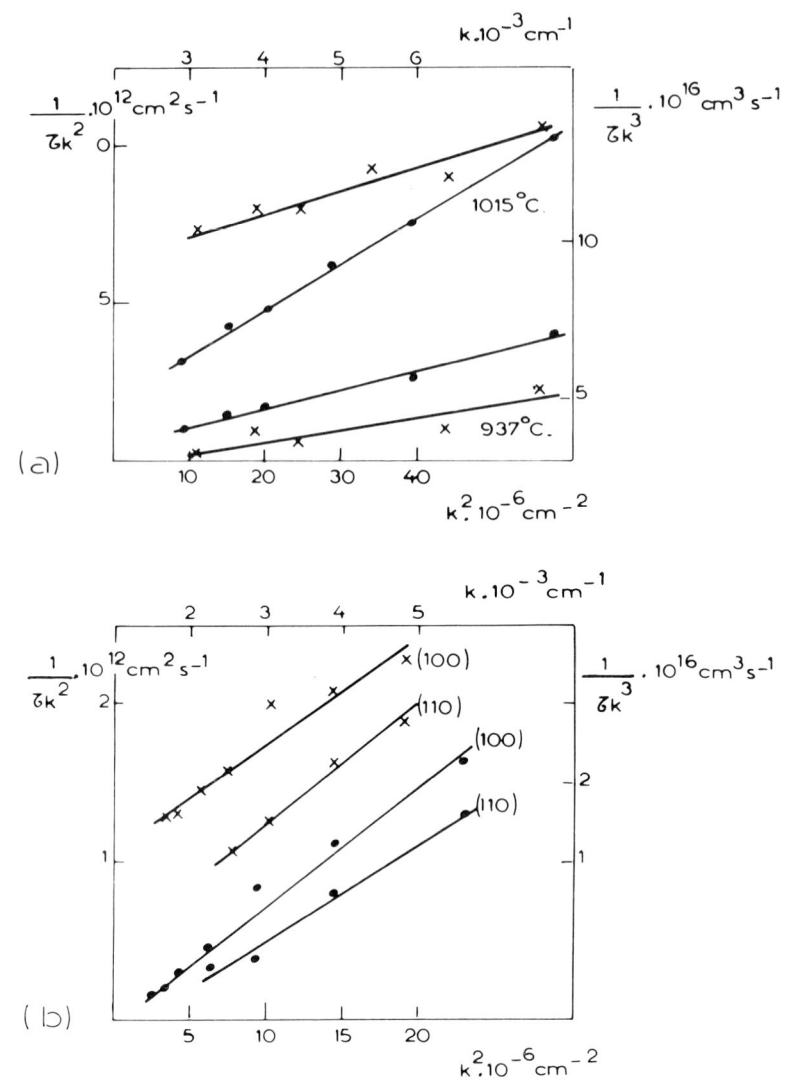

Fig. 3. Experimental dependence of $1/\tau$ versus k for: (a) (111)Cu surface [14]; (b) (100) and (110)Ni surfaces [15]. (+) Mullin's model: $1/\tau k^3$ versus k (right and upper scales respectively); (○) present model: $1/\tau k^2$ versus k^2 (left and lower scales respectively).

are certainly appropriate to Mullins' analysis. This proves that more precise experimental data, and a careful statistical analysis are necessary in order to distinguish between the two regimes.

(e) As stated before, this model is very crude indeed and neglects many complicating factors such as:

– The presence of steps of opposite sign. This second set of steps would be governed by an equation similar to eq. (7), both equations being now coupled by a chemical reaction term to account for the recombination of steps of opposite sign.
– The cooperative nature of step migration (if matter is to be conserved in the solid phase, each jump of step is accompanied by a jump of one neighbor step in the opposite direction).
– The configurational repulsion between steps on a two dimensional surface (2).

These two last questions could be studied by computer simulation.

5. Conclusion

We have proposed a crude model for the kinetics of the flattening of ondulations under conditions where the interaction between steps is the unique driving force.

This model predicts, for well choosen surface orientations, an exponential decay of the amplitude of the undulations, with an initial rate which is proportionnal to the second and fourth powers of the wave vector of the undulations.

A systematic study of this relaxation time should provide us with new data on step step interactions. Reliable data from the literature are better fitted by this model than by Mullins' standard model, although the conditions of validity of the present model are not fulfilled. This calls for more experimental and theoretical work.

Acknowledgments

The authors thank Professors Friedel and Lebowitz for their comments and Dr. Y. Adda for his continuous interest in this work.

References

[1] J. Friedel, Ann. Phys. (Paris) 1 (1976) 257.
[2] E.E. Gruber and W.W. Mullins, J. Phys. Chem. Solids 28 (1967) 875.
[3] L.D. Landau, Collected Papers (Pergamon, 1965) p. 540.
[4] P. Wynblatt, in: Interatomic Potentials and Simulation of Lattice Defects, Eds. P.G. Gehlen et al. (Plenum, New York, 1972) p. 633.
[5] J.M. Blakely and R.L. Schwoebel, Surface Sci. 26 (1971) 321.
[6] K.J. Matysik, Surface Sci. 46 (1974) 457.
[7] W.W. Mullins, J. Appl. Phys. 30 (1959) 77.
[8] W.W. Mullins, in: Metal Surfaces (Am. Soc. Metals, Cleveland, Ohio, 1963) p. 17.
[9] C.P. Flynn, Point Defects and Diffusion (Clarendon Press, Oxford, 1972) p. 467.
[10] J.W. Cahn, Acta Met. 9 (1961) 795.
[11] F.F. Abraham, Script Met. 10 (1976) 1.
[12] D.L. Holt, J. Appl. Phys. 41 (1970) 3197.
[13] E.g., G. Toulouse and P. Pfeuty, Introduction au Groupe de Renormalisation (Presse Universitaire, Grenoble, 1975) p. 55.
[14] K. Hoehne and R. Sizmann, Phys. Status Solidi (a) 5 (1971) 577.
[15] P.S. Maiya and J.M. Blakely, J. Appl. Phys. 38 (1967) 698.

PRECIPITATION AND RE-SOLUTION OF IMPURITIES AT THE SURFACE OF INDIUM ON TRAVERSING THE MELTING-POINT

M. GETTINGS and J.C. RIVIÈRE

Materials Development Division, AERE, Harwell, England

In the course of preparing a clean indium surface for solid-versus-liquid studies, changes in the surface concentrations of sulphur and oxygen were observed by AES and XPS while the metal was heated and cooled through its melting-point. Both impurities disappeared on melting, and reappeared on solidification, over a very narrow temperature range; the disappearance and reappearance were to a certain extent reproducible. The effect was found to be similar in characteristics to that observed for the behaviour of carbon on a nickel surface by Shelton et al., and the same Bragg–Williams model is invoked to explain the sharpness of the impurity concentration changes with temperature. Although the maximum temperature reached by the indium was only 200°C, traces of platinum were also observed on the indium surface after melting, in both the AES and XPS spectra, probably as a result of solution from the platinum boat.

1. Introduction

Differences between the bulk properties of the solid, liquid and, more recently, amorphous forms of a material have been the subject of much interesting work for many years, but the study of possible differences between the *surface* properties of the various forms has not been nearly so extensive. Most workers in surface physics and chemistry have preferred to restrict their experiments to the solid state and to study solid/gas reactions, for obvious practical reasons. However, there are several metals (and one or two semi-metals) that have both low melting-points, and sufficiently low vapour pressures at their melting-points, to be experimentally amenable within a conventional surface analysis system, without having recourse to special cells. It was in the course of studying one of these metals, indium, and in particular during the initial stages of surface cleaning, that the effects described in this paper were observed. A comparison paper will describe the differences observable by AES and XPS between the surfaces of the solid and liquid states of indium.

2. Experimental

Indium in the form of 2 mm diameter wire and with a purity of 99.99% was melted in a platinum boat using AC resistive heating. Temperature measurement was made via a chromel–alumel thermocouple spot-welded to the underside of the boat. The base pressure in the unbaked but clean system was 10^{-9} Torr, which deteriorated on the initial melting to 10^{-8} Torr but returned to 2×10^{-9} Torr after 20 min with indium in the molten state; on subsequent meltings the pressure never rose above 3×10^{-9} Torr.

Before melting, the indium was positioned accurately at the focal position for a Physical Electronics Industries (PHI) double-pass cylindrical mirror analyser (CMA) [1]. This analyser is designed for both AES and XPS measurements and the analysing position is the same for both techniques. The specimen-source-analyser geometry has already been described [2], but briefly the specimen surface normal is at 30° to the CMA axis and therefore also to the exis of the integral electron gun, and at 60° to the X-ray source. It is possible to change from one technique to the other within a minute. Typical operating conditions were: for XPS, Al K_α radiation, unfiltered, at a source power of 400 W, with band-pass energies of 100 eV for wide scans and of 50 eV for more detailed scans, and for AES, a primary beam of energy 4 keV and specimen current 10 μA, with modulations of 3 V p-p for wide scans and of 0.5 V p-p for detailed scans.

3. Results

Although the first AES analyses revealed a typically contaminated metal surface, a virtually clean indium surface was quickly produced by a combination of ion bombardment and melting. At that stage the surface was clean only when the metal was molten, for on solidifying, the sulphur, carbon and oxygen returned. A clean surface on solidified indium was obtained only after many repeated meltings and ion bombardments, and its Auger spectrum was then found to be identical with that for the molten state. In agreement with the AES observations, the XPS spectra also showed sulphur, carbon and oxygen reappearing on the surface during the early stages of cleaning. Once the surface of solid indium had been cleaned completely, then the XPS spectrum was almost identical to that of molten indium. The subtle differences observed between solid and liquid states will be the subject of another paper.

The first observations on the disappearance and reappearance of sulphur and oxygen at the indium surface during melting and solidifying were not sufficiently precise to be able to determine over what temperature range they were occurring. Accordingly the sulphur and oxygen Auger peaks were monitored continuously with a multiplexing unit (PHI Model 20-055) while the indium was heated and then cooled slowly. The results of this experiment are shown in figs. 1 and 2, for sulphur and oxygen, respectively.

On the first heating cycle above the melting-point (156.6°C) of indium, the sulphur Auger peak became undetectable. On cooling there was no reappearance of sulphur until the temperature had dropped to a few degrees below the melting-

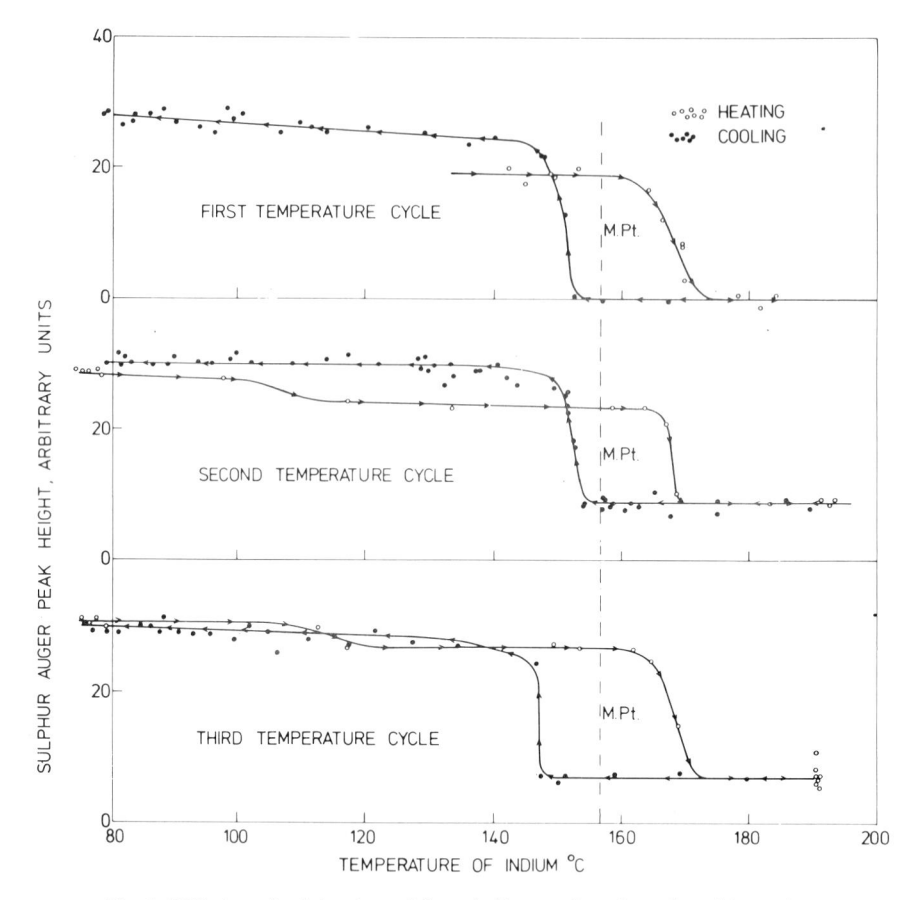

Fig. 1. Diffusion of sulphur to and from indium surface through melting point.

point, but at that point the magnitude of the sulphur peak rose suddenly to a value greater than that observed before heating. On subsequent temperature cycles the sulphur signal was again attenuated on heating through the melting-point but, unlike in the first cycle, did not disappear completely; the residual signal was about the same as the increase observed at the first cooling. There was no variation in the residual level with time above the melting-point, or with temperature in the molten state, at least up to ~193°C. An additional noticeable feature of the heating curve in the second and third cycles was an initial slight decrease in the sulphur level at ~110°C. The total hysteresis above and below the melting-point on heating and cooling, respectively, varied between 15 and 20°C, and did not appear to be closely related to rates of change of temperature near the melting-point. For example, the approximate temperature lags below the melting-point on cooling were

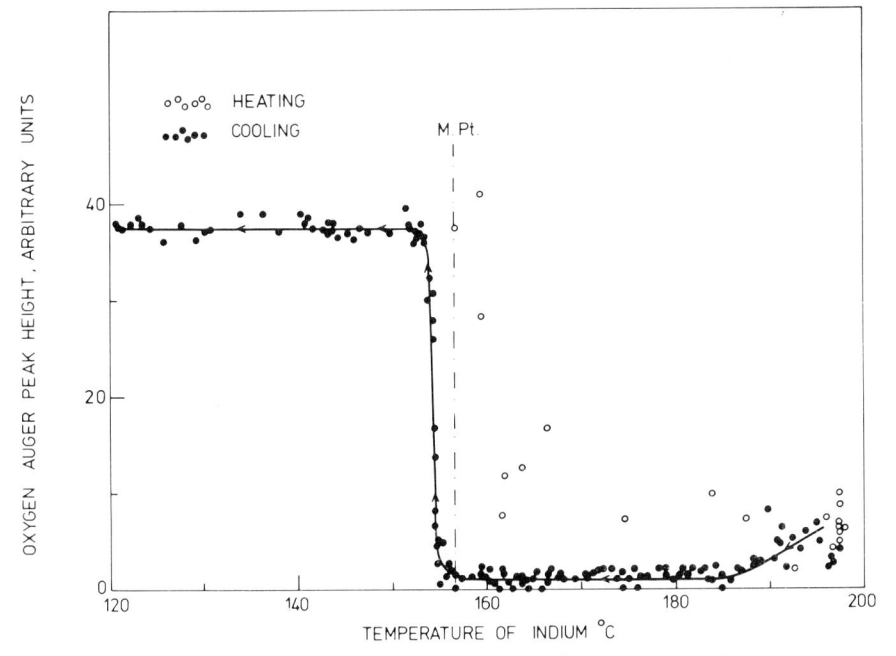

Fig. 2. Diffusion of oxygen to and from indium surface through melting point.

6, 5 and 13°C in the first, second and third cycles, respectively, compared with average cooling rates of 19, 7 and 21°C min^{-1}.

The behaviour of the oxygen peak was not quite the same as that of sulphur, since no trace of the element could be detected in the temperature range 160–185°C, but above 185°C it again became detectable and remained so as long as the temperature was maintained in that region. In addition, the oxygen level tended to decrease with each cycle on solidification, and the extent of total hysteresis was much less than for sulphur, being about 6°C.

Ion bombardment of the solidified indium at this stage removed the sulphur and most of the oxygen, but subsequent re-melting and solidification caused reversion to the behaviour shown in figs. 1 and 2. Eventually the sulphur and oxygen in the indium were exhausted and temperature cycling did not cause their reappearance. Within the experimental limitations there was no observable change in the behaviour of sulphur from that shown in fig. 1, as the level of sulphur was depleted. However, the absence of any change may have been due to the fact that even the initially observed amounts of sulphur on the surface were low, and it is possible changes might be seen if the indium contained deliberately added sulphur in much larger bulk quantities.

In both the AES and XPS spectra of solid indium, peaks were observed that could be ascribed to the presence of traces of platinum at the surface. Since the areas analysed in both techniques, 25 μm diameter in AES and 2 mm diameter in XPS, were much smaller than the surface area of the indium samples after melting, there could not have been any contribution to the spectra from the platinum boat itself. The thickness of the indium would also have precluded any such contribution. The conclusion is therefore that the molten indium must have dissolved some platinum from the boat, even though the maximum temperature reached was only 200°C.

4. Discussion

4.1. Precipitation and re-solution of sulphur and oxygen through the melting-point

It is interesting that the temperature dependence of the sulphur peak height shown in fig. 1 is very similar to that found by Shelton et al. [3] for the behaviour of the carbon peak on a nickel (111) surface (cf. their fig. 1). There is the same abrupt disappearance and reappearance in a narrow range about a characteristic precipitation temperature, with a hysteresis between heating and cooling, and the same slight change in peak height at another, lower, reproducible temperature. The principal differences between their observations and those reported here are (1) that whereas the precipitation temperature for the virtual disappearance of the carbon from the nickel surface was ~900°C, (i.e. *below* the melting-point of nickel) here the sulphur and oxygen disappeared from the indium surface as a result of melting, and (2) although the coverage of carbon on nickel reverted to at least a monolayer on cooling through the precipitation temperature, here the coverage of sulphur and of oxygen on re-solidified indium was only about 0.15. The coverage was estimated using the relative sensitivity factors for indium, sulphur and oxygen taken from the PHI Handbook [4]. Despite the differences, the similarities in behaviour of the two systems are so marked that it is believed that the same simple explanation for the abrupt changes put forward by Shelton et al. applies here.

The Bragg–Williams condensation model used by Shelton et al. attempts to describe adsorption (i.e. segregation) and subsequent condensation (or precipitation) in terms of the interaction of the adsorbate particles in two-co-existing phases in equilibrium. In their case the interacting particles were carbon atoms, and the two phases were graphite and the Ni–C solid solution. By equating the chemical potentials of carbon in graphite and in the solid solution reservoir, they were able to derive an expression for the variation of the equilibrium value of the coverage θ with absolute temperature T, and they deduced that for $T \ll T_c$, the critical temperature, only very low or very high values of θ occured at equilibrium. Thus slight changes in temperature on either side of the precipitation temperature produced large changes in coverage.

What is not clear from the work of Shelton et al. is the significance of the precipitation temperature for their Ni/C system, of ~900°C; they do not identify it in their paper with any feature of, say, the Ni/C phase diagram. Obviously it is the temperature at which the concentration of carbon at the surface exceeded the solubility of carbon in the surface region, but the specific precipitation temperature that they observed may not be unique. It may well be a function of the bulk level of carbon, *unless* the surface concentration after heating to sufficiently high temperatures is always the same, i.e. irrespective of the bulk concentration. In the present work, on the other hand, the precipitation temperature was clearly identified with the melting-point of indium, despite the hysteresis evident in figs. 1 and 2. (The latter effects were due partly to a lag between the temperature at the surface of the indium and that measured at the position of the thermocouple, and partly to a certain amount of genuine supercooling and superheating.) The interacting particles in this case were indium and sulphur (or oxygen) atoms, and the two phases were the In–S (or In–O) liquid solution and a sulphide (oxide) surface precipitate; since the system was indium-rich the sulphide may have approximated to InS (Ansell and Boorman [5]). Presumably the oxide was In_2O_3.

For graphite the critical temperature T_c calculated from the data of Shelton et al. is 4.4×10^4 K, but for InS it is bound to be lower because of the lower coordination in the orthorhombic structure [6]. However, a rough calculation shows that it is in the region of 6×10^3 K, which is still much greater than the melting-point, so that the requirement $T \ll T_c$ would certainly have held. The observation that the coverage θ *below* the melting-point was well below a monolayer (in contrast to carbon on nickel below 900°C) was due to the low level of sulphur within the indium, that is, there was simply not enough available to form a monolayer. *Above* the melting-point θ was zero on the first melting (cf. fig. 1) but during the subsequent heating cycles became constant at ~0.04 (which later on could be removed easily by ion bombardment). Since In_2S_3 is the only sulphide with measurable solid solution in indium [5], the appearance of a residual level of sulphur provides confirmation that some InS was formed at the surface during the first cycle. The composition of the sulphide represented by the additional sulphur that appeared on cooling is not known, but it was probably that of a non-stoichiometric compound incapable of separate bulk existence but thermodynamically possible on a surface, in which case the coordination must have been even lower than in InS.

Insufficient observations were made of the changes in the surface oxygen level back and forth through the melting-point to be sure that the same considerations applied, but the behaviour on cooling shown in fig. 2 suggests that they do. It would be interesting to repeat the experiment with indium specimens containing much higher levels of added sulphur and oxide.

4.2. Appearance of platinum peaks in the indium spectra

The solubility of platinum in liquid indium has been measured by Yatsenko and Dieva [7]; if their results can be extrapolated down to 200°C, which is doubtful, then the solubility is 4×10^{-3} at%. On the other hand, a rough estimate of the amount of platinum at the surface from XPS and using the relative sensitivities of Jørgensen and Berthou [8], gives a figure of ~3 at%. The difference of three orders of magnitude, if correct, may be due to various factors, viz. an enhancement of surface concentration of platinum over that in the bulk by segregation, surface diffusion of platinum over the surface of the molten indium from the sides of the boat, or greater solubility of platinum compounds, e.g. oxides, than of metallic platinum. Intermetallic compounds of indium and platinum are known to exist, but are unlikely to have been formed at such low temperatures.

5. Conclusions

During the initial stages of cleaning indium, the impurity elements sulphur and oxygen disappeared and reappeared at the surface, on melting and cooling, over a very narrow temperature range. The marked similarity in such behaviour to that observed for carbon on nickel by Shelton et al. [3] suggests a similar explanation, based on the Bragg—Williams condensation model. Even with the lower coordination in InS compared to graphite, the estimated critical temperature is far in excess of the melting-point of indium, so that the condition for very rapid changes in surface coverage on either side of the precipitation temperature (in this case the melting-point) was established. The observed behaviour is probably typical of many metal/non-metallic-impurity systems. In both spectra traces of platinum appeared after heating to no more than 200°C, and their appearance was attributed to dissolution of a small amount of platinum, possibly in the form of oxide, by the molten indium.

References

[1] P.W. Palmberg, J. Electron Spectrosc. 5 (1974) 691.
[2] M. Gettings and J.P. Coad, Surface Sci. 53 (1975) 636.
[3] J.C. Shelton, H.R. Patil and J.M. Blakely, Surface Sci. 43 (1974) 493.
[4] P.W. Palmberg, G.E. Riach, R.E. Weber and N.C. MacDonald, Handbook of Auger Electron Spectroscopy (Physical Electronics Industries, 1972).
[5] H.G. Ansell and R.S. Boorman, J. Electrochem. Soc.; Solid State Sci. 118 (1971) 133.
[6] W.J. Duffin and J.H.C. Hogg, Acta Cryst. 20 (1966) 566.
[7] S.P. Yatsenko and E.N. Dieva, Russian J. Phys. Chem. 47 (1973) 1658.
[8] C.K. Jørgensen and H. Berthou, Faraday Discussions 54 (1972) 269.

SEGREGATION OF TIN ON (111) AND (100) SURFACES OF COPPER

J. ERLEWEIN and S. HOFMANN

Max-Planck-Institut für Metallforschung, Institut für Werkstoffwissenschaften, Seestrasse 92, D-7000 Stuttgart-1, Fed. Rep. Germany

Surface segregation of Sn in Cu is measured at (111) and (100) surfaces by means of AES and LEED. In the case of at temperature measurements and no cosegregation of impurities occurring, equilibrium segregation is accomplished for Sn bulk concentrations between 40 and 4300 at ppm and temperatures of 800 to 1230 K. The maximum segregation level of Sn corresponds to a $(\sqrt{3} \times \sqrt{3})R30°$ structure for the (111) surface and a $p(2 \times 2)$ structure for the (100) surface. For theoretical analysis, the Langmuir–McLean equation has to be modified. No difference in segregation enthalpies for both surface orientations is found within the experimental error. The mean segregation enthalpy is determined to $\Delta H = -(53 \pm 5)$ kJ/g-atom.

1. Introduction

Segregation of solute atoms is a generally observed phenomenon in the study of interface or surface composition of alloys after temperature treatment. Many of the segregation experiments at surfaces have shown that complications in theoretical treatment may arise from analysis after quenching to room temperature [1,5–8], from contaminations [2,6,7,9] or from absence of true equilibrium composition due to the limiting effects of bulk diffusion and evaporation [8–11]. For a direct quantitive evaluation of equilibrium surface segregation, Auger electron spectroscopy (AES) has shown to be one of the most powerful techniques. To avoid any possible disturbance by structural imperfections like grain boundaries, experiments on single crystalline surfaces are expected to be most reliable. Using low index planes, auxiliary LEED studies have already shown the occurrence of definite segregation structures of foreign atoms on metal surfaces [1,12,13]. The quantitative thermodynamic data derived from carefully conducted segregation measurements can be readily compared with existing theories [15–18,24].

In this paper we report on equilibrium segregation studies of tin on (111) and (100) surfaces of copper by AES, which showed maximum segregation levels at ordered structures of tin revealed by LEED. Preliminary results on surface segregation kinetics have been published elsewhere [11] and will be discussed in detail in a subsequent paper [19].

2. Experimental

(111) and (100) surfaces of high purity (99.999%) copper single crystals doped with tin (0–0.5 at%, atom absorption analysis) were studied. Each sample of about 200 μm thickness was spot-welded to a NiCr/Ni thermocouple. Segregation experiments were performed in an UHV of ⩽30 nPa using CMA Auger analyzer and a threegrid LEED system (Physical Electronics Ind.). Due to the low residual gas pressure, it was possible to keep the sample surface free from any detectable contamination for a period of about one day. The surface of the sample, which could be heated by electron bombardment from the rear was repeatedly cleaned by bombardment with 1 keV argon ions followed by annealing. A typical Auger spectrum is shown in fig. 1. Only Cu and Sn signals are above the noise level. This proved to be essential for the reliability of segregation experiments.

Since the Auger signals were changed during electron bombardment heating in an unpredictable way, direct at temperature measurements gave no valuable results. Therefore, an indirect at temperature measurement was performed which is demonstrated in fig. 2. After heating the sample for a time long enough to establish equilibrium at the (constant) measured temperature, the electron beam heating was interrupted and the true Auger signal of Sn was obtained. By continuous monitoring of the Sn signal, the Auger peak to peak height (APPH) at temperature was

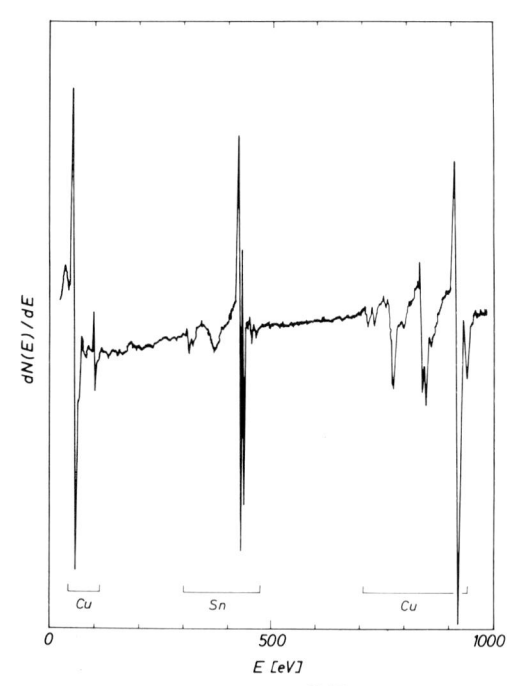

Fig. 1. Auger spectrum of a (111)Cu surface with $X_s^{max} = 0.33$ of Sn segregated on the surface.

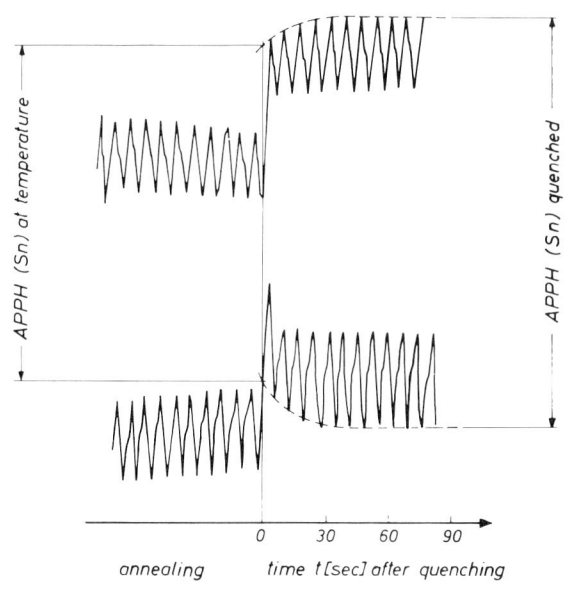

Fig. 2. Evaluation of true at temperature Sn APPH's. For details see text ($T_{\text{anneal}} = 1050$ K).

determined from extrapolation of the quenched value (at room temperature) as shown in fig. 2. To get the optimum time resolution, maximum and minimum of the Sn 430 eV Auger peak were monitored separately.

A crude test for the attainment of segregation equilibrium is a constant Sn signal with increasing annealing time. At low temperatures, however, this conclusion may be erroneous due to the suppressed foreign atom diffusion [8,11,20]. At high temperatures the surface concentration of the segregated species can be decreased by evaporation [8,20] and is likely to go through a maximum with time as shown recently by Lea and Seah [21]. In the system Cu–Sn no change of the Sn Auger signal with annealing time was detected up to 1230 K. Within the experimental errors, reversibility of Sn coverage with temperature was obtained between 800 and 1230 K.

3. Results and discussion

Only those measurements were taken for the evaluation of Sn segregation, where it was possible to completely avoid contamination of the specimen surface from residual gas atmosphere or from cosegregation of impurities [18,22]. Sulfur proved as the most obstinate impurity. An example of its influence on Sn segregation is given in fig. 3. At temperature measurements show a decrease of the Sn Auger signal with increasing S coverage (fig. 3, lower curve). A similar but more pro-

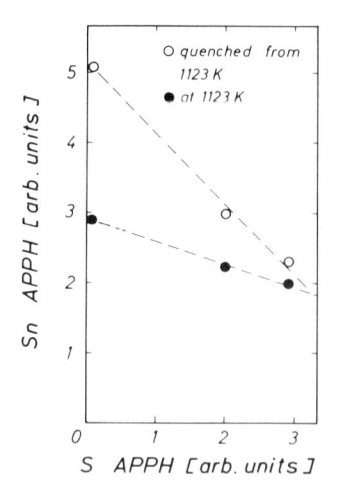

Fig. 3. Relation between tin and sulfur coverages (in APPH's) at temperature and quenched to room temperature.

nounced influence is obtained after quenching (upper curve). At temperature, an increasing number of Sn atoms in equilibrium surface concentration is replaced by the segregating sulfur. During quenching, the sulfur coverage remains constant because of its very low bulk concentration, which was estimated from segregation kinetics with the formalism given in ref. [11] to be $\leqslant 1$ atppm. In contrary, the tin coverage increases towards the equilibrium value at a lower temperature. If a certain amount of surface sites suitable for Sn segregation has been already taken by S atoms at temperature, the Sn segregation during quenching is impeded. Therefore, no simple correction seems to be possible by normalizing the Sn Auger signal at temperature with the Sn/S signal ratio in the quenched state [22].

Auger peak to peak heights (APPH's) of Sn on (111) and (100) surface of Cu with Sn bulk concentrations (in the solid solution region) between $X_b = 4 \times 10^{-5}$ and $X_b = 4.3 \times 10^{-3}$ (atom concentration ratios) were measured at temperatures from 800 to 1230 K. An example for the (111) surface is shown in fig. 4. For $X_b = 4.3 \times 10^{-3}$ a measureable deviation from the maximum attainable value X_s^{max} is obtained only above 1000 K. For $X_b = 2 \times 10^{-4}$, the decrease of the Sn signal is shifted by about 200 K to lower temperatures and X_s^{max} is reached at 800 K. LEED measurements at X_s^{max} (fig. 5) show a $(\sqrt{3} \times \sqrt{3})R30°$ supersturcture. Since it is known sputtering experiments that enrichment of Sn is confined to the first surface layer [11,23], the Sn/Cu atom number ratio at the (111)Cu surface is readily determined to $X_s^{max} = 0.33$ [24].

Corresponding experiments were performed at (100) surfaces which show comparable concentration and temperature dependence of X_s. In this case, however, LEED experiments reveal a p(2 × 2) structure for X_s^{max}, which points to $X_s^{max} = 0.25$. The X_s^{max} values obtained by LEED are in accordance with quantitative

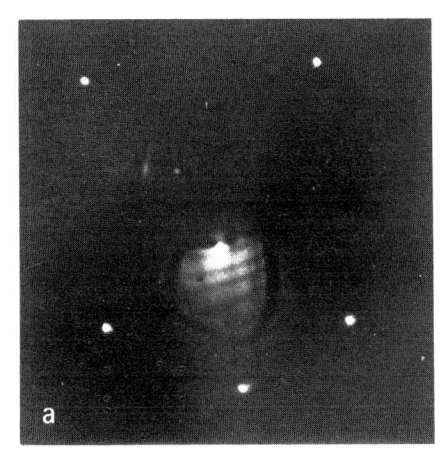

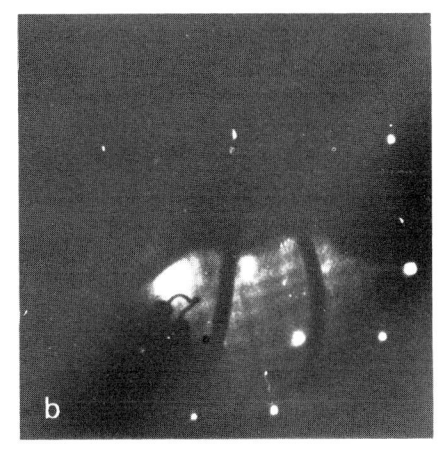

Fig. 4. LEED patterns of (a) Cu(111) surface with Sn segregation of $X_s < 0.01$ ((1 × 1) structure) and (b) Cu(111) surface with Sn segregation of $X_s^{max} = 0.33$ (($\sqrt{3} \times \sqrt{3}$)R30° structure) (Low quality of the pictures is mainly due to the multi-specimen sample holder with electron heating gun and thermocouples in front of the screen).

evaluation [19,23] by means of the measured 10% decrease of the Cu 920 eV Auger signal due to the segregated Sn at $X_s^{max}(111)$. Assuming an exponential decay of the Auger signal output with depth [23], an attenuation length $\lambda_T \cos 42°$ = 13 Å [29] for 920 eV gives a coverage of 0.5 at X_s^{max}, which by division with the ratio of atomic coverage per unit area of Cu and Sn (≈ 1.6) yields $X_s^{max}(111) \approx 0.3$.

Only absolute values of APPH of Sn can be set linearly proportional to Sn coverage [11,23] and not the often used signal ratios like APPH(Sn)/APPH(Cu). Small discrepancies between quantative Auger signal evaluation and LEED information at

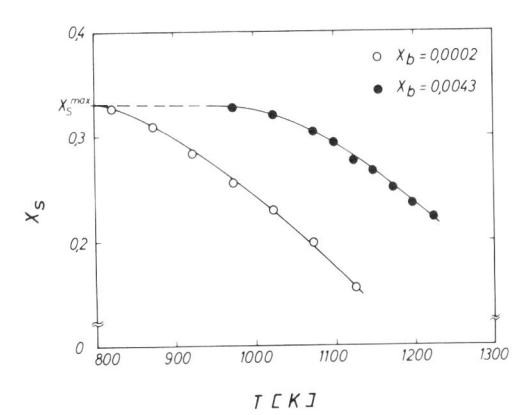

Fig. 5. Equilibrium surface coverage X_s of Sn at Cu(111) surfaces as a function of temperature for Sn bulk concentrations $X_b = 2 \times 10^{-4}$ and $X_b = 4.3 \times 10^{-3}$.

substantial monolayer fractions of the segregant [13] can be explained by this erroneous assumption [19,23,25].

The experimental results cannot be described adequately by the simple Langmuir–McLean equation [10,13,14]. In studies of the kinetics of surface segregation of Sn in Cu [11] a marked discontinuity of the $X_s(t)$ curves was observed at X_s^{max}. This behaviour is theoretically deduced [19] from the occurrence of an ordered segregation structure according to the LEED results. With the implication of site occupancy probabilities, the following modification of the Langmuir–McLean equation must be used for a proper description of $X_s(T, X_b)$ in thermodynamic equilibrium:

$$X_s/(1 - X_s/X_s^{max})^{X_s^{max}} = X_b/(1 - X_b)] \exp(-\Delta G/RT). \tag{1}$$

X_b must be small compared to the solubility limit of the segregant (~0.06 [26]). $\Delta G = \Delta H - T\Delta S$ is the free enthalpy (in terms of enthalpy ΔH and entropy ΔS) of surface segregation. Eq. (1) reduces to the simple Langmuir–McLean equation for $X_s^{max} \to 1$ [10].

Arrhenius plots of eq. (1) with the measured values of X_s and T for different X_b and surface orientations are shown in fig. 6, from which the segregation enthalpies and entropies are deduced and compiled in table 1. For comparison, the values obtained by applying the simple Langmuir–McLean equation are also included. Ob-

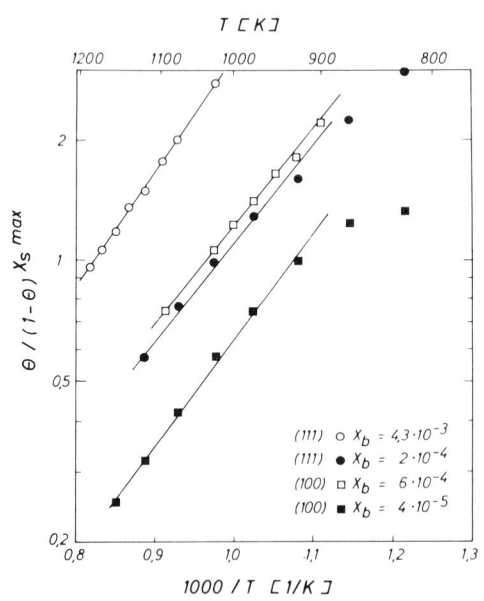

Fig. 6. Plot of $\theta/(1-\theta)^{X_s^{max}}$ versus $10^3/T$ ($\theta = X_s/X_s^{max}$). Straight lines according to eq. (1).

Table 1
Comparison of segregation parameters obtained by eq. (1) and by the simple Langmuir–McLean equation

Surface orientation	X_b	According to eq. (1):		After Langmuir–McLean eq.	
		$-\Delta H$ (kJ/g atom)	$-\Delta S/R$	$-\Delta H'$ (kJ/g atom)	$-\Delta S'/R$
(111)	4.3×10^{-3}	55	1	117	7
(111)	2×10^{-4}	48	2	80	1
(100)	6×10^{-4}	48	1	96	5
(100)	4×10^{-5}	53	1.5	82	1

viously this simplified evaluation leads to a considerable scatter in segregation enthalpy and to entropy values with unreasonable magnitude of $\Delta S'/R$ for the higher bulk concentrations of Sn. In the contrary, evaluation according to eq. (1) gives considerable agreement in the ΔH values and a reasonable order of magnitude for ΔS in all cases.

The deviations of the measured values form the straight lines in fig. 6 at low bulk concentrations and low temperatures are due to the fact, that in this case large mean diffusion lengths and hence too long annealing times will be requested for the attainment of equilibrium surface concentration [11,19]. A limiting experimental uncertainty of the measured data comes from the temperature determination which has been estimated to about ±10 K, resulting in a mean error for ΔH of ±3 kJ/g atom. Together with a small uncertainty in the measured APPH's, a mean value for the segregation enthalpy of Sn on Cu of $\Delta H = -53 \pm 5$ kJ/g-atom (-12.5 ± 1.5 kcal/g-atom) is determined including both surface orientations. It is expected, that the dependence of segregation enthalpy on surface orientation is of the order of the difference in the surface energies [17]. For copper, this difference between (111) and (100) orientations is about 5% [27] which lies within the mean error of ΔH.

Errors in the determination of the segregation entropy comprise those of the enthalpy. Additionally, local variations of bulk concentration in the sample are important. (The integral bulk concentration obtained by atom absorption analysis is a mean value over a relatively large volume of 10^{-2} cm^3.) The mean segregation entropy (in units of R) is determined to $\Delta S/R = -1 \pm 2$. Such a low value is expected from estimations of Ewing and Chalmers [28].

4. Conclusions

(1) Reliable equilibrium surface segregation measurements of Sn in Cu can only be performed if no detectable concentration of any other species is present at the surface (e.g., no sulfur co-segregation).

(2) Since segregating Sn forms an ordered structure on well-defined surface orientations, the Langmuir–McLean equation must be modified to evaluate the segregation enthalpies and entropies. This should be valid for all systems with ordered segregation structure.

(3) No appreciable difference is found between the segregation enthalpies of Sn on Cu(111) and Cu(100) surfaces. Their mean value is determined to $\Delta H = -53 \pm 5$ kJ/g atom.

References

[1] J. Ferrante, Acta Met. 19 (1971) 743.
[2] W.P. Ellis, J. Vacuum Sci. Technol. 9 (1972) 1027.
[3] S. Floreen and J.H. Westbrook, Acta Met. 17 (1969) 1175.
[4] A. Joshi and D.F. Stein, Met. Trans. 1 (1970) 2543.
[5] P. Douglas, J. Mater. Sci. 8 (1973) 1647.
[6] J.-P. Servais, H. Graas and V. Leroy, Centre Rech. Met. 44 (1975) 30.
[7] H.P. Bonzel and H.B. Aaron, Scripta Met. 5 (1971) 1057.
[8] J. Ferrante, Scripta Met. 5 (1971) 1129.
[9] M.P. Seah and C. Lea, Phil. Mag. 31 (1976) 627.
[10] S. Hofmann, G. Blank and H. Schultz, Z. Metallk. 67 (1976) 189.
[11] S. Hofmann and J. Erlewein, Scripta Met. 10 (1976) 857.
[12] L.C. Isett and J.M. Blakely, Surface Sci. 47 (1975) 645.
[13] H.J. Grabke, G. Tauber and H. Viefhaus, Scripta Met. 9 (1975) 1181.
[14] D. McLean, Grain Boundaries in Metals (Clarendon Press, Oxford, 1957).
[15] M.P. Seah and E.D. Hondros, Proc. Roy. Soc. (London) A335 (1973) 191.
[16] F.L. Williams and D. Nason, Surface Sci. 45 (1974) 377.
[17] G.A. Somorjay and S.H. Overbury, Faraday Discussions Chem. Soc. 60 (1975) 279.
[18] M. Guttmann, Metal Sci. 10 (1976) 337.
[19] S. Hofmann and J. Erlewein, to be published.
[20] J.J. Burton, C.R. Helms and R.S. Polizotti, J. Vacuum Sci. Technol. 13 (1976) 204.
[21] C. Lea and M.P. Seah, Phil. Mag. 35 (1977) 213.
[22] C. Lea and M.P. Seah, Surface Sci. 53 (1975) 272.
[23] S. Hofmann, Mikrochim. Acta Suppl. 9 (1977) 000.
[24] M. Prutton, Met. Rev. 152 (1971) 57.
[25] J.J. Grabke, W. Paulitschke, G. Tauber and H. Viefhaus, Surface Sci. 63 (1977) 377.
[26] M. Hansen, Constitution of Binary Alloys (McGraw-Hill, New York, 1958).
[27] W.L. Winterbottom, in: Surfaces and Interfaces, Vol. I, Eds. J.J. Burke, N.L. Reed and V. Weiss (Syracuse Univ. Press, Syracuse, NY, 1967) p. 133.
[28] R.H. Ewing and B. Chalmers, Surface Sci. 31 (1972) 161.
[29] C.J. Powell, Surface Sci. 44 (1974) 29.

SURFACE STUDIES WITH AN IMAGING ATOM-PROBE

A.R. WAUGH and M.J. SOUTHON

Department of Metallurgy and Materials Science, University of Cambridge, Pembroke St., Cambridge CB2 3QZ, UK

The imaging atom-probe (TAP) is an atom-probe field-ion microscope in which the conventional time-of-flight mass spectrometer is replaced by a nanosecond-gated channel-plate image intensifier which is used to record the arrival positions of ions of preselected mass-to-charge ratio which have been field-evaporated from the surface of a field-ion emitter. The evaporated metal ions have the same radial projection that is seen in the normal field-ion image and the recorded image displays the points of origin of the ions with a spatial resolution of 1 nm or better. The field of view covers a selected area of about 0.3 steradian at the curved specimen surface. The mass resolution, limited by the channel-plate gating time, is of order $\Delta m/m = 1/15$ to $1/25$. The wide field of view allows the rapid determination of the crystallographic distribution of differently-charged ions from those elements which evaporate as a mixture of charge species, and shows that this distribution is frequently a sensitive function both of the local surface geometry and of the presence or absence of field-adsorbed helium or neon at the specimen surface. The wide field of view of the IAP gives it a considerable advantage over the conventional atom-probe in the detection of segregation at grain boundaries.

1. Introduction

Since its introduction by Mueller and coworkers [1] the atom-probe field-ion microscope has become established as a useful microanalytical tool, capable of providing accurate analyses of field-ion microscope specimens from sample areas only a few atoms in extent. In the original conception of this instrument, metal ions field-evaporated from the specimen surface follow the same radial projection as the field-ionized gas ions in the simple field-ion microscope: metal ions from a very small area of the specimen pass through a small probe-hole in the microscope screen and are chemically identified in a time-of-flight mass spectrometer. Although this technique is useful in providing analyses of, for example, small precipitates [2,3], it is wasteful of information in that no analysis is obtained from the much larger specimen area lying outside the small probed region. By using a greatly enlarged probe-hole [4] it becomes possible to obtain accurate chemical analyses of very thin films on the specimen surface, but at the expense of decreased spatial resolution. Alternatively, by using a time-gated channel-plate image intensifier as a detector with spatial resolution in a time-of-flight spectrometer, in a configuration first successfully used by Panitz [5], it is possible to obtain micrographs showing the spatial distribution of preselected ions emitted from relatively large specimen areas (some tens of nanometres in diameter) with high spatial resolution ($\simeq 1$ nm). In this paper the application of an instrument of this type to a range of surface analysis problems is discussed.

2. Experimental

The "image-forming atom-probe" (IAP) used in the work discussed here differs in a number of respects [6] from the original design of Panitz [5] and will be described briefly here, and in more detail elsewhere. Essentially, the instrument (fig. 1) is simply a field-ion microscope in which the specimen is placed further (120 mm) than usual from the microscope screen. Metal ions are field-evaporated from the specimen, which is mounted on a goniometer stage, by applying 10 ns high-voltage pulses, generated using a mercury-wetted reed relay as a high voltage switch, at a repetition rate of 30 Hz. The ions drift at a speed determined by their mass-to-charge ratio to a microchannel-plate image intensifier, which may be switched on by electronically-delayed high-voltage pulses of 15 ns duration. By

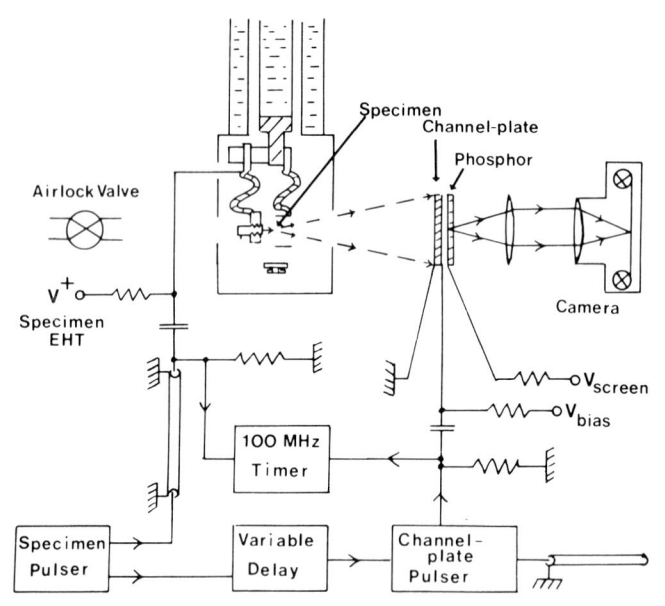

Fig. 1. Schematic diagram of the IAP. The ultra-high vacuum vessel containing the specimen and the channel-plate assembly, and the details of the mounting of the specimen on the goniometer stage, have been omitted for clarity. The specimen is cooled via an electrically-insulated flexible copper braid attached to the inner cryostat, and a radiation shield at 78 K surrounds the whole specimen stage.

varying the delay the on-time of the channel-plate may be made to coincide with the arrival of ions with a selected flight-time: the scintillations produced on the phosphor screen at the arrival positions of these ions are recorded photographically. A field-ion image of the specimen surface is also available simply by biasing the channel-plate continuously on, as in a normal field-ion microscope [7], and by admitting an imaging gas at a suitable pressure: the imaging gas pressure is reduced to 10^{-4} Pa or less, or it is removed entirely, during the recording of time-gated micrographs, so that the number of field-ionized image-gas ions arriving at the detector during the gating pulses is very much less than the number of field-evaporated metal ions arriving during this time. The mass resolution is limited to $\Delta m/m = 1/15$ to 1/25 by the differences in flight-distances across the (single) flat channel-plate used in this instrument, and by the 5 ns rise and fall-time of the gating pulse.

3. Applications

The most obvious applications of the IAP are in metallurgical investigations of segregation and of the initial stages of precipitation, for which the ability of the IAP to detect light elements (e.g. boron, carbon, oxygen, nitrogen) with high spatial resolution is especially useful, and in investigations of the field-evaporation process itself, which is at present imperfectly understood.

3.1. Field evaporation

The majority of metals field-evaporate as a mixture of differently-charged ion species, and it has been suggested [8] that a relatively simple relation exists between the average charge of the ions and the evaporation field of the metal concerned. The difference in local field which exists across the non-uniformly curved surface of a field-evaporated specimen may therefore be expected to cause differences in the crystallographic distributions of the different charge-species. The imaging atom-probe is ideal for detecting such differences in view of the wide angle of specimen crystallography ($\simeq 0.3$ steradian for the instrument used here) which is accessible for analysis at any one time, and has been used to show that the differences in the crystallographic distributions of differently-charged ions are more complex than is suggested by simple theory [8]. Panitz [5] obtained pictures showing the distribution of W^{3+} and W^{4+} from (110) tungsten evaporated in vacuum and later [9] published micrographs showing W^{3+} and He^+ desorbed from (111) tungsten, and Ir^{2+}, Ir^{3+}, He^+ and Ne^+ from (111) iridium. Waugh [6] obtained micrographs showing the crystallographic distributions of all types of ions desorbed from (110) tungsten and (111), (200) and (311) irridium in vacuum, in neon, and in helium, which demonstrated clearly the sensitivity of the crystallographic distributions to the gaseous environment of the specimen during evaporation. These results, and their origins, have been described and discussed in detail in a recent paper [10].

Attempts have been made to measure crystallographic distributions of charge-species using the small probed area of conventional atom-probes. For example, Kinoshita and coworkers [11] found differences in the relative abundances of Mo^{2+}, Mo^{3+} and Mo^{4+} from different crystallographic regions of molybdenum specimens, and Krishnaswamy and coworkers [12] have recently investigated a number of metals with a high-resolution energy-compensated atom-probe. The difficulties of obtaining reliable results using a small probehole are clearly illustrated by the failure of the latter authors to observe quadruply-charged ions in the evaporation products of rhenium, although a large number of planes were probed. With the IAP it is found (fig. 2) that considerable numbers of quadruply-charged ions are obtained from the environs of the $(10\bar{1}1)$ plane, but very few from its centre and from the $\langle 1\bar{2}10 \rangle$ zone-line; this is the case both in helium and in vacuum. Once the appro-

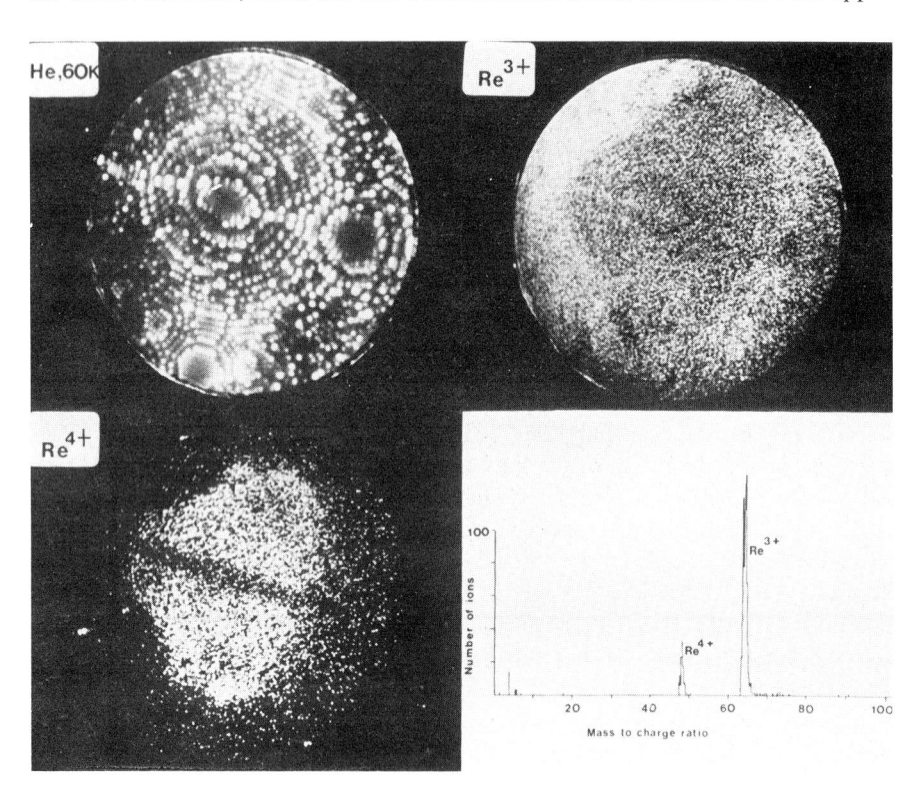

Fig. 2. Helium field-ion and rhenium field-desorption micrographs from $(10\bar{1}1)$ rhenium at 60 K. The Re^{3+} and Re^{4+} images were each produced by field-evaporating some tens of net planes in UHV: the differences in the spatial distributions, and in particular the relative scarcity of Re^{4+} from the (almost horizontal) $\langle 1210 \rangle$ zone-line, are striking. The mass spectrum was taken from the Re^{4+}-rich area in the conventional probe-hole atom-probe, at a background pressure of 6×10^{-10} Torr. This spectrum shows relative abundances of 783 Re^{3+}, 112 Re^{4+} and 2 Re^{2+} ions; 13 He^+ and 8 C^{2+} ions are attributable to the residual gases (He and CO).

priate regions have been identified with the IAP it is a simple matter to obtain accurate abundance ratios with the conventional atom-probe. In fact, for some materials the abundance ratio is such a sensitive function of the site chosen for the analysis and of the evaporation conditions that the graphical results obtained from the IAP are perhaps more useful; for example, for (110) molybdenum (fig. 3) the relative abundances of Mo^{2+}, Mo^{3+} and Mo^{4+} alter rapidly over short distances and are very

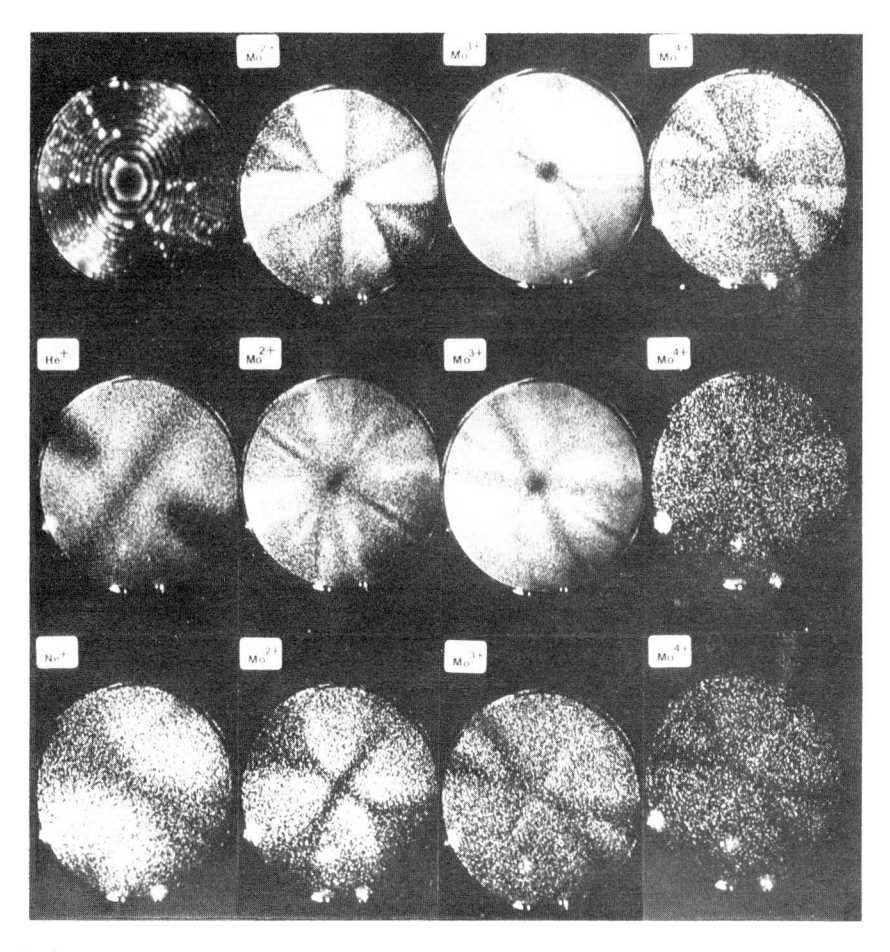

Fig. 3. A field-ion image and complex field-desorption images from (110) molybdenum at 60 K. The Mo^{2+}, Mo^{3+} and Mo^{4+} images were each produced by field-evaporating some tens of (110) net planes; for the top row, evaporation was in UHV; for the centre row, in helium; and for the bottom row, in neon. Also shown are He^+ and Ne^+ images produced by the pulsed desorption of field-adsorbed helium and neon; the differences between the He^+ and the Ne^+ images may be attributable to a crystallographic dependence of the amount of field-adsorbed helium or neon, or [12] to a crystallographic dependence of the formation of metal-helium complex ions, or to both.

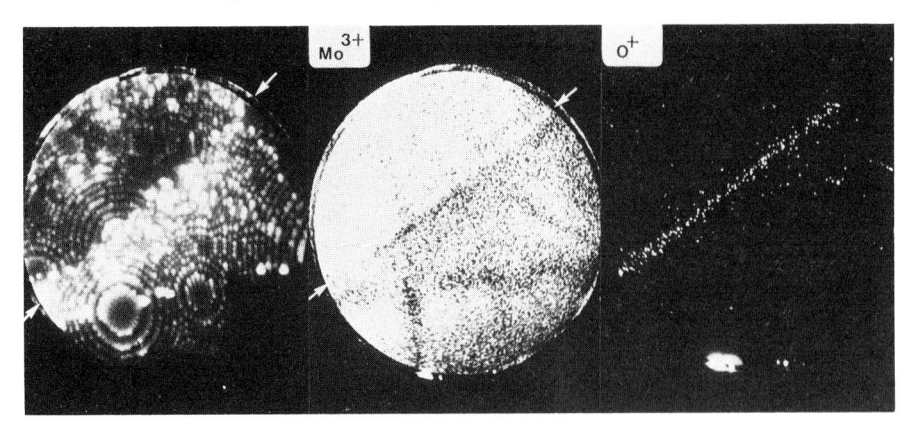

Fig. 4. Field-ion, Mo^{3+} and O^+ images from a molybdenum specimen containing a grain boundary (arrowed) at which segregation of oxygen has occurred. Some ten net planes were field-evaporated for the Mo^{3+} image, and thirty for the O^+ image. The majority of the O^+ ions lie within a band of 1 nm width centred on the grain boundary.

sensitive to the evaporation conditions. Merely to specify the source of a point analysis with a probe-hole atom-probe as a particular plane or on a particular zone-line is inadequate for molybdenum and for other metals; the precise location of the probed region has to be given if the results are to be meaningful.

3.2. Segregation

The ability of the IAP to contribute to segregation studies is illustrated here by fig. 4. This shows the helium field-ion image, the Mo^{3+} desorption image and the O^+ desorption image from a molybdenum specimen containing a grain boundary (arrowed). The Mo^{3+} image shows crystallographic modulations in intensity of the type seen in fig. 3, and also a modulation at the boundary, where the local radius of curvature and hence magnification of the specimen is disturbed. The O^+ image shows very clear evidence for segregation at the boundary; this was independently confirmed in the conventional atom-probe, where the mass-to-charge ratio of the segregant was established as 16 (O^+, or less probably S^{2+}). The segregant concentration at the boundary can be established by counting the number of segregant atoms detected in unit length of boundary when a known thickness of specimen material is field-evaporated: in the present example the concentration is approximately 1×10^{18} atoms/m^2. The segregant profile across the boundary can also be measured: in the present example it was found that under optimum conditions the great majority of segregant atoms were detected in a 1 nm wide band along the boundary; this could be broadened slightly (2 nm) by reducing the evaporation-pulse amplitude by comparison with the specimen standing-voltage and reducing the evaporation rate, probably because any exposed oxygen atoms then have an increased opportunity to migrate, under electrostatic polarization forces, before evaporation.

4. Conclusions

The IAP gives a "panoramic" view of the crystallographic dependence of the different ions formed during field-evaporation. This is especially useful in the detection of strongly-localized events such as segregation, as the tedious point-by-point analysis of the conventional atom-probe is eliminated. Quantitative results can be obtained directly from IAP micrographs, although for most purposes preliminary screening of specimens with the IAP followed by a quantitative point analysis with the conventional atom-probe is more convenient.

Acknowledgements

The authors thank the Science Research Council for support and wish to thank Mr. P. Mills for his assistance with the operation of the conventional atom-probe.

References

[1] E.W. Mueller, J.A. Panitz and S.B. McLane, Rev. Sci. Instr. 39 (1968) 83.
[2] S.S. Brenner and S.R. Goodman, Scripta Met. 5 (1971) 865.
[3] P.J. Turner and J.M. Papazian, Metal Sci. J 7 (1973) 81.
[4] B.J. Regan, C.D. Page, E.D. Boyes and M.J. Southon, 19th Field Emission Symposium, Urbana, Illinois, 1972.
[5] J.A. Panitz, J. Vacuum Sci. Technol. 11 (1974) 206.
[6] A.R. Waugh, 22nd Field Emission Symposium, Atlanta, Georgia 1975.
[7] P.J. Turner, P. Cartwright, E.D. Boyes and M.J. Southon, Advan. Electron. Electron Phys. 33B (1972) 1077.
[8] B.J. Regan, P.J. Turner and M.J. Southon, J. Phys. E9 (1976) 187.
[9] J.A. Panitz, J. Vacuum Sci. Technol. 12 (1975) 210.
[10] A.R. Waugh, E.D. Boyes and M.J. Southon, Surface Sci. 61 (1976) 109.
[11] K. Kinoshita, S. Nakamura and T. Kuroda, Japan J. Appl. Phys. 13 (1974) 1657.
[12] S.V. Krishnaswamy, S.B. McLane and E.W. Mueller, J. Vacuum Sci. Technol. 13 (1976) 665.

AN INELASTIC NEUTRON SCATTERING STUDY OF C_2H_2 ADSORBED ON TYPE 13X ZEOLITES

Josepth HOWARD and Thomas C. WADDINGTON

Chemistry Department, University of Durham, Durham DH1 3LE, England

The inelastic neutron scattering spectra of C_2H_2 and C_2D_2 adsorbed on a Ag^+ exchanged 13X zeolite (0–800 cm^{-1}) and of C_2H_2 on the Na^+ form (0–300 cm^{-1}) have been obtained. For the Na-13X system no distinct vibrational modes were observed, however for the Ag-13X systems the low frequency intramolecular modes of the adsorbed gas and some of the vibrations of the adsorbed gas relative to the surface have been assigned. From the deuteration shifts it appears that C_2H_2 and C_2D_2, adsorbed on Ag-13X, are non-linear.

1. Introduction

This study of adsorbed C_2H_2 follows naturally from our work with C_2H_4 adsorbed on the same zeolites [1]. Because there are so little IR and Raman data on complexes containing C_2H_2 π-bonded to a metal atom and because we have no inelastic neutron scattering (INS) data on such complexes we have used the results from our study of adsorbed C_2H_4 [1] to aid in the interpretation of the INS spectra of adsorbed C_2H_2 and C_2D_2.

There have been some recent infrared [2,3] and Raman [3] studies of alkali and alkaline earth ion exchanged zeolites containing C_2H_2. The authors [3] were able to observe five fundamentals for C_2H_2 on Na^+ and Ca^{2+} A type zeolites but only two on type X. These bands were assigned to intramolecular modes of the adsorbed gas and in no case was a vibration relative to the surface observed. From normal coordinate analyses (NCA) of the systems values of 46 cm^{-1} (NaA) and 41 cm^{-1} (CaA) were derived for the C_2H_2–cation vibrational frequencies [3], however, the accuracy with which these values can be predicted is necessarily low because the vibrational frequencies used in the NCA are high (>650 cm^{-1}).

There are at least two possible configurations for the adsorbed C_2H_2 involving either "side-on" interaction with a cation or "end-on" interaction with the oxide framework. Both configurations have been identified [4] for C_2H_2 adsorbed on alumina. The hydrogen-bonded form produced an increase in the $C \equiv C$ stretching frequency (2005 cm^{-1}) relative to the gas phase value (1974 cm^{-1}) while for the "side-on" form this vibration was found at lower frequency (ca. 1950 cm^{-1}). From an infrared study of the C–H vibrations of the adsorbed gas Tsitsishvili et al. [3]

have proposed the hydrogen-bonded configuration for C_2H_2 adsorbed on some A and X type zeolites. However some of their experiments have been repeated recently [3] and in all cases the ν_2 (C≡C stretching) has been observed in the infrared and always at lower frequency than the corresponding mode in the gas phase. These results favour the "side-on" form on interaction with the metal ions and this is in agreement with the available X-ray structures [5].

2. Inelastic neutron scattering

Neutron scattering techniques have been widely applied to the study of molecular motion and vibrations in solids, liquids and gases [6,7]. The cause of the scattering is the interaction between the incident neutrons and the nuclei of the scattering atoms. Neutrons are unique in that they can exchange both energy and momentum with the scattering nuclei. Thus thermal ($\lambda \sim 4$ Å) neutron scattering encompasses both the spatial measurements that can be made with X-rays and the vibrational measurements obtainable with optical spectroscopy. The neutron technique is analogous to Raman scattering but because of the finite mass of the neutron it can cover a much wider range of momentum transfer (typically 0.1 to 10 Å^{-1} compared to <0.01 Å^{-1} for light scattering). Furthermore as there are *no* optical selection rules to be obeyed, *all* modes of vibration are "active".

Neutrons can be scattered coherently and/or incoherently depending on the nuclear properties of the scatterer. We are concerned *only* with incoherent scattering, which yields information on the density of states of the scattering system [6,7]. In a neutron scattering experiment we normally measure the double differential cross section $d^2\sigma/d\Omega dE$ defined by

$$\frac{d^2\sigma}{d\Omega\,dE} = [\text{Number of neutrons/sec with energy between } E \text{ and } E + \delta E \text{ scattered into}$$

the solid angle $\delta\Omega$] × [Incident flux × δE × $\delta\Omega$]$^{-1}$.

It can be shown that $d^2\sigma/d\Omega dE$ is proportional to the incoherent scattering cross section (σ_{inc}) and the mean square amplitude of vibration of the protons (or the reciprocal mass of the scattering ligand) [8]. The scattering will therefore be particularly intense for nuclei which have large values of σ_{inc} and small M. Since σ_{inc} for hydrogen (80 barn) is larger by at least a factor of 10 than that of other nuclei, including deuterium and of course $M = 1$, only the incoherent scattering from the protons need be considered in a molecule which contains hydrogen. Similar arguments apply to the case of adsorbed (hydrogeneous) gases. Neutrons are not, in the usual sense, a surface probe. Applications of incoherent neutron scattering to the study of adsorbed species rely primarily on the relative transparency of the support material when compared to the large cross-section of the adsorbed species.

Consider for instance our work with hydrogen adsorbed on platinum black [9]. σ_{inc}/M for H is 79.7 whereas for Pt it is 0.004 in the same units. This large difference enables us to observe scattering from the H atoms even when the ratio, number of Pt atoms/number of H atoms, is considerably greater than unity.

3. Experimental

Experiments were carried out using the 6H time-of-flight (t-of-f) [10] and Beryllium Filter Detector (BFD) [11] spectrometers at AERE, Harwell. The 6H spectrometer is best suited to studying energy transfers in the region 0–300 cm^{-1} but we have obtained spectra up to 800 cm^{-1} using the BFD spectrometer. The t-of-f spectra are more difficult to interpret than the BFD spectra because they consist of incompletely resolved bands. In order to help in the problem of locating peak positions with accuracy the spectra from each of the nine angles of detection were fitted to a smooth curve using the Harwell routine "DOUGAL" [12].

The Na-13X was purchased from Linde (Lot No. 1976300) and ion exchanged with AgNO$_3$ solution [13]. The sample dehydration, degassing and adsorption were carried out exactly as described for our work on C$_2$H$_4$ + Ag-13X [1].

Deuterated gases were obtained from Merck, Sharpe and Dohme Ltd. and the cylinder ethylene and acetylene were >99.9% pure. Spectra were obtained at only one coverage, corresponding to a pressure of 40 Torr of C$_2$H$_2$, and at only one temperature on each spectrometer, i.e.: (a) 136 K on the 6H time of flight spectrometer; (b) ca. 90 K on the Beryllium Filter Detector spectrometer (BFD).

4. Discussion and assignments

4.1. C$_2$H$_2$ + Ag-13X

4.1.1. Intramolecular acetylene modes

By comparison with the IR and Raman results for solid C$_2$H$_2$ and C$_2$H$_2$ adsorbed on NaA and CaA zeolites [3] we can assign the bands at 794 and 673 cm^{-1} in the INS spectrum of C$_2$H$_2$ + Ag-13X (fig. 1a), to ν_4 and ν_5 respectively. The increase in frequency of these bands relative to solid C$_2$H$_2$ is greater than observed on type A zeolites [3]. Of the two possible explanations for this, (a) the change of cation and (b) the change of framework, the former is probably the more important since changing from an A to an X type zeolite increases ν_2 by only 1 cm^{-1} [3]. Although ν_4 and ν_5 are degenerate in the gas phase this degeneracy should be removed on adsorption and it is possible that we are observing some splitting at ca. 750 cm^{-1} (fig. 1a). The broad band, with a maximum at 584 cm^{-1} in the INS spectrum of C$_2$D$_2$ + Ag-13X, is unresolved ν_4 and ν_5; ν_5 has been assigned at 567 cm^{-1} and 2ν_4 at 1057.5 cm^{-1} in solid C$_2$D$_2$ [14].

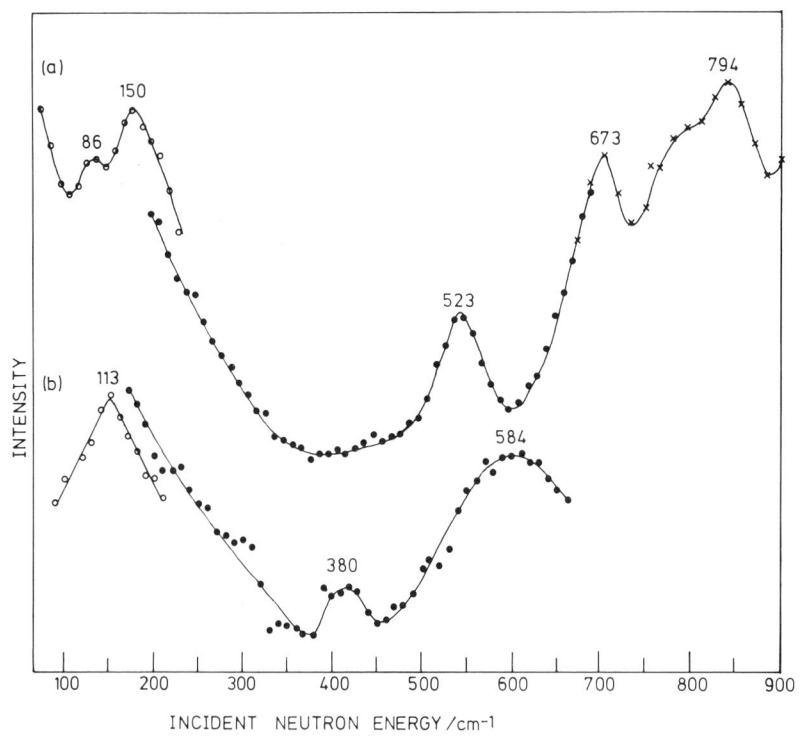

Fig. 1. Beryllium filter detector spectra of: (a) C_2H_2 + Ag-13X; (b) C_2D_2 + Ag-13X.

4.1.2. Acetylene-zeolite modes

There are no bands in the region 200 cm^{-1} → 550 cm^{-1} in the IR or Raman spectra of solid C_2H_2 or 200 → 400 cm^{-1} for C_2D_2 [14–18]. The peaks at 523 cm^{-1} (C_2H_2 + Ag-13X) and 380 cm^{-1} (C_2D_2 + Ag-13X) (figs. 1a, b) must therefore be vibrations of the acetylene molecule relative to the surface. (The six possible vibrations are shown in fig. 2.) It is reasonable to assume that they are the same mode which has shifted on deuteration. The ratio of their frequencies ($\nu C_2D_2/\nu C_2H_2$) is 0.727. This corresponds closely to a $1/\sqrt{2}$ shift (0.707) and this is only possible if the carbon atoms contribute very little (or nothing) to the moment of inertia of the mode. Clearly there is only one possible C_2H_2–zeolite vibration which can satisfy this criterion: τ_x (fig. 2). For a linear molecule τ_x is of zero frequency and so the adsorbed molecules must be non-linear. This differs from the conclusions reached for C_2H_2 on alkali and alkali metal zeolites [2,3], however, silver is able to use its filled "d" orbitals for back donation to C_2H_2 leading to stronger interactions with the adsorbed gas than are expected for Na$^+$ etc. The band at 523 cm^{-1} is also of higher frequency than all of the C_2H_4 to PtCl$_3$ modes in Zeise's Salt [19] except τ_x, and greater than any C_2H_4–Ag-13X mode [1] so that

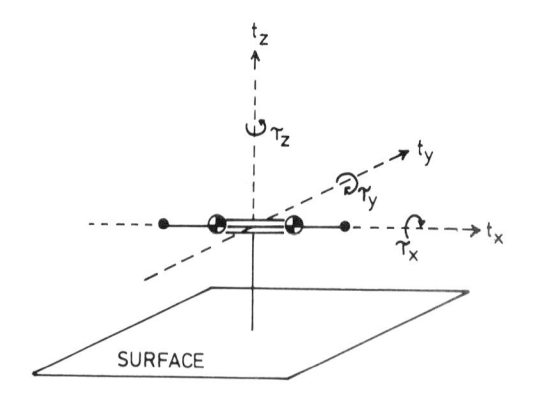

Fig. 2. C_2H_2 vibrations relative to the surface.

an assignment of this band to any vibration other than τ_x would be very difficult to justify. The higher frequency of this mode relative to that observed for adsorbed C_2H_4 [1] may be due to the large difference in their moments of inertia. It is interesting to note that the ratio $(I_{C_2H_2}/I_{C_2D_2})^{1/2}$ for mode τ_x is independent of the angle $\widehat{CCH}$ provided it is non-zero. With C–H as 1.06 Å and C≡C as 1.2 Å the ratio is 0.7338. The ratios $(I_{C_2H_2}/I_{C_2D_2})^{1/2}$ for modes τ_z and τ_y are shown in table 1. Uncertainty in the bond lengths of the adsorbed molecules and in the positions of the INS bands makes it impossible for us to estimate the angle $\widehat{CCH}$ from our data.

The peaks at 150 cm^{-1} (C_2H_2 + Ag-13X) and 113 cm^{-1} (C_2D_2 + Ag-13X) (figs. 1a, b) may also be assigned to the same mode. Their frequency ratio ($\nu C_2H_2/\nu C_2D_2$) is 0.75 and this is close to the value expected for τ_x, however, there is a *constant* uncertainty (± 14 cm^{-1}) in the location of the peak centres. At lower frequencies the frequency ratio becomes increasingly sensitive to this uncertainty and therefore in the lower energy region the predicted frequencies are a better guide to the accuracy of the assignments than are the frequency ratios. Table 1 lists the predicted frequencies for the C_2D_2 system calculated using the data from C_2H_2 + Ag-13X. τ_z and τ_y have almost identical moments of inertia if $\widehat{CCH} > 150°$. The frequency shift on deuteration indicates that the bands at 150 and 113 cm^{-1} should be assigned to one of them since the shifts predicted for the hindered translations are far too small. By comparison with our data for adsorbed C_2H_4 it would appear more reasonable to assign these bands to τ_y, i.e. the antisymmetric stretch is of higher frequency than the C_2 torsion.

The lower frequency bands at 53.5 and 27 cm^{-1} (C_2H_2 + Ag-13X) and 46 and 24 cm^{-1} (C_2D_2 + Ag-13X) (fig. 4) are more difficult to assign. The frequency shifts are subject to large errors because the centres of the peaks are difficult to determine and so any assignment must be regarded as tentative. Both peaks are, however, definitely shifted to lower frequencies on deuteration. Because the adsorbed mole-

Table 1
INS results for acetylene adsorbed on Ag-13X (cm^{-1})

C$_2$H$_2$ + Ag-13X		C$_2$D$_2$ + Ag-13X		C$_2$D$_2$ + Ag-13X Predicted from C$_2$H$_2$ + Ag-13X [a]	Assignments	Ratio of observed frequencies	Predicted ratio of frequencies	
BFD	t-of-f	BFD	t-of-f				Non-linear molecule [a]	Linear molecule
794 ± 14		584 ± 14			⎰ Intramolecular modes			
673 ± 14					⎱ v_4 and v_5			
523 ± 14		380 ± 14 (300?)		384 ± 10	τ_x	0.73 ± 0.05	0.734	Indeterminate
150 ± 14		113 ± 14	102 ± 6	127.5 ± 12	τ_y ?	0.75 ± 0.15	0.850	0.849
86 ± 14				83 ± 13	t_z ?		0.964	0.964
	53.5 ± 2.5		46 ± 2.5	45.5 ± 2	τ_z	0.86 ± 0.08	0.850	0.849
	27 ± 1.5		24 ± 1.5	23 ± 1.2	τ_z	0.89 ± 0.11	0.850	0.849

[a] C≡C as 1.2 Å, CCH = 170°, C–H as 1.06 Å.

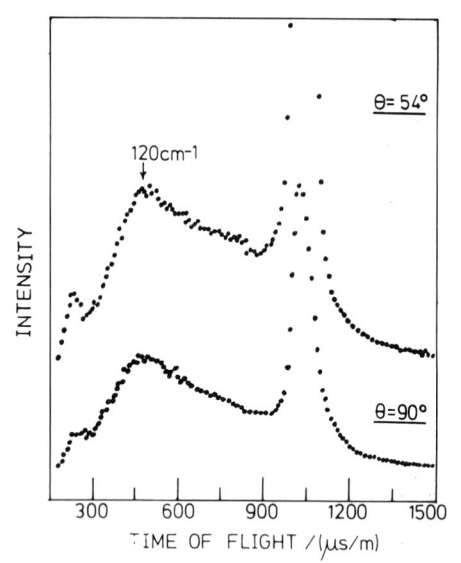

Fig. 3. Time-of-flight spectra of C_2H_2 + Na-13X.

cule is non-linear we would expect six C_2H_2–zeolite modes, i.e. τ_z, τ_y, τ_x and t_z, t_y, t_x (fig. 2). Of these our experience with model complexes and C_2H_4 + Ag-13X indicates that modes τ_z, τ_y and τ_x will be the more intense. It has also been possible to assign t_z in previous cases but it has been quite distinct from any other spectral feature. There are then at least three possibilities:

(a) From the frequency shifts and ratios (table 1) we could assign the 53.5 and 46 cm^{-1} peaks to τ_z and the lower energy peaks to one of the hindered translations (probably t_z).

(b) The peaks at ca. 50 cm^{-1} may be hindered translations and the lower frequency peaks may be τ_z.

(c) It is possible that we have a situation similar to that found for C_2H_4 + Ag-13X, i.e. the lower frequency peaks are both τ_z which has been split by steric hinderance within the supercage, i.e. we are observing the in and out-of-phase vibrations of pairs of coupled rotors. These modes have also been observed in cis-ethylene complexes [20].

We consider alternative (c) to be the correct one for two reasons. Firstly the radii of gyration of the protons in C_2H_4 and C_2H_2 are 1.56 and 1.66 Å respectively (gas-phase geometries) so that interaction is even more probable than for adsorbed C_2H_4. Secondly assignment (c) is more acceptable on intensity grounds. As explained earlier the mass sensitive modes are expected to be considerably less intense than the torsions.

Only one spectral feature remains to be assigned – the band at 86 cm^{-1} in the C_2H_2 + Ag-13X spectrum. There is no band corresponding to this in the spectrum

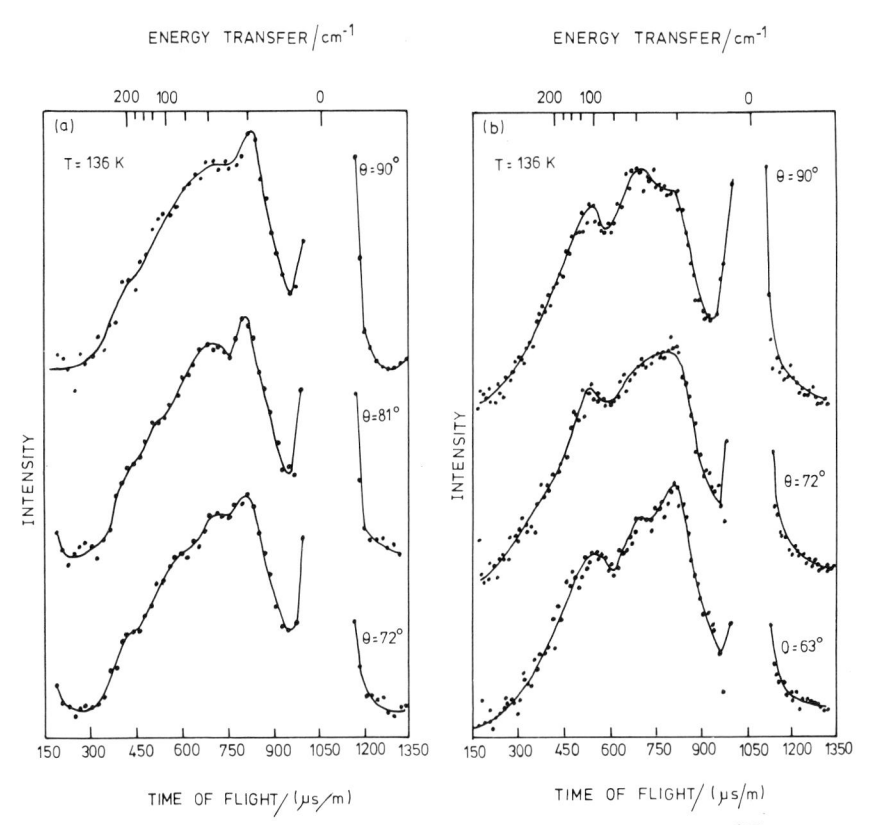

Fig. 4. Time-of-flight spectra of: (a) C_2H_2 + Ag-13X; (b) C_2D_2 + Ag-13X.

of adsorbed C_2D_2. If our previous assignments are correct then this must be a mode arising from a hindered translation. Its intensity relative to the τ_y mode (at 150 cm^{-1}) appears to be of the correct order, however, because we do not have any indication of the true background level we cannot make accurate quantitative measurements. The corresponding feature for C_2D_2 + Ag-13X is perhaps not resolved from the intense peak at 113 cm^{-1}.

4.2. C_2H_2 + Na-13X

We have obtained only time-of-flight data (fig. 3) for this system (120 K) and it can be seen that the spectrum is shifted to lower energies compared to that of C_2H_2 + Ag-13X. It is also very poorly resolved and there is just a single broad band which peaks at ca. 120 cm^{-1}. There are no distinct bands below 600 cm^{-1} and above this the intramolecular acetylene modes occur. This indicates that either τ_x has shifted (from 523 cm^{-1} in Ag-13X + C_2H_2) to less than 250 cm^{-1} or else the adsorbed molecule is linear. Because of the expected weaker interaction between C_2H_2 and Na^+ compared to Ag^+ we are inclined to the latter view.

In the quasi-elastic region the variation in intensity of the elastic peak with momentum transfer, relative to the inelastic region is far less marked than was found for C_2H_4 + Na-13X [1]. The elastic peak is broadened relative to the resolution function of the instrument and this indicates that some diffusive motion is present. This may also explain the broadness of the inelastic spectrum since the observed spectra are a convolution of the translational, rotational and vibrational modes. We do not have any structural data for Na-13X + C_2D_2 so that further analsysis of the broadening is not justified.

In general therefore the assignments from the time-of-flight data must be regarded as tentative, however, we hope to obtain better resolved INS spectra, in the range $0-200$ cm^{-1}, by using the IN4 spectrometer at the Institute Laue–Langevin in Grenoble.

Acknowledgements

One of us (J.H.) thanks the AERE (Harwell) for the provision of a research studentship and we thank C.J. Wright for assistance with one of the INS experiments.

References

[1] (a) J. Howard, Ph.D. Thesis, University of Durham (1976).
    (b) J. Howard, T.C. Waddington and C.J. Wright, to be published.
[2] G.V. Tsitsishvili, C.D. Bagratishvili and N.I. Oniashvili, Zh. Fiz. Khim. 43 (1969) 950.
[3] (a) Nguyen The Tam, R.P. Cooney and G. Curthoys, J.C.S. Faraday I, 72 (1976) 2577.
    (b) Nguyen The Tam, R.P. Cooney and G. Curthoys, J.C.S. Faraday I, 72 (1976) 2592.
[4] D.J.C. Yates and P.J. Luccesi, J. Chem. Phys. 35 (1961) 243.
[5] (a) A.A. Amara and K. Seff, J. Phys. Chem. 77(7) (1973) 908.
    (b) P.E. Riley and K. Seff, Inorg. Chem. 14(4) (1975) 714.
[6] B.T. M. Willis, Ed. Chemical Applications of Thermal Neutron Scattering (Oxford Univ. Press, 1973).
[7] P. Egelstaff, Ed., Thermal Neutron Scattering (Academic Press, 1965).
[8] W. Marshall and S. Lovesey, Theory of Thermal Neutron Scattering (Oxford Univ. Press, 1971).
[9] J. Howard, T.C. Waddington and C.J. Wright, J. Chem. Phys. 64 (1976) 3897.
[10] L.J. Bunce, D.H.C. Harris and G.C. Stirling, AERE Report R6246 (HMSO London 1970).
[11] P.H. Gamlen, N.F. Hall and A.D. Taylor, AERE Report RRL 74/693.
[12] R.E. Ghosh, AERE Report RRL 74/552.
[13] R.M. Barrer and W.M. Meier, Trans. Faraday Soc. 54 (1958) 1074.
[14] G.L. Bottger and D.F. Eggers, J. Chem. Phys. 40 (1964) 2010.
[15] M. Ito, T. Yokoyama and M. Suzuki, Spectrochim. Acta 26A (1970) 695.
[16] A. Anderson and W. Hayden Smith, J. Chem. Phys. 44 (1965) 4216.

[17] G. Hertzberg, Infrared and Raman Spectra of Polyatomic Molecules (Van Nostrand, New York, 1945).
[18] G. Glocke and M.M. Renfrew, J. Chem. Phys. 6 (1938) 340.
[19] J. Hiraishi, Spectrochim. Acta 25A (1969) 749.
[20] J. Howard, T.C. Waddington and C.J. Wright, J.C.S. Faraday II 72 (1976) 513.

THE BACKSCATTERING OF LOW ENERGY IONS AND SURFACE STRUCTURE

W. HEILAND and E. TAGLAUER

Max-Planck-Institut für Plasmaphysik, Euratom Association, 8046 Garching/München, Germany

Low energy ion scattering (<2 keV) in combination with LEED allows surface structure analysis in the case of gases adsorbed on single crystal surfaces. Strong shadowing and anisotropic effects observed are used to estimate the position of the adsorbed species. These techniques may also be extended to study adsorption on polycrystals. For the study of the faces of polar crystals multiple scattering effects in conjunction with shadowing offer a method of surface structure analysis. Surface relaxation effects may possibly be successfully measured by high energy (100 keV – 2 MeV) ion scattering making use of channelling and blocking effects.

1. Introduction

The determination of the position of surface atoms relative to the bulk atoms is still an unresolved problem, even though low energy electron diffraction (LEED) has been successful in some cases and has led to results which are in general agreement with physical or chemical expectations. However, such LEED experiments are cumbersome and the evaluation is time consuming and, in the case of the full-range dynamic treatment, prohibitively expensive. Any auxiliary technique which may narrow the field of possible structures should therefore be helpful. In this paper, we try to demonstrate that ion scattering is such a technique which is compatible with LEED experiments and indeed gives additional information about surface structures. The potential of ion scattering as a tool for surface structure studies has been recognized by Smith [1], who demonstrated that CO adsorbs on Mo with the C bound to the metal and the O sitting on top of the C, since the scattering from C is effectively screened by the presence of the oxygen.

2. Ion scattering

Low ion scattering was reviewed just recently in general [2] and for analytical purposes [3]. An excellent review of the whole field of ion–solid interaction has been given by McCracken [4]. For the purpose of this paper the elastic part of the ion–atom interaction is most important, i.e. the repulsive interaction between the

$$E = E_0 \frac{1}{(1+A)^2} [\cos\vartheta \pm \sqrt{A^2 - \sin^2\vartheta}\,]^2$$

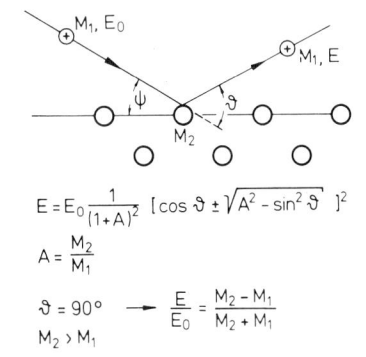

$$A = \frac{M_2}{M_1}$$

$$\vartheta = 90° \quad \rightarrow \quad \frac{E}{E_0} = \frac{M_2 - M_1}{M_2 + M_1}$$

$$M_2 > M_1$$

Fig. 1. Scheme of a binary collision between a projectile (mass M_1, energy E_0) and a surface atom (M_2) at rest before the collision; ψ = impact angle, ϑ = laboratory scattering angle.

nuclear masses. At high energies this interaction can be described by the Coulomb potential (Rutherford backscattering), while at lower energies the electron screening has to be taken into account. Each encounter between a projectile (ion) and a target atom can be described by a binary collision. Under special conditions multiple scattering and channeling/blocking become important, as will be discussed later. In the low energy range (ion energy $E_0 < 2$ keV) it can be ensured that each projectile scatters only once from one surface atom, and scattering occurs from the top monolayer only (fig. 1). The range of this single binary surface collison regime depends on the geometry (surface structure, impact angle ψ, laboratory scattering angle ϑ), on the mass ratio M_2/M_1 (M_2 = target mass, M_1 = projectile mass) and on the energy. Under such conditions extreme surface sensitivity is achieved and the interpretation of the energy spectra of the scattered ions becomes very simple. From the conservation of energy and momentum the energy E of the ions after the collisions is given by

$$\frac{E}{E_0} = \frac{1}{(1 + M_2/M_1)^2} \left\{ \cos\vartheta \pm \left[\left(\frac{M_2}{M_1}\right)^2 - \sin^2\vartheta \right]^{1/2} \right\}^2, \tag{1}$$

where the positive sign is for $M_2/M_1 > 1$ and both signs for $M_2/M_1 \leq 1$. The validity of this relation has been proved for a wide range of energies, mass ratios and scattering angles [1,2].

As a consequence of the single binary surface collisions any additions of adatoms to the top monolayer leads to screening or shadowing effects, i.e. the intensity of the scattering from the substrate atoms decreases with increasing coverage by adatoms (fig. 2). The intensity of the substrate I_s can be written as [5,6]

$$I_s \propto I_0 P_s (N_s - \alpha_a \sigma_a N_a) \Delta\Omega \, d\sigma_s/d\Omega, \tag{2}$$

where I_0 is the primary current density, P_s the neutralization factor, $N_s(\text{cm}^{-2})$ the surface density of the species s, α_a the shadowing coefficient, σ_a the scattering cross

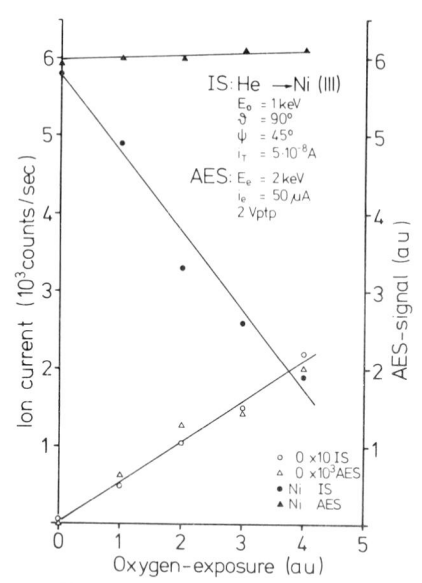

Fig. 2. Exposure dependence of ion scattering and AES signals (peak to peak) for O sorption on Ni. Maximum coverage is about 2/3 of a monolayer. AES signals are measured with a standard 4-grid retarding field system and an electron gun at grazing incidence (for further details see ref. [6]).

section of the adsorbate, N_a (cm^{-2}) the surface density of adatoms, $\Delta\Omega$ the solid angle seen by the detecting system and $d\sigma_s/d\Omega$ the differential scattering cross-section.

P_s, the probability that the ion is scattered as an ion, is in general a function of the electronic configuration of the atoms and of the velocity. The electron transfer from the solid to the ion occurs via Auger neutralisation or resonant charge exchange [7], which may lead in the latter case to oscillations in the ion yield as a function of energy (velocity) [8,9]. No "matrix effects" have been found in the case of the Auger neutralization [3,6]. Some details of the oscillatory ion yield depend on the chemical environment of the surface atom [8,9], but the main effect is still dominated by the charge exchange between the projectile and target atom only. This has been concluded from the angular dependence, where it has been shown that the effect does not depend on the impact angle, but only on the scattering angle via the velocity [10].

The shadowing effect is due to a geometrical factor α, which gives rise to an impact angle dependence [3,6,11,12] and a cross-section dependence, leading to a dependence on the energy and scattering angle [3,5,6,11–13]. The coverage with the adsorbate atoms N_a (cm^{-2}) is in general not known and has to be calibrated. If a linear relation between the signal and exposure can be established (fig. 2), calibration is not a severe problem.

Differential scattering cross-sections are derived from a screened Coulomb potential or a Born–Mayer potential [14]. Suitable values for the potential parameters are found in the literature [15]. A detailed discussion of the cross-sections and the models for the neutralization has been given earlier [2,3] and so we shall briefly summarize: In order to make use of ion scattering for surface analysis, especially structure analysis, measurements over some range of energies and scattering angles are useful in order to get an estimate of the energy (angular) dependence of the neutralization factor P, since the dependence of the cross-section on these quantities is relatively well established. If these functions $P(E)$, $P(\varphi)$ and $P(\vartheta)$ are well behaved, relative or comparative measurements become very useful, without actual knowledge of the absolute values. For example, the cross sections $\sigma(E)$ and $d\sigma(E, \vartheta)/d\Omega$ are independent of the angle of incidence ψ and also independent of the orientation of the plane of scattering relative to the surface, which is described by an azimuthal angle φ. It can also be safely assumed that P is independent of φ (if ψ and ϑ are not too small), and so measurements at constant E, ϑ and ψ with only φ being varied should allow estimates of structural parameters.

In the following sections we shall discuss some examples of the sorption of gases on single crystal surfaces, and the application of ion scattering to the study of the faces of polar crystals. We will conclude with a short look at the application of multiple scattering at low energies and channeling/blocking at high energies for the measurement of surface parameters.

3. Adsorption on metals

The adsorption of gases and other substances on single crystal surfaces has been widely studied with experiments combining LEED and Auger electron spectroscopy (AES) (for reviews, see, for example, refs [16–20]). The combination of LEED, AES and ion scattering has been used to study the sorption of O on Ni (110) [11,12], of O on W (110) [13], of O on Ag (110) [21] and of O on W (100) [22]. Further ion scattering studies have been done on O on Ni (110) [23] and S on Ni (100) [24], where the surfaces were prepared according to recipes elaborated in a different LEED experiment [25]. The results of the ion scattering experiments are generally in agreement with independent LEED studies, which include for some cases the elaborate calculations based on the dynamic theory. (For references: O/Ni(110), O/Ni(100), S/Ni(100) [26]; O/W(100) [27]; O/Ag(110) [28].)

The experimental efforts (UHV, target preparation, etc.) are, of course, the same for ion scattering as for a standard LEED experiment. The combination of the two methods offers rather simple approaches for the interpretation. For example, on the Ag(110) surface a (2 × 1) LEED pattern is formed by half a monolayer of oxygen. The ion scattering spectra (fig. 3) taken at constant ψ, ϑ and E immediately tell that the oxygen is adsorbed between the top (110) rows of the Ag, since nearly no O signal is observed if the plane of scattering is oriented along the

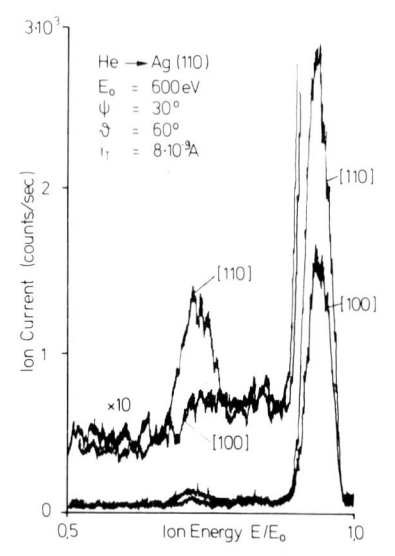

Fig. 3. He$^+$ scattering spectra from Ag covered with 1/2 monolayer of O forming a (2 × 1) LEED pattern. [110] means that the plane of scattering is parallel to the [110] surface chains [21].

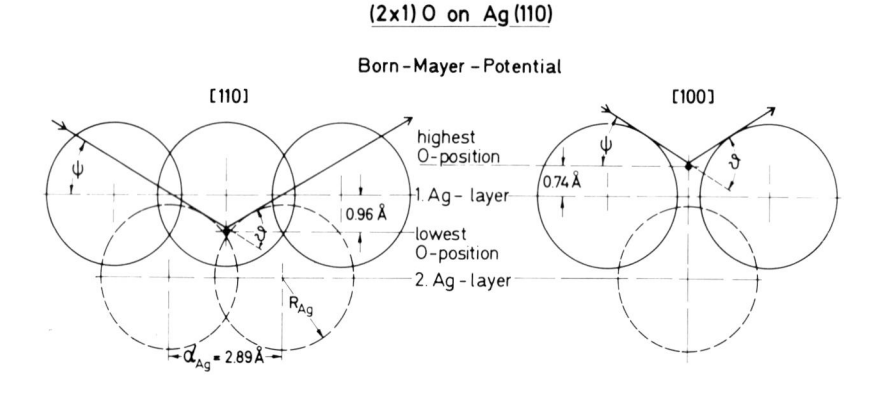

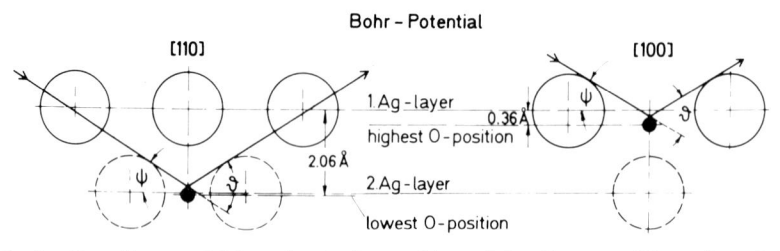

Fig. 4. Ag–O surface model to estimate the position of the O atoms. The radius of the Ag atoms (open circles) is equal to the impact parameter for $\vartheta = 1°$. The O atoms (black circles) have radii equal to the impact parameter for $\vartheta = 60°$. The limiting positions of the O atoms which fit the observed shadowing effect (fig. 3) are indicated [21].

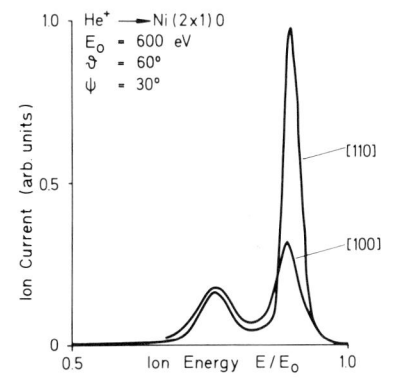

Fig. 5. As fig. 3, but for Ni(110)–O [11].

[100] direction. Furthermore, the very strong shadowing excludes the position on top of the Ag atoms of the second layer and favors the position between two Ag atoms of the top layer. An estimate of the actual height of the O relative to the Ag can be gained from a simple model (fig. 4), where the "size" of the atoms is the impact parameter taken for different scattering cross-sections, thus indicating the limiting values. For the best screened Coulomb potential, the Thomas–Fermi–

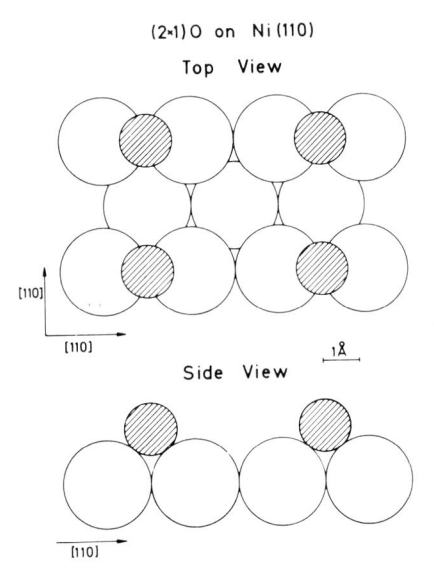

Fig. 6. Position of the O in the (2 × 1) structure on Ni(110). The sizes of the O atoms (hatched circles) and Ni atoms (Open circles) correspond to the "chemical" space occupied (Ni lattice constant $d = 2.48$ Å, O atomic radius $r = 0.73$ Å). [29].

Molière potential, the most probable position is obtained, i.e. the oxygen is located at the same height as the topmost silver layer, in agreement with other studies [28].

For the analogous (2 × 1) structure formed by half a monolayer of O on Ni(100) the ion scattering spectra obtained were strikingly different (fig. 5) [11]. Here the oxygen signal is independent of the azimuthal angle, but the Ni signal changes. This indicates the position of the oxygen to be on top of the outermost Ni layer. From a shadowing effect observed by changing the angle of incidence ψ it was concluded that the oxygen protrudes beyond the plane of the Ni atoms by about 0.4–0.8 Å. This is in good agreement with the LEED analysis (fig. 6) [26,29]. The "short

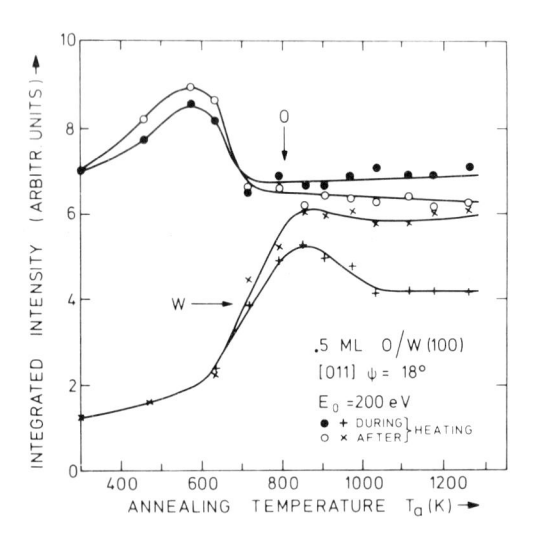

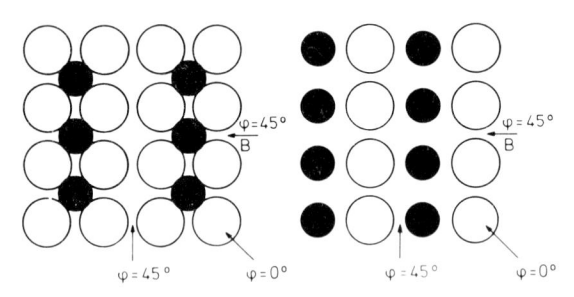

Fig. 7. Temperature dependence of the O and W signal on a (100)W surface covered with 1/2 monolayer of O forming a p(2 × 1) structure. The models of the low and high temperature modifications of the structure are shown in the lower half of the figure. Before heating (each heating period 60–80 s) the O atoms are assumed to take the fourfold symmetry positions above the W plane, after heating they are located in W atom sites in the topmost W plane. The O atoms are partly shadowed by the W atoms. ([011]) indicates the azimuth $\varphi = 45°$ B) [22].

bridge position" of fig. 6 has only recently been proposed [26,29], whereas in ref. [11] the "reconstruction model" [30] was favoured, without discussion of the short bridge position. On the basis of simple arguments (as presented in fig. 4) a distinction between the two models is not possible. On the other hand, the binding mechanism has been called in question by recent photoemission data [29,31], i.e. the binding in the short bridge position seems not to be fully understood.

Surface reordering has been observed in the case of O/W(100) and O/W(100). Fig. 7 shows ion scattering results indicating structural changes as a function of temperature at a constant coverage of 0.5 monolayer of oxygen on W(100) [22]. In the case of O/Ni(100), ion scattering results [23] were interpreted in agreement with LEED results [26], i.e. the oxygen forming on saturation a c(2 × 2) pattern takes the fourfold symmetry position 0.9 Å above the plane of the Ni atoms. The analogous result was obtained for S on Ni(100) [24].

Recently, the sorption of CO on Ni–Cu alloys was studied by ion scattering and SIMS (secondary ion mass spectrometry) [32]. The main conclusion from the shadowing effects observed by ion scattering were that the C is bound to the metal, and that at low coverages the CO is preferentially bound to the Ni-rich parts of the surface, i.e. with increasing coverage first a decrease of the Ni intensity is found before the Cu intensity decreases.

4. Polar crystals

A number of experiments have been done to measure the relative concentration of the constituents on the polar faces of II–VI crystals. The first results [33]

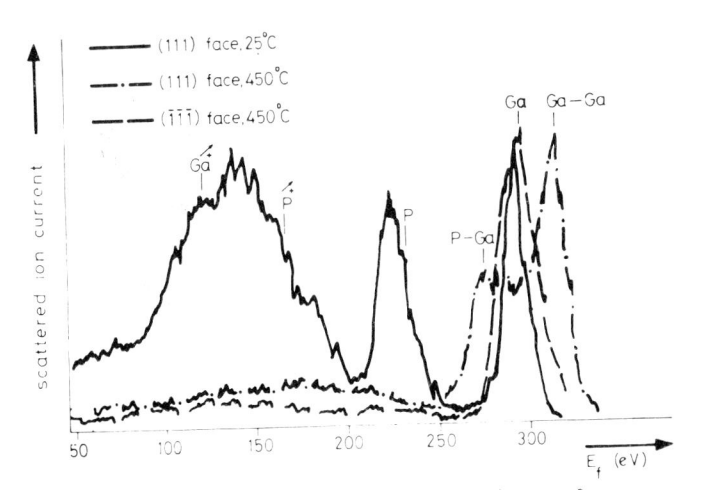

Fig. 8. Scattering of 350 eV Ne$^+$ from GaP surfaces $\varphi = 22,5°$, $\vartheta = 45°$. Double collisons (Ga–Ga and P–Ga) are observed after annealing at 450°C [36].

indicated a large concentration of Cd in the (111) face and of S in the ($\bar{1}\bar{1}\bar{1}$) face. Further experiments [34] then clearly showed that Cd forms the top layer of one face, and S that of the other. (These results again proved the extreme surface sensitivity of the technique as stated above.) Similar results were obtained for ZnTe [35] and CdSe [5].

So far only one study on a III–V crystal has been published, i.e. the (110), (111) and ($\bar{1}\bar{1}\bar{1}$) faces of GaP [36]. In all cases multiple scattering (see below) complicated the spectra, but again marked differences between the spectra from different surfaces indicate the potential of ion scattering for structure analyses (fig. 8).

5. Multiple scattering, channelling and blocking

Multiple scattering is observed with heavier projectiles scattered from single crystal surfaces [2]. It can be understood as a sequence of binary collisions from a chain of atoms (fig. 9). The energy retained by an ion in such an event is higher than for an ion scattered by a single event into the same scattering angle. Obviously, the energy becomes dependent on the lattice constant d and the impact angle, since the impact parameters p_i in the sequence depend on these quantities (fig. 9):

$$P_2 = d \sin(\psi - \vartheta_1) + P_1 . \tag{3}$$

The sensitivity to the surface structure is obvious, but a quantitative exploitation of the effect is mainly hampered by four facts: (i) the trajectories of the ions cannot be calculated analytically owing to the form of the interaction potentials involved, (ii) the parameters of the potentials are not sufficiently well known [2,37], (iii) the simple "chain model" (fig. 9) may be insufficient, while more realistic models lead to lengthy calculations [37] and (iv) the thermal vibrations of the surface atoms play an important role [37,38], but "surface" Debye temperatures are not too well known [39].

Nevertheless, the qualitative information obtained may be helpful, as demonstrated by fig. 8 [36] and by the changes of ion scattering spectra due to changes in the surface structure [40,41]. Fig. 10 demonstrates the changes in the spectra due

$$P_2 = P_1 - d \sin(\psi - \vartheta_1)$$

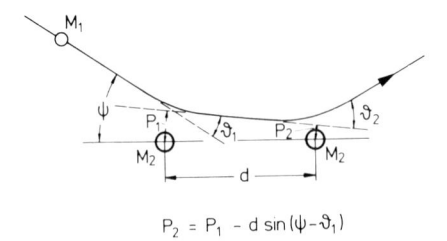

Fig. 9. Scheme of a double binary collision; P_1 and P_2 are the impact parameters, and d the lattice constant in a surface chain.

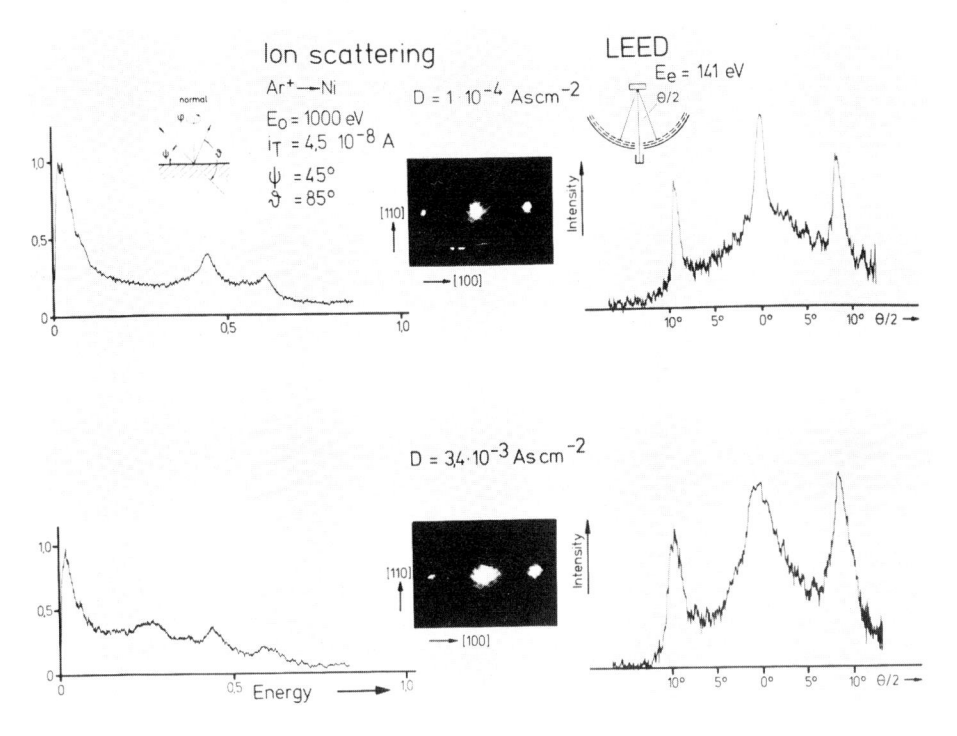

Fig. 10. Dose dependence of Ar$^+$ ion scattering and LEED structure at a bombardment energy of $E_0 = 1000$ eV, impact angle $\varphi = 45°$. The plane of scattering is parallel to the [110] surface direction and perpendicular to the surface. Note the decrease in the multiple scattering intensity (high energy peaks) and the increase of "single" scattering at about $0.23 \, E_0$ [41].

to Ar$^+$ bombardment of a Ni crystal. In good correlation with the change of the LEED pattern, the Ar backscattering spectra show a decrease of the multilple scattering events (high energy double peak) and an increase of a rather broad peak (at lower energy) due to more "single type" binary scattering from surface defects which are created without complete destruction of the crystalline symmetry, as observed for W in a field emission microscope [42].

In the higher ion energy ranges (from 100 keV to several MeV) surface scattering is less important and from single crystals channelling and blocking by atomic rows or planes is observed [4]. Even though the technique is mainly used to study lattice order, atomic positions etc. within the first 10^4 Å of a crystal, it was recently predicted [43–45] that by means of the channelling effect lattice relaxations within the first two layers of a single crystal can be observed. With ^{4}He$^+$ backscattering in the range from 0.5 to 2 MeV a 14% outward relaxation of the top monolayer of a gas covered Pt (111) surface was deduced [43]; these preliminary results have to be compared with "clean" LEED experiments [45], which showed no relaxation within 5%. Experiments [44] with 1 MeV He$^+$ on a (1 × 1) and reordered (5 × 20)

Au(001) surface indicate that only one layer is involved in the hexagonal reordering of the surface. In the 100–200 keV range Cu(110) and Ni(110) were studied with H^+ backscattering [45]. No significant surface relaxation is reported, but surface definition was a problem in this experiment (as in the others); in situ LEED or AES analysis was not yet available. Major efforts are under way in different laboratories to close this gap.

The general advantages of the high energy range (compared to low ion energies) are that the backscattering cross-sections and inelastic losses involved are well known, and that neutralization is no problem at all. It has been estimated that lattice relaxations can be measured to better than ±0.1 Å.

6. Conclusion

Ion scattering in combination with LEED offers the possibility of simplifying surface structure analysis. Further work is necessary (scattering cross-sections, neutralization effects, thermal vibrations) to obtain quantitative results. Low energy ion scattering is especially helpful to study adsorbed gases both on metals and on polycrystals, while high energy ion scattering will very probably help to solve the surface relaxation problem.

References

[1] D.P. Smith, Appl. Phys. 38 (1967) 340.
[2] W. Heiland and E. Taglauer, Nucl. Instr. Methods 132 (1976) 535.
[3] E. Taglauer and W. Heiland, Appl. Phys. 9 (1976) 261.
[4] G.M. McCracken, Rept. Progr. Phys. 38 (1975) 241.
[5] R.E. Honig and W.L. Harrington, Thin Solid Films 19 (1973) 43.
[6] E. Taglauer and W.Heiland, SurfaceSci. 47 (1975) 234.
[7] H.D. Hagstrum, Phys. Rev. 96 (1954) 336.
[8] R.L. Erickson and D.P. Smith, Phys. Rev. Letters 34 (1975) 291.
[9] T.W. Rusch and R.L. Erickson, J. Vacuum Sci. Technol. 13 (1976) 314.
[10] N.H.Tolk, J.C. Tully, J. Kraus, C.W. White and S.H. Neff, Phys. Rev. Letters 36 (1976) 747.
[11] W. Heiland and E. Taglauer, J. Vacuum Sci. Technol. 9 (1972) 620.
[12] W. Heiland, H.G. Schäffler and E. Taglauer, Surface Sci. 35 (1973) 381.
[13] H. Niehus and E. Bauer, Surface Sci. 47 (1975) 222.
[14] I.M. Torrens, Interatomic Potentials (Academic Press, New York, 1972).
[15] H.H. Andersen and P. Sigmund, Riso Report No 103 (1965);
   A.A. Abrahamson, Phys. Rev. 178 (1969) 76;
   M.T. Robinson, ORNL-Report 4556 (1970).
[16] J.B. Pendry, Low Energy Electron Difraction (Academic Press, New York, 1974).
[17] J.A. Strozier, D.W. Jepsen and F. Jona, in: Surface Physics of Crystalline Materials, Ed. J.M. Blakely (Academic Press, New York, 1975) ch. I.
[18] M.B. Webb and M.G. Lagally, in: Solid State Physics, Vol. 28, Ed. H. Ehrenreich (Academic Press, New York, 1973) pp. 301–405.

[19] S.Y. Tong, Progress in Surface Science 7 (1975) 1–48.
[20] C.B. Duke, in: Advances in Chemical Physics, Vol. 27, Eds. I. Prigogine and S.A. Rice, (Wiley, New York, 1974) pp. 1–210.
[21] W. Heiland, F. Iberl, E. Taglauer and D. Menzel, Surface Sci. 53 (1975) 383.
[22] H. Niehus, S. Prigge and E. Bauer, Verh. Deut. Physik. Ges. 9 (1976) 1008;
S. Prigge, H. Niehus, and E. Bauer, Surface Sci. 65 (1977) 141.
[23] H.H. Brongersma and J.B. Theeten, Surface Sci. 54 (1976) 519.
[24] J.B. Theeten and H.H. Brongersma, Rev. Phys. Appl. 11 (1976) 57.
[25] J.B. Theeten, J.L. Domange and J.P. Hurault, Surface Sci. 35 (1973) 145.
[26] P.M. Marcus, J.E. Demuth and D.W. Jepsen, Surface Sci. 53 (1975) 501.
[27] E. Bauer, H. Poppa and Y. Viswanath, Surface Sci. 58 (1976) 517.
[28] H.A. Engelhardt, Thesis, TU München (1975);
H.A. Engelhardt and D. Menzel, Surface Sci. 57 (1976) 591.
[29] J.E. Demuth, in: Proc. 50th Intern. Conf. on Colloids and Surfaces (Academic Press, New York, 1977).
[30] L.H. Germer and A.U. MacRae, J. Appl. Phys. 33 (1962) 2923.
[31] P.R. Norton, R.L. Tapping and J.W. Goodale, Surface Sci. 65 (1977) 13.
[32] D.G. Swartzfeger, private communication.
[33] W.H. Strehlow and D.P. Smith, Appl. Phys. Letters 13 (1968) 34.
[34] H.H. Brongersmaa and P. Mul, Chem. Phys. Letters 19 (1973) 217.
[35] R.F. Goff and D.P. Smith, J. Vacuum Sci. Technol. 7 (1970) 1.
[36] H.H. Brongersma and P. Mul, Surface Sci. 35 (1973) 393.
[37] W. Heiland, E. Taglauer and M.T. Robinson, Nucl. Instr. Methods 132 (1076) 655.
[38] B. Poelsema, L.K. Verhey and A.L. Boers, Nucl. Instr. Methods 132 (1976) 623.
[39] D.P. Jackson, Surface Sci. 43 (1974) 431.
[40] S.H.A. Begeman and A.L. Boers, Surface Sci. 30 (1972) 134.
[41] W. Heiland and E. Taglauer, Radiation Effects 19 (1973) 1.
[42] H. Vernickel, in: Trans. 3rd Intern. Vacuum Congr., Vol. 2 (Pergamon, Oxford, 1966) p. 109.
[43] J.A. Davies, D.P. Jackson, J.B. Mitchell, P.R. Norton and R.L. Tapping, Phys. Letters 54A (1975) 239.
[44] D.M. Zehner, B.R. Appleton, T.S. Noggle, J.W. Miller, J.H. Barret, L.H. Jenkins and O.E. Schow III, J. Vacuum Sci. Technol. 12 (1975) 454.
[45] W. Turkenburg, Thesis, Amsterdam (1976);
W. Turkenburg, W. Soszka, F.W. Saris, H.H. Kersten and B.G. Colenbrander, Nucl. Instr. Methods 132 (1976) 587;
W. Turkenburg, H.H. Kersten, B.G. Colenbrander, A.P. de Jongh and F.W. Saris, Nucl. Instr. Methods 132 (1976) 271.
[46] L.L. Kesmodel and G.A. Somorjai, Phys. Rev. B11 (1976) 630.

ANGULAR DEPENDENCE OF THE SCATTERED ION YIELDS IN ^{4}He$^+ \to$ Cu SCATTERING SPECTROMETRY

P. BERTRAND, F. DELANNAY [*], C. BULENS and J.-M. STREYDIO

Laboratoire de Physico-Chimie et de Physique de l'Etat Solide, Université Catholique de Louvain, 1 Pl. Croix du Sud, B-1348 Louvain-la-Neuve, Belgium

Assuming long range charge exchange mechanisms and neglecting shadowing effects, theory predicts the variation of the scattered ion yield with the scattering angle θ and the incidence angle ψ for some well defined experimental conditions. Such measurements were performed for ^{4}He$^+$ scattering on polycrystalline copper at incident energies ranging from 0.5 to 1.25 keV and at scattering angles from 20° to 130°. It is suggested that shadowing effects should be taken into account in order to explain the observed behaviour.

1. Introduction

Provided shadowing effects are neglected, the intensity of the scattered yield in ion scattering spectrometry for surface analysis is mainly governed by two mechanisms: the binary collision process between the incoming ion and the surface atom and the possible charge exchange between the ion and the surface.

The differential scattering cross section for the binary collision may be calculated by means of classical mechanics provided an adequate spatial variation of the interaction potential between the collision partners is accounted for.

The charge exchange mechanism has been studied namely by Hagstrum [1] in the case of ion–metal collision. This author evidenced a long range interaction due to resonant or Auger exchanges between the electronic levels of the ion and the conduction band of the metal. This process is described by a transition rate which depends only on the separation between the ion and the surface. It is assumed that this transition rate is insensitive to the velocity of the ion. From this theory, the probability that the ion remains ionized during its interaction with the surface is expressed by

$$P = \exp\left[-\left(\frac{1}{v_{0\perp}} + \frac{1}{v_{1\perp}}\right)v_c\right], \qquad (1)$$

[*] Aspirant FNRS.

where v_c is a characteristic velocity defined by

$$v_c = \int_{s_0}^{\infty} R(s) \, ds. \qquad (2)$$

$R(s)$ is the transition rate which depends on the ion-surface separation s, $v_{0\perp}$ and $v_{1\perp}$ are the normal components of the velocities of the ion before and after the scattering and s_0 is the distance of closest approach to the surface.

Since a long range neutralization mechanism is assumed, one may resonably write

$$v_c = \int_0^{\infty} R(s) \, ds, \qquad (3)$$

in which case v_c is constant whatever the type of experiment.

An oscillatory energy dependence of the scattered yield has recently been reported and interpreted in terms of a quasi resonant exchange between deep lying levels of the ion and the surface atom involved in the binary collision [2].

In this paper we will only focus on the $^4\mathrm{He}^+ \to \mathrm{Cu}$ scattering where this oscillatory behaviour has not been observed [3]. Our aim was to check experimentally the validity of the proposed Hagstrum's neutralization model, and, if found to be valid in our case, to determine the characteristic velocity (eq. (2)) by measuring the angular variations of the scattered ion yield. This characteristic velocity should be sensitive indeed to the electronic state of the surface and its accurate determination would be useful for surface characterization. $^4\mathrm{He}^+ \to \mathrm{Cu}$ scattering seemed to be an ideal case for preliminary measurements of this type. On the other hand such a study should contribute to our knowledge of the physical parameters responsible for the height of the scattering peaks in view of a quantitative analysis of surfaces.

2. Theory

In the elastic binary collision model (fig. 1a) the energy E_1 of the ions of mass M_1 scattered at some scattering angle θ after collision with a surface atom of mass M_2 is related to the energy E_0 of the ions before scattering by the well known relation [4]

$$E_1 = E_0 [F(\theta)]^2, \qquad (4)$$

where

$$F(\theta) = \frac{\cos \theta + (\gamma^2 - \sin^2 \theta)^{1/2}}{\gamma + 1}, \qquad (5)$$

with $\gamma = M_2/M_1$, this relation being valid for $\gamma \geq 1$.

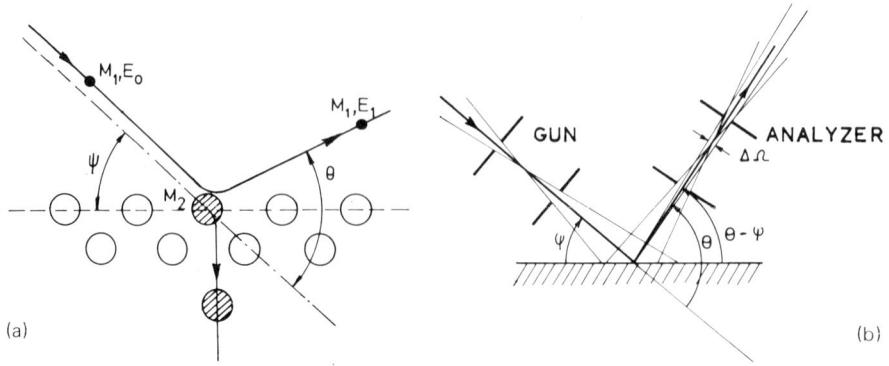

Fig. 1. (a) Diagram of the binary collision. θ is the scattering angle; ψ is the incidence angle; M_1 and M_2 are the masses of the ion and of the surface atom respectively; E_0 and E_1 are the kinetic energies of the ion before and after collision respectively. (b) Diagram of an ISS experiment. $\Delta\Omega$ is the solid angle accepted by the analyzer.

The total yield Y collected at the scattering angle by an energy analyser is the sum of the yields from each of the N scattering centers as long as they do not experience shadowing effects. One writes

$$Y = \sum_N J\sigma Pk \, \Delta\Omega , \qquad (6)$$

where J is the current density of ions incident on the scattering center considered, $\Delta\Omega$ is the solid angle emerging from this center and accepted by the analyser (fig. 1b), k is a dimensionless factor taking into account the transmission of the analyser and the sensitivity of the detection system, σ is the differential scattering cross section for the binary collision and P is the probability for ionic scattering, eq. (1). Owing to the high surface sensitivity of the ionic scattering, if the surface scattering centers do not experience shadowing, the eq. (6) may be written

$$Y = n\sigma P \int kJ \, \Delta\Omega \, dS, \qquad (7)$$

where n is the surface density of scattering centers in the first monolayer and dS is the surface element.

If the spot of the ion beam focused on the surface is much smaller than the portion of surface as viewed from the analyser, i.e. such as $\Delta\Omega \neq 0$ (focused beam), one can assume $\Delta\Omega$ and k to be constant throughout the bombarded area and write

$$Y = n\sigma Pk \, \Delta\Omega \, I/\sin\psi , \qquad (8)$$

where ψ is the incidence angle (fig. 1) and $I = \int J \sin\psi \, dS$ is the total ion current incident on the surface.

If, on the other hand, the surface portion accepted by the analyser is smaller than the bombarded surface area (unfocused beam, case of fig.1b) and if J is a constant on the accepted surface, looking at the scattering in a plane perpendicular to the surface, one may write

$$Y = n\sigma PKJ \, \omega/\sin(\theta - \psi), \qquad (9)$$

where $(\theta - \psi)$ is the analysis angle (fig. 1) and $\omega = \int \Delta\Omega \sin(\theta - \psi) \, dS$ is the input acceptance of the analyser. One sees that for angular measurements, it is very important to know the size of the beam relative to the acceptance of the analyser.

The easiest type of angular measurement consists in measuring the variation of the yield versus the incidence angle ψ at constant scattering angle θ and energy E_0. This may readily be done by rotating the specimen. If one assumes that n and σ remain constant during the experiment (no shadowing and pure binary collision), one may write

$$Y(\psi) \propto \frac{P}{\sin \alpha}, \qquad (10)$$

where α equals ψ or $\theta - \psi$ according to the experimental conditions. If the neutralization mechanism is only governed by Hagstrum's type processes, one obtains by using expression (1)

$$\ln[Y(\psi) \sin \alpha] = \text{constant} - v_c X(\psi), \qquad (11)$$

where

$$X(\psi) = \frac{1}{v_0} \left[\frac{1}{\sin \psi} + \frac{1}{F(\theta) \sin(\theta - \psi)} \right]. \qquad (12)$$

v_0 is the incidence velocity of the ion and $F(\theta)$ is given by (5).

A second type of measurement consists in varying the scattering angle at constant energy. Choosing the specular configuration ($\psi = \theta - \psi = \theta/2$) will make the yield follow the same angular variation whatever the experimental conditions (cf. relations (8) and (9)).

As it has been noted, it is possible to calculate the variation of σ with θ for a specific ion–atom combination. Assuming then that the density of scattering centers is constant during the experiment, one has

$$Y(\theta) \propto \frac{\sigma(\theta)}{\sin(\theta/2)} P, \qquad (13)$$

and similarily to (11)

$$\ln\left[\frac{Y(\theta) \sin(\theta/2)}{\sigma(\theta)}\right] = \text{constant} - v_c X(\theta), \qquad (14)$$

with

$$X(\theta) = \frac{1}{v_0 \sin(\theta/2)} \left(1 + \frac{1}{F(\theta)}\right). \qquad (15)$$

One must insist on the fact that the above relations (11) and (14) are only valid when only long range neutralization processes occur during the scattering.

3. Experimental

The experimental set up has been described in refs. [5] and [6]. The ion gun is based on the Colutron ion beam kit (Colutron MBK1D). The ion current reaches a few 10^{-7} A with a minimum spot diameter of 1 mm at energies between 0.3 and 2.5 keV. The energy dispersion is less than 1 eV. The experimental chamber is evacuated to 10^{-10} Torr by means of conventional sputter ion and titanium sublimation pumps. At the working conditions the pressure is maintained below a few 10^{-8} Torr of noble gases during the scattering experiments. The vessel is fitted with a rotating table allowing a 270° rotation of the analyser around the specimen. The analyser is of the spherical condenser type with a variable predecelerating lens system providing the energy scan [7]. The transmission is nearly constant throughout the investigated energy interval and has been carefully measured to allow correct estimation of the yield. The input acceptance, defined by means of two diaphragms, has a value of 2×10^{-4} mm^2 steradians.

The scattering and incidence angles may be adjusted within a precision of 0.25° and 1° respectively.

The sample is of carefully mechanically and electrolytically polished polycrystalline copper. The cleanliness was monitored by the observation of the disappearance of the ISS peaks associated with impurities as they were sputtered by the incident ion beam. The sample holder is fitted with a 1 mm diameter Faraday cup in the plane of the sample allowing regular control of the ion current and focussing.

The observed scattering peaks have a full width at half maximum of the order of 10 eV, which is much larger than the energy pass band of the analyser (0.6 eV). The yield measured at the energy corresponding to the maximum of the peak equals then the height of the scattered peak instead of its area. Eqs. (8) and (9) are strictly valid only when the area of the peak is considered. But experimentally we can take instead the height of the peak if its ratio to the area remains constant throughout the experiment. This applies in experiments where the only varying parameter is the incident angle. We found that the ratio increases slowly with the scattering angle [5], but we considered that this effect was sufficiently small to be neglected for the needs of our analysis. The experimentally determined energies correspond within less than 1% to the energies predicted by the binary collision model (eq. (4)).

4. Results

4.1. Variation of the yield with the scattering angle

We performed measurements within the interval $20° \leq \theta \leq 130°$ at the three following incident energies: 0.5, 0.75 and 1.25 keV. Fig. 2 shows the general behav-

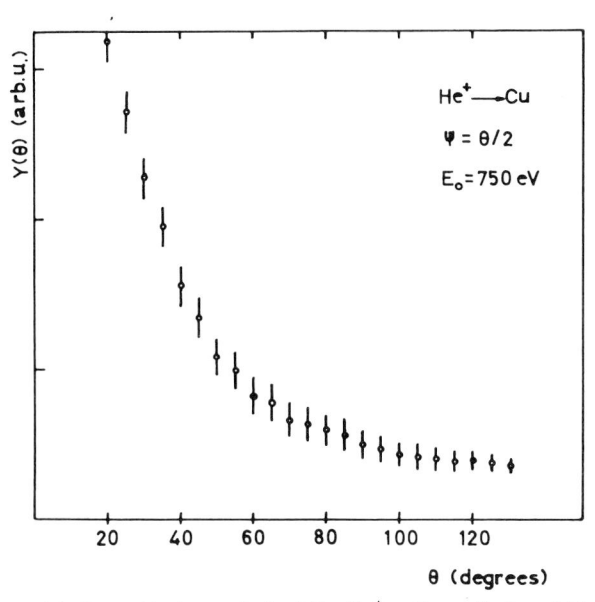

Fig. 2. Variation with the scattering angle θ of the He$^+ \to$ Cu scattering yield at specular configuration and 0.75 keV incident energy.

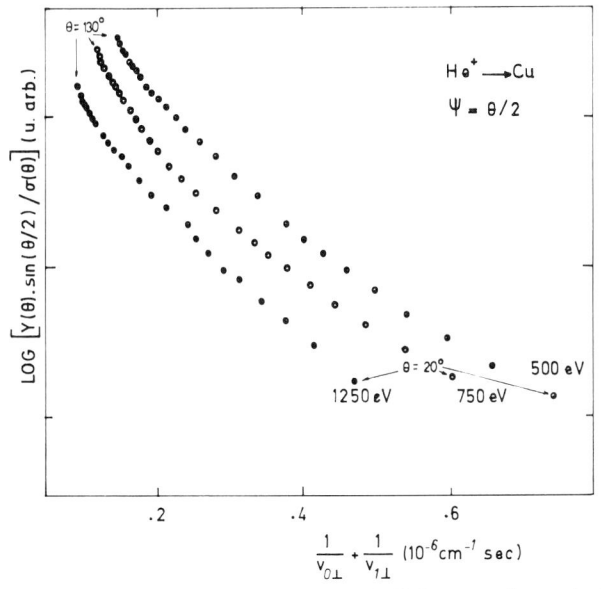

Fig. 3. The measured yield multiplied by the sine of half the scattering angle θ and divided by the scattering cross section $\sigma(\theta)$ is plotted on a logarithmic scale versus the sum of the inverse of the normal components of the velocities before and after the scattering.

iour exhibited by the variation of the yield with the scattering angle in the case where E_0 = 750 eV. One observes a nearly constant yield at large scattering angle ($90° \leq \theta \leq 130°$), and a steep increase of the yield at small angles.

Fig. 3 gives the plot of the results along the lines of the above discussion according to relation (14). The differential scattering cross section has been determined by interpolation on the tables of Robinson [8] using the Molière approximation of the Thomas Fermi interaction potential with Firsov's values of the screening length. Instead of a constant v_c, the slope increases monotonously with the scattering angle θ from a value of about 5.5×10^6 cm/sec for $\theta = 20°$ to a value of about 25×10^6 cm/sec for $\theta = 130°$. Within the experimental accuracy, one cannot observe marked differences between the results obtained at the three chosen energies.

4.2. Variation of the yield with the incidence angle

The relations (8) and (9) for the yields have been established for two ideal cases of focused and defocused beams. In order to interpret correctly the measurements, one must realize experimental conditions which comply with one of the two cases. With our experimental system, it is not possible to preserve the ideal focused case (relation (8)) when varying the incidence angle. We performed measurements of the ψ variation of the yield with a highly defocused beam, restricting the measurements to low incidence angle ($\psi \leq 25°$) in which case the hypothesis (J = constant) leading to relation (9) is most likely fulfilled.

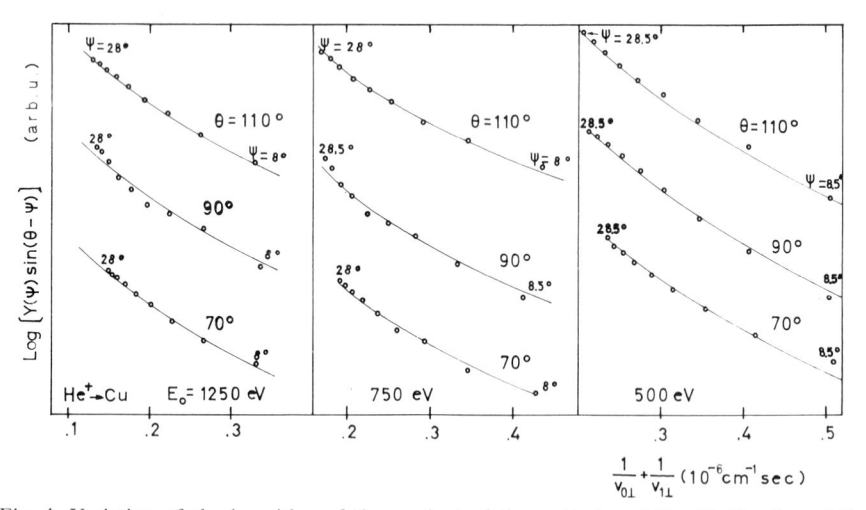

Fig. 4. Variation of the logarithm of the product of the scattering yield with the sine of the angle $(\theta - \psi)$ versus the sum of the inverse of the normal components of the velocities before and after the scattering for He$^+ \rightarrow$ Cu scattering at 110°, 90° and 70° incidence angle and 0.5, 0.75 and 1.25 keV incident energy.

Along the lines of relation (11), the plots of our results are presented on fig. (4) for the three energies 0.5, 0.75 and 1.25 keV and for the three scattering angles θ = 110°, 90°, 70°. Within the experimental accuracy, the behaviour is very much similar. The slopes of the curves calculated by linear regression on the experimental points remain in the interval between 5.5 and 6.5 × 10^6 cm/sec. One evidences again a regular increase of the slope as the incidence angle ψ increases from grazing angles. Experiments within the entire ψ domain showed that it is a general behaviour that the slope decreases if either the incidence angle ψ, or the analysis angle $(\theta - \psi)$ becomes more grazing. The maximum slope is in the vicinity of the specular configuration.

5. Discussion

Obviously the results disagree with our theoretical model. Instead of a straight line, one observes a regular increase of the slope as θ increases at specular configuration or as ψ increases from grazing at constant θ (figs. 3 and 4).

Although it remains up to now a controversial point [9], similar behaviour has been reported in the energy dependence of the scattered yield for a set of ion—surface combinations [10].

We verified that the behaviour of the curves sketched on fig. 4 for the θ variation of the yield is not very sensitive to the choice of the screening length in the interaction potential used to calculate the scattering cross section. On the other hand, it is known that multiple collision processes should not contribute much to the scattering yield in the case of helium scattering. Moreover, the observed behaviour is well marked in the large scattering angle domain where no multiple scattering has ever been reported.

The similar trends of the curves in fig. 4 suggests that the characteristic velocity should not vary much with the scattering angle or with the energy. It appears that the unexpected behaviour is mainly dependent on the particular configuration of the scattering with respect to the surface. One could tentatively assume that the long range Auger type charge exchange mechanism is not solely responsible for the total yield observed. Similarily to Brongersma et al. [11] one could involve a short range charge exchange mechanism occurring very close to the surface atom. Such a mechanism should be sensitive to the total velocity of the ion rather than to its normal component exclusively.

If this mechanism leads to an extra neutralization of the ions, the total probability that the ion remains ionized becomes $P_t = PP^*$, where P was previously defined (eq. (1)) and P^* is the probability associated with this short range mechanism. P^* will strongly depend on the distance of closest approach. Consequently it would vary with the scattering angle θ but remain constant when only the incidence angles vary. Thus this short range neutralization could not account for the deviation from linearity observed in fig. 4. Moreover it would cause a regular decrease of the slope with increasing θ in the plots shown in fig. 4 contrarily to what is observed [5].

On the other hand if one considers, as is the case at higher energies [12], a reionization at short range, the probability P_t becomes, according to Verhey [13]

$$P_t = P + (1 - P_{in})P^+ P_{out} \qquad (16)$$

where P^+ is the probability of reionization at short range. P_{in} and P_{out} are the probabilities that the ion remains ionized during the incoming or outgoing path respectively. This relation is less tractable, but if one can assume a reasonable value for v_c, it is possible to determine P^+ by fitting the experimental results. Such a procedure leads to physically inconsistent values of P^+. Furthermore Verhey [13] demonstrated that the probability P^+ is vanishingly small below 2.5 keV for $He^+ \rightarrow Cu$ scattering at $\theta = 30°$.

Another way is then to question the hypothesis of a constant number of scattering centers. This hypothesis neglects the shadowing effect exerted by the surface atoms on each other. It is known indeed that this shadowing effect can play a role in scattering spectrometry. It has namely been used to determine the exact location of adsorbed sulfur and oxygen in the nickel unit cell [14,15] and of adsorbed oxygen in the silver unit cell [16]. Furthermore, in spite of the careful polishing of the samples, microtopology and steps exist on our polycrystalline sample which could contribute to the shadowing at grazing angles.

The increase of the slopes with increasing scattering angles at specular configuration (fig. 3), or with increasing incidence angle at constant scattering angle (fig. 4) could indeed be interpreted as an increase of the density of scattering centers involved in the experiment.

One lacks a mathematical model to take account of such a shadowing effect. However, Harris [17] developed a theory about the influence of deeper lying atoms on the Auger yield in AES. When applied to our case such a theory predicts an angular variation of the number of scattering centers n taking the form

$$\left[\frac{1}{\sin \psi} + \frac{1}{\sin(\theta - \psi)}\right]^{-1}.$$

This treatment of our data gives no satisfactory results. It is obviously inadequate for ion scattering spectrometry owing to the very large surface sensitivity of the method. Nevertheless, one can verify that a correction to be brought to the plots of figs. 3 and 4 tends to vanish toward grazing angles.

Its seems then reasonable to assume that the slope of our experimental curves tends toward v_c at low incidence angles. In this case v_c should have a value of roughly 5 to 6×10^6 cm/sec. Such a value is consistent with the values found by Verhey [13] at higher energies (2 to 10 keV) and $30°$ scattering angle.

Further experiments on well defined monocrystalline copper surfaces and on other elementary samples seem necessary in order to gain a more comprehensive insight into the phenomenon of shadowing and in order to be able to propose a tentative theoretical treatment of this effect.

Acknowledgment

The authors are indebted to Professor J.-P. Issi for many fruitful discussions and to Messrs. P. Coopmans and L. Malcorps for technical support.

References

[1] H.D. Hagstrum, Phys. Rev. 96 (1954) 336.
[2] R.L. Erickson and D.P. Smith, Phys. Rev. Letters 34 (1975) 297.
[3] T.W. Rusch and R.L. Erickson, J. Vacuum Sci. Technol. 13 (1976) 374.
[4] E.g., D.P. Smith, Surface Sci. 25 (1971) 171.
[5] P. Bertrand and F. Delannay, Thesis, Louvain-la-Neuve (1976).
[6] F. Delannay, P. Bertrand and J.-M. Streydio, Phys. Rev. B, to be published.
[7] P. Bertrand, F. Delannay and J.-M. Streydio, J. Phys. E (Sci. Instr.) 10 (1977) 403.
[8] M.T. Robinson, Report ORNL 4556 (1970).
[9] H.H. Brongersma and T.M. Buck, Nucl. Instr. Methods 312 (1976) 559.
[10] E. Taglauer and W. Heiland, Surface Sci. 47 (1975) 234.
[11] H.H. Brongersma, N. Hazemindus, J.M. van Nieuwland, A.M.M. Otten and A.J. Smets, J. Vacuum Sci. Technol. 13 (1976) 670.
[12] W.F. van der Weg and D.J. Bierman, Physica 44 (1969) 177.
[13] L.K. Verhey, Thesis, Groningen (1976).
[14] H.H. Brongersma, J. Vacuum Sci. Technol. 11 (1974) 231.
[15] H.H. Brongersma and J.B. Theeten, Surface Sci. 54 (1976) 519.
[16] W. Heiland, F. Iberl, E. Taglauer and D. Menzel, Surface Sci. 53 (1975) 383.
[17] L.A. Harris, Surface Sci. 15 (1969) 77.

THE USE OF SECONDARY ION MASS SPECTROMETRY FOR STUDIES OF OXYGEN ADSORPTION AND OXIDATION

K. WITTMAACK

Gesellschaft für Strahlen- und Umweltforschung mbH, Physikalisch-Technische Abteilung, D-8042 Neuherberg, Germany

Thin oxide layers on silicon and tantalum as well as oxygen saturated silicon surfaces have been investigated by means of SIMS. Sputter depth profiling was achieved using a raster scanned 2 keV argon ion beam. The results clearly indicate that oxygen adsorption does not lead to on oxide-like structure. Moreover it is found that air-grown room temperature oxides on silicon are depleted of oxygen as compared to the composition of thick silicon dioxides. The results are discussed with reference to recently reported SIMS studies on oxygen covered surfaces.

1. Introduction

Secondary ion mass spectrometry (SIMS) is now a well-established analytical technique. This is true despite the fact that the mechanisms leading to secondary ion emission are presently not well understood [1]. Most commonly, SIMS has been used for depth profiling of impurity distributions and layer structures. E.g., range profiles of implanted ions could be determined with high accuracy [2–4] after minimizing profile broadening induced by energy dissipation of the primary ion [2,5].

Surface studies have also been carried out by means of SIMS. Most of the work has been done by Benninghoven and coworkers [6–14] who studied the interaction of oxygen with various metals and silicon using the so-called static method of SIMS. This simply means that the primary ion fluence required to monitor the intensity variation of characteristic secondary ions during progressive oxygen exposure was so small that bombardment induced changes in surface composition can be neglected. Based upon their SIMS measurements Benninghoven and coworkers [6–14] deduced a rather detailed model of oxidation. The main conclusions, however, are based upon the assumption that differences in the intensity variation of certain oxygen containing secondary ions as a function of oxygen exposure allow the identification of different oxide phases. The onset of oxidation has been deduced from the saturation behaviour at high oxygen exposure. These statements have not been confirmed by applying other experimental techniques such as Auger electron or photoelectron spectroscopy which would allow an unambiguous classification of oxide structures eventually produced.

In their analysis Benninghoven and coworkers did not take to account that surface coverage with oxygen and/or incorporation of oxygen will strongly change the energetics of the collision cascade [15] which in turn will affect the sputtering characteristics. Similarly the occurrence of certain types of cluster ions has not been discussed in terms of stability criteria [1]. Furthermore the strong dependence of the oxygen sticking coefficient upon the surface defect structure [16] has not been taken into consideration.

In trying to provide supporting evidence for the conclusions drawn from results of the oxygen uptake measurements, removal of oxygen enriched surface layers by ion bombardment has also been investigated by Benninghoven and coworkers. The "depth profiles" have been discussed by use of the so-called monolayer sputtering model [17] which is based upon the assumption that sputter erosion is completely governed by the statistics of primary ion impact. The resulting sputtering yields S and the corresponding cross sections for sputter removal, $\sigma = S/N_0$, where N_0 is the monolayer areal density, came out to be very large for certain secondary ion species. E.g., $S(VO_2^+) = 150$ atoms/ion [10] or $S(Ni^+) = 20$ atoms/ion [12] for 3 keV argon bombardment at an angle of incidence ϑ of $70°$ with respect to the surface normal [9–12]. Since these figures were calculated assuming a low areal density of only $N_0 = 5 \times 10^{14}$ atoms/cm^2 the corresponding cross sections are extremely large, $\sigma(VO_2^+) = 3 \times 10^{-13}$ cm^2 and $\sigma(Ni^+) = 4 \times 10^{-14}$ cm^2. A similarly large cross section has been reported by Dawson [18] by applying the same monolayer sputtering model to the removal of the "outer layer" of oxygen on aluminium by 500 eV argon ions ($\vartheta = 71°$).

These data become clearly unrealistic if comparison is made with recently reported cross sections for neon ion induced desorption of oxygen from nickel, as measured by low energy scattering. At 1.5 keV and $\vartheta = 60°$ Taglauer et al. [19] found a cross section of 10^{-15} cm^2. Extrapolation to 3 keV indicates that $\sigma(Ni^+)$ according for ref. [12] exceeds the cross section for oxygen desorption from nickel by an order of magnitude. (Differences between argon and neon bombardment can be estimated to be of minor importance.)

Note that the experiments reported by Benninghoven and coworkers [8–14] have not been carried out with constant primary ion current density across the area viewed by the mass spectrometer. From the mean bombarded area $A = 0.1$ cm^2 [8,10] a beam radius r of about 1 mm can be estimated (from $A = \pi r^2/\cos \vartheta$). Under such conditions the shape of the depth profiles must have been strongly affected by edge effects. Since the sputtering yields reported in refs. [9–12] have been deduced from measured depth profiles large errors might be introduced by non-uniform erosion rates across the field of view of the mass spectrometer.

Further arguments supporting the reservation with respect to the validity of the oxidation model proposed by Benninghoven and coworkers [6–14] are provided by supplementary studies reported by that group. In a recent investigation [20] the spectrum of secondary ions emitted from an oxygen saturated surface (silicon) has been compared with the pattern due to a thermal oxide (silicon dioxide). The data clearly show characteristic differences in the relative intensities of certain molecular ions. Analysis of the results of ref. [20] indicates [1] that the intensity of oxygen-rich cluster ions emitted from oxygen-saturated silicon is smaller by up to an order of magnitude or more than the intensity observed from silicon dioxide. This finding strongly suggests that saturation of SIMS signals during oxygen exposure does not provide evidence for oxide formation. This idea is supported by the observation that the saturation secondary ion intensities from oxygen covered surfaces increase considerably if oxygen exposure takes place during bombardment [1]. The effect is due to recoil implantation of adsorbed oxygen [1,21] (see section 3).

In an attempt to clarify the implications in the use of SIMS for studies of oxygen adsorption and oxidation we have carried out depth profiling measurements on samples with different oxygen coverage or oxide thickness. Most of the experiments have been performed with oxidized silicon samples. Some additional investigations concerned oxidized tantalum.

2. Experimental

The investigations were carried out using the quadrupole equipped DIDA scanning ion micorprobe [22,23]. The base pressure in the system was about 3×10^{-9} mbar. During bombardment the total pressure increased to 8×10^{-9} mbar at the most. Sputtering was performed with 2 keV argon ions hitting the target surface at normal incidence ($\vartheta = 0$). The ion beam was focussed to a spot of about 100 μm in diameter and raster scanned across areas between 0.5 and 2.5 mm^2. Edge effects in depth profiling could be avoided by use of an electronic aperture [24]. The relative gate width was 0.4 corresponding to a duty cycle of 16% [24].

The argon ions were not mass-analyzed. Sufficient beam purity could be achieved by adequate lay-out and operation of the ion source. E.g., the SiO$^+$ signal from a sputter cleaned silicon sample amounted to less than 10^{-3} of the signal observed with a silicon dioxide target.

The mean current densities $\langle j \rangle$ applied in raster scanning depth profiling ranged from 10^{-6} to 10^{-5} A/cm^2. The ratio of the primary ion flux to the residual gas molecule flux hitting the target surface was thus even better than the respective conditions provided in the investigations by Benninghoven and coworkers ($p \lesssim 10^{-10}$ mbar, $j = 10^{-9}$ A/cm^2 [9]). Despite the relatively large current densities applied in the present study depth profiles were measured in sufficiently small units of sputtered material. This was achieved by making use of the rapid peak switching capability of a quadrupole mass filter. The integration time per mass line was 10 s (print out time 2.8 s). As will be shown below this corresponds to sputtering a few percent of a monolayer (at $\langle j \rangle \simeq 10^{-6}$ A/cm^2). From this one can see that current densities as low as 10^{-9} A/cm^2 are by no means a necssary requirement for adequate surface studies.

Depth profiling of (insulating) thick silicon dioxide layers could only be done with the help of an additional electron beam which provided compensation of the primary ion charge. In this study a focussed 500 eV electron beam was used which hit the sample at an angle of about 60° with respect to the surface normal. The electron beam current density was adjusted such that no change in the setting of the energy filter preceding the quadrupole was required for optimum secondary ion transmission when switching from a silicon dummy to an insulating dioxide layer. The area covered by the electron beam ($\sim$10 mm^2) was markedly larger than the sputtered area. Electron induced desorption caused emission of H_3O^+ and spurious O^+. It was checked, however, that this did not affect the main results of this study.

Silicon dioxide layers on silicon were prepared either by thermal oxidation (at 1050°C in dry oxygen) or by anodic oxidation [25]. Anodic oxidation was also used to produce tantalum oxide layers on tantalum [25]. No attempt was made to dehydrate the anodic oxides. Comparison of the mass spectra of thermal and anodic oxide indicated, if any, only a small change in the relative intensity of the secondary ions investigated ($\leq$10%).

3. Results

Typical depth profiling results for silicon dioxide on silicon are presented in fig. 1. The intensity variation is shown for five ion species (out of a total of six) measured quasi-simultaneously in one run. To ease comparison the respective profiles are normalized to the intensity observed are having passed through the topmost layers of silicon dioxide. The absolute intensities are plotted in fig. 3.

According to fig. 1 characteristic features show up in the shape of the profiles for the different ion species. These features have been observed in repeated runs so that their is no doubt about their existence. One of the most astonishing facts is that Si^+ and SiO^+ start with a very small intensity. This effect is much less pronounced for the other ions. E.g., Si_2O^+ arrives at the 100% intensity level already after one measuring cycle. Curiously enough, Si_2^+ passes through a pronounced maximum at a depth where Si^+ and SiO^+ have only achieved the 50% level. Intercomparison of the different $Si_2O_n^+$ ions (n = 0, 1, 2) indicates, that the 100% level is approached the later the larger the number of oxygen atoms in the cluster. Note that $SiOH^+$ (not shown in fig. 1) starts with an intensity close to the 100% level.

Interesting phenomena are also observed in the vicinity of the interface between silicon dioxide and silicon (fig. 1). As might be expected the first indication for the fact that one approaches the silicon substrate comes from a decrease in the intensity of the oxygen-rich ion $Si_2O_2^+$. The ions Si^+ and SiO^+ differ from the clusters with two silicon atoms ($Si_2O_n^+$) in that they pass through a more or less pronounced maximum in the vicinity of the interface followed by a rapid decrease in intensity. A peculiar behaviour is observed with Si_2O^+ which exhibits a local maximum roughly at the half height position of Si^+ and SiO^+. Moreover one finds that Si_2O^+

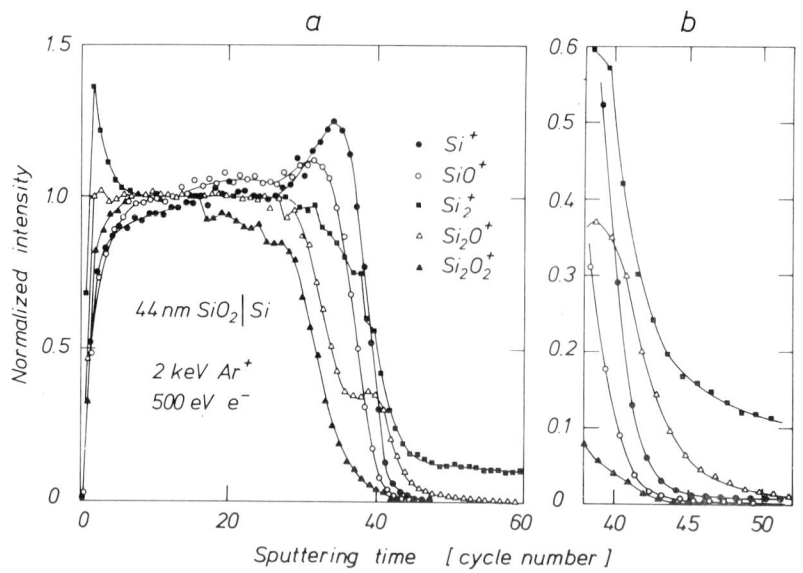

Fig. 1. (a) Normalized depth profiles of a silicon dioxide layer on (111) silicon, measured in one run by quasisimultaneous recording of the intensity of different mass lines. $\langle j \rangle \simeq 10^{-5}$ A/cm^2; 77 s/cycle. (b) Tails of the profiles of (a) on an enlarged scale. The thickness of the silicon dioxide has been determined by talysurf measurements.

provides the most sensitive indication for the presence of spurious oxygen, i.e. the normalized intensity of Si$_2$O$^+$ tails off most slowly as compared to the other oxygen containing ions and amounts to about 10% at a depth were SiO$^+$ and Si$_2$O$_2^+$ are already below the 1% level. Finally one should note the interesting observation that the intensity of Si$_2^+$ drops to as low as about 10% in passing from silicon dioxide to the silicon substrate.

Fig. 2 shows depth profiles of very thin silicon oxide. The mean current density was a factor of four smaller than in the measurements of fig. 1. The intensities of the different ions are normalized to the same reference levels as in fig. 1 (see fig. 3). Note the enlarged scale in fig. 2.

An interesting aspect of fig. 2 is that the maximum intensities observed with thin oxides are by far smaller than the respective 100% levels of the dioxide. This effect has already been deduced from previous studies [26]. The most pronounced reduction in intensity is found for SiO$^+$. The intensity of this ion passes through a maximum of only 10% for the room temperature oxide (fig. 2a) and about 1% for the sample etched prior to introduction into the target chamber (fig. 2b).

Comparison of fig. 2a with fig. 1b indicates striking similarities in the relative decrease of the various secondary ion intensities. This would mean that one can produce a layer with a composition similar to that of a thermally grown thin oxide by well-defined sputter erosion of a thick silicon dioxide.

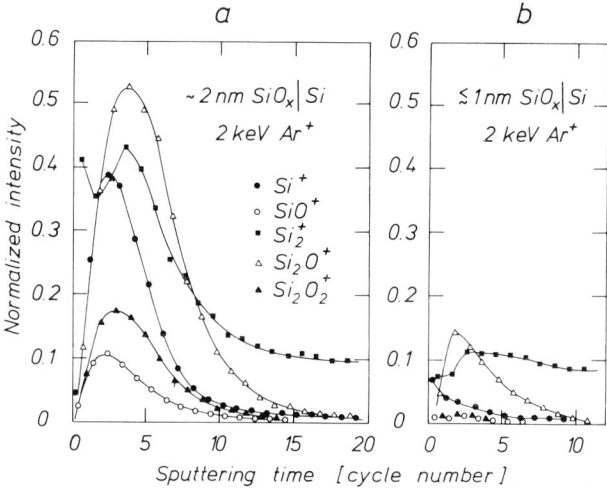

Fig. 2. Depth profiles of thin oxide layers on silicon. $\langle j \rangle \simeq 2.5 \times 10^{-6}$ A/cm^2; 77 s/cycle. The secondary ion intensities are normalized to the intensities characterizing a thick silicon dioxide (see figs. 1a and 3). (a) Room temperature oxide on (111) silicon (several years old). (b) Oxide grown in air at room temperature on (111) silicon within a few minutes after etching the sample in a 42% solution of NH$_4$F : HF (5 : 1) in water.

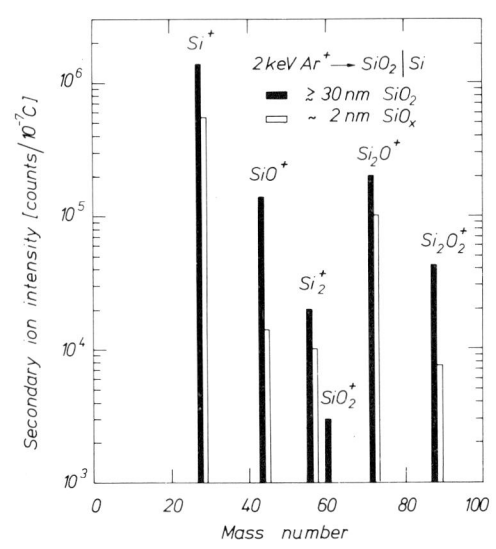

Fig. 3. Intensities of secondary ions emitted from silicon oxides of different thickness. The intensities due to thick silicon dioxide were recorded after having passed through the topmost layers of the respective sample (see fig. 1a), whereas the intensities due to a thin oxide represent the peak values of fig. 2a. Mass resolution $m/\Delta m \simeq 300$, energy band pass ~ 2 eV, energy window set at or close to the peak of the respective secondary ion energy distribution.

Results qualitatively similar to those of fig. 2b have been observed with samples which were sputter cleaned and oxygen saturated in the way used by Benninghoven and coworkers [8–14]. The maximum intensities observed after an oxygen exposure of 3000 L amounted to about half the intensities of fig. 2b (except for Si_2^+ which roughly followed the respective curve in fig. 2b).

The intensity variation introduced by raster scanning argon bombardment at elevated oxygen partial pressure is plotted in fig. 4. The left hand part of fig. 4 shows the effect due to build-up of oxygen in the target by recoil implantation whereas the right hand part represents the depth profile of the resulting oxide structure. The depth profile was measured after rapidly reducing the oxygen pressure from 7×10^{-6} to below 10^{-8} mbar without interrupting argon bombardment.

As already known from previous studies [27] the intensities of Si^+ and SiO^+ become very large under steady state bombardment at elevated oxygen pressure, in particular when compared with the intensities observed without simultaneous bombardment. Ions of the type $Si_2O_n^+$ exhibit a peculiar behaviour. The gain in intensity is relatively small during steady state oxygen recoil implantation. Depth profiling at low oxygen pressure, however, reveals pronounced maxima. In the tails of the depth distributions one observes similarities with the profiles in fig. 2a which may be partly masked due to the fact that the original base pressure had not been achieved in the experiments of fig. 4.

Depth profiles of tantalum oxides are presented in fig. 5. The intensity variation of the three most prominent mass lines observed at the early stage of sputtering of a thick (anodic) oxide is shown in fig. 5a. At the very surface the ratio of the abso-

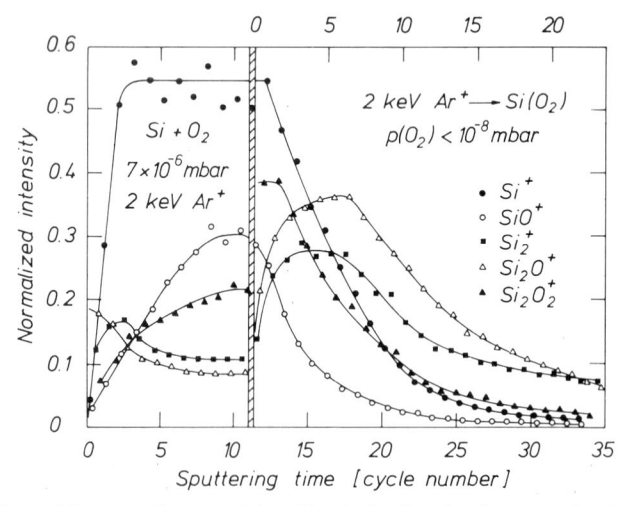

Fig. 4. Variation of the secondary ion intensities during bombardment at elevated oxygen partial pressure (left hand part) and after immediate return to a low pressure (right hand part). Raster scanning bombardment. $\langle j \rangle = 2.5 \times 10^{-6}$ A/cm^2; 77 s/cycle.

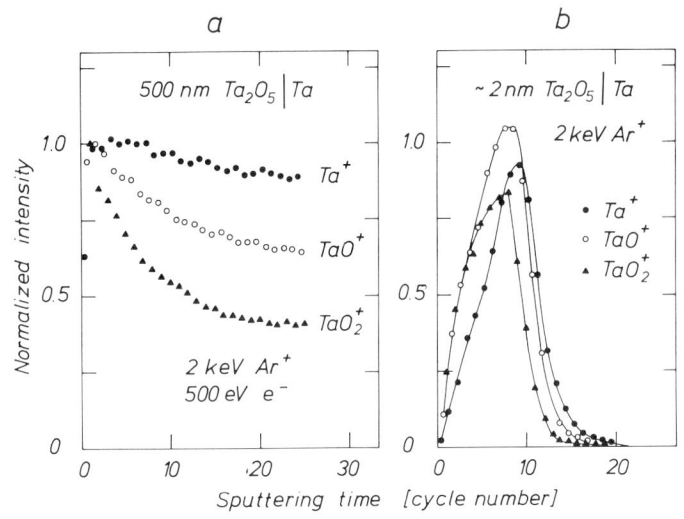

Fig. 5. Depth profiles of different tantalum oxide layers. $\langle j \rangle = 2.5 \times 10^{-6}$ A/cm^2; 38.5 s/cycle. (a) Thick tantalum oxide, surface region only. (b) Room temperature oxide on tantalum.

lute intensities of Ta$^+$, TaO$^+$ and TaO$_2^+$ is 1 : 3 : 0.5, i.e. TaO$^+$ is the most prominent line [1,10]. Sputtering produces a decrease in intensity which is the more pronounced the larger the number of oxygen atoms in the TaO$_n^+$ ion. This result may indicate that argon bombardment causes preferential sputtering of oxygen such that the surface becomes depleted of oxygen.

Fig. 5b shows a depth profile of a room temperature tantalum oxide layer on a tantalum foil (purity 99.98%). No cleaning treatment has been applied prior to SIMS analysis so that surface contamination with impurities cannot be excluded. The depth profile in fig. 5b may be considered as an example for routine material analysis of samples not specially prepared for the investigation of surface reactions under UHV conditions.

Note that the peak intensities in fig. 5b are close to or even slightly above the 100% level. This result strongly suggests that the composition of the room temperature oxide film on tantalum is close to the bulk oxide composition Ta$_2$O$_5$.

Finally we point out an interesting aspect of fig. 5b which concerns depth resolution in sputter depth profiling. After having passed through the maximum the intensity of all ions drops very rapidly. Taking the sputtering time required to pass from the 90% to the 10% intensity level as a measure of the depth resolution Δz one finds $\Delta z/z \simeq 0.25$ (for TaO$^+$) which corresponds to $\Delta z \simeq 0.5$ nm for the assumed oxide thickness of 2 nm. The upper limit of the oxide thickness can be estimated from the mean current density and by use of available sputtering yield data [25]. One finds $z \lesssim 4$ nm which would result in $\Delta z \lesssim 1$ nm. These depth resolution data are by far the best results ever reported. They may be due to a combination of favourable conditions such as low bombardment energy, narrow transition region from the oxide to the substrate and small sputtered depth. The latter fact might prevent pronounced atomic mixing.

4. Discussion

It is obvious from the complexity of the depth profiles that a detailed interpretation of both the absolute and the relative secondary ion intensities is far from being simple. Nevertheless one can state that much information about the composition of thin oxide layers can be achieved by a quantitative comparison with the secondary spectra of thick oxides.

Important quantities to be considered are the range of the argon projectile and the mean depth of origin of sputtered particles. These can be estimated to be about 3 nm [28] and 0.5 nm [29], respectively. On the other hand one has to take into account that the relevance of these quantities may be strongly modified by the mechanism leading to ionization. Since there are strong indications [1] that for oxide samples ionization mainly results from the breaking of ionic bonds one would expect the secondary ions to originate from the very surface. Information about the composition underneath the surface layer would than be contained in the ionization probability.

Under these assumptions the profiles shown in figs. 1a and 2a suggest that the original surfaces of both the thick and the thin oxide are rich in silicon. This can be deduced from the fact that the intensity of Si_2^+ becomes large as soon as adsorbed layers have been sputtered away (fig. 1a). Moreover we see that the thin oxides are depleted of oxygen throughout (fig. 2).

The SIMS results on thin silicon oxides are in qualitative agreement with low energy helium ion scattering (ISS) depth profiles reported by Harrington et al. [30]. E.g. it was found that very thin oxides (<1 nm) are strongly depleted of oxygen. However, it was concluded [30] that the stoichiometry of SiO_2 is already achieved for an oxide thickness of 1.5 to 2 nm. In trying to explain the discrepancy to the present results one should note that Harrington et al. [30] had to apply an empirical correction factor of 1.72 to the relative scattering cross section in order to arrive at the expected O/Si ratio of 2.0 for a thick dioxide. Bombardment induced changes in surface composition have not been taken into account [30]. Note that ISS probes the composition of the surface whereas SIMS reflects the composition of the sputtered material.

Excess silicon has also been detected in high energy backscattering (channeling) analysis of thin silicon oxides [31–33]. Although this has been interpreted in terms of a silicon-rich transition region between the assumed dioxide and the substrate [31] a stochiometry different from that of a dioxide cannot be excluded on the basis of channeling experiments.

The comparative SIMS studies reported above strongly support the idea that oxide layers with a thickness of up to 2 nm or more do not have the composition of SiO_2. This is clearly evident from a comparison of figs. 2 and 1b which demonstrates that the depth profiles of thin oxide layers exhibit the same features as the trailing edges of thick oxide profiles where one must expect a progressive reduction in oxygen concentration. The differences in profile shape observed for the different secondary ion species cannot be discussed in detail here. We only consider the intensity variation of Si_2O^+ in the interface region between a thick dioxide and the silicon substrate (fig. 1). It is not astonishing that the intensity of Si_2O^+ passes through a local maximum because one would expect the statistical probability for formation of Si_2O to pass through a maximum in going from SiO_2 to Si. Of course the simple statistical argument is not the whole truth because the ionization mechanism has to be taken into account as well. Note, however, that the occurrence of local maxima in the depth profile of secondary ions clearly rules out the possibility of describing through-oxide measurement by a simple superposition law [34,35].

It is evident from this study that the surface structure produced by saturation adsorption of oxygen on silicon reveals no oxide characteristics. Accordingly, one must conclude that the oxidation model developed and applied by Benninghoven and coworkers [6–14] in interpreting their SIMS measurements is unfounded in general. The onset of oxidation cannot be deduced simply from the saturation behaviour of oxygen sensitive secondary ion intensities. It is somewhat surprising that the large differences observed in the secondary ion spectra of silicon dioxide and oxygen saturated silicon, respectively [20], did not cause Benninghoven and coworkers to reconsider the validity of their oxidation model.

Structural differences between oxygen saturated silicon surfaces and silicon dioxide are also evident from Auger electron spectroscopy data. Room temperature oxygen adsorption merely causes a reduction in the 92 eV silicon Auger-peak [36,37], whereas the 75 eV silicon dioxide peak is observed only after high temperature oxidation [36].

One aspect related to quantitative SIMS measurements should be pointed out. The secondary ion yields reported for oxygen covered surfaces [7,14] have been considered to represent (absolute!) yields for oxidized metals [7,14,35]. The comparative study on silicon dioxide and oxygen covered silicon clearly indicates that this is an uncorrect statement (in general). E.g. the normalized intensity of Si^+ has been found to be a factor of about four larger for silicon dioxide than for oxygen saturated silicon [20]. From this one can estimate [14] a positive secondary ion yield for silicon dioxide of about 2.4. Using available sputtering yield data for silicon dioxide [38,39] the corresponding ionization probability is found to be larger than one. Since this is impossible we must conclude that *reliable* measurements of secondary yields are presently non-existing, neither for oxides nor for oxygen covered metals.

Finally we briefly discuss secondary ion yield enhancement due to bombardment at elevated oxygen partial pressure. This phenomenon has frequently been attributed to oxygen adsorption only [35,40,41]. The present study provides further evidence that secondary ion yield enhancement requires incorporation of oxygen which can be achieved by recoil implantation of adsorbed oxygen [1,21]. From fig. 4 one might conclude that the recoil-implanted oxygen distribution is strongly peaked at the instantaneous surface.

5. Conclusions

The results presented in this study demonstrate that rather detailed information can be obtained from SIMS investigations of surface layers. Compared to other surface analytical techniques SIMS has the advantage that even spurious modifications in surface composition show up in pronounced changes in the secondary ion mass spectra. (Semi-)quantitative interpretation of SIMS data can be achieved by comparison with the spectra of standards of well-known composition. Analysis of SIMS measurements without the use of standards will lead to erroneous results, at least at the present stage of understanding of secondary ion production [1].

Note added in proof

Very recent experiments concerning bombardment-induced uptake of oxygen have shown [42] that the oxygen incorporation yield (i.e. the number of oxygen atoms incorporated per incoming projectile) may be of the order of one or even larger. This result indicates that, in addition to recoil implantation, bombardment-enhanced diffusion may cause or assist oxygen incorporation.

References

[1] K. Wittmaack, in: Inelastic Ion-Surface Collisions, Eds. N.H. Tolk, J.C. Tully, W. Heiland and C.W. White (Academic Press, in press).
[2] F. Schulz, K. Wittmaack and J. Maul, Radiation Effects 18 (1973) 211.
[3] W.K. Hofker, D.P. Oosthoek, N.J. Koeman and H.A.M. de Grefte, Radiation Effects 24 (1975) 223.
[4] K. Wittmaack, F. Schulz and B. Hietel, in: Ion Implantation in Semiconductors, Ed. S. Namba (Plenum, New York, 1975) 193.
[5] J.A. McHugh, Radiation Effects 21 (1974) 209.
[6] A. Benninghoven, Surface Sci. 28 (1974) 541.
[7] A. Benninghoven, Surface Sci. 35 (1973) 427.
[8] A. Benninghoven and S. Storp, Appl. Phys. Letters 22 (1973) 170.
[9] A. Benninghoven and A. Müller, Surface Sci. 39 (1973) 416.
[10] A. Müller and A. Benninghoven, Surface Sci. 39 (1973) 427.
[11] A. Benninghoven and L. Wiedmann, Surface Sci. 41 (1974) 483.
[12] A. Müller and A. Benninghoven, Surface Sci. 41 (1974) 493.
[13] E. Stumpe and A. Benninghoven, Phys. Status Solidi (a) 21 (1974) 479.
[14] A. Benninghoven, Surface Sci. 53 (1975) 596.

[15] N. Andersen and P. Sigmund, Mat. Fys. Medd. Dan. Vid. Selsk. 39 (1974) No. 3.
[16] H. Ibach, K. Horn, R. Dorn and H. Lüth, Surface Sci. 38 (1973) 433.
[17] A. Benninghoven, Z. Physik 230 (1970) 403.
[18] P.H. Dawson, Surface Sci. 57 (1976) 229.
[19] E. Taglauer, G. Marin, W. Heiland and U. Beitat, Surface Sci. 63 (1977) 507.
[20] A. Benninghoven, W. Sichtermann and S. Storp, Thin Solid Films 28 (1975) 59.
[21] K. Wittmaack and P. Blank, in: Proc. 5th Intern. Conf. on Ion Implantation in Semiconductors and Other Materials (in press).
[22] K. Wittmaack, Rev. Sci. Instr. 47 (1976) 157.
[23] K. Wittmaack, in: Ion Beam Surface Layer Analysis, Vol. 2, Eds. O. Meyer, G. Linker and F. Käppeler (Plenum, New York, 1976) p. 649.
[24] K. Wittmaack, Appl. Phys. 12 (1977) 149.
[25] R. Kelly and N.Q. Lam, Radiation Effects 19 (1973) 39.
[26] K. Wittmaack, Intern. J. Mass Spectron. Ion Phys. 17 (1975) 39.
[27] J. Maul and K. Wittmaack, Surface Sci. 47 (1975) 358.
[28] J.F. Gibbons, W.S. Johnson and S.W. Mylroie, Projected Range Statistics, 2nd ed. (Dowden, Hutchinson and Ross, Stroudsbourg, PA, 1975).
[29] P. Sigmund, Phys. Rev. 184 (1969) 383.
[30] W.L. Harrington, R.E. Honig, A.M. Goodman and R. Williams, Appl. Phys. Letters 27 (1975) 644.
[31] W.K. Chu, E. Lugujjo, J.W. Mayer and T.W. Sigmon, Thin Solid Films 19 (1973) 329.
[32] W.F. van der Weg, W.H. Kool, H.E. Roosendaal and F.W. Saris, Radiation Effects 17 (1973) 245.
[33] G. Della Mea, A.V. Drigo, S. Lo Russo, P. Mazzoldi, S. Yamaguchi and G.G. Bentini, Appl. Phys. Letters 26 (1975) 147.
[34] H.W. Werner, H.A.M. de Grefte and J. van den Berg, Radiation Effects 18 (1973) 269.
[35] H.W. Werner, Surface Sci. 47 (1975) 301.
[36] A.B. Joyce, Surface Sci. 35 (1973) 1.
[37] J.J. Bellina, Jr., J. Vacuum Sci. Technol. 11 (1974) 1133.
[38] H. Bach, J. Non-Crystalline Solids 3 (1970) 1.
[39] H. Bach, I. Kitzmann and H. Schröder, Radiation Effects 21 (1974) 31.
[40] G. Blaise and M. Bernheim, Surface Sci. 47 (1975) 324.
[41] A.E. Morgan and H.W. Werner, Appl. Phys. 11 (1976) 193.
[42] W. Wach and K. Wittmaack, Nucl. Instr. Methods, in press.

THE APPLICATION OF SIMS TO THE STUDY OF CO ADSORPTION ON POLYCRYSTALLINE METAL SURFACES

M. BARBER, J.C. VICKERMAN and J. WOLSTENHOLME
Department of Chemistry, UMIST, Manchester, M60 1QD, England

Secondary ion mass spectroscopy (SIMS) was used to study the adsorption of carbon monoxide on polycrystalline nickel, copper, iron, palladium and tungsten foils. The results demonstrate the ability of SIMS to distinguish, qualitatively, between molecular and dissociative adsorption. A correlation between SIMS results and those obtained by infra-red spectroscopy for molecular adsorption is also suggested.

1. Introduction

The adsorption of carbon monoxide on metal surfaces is one of the most commonly studied aspects of surface science [1]. The aim of this work is to compare the results obtained from SIMS with those reported in the literature concerning carbon monoxide adsorption on polycrystalline nickel, copper, iron, palladium and tungsten.

More specifically, we shall be concerned with the extent of dissociation of the carbon monoxide molecule on the surface and a correlation between SIMS results and those obtained by infra-red spectroscopy is also suggested.

From their work using photoelectron spectroscopy, Roberts et al. [2,3] have suggested that there is a threshold heat of adsorption of carbon monoxide ($\sim$250 kJ mole^{-1}) below which molecular adsorption takes place and above which dissociative adsorption predominates at room temperature. Using this argument, molecular adsorption should take place on nickel, copper, palladium and the α-phase of adsorption on tungsten, while dissociative adsorption is expected for the β-phase on tungsten. The heat of adsorption of carbon monoxide on iron, however, is close to the threshold value and the expected form of adsorption is uncertain. We hope to show that SIMS is suitable to distinguish between these two types of adsorption.

2. Experimental

2.1. Apparatus

The apparatus used for these experiments was constructed by Vacuum Generators and has been described in detail elsewhere [4]. Briefly, the SIMS instrument consists of two, separately-pumped, UHV chambers, a preparation chamber and an analysis chamber, both bakable and capable of pressures less than 10^{-10} Torr (1 Torr = 133.3 Pa).

Fitted to the analysis chamber is a differentially-pumped, mass-filtered ion gun whose current is stabilised and is continuously variable from 10^{-12} to 5×10^{-8} A. The energy of the ion beam is variable from 0 to 5 keV and the beam may be focused into an area of about 1 cm^2. The ion beam strikes the sample at an angle of 20° to the surface plane. The secondary ions are mass analysed in a large quadrupole mass filter. Facilities have been added to enable this mass spectrometer to detect the largest possible energy range of secondary ions consistant with good mass resolution.

Each of the chambers is fitted with a residual gas analyser and a bakable leak valve to admit gases in a controlled manner.

2.2. Surface cleaning procedure

The metal surfaces were cleaned by argon ion bombardment. It was found, however, that a conventional cold-cathode discharge ion source was unsuitable for cleaning purposes since it was found to deposit the cathode material on the sample along with fairly large quantities of sodium and potassium. It was necessary, therefore, to clean the sample by means of the mass-filtered ion gun. The disadvantage of this method is the low current available which extends the time necessary to clean the sample and so puts stringent requirements on the quality of the vacuum. The cleanliness of the surfaces was checked by running SIMS spectra after the cleaning procedure. The ion current used for the analysis of each sample was chosen according to the sensitivity of SIMS to that sample. Thus, for nickel and iron a current density of 5×10^{-10} A cm^{-2} was used, for copper and palladium the current density was 2×10^{-9} A cm^{-2} while for tungsten it was 5×10^{-9} A cm^{-2}.

2.3. Sample purity

The metals used in this study were supplied by Goodfellow Metals and were of high purity (99.95% for palladium and tungsten and 99.99% for nickel, copper and iron).

The argon required for the primary ion gun was normal comercial grade provided by BOC but this was purified using a rare gas purifier. Despite the fact that the ion gun is mass filtered it is necessary to purify the argon since the differential pumping on the gun is not perfectly efficient. The pressure in the analysis chamber rises to about 10^{-8} Torr when the gun is in use and impurities in the argon would therefore contaminate the surface of the sample.

The carbon monoxide was spectroscopically pure and supplied by BOC. The purity of the gases was monitored using the residual gas analyser fitted to the SIMS chamber.

3. Results

3.1. Nickel

SIMS studies of the adsorption of carbon monoxide on nickel and copper have been described previously [5] and so will only be briefly reviewed here.

Adsorbed carbon monoxide on nickel at 77 K gives rise to peaks in the SIMS spectrum which were assigned to $NiCO^+$ and Ni_2CO^+ along with extremely weak peaks due to Ni_2C^+ and Ni_2O^+, there was no evidence for NiC^+ or NiO^+.

If carbon monoxide is admitted to a nickel surface at room temperature and 2×10^{-10} Torr peaks due to $NiCO^+$ and Ni_2CO^+ appear in the SIMS spectrum immediately and the intensity of each increases and eventually reaches a steady value after about 10^{-6} Torr s.

On heating the surface the $NiCO^+$ peak intensity decreased rapidly and was no longer visible in the spectrum at 395 K, the Ni_2CO^+ peaks remained.

Only prolonged exposure of the surface to carbon monoxide at 420 K caused a build up of carbon and oxygen on the surface, as shown by the increase of the intensities of Ni_2C^+ and Ni_2O^+ peaks with respect to the Ni_2CO^+ peaks.

3.2. Copper

Adsorption of carbon monoxide on copper at 77 K gave rise only to peaks due to $CuCO^+$, there was no evidence for Cu_2CO^+. On warming the sample the intensity of the $CuCO^+$ peaks decreased and a plot of its intensity against temperature showed a discontinuity at 195 K, the point at which desorption of one phase of adsorbed carbon monoxide is reported to be complete [6]. At room temperature no $CuCO^+$ could be detected but a very small amount of Cu_2CO^+ appeared. Again, only prolonged heating at 390 K at high pressures of carbon monoxide caused dissociation products to be observed.

3.3. Iron

Carbon monoxide was admitted to the sample at 195 K at a pressure of 10^{-8} Torr. A peak at mass 84 ($FeCO^+$) immediately appeared and its intensity increased to reach a steady state value in about $1\frac{1}{2}$ min, this is equivalent to an exposure of about 1 Langmuir and indicates that the sticking coefficient must be close to unity at this temperature. The spectrum recorded after equilibration is shown in fig. 1. It must be pointed out that the Fe_2CO^+ is very much smaller with respect to the other peaks in the spectrum than the equivalent species observed on nickel. Also to be noted are the very small peaks representing the breakdown products of carbon monoxide. Raising the pressure of carbon monoxide to 10^{-6} Torr did not cause any change in the appearance of the spectrum, indicating that the maximum coverage of carbon monoxide is achieved at or below 10^{-8} Torr at this temperature.

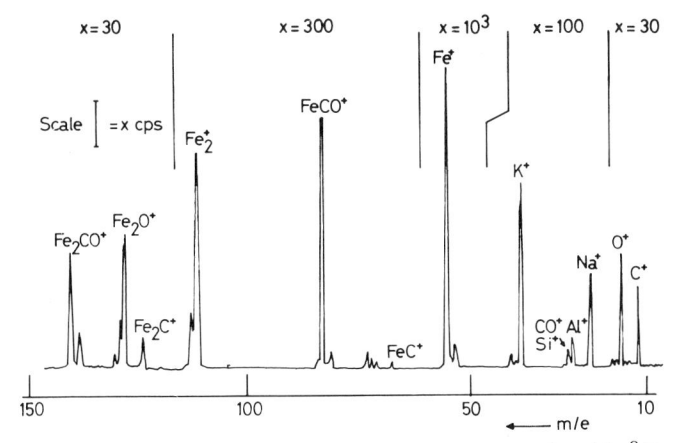

Fig. 1. SIMS spectrum recorded at equilibrium when iron is exposed to 10^{-8} Torr of carbon monoxide. Primary ion current density = 5×10^{-10} A cm^{-2}; energy = 3 keV.

A clean surface was again exposed to 10^{-8} Torr of carbon monoxide at 195 K and the spectrum recorded. The sample was then heated in steps to a series of temperatures up to 315 K and, at each temperature, time was allowed for a steady state to be achieved. Fig. 2a shows the effect of raising the temperature upon the intensities of the Fe$^+$ and FeCO$^+$ peaks. The Fe$_2$CO$^+$ peak intensity decreased rapidly and, at room temperature, could not be detected. Fig. 2b shows the variation of the FeCO$^+$/Fe$^+$ peak intensity ratios with temperature. It will be seen that the intensity of the FeC$^+$ increases over the same temperature range as the FeCO$^+$ intensity decreases.

Admitting carbon monoxide at 10^{-8} Torr and room temperature to a clean iron surface gave rise to SIMS spectra which were very similar to those recorded when an adsorbed layer is heated from 195 to 298 K in the presence of 10^{-8} Torr of carbon monoxide. Again, there was little difference in the spectrum when the pressure was raised to 10^{-6} Torr. The adsorption was found to be reversible since the FeCO$^+$ peak disappeared when the pressure of carbon monoxide was reduced to below 10^{-10} Torr.

3.4. Palladium

The clean palladium surface was exposed to carbon monoxide at 298 K and a pressure of 5×10^{-10} Torr. On admission of the gas, peaks due to PdCO$^+$ and Pd$_2$CO$^+$ appeared, these grew in intensity and reached a maximum value after about 30 min. No peaks which could be assigned to PdC$^+$, PdO$^+$, Pd$_2$C$^+$ or Pd$_2$O$^+$ were observed, suggesting that there is little, if any, dissociation of the carbon monoxide on palladium at room temperature. Increasing the pressure of carbon monoxide

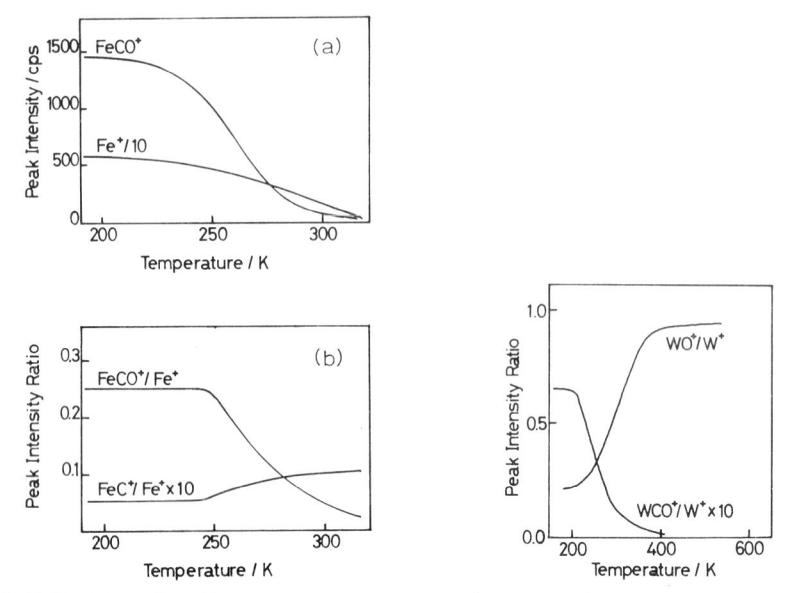

Fig. 2. (a) Variation of equilibrium peak intensities of Fe^+ and $FeCO^+$ with temperature; partial pressure of carbon monoxide is 10^{-8} Torr throughout. (b) Variation of relative peak intensities $FeCO^+/Fe^+$ and FeC^+/Fe^+ with temperature.

Fig. 3. Variation of relative peak intensities WCO^+/W^+ and WO^+/W^+ with temperature, in the presence of 10^{-6} Torr of carbon monoxide.

caused the $PdCO^+$ and Pd_2CO^+ peaks to increase their intensity reaching a maximum value at about 10^{-7} Torr, again with no evidence for dissociation.

3.5. Tungsten

A clean tungsten sample was cooled to 195 K and carbon monoxide was admitted to the sample at 10^{-6} Torr. This caused the appearance of peaks corresponding to WCO^+ but these were very weak in comparison with the MCO^+ peaks observed with other metals at this temperature. Much more intense than the WCO^+ were those due to WO^+. Although peaks due to W_2O^+ and $W_2O_2^+$ were also observed there was no evidence for W_2CO^+. Furthermore, the peaks due to WC^+ were also very small but because of the large WO^+ peaks it seems that some dissociation is taking place and therefore the low intensity of the WC^+ peaks may be due to its low secondary ion yield.

After adsorption of carbon monoxide at 195 K the sample was heated in the presence of 10^{-6} Torr of the gas to a series of temperatures. Time was allowed for a steady state to be established at each temperature and the intensities of the various peaks were measured. Fig. 3 shows the variation of the WCO^+/W^+ and WO^+/W^+

intensity ratios with temperature. It will be noted from this figure that, as the temperature is raised, the relative amount of WCO^+ decreases rapidly while that of WO^+ increases.

Adsorption of carbon monoxide at 10^{-6} Torr and room temperature gave rise to a spectrum which was very similar to that obtained when an adsorbed layer of carbon monoxide was heated from 195 to 298 K.

4. Discussion

Quantitative measurements of concentration are difficult using SIMS, partly because the probability for secondary ion formation is different for each ion and partly because changes in the work function of the sample, which accompany adsorption, will affect the ion yield for an individual secondary ion species. The effect of work function changes can be reduced by considering ratios of peak intensities rather than absolute peak intensities. If it is assumed, for example, that the ratio $FeCO^+/Fe^+$ is directly proportional to the coverage of carbon monoxide on iron this will lead to errors of about 10% (this figure was calculated using the formula proposed by Schroeer [7]). Nevertheless, it is possible to interpret our results in a manner which puts no quantitative significance on the peak intensities except that if a ratio X^+/M^+ increases this reflects an increase in the amount of the species corresponding to X on the surface relative to surface M.

On nickel at room temperature and below our results indicate little dissociation of carbon monoxide is taking place. Rasing the temperature results in an increase in the amount of dissociation, indicated by the increase in the intensities of the Ni_2C^+ and Ni_2O^+ peaks and the decrease in the intensity of the carbon monoxide containing secondary ions. Furthermore, the temperature at which we observe that dissociation becomes important is in agreement with that found using photoelectron spectroscopy [8].

For copper, a similar argument applies. Little, if any, dissociation is occurring at or below room temperature. At 390 K peaks were observed which could be assigned to dissociation products, although the possibility exists that the observed carbon-containing secondary ions arise from bulk impurities diffusing to the surface at this temperature. Even if this was the case, the fact that carbon-containing secondary ions were observed at high temperatures suggests that if dissociation were occurring at low temperatures similar species would have been observable in the SIMS spectra.

In the case of iron, the situation is shown clearly in fig. 2b. The $FeCO^+/Fe^+$ ratio decreases over the same temperature range as the FeC^+/Fe ratio increases. This suggests that the two effects are connected and indicates an increase in the amount of dissociated carbon monoxide as the temperature is raised. Thus, it may be concluded, there is little dissociation at low temperatures; at room temperature there is some dissociative adsorption which coexists with molecular adsorption, while at 350 K there is almost complete dissociation. These conclusions are precisely those drawn by Roberts and Kishi [3] from their work with photoelectron spectroscopy.

The results obtained on palladium are consistant with the adsorption being molecular at 298 K since there were no peaks which could be assigned to dissociation products.

The interpretation of our results concerning carbon monoxide adsorption on tungsten is a little more difficult than for the other metals. It is not clear whether WCO^+ should be assigned to the presence of virgin-CO on the surface, to the α-state of adsorption, or to a mixture of the two. If it is assigned exclusively to virgin-CO the problem arises that there is no obvious reason why the α-state should not be observed. However, it is known that on heating a virgin layer it transforms into a mixture of the α- and β-states [9,10], the β-state being due to dissociative adsorption [10]. This would explain why there is an increase in the WO^+/W^+ ratio as the WCO^+/W^+ ratio decreases. If, on the other hand, WCO^+ is assigned to the α-state the temperature at which desorption takes place is about 100 K below that which is reported in the literature. Desorption of the α-phase does not cause an increase in the amount of dissociation and so this would not be expected to give rise to an increase in the WO^+/W^+ ratio shown in fig. 3. It seems most logical to assign WCO^+ to virgin-CO, leaving open the problem about the absence of peaks which can be assigned to α-CO.

A further point of interest concerning carbon monoxide adsorption on these metals concerns a possible correlation between our results and those obtained by infra-red spectroscopy. When carbon monoxide is adsorbed on certain metals, including nickel [12] and palladium [13] two bands are observed in the ir spectrum whereas on others, including copper [6], iron [13] and tungsten [14], only one band appears. When SIMS spectra are recorded under similar conditions we observe two types of carbon monoxide-containing secondary ions, MCO^+ and M_2CO^+, for those metals which give rise to two ir bands and only one, MCO^+, for those which show only one ir band. It is tempting, therefore, to assign MCO^+ to a linearly bonded adsorbate molecule and M_2CO^+ to a bridge bonded molecule in much the same way as ir spectra have been assigned in the past.

Acknowledgements

We are grateful to the Science Research Council for a Research Studentship awarded to J.W. We would also like to thank R.S. Bordoli for his assistance with part of this work.

References

[1] R.R. Ford, Advan. Catalysis 21 (1970) 51.
[2] R.W. Joyner and M.W. Roberts, Chem. Phys. Letters 29 (1974) 447.

[3] K. Kishi and M.W. Roberts, J.C.S. Faraday I, 71 (1975) 1715.
[4] M. Barber and J.C. Vickerman, Surface and Defect Properties of Solids 5 (1976) 162.
[5] M. Barber, J.C. Vickerman and J. Wolstenholme, J.C.S. Faraday I, 72 (1976) 40.
[6] M.A. Chesters, J. Pritchard and M.L. Sims, Chem. Commun. (1970) 1454.
[7] J.M. Schroeer, T.N. Rhodin and R.C. Bradley, Surface Sci. 34 (1973) 571.
[8] R.W. Joyner and M.W. Roberts, J.C.S. Faraday I, 70 (1974) 1819.
[9] M.A. Chesters, B.J. Hopkins, A.R. Jones and R. Nathan, Surface Sci. 45 (1974) 740.
[10] J.T. Yates, Jr., T.E. Madey and N.E. Erickson, Surface Sci. 43 (1974) 257.
[11] C.G. Goymour and D.A. King, J.C.S. Faraday I, 69 (1973) 736, 749.
[12] A.M. Bradshaw and J. Pritchard, Surface Sci. 17 (1969) 372.
[13] A. Palazov, C.C. Chang and R.J. Kokes, J. Catalysis 36 (1975) 338.
[14] A.M. Bradshaw and J. Pritchard, Proc. Roy. Soc. (London) A316 (1970) 169.
[15] J.T. Yates, Jr., R.G. Greenler, I. Ratajczykowa and D.A. King, Surface Sci. 36 (1973) 739.

ATOMIC CHEMISORPTION ON SIMPLE METALS: CHEMICAL TRENDS AND CORE-HOLE RELAXATION EFFECTS

A.R. WILLIAMS and N.D. LANG

IBM Thomas J. Watson Research Center, Yorktown Heights, New York 10598, USA

We have obtained essentially exact numerical solutions for a simple model of atomic chemisorption on simple metals. The approximations constituting the model are the semi-infinite jellium simulation of the metal substrate and the self-consistent local density theory of exchange and correlation. The solutions provide a detailed picture of the electronic charge making up the chemisorption bond. The variation of this picture with the valence of the adatom exhibits in a direct and microscopic way the roles of electronegativity, charge transfer, and covalency. Predicted bond energies, bond lengths, and dipole moments are consistent with measurements and independent theoretical considerations.

1. Introduction

The mechanisms by which atoms and molecules bond to crystal surfaces have received much study in recent years. This interest stems in part from the natural desire to extend our growing understanding of bulk solids to surfaces and in part from the technological importance of catalysis. Heterogeneous catalysis often involves molecules interacting with the surface of a transition metal. Such systems contain several complicating features: the inter-atomic interactions within the molecule, the interaction of the molecular atoms with the mobile electrons of the substrate and the contribution of the localized d-electrons. We have taken what seems to be a necessary first step toward a fundamental understanding of chemical reactions on metal surfaces; we have developed and solved a realistic model for the interaction of individual atoms with the mobile electrons of a metal substrate. Even the study of such relatively simple systems required the development of elaborate analytical techniques. We have used these techniques to study two phenomena: the bonding of atoms to simple-metal surfaces and the metallic screening of core holes (created, for example, by X-ray photoemission). The results of these studies have been described separately elsewhere [1]; we summarize them below.

The characterizing feature of our study is that we make only two approximations, (1) the local-density treatment of exchange and correlation among the electrons and (2) the simulation of the simple-metal substrate by jellium [2] (the replacement of the discrete ion cores by a semi-infinite uniform distribution of positive charge). Because both approximations are widely used, their implications are relatively well understood. The absence of further approximations allows us to extract the full physical and chemical content of this simply stated and easily appreciated model.

The remainder of the review is organized as follows. In section 2 we discuss the atom-jellium model and the techniques we use to solve the equations which embody it. In section 3 we discuss results which exhibit the roles played by electronegativity, atomic shell filling and covalency in atomic chemisorption. In section 4 we summarize our more recent work in which the model was used to study chemical and extra-atomic relaxation shifts of deep-core-level binding energies. These shifts occur when an atom is allowed to interact with a metal.

2. The atom-jellium model

The atom-jellium model consists of a distribution of positive charges together with the neutralizing electron distribution $n(r)$, which minimizes the total energy of the combined system. The positive charges of the model are the semi-infinite uniform distribution of positive charge and the adatom nucleus. The dependence of the total energy E on the electron distribution is given, according to density-functional theory [3], by

$$E\{n\} = E^h\{n\} + E_{xc}\{n\}, \tag{1}$$

where $E^h\{n\}$ is the total energy of a system with electron density $n(r)$ in the Hartree self-consistent-field approximation [4] and $E_{xc}\{n\}$ is, by definition, the contribution of exchange and correlation. The local-density approximation which we use is simply

$$E_{xc}\{n\} \doteq \int d^3r \, n(r) \, \epsilon_{xc}^h(n(r)), \tag{2}$$

where $E^h\{n\}$ is the total energy of a system with electron density $n(r)$ in the Hartree self-consistent-field approximation [4] and $E_{xc}\{n\}$ is, by definition, the electron density is expressed in terms of one-electron orbitals,

$$n(r) = \sum_i n_i |\psi_i(r)|^2 \tag{3}$$

(the n_i give the occupation of each orbital), the minimization of $E\{n\}$ leads to Schrödinger-like Euler equations for the $\psi_i(r)$,

$$[-\tfrac{1}{2}\nabla^2 + V_H(r) + \mu_{xc}(r)]\, \psi_i(r) = \epsilon_i\, \psi_i(r). \tag{4}$$

The Hartree potential $V_H(r)$ [4] is the electrostatic potential due to the electrons and the positive charges; the exchange-correlation potential $\mu_{xc}(r)$ is given by

$$\mu_{xc}(r) \equiv \frac{\delta E_{xc}}{\delta n(r)} \doteq \frac{d}{dx}(x\epsilon_{xc}^h(x))_{x=n(r)}. \tag{5}$$

To relate our calculated state densities to excitation spectra, such as kinetic energy distributions of photoemitted electrons, we require a relationship between the ϵ_i's of eq. (4) and excitation energies. Although the following rigorous relationship exists,

$$\epsilon_i = \partial E/\partial n_i, \tag{6}$$

two complications prevent the straightforward identification of the ϵ's with excitation energies. First, excitations involve whole electrons, i.e.

$$E_{\text{final}} - E_{\text{initial}} = \int dn_i \, \epsilon_i(n_i), \tag{7}$$

and $\epsilon_i(n_i)$ is not necessarily a constant function of n_i. Second, the total energy has been shown to be a unique functional of the electron density only for the ground state.

Since both the Hartree and exchange-correlation potentials, which determine the orbitals $\psi_i(r)$ depend on the $\psi_i(r)$ through the electron density, the entire system of equations must be iterated until a self-consistent solution is obtained.

The important characteristics of this approach to electronic structure problems are: (1) the kinetic energy is treated wave-mechanically, in contrast to semi-classical Thomas–Fermi-like theories; (2) exchange and correlation are treated on an equal footing, in contrast to the Hartree–Fock approximation; (3) the quantitative understanding of exchange and correlation in the homogeneous interacting electron gas is exploited, and (4) the self-consistent-field structure of the theory allows polarization and charge transfer to freely develop to the extent that they lower the total energy.

The approach we have taken to the solution of eq. (4) (the Schrödinger equation) is scattering-theoretic. We transform eq. (4) to the equivalent Lippmann–Schwinger equation

$$\psi_k(r) = \phi_k(r) + \int d^3 r' \, G(r, r') \, \Delta(r') \, \psi_k(r'), \tag{8}$$

where $\Delta(r)$ is the difference between the effective potential ($V_H + \mu_{xc}$) for the combined metal–adatom system and that for the bare metal. The vector k indicates momentum parallel to the surface, a conserved quantity for the bare metal and a label for the scattering solutions of the combined metal-adatom system. This transformation allows us to exploit the fact that metallic screening greatly restricts the range of $\Delta(r)$ without making the unjustified assumption that the wave functions are perturbed only in a restricted region. The Green's function $G(r, r')$ and the incoming wave $\phi_k(r)$ correspond to the bare metal. The eigenvalue $\{\epsilon_i\}$ spectrum for the solutions of eq. (8) (ϵ_i is a fixed parameter in eq. (8)) contains both discrete and continuous parts. The discrete levels correspond to the core levels of the adatom. Atomic levels which after displacement by charge transfer etc. are degenerate with the band states of the metal become scattering resonances.

3. Ionicity and covalency in atomic chemisorption

Fig. 1 displays the atomic valence levels of Na, Si and Cl which, as mentioned above, have been broadened into scattering resonances. The state density shown in fig. 1 is the difference between the distribution of the $\{\epsilon_i\}$ appearing in eq. (4) for the combined metal-adatom system and that for the bare metal. The energy position of these resonances relative to the Fermi energy of the metal determines their occupation. We see that the 3s level of Na has lost its electron to the Fermi sea, while the 3p level of Cl has taken an electron from the metal. The large energy required to either fill or empty the 3p shell of Si forces the corresponding resonance to straddle the Fermi level. The determining factor in all three cases is the electronegativity of the adatom relative to that of the metal surface.

Fig. 2 displays contours of constant electron density for the same three systems represented in fig. 1. The upper three contour plots describe the total electron density in the vicinity of the adatom; the lower three describe the difference between the electron density of the combined metal-adatom system and the superposition of the bare-metal and atomic electron densities. The upper contours for Na and Cl are consistent with a simple ionic picture of the chemisorption bond. That is, the contours near the adatom are nearly circular, while those in the metal deflect according to the net charge on the ionized adatom. The upper contours in the case of Si show a projection of the atomic charge into the metal. We identify this projection with a covalent bond.

The contours of the density difference shown in the middle portion of fig. 2 show the charge displacements associated with the formation of the chemisorption bond. These difference contours provide a more detailed picture of the bond. We see, for example, that the electronic charge lost by Na does not simply disappear into the metal; it remains on the metal side of the atom. Similarly, the electron

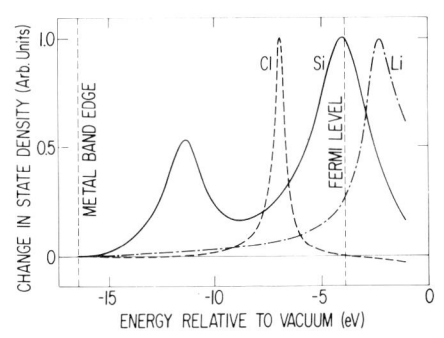

Fig. 1. Change in state density due to atomic chemisorption. Curves correspond to metal-adatom separations which minimize the total energy, 2.5, 2.3 and 2.6 Bohr for Li, Si and Cl respectively. The lower Si resonance corresponds to the 3s level; the corresponding level for Cl is a discrete level below the metal band edge.

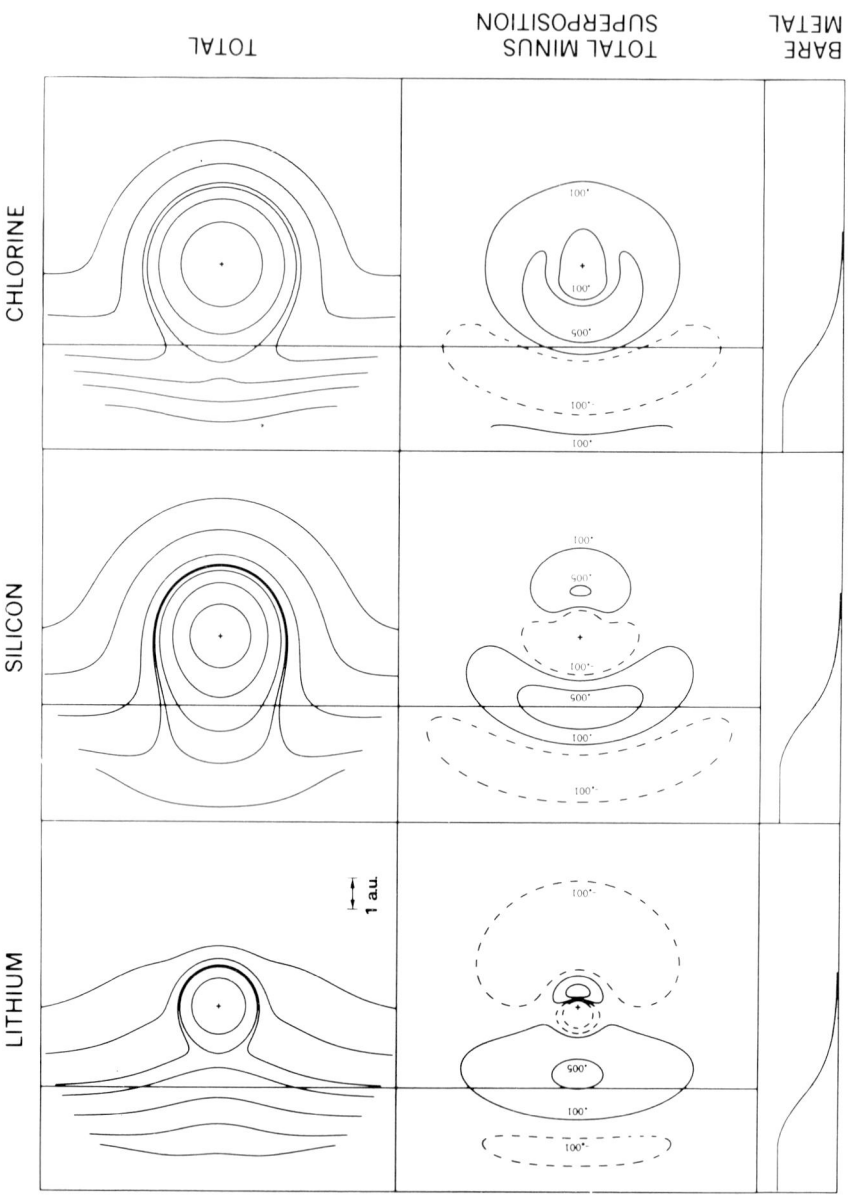

Fig. 2. Upper row: contours of constant electron density in (any) plane normal to the metal surface containing the adatom nucleus (indicated by +). Metal is to the left; solid vertical line is positive background edge. For computational convenience, contours are not shown outside the inscribed circle of each square. Contour values were selected to be visually informative. Center row: total electron density minus the superposition of atomic and bare-metal electron densities (electrons/Bohr³). The polarization of the core region, shown for Li, has been deleted for Si and Cl because of its complexity. Bottom row: bare metal electron density profile (shown to establish physical distance scale). (Bulk density is 0.03 electrons/Bohr³.)

captured by Cl comes primarily from the immediate vicinity of the adatom. The difference contours for Na and Cl are a vivid picture of the effectiveness of metallic screening.

The charge displacements resulting from the chemisorption of Si are more complicated, but are very similar to the displacements seen in the corresponding covalently bonded diatomic molecules [5]. Charge accumulates both in the bond region and on the vacuum side of the adatom.

4. Core-level binding energies

The atom-jellium model of chemisorption is an ideal tool with which to examine the change in the binding energy of core electrons which results when an atom is brought into contact with a metal [6]. Part of the change is due to initial state or chemical effects; the remainder is due to the metallic screening of the core hole. The latter is called the final-state or relaxation shift.

Changes in core-state binding energies are a common probe of extra-atomic chemical and relaxation shifts, but the effects originate in the valence states. The changes in the 3s level of Na shown in fig. 3 exhibit both types of effects. The chemical effects are those associated with the chemisorption process per se. In this case they are the broadening of the level and, because of its energy position above the Fermi energy of the metal, the transfer of the 3s electron to the metal. These are the effects discussed in the preceding section. When a core electron is removed,

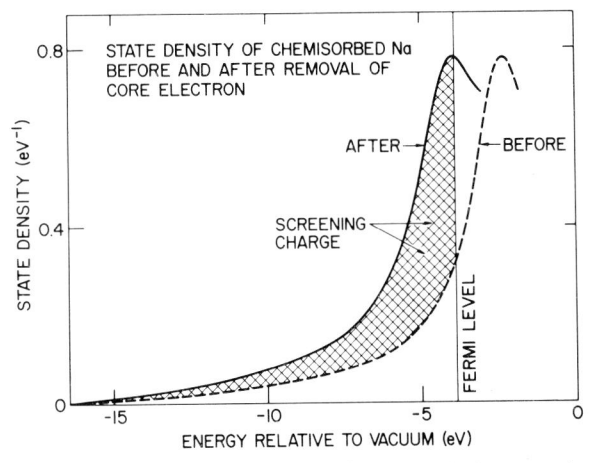

Fig. 3. State density of chemisorbed Na before and after removal of 2s core electron. State density of bare metal has been subtracted from that of metal–adatom system. High density ($r_s = 2$) substrate. Metal-adatom separation, 2.5 Bohr, minimizes total energy of ground state [7]. Discrete portion of spectrum not shown.

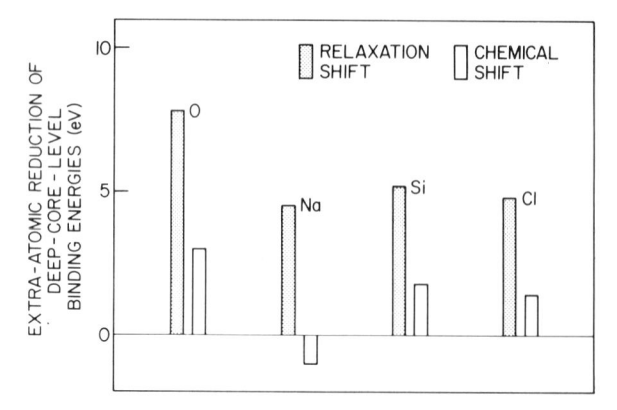

Fig. 4. Decomposition of extra-atomic reduction of deep-core-level binding energy. Single atoms of O, Na, Si and Cl chemisorbed on high density ($r_s = 2$) simple metal surface. Metal–adatom separations for Na, Si and Cl are 2.5, 1.6 and 1.5 Bohr; oxygen was placed inside the metal [7]. Hole state is 1s for O, 2s for Na, Si and Cl.

the reduced screening of the nucleus pulls the 3s resonance below E_f. In this case the relaxation, or final state screening of the core hole, consists of the reoccupation of the valence 3s level.

The changes in the core-level binding energies due to these valence electron rearrangements are shown in fig. 4 for O, Na, Si and Cl [7]. The chemical shift is seen to depend principally on charge transfer. Thus, for example, the chemical shift is negative only for Na. The relaxation shift depends primarily on the spatial extent of the orbital from which the screening charge is derived. The 2p orbital of oxygen is, for example, much more localized than the 3s orbital of sodium.

We have discussed the screening charge in terms of the atomic orbital from which it derives. Figs. 5 and 6 allow us to pursue further the atomic character of the screening charge. Shown in fig. 5 are contours of constant electron density for the extra-atomic screening charge. This quantity involves the self-consistent electron densities of four independent systems, (1) the isolated atom, (2) the isolated atom with a core hole, (3) the combined metal–adatom chemisorption system and (4) the metal–adatom system with a core hole. The difference between the first two is the atomic screening charge; it contains no net valence charge and results from the contraction of the atomic orbitals about the core hole. The difference between the third and fourth electron densities is the analogous quantity for the chemisorbed atom; it differs from the atomic screening charge in possessing a single net valence electron. The metal-adatom screening charge minus the atomic screening charge is the extra-atomic screening charge shown in fig. 5; it represents the spatial distribution of the extra-atomic screening electron supplied by the metal.

The spatial distribution of the extra-atomic screening electron shown in fig. 5 for Na is consistent with the discussion of fig. 3 above. The contours are approx-

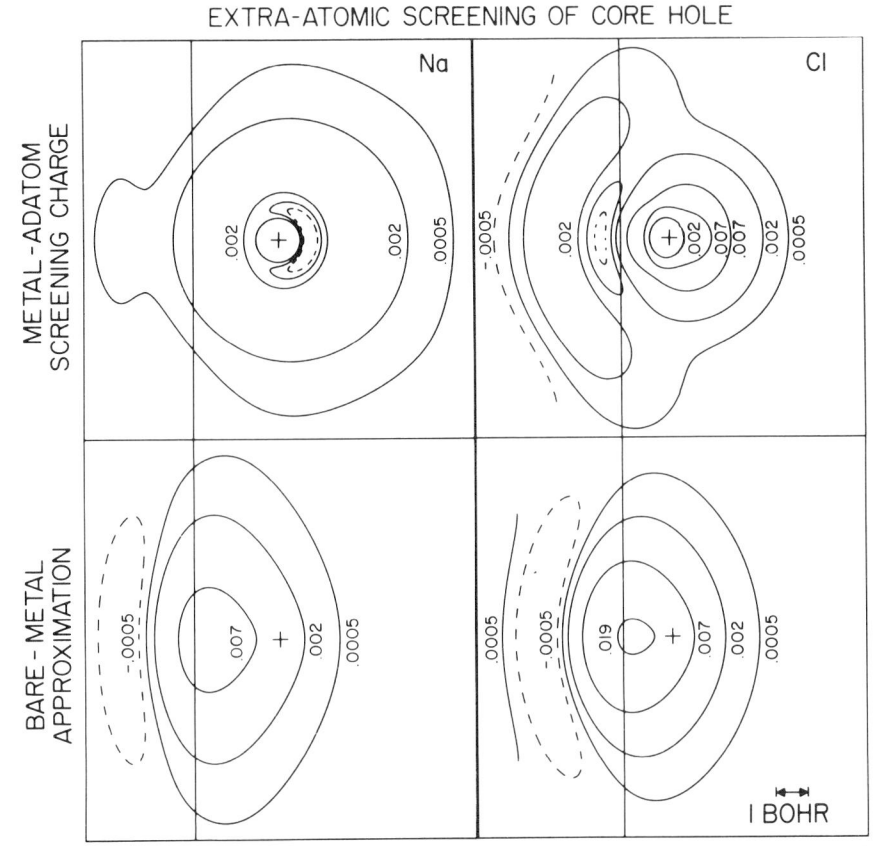

Fig. 5. Contours of constant electron density showing spatial distribution of extra-atomic screening charge for Na and Cl chemisorbed on high density ($r_s = 2$) simple metal surface. Upper plots: response of combined metal-adatom system. Lower plots: linear dielectric response of bare metal. Metal to left; vacuum to right. Adatom nucleus indicated by +. Solid vertical line indicates positive background edge. For computational convenience, contours not shown outside inscribed circle of each square. Contour values are 0.019, 0.007, 0.002 and 0.0005. For reference, the bulk metal density is 0.03 electrons/Bohr^3. Contours in core regions have been deleted for clarity. Hole state is 2s. Metal–adatom separations for Na and Cl are 2.5 and 1.5 Bohr [7].

imately circles surrounding the adatom nucleus. Fig. 5 reveals an interesting complication in the case of chemisorbed Cl. Here, initial-state charge transfer has made it impossible to accomodate the extra-atomic screening electron in the 3p shell. The contours encircling the Cl nucleus represent the contraction of the transferred charge about the core hole. The extra-atomic screening electron in this case lies in the metal (primarily in the kidney shaped contours).

The lower half of fig. 5 is concerned with the accuracy of an attractively simple

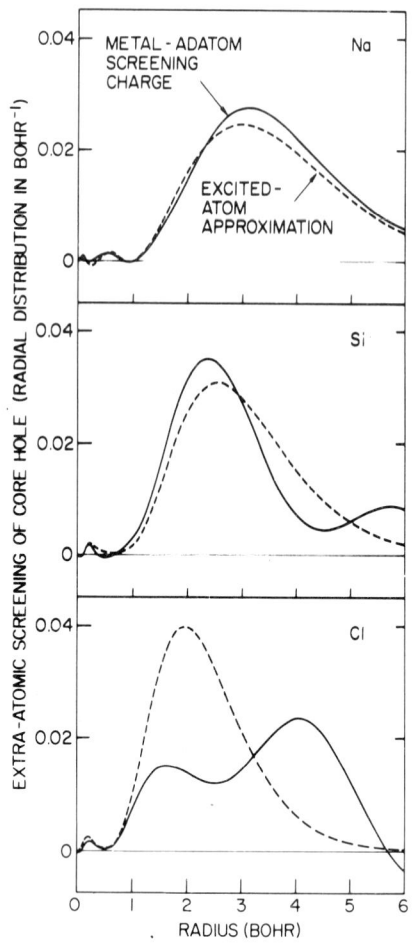

Fig. 6. Radial distribution of extra-atomic screening charge (charge between r and $r + \mathrm{d}r$) for Na, Si and Cl chemisorbed on high density ($r_s = 2$) simple metal surface. Hole state is 2s. Metal-adatom separations for Na and Cl are 2.5 and 1.5 Bohr [7]. Distance measured from adatom nucleus.

approximation to the extra-atomic screening charge distribution, the linear dielectric response of the bare metal to a point charge representing the core hole [8]. In this model the dependence of the extra-atomic screening charge distribution on the particular atom containing the hole is ignored; the only difference between the left and right portions of the lower half of fig. 5 is the separation of the point charge from the metal. In the case of Na the bare-metal screening model does not capture the atomic character of the extra-atomic screening charge distribution. Even in Cl, where the extra-atomic screening electron resides primarily in the metal, the description of the distribution given by the bare-metal-screening model is seen to be qualitatively incorrect. By examining the screening of fractional core holes in the atom-jellium model, we have determined that the inadequacy of the bare-metal-screening model stems not from the assumption of linearity, but from the absence of atomic-valence-orbital structure in the dielectric function of the bare metal.

Where fig. 5 compares the screening of the combined metal—adatom system to that of the bare metal, fig. 6 compares the combined-system screening to that of a purely atomic model. The latter, which we call the excited-atom model, consists of putting the screening electron in the lowest unoccupied orbital of the isolated atom. In other words, instead of removing the core electron to infinity, we excite it into the lowest unoccupied orbital [9]. Fig. 6 compares the radial distribution of the extra-atomic screening charge given by the atom-jellium model with that of the excited atom. (The process by which the four relevant electron densities are combined to form the extra-atomic screening charge distribution is described above in the discussion of fig. 5.) The excited-atom model provides an excellent description of the screening charge in Na and Si. As discussed above, the initial-state filling of the valence shell for chemisorbed Cl makes the screening in this case more complex. For Cl the extra-atomic-screening electron (the second peak in the radial distribution) can be thought of as arising from 4s—4p directional hybrids extending into the metal or from Friedel oscillations associated with metallic screening of the chemisorbed ion. As mentioned above, the first peak in the radial distribution of the extra-atomic screening charge for Cl is due to the contraction of the additional charge transferred to the 3p shell during the chemisorption process. (The negative contribution due to the inward contraction of the outer edge of the 3p shell is masked by the positive contribution in the metal.)

5. Conclusions

Our solutions of the atom-jellium model of chemisorption provide a detailed microscopic picture of the charge rearrangements associated with the chemisorption process and with the screening of core holes. For both processes the chemical identity of the adatom (size—valence—electronegativey) appears to be the decisive factor determining these rearrangements.

References

[1] N.D. Lang and A.R. Williams, Phys. Rev. Letters 34 (1975) 531;
    N.D. Lang and A.R. Williams, Phys. Rev. Letters 37 (1976) 212;
    N.D. Lang and A.R. Williams, Phys. Rev. B. (to be published);
[2] N.D. Lang, Solid State Phys. 28 (1973) 225.
[3] P. Hohenberg and W. Kohn, Phys. Rev. 136 (1964) B 864;

W. Kohn and L.J. Sham, Phys. Rev. 140 (1965) A 1133;
L. Hedin and B.I. Lundqvist, J. Phys. C (Solid State Phys.) 4 (1971) 2064.
[4] We include in our definition of the Hartree energy the electrostatic self-interaction of the electron density associated with each orbital.
[5] R.F.W. Bader, W.H. Henneker and P.E. Cade, J. Chem. Phys. 46 (1967) 3341;
R.F.W. Bader, I. Keaveny and P.E. Cade, J. Chem. Phys. 47 (1967) 3381.
[6] L. Ley, S.P. Kowalczyk, F.R. McFeely, R.A. Pollak and D.A. Shirley, Phys. Rev. B8 (1973) 2392;
P.H. Citrin and D.R. Hamann, Phys. Rev. B10 (1974) 4948;
R.E. Watson, J.F. Herbst and J.W. Wilkins, Phys. Rev. B14 (1976) 18.
[7] The equilibrium position of the adatoms Si and Na were calculated using the first-order pseudopotential perturbation theory described in ref. [2]. Adsorption was taken to be in a threefold site on the (111) face of Al. Let us define the distance from positive background edge to adatom nucleus along the surface normal as d. Then this procedure yields $d = 1.6$ bohr for Si (corresponding to an Al–Si bond length $b = 2.6$ Å) and $d = 2.5$ bohr for Na (corresponding to $b = 3.0$ Å). The calculated Na–Al bond length is in very reasonable agreement with the value found by B.A. Hutchins, T.N. Rhodin and J.E. Demuth, Surface Sci. 54 (1976) 419, using LEED analysis for Na on Al(100) (2.86 plus or minus 0.07 Å). As discussed in ref. [1b] the first-order theory does not appear to yield reliable results for Cl, and so a value for d was found by using fig. 3 of ref. [1b] in conjunction with the photoemission data of J.E. Rowe cited in ref. [9] of that paper ($d \doteq 1.5$ bohr). For oxygen the adatom was placed (in conjunction with the pseudopotential procedure described above) at an interstitial site inside the metal. This was done, because experimental studies (cf. K.Y. Yu, J.N. Miller, P. Chye, W.E. Spicer, N.D. Lang and A.R. Williams, Phys. Rev. B14 (1976) 1446) indicate that oxygen penetrates the crystal surface under the experimental conditions thus far explored.
[8] J.W. Gadzuk, J. Vacuum Sci. Technol. 12 (1975) 289;
G.E. Laramore and W.J. Camp, Phys. Rev. B9 (1974) 3270.
[9] R.E. Watson, M.L. Perlman and J.F. Herbst, Phys. Rev. B13 (1976) 2358;
R. Hoogewijs, L. Fiermans and J. Vennik, Chem. Phys. Letters 37 (1976) 87.

LOCALIZED ORBITAL APPROACH TO CHEMISORPTION: H AND O ADSORPTION ON Ni, Pt AND W(001) SURFACES

David W. BULLETT

Cavendish Laboratory, Madingley Road, Cambridge CB3 OHE, UK

Atomic orbitals and an atomic approximation to the potential are used in the chemical pseudopotential secular equation to calculate the covalent binding energy of an adsorbed monolayer on a transition metal slab. After showing that the method provides a realistic description of the bulk metal band structure we examine the changes which accompany adsorption in various surface arrangements. For H on W(001), maximum coverage is found to correspond to occupation of all bridge sites (β_1 phase). Bridge sites also provide maximum covalent binding at lower coverage but energy differences between alternative sites are small (~0.2 eV per H atom) and ionic effects may stabilise adsorption above atoms in the β_2 phase. On Ni and Pt(001) both adsorbates bind most strongly above 4-fold hole sites, but for hydrogen energy differences between alternative sites are particularly small (~0.1 eV on Pt). Surface densities of states are presented and discussed in relation to photoemission spectra and the mechanism of the chemisorptive bond.

1. Introduction

This paper summarises a few results of a parameter-free tight-binding approach for calculating the covalent adsorption energy and geometry of simple adsorbates on transition metal surfaces. Detailed results for other adsorbates will be published elsewhere [1].

2. Method

The approximations used are a simple extension of those already shown to be successful in describing the electronic and structural properties of covalent semiconductors in both crystalline and amorphous phases [2]. Briefly, the method proceeds as follows:

(i) Chemisorption energies are calculated by comparing total energies for a thin slab of transition metal with and without a periodic surface layer of adsorbate atoms. The slab extends infinitely in two dimensions so that adsorbate–adsorbate and solid state substrate–adsorbate effects are at least partially included, in an attempt to overcome the chief deficiency of cluster calculations.

(ii) The potential within any atomic cell is approximated by the Hartree–Fock–

Slater potential [3] of the isolated atom; valence electrons completely screen all but one of the ion cores. The boundaries of each cell are defined by the positions at which neighbouring atomic potentials are equal.

(iii) Valence level atomic functions ϕ_{il}^0 are used as approximate solutions of the localized pseudopotential equations [4]

$$H\phi_{il}^0 - \sum \phi_{jl'}^0 \langle \phi_{jl'}^0 | V_j^i | \phi_{il}^0 \rangle = \epsilon_{il} \phi_{il}^0 . \tag{1}$$

Suffixes i and l label the atomic site and angular momentum of each orbital and V_j^i is the local perturbation of the potential in cell j from that of the isolated atom V_i.

(iv) Assuming linear independence of the localized orbitals the crystal eigenvalues E_{nk} and wave functions ψ_{nk} are determined by diagonalising the usual secular equation

$$|D - E_{nk} I| = 0 , \tag{2}$$

where all of the matrix elements have been determined by direct integration using atomic wave functions and potentials. Off-diagonal elements of D are of the form

$$D_{rs} = \langle \phi_{jl'}^0 | V_j^i | \phi_{il}^0 \rangle$$

(where suffixes r and s now each label a particular orbital site and type) and diagonal elements are of the form

$$D_{rr} = \epsilon_{il} = \langle \phi_{il}^0 | H | \phi_{il}^0 \rangle - \sum_{jl'} \langle \phi_{il}^0 | \phi_{jl'}^0 \rangle \langle \phi_{jl'}^0 | V_j^i | \phi_{il}^0 \rangle = \epsilon_{il}^0 + \Delta\epsilon_{il} ,$$

where ϵ_{il}^0 is the corresponding energy level in the isolated atom. In practice we interpolate matrix elements between values calculated for steps of 1 au in the interatomic separation out to second or third neighbour distances and ignore all 3-centre contributions.

(v) Total energies are calculated by sampling a discrete set of k-points in the surface Brillouin zone and doubly occupying the resulting eigenvalues up to the Fermi level with the appropriate number of electrons.

(vi) Self-consistency effects are crudely incorporated by allowing diagonal energies ϵ_{il}^0 to vary linearly according to the charge on each atom. The proportionality constants can be estimated from free ion calculations.

(vii) For each arrangement of adsorbed atoms we calculate the total covalent energy as the difference between ΣE_{nk} over occupied states and the same sum for the isolated atoms in the same charge configuration. Subtracting the similarly calculated binding energy for the clean substrate (and chemisorbed molecule for non-atomic adsorbates) provides an estimate of the chemisorption energy per adsorbate atom.

3. Results

We now summarise a few of the results obtained for various situations. These calculations involved only s and d orbitals on the transition metal atoms. For convenience the metal d-level was fixed at an energy $E_d^0 = -6.5$ eV with the s-level 1 eV higher (chemisorption bond strengths were found to be relatively insensitive to the position of the metal s-level).

3.1. Bulk metals

As an example of a bulk metal band structure calculated by this technique, fig. 1 shows the result for tungsten (neglecting relativistic effects). There is reasonable

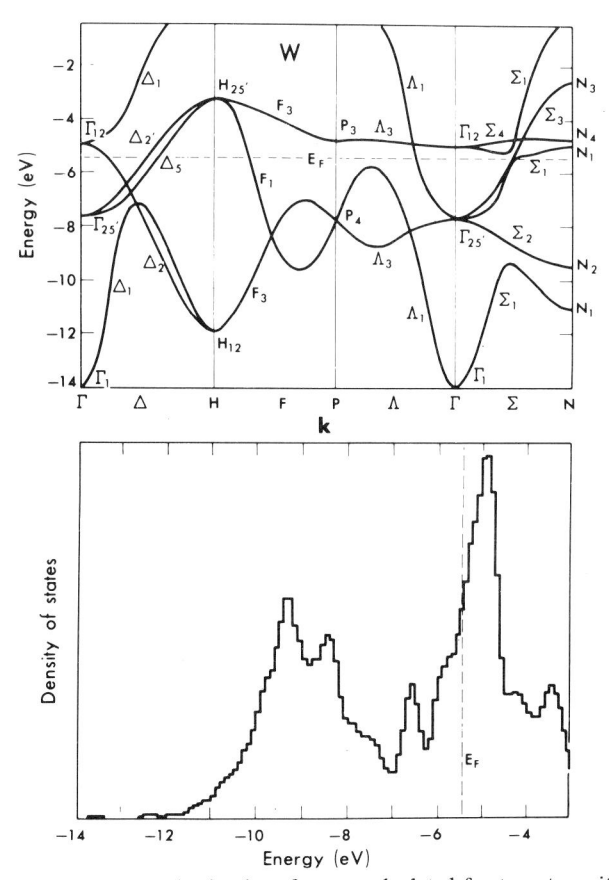

Fig. 1. (a) Band structure and (b) density of states calculated for tungsten with atomic levels $\epsilon_d^0 = -6.5$, $\epsilon_s^0 = -5.5$ eV.

agreement with APW calculations [5]. Transition metal d-bands tend to be about 20% too narrow because of insufficient hybridisation with the conduction band and of course the conduction band is not sufficiently free-electron-like but the overall picture is correct. For the same reasons our bulk configurations ($d^{5.8}s^{0.2}$ for W) contain too few conduction electrons and the Fermi level is rather high in the d-band. All of these errors are lessened when metal p-orbitals are included in the basis.

3.2. H adsorption on W(001)

The first model chemisorption system we consider concerns the interaction of atomic hydrogen with the outside faces of a five layer slab of tungsten. Three likely candidates for the adsorption site are above a surface atom (site A), in a bridging position between a pair of surface atoms (B) and in a four-fold site centred above a second layer atom (C). In each site adsorption energies were calculated for a c(2 × 2) lattice with $\theta = 0.5$ monolayer coverage and for a p(1 × 1) lattice with $\theta = 1.0$.

Fig. 2 displays the distance dependence of the adsorption energy for each site in the (1 × 1) lattices. The equilibrium distance is a little larger than that which maximizes valence shell covalent binding because of the rapidly rising repulsive force form W core electrons (calculated as the first order perturbation in the W 5s and 5p orbitals energies).

Calculated chemisorption energies for each surface arrangement are shown in table 1. In each case either 3 or 6 sample k-points were chosen from the set ($\frac{1}{4}, \frac{1}{4}; \frac{1}{4}, \frac{3}{4}; \frac{3}{4}, \frac{1}{4}; \frac{3}{4}, \frac{3}{4})\pi/a$ according to the surface lattice symmetry [6], but because of the large matrix size, calculations for the c(2 × 2) unit cell were performed only at the equilibrium distances found for the corresponding (1 × 1) structure. Covalent binding energies are essentially independent of coverage within this range; this suggests

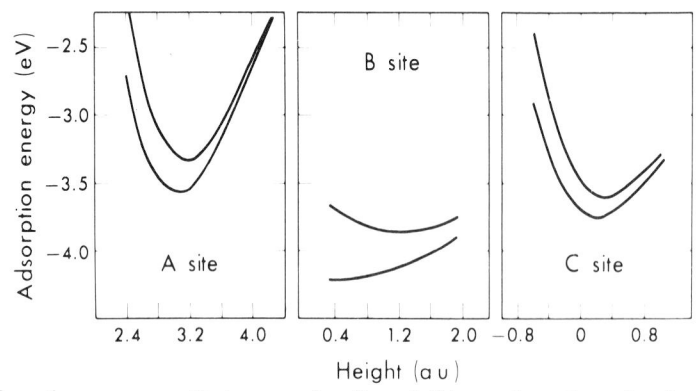

Fig. 2. Adsorption energy per H atom as a function of distance above the surface layer. Lower curves are without tungsten core repulsion, upper curves include W cores.

Table 1
Calculated (100) surface bond energies for H adsorption in various surface arrangements on a 5 layer slab of tungsten

Surface lattice	c(2 × 2)			p(1 × 1)			p(1 × 1)
Coverage θ	0.5			1.0			2.0
Adsorption site	A	B	C	A	B	C	B
Shortest W–H distance (au)	3.2	3.3	3.3	3.2	3.3	3.3	3.3
Negative charge on H (electrons)	0.57	0.46	0.43	0.57	0.48	0.38	0.46
Adsorption energy per H atom (eV)	3.5	3.8	3.7	3.4	3.8	3.6	3.8

that the bonding interaction is restricted almost entirely to near neighbours and gives us some confidence that the size of our k-point set is not introducing spurious effects. In both lattices the stability of adsorption sites decreases in the order B > C > A; in the B site the adsorption energy is ~3.8 eV per atom and the calculated energy differences between sites of order 0.1–0.2 eV are consistent with the small activation energy ~0.25 eV for diffusion of H on W measured by Gomer et al. [7] Results for higher coverages (θ = 2.0) indicated that the second set of B sites can be filled with essentially the same adsorption energy per hydrogen atom as the first set. At this stage adsorption appears to be saturated; for further adsorption on A sites after complete occupation of B sites the adsorption energy decreases by 0.9 eV per atom or 1.8 eV per molecule, which is considerably greater than the observed heat of desorption (~1.1 eV per H_2 molecule for the β_1 state).

An adsorption energy of ~3.8 eV per H atom is much too large when compared with the observed molecular dissociation energy of H_2 (4.75 eV). This method always overestimates binding energies if we ignore repulsive interactions resulting from the aspherical charge distribution in adjacent atomic cells. However if we compare these results with the binding energy of H_2 calculated by the same method, in order to maximize the cancellation of neglected terms, a reasonable value [8] ~1 eV per H_2 molecule is obtained for the desorption energy and this technique may be sufficiently accurate to allow studies of energy surfaces for dissociative adsorption of molecules on transition metals.

Experimental evidence [8,9] suggests that in the high coverage (θ = 2.0) β_1 state H atoms do occupy all the bridge sites but in the β_2 state ($\theta \lesssim 1.0$) adsorption occurs into A sites. The explanation for the change in site at low coverage perhaps lies in ionic effects. A-site adsorption is accompanied by a large (~0.57 e) shift of charge from the tungsten to the hydrogen atom which will then feel an attraction from its image charge in the metal. This may suffice to stabilize the A-site c(2 × 2) lattice whereas at higher coverages because of repulsive interactions between charged adsorbate atoms the bridged structure forms.

3.3. H on Pt(001)

On the (001) face of this fcc metal adsorption was found to be most stable in 4-fold hollow (C) sites but energy differences between alternative sites were again small and might be overwhelmed by ionic corrections: covalent chemisorption energies per H atom were 2.77, 2.84 and 2.94 eV in the A, B and C sites respectively. Calculated equilibrium spacings between H and nearest Pt atoms (3.2, 3.4 and 3.7 au respectively) appear reasonable in comparsion with the sum covalent and metallic radii (~0.6 au for H and 2.6 au for Pt). As expected, nearest neighbour distances increase with increasing coordination of the adsorbate by metal atoms, and the negative charge transferred to the hydrogen atom is again slightly larger for the A site (0.59 e) than for B (0.51 e) or C (0.49 e) sites.

3.4. O on Ni(001) and Pt(001)

Oxygen on Ni(001) represents one of the few overlayer structures for which LEED spectra have been analysed in sufficient detail to determine not only the surface lattice but also the adsorption site [10]. Oxygen atoms in the c(2 × 2) lattice sit in 4-fold (C) sites at a distance of 1.70 au above the surface Ni atomic layer, giving a nearest neighbour NiO distance of 3.74 au.

Table 2 shows the results of the present theory for c(2 × 2) adsorption in A, B or C sites above a three layer metal slab. Shortest metal-oxygen distances were fixed at 3.74 au for Ni and 4.1 au for Pt. For both metals the strongest adsorbate binding again occurs at C-sites. Binding energies are much larger and more site dependent than for hydrogen adsorption, now that chemisorption is taking place primarily through adsorbate p-orbitals. The charge transferred to the adsorbate is also larger (~0.7 e) but because adsorption takes place into the hollows of the surface metal

Table 2
Calculated binding energies and atomic charges for oxygen in different sites on (001) Ni and Pt for a c(2 × 2) overlayer with half monolayer coverage

Site	Nickel		Platinum	
	Adsorption energy (eV per O atom)	Negative charge transferred to O (electrons)	Adsorption energy	Negative charge on O
A (above atom)	2.9	0.69	3.8	0.77
B (bridge)	4.2	0.67	5.2	0.78
C (4-fold hollow)	4.7	0.67	5.8	0.77

layer this need not be incompatible with a small surface dipole and work function change. Differences between the two metals in table 3 are mainly an artifact of the higher Fermi level in Pt (−4.4 eV compared to −5.8 eV in Ni) which results when we use the same atomic levels for both metals. We can again estimate the heat of desorption into molecular oxygen using the O_2 dissociation energy determined either experimentally (5.2 eV) or by this type of simple calculation (8.1 eV).

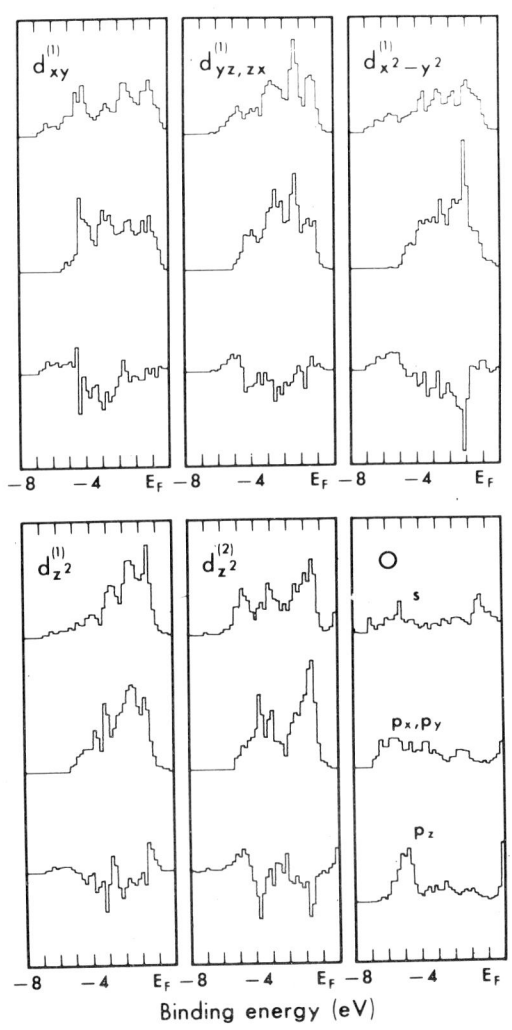

Fig. 3. Local densities of states projected on first layer d-orbitals and the second layer $d_{z^2}^{(2)}$ orbital immediately below the adsorption site for oxygen on platinum (001). Top line is after adsorption, second line is for the clean (3 layer) metal slab and the bottom line is the difference spectrum. Also shown are the densities of states on the adsorbate orbitals.

Results are in the range 1.3–4.2 eV for molecular desorption from Ni and 3.5–6.4 eV from Pt. Experimental determinations have given various values in a similar range, from 1.7 eV to over 6 eV [11,12].

Fig. 3 illustrates the oxygen induced modifications in the local density of states projected on to various orbitals. Strong bonding interactions obviously take place with the d_{yz}, d_{zx} and especially $d_{x^2-y^2}$ surface orbitals and with the second layer d_{z^2} orbital directly below the surface site, as expected for this bonding geometry. An interesting contrast between Ni and Pt occurs in the behaviour of the main oxygen induced feature. Whereas on Ni the spectral weight of all three oxygen p-orbitals is concentrated in a single peak just below the Ni d-band, on the wider d-band element the oxygen p_z orbital still contributes most of its weight below the metal d-band but the p_x and p_y peaks are now spread throughout the band. This difference seems to be confirmed in experimental photoemission spectra [11,13].

4. Comments and improvements

There are various simple improvements which can obviously be incorporated into the present scheme such as

(i) Inclusion of transition metal p-orbitals in the basis. This certainly improves the accuracy of the transition metal band structures but in the few preliminary cases considered the effect on adsorption energies has been slight.

(ii) A better method of assigning the atomic charges. Here we have used the simplest definition by which electrons in a molecular orbital $\psi = \Sigma a_i \phi_i$ are assigned to atoms in proportion to the values of $|a_i|^2$. A more appropriate prescription might assign overlap charges equally between atoms. Since our non-hermitian matrix D in eq. (2) is related to the basis orbital overlap S and Hamiltonian H matrices through [4]

$$D = S^{-1}H,$$

it is easy to show that corresponding left and right eigenvectors of D are related by

$$|\Psi_i^L\rangle = S|\Psi_i^R\rangle.$$

Expanding each in the localized basis

$$\Psi_i^R = \sum a_{ij}\phi_j, \quad \Psi_i^L = \sum b_{ij}\phi_j,$$

the normalisation condition on the molecular orbitals becomes

$$1 = \langle\Psi_i^R|\Psi_i^R\rangle = \sum_{jk} a_{ij}^* a_{ik} S_{jk} = \sum_j a_{ij}^* b_{ij},$$

and for this eigenvector the weight on an orbital j is just $a_{ij}^* b_{ij}$.

(iii) A fuller treatment of self-consistency, in which the s, p and d energy levels and wave functions on transition metal atoms vary independently according to the occupation of each angular momentum state, charge transfer effects appear in off-diagonal as well as diagonal matrix elements and ionic contributions to bonding are included.

(iv) For larger systems resolvent methods [14] will obviously have advantages over direct diagonalisation.

Acknowledgements

I should like to thank Professor Marvin Cohen and other members of the physics department at Berkeley, California for their hospitality while most of this work was performed. Financial support was kindly provided by the U.S. Energy Research and Development Administration, the National Science Foundation, the Royal Commission for the Exhibition of 1851 and Clare College, Cambridge.

References

[1] D.W. Bullett and M.L. Cohen, J. Phys. C (1977), to be published.
[2] D.W. Bullett, Phys. Rev. B14 (1976) 1683.
[3] F. Herman and S. Skillman, Atomic Structure Calculations (Prentice-Hall, Englewood Cliffs, NJ, 1963).
[4] P.W. Anderson, Phys. Rev. 181 (1969) 25.
   J.D. Weeks, P.W. Anderson and A.G.H. Davidson, J. Chem. Phys. 58 (1973) 1388.
[5] L.F. Mattheiss, Phys. Rev. 139 (1965) A1893;
   I. Petroff and C.R. Viswanathan, Phys. Rev. B4 (1970) 799;
   N.E. Christensen and B. Feuerbacher, Phys. Rev. B10 (1974) 2349.
[6] D.J. Chadi and M.L. Cohen, Phys. Rev. B7 (1973) 692; B8 (1973) 5747;
   S.L. Cunningham, Phys. Rev. B10 (1974) 4988.
[7] R. Gomer, R. Wortman and R. Lundy, J. Chem. Phys. 26 (1957) 1147.
[8] P.W. Tamm and L.D. Schmidt, J. Chem. Phys. 51 (1969) 5352; 52 (1969) 1150;
   T.E. Madey, Surface Sci. 36 (1973) 281.
[9] W. Jeland and D. Menzel, Surface Sci. 40 (1973) 295;
   H. Froitzheim, H. Ibach and S. Lehwald, Phys. Rev. Letters 36 (1976) 1549.
[10] J.E. Demuth, D.W. Jepsen and P.M. Marcus, Phys. Rev. Letters 31 (1973) 540;
   P.M. Marcus, J.E. Demuth and D.W. Jepsen, Surface Sci. 53 (1975) 501.
[11] D.M. Collins, J.B. Lee and W.E. Spicer, Surface Sci. 55 (1976) 389.
[12] D. Brennan, D.O. Hayward and B.M.W. Trapnell, Proc. Roy. Soc. (London) A256 (1960) 81;
   A.M. Horgan and D.A. King, Surface Sci. 23 (1970) 259.
[13] D.E. Eastman and J.K. Cashion, Phys. Rev. Letters 27 (1971) 1520;
   J.E. Demuth and D.E. Eastman, Japan. J. Appl. Phys. Suppl. 2, Pt. 2 (1974) 827.
[14] R. Haydock, V. Heine and M.J. Kelly, J. Phys. C5 (1972) 2845; C8 (1975) 2591.

THEORETICAL STUDIES OF ATOMIC ADSORPTION ON NEARLY-FREE-ELECTRON-METAL SURFACES *

H. HJELMBERG
Institute of Theoretical Physics, Chalmers University of Technology, Fack, S-402 20 Göteborg 5, Sweden

O. GUNNARSSON
Department of Physics, University of Pennsylvania, Philadelphia, PA 16174, USA

and

B.I. LUNDQVIST
Institute of Physics, University of Aarhus, DK-8000 Aarhus C, Denmark

A systematic theoretical study of atoms chemisorbed on free-electron like metal surfaces is being performed using an earlier published embedding scheme. Previously reported results have been supplemented in several essential ways. Improvements in the model, namely inclusion of part of the substrate pseudopotentials in a self-consistent way and only the remainder in perturbation theory, have been introduced to give a more realistic description of the surface. We find that (1) the position of the adatom resonance is largely determined by the location of the conduction band, (2) H is absorbed in Na and (3) the results of the improved model support the essential findings of the earlier used model, however, with important improvements of some obvious shortcomings.

1. Introduction

Most experimental studies of chemisorption systems are performed on transition-metal substrates due to their technical importance. We want to advocate that more attention will be given to free-electron like substrates as a detailed theoretical understanding is likely to come first for these systems. In addition such an understanding should be relevant for transition metals due to the important role of the sp-conduction electrons in chemisorption on these substrates [1].

We are performing an extensive and systematic study of atomic chemisorption on free-electron like substrate [2–4], from which we want to give a progress report with the following major results: (1) by varying the substrate parameters over the metallic range, we find that the position of the adatom resonance is largely determined by the location of the conduction band; (2) our model predicts that H is

* Supported by the Swedish Natural Science Research Council.

absorbed in Na as opposed to the adsorption occurring on, e.g. Al; (3) calculations with an improved model supports our earlier [3,4] predictions about the preferred configurations of H chemisorbed on low-indexed Al surfaces.

We have used our formalism [2] for solving the Kohn–Sham (KS) equations [5] to make self-consistent calculations without adjustable parameters on three successively more realistic models for chemisorption on simple metal surfaces. The first two models involve approximate descriptions of the metallic substrate by (a) a semi-infinite jellium system and by (b) adding effects of the substrate lattice as a first-order perturbation to the result of (a) [3,4]. The third model is an improvement of model (b) where part of the pseudopotential contribution is included in the self-consistent calculation.

Our study is aimed at calculating static properties, such as chemisorption energies, equilibrium configurations and electronic densities (charge transfers and dipole moments).

We believe that model (a) in a realistic way illustrates qualitative features, such as sorption character, charge transfer, and Friedel oscillations and that it gives gross numbers for, e.g., the binding energy. The introduction of the lattice in model (b) makes predictions about, e.g., preferred sites possible. These predictions should get improved using model (c). Similar calculations have independently been performed by Lang and Williams for H, Li and O [6] and Li, Cl and [7] on jellium ($r_s = 2$) and for O on jellium with perturbing Al-pseudopotentials (cf. model (b)) [8].

2. Theoretical method

The chemisorption problem is indeed complex from a theoretical point of view as it has both local and extended properties, a geometry made complicated by the loss of symmetry due to both the surface and the adatom and that correlation effects are important judging from molecular calculations. The Kohn–Sham (KS) scheme [5] is simple, includes exchange and correlation effects and have proved to give results with a useful accuracy, also in approximate versions, for atoms [9], molecules [10], solids [11] and metallic surfaces [12–14]. A shortcoming of the scheme is that it applies only to the ground state of the quantum system studied [15] and to certain excited states, i.e. the lowest state of each symmetry [9]. The scheme gives the density as

$$\rho(r) = \sum_i |\psi_i(r)|^2 , \qquad (1)$$

where the sum is over occupied orbitals. The orbital amplitudes are solutions to the Schrödinger-like equations

$$\left[-\frac{\hbar^2}{2m} \nabla^2 + V(r) + e^2 \int \frac{\rho(r')}{|r - r'|} \, d^3r' + v_{xc}(r) \right] \psi_i(r) = \epsilon_i \psi_i(r) . \qquad (2)$$

$v(r)$ is the external potential and $v_{xc}(r)$ is the exchange-correlation potential which we treat in the so-called local density approximation. This approximation, which was used in the applications mentioned above, is discussed in, e.g., ref. [9], together with an estimate of its accuracy. The energy parameter ϵ_i has no formal relation to the one-electron excitation spectrum [16,17].

In our method [2] we use the fact that due to the metallic screening the changes in the electron density when introducing an adatom are localized near the adatom. We therefore introduce a basis set $\{\varphi_n\}_{n=1}^N$ assumed to be complete in this region. Hence we can expand the induced density $\Delta\rho(r)$ as well as the orbital amplitudes $\psi_i(r)$ (the solutions of eq. (2)) in this region. The equation for the corresponding Green function is thus converted to a matrix equation due to the finite number N. The solution to this equation will give us properties inside the disturbed region like $\Delta\rho(r)$. All approximations assumes the basis set $\{\varphi_n\}_{n=1}^N$ to be complete in the disturbed region and the accuracy can therefore be improved by increasing N at the cost of calculational time. In our calculations we have assumed the disturbed region to be inside a sphere of radius R. The basis functions are on the form $F(r)Y_{lm}(\theta,\varphi)$ centered on the adatom and orthogonal in the sphere. The radial functions $F(r)$ are built from Slater-type orbitals. We have typically used $l = 0, 1, 2$ and for each of them 9-10 radial functions and a radius $R = 8$ a u ($r_s = 2, 3$) and $R = 10$ a u ($r_s = 4$).

In model (b) the effects of the ion lattice potentials have been reintroduced [12] through a lattice of weak pseudopotentials [18]. We have considered these effects to lowest order by adding (r_H is the position of the adatom)

$$\int \Delta V(r) \left[Z\delta(r - r_H) - \Delta\rho(r)\right] d^3r \tag{3}$$

to the jellium binding energy (model (a)). Here ΔV is the change in potential energy when replacing the positive background by the pseudopotentials. The adatom is here described by the potential of the bare nucleus. Calculations on many-electron adatoms, described by pseudopotential, have just started.

The obvious shortcoming of model (b) is that the pseudopotentials might not be weak enough to allow the use of first order perturbation theory. To do better we use the method of Perdew and Monnier [14] which makes it possible to include part of the pseudopotential contribution in the self-consistent calculation and treat only the remainder in perturbation theory. Their idea is to introduce a variational parameter c, whose optimum value depends on the crystal structure and the surface plane considered. This is done by adding a step potential $c\theta(-z)$ (z-axis pointing out of the surface) to the external potential in eq. (2). Then the KS equations are solved self-consistently giving a different electron density $\rho_c(z)$ for every c. The big advantage of the method is that the calculation is as easy to perform as for a pure jellium surface. The c was determined by minimizing the expression for the total surface energy (pseudopotentials included) using $\rho_c(z)$ as the electron density. With this method Perdew and Monnier managed to improve results for the surface energy especially for metals with relatively large pseudopotentials. The role of the step potential $c\theta(-z)$ is to mimic the real potential $\Delta V(r)$ in eq. (3). That this is done with succes can be concluded from the following features as well as from the successful application to the surface energy [14]. The c is found to be well correlated with the average value of $\Delta V(r)$. Also the first "peak" of $\rho_c(z)$ is well correlated with the first dip in $\overline{\Delta V(z)}$ (the average parallel to the surface of $\Delta V(r)$). We now proceed as follows. Using the $\rho_c(z)$ etc. of Perdew and Monnier with the c determined by them, we perform a calculation as in model (b) getting different induced densities for different surface planes. We then add the rest of the effect of the lattice pseudopotentials to lowest order by adding

$$\int (\Delta V(r) - c\theta(-z)) \, [z\delta(r - r_H) - \Delta\rho(r)] \, d^3r \qquad (4)$$

to the jellium-like binding energy. We have thus obtained an improved description of the variation of the density normal to the surface. Still lacking though in model (c) is the variation of the electron density parallel to the surface.

3. Results and discussion

The central quantity of the KS scheme is the electron density $\rho(r)$. In fig. 1 we show the total density for the case of H in the equilibrium position on jellium with $r_s = 2$ corresponding to Al ($4\pi r_s^3/3 = 1/n$, n = electron density). The figure illustrates the rapid screening of the perturbation caused by the adatom and the Friedel oscillations. While the oscillations due to the adatom are strong, best illustrated by the first oscillation in the figure, the oscillations caused by the surface are rather weak.

A secondary quantity of the KS scheme is the distribution of energy levels or the density of states (DOS). Fig. 2 shows the adsorbate induced DOS for H on jellium $r_s = 2$ (Al) as a function of the distance d from the surface [19]. The growing width of the resonance as d goes from 2 to 0 au can be understood in terms of the increasing adatom–metal coupling. Muscat and Newns [20] have recently in a one-parameter model analyzed the nature of these states.

So far results for model (a) and (b) have been published essentially only for one substrate, namely for $r_s = 2$ (Al) [6,7,3,8,4]. Our recent extension to other electron densities in the metallic range allows the study of various chemisorption quantities as a function of substrate parameters. In fig. 3 is shown how the resonance peak position of the adsorbate induced Kohn–Sham density of states (KS DOS) varies with the distance of an H atom from a simple metal surface. The result is shown for three values of r_s. It is striking how strongly the bulk parameters like the conduction band bottom, the band width and the effective potential of the free surface influence the peak position. Inside the metal surface the level is locked to the band bottom, lying just below for $r_s = 2-4$ as shown by bulk calculations [21,22]. This happens regardless of the fact that the energy of the band bottom measured from

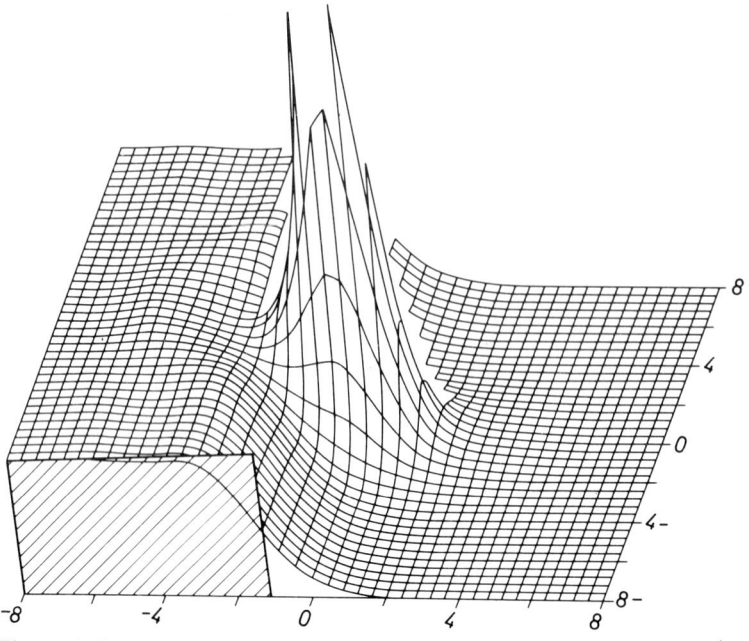

Fig. 1. The total electron density is plotted for H on a jellium $r_s = 2$ surface. The H atom is in its equilibrium position 1.1 au from the jellium edge. The shaded area indicates the positive background of the jellium. The electron density is cut near the adatom. Distances from the adatom position is given in atomic units.

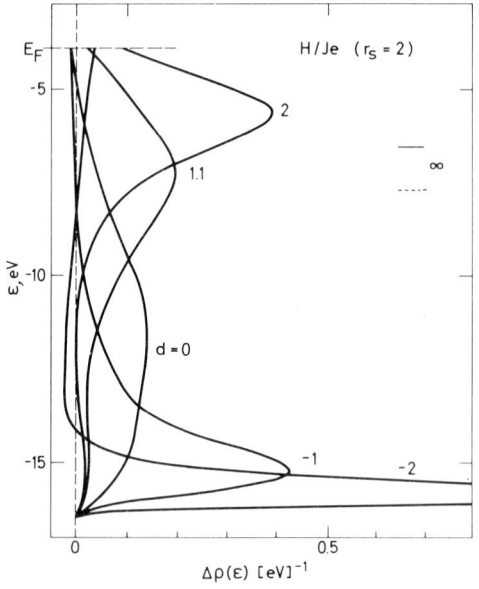

Fig. 2. Adsorbate induced Kohn–Sham density of states for H on a jellium ($r_s = 2$) surface is shown for 5 different distances from the jellium edge. The free H values are also given both from a spin-independent (full line) as well as a spin-dependent (dashed line) calculation.

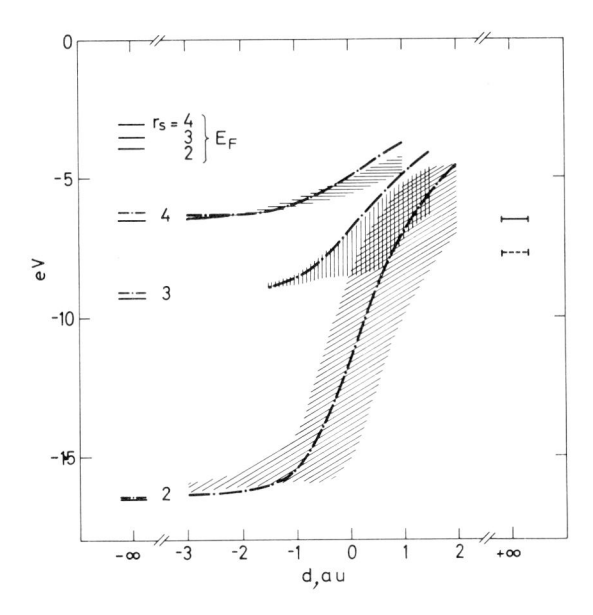

Fig. 3. The adsorbate induced resonance peak position of the Kohn–Sham density of states (shaded areas) as a function of distance d from the jellium edge is shown for three different r_s. The dash-dotted lines are the correponding values of the effective potential of the jellium surface. As a comparison bulk ($d = -\infty$) and free H values ($d = \infty$) are shown. In the bulk we have sharp levels (numerical values from [21,22]) as well as at $d = -3$ au for $r_s = 4$. For the free H two values, from a spin-independent (full line) and a spin-dependent (dashed line) calculation, are shown.

the vacuum level varies with a factor of 3 and that the bulk electron density varies with a factor of 8. Also there is a strong correlation, especially for $r_s = 2$, between the effective potential (the "local band bottom") of the substrate and the peak position. Such a correlation was first noted by Lang and Williams [7] for the 3s and 3p level of Si and the 3p level of Cl when adsorbed on an Al surface. Our extended study to other substrate parameters indicates strongly that it is a general phenomenon for adsorbate resonances.

One should not confuse the KS DOS with the one electron excitation spectrum. On the other hand, although there is no formal relation between the two, there are situations where the difference is small. Extended states with energies not too far away from the Fermi energy is such a situation [17]. The electron states building up the resonance extend quite far into the bulk. These states do not differ drastically from the extended bulk states [4]. This suggests that the difference between the KS DOS and the experimentally measurable one-electron spectrum is not very large for adsorbate resonances like that of H on free-electron like metals. When the adatom moves away from the surface approaching the free atom, the state should, however, get localized and the difference larger. The error of the local-density approximation should also get larger upon localization, as drastically illustrated by the free atom values in fig. 3.

When applied to the case of H on Na ($r_s = 4$) model (b) predicts that H is absorbed rather than adsorbed, in agreement with our earlier suggestion [4]. We find no minimum in the binding-energy curves when the H atom is moved perpendicular to the surface in a bridge (B) or a center (C) position. We have tested this for distances between 1 and −1 au, measured from the jellium edge. The H atom prefers to enter in the bridge (B) configuration for the (100) and (111) surfaces while all the symmetrical configurations are possible for the (110) surface. It is however appropriate to point out that our calculated energy gain for H absorbed by sodium is twice the cohesive energy of the sodium metal. One could thus expect reconstruction of the surface or even the formation of hydrides, which has not been accounted for in our model.

The predictions of model (b) about preferred adsorption sites of H on Al are confirmed by the results of model (c). In fig. 4 are shown some preliminary results for the binding energy curves for H on Al(111) both from the (b) and the (c) model. When calculating the results of model (b) [3,4] we found one deficiency [4] which was most apparent for the (111) surface. Due to charge neutrality the first layer of ions is placed at different distances from the jellium edge for different surface planes. Still, in model (b), the same electron density and effective potential

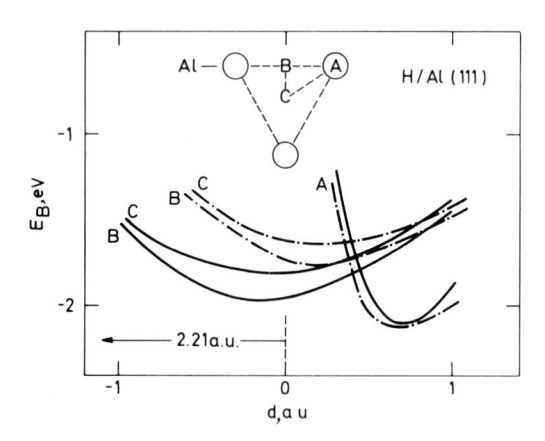

Fig. 4. Calculated binding-energy curves for H chemisorbed on a (111) surface of Al in the atop (A), bridge (B) and centered (C) position. The full lines represent results of model (c) and the dash-dotted lines the results of model (b). The C position shown is the one which is A with respect to the second ion layer. In the figure is also indicated the position of the first ion layer with respect to the jellium edge ($d = 0$).

were used. One of the bond-determining mechanisms is the increase in kinetic energy upon entering the surface, an increase due to the reduced volume available for the adatom electron as can be seen from the bare jellium result (model (a)) [3,4]. This effect tends to place the equilibrium H positions for the (111) surface outside the true ones, as the first ion layer there is farthest from the jellium edge. (2.21 au). Fig. 3 shows that this defect of model (b) is significantly reduced in model (c) for the bridge (B) and center (C) position. The bond length for the atop (A) is mainly determined by the repulsive core of the ion and should not change appreciably. For the (100) surface the changes are small (1.91 au from jellium edge to first ion layer), while for the (110) surface the equilibrium distances are pushed outwards (1.35 au from jellium to first ion layer), in accord with our expectations.

Acknowledgements

We are especially grateful towards J.P. Perdew and R.Monnier who have provided us with data from their surface calculations. One of us (H.H.) acknowledges with thanks the hospitality of the Institute of Physics, University of Aarhus.

References

[1] G. Blyholder, J. Chem. Phys 62 (1975) 3193;
    I.P. Batra and O. Robaux, Surface Sci. 49 (1975) 653;
    D.J.M. Fassaert and A. van der Avoird, Surface Sci. 55 (1976) 291, 313;
    D.E. Ellis, H. Adachi and F.W. Averill, Surface Sci. 58 (1976) 497;
    C.F. Melius, J.W. Moskowitz, A.P. Mortola, M.B. Baille and M.A. Ratner, Surface Sci. 59 (1976) 279.
[2] O. Gunnarsson and H. Hjelmberg, Phys. Scripta 11 (1975) 97.
[3] O. Gunnarsson, H. Hjelmberg and B.I. Lundqvist, Phys. Rev. Letters 37 (1976) 292.
[4] O. Gunnarsson, H. Hjelmberg and B.I. Lundqvist, Surface Sci. 63 (1977) 263.
[5] W. Kohn and L.J. Sham, Phys. Rev. 140 (1965) A1133.
[6] N.D. Lang and A.R.Williams, Phys. Rev. Letters 34 (1975) 531.
[7] N.D. Lang and A.R. Williams, Phys. Rev. Letters 37 (1976) 212.
[8] K.Y. Yu, J.N. Miller, P. Chye, W.E. Spicer, N.D. Lang and A.R. Williams, Phys. Rev. B14 (1976) 1446.
[9] O. Gunnarsson and B.I. Lundqvist, Phys. Rev. B13 (1976) 4274.
[10] O. Gunnarsson, P. Johansson, S. Lundqvist and B.I. Lundqvist, Intern. J. Quantum Chem. 9S (1975) 83;
    O. Gunnarsson and P. Johansson, Intern. J. Quantum Chem. 10 (1976) 307.
[11] O. Gunnarsson, B.I. Lundqvist and J.W. Wilkins, Phys. Rev. B10 (1974) 1319;
    J. Janek, V.L. Moruzzi and A.R. Wiliams, Phys. Rev. B12 (1975) 1257;
    V.L. Moruzzi, A.R. Williams and J.F. Janak, to be published;
    J.F. Janak and A.R. Williams, to be published.
[12] N.D. Lang and W. Kohn, Phys. Rev B1 (1970) 4555; B3 (1971) 1215.
[13] N.D. Lang, Solid State Phys. 28 (1973) 225.
[14] J.P. Perdew and R. Monnier, Phys. Rev. Letters 37 (1976) 1286.
[15] P. Hohenberg and W. Kohn. Phys. Rev. 136 (1964) B864.
[16] L.J. Sham and W. Kohn, Phys. Rev. 145 (1966) 561.

[17] L. Hedin and B.I. Lundqvist, J. Phys. C4 (1971) 2064.
[18] We use the pseudopotentials of N.W. Ashcroft, Phys. Letters 23 (1966) 48.
[19] The small bump seen for low energies in fig. 2, mainly for the distancies 2 and 1.1 au, may be due to an approximate correction. This correction, which does not enter in the calculation of $\Delta\rho(r)$ or the energy, is due to the finite value of R.
[20] J.P. Muscat and D.M. Newns, Phys. Letters 60A (1977) 348, 61A (1977) 481.
[21] C.O. Almbladh, U. von Barth, Z.D. Popović and M.J. Stott, Phys. Rev. B14 (1976) 2250.
[22] U. von Barth, private communication;
J.K. Nørskov, private communication.

THEORETICAL STUDIES OF THE ELECTRONIC STRUCTURE OF SEMICONDUCTOR SURFACES

D.R. HAMANN

Bell Laboratories, Murray Hill, New Jersey 07974, USA

Methods developed by the author and J.A. Appelbaum have been used to examine the electronic structure of semiconductor surfaces in a variety of situations. These essentially first principles calculations self-consistently determine the potential in a surface region several atom layers thick which continuously join to a semi-infinite bulk region whose self-consistent potential is previously determined. Model potentials adjusted to fit bulk and atomic levels represent the ion cores, and a local functional of the density represents the exchange and correlation potential. Electronic states are found in a mixed real-space Fourier-space representation using a transfer matrix technique. A primary emphasis in the problems studied has been to relate features in experimentally accessible surface spectra to hypothesized rearrangements in the atomic geometry and chemical bonding at the surface. Studies of the unreconstructed and reconstructed clean Si(100) surface and of H chemisorption on several Si surfaces will be described as examples of the utility of the methods.

1. Introduction

Theoretical calculations of electronic structure are playing an increasing role in surface science today. I would like to selectively review the results of some of these calculations, examining in the process some seldom discussed questions of motivation. The possible motivations for performing electronic structure calculations can be roughly grouped into three categories: (1) to fit experimental data, (2) to test existing models, and (3) to generate new ideas. While (1) is certainly a valid motivation, the fact that an established model and set of calculational techniques produce, say, a spectrum that agrees with a measured spectrum to the anticipated accuracy does not increase our understanding. (A significant disagreement would, in fact, be a more stimulating result.) When past experience or independent measurement have not established a model, or when conflicting models have been proposed, the motivation falls into category (2). The selection of an adequate model through intercomparisons of theoretical results and experiment represents a real increase in our understanding of the system being studied. The most significant achievement that can be expected from the theoretical investigation of electronic structure is the generation of a new idea, category (3). By this, I mean the identification of a qualitatively new effect or mechanism which may be of consequence beyond the scope of the particular problem being investigated.

Electronic structure calculations vary considerably in their complexity, and in the extent to which they are empirically adjusted. While calculations from a considerable range of these variables can make significant contributions, I would maintain that those which stay closer to first principles and retain at least a moderate degree of complexity are more likely to stimulate new ideas or convincingly discriminate among models. The qualitative results of highly simplified models are usually anticipated when the model is constructed. A highly adjustable theory runs the risk of incorrectly identifying the cause of an observed effect.

A set of calculational techniques have been developed by the author and Joel A. Appelbaum to treat the surface electronic structure of simple metals and semiconductors [1,2]. A semi-infinite geometry is utilized. The self-consistent potential is calculated in the surface region and is joined continuously to the self-consistent bulk potential 2 to 5 atom planes inside the surface. The ion cores are represented by pseudopotentials, and the exact Hartree electrostatic potential produced by the valence elctrons is calculated. The exchange and correlation potential is approximated by a local function of the valence electron density. The potential and the wave functions are expressed in a mixed representation, utilizing a Fourier expansion parallel to the surface and a real space coordinate mesh normal to the surface.

The surface to be studied is specified by supplying the self-consistent bulk potential, the positions of the surface region ion cores, and their (previously fitted) pseudopotential. An approximate valence electron potential is initially formed for the surface region, and the Schroedinger equation is solved for a sampling of the occupied states. The valence charge density found from these wave functions is used to compute a new potential, and the process is iterated until the input and output potential agree.

The model potentials describing the ion cores are smooth and physically sensible real space functions, and are determined by fitting bulk band structures or performing self-consistent pseudo-atom calculations. From the latter calculations, we have verified that a well chosen pseudopotential has wave functions which are identical in shape and magnitude to "real" atom wave functions outside the core region. The pseudo-atom wave function has no nodes in the core, but its integrated charge in this region is equal to that of the rapidly oscillating real wave function. This quality is important for valid self-consistency. Once the ion model potentials are determined, no adjustments are made in the course of calculating the surface electronic structure.

2. Clean Si(111)

The first semiconductor surface to be studied using self-consistent methods was the unreconstructed (111) surface of Si. The geometry of this surface is such that one bond per surface atom is broken to form the surface, and that bond lies in the surface normal direction. A band of surface states lying in the gap between the valence and conduction bands was first predicted by Shockley [3]. While these were intuitively identified as "dangling bond" surface states, the validity of this interpretation was disputed as subsequent model calculations showed varying degrees of charge localization for this state [4]. Our calculations show that the gap surface state is, in fact, highly localized in the region of the broken bond, in keeping with the intuitive picture [5].

Analogy with empirical ideas about bond lengths in molecules [6] led us to expect that the breaking of one bond of the surface atom would cause its three remaining bonds to be stronger and therefore shorter, causing the surface layer to relax inward toward the bulk. We explored the consequences of such relaxation, and found that for a relaxation of ~.6 au, two previously unanticipated bands of surface states formed. These split off below two of the bulk valence bands, having energies 2 to 4 eV and 11 to 13 eV below the valence band maximum, respectively. The charge in these surface states is highly localized in the first to second layer bonds, so they are clearly associated with the strengthening of these bonds [5].

While the $1 \times 1(111)$Si surface exists only as a high temperature phase, and has not been studied spectroscopically, LEED evidence indicates that the 7×7 room temperature structure is not too different [7]. Inelastic low energy electron scattering and ultraviolet photoelectron spectroscopy of this surface were being undertaken by Rowe and Ibach at the same time our calculations were being performed, and revealed sharp features which could clearly be identified with the back-bond surface states we found [8]. The fact that the Si(111) surface relaxes enough to generate additional deep lying surface states is a result we consider in the "new idea" category.

3. H chemisorption on Si(111)

The chemisorption of atomic H on Si(111) should be chemically simple, with the H attaching itself to the dangling bond. We explored this and studied the force on the H as a function of distance. The equilibrium position and bond force constant were found to be in good agreement with molecular values and measurements of surface vibrational modes [9,10]. The charge distribution at equilibrium is shown in fig. 1.

The spectrum of this surface showed two distinct peaks, at -4 and -7 eV. Existing data did not show this structure, and the calcalations motivated new measurements by Hagstrum and Sakurai. These turned out to give two peaks, in good agreement with the theory, apparently completing the story and confirming that H on Si(111) does everything as anticipated [11].

In the course of their experiments, however, Hagstrum and Sakurai observed that the spectrum of submonolayer H depended in its shape on the substrate temperature at the time of deposition. For deposition at room temperature, the -4 eV peak grew first, with the -7 eV feature present only near saturation, while for $T =$

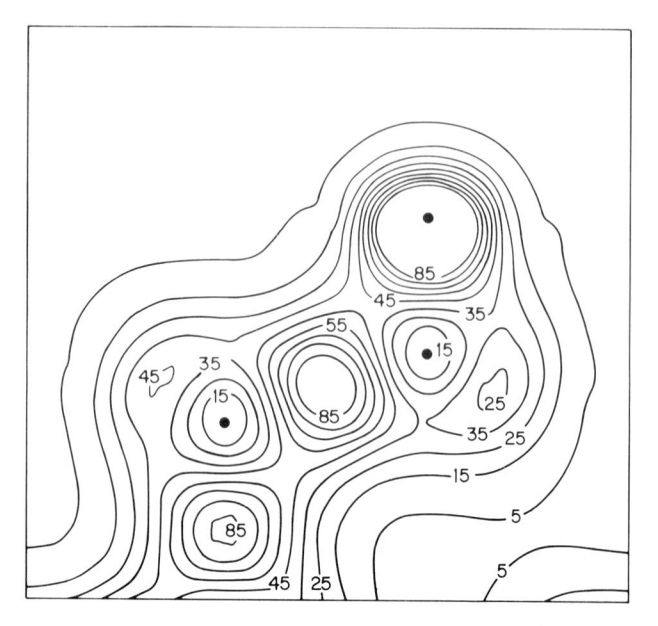

Fig. 1. Contour plot of the total valence charge density for H bonded to Si(111). The plot is in a plane normal to the surface passing through the H, first layer Si, and second layer Si, indicate by heavy dots. Vacuum is at the top of the plot. The density is in atomic units $\times 10^3$, the average density of bulk Si being 30. From ref. [28].

150°C, the two peaks grew proportionally [12]. It was speculated that the H was immobile at room temperature, leading to random coverage, but that it grew in islands of saturated coverage at 150°C.

In trying to decipher the discrepancies in the spectra in the two situations, Appelbaum and I noted that the two peaks came from different parts of the surface Brillouin zone. The −4 eV peak corresponded to a critical point structure of a surface state ϵ versus $k_\parallel$ relation near the outer edge of the SBZ, while the −7 eV peak represented resonance with a bulk feature near the SBZ center. We constructed a simple planar model with one state per Si–H bond, and adjusted strong first neighbor and weak second and third neighbor matrix elements to mimic the saturated-layer behavior found in our detailed calculations. Randomly introducing vacancies into this model removed the lower peak much more quickly than the upper, confirming the interpretation of the data [12]. The qualitative behavior of the model with random vacancies is also clear. Near the SBZ boundary, the phase of the wave function on neighboring atoms alternates, and the effects of vacancies tend to cancel. Near the zone center, however, everything is in phase and the vacancies raise the energy in proportion to their number, removing the Si–H bond from resonance with the bulk peak. We believe that the role of wave function coherence in changing spectra with coverage is a new idea which may prove important in a wide variety of situations.

4. H chemisorption on Si(100)

The Si(100) surface has two broken bonds per surface atom [13]. There are four possible positions for an H monolayer over this surface which preserve its point group symmetry and we are in the process of exploring all of these.

The geometry with H directly above the surface atom is unconventional because it lies at a nontetrahedral position midway between the two broken bonds. Nevertheless, we found an equilibrium position and force constant essentially identical to that on the (111) [14]. A wide range of normal displacements was explored, and the formation of the bond is shown for several of these in fig. 2. Note that the charge at equilibrium is very similar to that in fig. 1. The force on the H varied in a smooth manner over the entire range, and we found, to our surprise, that it could be fit essentially exactly by a Morse potential [14] as shown in fig. 3.

In the position midway between the surface atoms in the direction of the broken bonds, the H experience a relatively weak attractive force at large separations, and achieve equilibrium at 0.75 au above the surface layer with a relatively weak force constant. The bond lengths are too long for this to be a real chemical bond, and it is perhaps better described as a combination of a chemical bond and polarization, or Van der Waals bond.

The third geometry explored placed the H over the second Si layer. In this position, it would pass between the surface atoms at right angles to the broken bond direction. The force at large distances was weak, consistent with a polarization force, as anticipated. The curve of force versus distance extrapolated to zero at 2.2 au above the surface layer. Calculations at this distance and smaller distances, how-

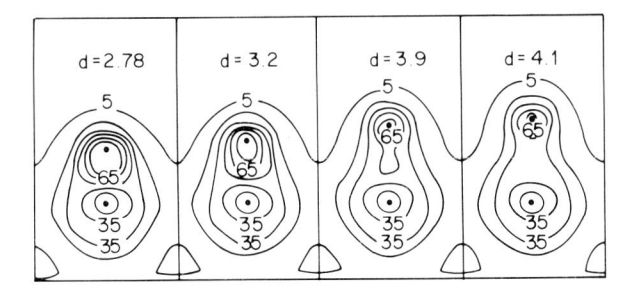

Fig. 2. Charge density contour plot showing the progressive formation of the Si–H bond on the Si(100) surface. The plane of the plot is through the H and Si (indicated by the upper and lower dots) in the direction of the broken bonds. The Si–H distance d is in atomic units, and the density is in au $\times 10^3$. From ref. [14].

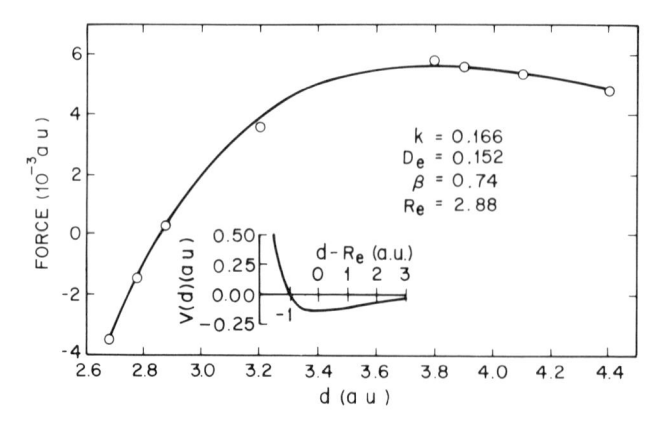

Fig. 3. Force as a function of Si–H separation for bonding over the surface atom on (100), and fit derived from Morse potential. Inset shows the Morse potential itself, $V_M(d) = D_e \{\exp[-\beta(d - R_e)] - 2\exp[-\beta(d - R_e)]\}$, and k is the force constant in au. From ref. [14].

ever, showed a dramatic increase in the inward force, followed by a rapid decrease and an equilibrium almost in the plane of the surface layer. The charge contours in this situation reveal the formation of a three center bond to the bond charges of the first to second layer Si bonds [15].

This is certainly an unconventional type of chemical bond. The force law near equilibrium is similar to the previously fitted Morse potential, only the second layer is the origin. Yet the second layer Si already has its 4 tetrahedral bonds saturated by 4 other Si atoms. These bonds are certainly weakened by the H, since the charge density at the bond center is reduced by ~20%, and the surface Si layer should presumably relax outward to some extent. We speculate that this unusual bond may be a precursor to the corrosive attack experienced by appropriately prepared surfaces under prolonged exposure to atomic H, and that it may be a prototype for an interstitial bond of H in crystalline or amorphous bulk Si [15].

5. Si(111)2 × 1 reconstruction

Perhaps the most studied semiconductor surface is the metastable 2 × 1 form of the Si(111) surface prepared by cleavage. A buckling model for this surface, originally proposed by Haneman [16], has become generally accepted after several recent theoretical studies of this model and variations on it [17–19]. Alternate rows of atoms are raised and lowered, splitting the dangling bond surface state band into two nonoverlapping bands. The transitions between these bands produce absorption in the 0.3 to 0.5 eV range, calculated in detail by Schluter et al. [19] and found to be in good agreement with infrared absorption and inelastic electron scattering experiments [20].

The band splitting and its magnitude is not a surprise. What is more surprising is that the states of the occupied band are located almost completely on the raised atoms, while those of the unoccupied band are almost completely on the lowered atoms. This charge separation means that the surface in fact has an ionic character, and may explain the fact that the strong surface phonon peak seen in inelastic electron scattering from the 2 × 1 surface has not been reported on the 7 × 7 surface, which is presumably nonpolar and lacks the requisite dipole to couple to the scattering electron [21].

The split dangling bond bands do not explain all the distinctive features of this surface, however. The occupied band is centered at essentially the valence band maximum, but a surface sensitive feature in the UPS spectrum originally observed by several groups [22], extends to ~ -1 eV, and is peaked at ~ -0.5 eV. This feature was originally interpreted as the dangling bond surface state, but this interpretation is clearly inconsistent with the recent calculations. The explanation of this seeming paradox lay in the observation that the originally observed feature in fact

Fig. 4. Contour plot of the charge density in the surface state associated with the back bonds of the raised atoms on the Si(111)2 × 1. The plane of the plot is parallel to the surface, midway between the first and second layer. The positions of the first and second layer atom above and below the plane are shown by the crosses at the center and edges, respectively. From ref. [17].

has two physically distinct components. Experimentally, this was seen by angular-resolved photoemission measurements by Rowe, Traum and Smith [23]. These results indicated that the emission in the upper part of the "surface state peak" was concentrated near the surface normal, and independent of azimuthal angle, as might be expected of the dangling bond. The lower portion of the peak, however, was concentrated at larger polar angles, and displayed a strong three-lobed intensity variation with azimuthal angle. The theoretical calculations reveal another band of surface states in this energy range associated not with the dangling bond but with the stretched back bonds of the raised atoms, shown in fig. 4. This is consistent with the three-lobed pattern, and is, in a sense, a complement of the surface states associated with shortened bonds on the 7×7 surface discussed in section 2. The present back-bond states are split from the top of the same valence band from whose bottom those states were split.

6. Si(100)2 × 1 reconstruction

Another clean-surface reconstruction which has received extensive study is the 2×1 reconstruction of the (100)Si surface. Unlike that of the (111), this reconstructed phase is the stable form of the clean surface at all temperatures below those which induce faceting. Attention has centered on two models for this surface. In the Schlier–Farnsworth–Levine model [24], rows of surface atoms are moved together in pairs, with the aim of rebonding one of the two broken bonds on each atom. In the Lander–Phillips model [25], alternate rows of surface atoms are removed to allow the remaining atoms to form double bonds to the second layer. Comparison of the surface density of states for the two models with UPS data indicated considerably better agreement for the pairing model [26]. The pair bond in this model has a charge distribution essentially identical with that of the bulk, despite the fact that it is aligned parallel to the surface, at a different angle from the bulk bonds. The local density of states at this bond, consisting of one sharp peak in the p-band region at -2.5 eV, and another in the s-band region at -10 eV, may be qualitatively explained by a lessening of hybridization with the bulk bonds due to this misalignment.

The surface atoms in the pairing model possess three bonds in nearly perfect tetrahedral configuration, suggesting the fourth tetrahedral site as a natural location for chemisorbed H. It was observed that atomic H would bond to this surface without destroying the reconstruction [27], and a theoretical study of this system seemed a natural extension of our studies [28]. The charge distribution calculated for this system is shown in fig. 5. The Si–H bonds are only slightly distorted from their configuration on the (111) surface. The pair bond remains bulk-like, and essentially undisturbed.

The calculated spectrum on the chemisorbed H is considerably more complex than that found on the (111), showing five peaks as indicated in fig. 6. Decomposi-

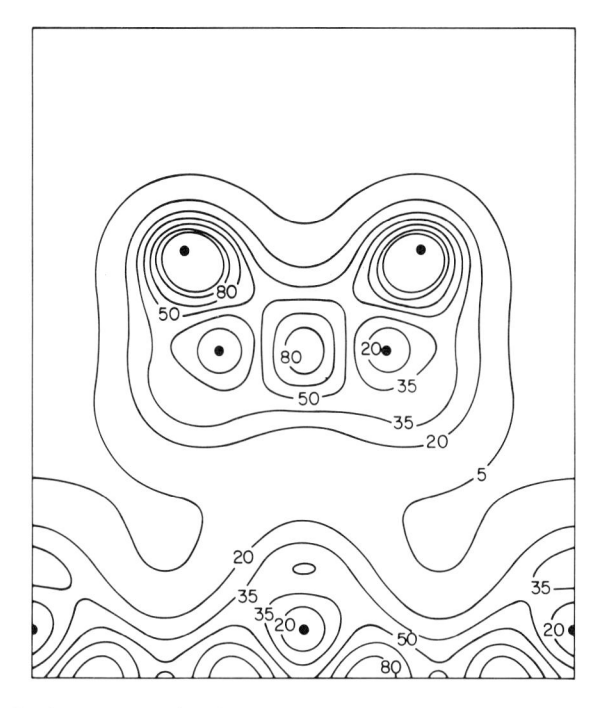

Fig. 5. Charge density contour plot for H bonded on Si(100)2 × 1 with pairing model reconstruction. The H, paired first layer Si, and third layer Si are shown. The second layer Si atoms are out of the plane of the plot. Units are as in fig. 1. From ref. [28].

tion of the spectrum into components arising from states of even and odd symmetry with respect to a plane bisecting the pair bond shows that the upper split pair is, in fact, an even-odd doublet split by an indirect interaction of the two Si-H bonds. The lower doublet is associated only with the odd states, however, Closer examination shows that it arises from a band of surface states and resonances whose ϵ versus $k_\parallel$ relation is essentially one-dimensional, and has its critical points at its upper and lower extrema. The fifth feature, lowest in energy, is associated with the lower peak of the pair bond spectrum, which is apparently resonantly coupled into the Si–H bond.

It would be gratifying to report that these characteristic five features have been seen in experimental spectra of the Si(100)2 × 1–H surface. Unfortunately, only rather broad changes are seen with H adsorption [28]. This may be due to the existence of a mixed surface phase, since a 1 × 1 phase has recently been shown to be stable at room temperature with prolonged H exposure [29]. It may also simply indicate that the pairing model for the reconstruction is incorrect. The five calculated features have been seen, however, in H adsorption on Ge(100)2 × 1 surface by UPS and ion neutralization spectroscopy [28]. The electronic structure of Si and

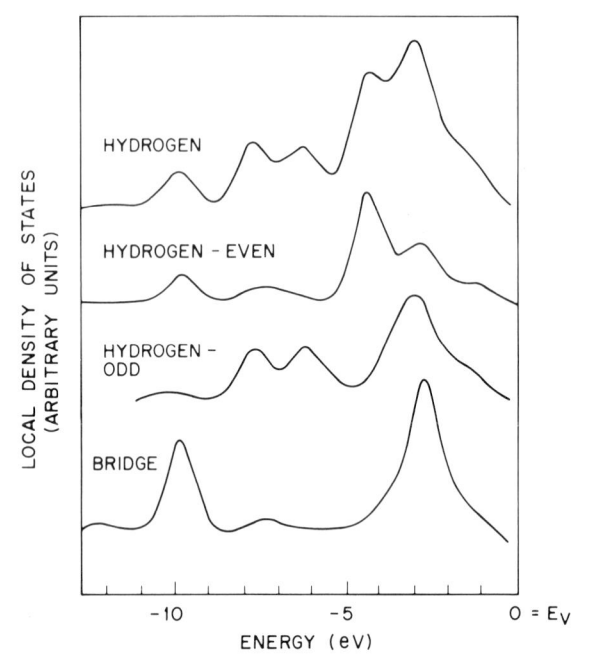

Fig. 6. Local density of states versus energy measured from the valence band maximum at the hydrogen and bridge-bond centers for H on Si(100)2 × 1. The hydrogen dos is decomposed into components contributed by even and odd symmetry states with respect to the plane bisecting the pair or bridge bond. From ref. [28].

Ge is very similar in the valence band region, and the Si calculations should be valid for this system as well.

7. Summary and conclusions

I have tried to illustrate the thesis proposed in the Introduction — that detailed electronic structure calculations can lead to novel ideas and interpretations of phenomena in surface science. Among the new ideas discussed I would include the following: that relaxation-induced back-bond surface states exist, that disorder can change the spectrum of a partial monolayer, that saturated subsurface atoms can bond H, that reconstruction polarizes the Si(111)2 × 1 surface, and that the classic UPS "surface state peak" of this surface is partially due to deformed back bonds. Two ideas which perhaps fit better in the model-choosing category are the validity of a simple force law over a wide range for some chemisorption situations, and the strong support for the pairing model of the Ge(100)2 × 1 provided by the H adsorption studies.

While the major advances in theoretical surface studies in the past few years have been in the area of semiconductor surfaces, I believe that we can look forward to a comparable growth in the extremely important area of transition metal surfaces in the near future. I hope that the semiconductor work has set a example for these anticipated developments, and that the value of studying realistic models of specific systems will not be forgotten.

References

[1] J.A. Appelbaum and D.R. Hamann, Phys. Rev. 86 (1972) 2166.
[2] J.A. Appelbaum and D.R. Hamann, in: Proc. of the 12th Intern. Conf. on the Physics of Semiconductors, Ed. M.H. Pilkuhn (Teubner, Stuttgart, 1974) p. 675.
[3] W. Shockley, Phys. Rev. 56 (1939) 317.
[4] R.O. Jones, in: Surface Physics of Semiconductors and Phosphors, Eds. C.G. Scott and C.C. Read (Academic Press, London, 1975), p. 96.
[5] J.A. Appelbaum and D.R. Hamann, Phys. Rev. Letters 31 (1973) 106; 32 (1974) 225.
[6] L. Pauling, The Nature of the Chemical Bond, 3rd ed. (Cornell Univ. Press, Ithaca, NY, 1960).
[7] J.E. Florio and W.D. Robertson, Surface Sci. 24 (1971) 17;
H.D. Shih, F. Jona, P.W. Jepsen and P.M. Marcus, Phys. Rev. Letters 37 (1976) 1622.
[8] J.E. Rowe and H. Ibach, Phys. Rev. Letters 31 (1973) 102; 32 (1974) 421.
[9] J.A. Appelbaum and D.R. Hamann, Phys. Rev. Letters 34 (1975) 806.
[10] G.E. Becker and G.W. Gobeli, J. Chem. Phys. 38 (1963) 2942;
H. Froitzheim, H. Ibach and S. Lehwald, Phys. Letters 55A (1975) 247.
[11] H.D. Hagstrum and T. Sakurai, Phys. Rev. B12 (1975) 5349.
[12] J.A. Appelbaum, H.D. Hagstrum, D.R. Hamann and T. Sakurai, Surface Sci. 58 (1976) 479.
[13] J.A. Appelbaum, G.A. Baraff and D.R. Hamann, Phys. Rev. B11 (1975) 3822; B12 (1975) 5749.
[14] J.A. Appelbaum and D.R. Hamann, Phys. Rev. B15 (1977) 2006.
[15] J.A. Appelbaum and D.R. Hamann, to be published.
[16] D. Haneman, Phys. Rev. 121 (1961) 1093.
[17] J.A. Appelbaum and D.R. Hamann, Phys. Rev. B12 (1975) 1410.
[18] K.C. Pandey and J.C. Phillips, Phys. Rev. Letters 34 (1975) 1450.
[19] M. Schluter, J.R. Chelikowsky, S.G. Louie and M.L. Cohen, Phys. Rev. Letters 34 (1975) 1385.
[20] G. Chiarotti, S. Nannarona, R. Pastore and X. Chiaradia, Phys. Rev. B4 (1971) 3398;
H. Froitzheim, H. Ibach and D.L. Mills, Phys. Rev. B11 (1974) 4980.
[21] H. Ibach, Phys. Rev. Letters 27 (1971) 253.
[22] D.E. Eastman and W.D. Grobman, Phys. Rev. Letters 28 (1972) 1378;
L.F. Wagner and W.E. Spicer, Phys. Rev. Letters 28 (1972) 1381.
[23] J.E. Rowe, M. Traum and N.V. Smith, Phys. Rev. Letters 33 (1974) 1335.
[24] R.E. Schlier and H.E. Farnsworth, J. Chem. Phys. 30 (1959) 917;
J. Levine, Surface Sci. 34 (1973) 90.
[25] J.J. Lander and J. Morrison, J. Chem. Phys. 33 (1962) 729;
J.C. Phillips, Surface Sci. 40 (1973) 459.
[26] J.A. Appelbaum, G.A. Baraff and D.R. Hamann, Phys. Rev. Letters 35 (1975) 729;
Phys. Rev. B15 (1977) 2408.
[27] H. Ibach and J.E. Rowe, Surface Sci. 43 (1974) 48.
[28] J.A. Appelbaum, G.A. Baraff, D.R. Hamann, H.D. Hagstrum and T. Sakurai, Surface Sci., to be published.
[29] T. Sakurai and H.D. Hagstrum, Phys. Rev. B14 (1976) 1593.

PHOTOEMISSION STUDIES ON ZnO(0001), (000$\bar{1}$) and (10$\bar{1}$0)

F. HUMBLET, H. VAN HOVE and A. NEYENS

Fysico-chemisch Laboratorium Celestijnenlaan 200 G, 3030 Heverlee, Belgium

Ultra violet photoemission experiments (up to LiF cut-off frequencies, incidence angle 45°) with subsequent low acceptance angle (about 1°) energy analysis of the perpendicularly emitted electrons, were carried out on the principal low index faces of ZnO.

Different surface conditions were explored: (i) argon bombarded and subsequent vacuum annealed surfaces at increasing temperatures till 650°C, (ii) bombarded but 10^{-6} Torr oxygen annealed surfaces and (iii) cesium covered surfaces, heated till temperatures where one may assume that ($\sqrt{3} \times \sqrt{3}$) superstructures could readily be formed at least at (000$\bar{1}$) surfaces [1].

After the bombardement and annealing cycles three pieces of structure were observed on all the faces: two main bulk peaks that were identified as due to valence and conduction band density of states in accordance with Powell et al. [2]; a third prominent feature appeared as a more or less developed shoulder on the high energy side of the EDC at a final energy of about 9 eV above the valence band maximum. A possible extrinsic surface state origin for this peak is not to be ruled out.

Apart from the extension of the EDC to the low energy side, no single cesium induced photoemission feature could be observed up to now even not on the (000$\bar{1}$) face [1], moreover, on all the surfaces the pre-cesium situation could easily be restored by moderate heating.

The possible influence of the degree of stabilisation of one of the polar terminations on the adsorption or on other surface energy related phenomena on the other polar surface, is under study.

References

[1] R. Leysen, P. Taylor and B.J. Hopkins, J. Phys. C.8 (1975) 907.
[2] R. Powell, W. Spicer and J. McMenamin, Phys. Rev. B6 (1972) 3056.

THE INTERACTION OF Ag WITH Si(111)

M. HOUSLEY [*], R. HECKINGBOTTOM and C.J. TODD
Post Office Research Centre, Martlesham Heath, Ipswich IP5 7RE, England

Deposits of Ag on Si(111), at room temperature, have yielded a linear Auger signal–time characteristic to a gradient break point at $(7.6 \pm 0.9) \times 10^{14}$ atoms of Ag cm^{-2}, which is very close to the Si surface state density of $(8 - 10) \times 10^{14}$ cm^{-2}, and which supports a Stranski-Krastanov growth mechanism. Analysis of the Auger spectra at the monolayer end point revealed a new peak at 82 ± 1 eV. This peak is believed to arise from an Auger process involving an induced Ag–Si interface state. A model is proposed for this state arising from the chemisorption of Ag on Si.

1. Introduction

The adsorption of Ag on Si is of theoretical interest as an example of a metallic–covalent interface, and is of some importance in semiconductor metallisation technologies. Growth of Ag on Si has been studied by Spiegel [1], Bauer and Poppa [2], and recently by Le Lay et al. [3]. Since Ag has a cohesive energy of 2.9 eV/atom but a desorption energy from Si of 2.5 eV [2] an island growth mode would be anticipated. However Bauer and Poppa argued that their flash desorption results and Spiegel's LEED data implied a Stranski-Krastanov growth mechanism, i.e. initial layer growth then island formation. Unfortunately expected gradient break points were not evident in their Auger signal–time (AS–T) characteristics. The purpose of this work has been to further characterise the growth mode at room temperature and to probe the nature of any Ag–Si interfacial states.

2. Experimental

The measurements were carried out in a standard ion pumped 10^{-11} Torr UHV system fitted with a CMA (Physical Electronics), a specimen manipulator with 25 μm linear drives and a claimed ±12' rotary drive (Vacuum Generators) an Ag evaporator and an ion gun (Physical Electronics). The substrate p-type Si (carrier concentration ca. 10^{16} cm^{-3}) was cut and polished parallel to (111) and mounted

---

[*] Present address: Centre de Cinétique Physico Chimique CNRS, Route de Vandoeuvre, 54 Villiers–Nancy, France.

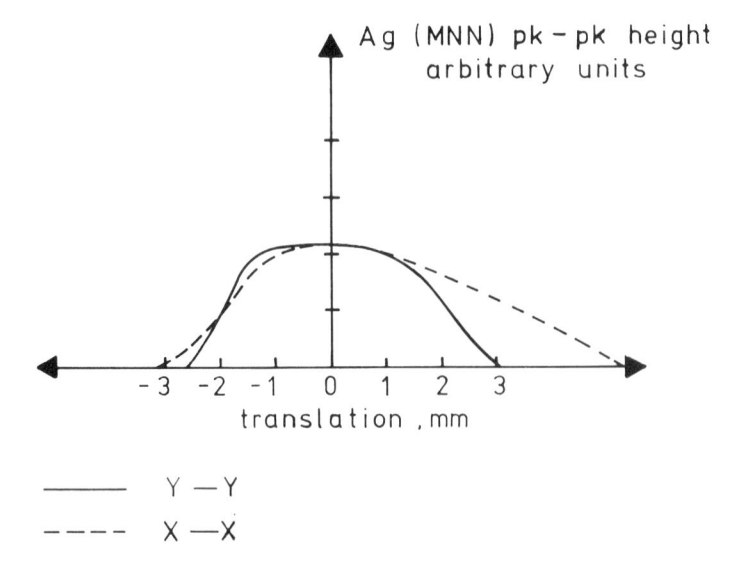

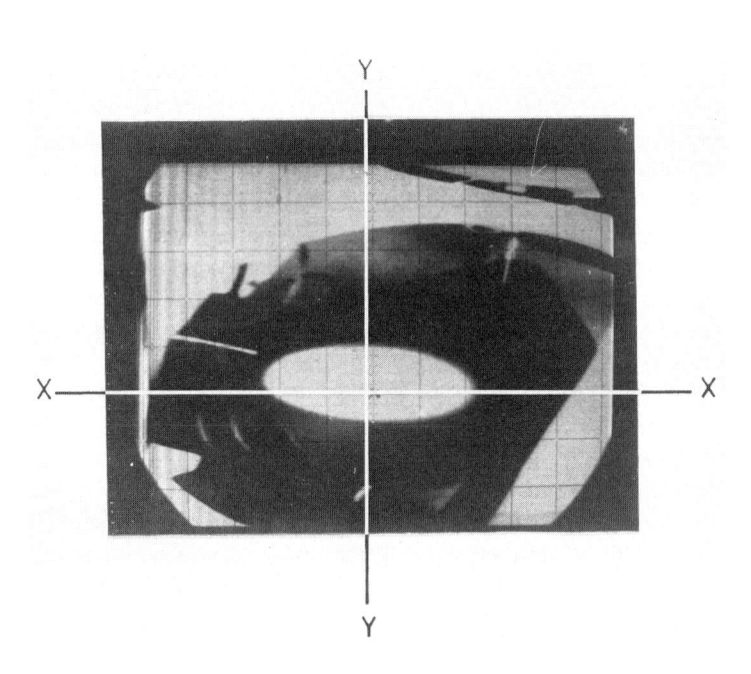

Fig. 1. Extent of Ag deposit; the peak-to-peak height of the Ag-MNN transitions plotted along the two directions shown on the low resolution SEM micrograph.

off molybdenum supports via 127 μm OD tungsten wires, to facilitate uniform heating by electron bombardment of the back face. A W/W–26% Re thermocouple inserted into the body of the crystal provided feedback temperature control to the electron bombardment supply of ±1°C. An identical control system was used to heat a molybdenum Knudson cell, containing 6N Ag, providing flux control to ±5%. Attempts to calibrate absolute flux rates in ancillary experiments were thwarted by the movement of collimating apertures during refilling and mounting on the main UHV chamber. The Si substrate was mounted 10 cm off axis to enable both near normal incidence deposition and normal incidence Auger by rotation from port to port, and to avoid other geometrical conflicts. However it is clear from fig. 1 that to maintain a <5% reproducibility of Auger signal strength with this design, a rotational reproducibility of <30' is required; unfortunately backlash in the rotary drive exceeded this figure. This problem was overcome by driving SE3K5U electron guns (Superior Electronics), mounted on each port, in the SEM mode [4] which permitted absolute substrate location to ≃200 μm.

3. Deposition of Ag and film growth

The depositions were carried out in stages of equal duration at the same Ag source temperature, onto a room temperature Si(111) substrate previously cleaned by argon ion bombardment and heating to 1100°C to reduce remnant carbon and oxygen to ≤1% monolayer; during deposition $P < 10^{-9}$ Torr. AS–T plots were constructed by monitoring the background to maximum negative slope point of the Si-$L_{2,3}$VV and Ag-$M_4N_{4,5}N_{4,5}$ transitions in the differential spectra. (The presence of new structure at 82 eV (section 4) did not affect the validity of this measurement of $L_{2,3}$VV.)

The AS–T characteristic for this system, emphasising low coverage, is shown in fig. 2. The occurrence of an initial linear section and an abrupt gradient break in the Ag characteristic followed by a pseudolinear region and slow exponential climb (not shown) to within 5% of the bulk signal level after 100 min, and the persistance of the Si signal, are strong evidence for a Stranski–Krastanov growth mode [5]. Spiegel [1] has shown that a strong background grows, during the deposition of Ag onto Si at room temperature, and the 7×7 LEED pattern fades gradually; with increasing thickness LEED shows the development of a Ag(111) texture pattern. These results were confirmed in essence by the RHEED data of Le Lay et al. [3].

To establish the number of interface states cm^{-2} it is important to compute the atom density at the "monolayer end point" or gradient break; this can be estimated from the data in fig. 2. If I_n is the signal from n layers and I_∞ the bulk signal

$$I_n = I_\infty \left[1 - \exp(-nd/\lambda \cos\theta) \right],$$

where d is the layer thickness, λ the escape depth and θ the collection angle [6]. Now I_1 and I_∞ are measured and $\cos\theta = 0.74$ for the normal incidence CMA [7],

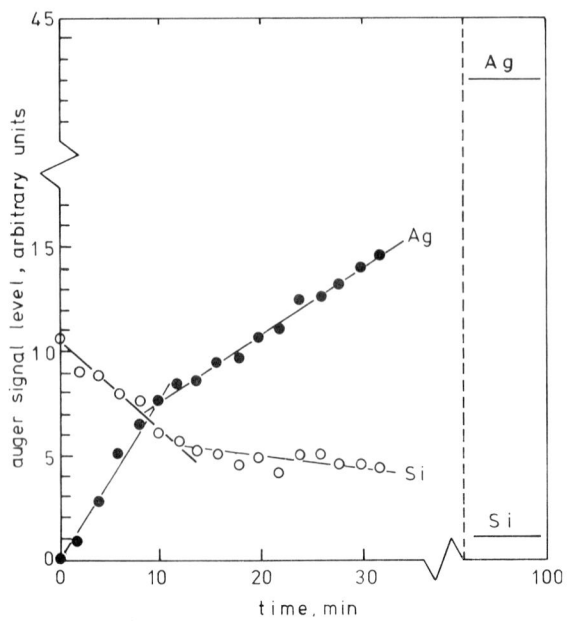

Fig. 2. Auger signal–time characteristic for Ag deposited onto Si(111) at room temperature. The displacement of the Ag and Si break points is not considered significant in view of the scatter in the Si data.

but d and λ are subject to some uncertainty. However the value of λ (MNN-Ag) is perhaps better known that most; several independent determinations have been made, (see e.g. ref. [7]) yielding $8 \leqslant \lambda \leqslant 9$ Å. In fig. 1, the Ag deposit shows an increased secondary electron yield compared with the surrounding Si. In the case of a monolayer of Ag a higher yield must imply a lowering of work function upon interaction of Ag with Si. To allow for charge transfer an estimate of d would require 2.52 (Ag^+) $\leqslant d \leqslant$ 2.88 (Ag^0) Å (this range includes the average layer thickness (2.58 Å) computed from the atomic volume). Taking the worst cases these data imply a ratio $I_1/I_\infty = 0.36 \pm 0.04$ for a single layer having bulk density. This is to be compared with the value of $I_1/I_\infty = 0.32 \pm 0.04$ measured by Jackson et al. [8], for Ag deposited on Ni. However, if I_b is the signal strength measured at the break in the Ag AS–T characteristic then $I_b/I_1 = 0.51 \pm 0.05$ which implies a density of Ag at the break point of $7.6 \pm 0.9 \times 10^{14}$ atoms cm^{-2}. In the case of Si, larger scatter obscures the break point, but using the same d range as above and $5 \leqslant \lambda$ (LVV) $\leqslant 9$ Å the monolayer end point can be predicted to occur at $I_b/I_\infty = 0.57 \pm 0.11$; this compares well with the measured value of $\simeq 0.55$. Using these data a density $\leqslant 4 \times 10^{12}$ islands cm^{-2} having a hemispherical shape and idential size can be computed to yield the observed signal levels after 100 min of deposition. In practice, the (111)Ag islands, exhibiting a variety of azimuthal orientations, will encompass a range of sizes and shapes, but the model yields a useful magnitude

The surface state density on cleaved Si(111) has been measured by photoemission and ellipsiometry [9] to be in the range $8-10 \times 10^{14}$ electrons cm^{-2}. This is in good agreement with the simple model that each of the 7.85×10^{14} surface atoms has one occupied and one unoccupied surface state. Although a number of different models have been proposed for the annealed Si(111) 7 × 7, the number of exposed Si surface atoms is identical to the cleaved surface and the density of surface states is taken to remain unchanged. (This is not in conflict with the evidence that the energy distribution of surface states is dependent on the surface type). The experimentally determined Ag density at the break point is therefore in good agreement with the expected filled surface state density. A model is proposed for the strong Ag–Si chemisorbed bond ($E_p \simeq 2.5$ eV [2]), involving overlap of the 5s electron from Ag with the $3p_z$ from Si; distortion of the Ag 5s through formation of this state could be compatible with a net charge transfer in the sense predicted by the lowering of work function. The applicability of this chemisorption model is probably not strongly dependent upon the exact site location of the chemisorbed Ag, although the magnitude of the binding energy may well be. The Lander model of the Si(111) 7 × 7 is preferred here since as argued by Philips [10] there is the suggestion that large perturbation have occurred in Si atomic positions. It is proposed that at room temperature the arriving Ag populates valley sites A, B and C of fig. 3 in a random manner. The presence of Ag chemisorbed on site A in particular will hinder the relaxation of the Si 7 × 7 to a 1 × 1 structure. According to this model as Ag arrives it will contribute to an increasing background in LEED and the 7 × 7 pattern will be obscured but not relax, in accord with Spiegel's observations [1]. The infilling by Ag will lead to a smoother surface but it is likely that an end point will not involve occupation of all A, B or C sites due to interferences, nor is

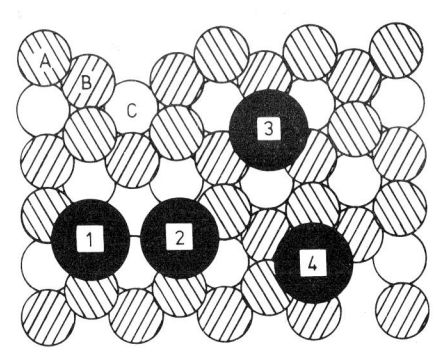

Fig. 3. Proposed model of location of Ag on a Lander type Si(111) 7 × 7 surface. A, B and C refer to 1st, 2nd and 3rd layer Si atoms respectively. In the Lander model Si of type A are systemmatically absent. Ag atoms (drawn large for clarity) are shown in a variety of possible valley sites: 1 and 2 on B valleys, 3 on an absent A valley site and 4 on a C valley with an adjacent A absent.

it likely that island growth will only begin when the optimum "monolayer" filling has been achieved, rather the break point will mark the changeover from the dominance of infilling to the growth of observable islands. It follows that this model presumes a higher mobility for Ag on the Ag–Si chemisorbed phase than on bare Si. The nucleation mechanism of the islands remains unknown; they could grow on top of the "monolayer raft" or interface directly with the bare Si. Comparison of this work and that of Bauer and Poppa, and Le Lay et al. suggests a degree of variability of end points of different phases with measured coverage. Henzler [11] showed that the sticking probability of oxygen on Si was a strong function of the surface defect density. In the case of Ag interacting with Si, the Si surface defect density may well effect the island nucleation rate and hence, e.g., the occurrence of gradient breaks in AS–T plots. Future work should include careful control and monitoring of defect levels. It should be noted that work to date on this system indicates a uniform Ag sticking probability, which is likely to be close to unity at room temperature.

4. Auger spectroscopy of Ag–Si system

The occurrence of Ag–Si interface states could lead to new structure in the Auger spectrum. Fiebelman et al. [12] have suggested that the presence of a monolayer on Si(111), effecting double occupancy of the surface state, could yield additional structure to modify the $L_{2,3}VV$ line shape. The spectral region below 100 eV has been carefully monitored and a new peak of low intensity has indeed been identified at 82 ± 1 eV arising from a monolayer of Ag on Si(111). A similar study of the Ag-$M_4(M_5)N_{4,5}N_{4,5}$ showed no observable change in line shape or position with Ag thickness; however in this spectral region (340–360 eV) the CMA resolution is poor.

The 20–100 eV spectrum is complex, both Ag and Si separately contribute several Auger features. To identify new features not associated with either bulk Ag or Si difference spectra were computed. In an attempt to avoid artifacts arising from inadequate background assessment, two approaches were adopted and the results compared: (1) backgrounds were fitted and removed from the individual differential spectra which were then integrated and a scaled Si spectrum was subtracted from the monolayer spectrum; (2) the differential spectra after and before deposition of an Ag monolayer were subtracted and then integrated, no background assessment was made, i.e. here the presence of the Ag was assumed not to affect the background significantly. In all cases the spectra were normalized by E^{-1} to account for the transmission characteristics of the CMA, and the first dynode of the multiplier was biased to 250 V. The results are shown in fig. 4. The curves (a) and (c) contain bulk Ag peaks and the additional weak feature at 82 ± 1 eV. As the thickness of Ag increases and island growth blocks out interface signal, there is a tendency for the strength of the 82 eV peak to diminish further (b); unfortunately

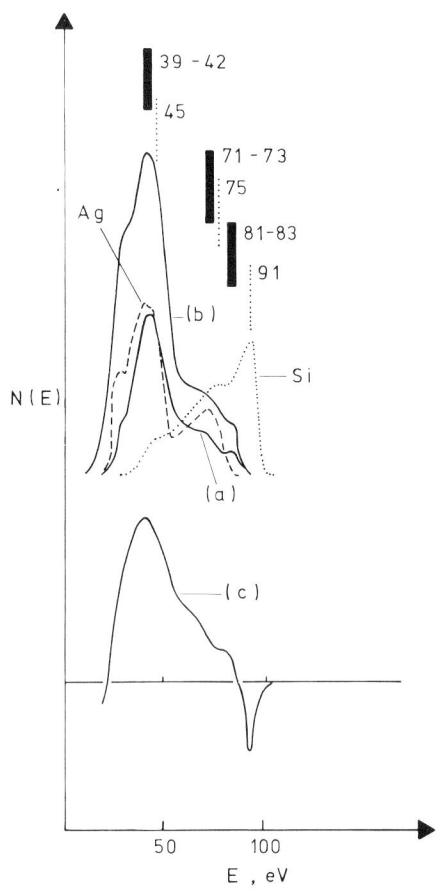

Fig. 4. Difference curves in the 20–100 eV region. Curve (a) and (c) at the monolayer end point, (b) after 30 min deposition; (a) and (b) extracted by procedure (1), (c) by procedure (2). The negative excursion in (c) at 91 eV results from oversubtraction of the clean Si spectrum, an artifact of procedure (2). The bulk Si (÷2) and bulk Ag (÷3) are shown for comparison. The dotted energy assignments refer to bulk Si, the full blocked outlines to (a), (b), (c) and bulk Ag; the range shown gives an estimate of error arising from the data manipulation.

a quantitative assessment of this decrease with increasing Ag is not feasible due to the growing high energy edge of the peak at 72 ± 1 eV. Peak energies were assigned assuming a 4 eV spectrometer work function correction; this yielded a dominant Si-LVV at 91 eV, in close agreement with the work of Amelio [13].

To proceed to a plausible assignment for the Ag–Si interface peak at 82 eV it is necessary to estimate the contribution that arises from the final state "hole–hole" interaction (H) and the polarization of lattice electrons (P). Following Matthew

[14] the energy of an ABC Auger transition may be written

$$E_{ABC} = E_A - E_B - E_C - (H - P) . \quad (1)$$

Kowalczyk et al. [15] have calculated $(H - P)$ for bulk Ag (using a different notation) but difficulties arise in interpreting their numerical data and in applying their quasi atomic state calculations to solid state environments. An experimental approach is adopted here. First of all the bulk Si and Ag peaks must be properly assigned. In the case of Si, following Amelio [13] and using the terminology of Ley et al. [16], the process is straightforward yielding $L_1L_{2,3}X_4$ (45 eV), $L_{2,3}\mathcal{L}_2\mathcal{L}_2$ (75 eV) and $L_{2,3}X_4X_4$ (91 eV); X_4 is the 3p like maxima (E_{X_4} = 2.5 eV) and $\mathcal{L}_2$ the lower 3s like maxima ($E_{\mathcal{L}_2}$ = 9.8 eV) in the valence band density of states. Little help is available from the literature to assign the low energy Ag peaks. The best fit appears to be $N_1N_{4,5}N_{4,5}$ (72 eV), $N_2N_{4,5}N_{4,5}$ (40 eV) and $N_3N_{4,5}N_{4,5}$ (31 eV, shown but not marked in fig. 4). Since each of these proposed transitions involves the same final state configuration a comparison between initial filled state binding energies and differences in measured peak positions, detailed in table 1, emphasizes the plausibility of this assignment. (The discrepancy in the case of N_3 is not fully understood but it is noted that the method of computing difference curves implies an increasing experimental error with decreasing energy.) In table 2, the value of $(H - P)$ required to fit the absolute peak energies are listed. The marked difference between Ag and Si is probably related to the degree of hole delocalization, being larger in the case of Si. Since the environment of the Ag–Si bond is intermediate between bulk Si and Ag, it is argued that transitions involving final state holes in the interface state will have an $(H - P)$ magnitude bounded by $3.5 \leqslant (H - P)_{Ag-Si} \leqslant 11$ eV, and further that possible transitions involving a single hole in the interface state will be better represented by an $(H - P)$ term in the lower half of this bound, whereas those transitions leaving two holes in the interface state will have an $(H - P)$ term in the upperhalf of the bound. Within this framework five possible transitions must be considered: $L_{2,3}\mathcal{L}_2 I$ and $N_1N_{4,5}I$ imply an interface state binding energy E_I within 3 eV of E_F, $L_{2,3}\mathcal{L}_1 I$ with $1.5 \leqslant E_I \leqslant 5$ eV, $N_1I_1I_2$ with $1 \leqslant (E_{I_1} + E_{I_2}) \leqslant 5$ eV, $L_{2,3}I_1I_2$ with $6.5 \leqslant (E_{I_1} + E_{I_2}) \leqslant 10.5$ eV. Now the Ag centred

Table 1
Differences in binding energy (BE) compared with differences between experimental peak positions assuming an (N_1, N_2, N_3)–$N_{4,5}N_{4,5}$ assignment for the Ag low energy Auger peaks

	BE data	From exp.
$E_{N_1} - E_{N_2}$	33	32 ± 2
$E_{N_1} - E_{N_3}$	39	41 ± 2
$E_{M_5} - E_{N_1}$	272	275 ± 2
$E_{M_5} - E_{N_2}$	305	307 ± 2
$E_{M_5} - E_{N_3}$	311	316 ± 2

Table 2
Experimentally determined values of $(H - P)$ term for Ag and Si, using BE data and $E_{N_{4,5}}$ = 5.5 eV.

	Transition	$(H - P)_{exp}$ (eV)	Mean $(H - P)$ (eV)
Ag	$M_5N_{4,5}N_{4,5}$	9 ± 1	$\simeq 11$
	$N_1N_{4,5}N_{4,5}$	12 ± 1	
	$N_2N_{4,5}N_{4,5}$	11 ± 1	
Si	$L_{2,3}X_4X_4$	3.5 ± 1	$\simeq 3.5$
	$L_{2,3}\mathcal{L}_2\mathcal{L}_2$	4.9 ± 1	
	$L_1L_{2,3}X_4$	2.0 ± 1	

transitions will be in competition with $N_1N_{4,5}N_{4,5}$. Because of the high occupancy of the Ag 4d levels these Ag centred transitions will therefore be unlikely candidates. The states involved in the interface orbital are expected to shift as a result of the interaction. Recently Wehking et al. [17] examined the Ag–Si system with UPS and found additional weak broad structure within 4 eV of E_F at low Ag coverage. Thus $L_{2,3}\mathcal{L}_1$I, $L_{2,3}\mathcal{L}_2$I and $L_{2,3}$II remain plausible designations for the 82 eV peak which are in accord with Fiebelman et al. [12] suggestion and with the 3p–5s model proposed for the Ag–Si chemisorption bond.

5. Conclusions

Further evidence from the Auger signal–time characteristic supports the Stranski–Krastanov model of growth of Ag on Si; although the changeover from monolayer to island growth may be affected by substrate defect density. It is believed that the Ag–Si chemisorbed bond has (Si)3p (Ag)5s character and a new Auger transition is found arising from this interface state.

Acknowledgements

The authors gratefully acknowledge the Director of Research of the Post Office for permission to publish this paper, and Dr. J.A.D. Matthew for several valuable discussions during the course of this work.

References

[1] K. Spiegel, Surface Sci. 7 (1967) 125.
[2] E. Bauer and H. Poppa, Thin Solid Films 12 (1972) 167.

[3] G. Le Lay, G. Quentel, J.P. Faurie and A. Masson, Thin Solid Films 35 (1976) 273.
[4] G.W.B. Ashwell, C.J. Todd and R. Heckingbottom, J. Phys. E6 (1973) 435.
[5] G.E. Rhead, J. Vacuum Sci. Technol. 13 (1976) 603.
[6] C.J. Todd, Vacuum 23 (1973) 195.
[7] M.P. Seah, Surface Sci. 32 (1972) 703.
[8] D.C. Jackson, T.E. Gallon and A. Chambers, Surface Sci. 36 (1973) 381.
[9] Reviewed by F. Meyer and M.J. Sparnaay, in: Surface Physics of Phosphors and Semiconductors, Eds. C.G. Scott and C.E. Reed (Academic Press, 1975) p. 321.
[10] J.C. Philips, Surface Sci. 40 (1973) 459.
[11] M. Henzler, Surface Sci. 36 (1973) 109.
[12] P.J. Feibelman, E.J. McGuire and K.C. Pandey, Phys. Rev. Letters 36 (1976) 1154.
[13] G.F. Amelio, Surface Sci. 22 (1970) 301.
[14] J.A.D. Matthew, Surface Sci. 40 (1973) 451.
[15] S.P. Kowalczyk, L. Ley, F.R. McFeely, R.A. Pollack and D.A. Shirley, Phys. Rev. B9 (1974) 381.
[16] L. Ley, S. Kowalczyk, R. Pollack and D.A. Shirley, Phys. Rev. Letters 29 (1972) 1088.
[17] F. Wehking, H. Beckermann and R. Niedermayer, Thin Solid Films 36 (1976) 265.

ADSORPTION STUDIES OF Cs ON Si (111)

P. WAGNER, K. MÜLLER and K. HEINZ

Institut für Angewandte Physik, Lehrstuhl für Festkörperphysik, Erwin-Rommel-Str. 1, D-8520 Erlangen, West Germany

Cs adsorption on the Si(111)-7 × 7 surface has been investigated using the combined methods of LEED, work function measurements and thermal desorption. LEED patterns are only affected with respect to the intensities indicating that no new superstructure is developed by Cs adsorption. The thermal desorption spectra reveal two distinct peaks. The first one arises already at low coverages and shifts in desorption energy from about 2.15 eV to 2.4 eV with increasing coverage. Then the second peak comes up showing a similar coverage dependence. It finally unites with the first one. There is a clear correspondence between the general decrease of the intensity of integer order spots in the 7 × 7 LEED pattern during adsorption and the behaviour of the desorption peaks. More specifically, pronounced peaks in the intensity spectrum of the 01 spot have been recorded as a function of coverage for comparison with the relative hights of the desorption peaks. Moreover, a similar correspondence could be observed with respect to the coverage dependence of the work function decrease and surface conductivity increase. Saturation of coverage was already achieved at a Cs coverage of about 2×10^{14} atoms cm^{-2} which could not be exceeded even with prolonged exposures. Two different adsorption sites, their relative population as a function of coverage as well as the low saturation coverage can be well understood on the basis of the Lander model for the Si(111)-7 × 7 surface.

ADSORPTION–DESORPTION STUDIES OF Zn ON GaAs

G. LAURENCE *, B.A. JOYCE and C.T. FOXON

Philips Research Laboratories, Redhill, Surrey, England

and

A.P. JANSSEN, G.S. SAMUEL and J.A. VENABLES

School of Mathematical and Physical Sciences, University of Sussex, Falmer, Brighton, Sussex, England

By employing a wide range of techniques to study adsorption–desorption behaviour in the ostensibly simple system of the metal Zn on the semiconductor GaAs it has been found that many complicating factors can occur, and reliance on any one of the techniques would have given a totally misleading picture. The methods used were temperature programmed thermal desorption, modulated atomic beam adsorption measurements, AES, high resolution UHV SEM combined with AES and high resolution (300 Å) AES, and RHEED. GaAs substrate surfaces were cleaned in-situ by thermal treatment or inert gas ion bombardment followed by annealing. It was established by SEM and RHEED observations that surface topographic and compositional changes could occur at this stage. Zn sticking coefficient measurements by modulated beam and AES techniques showed that it could vary widely for minor changes in surface composition, and that it was also a strong function of deposition time and substrate temperature. However, initial growth of the Zn deposit was always two dimensional (within the resolution limits of the SEM) and epitaxial, for the range of substrate conditions used. Thermal desorption spectra were also found to depend rather critically on the substrate surface, with very pronounced difference between the Ga and As stabilized forms. An attempt has been made at a systematic interpretation of the kinetic data based on reasonably simple models, and also to relate it to previously published work on this system, but the profound influence of substrate surface effects makes a fully quantitative evaluation extremely difficult. Nevertheless, the value, and perhaps the necessity, of employing a wide range of techniques to investigate metal–semiconductor systems is clearly demonstrated.

1. Introduction

There is considerable interest at the present time in obtaining a more fundamental understanding of doping effects in molecular beam epitaxy (MBE). This is the process, recently reviewed by Cho and Arthur [1], in which epitaxial thin films of compound semiconductors are prepared from molecular beams of their constituent

---

* Permanent address: Laboratoires d'Electronique et de Physique Appliquée, BP 15, 3, Avenue Descartes, 94450 Limeil-Brévannes, France.

elements. Zn has been used successfully as a p-type dopant in GaAs films prepared by liquid phase and vapour phase epitaxial deposition processes, and in principle it could be useful in MBE if, by directing a Zn beam at the growing film, the Zn atoms were to be incorporated on electrically active (acceptor) sites. Previous work [2,3] has suggested that the interaction between Zn and a GaAs surface may be too weak to permit its incorporation at realistic growth temperatures ($\geqslant$750 K), but there are sufficient discrepancies between the two sets of reported data, even though very similar experimental techniques were used to make a further study worthwhile. The proposed adsorption–desorption models were also significantly different. Arthur [2], using pulsed beam techniques combined with thermally stimulated desorption and electron diffraction, proposed a model in which epitaxial two-dimensional Zn islands formed from a mobile precursor state, the adsorption rate being controlled by interaction between Zn atoms in this state and island boundaries. The desorption energy of Zn was equal to its enthalpy of vaporization (1.37 eV), with a very small population ($\sim 1 \times 10^{13}$ atoms cm^{-2}) in a more tightly bound state, the exact number depending on surface composition. Harvey [3] using the same experimental approach, except for the replacement of electron diffraction by SEM observations in a separate instrument outside the UHV chamber, found a coverage dependent desorption energy (0.75–0.26 eV) which was always less than the enthalpy of vapourization, a negligible tightly bound population on clean surfaces, and a three-dimensional island growth mode. She suggested that desorption occurred from the peripheries of these islands.

The present work was undertaken to attempt to resolve these differences, principally by extending the range of techniques used. We believed that this approach would have two advantages: (i) if the same parameter could be measured by more than one method, not only would agreement strengthen our confidence in the values obtained, but it would also help to substantiate the validity of the application of the method in question to this and similar systems: (ii) additional information on surface composition, structure and rate processes, not available in previous work, might help resolve the differences.

2. Experimental

The techniques employed included temperature programmed thermal desorption (TPTD), modulated atomic beam adsorption/desorption measurements, AES, high resolution UHV SEM combined with AES and high resolution (300 Å) AES, and RHEED. Two experimental systems were used, one principally designed for surface kinetic studies, and one for structural, compositional and topographic analysis of the surfaces. However, it was established that measurements by two techniques (TPTD and AES) were reproducible, at least semiquantitatively, from one system to the other. The basic instrumentation and methods of data acquisition and signal processing of each system have been described previously [4–6]. The first system had a base pressure after bakeout of $\sim 10^{-10}$ Torr and the second $<10^{-9}$ Torr.

The substrates were GaAs bars 20 mm $\times$ 5 mm $\times$ 1 mm thick, most of which were oriented such that the widest face was {100}, although some experiments were also carried out on {110} surfaces. They were polished using bromine–methanol on a felt pad, free-etched in an $H_2O_2/H_2SO_4/H_2O$ mixture and cleaned in-situ either thermally or by argon-ion bombardment. The only contaminants then detectable by AES were C and O. It was always possible to reduce oxygen below the detection limit (<1% monolayer) by continued treatment, but on some substrates small amounts (1–5% monolayer) of carbon remained.

Atomic beam of Zn were produced from Knudsen cells which gave fluxes at the substrate from 10^{11} to 10^{15} atoms cm^{-2} s^{-1}. The Zn was of semiconductor grade purity.

3. Results

For simplicity of presentation, results will be detailed under four separate headings, modulated beam measurements, AES, TPTD and microstructure determination but we would emphasize that particular sets of data cannot be analysed in isolation, and it is important to consider the results as a whole.

3.1. Modulated beam measurements

The basis of this technique has been described previously in considerable detail [4], and it has been used in the present work to obtain values of the sticking coefficient and surface lifetime of Zn on {100} oriented GaAs surfaces over a fairly wide range of conditions. In these experiments sticking coefficients were determined by modulation of the desorbing flux, which ensured that only Zn atoms desorbing directly from the substrate were detected. However, because a signal averaging technique was used for data acquisition, instantaneous values of the Zn sticking coefficient (S_{Zn}) at time $t = 0$ could not be obtained.

Before attempting to measure S_{Zn} as a function of t (and therefore coverage θ) and substrate temperature, T_s, it was important to determine the effect of GaAs surface conditions on the condensation behaviour of Zn at room temperature (295 K). The results can be summarized as follows:

(i) $S_{Zn} = 0$ if the surface concentration of oxygen (as determined by AES) exceeds a few percent of a monolayer.

(ii) For all other known surface conditions, including varying amounts of contamination, and on both gallium and arsenic stabilized surfaces [7,8], $S_{Zn} = 1$.

All subsequent work was therefore carried out using surfaces on which oxygen contamination was below the detection limit of AES ($\sim$0.01 monolayers), and the results for S_{Zn} as a function of time for three different temperatures (293, 305 and 319 K) are shown in fig. 1. By 373 K, $S_{Zn} = 0$ for all measurable t ($t \geqslant 300$ s for a

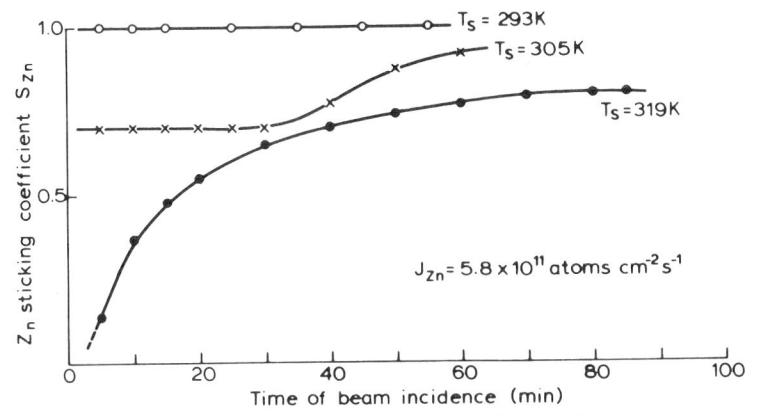

Fig. 1. Sticking coefficient as a function of time for different substrate temperatures.

beam flux of $\sim 3 \times 10^{11}$ atoms cm^{-2} s^{-1}). It is evident from these curves that for $0 < S_{Zn} < 1$, S_{Zn} increases as a function of time, in qualitative agreement with previous observations [2,3].

We attempted to measure the surface lifetime, τ, of Zn by modulating the incident flux and analysing the frequency dependent attenuation and phase shift of the signal produced by the desorbing Zn [4]. We were only able to set limits however, since direct values were experimentally inaccessible. The upper limit, τ_{max}, was set by the minimum modulation frequency which can be used (0.5 Hz, corresponding to $\tau \approx 1$ s), while the lower limit was fixed by the signal to noise ratio of the detected signal, which in this case corresponded to a minimum lifetime, $\tau_{min} \approx 10^{-5}$ s. Somewhat surprisingly, we could find no substrate temperature where $10^{-5} < \tau < 1$ s; either the detected signal was totally demodulated (i.e. $\tau > 1$ s) or there was no measurable frequency dependent phase shift and attenuation ($\tau < 10^{-5}$), the change occurring for a change in T_s of ~ 20 K, i.e. from ~ 320 to 340 K. Evaluation of these data, using the limit values of τ at the two temperatures, in terms of desorption obeying the simple Frenkel equation

$$\tau = \tau_0 \exp(E_D/kT), \tag{1}$$

where E_D is the activation energy for desorption and τ_0 a surface vibration period, gives an impossibly large value for E_D (>5 eV). In section 3.2, we show, using TPTD techniques, that E_D is coverage dependent, with a maximum value of <1 eV; in section 4 we discuss alternative explanations of the surface lifetime behaviour.

3.2. Temperature programmed thermal desorption (TPTD)

Desorption of Zn films condensed at room temperature (295 K) on {100} GaAs surfaces was investigated by TPTD, using a computer generated temperature ramp

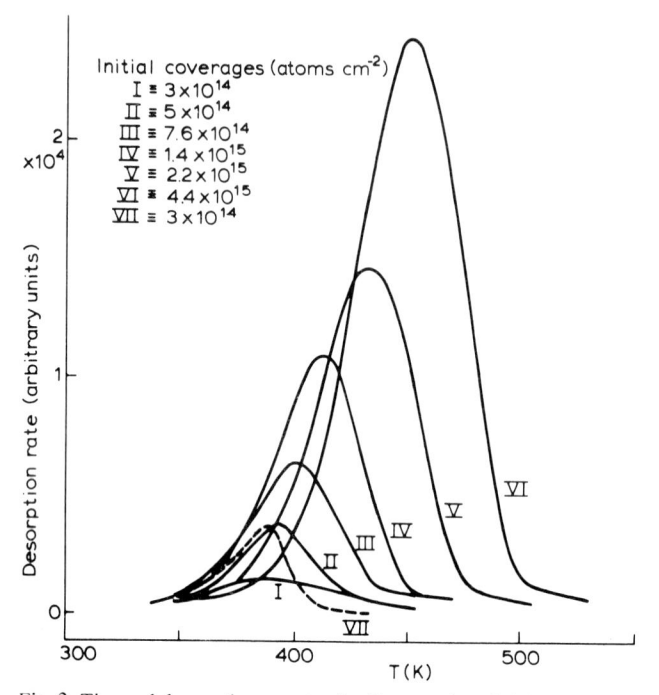

Fig. 2. Thermal desorption spectra for Zn at various initial coverages.

Initial coverages (atoms cm⁻²)
I ≡ 3×10¹⁴
II ≡ 5×10¹⁴
III ≡ 7.6×10¹⁴
IV ≡ 1.4×10¹⁵
V ≡ 2.2×10¹⁵
VI ≡ 4.4×10¹⁵
VII ≡ 3×10¹⁴

covering the range $1-15$ K s^{-1}. Desorption spectra could be obtained simultaneously in digital and analogue form, but all analyses were carried out on digital data. The initial state of the substrate surface, prior to Zn deposition had a major effect on desorption behaviour, and it was only possible to obtain reproducible results from a surface prepared as "Ga stabilized" [7] before depositing Zn. Desorption spectra obtained for increasing Zn coverage on such a surface are shown in fig. 2 (curves I–VI); curve VII was produced by desorption from a surface exposed to Zn for the first time; results from such surfaces do not fit into the general pattern and are discussed further below. It is immediately apparent that spectra I–VI are not compatible with a simple integral order process, since the maximum desorption rate moves to higher temperatures with increasing coverage.

We therefore used two analytical procedures, both based on the Polanyi–Wigner desorption model, in which it is assumed that the rate of desorption from a state i is given by:

$$-dN_i/dt = \nu_i N_i^{x_i} \exp(-E_{D_i}/kT) \qquad (2)$$

where x_i is the reaction order for desorption from state i, ν_i is a frequency factor, N_i is the population of state i and E_{D_i} is the activation energy for desorption. In the first method of treatment, only the initial part of the desorption curve is considered, and in this region, where the surface coverage is not changing significantly, the desorption rate is approximately proportional to the exponential term, especially for small values of x_i. Thus if the desorption energy is not a strong function of coverage, Arrhenius plots of ln (initial desorption rate) as a function of reciprocal temperature should produce straight lines, from the slopes of which the desorption energy for this first stage of the process can be calculated. The result of analysing the desorption curves in this way is shown in fig. 3, from which it is clear that good straight lines are obtained, suggesting a desorption energy independent of coverage for any specific coverage. However, E_{D_i} is not independent of initial coverage, but increases with increasing coverage, asymptotically approaching a value of ~0.85 eV for deposits several tens of monolayers thick, as indicated in fig. 4.

The second approach involved the so-called "complete analysis" described by King [9]. Here, the general form of eq. (2) is used without assuming any term to be coverage independent. The analysis is performed by integrating the desorption curves from the high temperature end until a value of the integral corresponding to a particular coverage still remaining on the surface is reached on each curve. Thus from each curve a single rate–temperature point is obtained for a specific coverage. An Arrhenius plot can therefore be produced for any predetermined value of cover-

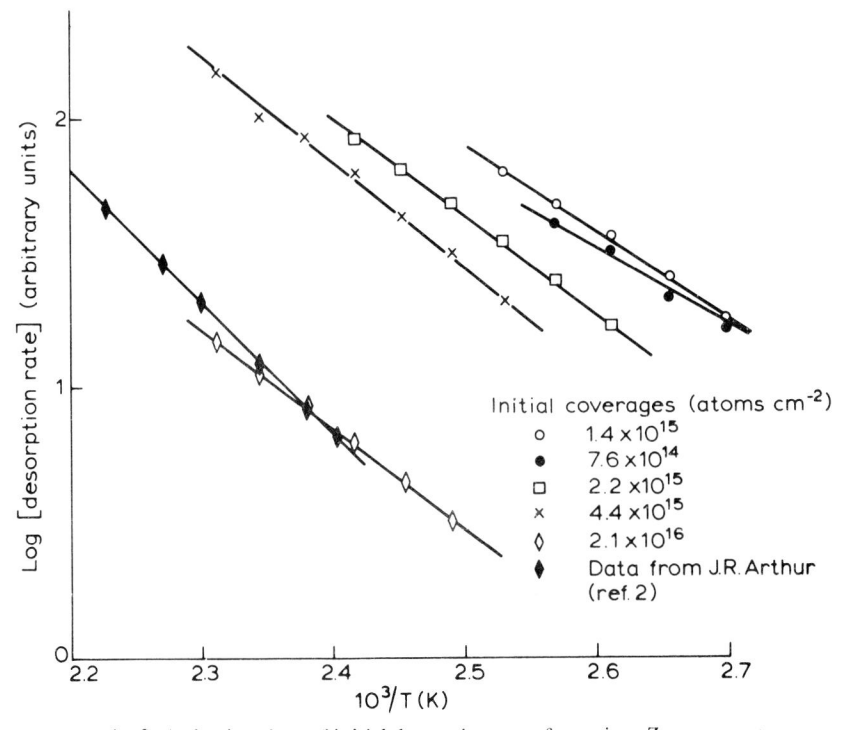

Fig. 3. Arrhenius plots of initial desorption rates for various Zn coverages.

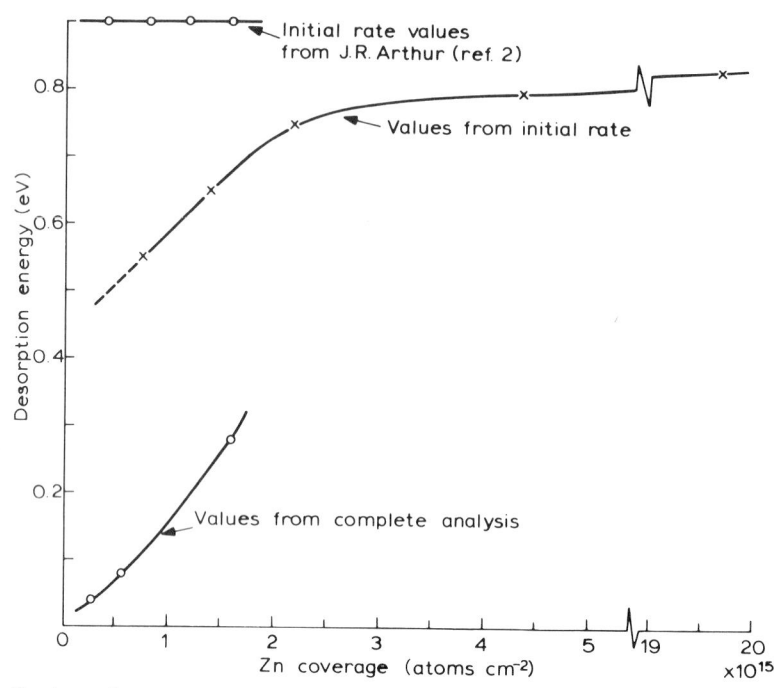

Fig. 4. Zn desorption energy as a function of coverage, desorption energies determined from both initial rates and complete analysis.

age, from which the desorption activation energy can be obtained as a function of coverage. The results of analysing the desorption spectra of fig. 2 in this way are presented in fig. 4, which shows that again the desorption energy increases with increasing coverage, although the actual values are very much lower than from the initial rate analysis. These very low values of E_D, coupled with correspondingly low values of the pre-exponential term, suggest a low binding energy but make it difficult to propose a physically meaningful model for the final stages of desorption. It cannot be expected, of course, that the same value of E_D will be obtained from the two methods of analysis, since in the first case we are making an initial state evaluation, whereas the complete analysis is a final state determination, obtained near the end of the temperature ramp. We attach more significance to the initial rate measurements, however, if only because experimental errors are much less in that case, and measured rates are over a wider temperature range.

Although sets of spectra from a Ga-stabilized surface are in general reproducible and self-consistent, in that areas under the peaks,

$$\int_{T_1}^{T_2} \frac{dN}{dt} \frac{dt}{dT} dT,$$

are proportional to coverage, the first adsorption–desorption cycle on a new substrate (i.e. one not previously exposed to a Zn flux) always produced an anomalous result, which did not fit into the pattern of subsequent spectra. This is shown by curve VII in fig. 2, produced from the same initial coverage as curve I, but clearly corresponding to a much larger amount of desorbed material. This anomaly occurred in spite of the fact that before a Zn film was deposited, the substrate was in every case checked by AES to ensure that no Zn remained from previous adsorption–desorption processes. It was found to be necessary to heat the substrate to ~850 K to ensure that Zn in the surface region was below the detection of limit of AES, even though no desorption of Zn could be detected above ~500 K from a Ga-stabilized surface.

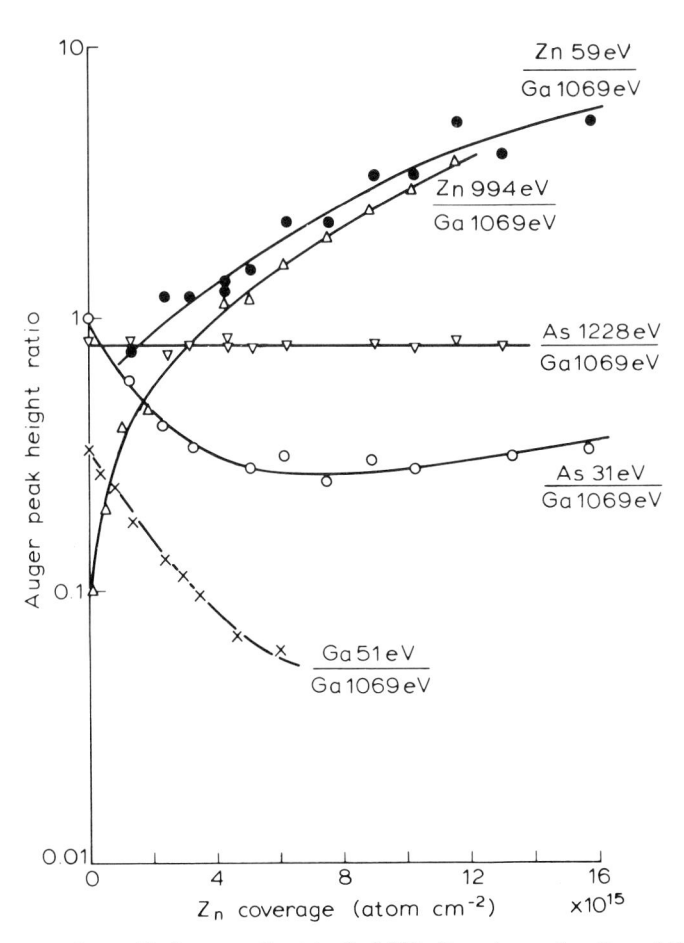

Fig. 5. Auger peak amplitudes normalized to Ga 1069 eV peak as a function of Zn coverage.

The only extraneous surface impurity detected (by AES) after Zn desorption was small amounts of carbon, but by deliberately varying the quantity present over a large range (~5 ×) we established that it had no observable effect on desorption behaviour.

Desorption spectra from "As-stabilized" surface were quite irreproducible, to the extent that sometimes one and sometimes two desorption peaks were produced from apparently identical substrate surface conditions, deposition procedures, coverages and desorption conditions. Since we were unable to identify the parameter(s) responsible for the irreproducibility, however, we did not make any other measurements on As-stabilized surfaces, nor have we attempted to analyse desorption spectra from surfaces in this state.

3.3. Auger spectroscopy

By monitoring the Auger electron currents from specific transitions in Zn, Ga and As as a function of Zn deposition time, it should in principle be possible to obtain information on the growth mode of Zn, and also of any interdiffusion between Zn and GaAs. For each element at least one low energy (<100 eV) and one high energy (>994 eV) transition occurs, and because of the significantly different information depths associated with the two ranges of electron energies (~5 Å and 10–15 Å respectively) any interface interpenetration effects should be revealed. The following Auger transitions were monitored, the intensity being measured as the peak-to-peak amplitude in the $N'(E)$ versus E spectrum:

Ga: $M_3M_4M_4$, 51 eV; $L_3M_4M_4$, 1069 eV;

As: M_4VV, 31 eV; $L_3M_4M_4$, 1228 eV;

Zn: $M_2M_4M_4$, 59 eV; $L_3M_4M_4$, 994 eV.

The Ga 51 eV peak was used in preference to the 53 eV $M_2M_4M_4$ peak, which is larger, to reduce interference from the Zn 55 eV peak.

In fig. 5, the Auger peak amplitudes, normalized to the Ga 1069 eV peak, are plotted as a function of Zn coverage. These results were obtained in the SEM, but qualitatively similar behaviour was observed in the kinetic apparatus.

The essential features revealed by the results are (i) neither the low nor high energy Zn peaks reach a constant value even when the deposit thickness greatly exceeds the escape depths of the Auger electrons. (ii) The As peaks, while being initially attenuated as Zn is deposited either stay at constant amplitude or increase with further deposition of Zn. (iii) The Ga peaks are not completely attenuated even for very large coverages. These results appear to indicate that significant interdiffusion is occurring, not simply of Zn into GaAs, but with some segregation of As to the Zn surface (cf. As and Ga ratios). These results were obtained after previous Zn deposits had been desorbed from the substrates, but no Zn was detectable by AES after this desorption.

3.4. Microstructure

This work was carried out in the UHV SEM which included facilities for RHEED, AES and high resolution AES, the details of which have been described previously [5,6]. The instrument also contained a Knudsen cell identical in design and operation to those used in the kinetic measurements. The experiments were designed principally to determine the surface topography and uniformity of the Zn deposits

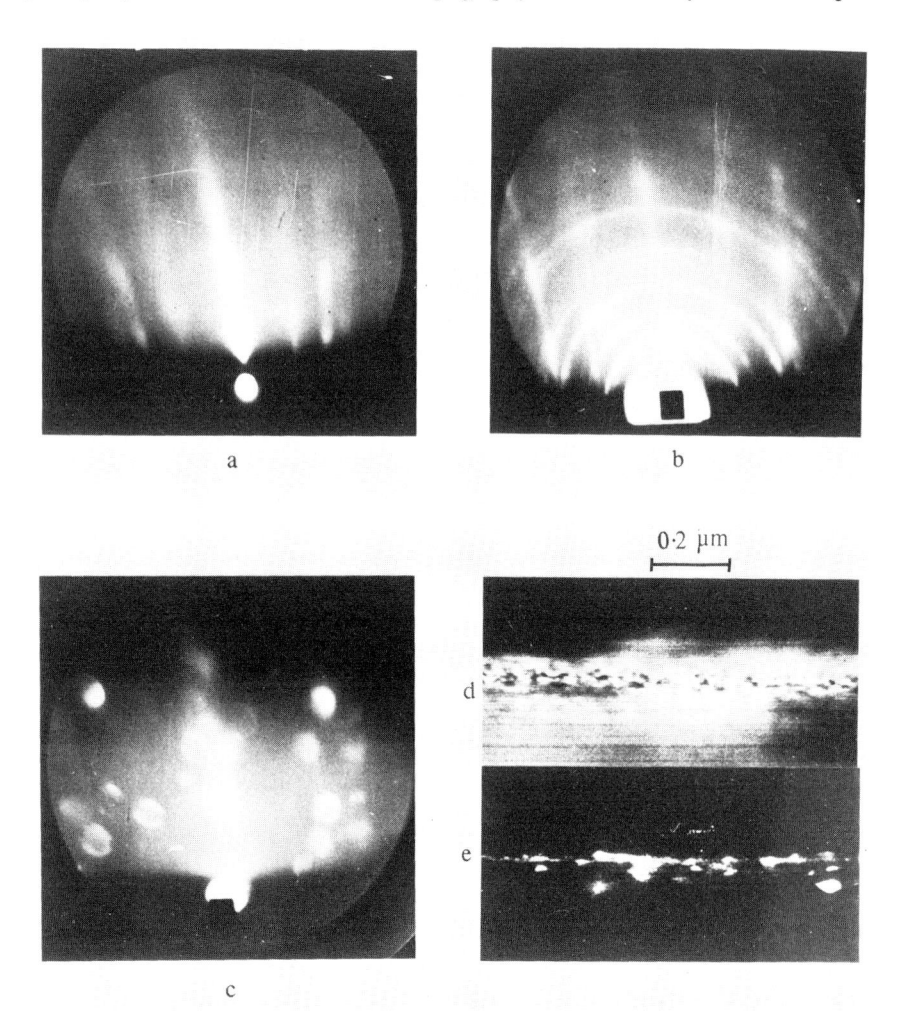

Fig. 6. RHEED (60 eV) patterns of Zn on (001)GaAs along ⟨100⟩ azimuth. (a) 2.7×10^{15} atoms cm^{-2}; (b) 1.4×10^{16} atoms cm^{-2}, polycrystalline layer; (c) selected area transmission pattern from several Zn crystallites, 1.4×10^{16} atoms cm^{-2}; (d) secondary electron image at glancing incidence; (e) dark field transmitted electron image, using one of the Zn spots in (c).

as a function of coverage. We found that with the substrate at room temperature the films were two-dimensional and epitaxial, with a lattice spacing close to that of bulk Zn up to coverage of $\sim 6 \times 10^{15}$ atom cm^{-2} (assuming $S_{Zn} = 1$), with no evidence of island growth, and were of uniform thickness, as determined by high resolution AES. Pronounced streaking in the RHEED pattern showed the two-dimensional nature of the film (fig. 6a). For deposits $>6 \times 10^{15}$ atoms cm^{-2}, diffraction patterns showed evidence of polycrystallinity (fig. 6b), and by 1.4×10^{16} atoms cm^{-2} crystallites having dimensions of $\sim 100-200$ Å could be seen by SEM. Fig. 6c shows the crystallite distribution over the surface, imaged using a grazing incidence electron beam, while the selected area diffraction pattern from several individual Zn crystallites (fig. 6d) can be used to generate a dark field image (fig. 6e) which shows the size of the Zn crystallites more clearly.

The topography of the Zn deposit was monitored continuously during the complete desorption process, but we could observe no change from the two-dimensional state for low coverages. Using a mass spectrometer in the system we were able to show that at least qualitative agreement could be obtained between TPTD spectra from the two systems. Both {100} and {110} surfaces were studied with very similar results.

4. Discussion

Before attempting to formulate a quantitative model it is important to consider the overall physical significance of the data, since any model must obviously be compatible with results from all techniques. Firstly the early stages of growth are two-dimensional and ordered, and no detectable topographic changes occur during desorption. Not only is growth two-dimensional, however, but there is also a considerable amount of evidence of interdiffusion and segregation processes. The most direct indication comes from the Auger peak height ratios (fig. 5) which suggest that arsenic is tending to segregate and/or combine with Zn in the surface of the film, i.e. the deposit cannot simply be treated as a Zn film on a GaAs substrate. Additional evidence for interdiffusion is provided by (i) the very high temperature (~ 850 K) to which the substrate must be heated to reduce the Zn Auger signal below the detection limit, even though no desorbing Zn can be detected >500 K. Apparently Zn is diffusing out from the substrate in this temperature interval. (ii) The anomalously large amount of Zn which desorbs the first time a substrate is used (c.f. curves I and VII, fig. 2). It must be concluded that in subsequent desorption experiments a substantial amount of Zn diffuses into the bulk, (which we have confirmed by atomic absorption analysis) in a way which is strongly influenced by the first Zn–GaAs interaction. (iii) The observation that even at coverages of several tens of monolayers the activation energy for desorption of Zn (as measured from the initial rate) does not reach the value (1.37 eV) corresponding to the Langmuir evaporation of pure Zn [10]. Arthur [2] claimed agreement with this value from his initial desorption rate data, but it is not consistent with his published spectra, from which we calculate an energy of 0.9 eV. We would suggest that if these comparatively thick films were pure Zn, then indeed the desorption energy must reach that for bulk Zn. That it does not can be taken to suggest that we are not dealing with a pure Zn film, but one in which surface Zn atoms have As and perhaps Ga, as well as Zn nearest neighbours, where Zn–As and Zn–Ga interactions are weaker than Zn–Zn. The diffusion of Zn in GaAs has been studied extensively (e.g. ref. [12], and references quoted there), but it is difficult to make direct comparison with the present results, because diffusion measurements have all been made at substantially higher temperatures ($\geqslant 1000$ K) than we used. Nevertheless, the low temperature behaviour is not incompatible with the substitutional–interstitial diffusion model [11,12], in which the Zn atom is incorporated in the lattice at the surface, goes interstitial and diffuses very rapidly. It eventually combines with a Ga vacancy and becomes substitutional again, but the surface and dislocations in the substrate can act as further vacancy sources [13].

Having discussed the structure and possible composition of the deposit we can now turn our attention to the adsorption–desorption kinetics. The total inhibition of Zn condensation by sub-monolayer (<10%) quantities of oxygen strongly suggests that specific surface sites, perhaps steps, are required for nucleation, but they are easily blocked by oxygen. On a clean surface, the increase of S_{zn} with time (i.e. coverage), and the corresponding increases of desorption energy with coverage, particularly as determined from the complete analysis, both point to a stronger mutual interaction between Zn atoms than between Zn and GaAs. The behaviour of desorption energy with coverage requires further comment however. The constancy of E_D during the first stages of desorption (as determined from the initial rate) for a specific initial coverage contrasts with its variation as the initial coverage is changed, especially as we are considering coverages >1 monolayer. We would interpret this to mean that different surface phases are formed for different initial coverages, from which Zn has varying desorption energies. For the coverages used the initial rate analysis does not provide information on Zn atom desorption from sites on GaAs surfaces. In the complete analysis, however, the final stages of desorption are being considered, so Zn–GaAs interactions become more important, with the reservation that diffusion effects could introduce complicating features. If we assume, however, that the values are at least relatable to the simple Zn–GaAs system, it is clear that the interaction is very weak.

The surface lifetime values provide the final limitation to the system model. We have already pointed out that they cannot be understood in terms of a single process, but at least two parallel processes are required to produce the observed effect. A Zn atom incident on the surface can either desorb with a surface lifetime $<10^{-5}$ s (quite compatible with the measured desorption energies), or it can be transferred to a more tightly bound state with a very long surface lifetime in relation to the modulation period (i.e. $\tau \gg 1$ s). Atoms cannot have long surface lifetimes prior to desorption, so if a precursor state is assumed [2], transfer to the bound state and desorption must both be rapid. Atoms could of course return from the bound state to the vapour phase, but with an extremely long time constant.

Although the complex nature of the system precludes quantitative modelling, it is possible to write a simple mass balance equation in which the importance of the various terms can be demonstrated. This equation must be of the form:

$$\frac{dn}{dt} = J_{Zn} - \frac{n}{\tau_1} - \frac{n}{\tau_2} - \sigma Dn(n_x) - J_{Zn}n_x a_x , \qquad (3)$$

where n is the surface concentration of single Zn atoms, τ_1 the liftime for desorption, τ_2 the time constant for diffusion into the bulk, D the surface diffusion constant, n_x the surface concentration of stable islands of area a_x, and σ the capture number discussed in nucleation theories (see e.g. Venables [14]). J_{Zn} is the normalized Zn flux, i.e. number of atoms/surface site/s. The fourth term of eq. (3) then represents capture of single atoms by existing growth centres and the fifth term direct impingement onto these islands, which becomes increasingly important with increasing coverage. The loss rate of single atoms by diffusion into the bulk can be represented by a term $(-n/\tau_2)$ [15] exactly analogous to the term for loss by desorption $(-n/\tau_1)$. Without the fourth and fifth terms, $dn/dt \to 0$, and the desorption rate would become constant and finite, independent of time. This is not the observed behaviour (see fig. 1, where $S_{Zn} = 1 -$ desorption rate) and such terms are needed to account of the desorption rate decreasing with time. Because of the limited temperature range available for measurements (S_{Zn} changes from 1 to 0 over <50 K) it is not possible to collect sufficient data to enable the relative importance of the various processes to be assessed.

Although we have not produced a more quantitative model than previous investigators, the physical complexities of this system have been demonstrated by use of a wide range of techniques. The use of any of these alone would have resulted in an inadequate physical picture of the Zn–GaAs surface interaction behaviour.

References

[1] A.Y. Cho and J.R. Arthur, Progr. Solid State Chem. 10 (1975) 157.
[2] J.R. Arthur, Surface Sci. 38 (1973) 394.
[3] J.A. Harvey, Ph.D. Thesis, University of Surrey, England (1975).
[4] C.T. Foxon, M.R. Boundry and B.A. Joyce, Surface Sci. 44 (1974) 69.
[5] J.A. Venables in: Developments in Electron Microscopy and Analysis, Ed. J.A. Venables (Academic Press, 1976) p. 23.
[6] J.A. Venables, A.P. Janssen, C.J. Harland and B.A. Joyce, Phil. Mag. 34 (1976) 495.
[7] A.Y. Cho, J. Appl. Phys. 42 (1971) 2074.
[8] A.Y. Cho J. Appl. Phys. 47 (1976) 2841.
[9] D.A. King, Surface Sci. 47 (1975) 384.
[10] R.W. Mar and Searcy, A.W. J. Chem. Phys. 53 (3076) 1970.

[11] B. Tuck and M.A.H. Kadhim, J. Mater. Sci. 7 (1972) 581.
[12] L.L. Chang and G.L. Pearson, J. Appl. Phys. 35 (1964) 1960.
[13] J. Blanc, J. Appl. Phys. 45 (1974) 1948.
[14] J.A. Venables, Phil. Mag. 27 (1973) 697.
[15] J.A. Venables, Thin Solid Films 18 (1973) S 11.

PAST AND FUTURE SURFACE CRYSTALLOGRAPHY BY LEED

F. JONA

Department of Materials Science, State University of New York, Stony Brook, New York 11794, USA

The first eight years of surface crystallography by low-energy electron diffraction (LEED) have provided much novel and interesting information about atomic arrangements on crystalline surfaces. A brief review of methods, procedures and instrumentation shows, however, that many improvements are needed or desirable, particularly in the areas of: LEED diffractometers for structure analyses and devices for rapid data collection and display; characterization of surfaces; theoretical and computational methods; search procedures for the correct structural model; objective and quantitative criteria for evaluation of the reliability of structural models. A few of the most pressing problems in LEED crystallography are discussed.

1. Scope

Surface crystallography, i.e., the determination of atomic arrangements within the first two or three layers on the surfaces of crystalline solids, is a relatively new branch of science that has been growing rapidly in recent years. The oldest tool for this work, and the one that has been most successful so far, is low-energy electron diffraction (LEED), but several interesting alternatives are being actively and successfully pursued, viz., ion scattering, angle-resolved photo-emission, angle-resolved Auger electron spectroscopy (AES), etc.

This brief review is concerned exclusively with what may be called LEED crystallography, and in particular with that part of LEED crystallography that makes use of the dynamical theory of diffraction. We are not going to include, in other words, discussions of the so-called data-reduction or data-averaging methods and of the Fourier-transform procedures that are used by a few workers in LEED crystallography. The reasons for this neglect are twofold. First, discussions of either method are already available in the literature in comprehensive form [36–38]. Second, the accomplishments of either method in terms of real surface structure determinations are still rather limited.

LEED crystallography by dynamical theory was started in 1969–70 with the first detailed measurements and the first calculations of intensity spectra from three low-index surfaces of aluminum. Much has been accomplished in the intervening 7 or 8 years. A review of these accomplishments is presented in section 2 but is kept rather brief because more comprehensive discussions of the subject matter have already appeared recently in the literature [1,2,22,23]. In section 3, we attempt to review how those accomplishments were achieved, i.e., to discuss the present methods of LEED crystallography in its three major aspects: experimental, theoretical or computational, and analytical or evaluational. In this discussion we bring out some of what we believe are the most serious defects and limitations of the field, particularly in the experimental area and in the procedures followed almost universally for evaluation of structural models. In section 4, we discuss some of the proposed solutions for the problems at hand and we identify some of the topics that are likely to attract most attention in the years immediately ahead.

2. Accomplishments

To date, the exploration of surface structures has been limited, with only a few exceptions, to low-index surfaces of solid elements, either clean of after reactions with minute amounts of gaseous agents. Such gaseous agents have been also mostly elemental (oxygen, sulfur, nitrogen, etc.), the surface coverage has been always smaller than, or at most equal to, one monolayer, and the distribution of the chemisorbed atoms has been always ordered.

Within these limits, a few conclusions of general validity have begun to emerge in surface crystallography. The first is that most low-index surfaces of metals are essentially simple truncations of the corresponding bulk structures – the atomic arrangement on the surface has the same registry and approximately the same interlayer spacing as that in the bulk. The more open, i.e., less close-packed, surfaces (such as the {001} surfaces of the body-centered cubic (bcc) and the {110} surfaces of the face-centered cubic (fcc) lattice) exhibit compressions of the first interlayer spacings, varying from 1.5% for Fe{001} to 6% for W{001}, about 10% for Ag{110} and 11.5% for Mo{001}.

The second conclusion is that hard-sphere models, already so successful in bulk crystallography, are often also adequate and convenient for describing surface structures, particularly when chemisorbed foreign atoms are involved. Accordingly, each type of adsorbate atoms may be assigned a characteristic "effective radius" r_{eff}, which can be calculated by subtracting the atomic radius of the substrate atom from the adsorbate–substrate bond length determined by the surface structure analysis. Table 1 lists the effective radii of a few selected adsorbate atoms, and shows that these radii are often somewhat different from the single-bond covalent radii and vary from substrate to substrate. It is perhaps too early to say to what extent these variations are meaningful and to what extent they are merely consequences of inaccuracies in the corresponding structure analyses.

The third conclusion is that on any surface there are indeed preferred sites for chemisorption of foreign atoms – these sites are usually (but not always, as we will discuss below) those which would be occupied by indigenous atoms if the substrate were to grow. Such sites, which may conveniently be called "expected" locations

Table 1

Effective radii of chemisorbed atoms on selected substrates; the single-bond covalent radii and the metallic radii are given in parentheses; all numbers are in Ångstrom units

Substrate (metallic radius)	Chemisorbed atom (single-bond covalent radius)					
	N (0.70)	O (0.66)	Si (1.17)	S (1.04)	Se (1.13)	Te (1.39)
Ti {0001} (1.47)	0.62					
Fe {001} (1.26)		0.78		1.06		
Ni {001} (1.24)		0.74		0.94	1.03	1.34
Mo {001} (1.36)			1.15			
W {110} (1.39)		0.72				

for chemisorption [1,2], are indicated schematically in fig. 1 for the fcc and the bcc {001}, the fcc {111} or the hcp {0001}, the fcc {110} and the bcc {110} surfaces, respectively. Of course, not all the "expected" locations are always occupied: on fcc {001}, for example, in an overlayer structure of the p(2 × 2) type only 1/4 of the "expected" sites may be occupied, in a structure of the c(2 × 2) type only 1/2

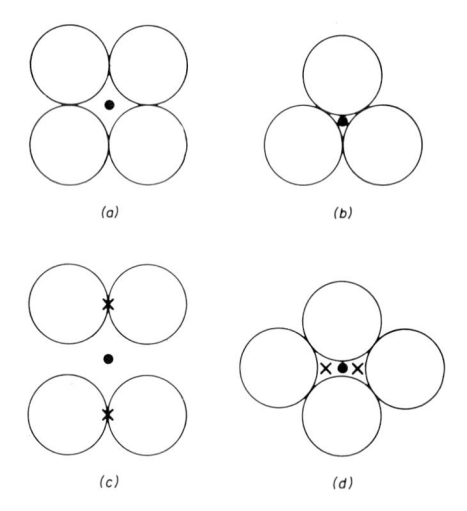

Fig. 1. Hard-sphere models of the nearest neighbors (open circles) of chemisorbed atoms on low-index surfaces of metals. The "expected" adsorption sites are indicated with black dots. (a) fcc {001} and bcc {001} (the latter would of course have the atoms less close-packed than indicated). (b) fcc {111} and hcp {001}. (c) fcc {110}: the crosses indicate the positions of oxygen atoms in the Ni {110} 2 × 1–O structure. (d) bcc {110}: the crosses indicate the positions of oxygen atoms in the W {110} 2 × 1–O structure. Only one of the crossed positions may be occupied by oxygen.

of them are occupied, while in a 1 × 1 structure all are occupied. The point is that, in general, whatever atoms are chemisorbed reside in "expected" locations.

No exception to this rule is known for the {001} surfaces of either the bcc or the fcc lattice, but an interesting modification of it is encountered in the Fe{001}-1×1-O and in the Ni{001}2×2-C structure. In the former, the O atoms "sink" into the four-fold symmetrical hollows (fig. 1a) until they come to rest on top of iron atoms in the second layer. Perhaps as a consequence of this position and of a tendency toward formation of a planar mixed layer in the top atomic plane, the first substrate layer is lifted upwards to increase the first substrate interlayer spacing 7.5% over the bulk value [3]. Similarly, in the Ni{001}2×2-C structure, the presence of carbon seems to cause an expansion of the Ni{001} first interlayer spacing by about 8.5% with respect to the bulk [4].

On close-packed atomic planes, such as those encountered on fcc{111} or hcp{0001}, the "expected" locations are the three-fold symmetrical sites, one of which is indicated schematically in fig. 1b. Not all such sites are equivalent, however. On fcc{111} the "expected" locations are those below which there *is not* an atom in the second substrate layer, on hcp{0001} they are those below which there *is* an atom in the second substrate layer. Two cases are known in which chemisorption does *not* occur on the "expected" locations, viz., (i) the case of Ti{0001}-1×1-Cd, where the registry of the Cd atoms corresponds to a stacking fault in the normal ABAB .... type of stacking sequence characteristic of the hcp structure [5], and, (ii) the case of Ti{0001}1×1-N, where the small N atoms penetrate into the octahedral holes between the first and the second layer of the Ti structure to form an "underlayer" of nitrogen atoms [6].

On either fcc{110} or bcc{110}surfaces only one example of chemisorbed structures is known. In both cases, the adsorbed atoms are *not* found in "expected" locations. In the Ni{110}2×1-O structure, the O atoms reside on the bridge sites (indicated with crosses in fig. 1c) across two adjacent substrate atoms [7,8]. In the W{110}2×1-O structure, the O atoms are located on either one of the two quasi-three-fold symmetrical saddles (indicated by crosses in fig. 1d) in contact with three W atoms [34].

Comparatively little work has been done on reconstructed surfaces of solid elements or on surfaces of compound crystals, primarily because the structure analysis of such surfaces is computationally much more complicated than that of the simple metal surfaces discussed above. On a reconstructed surface, the problem is to find the correct structural model by taking into account possible different distortions from the bulk structure in any number of layers. On a surface of a compound crystal, the problem is to develop computer programs capable of handling a unit cell with an arbitrary basis throughout a semi-infinite bulk. One also wishes to choose systems in which the stoichiometry of the top layers is constant and, hopefully, known. At the present time, advances are being made in the problem of surfaces of silicon [9–11] and of compound semiconductors [12,13]. Satisfactory results have been reported for the structures of the {001} surfaces of MgO [14] and MoS_2 [15], both of which were found to be little distorted from simple terminations of the bulk structures.

3. Methods of LEED structure analysis

3.1. Experimental procedures

Preparation of the surface structure to be studied should be carefully executed and documented. The task is relatively easy for clean surfaces, where the degree of cleanness can be monitored by such analytical techniques as Auger electron spectroscopy (AES), while the degree of crystallinity can be tested by the low background and high contrast of the LEED pattern. More difficult is the preparation of a superstructure, i.e., a structure producing both integral- and fractional-order beams, where sharpness and intensity of the fractional-order beams are the only available criteria for good crystallization and completeness of the structure at hand. Even more difficult is the preparation of surface structures involving a full, ordered and epitaxial monolayer of chemisorbed atoms (the so-called 1 × 1 structures), because in these cases there are no fractional-order beams at all — the geometry of the LEED pattern is the same with or without the chemisorbed monolayer. Rules for the recognition and preparation of this type of structures have been given elsewhere [5,6,16].

Before measurements of the diffracted intensities can begin, the magnetic field in the vicinity of the sample must be substantially reduced, and the sample itself must be aligned in such a way that the angle of incidence θ and the azimuthal angle ϕ can be measured accurately [17]. Most sample holders currently in use do not allow variations in situ of the azimuth ϕ (in which case this angle may be measured, e.g., on a suitable photograph of the LEED pattern [18]) and practically none allow simultaneous in situ control of both the azimuth ϕ and the tilt ψ.

In most cases, the diffracted intensities are measured as functions of the incident electron energy for fixed values of θ and ϕ (so-called LEED spectra, or $I-V$ curves, or intensity profiles, etc.). The measurement is done either directly with a movable Faraday cage or indirectly with a brightness meter ("spot photometer") aimed at the LEED "spots" on the fluorescent screen of display-type equipment. In the former case, the measurement yields directly absolute intensities, but only the specular beam is easy to record — the non-specular beams require much more skill and attention (which may explain why the data sets collected with a Faraday cage often show considerable predilection for the specular beam at different angles of incidence). In display-type LEED equipment, measurement of the specular spectrum is also easier to carry out, but the measurements of both the specular and the non-specular spectra are much easier and faster than with the Faraday cage. Absolute intensities, however, can be determined only by calibrating the response of the fluorescent screen and are, therefore, less precise than those measured with the Faraday cage. In either case, the acceptance angle of the detector is a critical quantity that should be controlled and known. Again, the spot-photometer lends itself more easily than the Faraday-cage to the task because test and control of the acceptance angle can be achieved by changing the optical lenses and the distance between detector and screen, all of which is more easily done than changes of the aperture of the Faraday cage inside the vacuum system. On balance, since absolute intensities are not (yet) important in the comparison between calculations and observations that is required for a structure analysis (see below), it appears that use of the spot-photometer technique is more convenient and perhaps even more precise than that of a Faraday cage.

The disadvantage of either method is that the time required for collection of a set of intensity data is so long that it may interfere with the precision and the reliability of the data themselves. Recording a single LEED spectrum over an energy range of, say, 300 eV requires typically about 10–15 min, so that it is not uncommon to spend about 1 h in collecting a set of 4 or 6 beams at a given angle of incidence. Some of the more reactive surfaces cannot be kept clean for much longer than about 1 h even at pressures of the order of 10^{-10} Torr (see, e.g., ref. [19]). Consequently, the surface must be re-cleaned frequently. The cleaning process (usually ion bombardment followed by annealing treatments) is likely to upset the alignment of the sample, which must, therefore, be checked before data collection can be resumed. All things considered, recording of a data set involving, say, 15 LEED spectra distributed over three angles of incidence may require 5 to 6 h. Under these conditions, even if the surface structure under study is not very sensitive to contamination, there is always some question whether, e.g., the spectrum collected first and that collected last pertain exactly to the same structure. In addition, there are adsorption systems, particularly those involving organic molecules [20], which are destroyed within a few seconds by electron beams of the intensity commonly used in LEED analysis. For such systems, data collection at the rates mentioned above is just out of the question.

It is obvious, therefore, that schemes for rapid and reliable collection of LEED intensities must be developed. The photographic technique developed by the Berkeley group is an acceptable answer to this problem [21] and represents a remarkable improvement over the "manual" procedure described above. It is, however, still not as rapid (recording of a data set for *one* angle of incidence in the energy range up to 150 eV involves times of the order of several minutes) as may be required to observe dissociating chemisorbed molecules, and it has the disadvantage that it is not "on-line". Electronic techniques are potentially much faster, but more difficult to realize. We shall discuss some possibilities in section 4.1.

Before they can be used for structure analysis, the intensity data must be "normalized" to constant incident current and corrected for contact potential differences. Normalization can, of course, be done after the intensities have been recorded by dividing point by point the intensity recorded at a given electron energy by the value of the incident electron current at the same energy. The incident current can be measured either with a Faraday cage or by monitoring the current to ground through the sample after this has been biased positively to avoid loss of the emitted secondary electrons. Normalization can also conveniently be done electronically on line: in this case the incident electron current I_0 is monitored continuously by measuring the total current to ground through sample, grids, container walls, etc. This current is fed to a linear analogue divider whose output is the ratio of spot-photometer (or Faraday-cage) current to I_0.

3.2. Intensity calculations

Once the experimental intensity data are collected, the process of solving a structural problem consists of three steps: (i) postulation of a structural model; (ii) calculation of intensities expected for such a model, (iii) comparison between calculated and observed intensities. We discuss here some of the aspects of step (ii).

Several theoretical treatments [22,23] and computer programs are being used to calculate LEED intensities of postulated structural models. Following the most common experimental format, the intensities are calculated as functions of the incident electron energy, usually every 2 eV. The energy range may vary, and may be limited to about 100 eV at the upper end, in order to reduce computer time and storage requirements. We found it satisfactory to standardize on the range 30–150 eV and occasionally extend it to about 200 eV. Below 30 eV the experimental data taken with most of the available LEED apparati are unreliable [24] and the theoretical results calculated with most of the available multiple-scattering programs are doubtful because of inaccuracies in the scattering potential. About 150–180 eV the number of beams required to attain acceptable accuracy becomes very large and hence increases both the computer storage size and the computation time considerably.

In the range 30–150 eV we always use 8 phase shifts and often a minimum of 29–30 beams [41] for a clean surface structure or a 1 ×1 overlayer structure, and 48–60 beams for superstructures that involve one fractional-order beam per reciprocal unit mesh, such as a c(2 ×2) or a p(2 ×1) superstructure. These choices are satisfactory for exploratory purposes as long as the surface interlayer distances do not get very small (in practice, not smaller than about 0.5 Å). If they do, then the number of beams must be increased accordingly. The same is true if the substrate's bulk involves short interlayer distances throughout (as is the case, e.g., for {111} surfaces of the bcc lattice). In any case, after the "correct" structural model has been found it is advisable to make at least one final calculation with a larger number of beams, typically 58–61 beams for a 1 ×1 structure and 100–130 beams for a c(2 ×2) or p(2 ×1) structure.

The search for the correct structural model requires computer programs that allow arbitrary choices of atomic coordinates in one or more layers on the surface. However, the complexity of surface structures that can arise is so large that only a limited number of choices have been taken into consideration so far. We have been using for several years four basic computer programs that were written and tested by D.W. Jepsen and P.M. Marcus of the IBM Research Center. Two of these programs (called SUB and BASE) are used for substrates with Bravais lattices, the other two (called HEX and HEXBAS) for substrates involving double layers as in the hcp or in the diamond structure. SUB allows treatment of a Bravais surface layer whose registry with and distance from the substrate can be varied, while the substrate has a Bravais lattice whose first interlayer spacing can also be varied. BASE allows treatment of a surface layer with a basis of two atoms, which can be different from one another and from the substrate's and need not lie in the same plane parallel to the surface, while the substrate has the same attributes as in SUB. In BASE the multiple scattering between atoms in the layer is handled in a spherical-wave basis using an Ewald procedure. HEX is analogous to SUB but the substrate is built of double layers as indicated above. HEXBAS is analogous to BASE but, again, the substrate involves double layers. A few examples of the uses of these programs are the following: Simple overlayer systems such as Ag{001}c(2×2)–Cl [18], Mo{001}1×1–Si [16], or Fe{001}c(2×2)–S [40] can be handled adequately by SUB; diatomic surface structures such as Mo{001}1×1–CO [25] or those involving chemisorption of oxygen molecules on molybdenum require the use of BASE, and so does the Fe{001}1×1–O structure, in which the first O–Fe interlayer distance is very small [3]; the clean {0001} surfaces of Ti or Be call for the use of HEX, and so does the single layer of Cd on Ti(0001) [5] and the unreconstructed {111} surface of silicon [9]; the "underlayer" structure Ti{0001}1×1–N [6], the structure of a two-layer film of Cd on Ti{0001} [26] or the ordered structures of CO and NO on Ti{0001} [27] require the use of HEXBAS.

In general, the computation times are considerable. They depend, of course, on the type of program, the number of beams and phase shifts, and the computer used. A comparison between the efficiencies of two different programs is difficult to make unless the two programs are run on the same computer with exactly the same input data — a hardly realizable situation. Nevertheless, in order to provide some pertinent information, we have carried out a survey of the characteristics of the computer programs used by the groups involved in LEED crystallography. The results, for those groups which chose to answer the survey questions, are summarized in table 2. Note that, to simplify matters, the survey was limited to the calculations of a c(2×2) or a p(2×1) structure on fcc{001}, i.e., a structure with one integral- and one fractional-order beam per reciprocal-net mesh.

It is obvious that considerable savings are possible by taking advantage of the high symmetry (i.e., beam degeneracies) at normal incidence. Also, it appears that there are no great differences between different programs, if the input data are equal of similar. All groups take advantage of the fact that a relatively modest investment of additional time (5–10%) allows the calculation of structural models with a different interlayer spacing or a different geometry.

3.3. Evaluation of structural models

The third step in LEED structure analysis consists of the comparison between calculated and observed intensity spectra. This comparison has been done almost

Table 2
Typical LEED programs for a c(2 × 2) or p(2 × 1) overlayer structure on fcc {001} (energy range approx. 20–150 eV)

Program by	Total number of beams	Number of phase shifts	Computer	Storage size	Time per point (sec)	Time per additional interlayer spacing (or geometry)
Pendry [a]	25	5	IBM 165	200K [b]	Sym. 0.15 RFS 3 layer doubl. 5	
Kesmodel and Baetzold [c]	33	5	CDC 7600	260 K words [d]	9	5.5%
Van Hove and Tong [e]	13 max. [f] 49 max.	8 max. [f] 8 max.	UNIVAC 1110 UNIVAC 1110	40 K words [g] 80 K words	{15 at 60 eV, 19 at 100 eV, 23 at 150 eV} (h)	~10%
Prutton [i]	40 max. [j]	7 max. [j]	DEC 10	88 K words	90 [j]	
Jepsen and Marcus [k] (SUB program)	24	5	IBM 168	(700K est.) [l]	9.9 [m]	
	26	5	IBM 168		10.8	4%
	34	5	IBM 168		14	4.2%
	58	8	IBM 168		63	7%
	102	8	IBM 168		178	8%

a   J.B. Pendry, private communication (1977). See also Pendry, Low Energy Electron Diffraction (Academic Press, 1974).
b   K = kilobyte = 10^3 bytes (IBM).
c   L.L. Kesmodel, private communication (1977).
d   1 word CDC = 60 bits = 7.5 bytes IBM. The core size quoted assumes a maximum of 61 beams.
e   M. Van Hove and S.Y. Tong, private communication (1977).
f   The program involves a variable number of beams and phase shifts, depending on the accuracy required, the energy and the symmetry. At normal incidence and 150 eV, 13 beams and 8 phase shifts are included. At arbitrary angles of incidence and 150 eV, 49 and 8 are the corresponding maximum numbers of beams and phase shifts.
g   1 word Univac = 4 bytes IBM. The 40 Kwords core applies to the case of normal incidence, the 80 Kwords core to arbitrary angles of incidence.
h   Since the numbers of beams and phase shifts are variable with energy, so is the time. The numbers quoted in the table apply to the case of normal incidence only. For completely arbitrary angles of incidence the times are multiplied approximately by 4.
i   M. Prutton, private communication (1977).
j   For arbitrary angles of incidence, energy range 100–250 eV.
k   D.W. Jepsen and P.M. Marcus, in: Computational Methods in Band Theory, Eds. P.M. Marcus, J.F. Janak and A.R. Williams (Plenum, New York, 1971) pp. 416–43.
l   The pertinent part of the program is written in PL/I (not in FORTRAN), hence the dimension statements are not fixed and the storage size varies with the input data, i.e., it is smaller for, say, 34 than for 58 beams.
m   These times apply exclusively to the case of arbitrary angles of incidence.

exclusively by eye, so far, and models have been called correct or incorrect on the basis of visual evaluation of the correspondence between calculated and observed intensity spectra.

In this evaluation, no attention is paid to the magnitudes of the absolute intensities. This action is justified on several grounds. On the theoretical side, the calculation of reliable absolute intensities probably requires some consideration of the fact that the surface of the crystal is not made of ideally flat planes, i.e., consideration of the analogue of mosaic effects. On the experimental side, the magnitude of the diffracted intensities depends on the aperture of the detector used for the measurement, an again on the nature and the extent of (usually unknown) morphological and crystallographic "roughness" of the surface studied. Practically nothing is known about the effect of surface roughness upon LEED intensities [28,29]. All studies carried out to date indicate that the observed intensities are smaller (by factors varying from 1 to 100) than the calculated intensities.

Comparison is, therefore, limited to the overall shape of the intensity spectra, the relative intensities of peaks and, especially, the peak positions. A reliable analysis requires consideration of several structural models and, for each model, consideration of several values of the interlayer spacings on the surface. It turns out that most intensity spectra do not change much with, i.e., are insensitive to, small variations of the first or second interlayer spacing on the surface, while other spectra are indeed sensitive to such variations. It would be useful, of course, to know a priori which beams are most affected by interlayer spacings in any given structure, so that one could concentrate attention on them. Unfortunately, we have not been able to make such predictions or to develop rules of general validity for all structures. As an initial attempt at documenting our experience in this area, we present in tabel 3 a list of those beams from several surface structures that we have found to be particularly sensitive to variations of the first interlayer spacing. Perhaps the only guide that may appear from this table is that the 11-type beams occur more often than others, followed by the 00 beam.

Identification of the correct structural model may be ambiguous if only very few beams are tested. It is not uncommon for models involving different atomic geometries to produce two or three intensity spectra that are almost equally acceptable in terms of the agreement with experiment (see, e.g., refs. [16] and [18]. The analysis must, therefore, include more than three beams and more than one angle of incidence. We have often considered 4 to 8 beams at three different angles (usually $\theta = 0°$, $\theta = 10°$ and $\theta = 20°$ at a single value of the azimuth ϕ) for a grand total of 16 to 24 spectra. An adequate analysis can be carried out with 4 to 5 beams at two angles (say, $\theta = 0°$ and $\theta = 20°$) for a total of 9 to 10 spectra, which should probably be considered a minimum. The choice of angles of incidence θ is unfortunately limited in most display-type systems not provided with a Faraday cage, as mentioned above, to the range 0° to about 24–25°. Very few calculations and comparisons with experiment have been done for larger values of θ, although it is likely that many spectra would show much more sensitivity to structural parameters in the high range than in the low range of θ values.

Table 3

LEED beams that were found to be particularly sensitive to variations of the first interlayer spacings in the structures indicated

Surface structure	Beams and angles	Remarks
Fe {001}	21 at $\theta = 0°$ 21 at $\theta = 20°, \phi = 0°$	Changes in range 80–120 eV Peak positions change in range 120–150 eV
Mo {001}	10 at $\theta = 8°, \phi = 0°$ 11 at $\theta = 8°, \phi = 0°$	Moderate changes for 0.1 Å, large changes for 0.2 Å compression
Co {001}	11 at $\theta = 10°, \phi = 62.5°$	
Fe {111}	11 at $\theta = 0°$ 00 and 10 at $\theta = 0°$	Large changes beyond 0.1 Å compression
Ti {0001}	11 at $\theta = 0°$ 00, $\bar{2}2$, $\bar{3}2$ at $\theta = 20°, \phi = -30°$	
Ag {001} c (2 × 2) –Cl	$\bar{1}\bar{1}$ at $\theta = 20°, \phi = 129.5°$	
Fe {001} c (2 × 2) –S	$\bar{2}0$, 1/2 1/2 at $\theta = 10°, \phi = 0°$	Changes in range 80–140
Mo {001} 1 × 1 – Si	00, 11 at $\theta = 8°, \phi = 0°$	
Ti {0001} 1 × 1 – Cd	11 at $\theta = 0°$ 00, 11, $\bar{2}2$ at $\theta = 7°, \phi = -30°$	
Ti {0001} 1 × 1 – N (underlayer)	11 at $\theta = 0°$ 00, 11, $\bar{2}2$ at $\theta = 8°, \phi = -30°$	

4. Improvements

The brief review presented above indicates that several features in LEED crystallography need improvement. In some cases, the improvement may be only desirable, in other cases, it appears to be necessary. We discuss in the following some of the developments that are likely to occur in the field in the near future.

4.1 Experimental

There are, as we mentioned above, distinct advantages of the display-type LEED apparatus over the Farady-cage option. It is well known, however, that the most popular display-type machines in use today were designed only for qualitative observations of LEED pattern and are, therefore, not suitable for the precise quantitative measurements that are required in LEED crystallography. Some of the most important limitations of this type of equipment, mentioned in part above, are summarized in the following: (i) The angle of incidence θ is limited to the range 0–25°. (ii) The visibility of the fluorescent screen is substantially reduced if the sample holder is as large and bulky as may be required by the need for precise adjustments of the sample orientation. (iii) Most sample holders do not allow in situ selection of a given value of the azimuth ϕ, and almost none allow in situ adjustment of both the azimuth and the tilt of the surface plane. (iv) Unless suitable protective shields are introduced, the fluorescent screen is easily contaminated during experiments in which condensable vapors are deposited on the sample surface.

It would be desirable to remove these limitations in the existing LEED systems, but is is unlikely that this goal can be achieved without redesigning the systems completely. Several solutions are possible, of course. One that has the advantage that it has been shown to work satisfactorily is the LEED goniometer designed and built by De Bersuder [30] and used routinely by the Grenoble group [31]. At the present time, the scheme developed by De Bersuder seems to be the best for precise LEED crystallographic work.

Whatever the design, however, it is the collection of intensity data that needs drastic improvements. There are at least three reasons for these improvements: (i) To insure that the surface is not contaminated by the residual atmosphere during data collection. (ii) To enable one to study adsorbates that are (relatively slowly) dissociated by the incident electron beam. (iii) To allow studies of transient phenomena on surfaces. We already mentioned above that the photographic technique developed by the Somorjai's group [21] is a considerable advance over the manually operated spot-photometer method. The ideal solution, however, must be much faster and, especially, on-line, i.e. the intensity spectra must be available while the experiment is in progress. Image-analyzing schemes involving Vidicon cameras exist and are available commercially (at very high prices). Their application to LEED crystallography has been described by at least two groups [32,33], but the present systems do not seem to be sufficiently fast to satisfy the requirements. One of the problems is that considerable time is wasted in digitizing portions of the LEED pattern (i.e., the background) that are not needed in present-day LEED crystallograhy — in fact, at any incident electron energy between 30 and, say, 300 eV the total area occupied by the diffracted beams is only a small portion of the total. The goal is to record the intensities, of say, 9 or 10 beams simultaneously over the range of energies from 30 to 200 eV in times of the order of seconds or less (currently available image-analyzing systems may accomplish this task in about 30 min). It may be necessary, for this purpose, to increase the overall intensity of the LEED pattern by means of suitable channel plates or other image-intensifying devices.

Finally, more attention will have to be devoted to the morphological, chemical and crystallographic characterization of the surface under study. To be sure, the lack of such characterization has not hindered structure analysis in the past, but it may have affected the accuracy of results that could have been more precise. In addition, quantitative correlation between, e.g., surface roughness and LEED intensities is likely to open the way toward a much needed study of surface defects, cluster sizes etc. Despite early theoretical attempts by Laramore et al. [28] and recent applications of that theory to a specific case [29], practically nothing is known about the effects of surface steps on LEED intensities.

Chemical characterization of the surface is in better shape than morphological characterization, thanks primarily to Auger electron spectroscopy (AES). However, it would be desirable to make AES much more reliably quantitative. In many cases, information about the precise concentration of adsorbates on a surface would be an invaluable help in the work toward solution of a structural problem. Finally, characterization of the degree of crystallinity achieved on a surface would help establishing the effects of amorphous patches on LEED beams' and background intensities.

4.2. Theoretical

The most urgently needed developments in the theoretical and computational areas can perhaps be merged in four groups:

(1) Computer programs should be available that allow arbitrary distortions in each of, say, the first four or five layers of a surface, with one or more atoms per unit mesh. In other words, the flexibility that programs such as BASE and HEXBAS offer for the first layer (see section 2.2) should be extended to the bulk.

(2) Systematic and rational search procedures should be developed to facilitate the solution of structural problems and thus reduce the demands on computation time. For example, given a model which produces S beams in satisfactory agreement with experiment and U beams with unsatisfactory agreement with experiment, how should the model be modified to increase S and decrease U? Can we learn something about the next step in the analysis even if $S = 0$? Can we infer something about the structural model of a superstructure by comparing the experimental spectra of integral-order beams for the clean surface with the corresponding spectra of the superstructure? Can we identify a priori those ranges of electron energy, angle of incidence θ or azimuth ϕ, and those beams which are particularly suitable and sensitive to structural parameters?

(3) New ideas aimed at substantial decrease of the computation time should be developed. Also, the model of the solid used for calculation, e.g., the treatment of phonon effects, could be improved, thus improving the agreement with experiment.

(4) The treatment of LEED on disordered (amorphous) structures should be pursued quantitatively and the effects of mono- or poly-atomic surface steps on LEED intensities should be investigated.

4.3 Evaluation

Several workers have been trying for some time to introduce objectivity and quantification in the comparison between calculated and observed LEED spectra. The task is complicated by a number of requirements: comparison of shapes, curvatures and relative peak heights without consideration of absolute intensities; neglect of the background of the experimental spectra; emphasis on peak positions, on high and narrow peaks over low and broad peaks, etc.

Several reliability factors have been introduced by a number of workers, each of which attempts to evaluate one given feature of the curves to be compared [34]. Many or all of them can be used to reach a decision about the probability of correctness of a given structural model. Recently, a reliability factor has been defined that purports to evaluate and weigh in a single formula *all* important features of a curve as compared to another [35]. Systematic use of this reliability factor (and others) is just beginning.

It is, therefore, too early to draw conclusions about advantages and disadvantages of any specific reliability factor. If none of those that have been proposed works out satisfactorily, others will have to be invented. There is no doubt, in any case, that quantitative and objective measures of reliability must be introduced and used to assess the reliability of structural models, to choose the most probable ("correct") model and to refine it with respect to a variety of structural and non-structural parameters.

4.4. Problems

While work similar to that done in the past is likely to go on and to continue producing interesting examples of atomic arrangements on solid surfaces, several problems, both old and new, should be studied and solved.

One of the most difficult "old" problems is that of reconstructured surfaces. After much speculation during the past twenty years some significant progress is now being made in determining atomic arrangements on reconstrcuted Si surfaces [9–11], although the complete solution is probably still far.

In the area of chemisorption on low-index metallic surfaces two types of developments are awaited with interest. One is the study of superstructures with multiplicities larger than 2 (e.g., 3×3, 5×1, etc.). The other is the study of surface structures involving more than one layer different from the bulk, e.g., two, three or four layers. As far as we know, only one such study has been carried out so far [5,26]. More work in this direction is likely to contribute substantially to the understanding of epitaxy and of the intermediate stages of oxidation.

Surfaces of binary and ternary compounds are also likely to be more intensely studied than heretofore. Only a few binary compounds have been scrutinized so far [12–15,39], but more work is expected on the surfaces of III–V and II–V semiconductors. Finally, it would be interesting and productive to pursue the problem of amorphous surface films, if both the theoretical and experimental tools of LEED crystallography can be adapted to the task.

Acknowledgements

The work of the author in the field of LEED crystallography has been done in collaboration with a number of people over a period of several years. These people are, at the IBM Research Center in Yorktown Heights, D.W. Jepsen and P.M. Marcus, and at Stony Brook (in alphabetical order). A. Ignatiev, K.O. Legg, H.D. Shih, J.A. Strozier, Jr. and E. Zanazzi, The author is also indebted to three agencies for partial support of LEED research: the National Science Foundation, the Army Research Office and the Air Force Office of Scientific Research. For the compilation of table 2, the author is grateful to the people who graciously answered his survey questions and allowed him to use the answers for this article, i.e., J.B. Pendry, L.L. Kesmodel and G.A. Somorjai, M. Van Hove and S.Y. Tong, and M. Prutton.

References

[1] P.M. Marcus, in: Advances in Characterization of Metal and Polymer Surfaces, Ed. L.H. Lee (Academic Press, 1976).
[2] F. Jona and P.M. Marcus, Comments on Solid State Physics 8 (1977) 1.
[3] K.O. Legg, F. Jona, D.W. Jepsen and P.M. Marcus, J. Phys. C8 (1975) L492.
[4] M.A. Van Hove and S.Y. Tong, Surface Sci. 52 (1975) 673.
[5] H.D. Shih, F. Jona, D.W. Jepsen and P.M. Marcus, Communications on Physics 1 (1976) 25; and submitted to Phys. Rev. B (1977).
[6] H.D. Shih, F. Jona, D.W. Jepsen and P.M. Marcus, Phys. Rev. Letters 36 (1976) 798; Surface Sci. 60 (1976) 445.
[7] P.M. Marcus, J.E. Demuth and D.W. Jepsen, Surface Sci. 53 (1975) 501.
[8] The structure proposed by W. Heiland, F. Iberl and E. Taglauer, Surface Sci. 53 (1975) 383, for Ag{110}2 × 1−O does not seem to be convincing from the point of view of LEED analysis (M. Maglietta, E. Zanazzi, F. Jona, D.W. Jepsen and P.M. Marcus, to be published).
[9] H.D. Shih, F. Jona, D.W. Jepsen and P.M. Marcus, Phys. Rev. Letters 37 (1976) 1622.
[10] S.J. White and D.P. Woodruff, J. Phys. C9 (1976) L451.
[11] F. Jona, H.D. Shih, A. Ignatiev, D.W. Jepsen and P.M. Marcus, J. Phys. C10 (1977) L67.
[12] A.R. Lubinsky, C.B. Duke, B.W. Lee and P. Mark, Phys. Rev. Letters 36 (1976) 1058.
[13] C.B. Duke and A.R. Lubinsky, Surface Sci. 50 (1975) 605;
A.R. Lubinsky, C.B. Duke, S.C. Chang, B.W. Lee and P. Mark, J. Vacuum Sci. Technol. 13 (1976) 189;
C.B. Duke, A.R. Lubinsky, B.W. Lee and P. Mark, J. Vacuum Sci. Technol. 13 (1976) 761.
[14] K.O. Legg, M. Prutton and C. Kinniburgh, J. Phys. C7 (1974) 4236;
C.G. Kinniburg, J. Phys. C8 (1975) 2382; 9 (1976) 2695.
[15] B.J. Mrstik, S.Y. Tong, R. Kaplan and A.K. Ganguly, Solid State Commun. 17 (1975) 755.
[16] A. Ignatiev, F. Jona, D.W. Jepsen and P.M. Marcus, J. Vacuum Sci. Technol. 12 (1975) 226; Phys. Rev. B11 (1975) 4780.
[17] F. Jona, IBM J. Res. Develop. 14 (1970) 444.
[18] E. Zanazzi, F. Jona, D.W. Jepsen and P.M. Marcus, Phys. Rev. B14 (1976) 432.
[19] H.D. Shih, F. Jona, D.W. Jepsen and P.M. Marcus, J. Phys. C9 (1976) 1405.

[20] L.L. Kesmodel, P.C. Stair, R.C. Baetzold and G.A. Somorjai, Phys. Rev. Letters 36 (1976) 1316.
[21] P.C. Stair, T.J. Kaminska, L.L. Kesmodel and G.A. Somorjai, Phys. Rev. B11 (1975) 623.
[22] J.A. Strozier, Jr., D.W. Jepsen and F. Jona, in: Surface Physics of Materials, Vol. 1 (Academic Press, New York, 1975) p. 1.
[23] S.Y. Tong, in: Progress in Surface Science, Vol. 7 (Pergamon, Oxford, 1975).
[24] A. Ignatiev, F. Jona, M. Debe, D.C. Johnson, S.J. White and D.P. Woodruff, J. Phys. C10 (1977) 1109.
[25] F. Jona, A. Ignative, D.W. Jepsen and P.M. Marcus, Bull. Am. Phys. Soc. (II) 20 (1975) 407.
[26] H.D. Shih, F. Jona, D.W. Jepsen and P.M. Marcus, Phys. Rev. B (1977), in press.
[27] H.D. Shih, Ph.D. Dissertation, State University of New York at Stony Brook (1976); H.D. Shih, F. Jona, D.W. Jepsen and P.M. Marcus, to be published (1977).
[28] G.E. Laramore, J.E. Houston and R.L. Park, J. Vacuum Sci. Technol. 10 (1973) 196.
[29] E. Zanazzi, F. Jona, D.W. Jepsen and P.M. Marcus, J. Phys. C10 (1977) 375.
[30] L. De Bersuder, Rev. Sci. Instr. 45 (1974) 1569.
[31] D. Aberdam, R. Baudoing and L. De Bersuder, Rev. Sci. Instr. 45 (1974) 1573.
[32] P. Heilmann, E. Land, K. Heinz and K. Muller, Appl. Phys. 9 (1976) 247.
[33] D.C. Frost, K.A.R. Mitchell, F.R. Shepherd and P.R. Watson, J. Vacuum Sci. Technol. 13 (1976) 1196.
[34] M.A. Van Hove and S.Y. Tong, Phys. Rev. Letters 35 (1975) 1092.
[35] E. Zanazzi and F. Jona, Surface Sci. 62 (1977) 61.
[36] M.B. Webb and M.G. Lagally, Solid State Phys. 28 (1973) 301.
[37] T.A. Clarke, R. Mason and M. Tescari, Proc. Roy. Soc. (London) A331 (1972) 321.
[38] D.L. Adams and U. Landman, Phys. Rev. Letters 33 (1974) 585;
U. Landman and D.L. Adams, J. Vacuum Sci. Technol. 11 (1974) 195;
D.L. Adams, U. Landman and J.C. Hamilton, J. Vacuum Sci. Technol. 12 (1975) 260.
[39] C.G. Kinniburg and J.A. Walker, Surface Sci. 63 (1977) 274.
[40] K.O. Legg, F. Jona, D.W. Jepsen and P.M. Marcus, Surface Sci. 66 (1977) 25.
[41] To avoid misunderstandings it may be necessary to point out that, in this context, "beam" means a reciprocal-lattice vector of the surface net.

LEED CALCULATIONS OF EXCHANGE REFLECTIONS FROM ANTIFERROMAGNETIC NiO(100)

J.A. WALKER, C.G. KINNIBURGH * and J.A.D. MATTHEW

Department of Physics, University of York, Heslington, York YO1 5DD, England

Multiple scattering LEED calculations have been performed for the intensities of the half-order features of the NiO(100) surface, assuming different exchange potentials on the different magnetic sublattices. Approximate methods have been tested on a model NaCl like structure, in which the different sublattices have opposite spin, and for which exact calculations can be performed. These techniques, thus validated, have been applied to the NiO(100) surface, and comparisons are made with the limited experimental data currently available.

1. Introduction

The existence of half-order features in the diffraction pattern of the NiO(100) surface has been reported by Palmberg et al. [1] and others. A recent paper [2] has shown that this surface is a simple termination of the bulk accompanied by relatively small relaxation and rumpling, therefore ruling out the possibility that the extra beams are dominantly geometrical in origin. These extra beams, which are observed for energies less than 100 eV and for temperatures below the Néel temperature (T_N = 520 K), are thought to be associated with magnetic ordering of the crystal. Below the Néel temperature the exchange interaction between the spin of an incoming electron and the spins of nickel ions on the two magnetic sublattices will be different, hence causing a doubling in the periodicity of the lattice.

At the present time only very limited calculations exist for the exchange scattering. Palmberg et al. used the first order Born approximation, an approximation which is not valid in the energy range considered, to calculate the ratio of the forward exchange and Coulomb scattering for a single nickel ion. This contrasts with a LEED experiment where it is the backscattered wavefield which is actually observed. Namikawa et al. [3] have investigated the coupling between the half-order and integer beams by solving a three beam problem, but it is well known that many beams are necessary to represent the wavefield in a LEED calculation.

In applying LEED theory to antiferromagnetic NiO, we were faced with two problems. The first of these is how to accommodate the exchange scattering on differ-

---

* Cavendish Laboratory, Madingley Road, Cambridge, CB3 0HE, England

ent nickel sublattices within the muffin-tin approximation, and this is discussed in section 2. The second problem is that antiferromagnetic NiO has 4 atoms per unit cell, and to calculate the layer scattering matrices for such a structure would be prohibitive in terms of the amount of computer time necessary. Because the difference in the t-matrix for nickel ions on the different magnetic sublattices is small it was thought that it should be possible to find an approximate method of including the exchange scattering within the framework of a two atom per unit cell calculation. Such an approximation is discussed and tested in section 3 and, thus validated, applied to NiO in section 4.

2. Model of the potential

The free Ni^{2+} ion is in the spectroscopic state $^3F_4$, the one electron orbitals having occupations $(1s)^2$ $(2s)^2$ $(2p)^6$ $(3s)^2$ $(3d)^8$. Assuming the 3d electrons in NiO are localised with two unpaired spins, the crystalline field would be expected to quench the orbital contribution to the magnetic moment, and the resultant should essentially be 2 μ_B, the spin value only. This is confirmed by the neutron diffraction studies on NiO of Roth [4,5].

Electrons in a particular angular momentum state l can have $(2l + 1)$ values of the z component of angular momentum m_1, with two spin states for each. In order to preserve spherical symmetry the muffin-tin approximation places the electrons in a partially filled sub-shell equally amongst the $2(2l + 1)$ orbitals. There the Ni^{2+} ion in NiO will have 0.8 electrons in each of the ten 3d orbitals. If the exchange potential is calculating using the Slater formula,

$$V_{ex}(r) = -3\alpha[\tfrac{3}{4}\pi\rho_{11}(r)]^{1/3},$$

where $\rho_{11}(r)$ is the density of electrons with spins parallel to that of an incident electron, then there will normally be four of the 3d electrons contributing to the exchange potential. A more detailed discussion on the construction of the muffin-tin potential is given in an earlier paper [2].

The way we have chosen to represent the magnetic interaction is to adjust the exchange potential so that five 3d electron contribute to the exchange potential of the Ni^{2+} ions on one magnetic sublattice, and three 3d electrons on the other. This has been done in two different ways:
(i) By including five 3d electrons in the electron density $\rho_{11}(r)$ for the Ni^{2+} ion on one magnetic sublattice and three on the other.
(ii) By adding a constant potential U_{ex} to the Ni^{2+} muffin tin sphere on one magnetic sublattice, and subtracting U_{ex} from the muffin tin sphere of the other, where U_{ex} is equal to the average extra echange potential seen by a free electron.
Following Mattheiss [6],

$$U_{ex} = \tfrac{1}{3}\langle k|(\delta\rho_{11}/\rho_{11})V_{ex}|k\rangle,$$

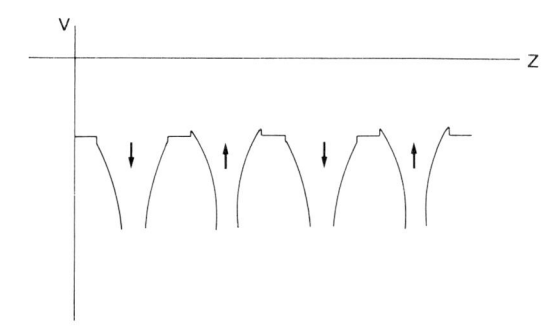

Fig. 1; Muffin tin potential including different exchange potentials for the nickel ions on different magnetic sublattices.

where $2\delta\rho_{11}$ is the difference in magnetic charge density on the two sites. These are shown diagrammatically in fig. 1. By averaging the exchange potential over the Ni^{2+} muffin tin sphere in NiO it is found that $U_{ex} = 2.9$ eV.

3. Approximate methods and their application to a model structure

To reduce computational time it was felt necessary to try and find an approximate method of calculation within the framework of a two atom per unit cell structure. This was found to be possible if the two following assumptions are made:
(i) In the calculation of the intralayer multiple scattering, to determine the amplitude of the wavefield at the position of each atom in the unit cell, the difference in the t-matrix for nickel ions on different magnetic sublattices is neglected.
(ii) The scattering from half order to half order beams and from half order to integer beams is treated kinematically.

In the usual notation the intralayer scattering can be written,

$$[I - (t + \Delta t)G]^{-1} (t + \Delta t) \simeq (I - tG)^{-1} (t + \Delta t)$$

$$+ (I - tG)^{-1} \Delta t G (I - tG)^{-1} t,$$

to first order in Δt.

The first approximation corresponds to considering only the first term in the expansion, since the second term cannot be readily evaluated without treating the full four atom per unit cell problem. The second approximation arises from expanding G into a form compatible with a two atom per unit cell calculation.

At first sight these seem rather drastic approximations, and it is necessary to test their validity on a simpler structure for which exact calculations can be performed. This was done on a NaCl type structure which consists of Ni^{2+} ions on two magnetic sublattices, the spin on the ions being reversed on each. The value of the lattice constant used in this calculation was 4.168 Å. This value was adopted for two reasons, firstly it is the lattice constant for NiO [7], and secondly in terms of the multiple scattering between the nickel ions it is a less dilute lattice than NiO. Therefore, validation of the approximations on this structure would give confidence in their application to NiO.

The (10) beam (the first "half order beam") for this model calculation is shown in fig. 2, the potential model (i) being used. Curve (a) corresponds to an exact calculation, curve (b) includes approximation (i), and curve (c) includes approximations (i) and (ii). The calculations were performed at normal incidence and for zero relaxation of the surface layer. A layer multiple scattering approach was used as described elsewhere [2], with a maximum of seven phase shifts and twenty symmetrised beams. There is clearly good agreement between the exact and approximate calculations both in terms of peak shapes and positions, particularly for the low energy peak around 35 eV. The absolute intensities of the approximate calculations are also comparable with those of the exact calculation. The effect of the approximations on the integer beams was found to be a reduction in intensity rather than a change in peak shape and position.

Fig. 3. shows the results of a similar calculation for potential model (ii), for which $U_{ex} = 3.4$ eV. The same level of agreement is obtained as in the previous calculation, both in terms of peak positions, peak shapes, and absolute intensities. Calculations were also performed using this potential model for a series of values of

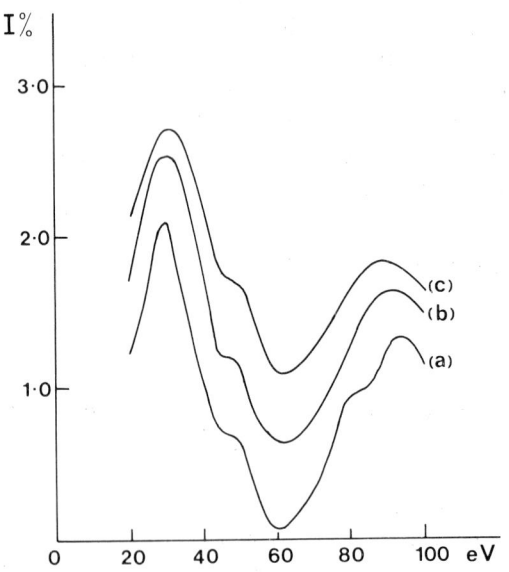

Fig. 2. Calculated spectra for the (10) beam at normal incidence for model potential (i). Curve (a) is an exact calculation, curve (b) includes approximation (i) and curve (c) includes approximations (i) and (ii). Curves (b) and (c) have been displaced vertically.

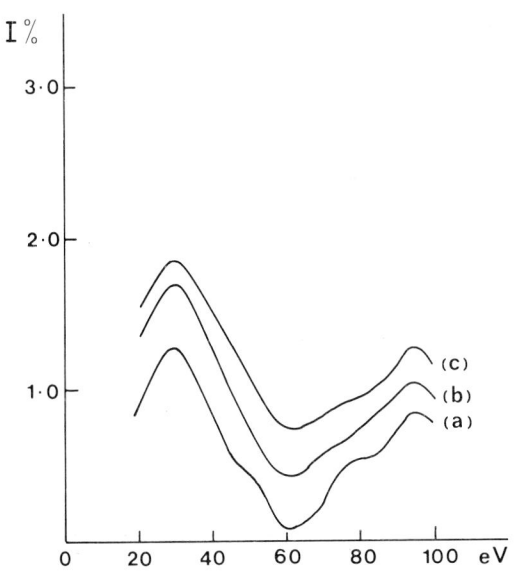

Fig. 3. Calculated spectra for the (10) beam at normal incidence for model potential (ii) with $U_{ex} = 3.4$ eV. Curve (a) is an exact calculation, curve (b) includes approximation (i) and curve (c) includes approximations (i) and (ii). Curves (b) and (c) have been displaced vertically.

U_{ex} ranging from zero to 27 eV. For values of U_{ex} of up to 3.5 eV the intensity of the main peak scales as the square of U_{ex} as would be expected from first order perturbation theory. For values of U_{ex} greater than this the intensity increases less rapidly with U_{ex}. For values of U_{ex} greater than 16 eV the approximations start to break down, particularly at low energies where the approximate calculations cease to converge.

Comparison of fig. 2 and fig. 3 shows that the spectra obtained using the two different potential models are very similar, the main difference being the values of the absolute intensities. The close agreement between the exact and approximate calculations imply that within the range of the differnce in exchange potentials required to represent the exchange scattering from NiO the dominant processes are the scattering of the integer beams into the half order beams, and that half order–half order scattering is of secondary importance. For both potential models used the main peak in the (10) beam corresponds closely with a peak in the (00) beam, therefore indicating strong coupling between the two.

4. Application to NiO

Because of the level of agreement obtained between the approximate and exact calculations discussed in the last section we felt confident in applying the approxi-

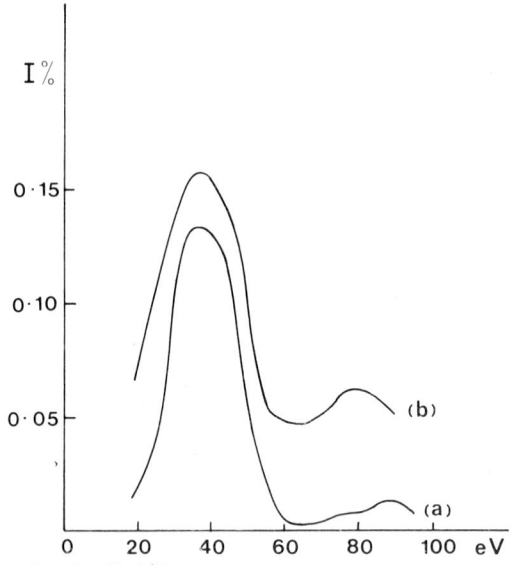

Fig. 4. The spectra for the (0 1/2) beam of the NiO(100) surface calculated at normal incidence. (Curve (a) for potential (i) and curve (b) is for potential (ii).)

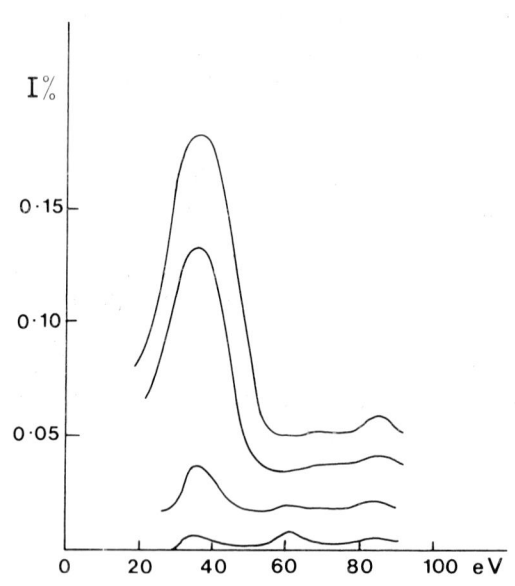

Fig. 5. Calculated spectra for the (0 1/2) beam for $\phi = 0°$ and θ ranging from $\theta = 0°$ (upper curve) to $\theta = 30°$ (lower curve).

mations to NiO. Fig. 4 shows the (0 1/2) spectra for NiO at normal incidnce and for zero relaxation of the surface layer. Curve (a) is for potential model (i) and curve (b) is for potential model (ii). The parameters of the potential are those of the ionic potential described in an earlier paper [2]. Both curves show a main peak in the 30–40 eV energy range, and this is in agreement with the limited experimental data currently available [1,8,9]. The ratio of the intensities of the (0 1/2) and (10) beam is of the order of 8%, whereas the experimentally measured ratio is of the order of 1–4%. Because the surface region covered by the incident beam is a multidomained structure the surface area contributing to the magnetic beam is only 50% of the area contributing to the integer beam intensity. Hence for a single domain the intensity of the magnetic beam is actually 2–8% of the intensity of the integer beam, in close agreement with the calculations.

Fig. 5 shows the (0 1/2) beam calculated using the potential model (i) for a range of angles of incidence and for zero relaxation of the surface layer. As the angle of incidence is decreased from $\theta = 30°$ to $\theta = 0°$ the peak height at 35 eV in the (0 1/2) beam can be seem to increase. This corresponds to similar increases in intensity in peaks in the (00) and (10) beams at the same energy, again demonstrating the coupling between the half order and integer beams.

The calculations discussed are in agreement with the very limited data currently available, and in order to test the potentials and approximations fully, it is necessary to compare the results with experimentally measured $I(V)$ spectra.

5. Discussion

In this paper we have discussed different theoretical models for the magnetic interaction of an electron beam within the framework of a muffin-tin model with inclusion of multiple scattering, and have tested approximate methods for the calculation of the intensity of the half order beams. The results are in agreement with the limited data currently available and confirm the importance of the coupling of the integer and half order beams in producing the intensity of the half order features.

The small value of the magnetic moment of the Ni^{2+} ion results in the validation of the approximations for NiO, but for a system with much larger magnetic moments it would be necessary to revert to an exact calculation of the layer scattering matrices.

The experimental results of Namikawa [9] have indicated that maxima in the half order beams can be observed under conditions of surface resonance, and this could provide scope for further work.

Until $I(V)$ spectra can be measured for the half order beams the usefulness of this type of calculation remains limited, but it is felt that, if the $I(V)$ spectra can be obtained, LEED could prove to be a useful tool in the investigation of the state of magnetisation of the surface.

Acknowledgements

The authors would like to thank Professor M. Prutton for some useful discussions concerning this work. We are also grateful for the support of the Science Research Council.

References

[1] P.W. Palmberg, R.E. de Wames and L.A. Vredovoe, J. Appl. Phys. 40 (1969) 1158.
[2] C.G. Kinniburg and J.A. Walker, Surface Sci. 63 (1977) 274.
[3] K. Namikawa, K. Hayakawa and S. Miyake, J. Phys. Soc. Japan 37 (1974) 733.
[4] W.L. Roth, Phys. Rev. 110 (1958) 1333.
[5] W.L. Roth, Phys. Rev. 111 (1958) 772.
[6] L.F. Mattheiss, Phys. Rev. B10 (1974) 995.
[7] R.W.G. Wyckoff, Crystal Structures, Vol I (Interscience, New York, 1960).
[8] K. Hayakawa, K. Namikawa and A. Miyake, J. Phys. Soc. Japan 31 (1971) 1408.
[9] K. Namikawa, private communication.

SPIN-POLARIZED LEED FROM LOW-INDEX SURFACES OF PLATINUM AND GOLD

Roland FEDER

Institut für Festkörperforschung, KFA Jülich, D-5170 Jülich, Germany

A relativistic theory of LEED has been applied, in the energy range from 0 to 200 eV, to (111) and (001) faces of Pt and Au. The calculated intensity versus energy profiles are in good agreement with experimental data, which, in particular, provides support for interpreting the observed surfaces as unreconstructed. As a consequence of strong spin-orbit coupling, large spin-polarization peaks are found, partly in conjunction with intensity maxima. As a rough guide to diffraction conditions, at which both polarization and intensity are sizeable, a simple kinematic model might be of some use.

1. Introduction

The agreement between recently observed spin polarization effects in low-energy electron diffraction (LEED) from W(001) [1] and calculations using a relativistic LEED theory appears, in spite of some discrepancies, to be very encouraging [2,3]. The calculations [2,3] further revealed a strong sensitivity of parts of the polarization profiles to a contraction of the top layer spacing, which suggests that spin-polarized LEED could be of use in surface structure determination. It therefore seems pertinent to extend previous theoretical work [2–4] on tungsten to other large-Z materials, for which spin-polarization effects can be expected to be large.

The present work focusses on low-index surfaces of Pt ($Z = 78$) and Au ($Z = 79$), for which experimental intensity data are already available and either not quite satisfactorily or not at all theoretically interpreted. In section 2, the theoretical model assumptions are outlined. In section 3, calculated intensity and polarization results are presented and discussed, and an attempt is made to retrieve some prominent LEED polarization features from atomic polarization results in a kinematic framework.

2. Theoretical model

Spin polarization and intensity spectra have been obtained by means of a relativistic theory of elastic LEED, which has been presented elsewhere [2,5]. With specific regard to Pt and Au, the following assumptions have been made. Relativistic ion-core scattering phase shifts were included upto $l = 7$. The actual spin-up and spin-down phase shifts $\delta_l^\pm$ were calculated from a muffin-tin band-structure potential obtained by overlapping relativistic atomic charge densities [6,7]. Temperature-corrected phase shifts were determined from the $\delta_l^\pm$ according to ref. [2] for $T = 300$ K using effective Debye temperatures of 178 K for Pt [8] and 116 K for Au [9]. The inner potential was assumed as constant, its real part being 14 eV and its imaginary part 4 eV, a choice used with success recently [10]. Both real and imaginary part of the surface potential barrier were taken as exponential-type smooth functions. The surface reciprocal lattice vectors included in the computations were automatically selected such as to ensure convergence. As for the number of monatomic layers, six (eight) were taken at lower (higher) energies. The primary beam is assumed as unpolarized and normalized to unit intensity. By "polarization" of a diffracted beam we mean, in the following, the projection of its spin-polarization vector on to the normal to the scattering plane.

3. Computational results and discussion

Since there are several discrepancies (with regard to relative peak heights) between recent experimental intensity spectra from Pt(111) and nonrelativistically calculated ones [10,11], it seemed worthwhile to apply the present theory to exactly the same diffraction conditions. As a consequence of using a relativistic formalism and a continuous surface barrier (as opposed to a "non-reflecting" abrupt barrier in ref. [10]), the agreement with the experimental data could be substanti-

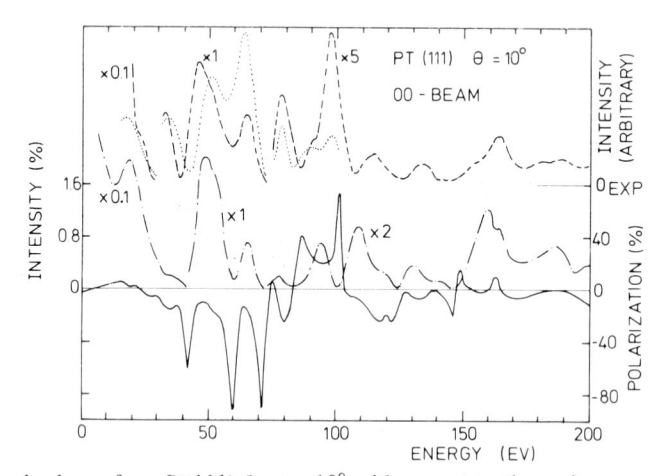

Fig. 1. Specular beam from Pt(111) for $\theta = 10°$ with respect to the surface normal. Intensity profiles from experiment [11] (– – –), a nonrelativistic calculation (· · · ·) and the present calculation (– · – · –), and polarization profile (———).

ally improved. Fig. 1. shows, as a typical example, specular beam results for $\theta = 10°$ (with respect to the surface normal) and azimuthal angle $\phi = 0$ (cf. ref. [10], fig. 3). Below 100 eV, relative intensity–peak heights from the present calculation can be seen to agree well with the data, in particular for the peaks near 50 and 65 eV as well as the two peaks near 78 and 95 eV. In the first case, an additional calculation using a non-reflecting barrier showed only a minor change, whilst in the second case it produced a result very close to the one calculated in ref. [10]. It thus seems that relativistic effects are responsible in the first case and properties of the surface barrier in the second. Above 100 eV, the present intensity spectrum also checks very well with the experiment. The ratios of the heights of peaks below 100 eV to the heights of peaks above 100 eV are, however, at variance by approximately a common factor. As for the spin polarization of the specular beam, peaks of about 90% are found near intensity minima and some minor peaks near intensity maxima (near 116 and 164 eV).

Recent LEED experiments on Pt(001), which normally exhibits a complicated superstructure, indicate that an unreconstructed [12] (meta)stable phase can be obtained by rapid cooling [8,13]. A LEED spectrum taken at this phase is shown, in fig. 2, to be reproduced accurately by a calculation on an unreconstructed [12] Pt(001) face [14]. In particular, the peak near 130 eV, the origin of which could not be explained by the "Bragg-type arguments" in ref. [8], is present in our results. It is therefore likely to be due to higher order multiple scattering processes. The close agreement of the calculated spectrum with the data from the one phase together with its clear distinction from the data from other phases (cf. ref. [8]) pro-

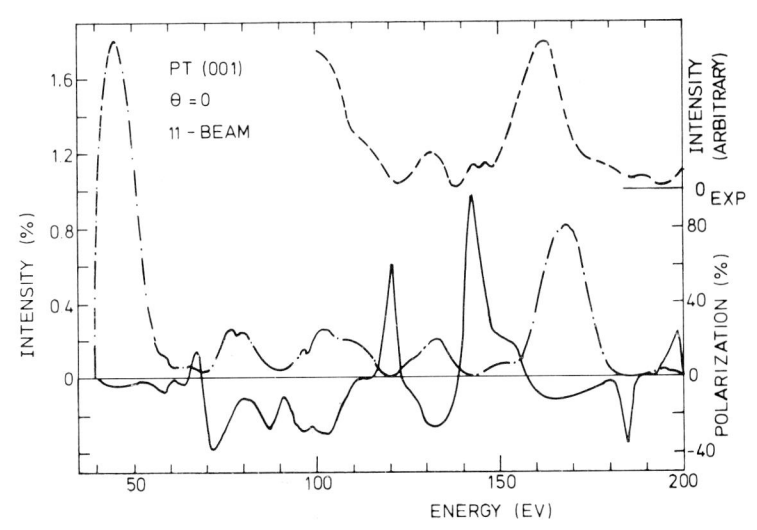

Fig. 2. 11 beam from Pt(001) for normal incidence. Experimental [8] (– – –) and theoretical (– · – · –) intensity profiles and (theoretical) polarization profile (———).

vides strong support that the phase under consideration was indeed unreconstructed [12]. The polarization profile obtained in the same calculation can be seen to exhibit a number of large peaks, out of which two fairly sizeable ones, around 103 and 132 eV, are correlated with intensity maxima. The occurrence of an unreconstructed phase of the (001) face has also been reported for Au [15]. Fig. 3 shows the agreement between an experimental intensity profile and its theoretical counterpart calculated for the unreconstructed face [16]. Again, large polarization features are found. In particular, a peak of 35% occurs in conjunction with an intensity maximum, which can be associated with a primary Bragg peak in a kinematic treatment.

Since high polarization together with high intensity is essential in view of using a LEED process for detecting spin-polarization (cf. ref. [17]) and for producing polarized electrons, a systematic theoretical procedure for locating "favourable" energy and beam orientation combinations would be very desirable. The obvious way, namely a scan over energy and angles using the present relativistic LEED theory, is however rather (computing-)time-consuming. It seems therefore worthwhile to investigate, how much information could already be obtained from a much simpler kinematic treatment. Within the framework of a kinematic theory, spin-polarization in LEED, as a function of energy and polar angle of incidence, is the same as for a single atom (in its solid environment) [18]. A favourable energy–angle combination would then simply be one, where a kinematic intensity peak coincides with an atomic polarization maximum. In order to illustrate and exploit this idea, fig. 4 shows an energy–angle contour plot of the single Au "muffin-tin atom" spin polarization together with primary Bragg peak locations for Au(001) and Au(111). Both angle of incidence (with respect to the surface normal) and energy have been

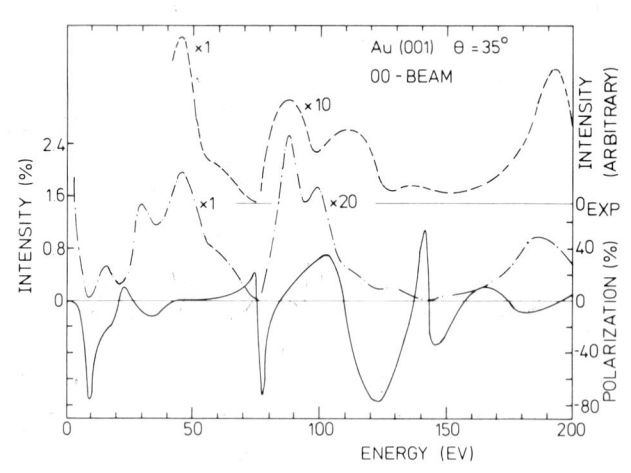

Fig. 3. Specular beam from Au(001) for $\theta = 35°$ and 01 azimuth. Experimental [15] (– – –) and theoretical (– · – · –) intensity profiles and polarization profile (———).

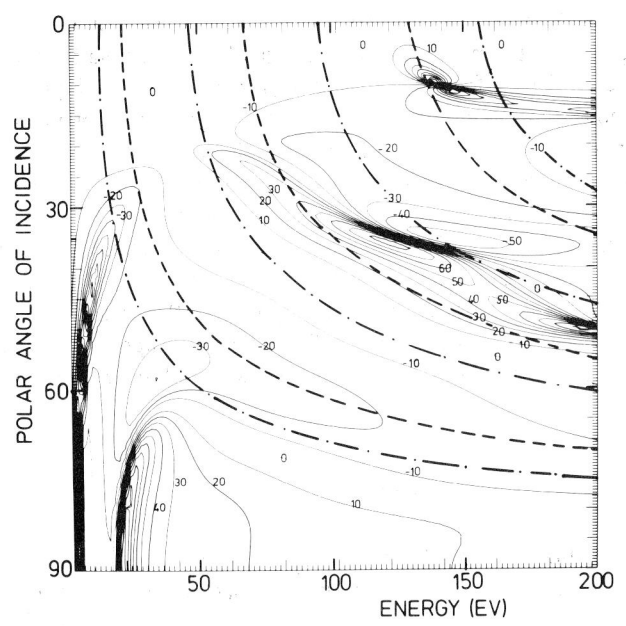

Fig. 4. Contours of constant polarization – in %, at 10% intervals – for electrons scattered from a single Au "muffin-tin atom". The dashed and dash-dotted lines indicate the position of kinematic Bragg peaks for Au(001) and Au(111), respectively. The angle and energy coordinates have been corrected for refraction by an inner potential of 14 eV.

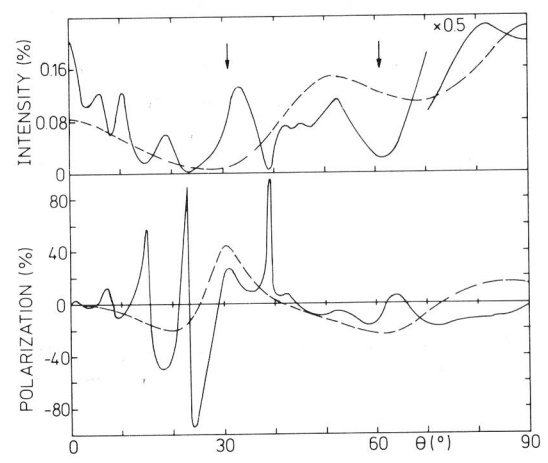

Fig. 5. Scattering cross section and spin polarization for a single Au "muffin-tin atom" (corrected for an inner potential of 14 eV (– – –) and intensity and polarization versus θ of the specular beam from Au(001) for $E = 90$ eV and the 01 azimuth (———). The two vertical arrows indicate kinematic peak positions.

corrected for an inner potential of 14 eV. Comparing a scan along the $\theta = 35°$ horizontal in fig. 4 with the polarization profile in fig. 3, the polarization at the Bragg positions at 34 eV and 100 eV in fig. 4 is seen to agree well with polarization peaks in fig. 3 associated with intensity maxima. An angle scan through fig. 4 at $E = 90$ eV is shown in fig. 5, together with the atomic scattering cross section and the intensity and polarization profiles obtained by means of the present LEED theory. In the vicinity of the Bragg peak location at 31 eV, at which the atomic polarization attains its maximum of 44%, the LEED intensity exhibits a peak associated with a polarization peak of 26%, i.e. "kinematic expectation" is still qualitatively confirmed. For the Bragg location at 61 eV, on the other hand, multiple scattering effects are dominant. Comparing in fig. 5 the overall shapes of atomic and crystal profiles, the latter can be viewed as atomic ones modulated by very strong "rapid" oscillations.

4. Conclusion

In summary, the present relativistic LEED calculations for Pt and Au have produced intensity versus energy profiles in good agreement with experimental data [19], thus providing theoretical evidence that the observed Pt(111), Pt(001) and Au(001) faces were indeed unreconstructed [12]. In the case of Pt(111), where a nonrelativistically calculated intensity profile was available for comparison [10], the present relativistic result shows significantly better agreement with experiment [20]. In accordance with the large spin-orbit coupling in Pt and Au, large spin polarization peaks were found, several of which are associated with intensity maxima [21]. Diffraction conditions, for which a kinematic model predicts high polarization and high intensity, were found to be "favourable" in the full multiple scattering theory as well in the cases of two Bragg maxima on Au(001). Though the outlined kinematic approach for finding "favourable" diffraction conditions is a priori restricted, it could still have its merits in providing — at virtually zero computing cost — some rough guideline for setting up experiments. Further calculations are in progress.

Acknowledgement

I would like to thank Dr. J. Kirschner for stimulating discussions, and Dr. O.K. Andersen for making the Pt and Au muffin-tin potentials available.

References

[1] M.R. O'Neill, M. Kalisvaart, F.B. Dunning and G.K. Walters, Phys. Rev. Letters 34 (1975) 1167.

[2] R. Feder, Phys. Rev. Letters 36 (1976) 598.
[3] R. Feder, Surface Sci. 63 (1977) 283.
[4] R. Feder, Phys. Status Solidi (b) 62 (1974) 135.
[5] R. Feder, Phys. Status Solidi (b) 49 (1972) 699.
[6] O.K. Andersen, private communication.
[7] The adequacy of such a superposition potential as compared to a more sophisticated self-consistent potential was explicitly demonstrated, in the case of Ni, by S.Y. Tong, J.B. Pendry and L.L. Kesmodel, Surface Sci. 54 (1976) 21.
[8] G.G. Waldecker, Dissertation, Erlangen (1976).
[9] D.P. Jackson, Surface Sci. 43 (1974) 431.
[10] L.L. Kesmodel and G.A. Somorjai, Phys. Rev. B11 (1975) 630.
[11] P.C. Stair, T.J. Kaminska, L.L. Kesmodel and G.A. Somorjai, Phys. Rev. B11 (1975) 623.
[12] The term "unreconstructed" is used in the present work in the sense that the topmost ordered metal layer is identical to a bulk layer both with regard to its two-dimensional structure and its position relative to adjacent layers.
[13] A clean Pt(001)-(1 × 1) surface is also reported by C.R. Helms, H.P. Bonzel and S. Kelemen, J. Chem. Phys. 65 (1976) 1773. Since these authors have not measured LEED intensities, it is however not possible to extract from their work any information regarding the position of the top (1 × 1) layer relative to the adjacent layers.
[14] The slight discrepancy with regard to peak positions above 140 eV can be easily remedied by using a reduced inner potential (rather than the static value of 14 eV).
[15] D. Wolf, Dissertation, München (1972).
[16] The otherwise excellent agreement is somewhat marred by the merging of the two theoretical peaks near 100 eV and 110 eV into a single experimental peak. The discrepancy could be due to slight deviations of the experimental angle of incidence from the nominal value $35°$.
[17] R. Feder, Surface Sci. 51 (1975) 297.
[18] For details, I refer to J. Kessler, Polarized Electrons (Springer, Berlin, 1976) p. 172.
[19] The slight misalignment of the intensity profiles above about 90 eV could be easily remedied by replacing the static inner potential value by a more realistic inner potential taking into account the energy-dependence of the exchange-correlation contribution (cf., e.g., D.W. Jepsen, P.M. Marcus and F. Jona, Phys. Rev. B5 (1972) 3933).
[20] It thus seems that, for heavy metals, relativistic effects are of importance for surface structure determination by means of LEED intensity analysis. A quantitative assessment of their overall significance requires, however, further relativistic and nonrelativistic computations. From a point of view of computational convenience, it is worthwhile to investigate in this context to what extent relativistic effects on LEED intensities can be reproduced by a nonrelativistic LEED calculation using spin-averaged relativistic phase shifts instead of the customary nonrelativistic phase shifts.
[21] An investigation of the sensitivity of the polarization features to variations in the model assumptions, in particular the phase shift input, the imaginary potential, the shape of the surface barrier and the Debye temperature, is currently in progress. Preliminary results indicate that the variations are fairly mild — except for specular beam sensitivity to the shape of the surface barrier — if the assumptions are varied in a physically reasonable way.

THE ROLE OF SURFACE SCIENCE EXPERIMENTS IN UNDERSTANDING HETEROGENEOUS CATALYSIS

H.P. BONZEL

Institut für Grenzflächenforschung und Vakuumphysik, Kernforschungsanlage Jülich, D-5170 Jülich, Germany

Heterogeneous catalysis is of enormous industrial importance for the large scale production of chemicals. A continuous research effort is devoted to improving the activity and selectivity of catalysts. Research is also expected to replace gradually empiricism by a more fundamental understanding of the governing factors of catalysis. Surface science in particular is capable of providing this fundamental information because of the advent of a multitude of novel surface characterization techniques. These new surface analytical tools operative under vacuum conditions are used to characterize catalytic surfaces with respect to composition, crystallographic and electronic structure. At the same time catalytic reactions are studied on these well characterized surfaces. A comparison of results from these studies with those obtained by more traditional catalysis research, e.g. at high pressure and on supported catalysts, turns out to be most interesting and informative.

1. Introduction

Heterogeneous catalysis as a chemical process plays a key role for the large scale production of plastics, liquid fuels, fertilizers, pharmaceuticals and other every-day chemicals. More recently heterogeneous catalysis has also been applied in the area of pollution control, in particular in the automotive industry. In the future one expects this process to play a major role in the gasification and liquefaction of coal in order to supplement the supply of natural gas and crude oil. Therefore heterogeneous catalysis will most likely continue to be a technology of central importance in the chemical and energy industries for a long time to come. In this prospect we see the major justification for a continuation of intense basic and applied research in this field.

The widespread industrial utilization of heterogeneous catalysis in the present and past was already responsible for many decades of successful research. Much progress has been made in recent years in the phenomenological description of heterogeneous catalysis but the whole subject has to a large extent remained in an empirical state. However, a more basic understanding of catalysis has always been a specific research goal of the catalytic chemist [1,2]. That goal has been very difficult to come close to because of the high degree of dispersion of the catalyst particles. For this reason the active surface of the catalyst is extremely difficult to characterize.

It has been recognized for a long time that the process of heterogeneous catalysis is governed by a multitude of surface parameters [3,4], such as crystallographic structure, surface composition, adsorption energies, migration energies of adsorbed species, dissociation energies, electronic structure and molecular orbital configurations, vibrational properties of adsorbates and others. In order to come to a more systematic description of heterogeneous catalysis, it would be desirable to measure as many of these surface properties as possible and thus to characterize the catalytic surface to a higher degree of completeness than heretofore feasable.

New surface analytical tools [5–8] may be utilized to provide new information on catalytic surfaces and processes. The corresponding approach to studying heterogeneous catalysis is via model experiments: The catalyst is a small area single crystal (metal) or a polycrystalline foil, the characterization of the surface and catalytic reactions are carried out under vacuum conditions. Thus, the desired new information is gained under conditions different from those prevalent in technical catalysis, and hence their validity is sometimes questioned by the catalytic chemist.

Despite recognized shortcomings it is important to take note of the surface scientist's primary objectives in this endeavor. These can be grouped into four categories: (1) To carry out basic research on metal surfaces by means of new surface analytical tools, such as LEED, Auger electron spectroscopy (AES), photoemission spectroscopy (UPS and XPS), and others. (2) To select catalytically relevant problems and study them with these new techniques, gain new information and complement the work of the catalytic chemist. (3) To study catalytic reactions (of industrial importance) in combination experiments at low pressure (vacuum) and at high pressure in order to provide a more direct basis for comparison of ideal and technical catalysts. (4) To propose new mechanisms and concepts in heterogeneous catalysis.

In this paper we will review some of the more recent surface science experiments which more or less fulfill one or several of the above directives. At first we discuss what one may call the roots of the surface scientist's interest in heterogeneous catalysis: the more fundamental problems in that area which have to do with the atomistic detail of heterogeneous reactions. Then we present examples of investigations of surface structure, adsorption and surface reactions. Hopefully this review will complement the previous review papers and supplementary discussions on this subject [9–13].

Parallel to the development of new surface instrumentation and numerous experiments there has also been considerable progress made in the theoretical description of surface processes [14–16]. In particular adsorption on small metal clusters has been the focus of attention by several theorists [17–19].

2. Fundamental problems in heterogeneous catalysis

The most fundamental problems in heterogeneous catalysis can often be traced back to an information gap between phenomenological data, e.g. reaction rates, order of reaction, apparent activation energies, dependence of rate on external parameters, and surface characterization with respect to the three most important surface properties: crystallographic *structure, composition* of the surface and the adsorbed layer, and *electronic* structure. Let us illustrate this statement by a brief discussion of the most intriguing subjects in heterogeneous catalysis.

The first subject is the structure sensitivity in chemisorption and catalysis [20]. Here the phenomenological observations are such that the specific activity (reaction rate per surface site of catalytically active material) of a given reaction depends on the average particle size of the catalyst. Since the surface structure of small metal particles is presumably a function of their size [21,22], one has indirect evidence for the dependence of catalytic activity on crystallographic orientation. Early experiments with single crystal specimen have supported this point of view [23–26].

An important new facet in experimentation was introduced by Rhead [27,28] and Somorjai [29] when they began to study gas adsorption on vicinal (single crystal) surfaces of Cu and Pt. These surfaces annealed at high temperature exhibit well ordered arrays of monatomic steps [29] and kinks.

Their surface structure is presumably not unlike the crystallography of small metal particles of supported catalysts. This simulation of small particles by stepped surfaces is a new dimension in single crystal catalysis research.

Another way of experimentally demonstrating structure sensitivity in catalysis is by studying reconstructed single crystal surfaces. An example is the Pt(100) surface which can exist in a reconstructed state [30] and a non-reconstructed state [31]. Effects of surface reconstruction are also suspected to play a role for dispersed catalysts [32,33] although direct proof is difficult to obtain.

Another subject of interest is the "electronic factor" in heterogeneous catalysis. This electronic factor is because of its diverse phenomenology difficult to describe. One may define it as the dependence of catalytic activity on variables which characterize the electronic structure, such as the work function, the density of electrons in the valence band, general features of the band structure, or in a more local picture: the steric configuration of surface molecular orbitals. These variables themselves depend strongly on the chemical composition of the surface, i.e. they vary from metal to metal, with alloy composition, etc. Of course, changes in the chemical surface composition are connected with changes in the surface structure. Thus it is evident that a clean separation between structure and electronic factor is not possible. The task of characterizing the electronic structure of a metal surface is a difficult one for extended surfaces but it is nearly impossible for dispersed metal particles in a supported catalyst. Experimentalists experience also great difficulty in trying to extract data on the electronic nature of unsupported small particles

[35–36]. This work is guided by theoretical calculations of the electronic properties of small clusters [17–19] of up to 13 atoms in size. However, it remains to be seen to what extent the properties of these ideal clusters resemble those of supported particles in a catalyst. Another problem in context with small, supported particles, which one would like to see solved in the framework of theory, concerns the "equilibrium shape" of such particles, particularly in the presence of adsorbates or in the case of surface segregation. This kind of knowledge would also be quite valuable for the interpretation of catalytic experiments.

Finally there is the selectivity of a catalyst which according to Sinfelt [37] is perhaps the most intriguing and important aspect of heterogeneous catalysis. Here selectivity refers to the ability of a substance to catalyze preferentially a particular type of chemical reaction and thus produce selectively a certain product out of a possible multitude. There are few, if any, examples where the factors determining catalytic selectivity are understood in detail [37]. Yet the manipulation of selectivity is one of the most important tasks of the industrial chemist and is largely accomplished by empirical means. The selectivity depends on many parameters, e.g. particle size, support, alloy composition, presence of promoting elements and others. Surface science can clearly lead the way in selectivity research because physical surface characterization and reaction rate as well as product distribution measurements are feasable [38,39] under vacuum conditions.

3. Contributions of surface science

3.1. Surface structure and chemisorption

Perhaps the most fundamental information which was obtained from surface science experiments concerns the structure of metal surfaces. It was learned from early investigations by field electron and particularly by field ion microscopy (FIM) [40] that metal surfaces were crystalline and exhibited crystallographic defects both expected from their bulk structure. These findings were confirmed by low-energy electron diffraction (LEED) in that the surface structure of most metals corresponded to the structure of the bulk, with the exception of Ir, Pt and Au [41]. These exceptions are quite interesting from a catalysis point of view, and we will see lateron how the structure of Pt surfaces, for example, has a special influence on the rate of chemisorption.

The investigation of surface structure in connection with an adsorption or catalytic reaction experiment is extremely important in order to elucidate the question of structure sensitivity in catalysis. This is the reason why in surface science an adsorption or reaction study is always combined with the characterization of surface structure by LEED, FIM or an equivalent technique. At the same time the cleanliness (or composition) of the surface is investigated by AES, ESCA, SIMS or

Table 1
Adsorption parameters on single crystal surfaces

Surface	Gas	T(K)	s_0	$\Delta\phi$ (eV)	E_{ad} (kcal/mole)	Reference
W(110)	N_2	300	0.004	0	79	42, 43
W(110)	N_2	300	$<10^{-3}$			44
W(110)	N_2	300	< 0.01			44a
W(100)	N_2	300	0.25–0.59	−0.4	75–79	42, 44b, 45
W(111)	N_2	300	< 0.04	0.17	75	42
W(111)	N_2	300	0.08			44a
W(310)	N_2	300	0.25–0.72	0.2	75	46, 46b
W(210)	N_2	300	0.28	0.27	75	46
W(320)	N_2	300	0.73			46b
W(111)	N_2	300	0.4			46b
W(110)	H_2	300	0.07	−0.4	33	47
W(110)	H_2	80	$<10^{-4}$			48
W(100)	H_2	300	0.18	1.0	32	47
W(111)	H_2	425	0.24	0.3	37	47
W(211)	H_2	300		0.6	38	47
Re(0001)	N_2	300	$<10^{-5}$	0.08		49
Re	N_2	300	0.002	0.25		50
Fe(110)	N_2	300				51
Fe(100)	N_2	300	10^{-7}			52, 53
Fe(111)	N_2	300	10^{-4}			53
Pt(100)						
5 × 20	O_2	300	4×10^{-4}			54
1 × 1	O_2	300	0.1	0.4		55, 56
Pt(110)	O_2	300	0.4			57
Pt(111)	O_2	550	0.02			58, 59
Pt(100)						
5 × 20	H_2	300	0.17			60
5 × 20	H_2	125	0.07			61
1 × 1	H_2	300		−0.5		56
Pt(110)	H_2	125	0.33			61
Pt(111)	H_2	125	0.016		17.5	61
Pt(111)	H_2	150	0.1		9.5	62
Pt(211)	H_2	125	0.14			61
Pd(111)	H_2			0.18	20.8	63
Pd(110)	H_2			0.36	24.4	63
Pd(111)$_{step}$	H_2			0.23	23.8	63

ISS *. In this way a large number of gas adsorption studies on well-characterized surfaces has already been carried out. Some examples of systems where significant structure effects were found are listed in table 1.

The data in table 1 show that large differences are observed in kinetic parameters (sticking coefficient), not so much in the energetic parameters (adsorption energy). A striking example are the sticking coefficients for N_2 on W surfaces. The initial sticking coefficient is by a factor of at least 100 larger on the W(100) surface than on W(110) whereas the value for W(111) appears to be intermediate. The stepped W surfaces generally exhibit a high sticking coefficient. Adams and Germer [46a] correlate the behavior along the [01$\bar{1}$] zone with the density of fourfold coordinated sites. For the system H_2 on W, s_0 is again much lower for W(110) than for the other surfaces. In a more recent investigation of H_2 and N_2 adsorption on W(110) Polizzotti [48] finds that the sticking coefficient for these gases is extremely low; he suggests that the W(110) surface is filled by surface diffusion of adsorbed atomic species from neighboring atomically rough planes which occurs readily at 300 K and above. Such a process was also suggested for N_2 adsorption on Re(0001) by Liu and Ehrlich [49] who quote a sticking coefficient $<10^{-5}$ at 300 K for this system while on other Re surfaces $s_0 = 0.002$ has been measured [50].

Another interesting case is N_2 adsorption on various Fe surfaces. In an early investigation by Brill et al. [64] it was stated that N_2 adsorbs much faster on Fe(111) than on other low-index planes. More recent studies basically confirmed this result. N_2 adsorption on Fe(110) at and above room temperature could not be measured [51] while the sticking coefficient for N_2 adsorption on Fe(100) was reported as 10^{-7} by Ertl et al. [52]. Adsorption of N_2 on the rougher Fe(111) surface is characterized by $s_0 = 2 \times 10^{-4}$ [53].

It is interesting to note at this point that surfaces with very low sticking coefficients for dissociative adsorption of a particular gas generally exhibit rather high values of the work function [51]. For example the work functions for W(110), Re(0001) and Fe(110) are 5.43–6.0 eV [48,65], 6.4 eV [49] and 5.0 eV [51], respectively. Also the Pt(100) surface in its clean state has a high work function of 5.8 eV [66] and is according to table 1 another candidate for low sticking of O_2 and H_2.

The Pt(100) surface is one of the special cases where reconstruction has been noted [30]. The LEED pattern of a well annealed, clean Pt(100) surface shows a 5 × 20 symmetry, fig. 1a, for which a quasi-hexagonal structure similar to that of Pt(111) was proposed [67]. Adsorption rates of O_2 and H_2 on this 5 × 20 structure are very low [68]. On the other hand, it is possible to create the non-reconstructed version of a Pt(100) surface [31] which shows a clear 1 × 1 LEED pattern, fig. 1b.

This surface is clean by Auger spectroscopy standards, its surface structure is metastable and transforms back to the 5 × 20 version at temperatures greater than 125°C. The two surface structures of the Pt(100) surface were also investigated by

---

* SIMS = secondary ion mass spectroscopy; ISS = ion scattering spectroscopy.

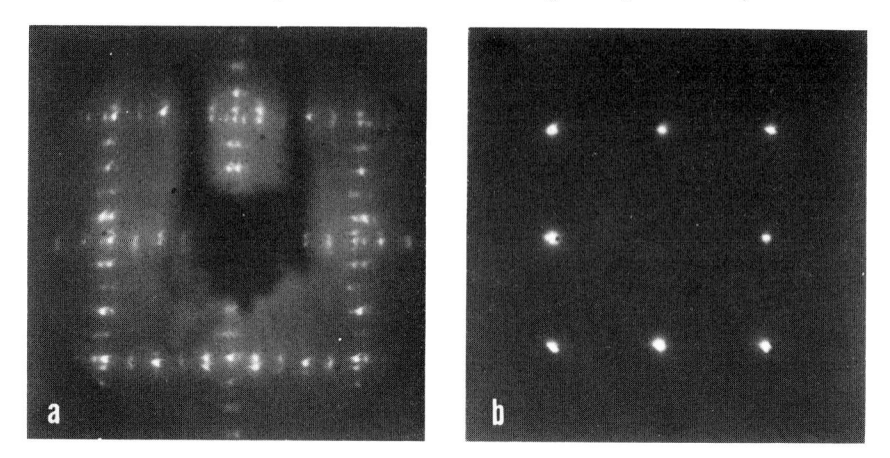

Fig. 1. LEED patterns of the clean Pt(100) surface: (a) reconstructed 5 × 20 surface, E = 83 V; (b) unreconstructed 1 × 1 surface, E = 105 V.

photoemission spectroscopy (UPS) [31]. Fig. 2 shows spectra for $h\nu$ = 21.2 eV for both the clean 5 × 20 and 1 × 1 surface. The difference curve "1 × 1 − 5 × 20" which is also presented in fig. 2 is characterized by a sharp emission peak at 0.2 eV below the Fermi level. This peak is believed to represent a surface state or surface resonance only present on the 1 × 1 surface [31]. This surface is also quite active for the adsorption of O_2 and H_2 at 300 K [56]. In addition it was demonstrated

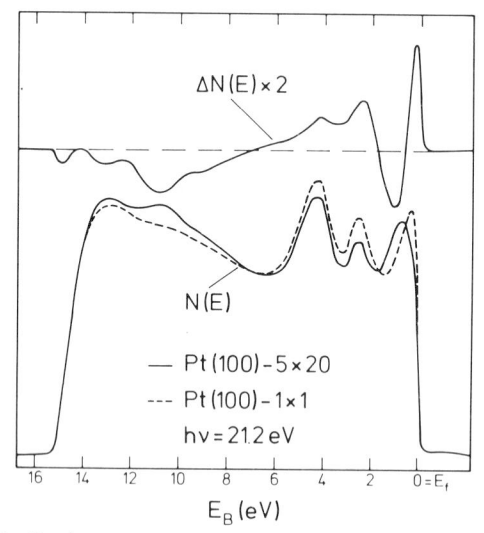

Fig. 2. UPS energy distribution curves for the clean Pt(100) 5 × 20 and 1 × 1 structures and their difference 1 × 1−5 × 20 which shows a characteristic peak near the Fermi level.

recently [55] that the 1×1 surface is able to adsorb NO dissociatively at 420 K whereas the 5×20 surface is quite inert to NO at $T > 370$ K.

The adsorption of O_2 was also investigated on Pt(110) and Pt(111) surfaces [57–59]. There was never any doubt about a high sticking coefficient of O_2 on Pt(110) [69], and it was in fact measured by Ducros and Merrill as 0.4 [57]. On the other hand, the sticking coefficient of O_2 on Pt(111) was a controversial issue for some time mainly due to the initial statement by Lang et al. [29] that "O_2 does not adsorb on a smooth Pt(111) surface". An investigation by Weinberg et al. [70] also indicated an extremely low initial sticking coefficient of O_2 on Pt(111). Separate studies by Bonzel and Ku [58] and Hopster et al. [59] showed that 0.02–0.10 was indeed a realistic value for s_0. Hopster et al. [59] used a cylindrical Pt sample for their studies which had in addition to the (111) orientation a surface region with a continuously varying step density. In this way they were able to obtain the sticking coefficient of O_2 at different step densities and as a function of coverage. Fig. 3 shows a plot of s versus coverage for the smooth (111) orientation and a stepped Pt[14(111)×(11$\bar{1}$)] crystal. It can be seen that the functional dependence is quite similar for these two orientations while the initial sticking coefficient is about twice as large for the stepped crystal as it is for Pt(111).

The kinetics of H_2 adsorption on Pt single crystal surfaces was also studied fairly extensively (table 1). The situation for Pt(100)-5×20 is still somewhat uncertain since Lu and Rye [61] determined $s_0 = 0.07$ at 125 K whereas Helms et al. [56] found no evidence of H_2 adsorption at 300 K on that surface. Netzer and Kneringer [60] measured even $s_0 = 0.17$ at 300 K which is in flagrant disagreement with the observations by Helms et al. [56] as well as our own more recent work [55]. Relatively high sticking coefficients of 0.02–0.10 were determined for H_2 on Pt(111)

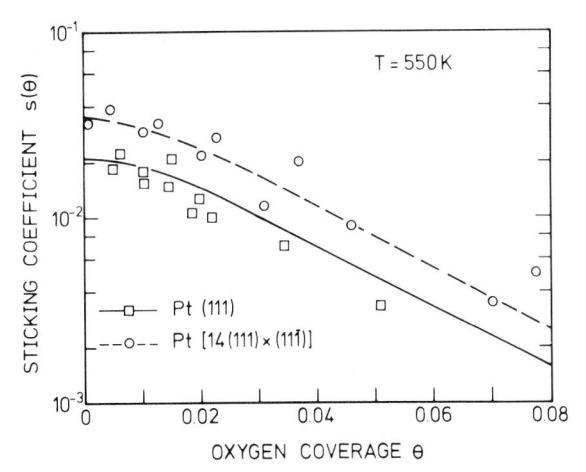

Fig. 3. Sticking coefficient of O_2 versus coverage for a smooth Pt(111) and a stepped Pt[14(111) × (11$\bar{1}$)] surface [59].

[61,62] and also on the rougher surfaces Pt(110) and Pt(211) [61].

Although not all of the sticking coefficient data in table 1 are equally reliable, an interesting pattern in general behavior seems to be emerging: Extremely smooth surfaces, i.e. surfaces with the highest density of surface atoms and a high work function, seem to be characterized by very low sticking coefficients for the dissociative adsorption of N_2, O_2 and H_2. Rougher surfaces of these same metals show higher sticking probabilities, regardless whether this roughness is homogeneous or of a local nature, such as caused by steps. This result is in agreement with the old idea that more exposed surface atoms, i.e. those with a lower coordination number, were preferred sites for adsorption in a kinetic and energetic sense. The connection between kinetics and energetics can be thought of in terms of a Brønsted or the more general Hammet–Taft relationship [4] which in the case of activated chemisorption relates change in the activation energy of adsorption to changes in the energy of adsorption

$$\Delta Q = \rho \, \Delta E_{ad}, \qquad \rho < 1. \tag{1}$$

Changes in entropy are neglected. This relationship means that for a given adsorption system a change in the adsorption energy, ΔE_{ad}, caused by a different surface orientation or steps, will lead to a change in the activation energy, ΔQ.

However, this relationship is not really supported by the experimental data in table 1. For instance, the rate of adsorption of N_2 on W(110) is very low compared to W(100) [42] but there is hardly any difference in E_{ad} for the two orientations. The same is true for H_2 on W(110) and W(100), and the two sets of data for H_2 on Pt(111) are the wrong way around, since Lu and Rye [61] measured a low s_0 and high E_{ad} while Ertl et al. [62] measured a high s_0 and low E_{ad}. Better agreement between this formalism and adsorption experiments on stepped surfaces was obtained by Ibach [71] for O_2 on Si(111) where he derived a relationship between changes in the activation energy of adsorption and changes in the ionization potential and electron affinity of the substrate. For metals this relationship simplifies to

$$\Delta Q = \text{const.} \, \Delta \phi, \tag{2}$$

with $\Delta \phi$ being a change in the work function. This equation permits also a convenient rational for the effect of steps on the rate of chemisorption. Experimentally a decrease in work function is observed for stepped W, Au and Pt surfaces [72,73],

$$\Delta \phi = an, \tag{3}$$

with n = step density. Combining eqs. (2) and (3) Ibach pointed out the relationship between ΔQ and step density n. Thus steps which lower the work function of a metal surface also lower the activation energy of adsorption, increasing the rate of adsorption. In addition there is some evidence that E_{ad} is larger on stepped surfaces than on smooth surfaces [63,75].

Other techniques for studying the chemisorption behavior of gases on transition metal surfaces are suited to yield more information on the electronic, vibrational

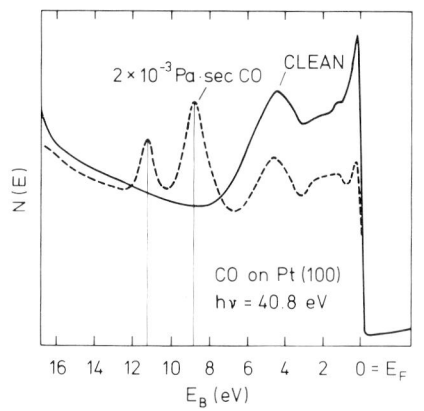

Fig. 4. UPS energy distribution curves for CO adsorbed on Pt(100); molecular CO orbitals emerge at 8.8 and 11.3 eV below E_f.

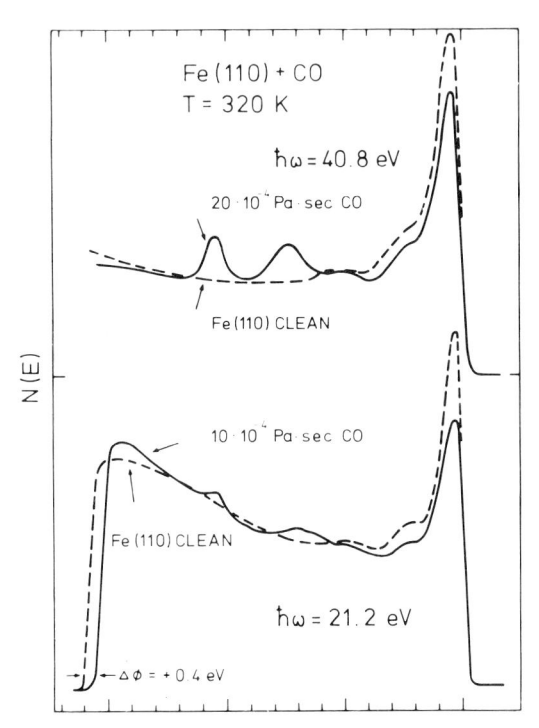

Fig. 5. UPS energy distribution curves for CO adsorbed on Fe(110) at 320 K; molecular CO orbital resonances are clearly visible at 7.0 and 10.5 eV below the Fermi level [51].

and site specific properties of the adsorbate. These techniques are in particular UPS and electron energy loss spectroscopy (ELS). The study of the valence band region by UPS yielded in many cases information on the electronic binding energies of molecular orbitals of the adsorbed gas. In such instances it could be fairly well decided, for example, whether a gas was dissociatively or molecularly adsorbed. As an example, we see in fig. 4 the spectrum of CO adsorbed on a Pt(100) surface taken with $h\nu = 21.2$ eV and $h\nu = 40.8$ eV. This CO spectrum is characterized by essentially two resonances, one at 11.3 eV and the other at 8.8 eV below the Fermi level of Pt. It is now fairly well accepted that these peaks represent the 4σ, 1π and 5σ molecular orbitals of CO, respectively [76,77].

In contrast to Pt(100), CO adsorbs molecularly as well as dissociatively on Fe(110) [51]. This behavior is illustrated in fig. 5 by UPS data for CO adsorbed at $T = 320$ K on a clean Fe(110) surface, $h\nu = 21.2$ eV and $h\nu = 40.8$ eV. This spectrum shows again the features typical of molecular CO. On the other hand, adsorbing CO at $T = 385$ K gives rise to the spectrum in fig. 6 taken with $h\nu = 21.2$ eV; peaks at 3.7 eV and 5.4 eV below the Fermi level are noted which are characteristic of atomic species C and O [51]. Thus CO dissociates quite easily on Fe(110), a fact which is presumably significant for the hydrogenation of CO on Fe catalysts. The bonding of CO to the metal surface is believed to occur mainly via donation of electrons from the 5σ level into the metal d-band and back-donation of metal electrons

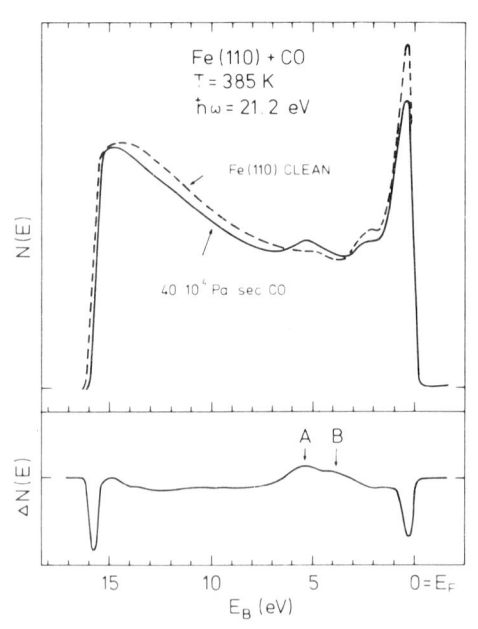

Fig. 6. UPS energy distribution curve for CO dissociatively adsorbed on Fe(110) at 385 K; the difference spectrum shows peaks at 3.7 and 5.4 eV below E_f [51].

into the anti-bonding $1\pi^*$ level of CO [77]. Thus one expects a weakening of the C–O bond within the adsorbed molecule. This idea was carried on further by Brodén et al. [79] in that they linked the stretching of the molecule with energetic shifts of the 4σ and 1π orbitals which are involved in the C–O bond. Brodén et al. [79] applied this idea to all the CO–metal systems investigated by UPS and found that the energy separation between 4σ and 1π level varied from 2.60 eV for Pt to 3.5 eV for Mo and Fe, compared to 2.75 eV for gaseous CO. In other words, adsorbed CO on Pt is hardly deformed but on Fe and Mo it experiences considerable stretching, or dissociation into carbon and oxygen. As a matter of fact, a dissociated state of CO is found on W, Mo and Fe at room temperature whereas molecular CO is only found at lower temperature. Brodén et al. [79] suggested on the basis of these data that the tendency for CO dissociation on transition metals increases from right to left and bottom to top of the periodic table. A border line between molecular and dissociated CO at room temperature runs to the right of Fe, Tc and W. For these borderline elements the adsorption characteristics seem to be particularly structure sensitive [79]. A similar behavior was proposed for the adsorption of N_2 and NO.

Examples, where the deformation of a molecule upon adsorption on a metal was postulated on the basis of UPS data, are acetylene and ethylene on Ir(100) [80] and acetylene on Pt(111) [81] and Pt(100) [82]. Thus useful information on the bonding configuration of adsorbed molecules may be obtained from angle integrated UPS but no information on the geometry of the adsorption site.

With regard to the identification of adsorption sites ELS has proven to be a powerful technique. For H_2 on W(100) it was found that at low coverage the on top position is the preferred site for hydrogen atoms whereas at higher coverage ($\theta > 0.5$) hydrogen occupies the bridge site [83]. This information stems from ELS spectra of H on W(100) at various coverages, shown in fig. 7. Energy loss peaks at 155 and 130 meV were detected at low and high coverage, respectively. On the basis of a central force potential it is argued that a singly coordinated bond between H and the metal surface would have a higher frequency (= energy loss) than a bond with coordination two, three etc .... Thus the higher energy loss of 155 meV represents the on top site while the 130 meV loss represents the bridge site. This assignment is also consistent with the coverage scale, LEED pattern and electron stimulated desorption behavior [83].

Other systems which were investigated by ELS are CO and O_2 on W(100) [84], O_2 and CO on Ni(100) [85,86], CO and C_2H_2 on Pt(111) [87]. A particularly interesting system is CO on Pt(111) [87] which was also studied by infra-red spectroscopy (IRS) by Shigheishi and King [88] and Horn and Pritchard [89]. Whereas IR data show a single absorption band at 2060–2100 cm^{-1}, the energy loss spectrum shown in fig. 8 reveals two adsorption states characterized by pairs of loss peaks at 258 and 58 meV, and 232 and 45 meV, respectively. The latter peaks are lower in intensity and broader than the first pair of peaks; for this reason the 232 meV (1870 cm^{-1}) peak may not be detectable with IRS [87]. Froitzheim et al.

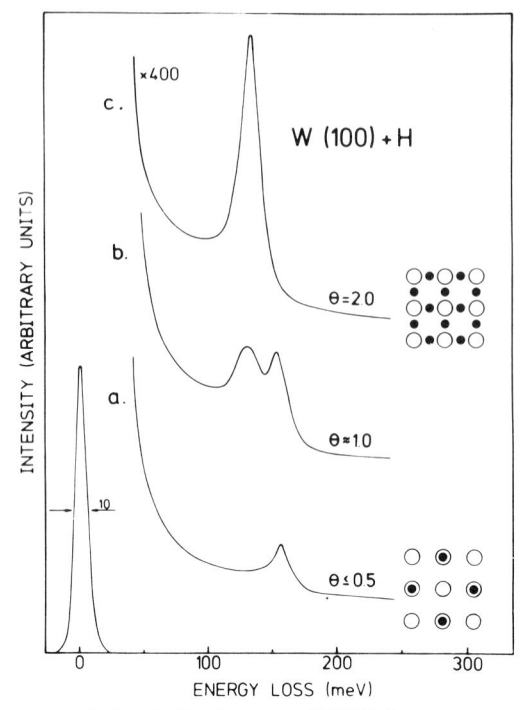

Fig. 7. Energy loss spectra of adsorbed hydrogen on W(100) for various coverages of hydrogen: (a) at $\theta \leq 0.5$ only top sites are occupied by hydrogen; (b) at $\theta = 1.0$ both top and bridge sites are occupied; (c) at saturation coverage all bridge sites are filled with hydrogen [83].

[87] interpret their ELS spectra in such a way that CO initially sticks to on top sites and at higher coverage also on bridge sites. This assignment permits a straightforward understanding of the two LEED patterns observed for CO on Pt(111) by Neumann et al. [90]. The interpretation by Froitzheim et al. is reminiscent of the original work by Eischens et al. [91] who proposed the "linear" and "bridge" states of adsorbed CO on aluminia supported Pt catalysts.

3.2. Heterogeneous reactions

The most direct relationship between surface science and heterogeneous catalysis is achieved by studying a particular catalytic reaction on a certain catalyst. Here the surface scientist investigates this reaction on a small area single crystal in a vacuum chamber, and the catalytic chemist performs the corresponding experiment on a dispersed catalyst at high pressure, e.g. one atmosphere. The question arises whether information obtained at low pressure, say 10^{-5} Pa, can still be meaningfully extrapolated and compared with data gathered at 10^5 Pa (= 1 atm) since we are dealing with a pressure differential of 10 orders of magnitude. It is conceivable

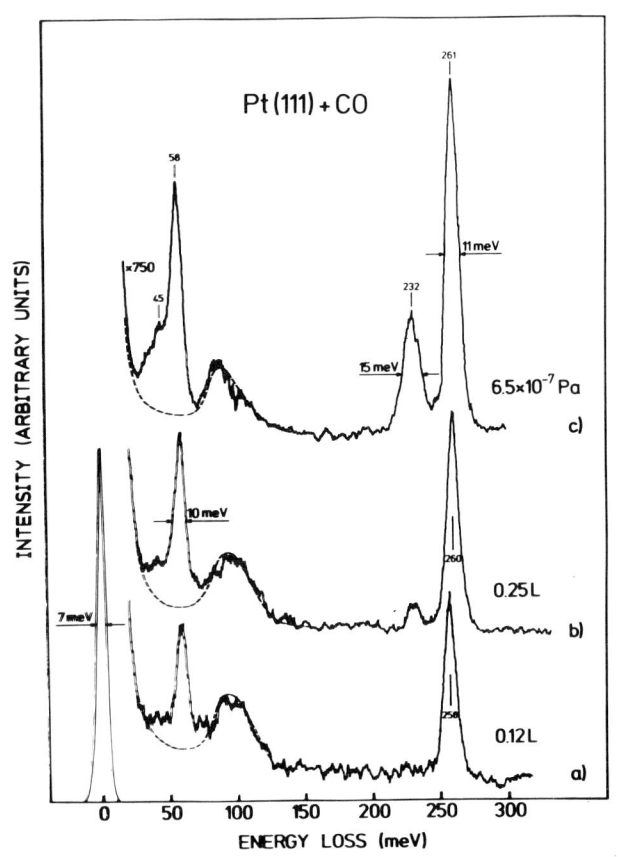

Fig. 8. Energy loss spectra of CO adsorbed on Pt(111) at 320 K with increasing CO coverage. Spectra (a) and (b) were obtained after CO exposures of 10^{-5} Pa sec and 2×10^{-5} Pa sec, respectively. Spectrum (c) was measured in ambient CO of 6.5×10^{-7} Pa and at 350 K [87].

that the reaction mechanism or the surface structure of the catalytic material, for example, could change with pressure — so the question concerning the relevance of reaction measurements under such idealized conditions is justified. Therefore the surface scientist's task is really a dual one: To get new information on the atomistics of catalysis with his tools under the chosen conditions *and* to demonstrate how this information can be useful at the conditions of technical catalysis.

Fig. 9 is an attempt to illustrate the extent of the "pressure gap" between surface science and technical catalysis. The ordinate of the figure gives the pressure (in Pascal) at which typical experiments are carried out; the abscissa is labelled "structural complexity" of the catalyst surface. By this term we mean to imply that a smooth single crystal surface of low-index orientation has a rather isotropic surface and thus very low structural complexity. A vicinal single crystal surface containing

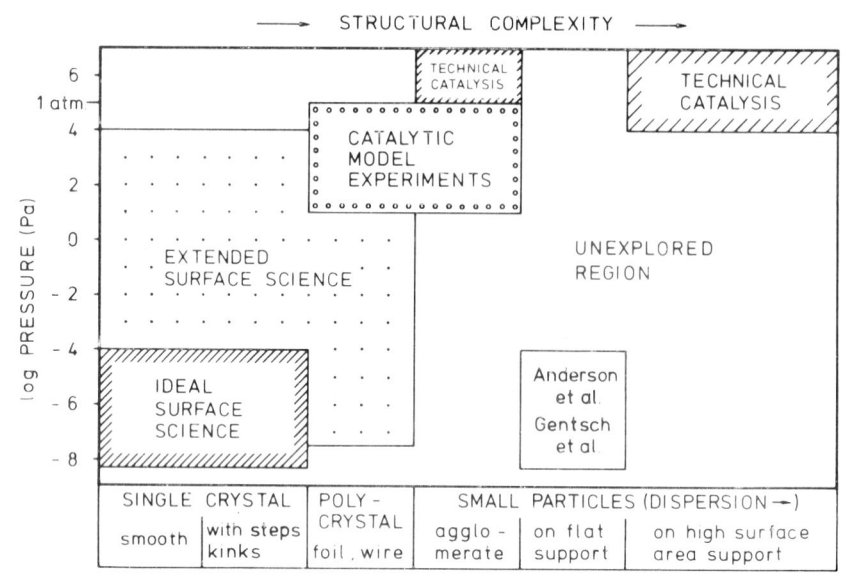

Fig. 9. Schematic diagram illustrating the "pressure gap" between surface science and technical catalysis. The variable "structural complexity" refers to the surface structure of the catalytic material and is also proportional to the "dispersion".

monatomic steps and perhaps kinks has a higher but well defined structural complexity. These samples are the primary research objects of surface science and usually studied in the $10^{-8}-10^{-7}$ Pa regime. Polycrystalline foils and wires have because of various exposed crystallographic orientations and due to grainboundaries a considerably higher structural complexity and have been studied at low as well as high pressures in catalytic model experiments, such as the one by Yao [92]. Such samples are also being studied by surface science techniques at low pressures, e.g. in context with CO oxidation [38,39,69,93], NO reduction [94] and most recently CO hydrogenation [95]. An even higher structural complexity is expected for small particle agglomerates, such as in the study by Carter et al. [96] on the hydrogenolysis of n-heptane. However, this study was only carried out at high pressure. Very few catalytic reactions were as yet measured for model dispersed catalysts [97,98], such as used by Wynblatt and Gjostein [99] for sintering studies of Pt particles. Finally there are the industrial supported catalysts where the active metal particles are highly dispersed and have a high structural complexity which in detail is not known at all. These catalysts have only been studied at high pressure so that a large unexplored area exists for research on structurally more complex catalysts at medium and low pressures.

Fortunately there is a noticeable trend in surface science to extend catalysis related research on single crystal and poly-crystal materials to high pressures. Various types of complex machinery has been designed to accomplish a combination of

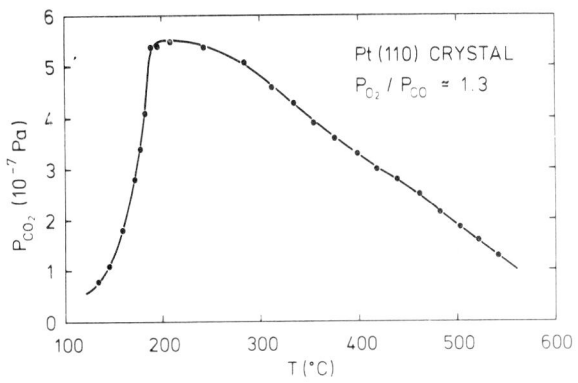

Fig. 10. Rate of CO oxidation on a clean Pt(110) crystal at a total pressure of 1.5×10^{-5} Pa as a function of crystal temperature [40].

low and high pressure studies [100] without sacrificing the use of surface characterization techniques. On the other hand, wherever experimental data at high pressures are already available, the surface scientist should make an effort to compare them with his low pressure data and try to come to a unifying picture of understanding.

Let us illustrate this point by giving a few examples. The first example deals with the oxidation of CO on Pt which was studied on Pt(110) at low pressure [39,101] and on Pt foil at high pressure [92]. This particular reaction has been studied since the time of Langmuir [102] and is today fairly well understood. At a pressure of about 10^{-5} Pa and a nearly stochiometric ratio of O_2 to CO the reaction rate exhibits a maximum near 250°C as shown in fig. 10 [39]. Below the maximum temperature the rate of CO_2 production is influenced to a large extent by adsorption of CO which inhibits the reaction. This inhibition can also be noted from the kinetic rate law:

$$\text{Rate} = k P_{O_2}^m P_{CO}^n , \qquad (4)$$

where k is the rate constant at a given temperature. The exponents m and n were found to be nearly identical at low and high pressure and, namely $m = 1.1-1.2$ and $n = -(0.5-0.6)$.

The maximum rate at 264°C was measured on a Pt(110) crystal for the following pressures [101]:

$$P_{CO} = 6.0 \times 10^{-6} \text{ Pa}, \quad P_{O_2} = 4.0 \times 10^{-5} \text{ Pa}, \quad P_{CO_2} = 8.8 \times 10^{-7} \text{ Pa}.$$

With a pumping speed of about 45 ℓ/sec and a sample area of 0.6 cm² a rate of CO_2 production of 1.6×10^{13} molecules/cm² sec is calculated. In the high pressure oxidation of CO by Yao [92] the measured rate was 30 mℓ CO_2/cm² at 264°C or 1.3×10^{15} molecules/cm² sec; in this experiment the partial pressures were:

$$P_{CO} = 910 \text{ Pa}, \quad P_{O_2} = 1530 \text{ Pa}.$$

We can now compute the turnover number in both cases as well as the yield of the reaction which we define as

$$\text{yield} = \frac{\text{rate of CO}_2 \text{ production}}{\text{rate of incident CO molecules}}. \quad (5)$$

For the calculation of the turnover number we assume in both cases a site density of 10^{15} cm^{-2}. We note that at 10^{-6} Pa the turnover number is 0.016 and at 10^3 Pa (total pressure was 10^5 Pa) it is 1.3. However, the more interesting figure is the yield which at low pressure is *unity* and at high pressure 6×10^{-7}. This about six orders of magnitude difference implies that the reaction efficiency is much higher on the clean single crystal surface at low pressure than on the foil at high pressure. Presumably the high pressure run was heavily inhibited by adsorbed CO while CO inhibition was negligible in vacuum. The intrinsic reaction rate k defined by eq. (4), on the other hand, can also be calculated for the two cases. Using $m = 1.1$ and $n = -0.5$ one obtains 8×10^{16} and 2.5×10^{15} (molecules cm^{-2} sec^{-1} Pa$^{-0.6}$) for low and high pressure, respectively. As expected, these values lie fairly close together suggesting that the reaction mechanism is the same at low and high pressure.

Thus the outcome of this research on CO oxidation on Pt at low pressures is substantial: Besides a clarification of the mechanism of the reaction [69] there have been studies on the influence of steps on reactivity [59] as well as the poisoning by sulfur [103]. The demonstrated correspondence between the low and high pressure behavior justifies a generalization of the results obtained for well characterized surfaces into the range of dispersed catalysts. The high value of one of the reaction yield at 10^{-6} Pa also implies that there are no special "active sites" for this particular reaction.

Another example of a heterogeneous reaction where good correspondence between low and high pressure data was observed is the reduction of NO by N_2 on Pt [94]. A low pressure steady state reaction on polycrystalline Pt foil was run in the temperature range of room temperature to 600°C at a total pressure of 2×10^{-5} Pa. Fig. 11 shows a product distribution for a partial pressure ratio of $P_{H_2}/P_{NO} = 5$ as a function of temperature. The general behavior is quite similar to that found for dispersed Pt on Al_2O_3 at high pressure [104–106] and also for Rh on Al_2O_3 [106] which behaves analogously to Pt. In particular a significant amount of NH_3 is produced by this reaction at temperatures around 170–300°C, with a peak at 220°C. At this temperature less than 5% N_2 are being formed such that the selectivity of the reaction is shifted towards the production of NH_3. This result agrees very well with high pressure studies of this reaction on Pt/Al_2O_3 catalysts by Shelef and Gandhi [104] who reported a 8% selectivity minimum for N_2 production at 300°C and with data by Klimish and Taylor [105] who found a NH_3 maximum at 310°C for a 3 : 1 ratio of P_{H_2}/P_{NO}. Kobylinski and Taylor [106] measured a temperature of 220°C for the NH_3 maximum for RH/Al_2O_3 and a partial pressure ratio of 4 : 1.

Good agreement is also found for the amount of NO reduced (in percent) as a

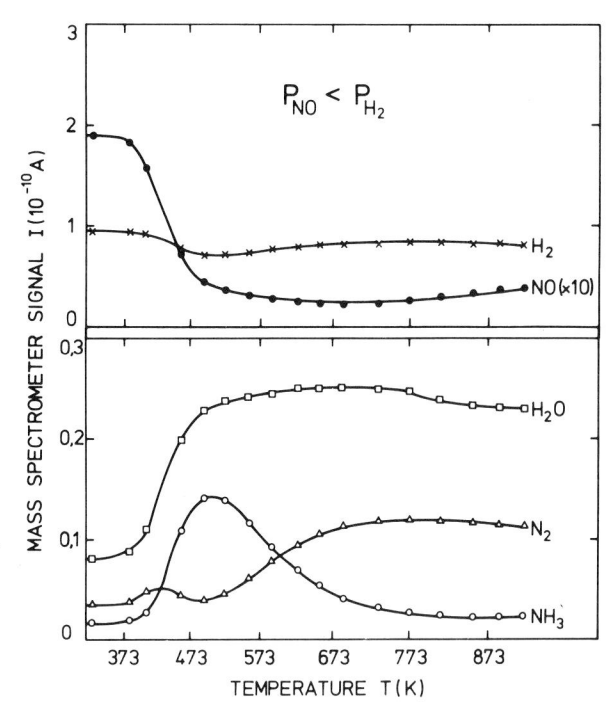

Fig. 11. Temperature dependence of the NO + H_2 reaction on polycrystalline platinum at a total pressure of 2×10^{-5} Pa. The upper portion shows the reactants while the lower portion illustrates the product selectivity [94].

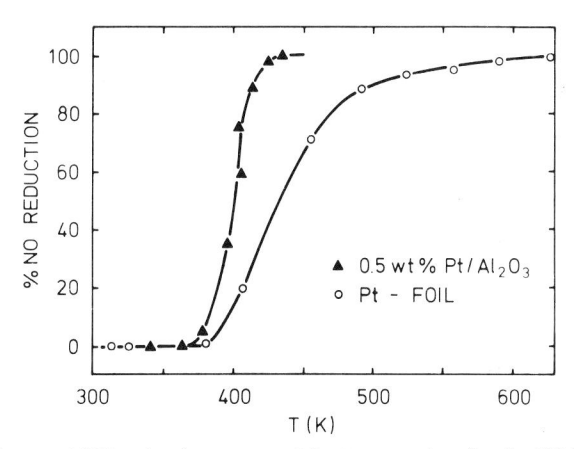

Fig. 12. Plot of percent NO reduction versus catalyst temperature for the NO/H_2 reaction. The data for the 0.5 wt% Pt/Al_2O_3 catalyst were taken under the following conditions: 0.5% NO, 2% H_2, 97.5% Ar, 24000 GHSV. The data for the Pt-foil were obtained for a 5 : 1 partial pressure ratio of H_2 to NO and at a total pressure of ~10^{-5} Pa [94,106].

function of catalyst temperature, shown in fig. 12 for the Pt-foil at 10^{-5} Pa total pressure [94] and for a Pt/Al$_2$O$_3$ catalyst at 10^5 Pa total pressure [106]. Whereas the starting temperatures of the reaction are nearly equal in both cases, the rate of increase in percent NO reduction is higher for the supported catalyst than for the Pt-foil. The reason for this difference is presumably that the Pt-foil in a vacuum system resembles a true differential reactor while the catalyst at high pressure is more typical of an integral reactor where the gas composition varies strongly over the catalyst volume.

The mechanism of the NO reduction reaction was elucidated by surface science experiments dealing with the details of NO adsorption of Pt [66]. It was found that NO adsorbs in molecular form at temperatures below 80°C and dissociatively at $\gtrsim$180°C on polycrystalline Pt. The onset temperature of dissociation is likely to be lowered in the presence of adsorbed hydrogen. For the understanding of the reaction data in fig. 11 the dissociation of NO seems to be of crucial importance since essentially no NO is being reduced below 130°C, i.e. molecular NO is not reactive towards H$_2$. With the beginning of NO dissociation the reaction products H$_2$O, N$_2$ and NH$_3$ appear. Therefore these products are most likely being formed via the dissociated atomic species, N, O and H [94].

The third reaction for which data at low and high pressure and on extended surfaces as well as supported catalysts are available is the hydrogenation of CO on rhodium. Sexton and Somorjai [95] studied this reaction on a Rh foil of 1 cm^2 area at 10^{-2} Pa and at 9.3×10^4 Pa ($\approx$700 Torr) in the temperature range 250–350°C. At low pressure there were no detectable reaction products but at 9.4×10^4 Pa total pressure and a ratio of H$_2$ to CO of 3 : 1 significant amounts of hydrocarbons were produced. For example, at 300°C the turnover number was 0.13 sec^{-1} with 90% methane, 5% ethylene, 2% ethane, 3% butane and the rest higher molecular weight species [95]. The apparent activation energy for methane formation was 24 kcal/mole. A comparison of these data with those of Vannice [107] who studied the CO + H$_2$ reaction on 1% Rh/Al$_2$O$_3$ supported catalyst at 10^5 Pa is quite favorable. In the latter case the turnover number was 0.034 sec^{-1} for methane and the product distribution similar to that from Rh foil, both at 300°C, although no ethylene was reported by Vannice [107]. The apparent activation energy for methane formation was also 24 kcal/mole for supported Rh.

The observation that Rh is only active for methane formation at high pressure was also made with Ni as a catalyst. Thus CH$_4$ formation is easily observed from CO and H$_2$ on Ni at a total pressure of 70 Pa [108] whereas low pressure studies failed to produce any methane [109]. It is interesting to note, however, that the formation of CH$_4$ from carbon and H$_2$ on Ni was observed under vacuum conditions by Gentsch et al. [98]. Here Ni particles in the sub-monolayer range were deposited onto a smooth carbon film which itself was sitting on the inside wall of a glass bulb at a base pressure of 10^{-8} Pa. After exposure of the Ni on carbon to 8×10^{-5} Pa H$_2$, the formation of CH$_4$ was measured at 47°C and 6×10^{-8} Pa partial pressure [98]. Similar experiments with Pd and Pt [110] as catalytic materials also showed

CH_4 formation but with considerably higher activity. These experiments suggest that CO dissociation may be a necessary step for the formation of CH_4, in analogy to the mechanism of NH_3 formation from NO and H_2 on Pt [94]. CO dissociation is not known to occur on Rh or Ni at low pressures but on Ni, for example, it is a well known phenomenon at high pressure [111].

From the brief discussion of these examples it can be seen that the investigation of catalytic reactions at low and high pressures on well characterized surfaces is well underway. The first results of this work are promising in the sense that new information about the fundamental steps of heterogeneous catalysis is becoming available and that new ideas about reaction mechanism are being deduced. The agreement between surface science results and those of conventional catalytic chemistry is encouraging and supporting the credibility of the surface science approach. Based on this agreement it is often possible to conclude that structural effects and even support effects for some catalytic reactions are of secondary significance. This was also observed in a study of the hydrogenolysis of cyclopropane on a Pt[6(111) $\times$ (100)] stepped crystal [112] at 1.8×10^4 Pa where the measured turnover numbers as well as activation energies for propane formation agreed very well with those measured for dispersed supported catalysts.

4. Conclusion

New surface analytical techniques in the area of surface science are capable of providing important information on the structure, composition and electronic nature of solid surfaces in an atomically clean state. Many of these techniques can also be used to study alloy surfaces, adsorption layers and deliberately contaminated surfaces. New data on bonding, molecular orbitals and vibrational levels of adsorbates are being produced. The combination of such extensive surface characterization with kinetic gas adsorption and reaction experiments can be called the surface science approach to research in heterogeneous catalysis. Although the spectrum of information thus obtained is extremely broad, it is now evident that a continuous linkage between the two disciplines is emerging. To a first order the surface scientist's task is to complement research efforts of the catalytic chemist with fundamental data on surface processes. Such data will hopefully assist the catalytic chemist in understanding the phenomenology of catalysis as well as provide some guidance in designing future experiments.

Acknowledgement

The author is grateful to G. Brodén, G. Comsa, H. Ibach and G. Pirug for valuable comments on the manuscript.

References

[1] Consult, for example the serial publications: Journal of Catalysis, Advances in Catalysis, Catalysis Reviews.
[2] G.C. Bond, Catalysis by Metals (Academic Press, London, 1962).
[3] A. Clark, The Theory of Adsorption and Catalysis (Academic Press, New York, 1970).
[4] E.G. Schlosser, Heterogene Katalyse (Verlag Chemie, Weinheim, 1972).
[5] J.T. Yates, Jr., Chem. Eng. News (Aug. 26, 1974).
[6] G. Ertl and J. Küppers, Low Energy Electrons and Surface Chemistry (Verlag Chemie, Weinheim, 1974).
[7] E.W. Plummer, in: Interactions on Metal Surfaces, Ed. R. Gomer (Springer, Berlin, 1975).
[8] H. Froitzheim, in: Topics in Current Physics, Vol. 4, Ed. H. Ibach (Springer, Berlin, 1977).
[9] T.E. Fischer, Phys. Today 27 (5) (1974) 23.
[10] J.J. McCarroll, Surface Sci. 53 (1975) 297.
[11] T.E. Fischer, CRC Crit. Rev. Solid State Phys. 6 (1976).
[12] Proc. Battelle Conf. on The Physical Basis of Heterogeneous Catalysis, Eds. E. Drauglis and R.I. Jaffee (Plenum, New York, 1975).
[13] H.P. Bonzel, Physik. Bl. 32 (1976) 392.
[14] T.B. Grimley, in: Molecular Processes on Solid Surfaces, Eds. E. Drauglis, R.D. Gretz and R.I. Jaffee (McGraw-Hill, New York, 1969).
[15] J.R. Schrieffer and P. Soven, Phys. Today 28 (4) (1975) 24.
[16] J.R. Smith, in: Interaction on Metal Surfaces, Ed. R. Gomer (Springer, Berlin, 1975).
[17] J.C. Slater and K.H. Johnson, Phys. Today 27 (10) (1974) 34.
[18] R.C. Baetzold and R.E. Mack, J. Chem. Phys. 62 (1975) 1513.
[19] R.P. Messmer, S.K. Knudson, K.H. Johnson, J.B. Diamond and C.Y. Yang, Phys. Rev. B13 (1976) 1396.
[20] M. Boudart, J. Vacuum Sci. Technol. 12 (1975) 329.
[21] J.J. Burton, Catalysis Rev. Sci. Eng. 9 (1974) 209.
[22] J.J. Burton, E. Hyman and D. Fedak, J. Catalysis 37 (1975) 106.
[23] H. Leidheiser and A.T. Gwathmey, J. Am. Chem. Soc. 70 (1948) 1206.
[24] A.T. Gwathmey, Chem. Progr. 14 (1953) 117.
[25] R.E. Cunningham and A.T. Gwathmey, Advan. Catalysis 9 (1957) 25.
[26] J. McAllister and R.S. Hansen, J. Chem. Phys. 59 (1973) 414.
[27] G.E. Rhead and J. Perdereau, in: Structure et Propriétés des Surfaces des Solides, Colloq. Intern. CNRS Nr. 187, (CNRS, Paris, 1970).
[28] J. Perdereau and G.E. Rhead, Surface Sci. 24 (1971) 555.
[29] B. Lang, R.W. Joyner and G.A. Somorjai, Surface Sci. 30 (1972) 440, 454.
[30] S. Hagstrom, H.B. Lyon and G.A. Somorjai, Phys. Rev. Letters 15 (1965) 491.
[31] H.P. Bonzel, C.R. Helms and S. Kelemen, Phys. Rev. Letters 35 (1975) 1237.
[32] K.C. Taylor, R.M. Sinkevitch and R.L. Klimisch, J. Catalysis 35 (1974) 34.
[33] M. Boudart and J.A. Dumesic, in: The Physical Basis of Heterogeneous Catalysis, Eds. E. Drauglis and R.I. Jaffee (Plenum, New York, 1975).
[34] M.G. Mason and R.C. Baetzold, J. Chem. Phys. 64 (1976) 271.
[35] K.H. Johnson and G. Hochstrasser, Agenda Discussion in: The Physical Basis of Heterogeneous Catalysis, Eds. E. Drauglis and R.I. Jaffee (Plenum, New York, 1975) p. 373.
[36] W. Schulze, D.M. Kolb and H. Gerischer, J. Chem. Soc. Faraday Trans. II 71 (1975) 1763.
[37] J.H. Sinfelt, Progr. Solid State Chem. 10, Part 2 (1975) 55.
[38] G. Ertl and P. Rau, Surface Sci. 15 (1969) 443.
[39] H.P. Bonzel and R. Ku, J. Vacuum Sci. Technol. 9 (1972) 663.
[40] E.W. Müller and T.T. Tsong, Field Ion Microscopy (Elsevier, New York, 1969).

[41] G.A. Somorjai, Principles of Surface Chemistry (Prentice-Hall, Englewood Cliffs, NJ, 1972).
[42] T.A. Delchar and G. Ehrlich, J. Chem. Phys. 42 (1965) 2686.
[43] P.W. Tamm and L.D. Schmidt, Surface Sci. 26 (1971) 286.
[44] (a) T.E. Madey and J.T. Yates, Nuovo Cimento Suppl. 5 (1967) 483.
 (b) D.A. King and M.G. Wells, Surface Sci. 29 (1972) 454; Proc. Roy. Soc. (London) A339 (1974) 245.
[45] L.R. Clavenna and L.D. Schmidt, Surface Sci. 22 (1970) 365.
[46] (a) D.L. Adams and L.H. Germer, Surface Sci. 27 (1971) 21.
 (b) S.P. Singh-Boparai, M. Bowker and D.A. King, Surface Sci. 53 (1975) 55.
[47] L.D. Schmidt, in: Interactions on Metal Surfaces, Ed. R. Gomer (Springer, Berlin, 1975) p. 63.
[48] R.S. Polizzotti, Structure-Sensitive Chemisorption: Hydrogen and Nitrogen on Single Crystal Planes of W and Rh, Ph.D. Thesis, Univ. of Illinois, Urbana (1974).
[49] R. Liu and G. Ehrlich, J. Vacuum Sci. Technol. 13 (1976) 310.
[50] A. van Oostrom, in: Proc. 8th Intern. Conf. on Phenomena in Ionized Gases (Intern. Atomic Energy Agency, Vienna, 1968).
[51] G. Brodén, G. Gafner and H.P. Bonzel, Appl. Phys., 13 (1977) 333.
[52] G. Ertl, M. Grunze and W. Weiss, J. Vacuum Sci. Technol. 13 (1976) 314.
[53] F. Bozso, G. Ertl, M. Grunze and M. Weiss, J. Catalysis, to be published.
[54] G. Kneringer and F.P. Netzer, Surface Sci. 49 (1975) 125.
[55] G. Brodén, G. Pirug and H.P. Bonzel, to be published.
[56] C.R. Helms, H.P. Bonzel and S. Kelemen, J. Chem. Phys. 65 (1976) 1773.
[57] R. Ducros and R.P. Merrill, Surface Sci. 55 (1976) 227.
[58] H.P. Bonzel and R. Ku, Surface Sci. 40 (1973) 85.
[59] H. Hopster, H. Ibach and G. Comsa, J. Catalysis 46 (1977) 32.
[60] F.P. Netzer and G. Kneringer, Surface Sci. 51 (1975) 526.
[61] K.E. Lu and R.R. Rye, Surface Sci. 45 (1974) 667.
[62] K. Christmann, G. Ertl and T. Pignet, Surface Sci. 54 (1976) 365.
[63] H. Conrad, G. Ertl and E.E. Latta, Surface Sci. 41 (1974) 435.
[64] R. Brill, E.L. Richter and R. Ruch, Angew. Chem. 79 (1967) 905.
[65] E.W. Müller, J. Appl. Phys. 26 (1955) 732.
[66] H.P. Bonzel and G. Pirug, Surface Sci. 62 (1977) 45.
[67] D.G. Fedak and N.A. Gjostein, Surface Sci. 8 (1967) 77.
[68] A.E. Morgan and G.A. Somorjai, Surface Sci. 12 (1968) 405.
[69] H.P. Bonzel and R. Ku, Surface Sci. 33 (1972) 91.
[70] W.H. Weinberg, R.M. Lambert, C.M. Comrie and J.W. Linnett, in: Proc. 5th Intern. Congr. on Catalysis, Ed. J.W. Hightower (North-Holland, Amsterdam, 1973).
[71] H. Ibach, Surface Sci. 53 (1975) 444.
[72] K. Besocke and H. Wagner, Phys. Rev. B8 (1973) 4597.
[73] K. Besocke, B. Krahl-Urban and H. Wagner, Surface Sci. 68 (1977) 39.
[74] K. Christmann and G. Ertl, Surface Sci. 60 (1976) 365.
[75] B. Feuerbacher, in: Topics in Current Physics Vol. 4, Ed. H. Ibach (Springer, Berlin, 1977).
[76] H.P. Bonzel and T.E. Fischer, Surface Sci. 51 (1975) 213.
[77] T. Gustafsson, E.W. Plummer, D.E. Eastman and J.L. Freeouf, Solid State Commun. 17 (1975) 391.
[78] G. Blyholder, J. Phys. Chem. 68 (1964) 2772.
[79] G. Brodén, T.N. Rhodin, C.F. Brucker, R. Benbow and Z. Hurych, Surface Sci. 59 (1976) 593.
[80] G. Brodén and T.N. Rhodin, Chem. Phys. Letters 40 (1976) 247.

[81] J.E. Demuth, Chem. Phys. Letters 45 (1977) 12.
[82] T.E. Fischer, S. Kelemen and H.P. Bonzel, Surface Sci. 64 (1977) 157.
[83] H. Froitzheim, H. Ibach and S. Lehwald, Phys. Rev. Letters 36 (1976) 1549.
[84] H. Froitzheim, H. Ibach and S. Lehwald, Phys. Rev. B14 (1976) 1362.
[85] S. Anderson, Solid State Commun. 20 (1976) 229.
[86] S. Anderson, Solid State Commun. 21 (1977) 75.
[87] H. Froitzheim, H. Hopster, H. Ibach and S. Lehwald, Appl. Phys. 13 (1977) 147.
[88] R.A. Shigeishi and D.A. King, Surface Sci. 58 (1976) 379.
[89] K. Horn and J. Pritchard, to be published.
[90] G. Ertl, M. Neumann and K.M. Streit, Surface Sci. 64 (1977) 393.
[91] R.P. Eischens and W.A. Pliskin, Advan. Catalysis 10 (1958) 1.
[92] Y.F. Yu-Yao, unpublished research: see also: J. Catalysis 39 (1975) 104.
[93] H.P. Bonzel and J.J. Burton, Surface Sci. 52 (1975) 223.
[94] G. Pirug and H.P. Bonzel, J. Catalysis, to be published.
[95] B.A. Sexton and G.A. Somorjai, J. Catalysis 46 (1977) 167.
[96] J.L. Carter, J.A. Cusumano and J.H. Sinfelt, J. Catalysis 20 (1971) 223.
[97] J.R. Anderson and R.J. MacDonald, J. Catalysis 19 (1970) 227.
[98] H. Gentsch, V. Härtel and M. Köpp, Ber. Bunsenges, Physik. Chem. 75 (1971) 1086.
[99] P. Wynblatt and N.A. Gjostein, Scripta Met. 9 (1973) 969.
[100] D.W. Blakely, E.I. Kozak, B.A. Sexton and G.A. Somorjai, J. Vacuum Sci. Technol. 13 (1976) 1091.
[101] H.P. Bonzel, unpublished research.
[102] I. Langmuir, Trans. Faraday Soc. 17 (1922) 621.
[103] H.P. Bonzel and R. Ku, J. Chem. Phys. 59 (1973) 164.
[104] M. Shelef and H.S. Gandhi, Ind. Eng. Chem., Prod. Res. Develop. 11 (1972) 393.
[105] R.L. Klimisch and K.C. Taylor, Ind. Eng. Chem., Prod. Res. Develop. 14 (1975) 26.
[106] T.P. Kobylinski and B.W. Taylor, J. Catalysis 33 (1974) 376.
[107] M.A. Vannice, J. Catalysis 37 (1975) 449.
[108] M. Araki and V. Ponec, J. Catalysis 44 (1976) 439.
[109] W. Erley and H. Wagner, unpublished research.
[110] N. Guillen, Ph.D. Thesis, University of Hannover (1973).
[111] H.E. Grenga and K.R. Lawless, J. Appl. Phys. 43 (1972) 1508.
[112] D.R. Kahn, E.E. Petersen and G.A. Somorjai, J. Catalysis 34 (1974) 294.

ELECTRON ENERGY LOSS SPECTRUM OF CYANOGEN ON Pt (100)

R.A. WILLE and F.P. NETZER

Institut für Physikalische Chemie der Universität Innsbruck, Innsbruck, Austria

and

J.A.D. MATTHEW

Department of Physics, University of York, Heslington, York, England

Electron energy loss spectra for adsorbed cyanogen on Pt(100) are presented, and discussed in terms of possible models suggested by other techniques. Adsorbate induced loss features are found at 11 and 14 eV, and these are associated with levels below the Fermi level as observed in ultra-violet photoemission. As in the case of CO adsorption losses in the adsorbed phase occur at energies significantly larger than in the gas phase, indicating an upward shift of the final $2\pi^*$ state due to mixing with metal orbitals. Thermal desorption studies of C_2N_2 on Pt clearly resolve α and β phases, but there is some controversy over whether the β phase involves mainly single CN units bonding to the metal, whether C_2N_2 is molecularly adsorbed, or whether paracyanogenlike structures form at the surface. The electron spectroscopic evidence is examined, and is shown on balance to support adsorption in some molecular form.

1. Introduction

Electron spectroscopy is currently widely being used for the investigation of adsorbate—metal surface interactions and many of the finer details of the chemisorptive bond have been revealed by suitable combinations of various electron spectroscopic techniques. In ultraviolet photoelectron spectroscopy (UPS) the emphasis is mainly concentrated in the recognition and characterization of filled adsorbate induced energy levels. Therefore, final state effects are usually neglected in qualitative interpretations of UPS data. Knowledge of the position of the final states is crucial, however, in interpreting electron energy loss spectra (ELS) of adsorbed particles, where electronic transitions from filled adsorbate induced states to empty adsorbate or metal states are involved. In combining ELS data with UPS evidence it is therefore possible to probe both occupied and unoccupied energy levels and to obtain useful information about differential chemical shifts of filled and empty chemisorption induced states. In this paper ELS spectra for adsorbed cyanogen on Pt(100) are presented and discussed by comparison with UPS results of adsorbed C_2N_2 [1] and with spectroscopic evidence from transition metal cyanogen complexes [2]. Possible models of C_2N_2 adsorption states as suggested by other techniquesques are taken into consideration.

The adsorption of C_2N_2 on a Pt(100) surface has been studied previously by LEED, Auger spectroscopy and flash desorption mass spectrometry (FDMS) [3,4]. C_2N_2 exposure causes first the (5 × 1) LEED pattern of the clean Pt(100) surface to change into a diffuse (1 × 1) pattern, but at saturation almost no diffraction features are discernible indicating a densely packed adsorbate layer out of registry with the substrate. At saturation, a work function increase of about 0.5 eV was observed [1]. Two main desorption states were detected by FDMS with desorption maxima at 140°C (α) and 410–480°C (β_1, β_2), the total saturation coverage being close to a monolayer. Desorption from the surface took place as molecular C_2N_2 only, following first order desorption kinetics for both low and high temperature desorption states. A large difference in the initial sticking coefficient of α and β was noted with initial values of 0.06 and 0.9, respectively. The kinetic results mentioned briefly above favour the assumption that C_2N_2 in the α state is molecularly adsorbed with undissociated C_2N_2 present at the surface. It is more difficult to incorporate the kinetic evidence for the β states into a suitable adsorption model. It has been proposed that the β phases involve mainly isolated CN units formed by dissociative adsorption of C_2N_2 either in a mobile form [5] or in an immobile adsorbate layer without surface diffusion during desorption [4]. It was felt, however, that the analogy to adsorbed CO should be more striking with isolated CN particles at the surface than in fact it is. In an alternative model [4] it was therefore proposed that, by analogy with the formation of para-cyanogen from gaseous C_2N_2 attractive lateral interactions between C_2N_2 molecules in the adsorbed layer induce association into some sort of polymeric, paracyanogen-like structures at the surface. As will be discussed below, electron spectroscopic evidence seems to support adsorption in some associated, molecular form.

2. UPS evidence

Ultraviolet photoemission spectra of C_2N_2 adsorbed on a Pt(100) surface have been recorded recently by Conrad et al. [1]. The He II spectra of the clean and C_2N_2 saturated surface with a corresponding difference spectrum taken from their work are reproduced in fig. 1 together with a schematic C_2N_2 gas spectrum. Apart from a strong decrease of the d-band emission intensity near the Fermi level three adsorbate induced features were recognised at 3, 6, and around 16 eV below E_F. Due to the lack of theoretical treatments of C_2N_2 in the adsorbed state tentative assignments were given by comparison with known gas phase data taking into account the expected relatively strong structural changes upon adsorption. The prominent, asymmetrical peak 6 eV below E_F was ascribed to the overlapping π and nitrogen lone-pair orbitals, the weak, broad feature around 16 eV assigned to C–C σ bonds present in the adsorbate layer. The feature within the Pt d-band region at 3 eV was

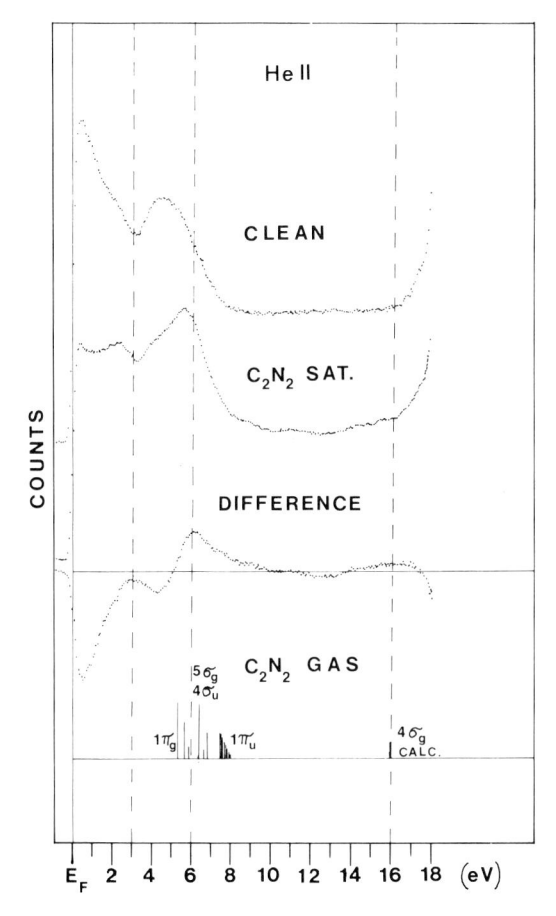

Fig. 1. UPS data of Conrad et al. [1] for cyanogen chemisorbed on Pt(100).

associated with ionisation from levels formed by the coupling of filled metallic d-band states and empty $2\pi^*$ states of the adsorbate.

Although the interpretations of the C_2N_2 induced UPS features given by Conrad et al. have to be considered as preliminary ones, the data seem to support the view of an associated, possibly polymeric adsorbate layer. In this paper we see how the addition of ELS data helps to distinguish between dissociated and non-dissociated forms of cyanogen adsorption.

3. Experimental

The experimental apparatus and procedure were similar to that of previous studies [7,8]. Electron energy loss spectra were taken in $N(E)$ and dN/dE form at normal incidence of the primary beam using the LEED optics as retarding field analyser. No electron induced desorption or dissociation effects on adsorbed C_2N_2 were observed throughout the experiments unlike the case of adsorbed CO, where strong deteriorations of CO induced loss features were detected [7]. Cyanogen was generated in a glass cell attached to the UHV chamber by thermal decomposition of high purity AgCN as described previously [3,4].

4. Results

ELS spectra of the clean and C_2N_2 saturated Pt(100) surface in $N(E)$ and differentiated dN/dE form are shown in figs. 2 and 3. In order to observe the general appearance of the loss region up to 30 eV loss energy $N(E)$ spectra had to be recorded at lower sensitivity due to the strongly varying secondary electron background. Upon C_2N_2 adsorption a broad loss feature shows up in the 10–20 eV loss region, peaked around 14 eV, with some structure suggested on the low energy side (fig. 2). This suggestion is confirmed by the differentiated spectra in fig. 3, which were taken at enhanced sensitivity. Here, two clearly resolved C_2N_2 induced loss peaks are recognized at 11 and 14 eV loss energy. The appearance of the C_2N_2 induced loss features is accompanied by a noticeable enhancement in the reflected elastic current (see also fig. 4). This effects has been investigated in a separate study

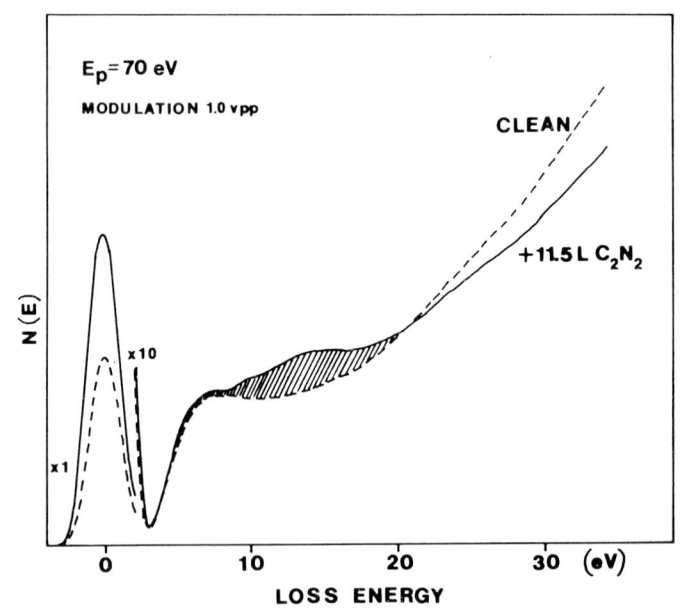

Fig. 2. Electron loss spectra of the clean and C_2N_2 saturated Pt(100) surface in $N(E)$ form.

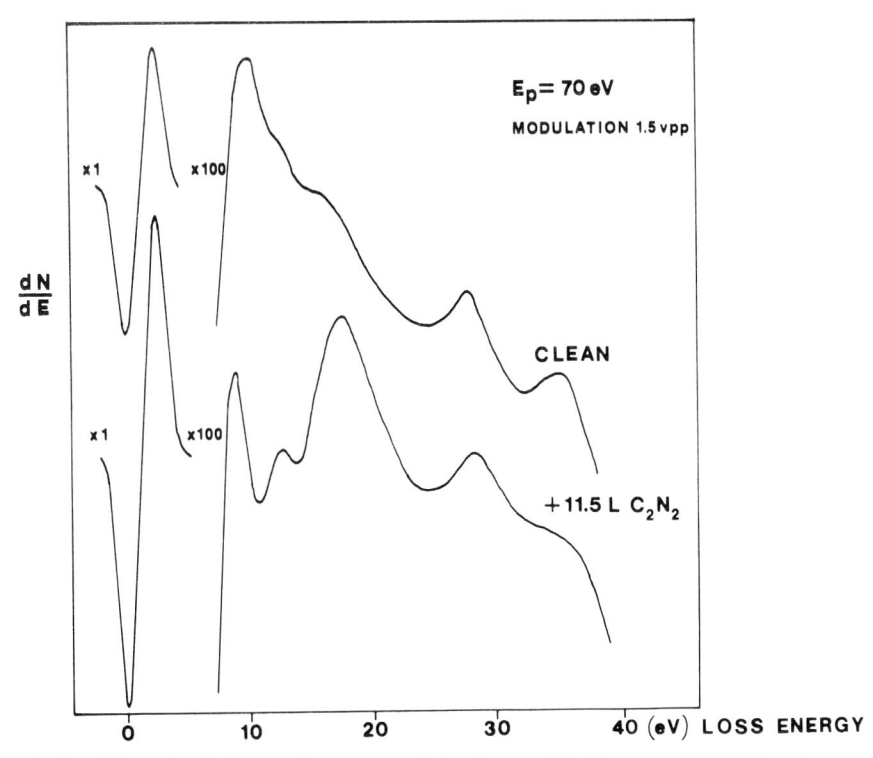

Fig. 3. Differentiated electron loss spectra of the clean and C_2N_2 covered Pt(100) surface.

[9] and shall not be discussed further here. The build-up of the new extrinsic losses as a function of surface coverage is shown fig. 4, where $N(E)$ difference spectra are displayed for various C_2N_2 exposures. The areas below the characteristic adsorbate loss features in the difference spectra can be used as monitors of surface coverage and closely follow the coverage dependence of corresponding Auger signals [9]. Although the resolution is poor in the difference spectra, it has to be noticed that there is no strong extrinsic loss feature in the lower energy region below 10 eV as was observed with adsorbed CO [8].

In order to investigate the differential effects of α and β-C_2N_2 on the loss spectra the surface was saturated with C_2N_2 at room temperature, then heated up to above the desorption temperature of the α state and loss spectra recorded. A sequence of loss spectra in differentiated form according to this procedure is shown in fig. 5. As indicated in the figure desorption of the α state occurs at about 140°C, the loss spectra, however, remain almost unchanged up to 400°C, where desorption of the β states sets in. The spectrum of the clean surface is regenerated after heating to 700–800°C (not shown in fig. 5). We therefore conclude that the α state has only a minor quantitative effect on the loss spectra, which has to be taken into account in

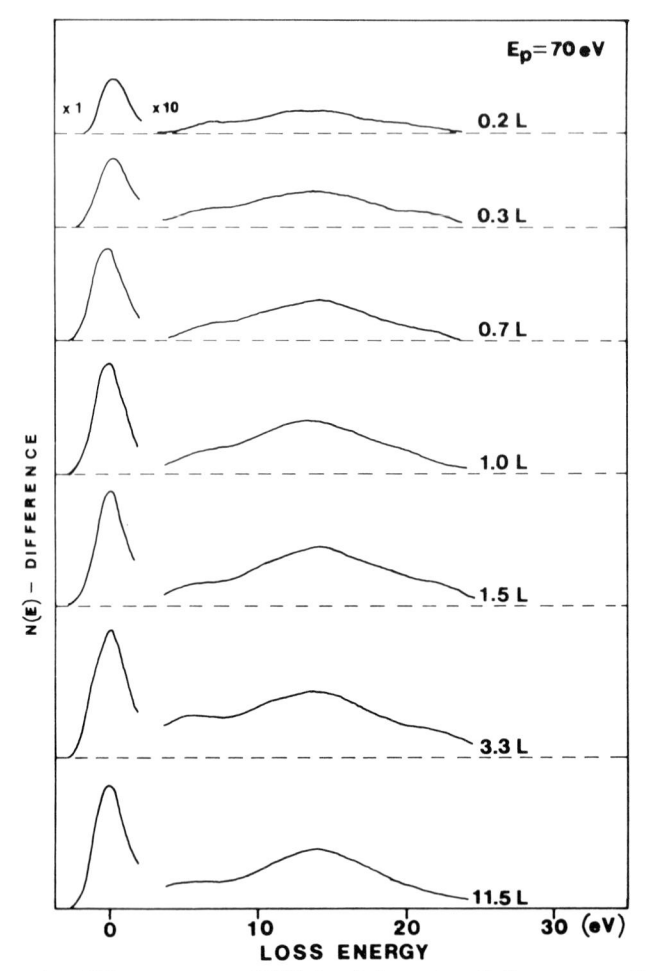

Fig. 4. Electron loss difference spectra ($N(E)$ form) for various C_2N_2 exposures (in Langmuir) on Pt(100) showing the build-up of the C_2N_2 induced loss features as a function of surface coverage.

the interpretation. These findings closely parallel with the results of the UPS study was described by Conrad et al. [1].

5. Discussion

Interpretation of the ELS of C_2N_2 on Pt(100) in relation to other observations on the system poses two main questions:

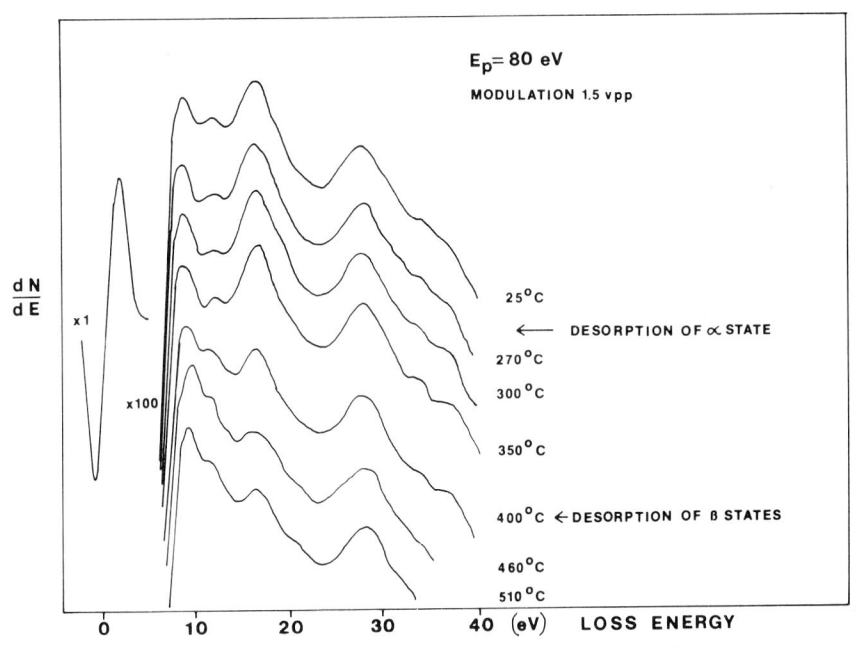

Fig. 5. Electron loss spectra in dN/dE form taken after heating cycles of the room temperature C_2N_2 saturated Pt(100) surface (see text). Desorption processes are schematically indicated.

(a) Does the data help us to distinguish between the very different possibilities of dissociated adsorption and adsorption in some molecular form?
(b) Is it possible to assign initial and final states to the loss features?

If dissociated CN units were bonded to the surface one might expect close correspondence between the properties of Pt(100)/C_2N_2 and those of platinum cyanides and cyanide complexes in solution, by analogy with the similarities observed between metal carbonyl compounds and CO adsorbed on transition and noble metals: ultra violet photoemission, infrared absorption [10] electron losses, and ultra violet absorption [8] all show strong parallels in these systems.

It is interesting therefore to compare UPS from Pt(100)/C_2N_2 [1] with the X ray photoemission spectra of several cyanides including $Li_2Pt(CN)_4$ obtained by Prins and Biloen [11]. In the former case there are adsorbate features 3 and 6 eV below the Fermi level, while in the latter peaks associated with CN units appear at 6 and 9 eV relative to a zero energy level referenced by setting the C_{1s} level to 285.0 eV. Because of the different photon energies used in the two experiments, no intensity comparisons can readily be made, and because the reference levels in the two experiments are not consistent, no firm conclusions can be drawn.

If dissociated adsorption occurs, the electron loss spectrum presented in this paper should show similarities to the ultra-violet absorption spectrum of cyanide complexes. Alexander and Gray [12], Moreau-Colin [2] and others have made very detailed studies of the electronic spectra of Pt cyanides; at around 4.9 and 5.9 eV (the exact energies depending on the precise environment) there are absorption bands of very high extinction coefficient, characteristic of charge transfer between metal and ligand. Such an interpretation is confirmed by Interrante and Messmer [13] in their Xα cluster calculations: though some controversy remains over the extent of metal–ligand mixing in the filled states nominally associated with the cyanide units, the theory predicts ultra violet absorption behaviour similar to that of metal carbonyls. This strongly suggests that for CN adsorbed on a metal there should be a prominent ELS feature at around 5 eV, as is observed in many cases of CO adsorption. The marked contrast between the ELS of Pt(100)/CO and Pt(100)/C_2N_2 in this lower energy loss region provides evidence against dissociated adsorption.

While neutral metal carbonyl complexes are often stable, cyanide complexes e.g. $Pt(CN)_4^{2-}$ are always negatively charged, emphasising the highly ionic character of the CN unit. When CN is bonded directly to a metal surface, one might expect rather large work function changes, contrary to the small shifts noted by Conrad et al. [1]. While none of these comparisons between the properties of cyanide complexes and adsorbed cyanogen is conclusive in itself, taken together they tend to argue against the dissociated adsorption model. However, is it possible to reconcile the observations with the likely properties of C_2N_2 adsorbed, possibly in paracyanogen like units, on the surface?

A useful starting point is examination of the ultraviolet absorption of gas phase cyanogen, which has been studied in detail by Connors, Roebber and Weiss [14]. The most prominent bands are around 9.5 and 13 eV, corresponding to $1\pi_g \rightarrow 2\pi_u^*$ and $1\pi_u \rightarrow 2\pi_g^*$ transitions respectively. A striking feature of free C_2N_2 is the clean splitting between the filled even and odd π states (see fig. 1), and the difference in energies of the two strong absorptions mainly reflects this initial state splitting. However, for molecularly adsorbed C_2N_2, whether adsorbed as single molecules or as associated groups of molecules, the symmetry will be lowered, and the two discrete states are likely to be replaced by a single "band" of states, probably overlapping the σ like N lone pair states, as suggested by Conrad et al. [1]. In the absence of differential shifts of the filled and unfilled levels on adsorption, one might then expect a broad adsorbate induced loss centred at around 11 eV; however, as in the case of CO adsorption, it is possible that, through interaction with the substrate, the unfilled $2\pi^*$ levels will be raised relative to the filled levels [8]. With a differential shift of 3 eV (comparable to that implied in CO adsorption), one can mainly associate the strong 14 eV loss with $\pi \rightarrow \pi^*$ transitions from the states observed in UPS at 6 eV below the Fermi level. The weaker 11 eV loss may then arise by electron promotion from the adsorbate related level 3 eV below the Fermi level to the same excited π^* states. One may note that the final state of the loss process must be placed approximately 8 eV above the Fermi level in contrast to the final states about 4–5 eV above the Fermi level that would be expected for metal–ligand charge transfer transitions in dissociatively adsorbed CN, assuming initial states with energies at the main peak of the Pt d states.

The precise nature of the UPS peak at 3 eV is still not clear. It possibly corresponds to Pt d orbitals strongly perturbed by the adsorbate. In contrast to CN adsorption, the C_2N_2 units are likely to lie approximately parallel to the surface plane, and the interactions would be very different from the relatively ionic bonding likely for a single CN unit adsorption. Theoretical calculations of C_2N_2 incorporated in a cluster of Pt atoms are needed to clarify these ideas: the Xα method could probably be applied to a cluster of sufficient size to give some further insight into the adsorption process.

The ultimate verification of molecular rather than dissociated adsorption lies in the direct identification of C–C bonds. In the UPS work of Conrad et al. [1] there is some evidence to support their presence; electron loss features from such deeper levels are generally rather weak, and the data presented here give no further indication on this question. On the other hand measurement of very low energy vibrational losses in the manner of recent experiments by Backx et al. [15] for C_2H_2 on W could provide further important information.

5. Conclusions

(a) Adsorbate induced electron losses at 11 and 14 eV are observed when C_2N_2 is adsorbed on Pt(100).

(b) The absence of a prominent loss at around 5 eV tends to give support to the hypothesis of adsorption in the β phase in some molecular form.

(c) The losses are likely to be associated with transitions from initial states 3 eV and 6 eV below the Fermi level to unfilled π^* levels shifted upwards relative to the filled levels by about 3 eV due to interactions with the substrate.

Acknowledgements

The Fonds zur Förderung der Wissenschaftlichen Forschung of Austria is acknowledged for supporting the experimental programme, and one of us (J.A.D.M.) wishes to thank the Royal Society of Great Britain for a travel grant under the European Science Exchange Programme.

References

[1] H. Conrad, J. Küppers, F. Nitschke and F.P. Netzer, Chem. Phys. Letters 46 (1977) 571.
[2] M.L. Moreau-Colin, in: Structure and Bonding, Vol. 10 (Springer, Berlin, 1972) p. 167.
[3] F.P. Netzer, Surface Sci. 52 (1975) 709.

[4] F.P. Netzer, Surface Sci. 61 (1977) 343.
[5] M.E. Bridge, R.A. Marbrow and R.M. Lambert, Surface Sci. 57 (1976) 415.
[6] M.E. Bridge and R.M. Lambert, Surface Sci. 63 (1977) 315.
[7] F.P. Netzer and J.A.D. Matthew, Surface Sci. 51 (1975) 352.
[8] F.P. Netzer, R.A. Wille and J.A.D. Matthew, Solid State Commun. 21 (1977) 97.
[9] J.A.D. Matthew, R.A. Wille and F.P. Netzer, Surface Sci. 67 (1977) 269.
[10] H. Conrad, G. Ertl, H. Knözinger, J. Küppers and E.E. Latta, Chem. Phys. Letters 42 (1976) 115.
[11] R. Prins and P. Biloen, Chem. Phys. Letters 30 (1975) 340.
[12] J.J. Alexander and H.B. Gray, J. Am. Chem. Soc. 90 (1968) 4260.
[13] L.V. Interrante and R.P. Messmer, Chem. Phys. Letters 26 (1974) 225.
[14] R.E. Connors, J.L. Roebber and K. Weiss, J. Chem. Phys. 60 (1974) 5011.
[15] C. Backx, B. Feurbacher, R.J. Willis and B. Fitton, Surface Sci. 63 (1977) 193.

Surface Science 68 (1977) 269-276
© North-Holland Publishing Company

THE INTERACTION OF OXYGEN WITH ALUMINIUM (111)

A.M. BRADSHAW, P. HOFMANN and W. WYROBISCH
Fritz-Haber-Institut der Max-Planck-Gesellschaft, Faradayweg 4-6, D-1000 Berlin 33, Germany

The initial uptake of oxygen by an aluminium (111) surface at 300 K has been studied by several experimental techniques including XPS, AES, $\Delta\phi$, ellipsometry and surface plasmon spectroscopy. The interaction falls into two stages. Oxygen exposure up to 30-50 L in the first stage results in a chemisorbed layer with oxygen atoms located in the immediate surface region. In the second stage increased exposure produces a thin, relatively homogeneous oxide film.

1. Introduction

The interaction of oxygen with polycrystalline aluminium surfaces at room temperature has been thought by many authors to lead directly to incorporation or oxidation without an inital chemisorption phase [1-4]. Experimental evidence for this conclusion lies mainly in the decrease in work function brought about by oxygen sorption [1-3,5-7]: in the majority of oxygen/metal systems work function increases are observed. Recent electron spectroscopic [8] and Auger studies [7,9] offer little insight into this problem. With two very recent exceptions [10,11] no new work on single crystal aluminium surfaces has appeared since the LEED investigations of Jona [12], who found no evidence for ordered adsorption. Prompted by our interest in the inital stages of corrosive oxidation we have studied the interaction of oxygen with aluminium (111) using a variety of experimental techniques. The results indicate that chemisorption precedes oxide film formation.

2. Experimental

The XPS measurements were carried out with an AEI 200 A photoelectron spectrometer with a base pressure of $2-3 \times 10^{-10}$ Torr. LEED, Auger and electron energy loss investigations were performed in a separate UHV system in which work function changes, $\Delta\phi$, could be measured with a Kelvin probe. Suitable window ports also permitted ellipsometric investigations with a Rudolph null ellipsometer. Pressure was measured with the same nude ionisation gauge and control unit at approximately the same distance from the crystal in the two systems.

The aluminium single crystals (Metals Research, 6 N) were oriented to within 1/2° of the (111) plane, cut by spark erosion, re-oriented on the polishing jig and then polished with successively finer diamond pastes. One crystal was then electropolished in a perchloric acid/acetic anhydride solution which left a bright mirror finish but with several isolated etch pits. The other crystal was electropolished in perchloric acid/ glacial acetic acid using a rotating electrode arrangement. The bloom caused by the latter solution was removed with a felt polishing cloth. The XPS investigations were carried out with the first crystal; the remaining measurements with both crystals. Only in the case of the work function change was a difference in behaviour between the two crystals observed. In situ cleaning consisted of argon-bombardment– anneal cycles lasting for several days (anneal temperature 450°C). Between oxygen sorption experiments the crystal was subjected to two or three such cycles; residual oxygen was checked with XPS or AES. Oxygen exposures under 300 L were carried out at $\leqslant 2 \times 10^{-7}$ Torr.

3. Results and discussion

As shown in a previous paper [13] the aluminium surface plasmon (~11 eV) couples to the 1s level of oxygen adsorbed at low coverages on aluminium (111). This is shown in fig. 1a. The coupling parameter was shown to be strongly coverage dependent, decreasing rapidly up to 30 L (1 L = 1 Langmuir = 1×10^{-6} Torr-sec). At higher exposures the plasmon satellite shifts to lower energies corresponding to the interface plasmon Al/Al_2O_3 at ~8 eV [14–17] (see fig. 1b). The fact that coupling occurs only to the surface plasmon and not to the volume plasmon indicates that the oxygen atoms are adsorbed on the surface or in the immediate surface region [13] during the early stages of the interaction. As Šunjić and Šokčević [18] show, the coupling to the surface plasmon can still take place when photoionization occurs in the bulk, but then the probability of coupling to the volume plasmon should become non-zero. At exposures greater than 50 L the O 1s peak, initially at 531.3 ± 0.2 eV binding energy with fwhm 1.4 eV, shifts to higher binding energies and broadens considerably. Thus after 10^3 L the binding energy is 531.7 eV with fwhm 1.9 eV and after 10^5 L, 532.4 eV with fwhm 2.2 eV. This is evidence for the formation of oxide after an exposure of 30–50 L. The integrated intensity of the O 1s peak as function of exposure is plotted in fig. 2a. Because of the stability of the X-ray source and the accurate re-positioning of the sample the data from five different runs could be plotted without normalizing. By assuming that 50 L corresponds to a chemisorbed monolayer an initial sticking coefficient of ~0.02 is estimated. A more detailed kinetic analysis is not possible because of the difficulty of finding the exact point at which oxide formation begins. By heating the surface to 450°C after exposure to 30 L oxygen the O 1s peak also shifted to the binding energy 532.4 eV and broadened considerably (fwhm 2.2 eV) without any loss of integrated intensity. This indicates complete conversion of chemisorbed

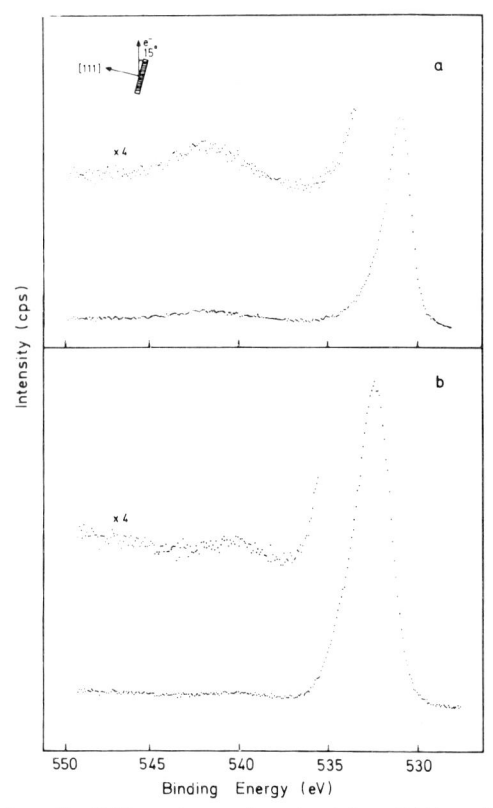

Fig. 1. (a) O 1s peak from the XPS spectrum of an aluminium (111) surface after exposure to 5 L oxygen (~0.17 monolayer); source radiation MgK$_\alpha$. (b) O 1s peak after 1000 L; intensity scale reduced by factor 0.5 compared to (a).

oxygen into "oxidic" oxygen at this temperature. The difference between the two forms of bound oxygen on aluminium (111) is also shown in recent XPS work by Flodström et al. [10]. Using synchrotron radiation this group has been able to separate two distinct substrate chemical shifts on the Al 2p level. On aluminium (100) and (110), however, no initial chemisorption phase could be distinguished.

The corresponding development of the oxygen KLL Auger peak as function of exposure is shown in fig. 2b. The spectra were recorded in the $N(E)$ mode and the results from three runs normalized to an integrated intensity at 150 L. The shape of the curve is somewhat different to that obtained in XPS although it agrees well with the result of Martinsson et al. [10]. At higher exposures a levelling-off is apparent much earlier in the Auger case, probably because of a lower effective sampling depth. More pertinent to the present paper is the fact that the peak maximum shifts from about 506 to 504 eV over the range 0—50 L, thereafter changing only minimally. The significance of this exposure range for the interaction of oxygen with

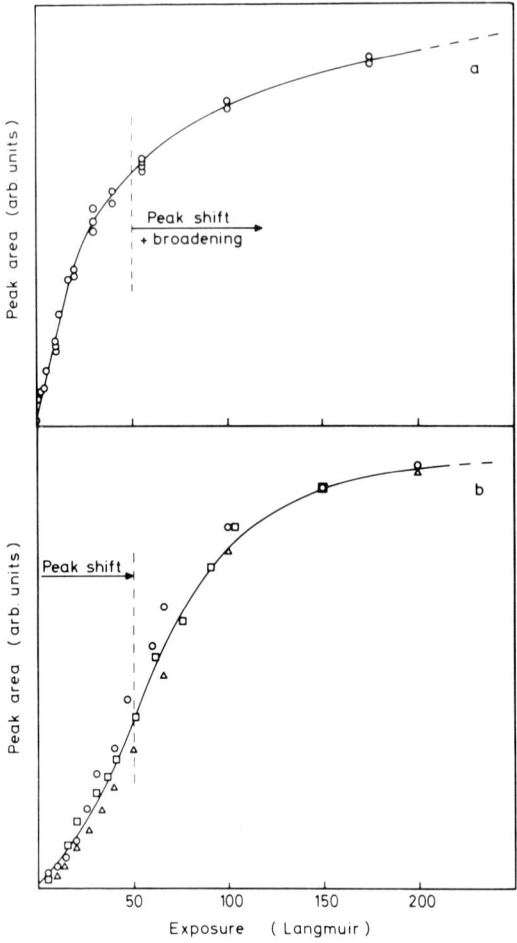

Fig. 2. (a) Oxygen 1s integrated intensity in XPS as function of exposure. (b) Oxygen Auger KLL integrated intensity as function of exposure. As in all the measurements reported here in the range 0–50 L no pressure dependence was observed for exposures at and below 2×10^{-7} Torr.

aluminium (111) is also seen in electron energy loss measurements on the elastic peak. From the curves of fig. 3 we see that the intensity of the surface plasmon peak decreases by 70% over the range 0–50 L as observed in the XPS coupling experiment described above. Its position, however, remains virtually constant at approx. 10.7 eV until 30 L where its energy starts to decrease, reaching about 9 eV after 10^2 L. Higher exposures bring about a more gradual energy shift: after 10^3 L the peak is found at 8.8 eV loss energy and after 5×10^4 L at 7.5 eV. The latter value corresponds to the interface plasmon first observed by Powell and Swan [14]

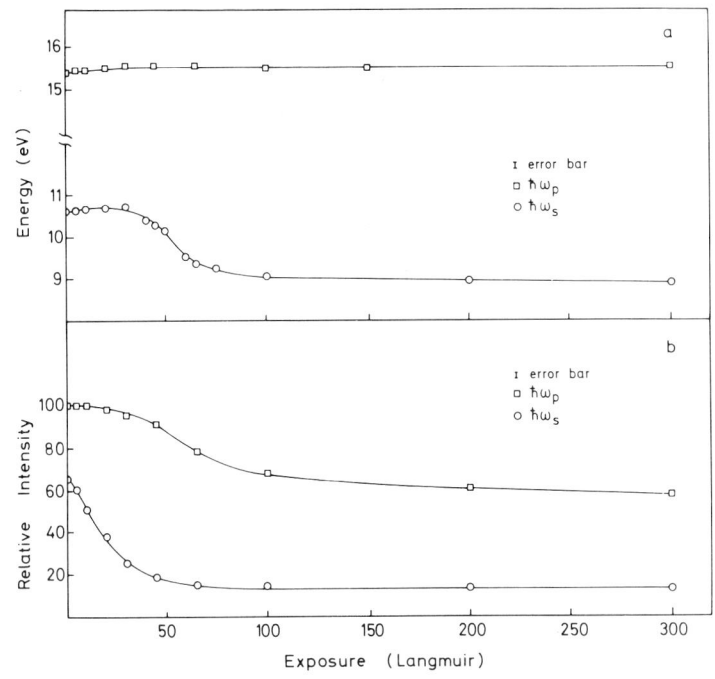

Fig. 3. (a) Energy, and (b) peak area (intensity) of volume and surface plasmon peaks in the electron energy loss spectrum as function of oxygen exposure. Primary beam energy 350 V.

for thick oxide layers on evaporated aluminium films. The continuous shift would appear to be characteristic for the nature of the oxidation process on aluminium (111). Several investigations on polycrystalline material [14,16,17] have indicated the extinguishing of the surface plasmon peak concomitant with the growth of the interface plasmon peak. Kunz [15] has argued that the two peaks can occur simultaneously in measurements which integrate over various scattering angles and thus produce a weighting of certain plasmon k vectors. In most integrating experiments, however, only small k vectors will play a role because of the inverse cubic dependence on scattering angle. Murata and Ohtani [16] obtained both types of behaviour depending only on the oxidation conditions. It seems therefore more likely that the nature of the oxidation process is responsible. When nucleation in the oxide phase leads to comparatively large areas of surface which are oxidised as well as areas where only chemisorbed oxygen is present then two peaks may be expected. This is probably the case on polycrystalline films. For the more homogeneous oxide layers expected on a smooth single crystal surface, the continuous shift predicted by Stern and Ferrell [19] will be observed, despite the integration over scattering angle. An additional effect not to be neglected is the change in the dielectric constant of the oxide layer during its growth. Because perfect homogeneity is very unlikely the dielectric constant will be below the bulk value particularly during the early stages of oxidation.

The work function change observed for oxygen on aluminium (111) was found to depend strongly on the preparation conditions. The crystal which was polished in perchloric acid/glacial acetic generally gave larger changes than the one polished in perchloric acid/acetic anhydride. A dependence on bombardment and annealing times was also observed. The standard two cycles of one hour argon ion bombardment followed by 20 min anneal at 450°C, which was found in XPS and AES to produce a virtually oxygen-free surface, gave the result shown in fig. 4. The decrease in work function is unexpected in view of the conclusions above concerning the nature of the interaction in the initial stages. A chemisorbed layer is expected to produce an increase in work function, whereas incorporation or oxidation would result in a decrease. The result is also at variance with the photoelectric measurements of Gartland [11], who observed initially no change in work function and then after 30–50 L a slight increase. The discrepancy here could be due to a number of factors. The photoelectric method measures for insulators and metal/insulator systems a somewhat different quantity compared to the Kelvin method and this could be important during the later stages of the interaction. In fact discrepancies between the photoelectric and Kelvin methods occur quite often, for instance in the system Xe/Ag [20], which would not be expected to offer the same complexity as oxygen/Al. Alternatively, our results could be influenced by adsorption on the flame-annealed stannous oxide coating of the reference electrode or even by sputtering of aluminium onto it during argon ion bombardment. However, our results on copper surfaces [21] obtained with the same arrangement yield good agreement with previous investigations [22]. Although the extent of the decrease in work function in the present results depended sensitiviely on preparation conditions, the form of the curve remained the same: the rate of decrease is initially higher, namely, in the exposure range 0–50 L. Some recent results of Ramsay and Pritchard [23] on Al (111) also show a decrease in work function, but preceded by an initial increase. Despite the difference in behaviour between the (111) surface and polycrystalline material, the result reported here resembles paradoxically that of Agarwala and Fort [3].

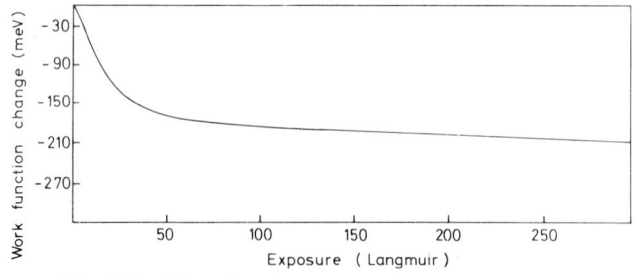

Fig. 4. Work function change $\Delta\phi$ against oxygen exposure.

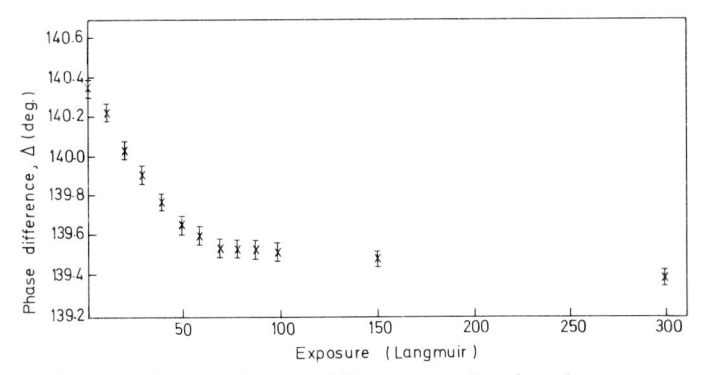

Fig. 5. Change in ellipsometric phase difference, Δ as function of oxygen exposure.

The preliminary ellipsometric data from this system is given in fig. 5, which illustrates the change in phase difference, Δ, as a function of oxygen exposure at $\lambda = 546.1$ nm. In the Drude–Tronstad approximation [24] Δ is directly proportional to the optical film thickness, $\bar{L}$ for a film of constant refractive index. Because the changes in azimuth, ψ, at this wavelength, are of the same order of magnitude as the accuracy of the ellipsometer, it is not possible to determine both the refractive index (real, at this wavelength) and the film thickness. Using the refractive index of bulk Al_2O_3 ($n_L = 1.635$ [25]) only seems sensible at higher film thicknesses. After 10^4 and 10^5 L we obtained 0.77 nm and 0.94 nm respectively. After prolonged exosure to air a thickness of 1.6 nm was estimated. The optical constants of the clean aluminium (111) surface at $\lambda = 546.1$ nm were found to be $1.14 - i\, 7.42$. The ellipsometric parameters for the clean surface and thus the refractive index depended sensitively on the preparation conditions. For bombarded but unannealed surfaces the real part of the refractive index could be as much as 20% lower than for the smooth annealed surface. The corresponding decrease in the imaginary part was an order of magnitude smaller. By measuring at shorter wavelengths it will be possible to observe larger values of $\delta\psi$ which will make possible a complete ellipsometric characterisation. (For instance, we estimate using literature values for the optical constant [25] that $\delta\psi/\sin 2\bar{\psi}; \bar{L}$ is 0.4 and 0.7 deg nm^{-1} at 350 and 250 nm respectively, compared to values under 0.2 deg nm^{-1} in the visible.) The curve of fig. 5 also shows a "break" but slightly higher than 50 L exposure. This discrepancy could be due to the changing dielectric constant.

4. Summary

The present results show that the initial stages of corrosive oxidation on aluminium single crystal surfaces can only be successfully investigated with a wide range of surface sensitive techniques. We show for the (111) surface with XPS, AES and surface plasmon spectroscopy that a chemisorbed layer results in the exposure range 0–50 L. Thereafter oxidation or incorporation takes place producing an oxide layer more homogeneous than that encountered on polycrystalline films. The $\Delta\phi$ results for this system do not agree with those obtained with the photoelectric method and are somewhat surprising in view of the conclusion that a chemisorbed monolayer is formed. Further work is required to account for this discrepancy. Further characterisation of the oxide layer with ellipsometry awaits measurements at shorter wavelengths.

Acknowledgements

The authors would like to thank the referee for pointing out the implications of ref. [18]. This project has been supported financially by the Deutsche Forschungsgemeinschaft.

References

[1] M.W. Roberts and B.R. Wells, Surface Sci. 15 (1969) 325.
[2] W.H. Krueger and S.R. Pollack, Surface Sci. 30 (1972) 263.
[3] V.K. Agarwala and T. Fort, Surface Sci. 45 (1974) 470.
[4] K. Yu, J.N. Miller, P. Chye, W.E. Spicer, N.D. Lang and A.R. Williams, Phys. Rev. B14 (1976) 1446.
[5] E.E. Huber and C.T. Kirk, Surface Sci. 5 (1966) 447.
[6] V.K. Agarwala and T. Fort, Surface Sci. 48 (1975) 527; 54 (1976) 60.
[7] C. Benndorf, H. Seidel and F. Thieme, Surface Sci., in press.
[8] J.C. Fuggle, L.M. Watson, D.J. Fabian and S. Affrossman, Surface Sci. 49 (1975) 61.
[9] T.H. Allen, J. Vacuum Sci. Technol. 13 (1976) 112.
[10] C.W.B. Martinsson, L.-G. Petersson, S.A. Flodström and S.B.M. Hagström, in: Proc. Intern. Photoemission Symposium, Noordwijk (ESA, 1976) p. 177;
S.A. Flodström, private communication.
[11] P.O. Gartland, Surface Sci. 62 (1977) 183.
[12] F. Jona, J. Phys. Chem. Solids 28 (1967) 2155.
[13] A.M. Bradshaw, W. Domcke and L.S. Cederbaum, Phys. Rev. B15 in press.
[14] C.J. Powell and J.B. Swan, Phys. Rev. 118 (1960).
[15] C. Kunz, Z. Physik 196 (1966) 311.
[16] Y. Murata and S. Ohtani, J. Vacuum Sci. Technol. 9 (1972) 789.
[17] C. Benndorf, G. Keller, H. Seidel and F. Thieme, J. Vacuum Sci. Technol., in press.
[18] M. Sunjić and D. Sokcević, Solid State Commun. 18 (1976) 373.
[19] E.A. Stern and R.A. Ferrell, Phys. Rev. 120 (1960) 130.
[20] J. Pritchard, private communication.
[21] P. Hofmann, R. Unwin, W. Wyrobisch and A.H. Bradshaw, Surface Sci., submitted.
[22] T. Delchar, Surface Sci. 27 (1971) 11.
[23] J. Ramsay and J. Pritchard, to be published.
[24] A. Vasicek, NBS Miscellaneous Publication No. 256 (1964) 25.
[25] H.J. Hagemann, W. Gudat and C. Kunz, J. Opt. Soc. Am. 65 (1975) 742, and references therein.

THE ADSORPTION OF CARBON MONOXIDE ON RHENIUM: BASAL (0001) AND STEPPED |14 (0001) $\times$ (10$\bar{1}$1)| PLANES

M. HOUSLEY, R. DUCROS, G. PIQUARD and A. CASSUTO

Centre de Cinétique Physique et Chimique, CNRS, BP 104, 54600 Villers-les-Nancy, France

The influence of surface defects on the adsorption of CO by rhenium is investigated using LEED, AES and linear temperature programmed desorption. On both surfaces, thermal desorption reveals two adsorption states, the lower temperature α state being resolved into two substates, and one β state, all desorbing with first order kinetics. The α state is unaffected by the surface texture, its maximum population being the same on both surfaces, around 4×10^{14} molecules cm^{-2}, similar to the value found for polycrystalline rhenium. On the other hand, the β state is strongly dependent on surface structure. On Re(0001) a maximum of 4×10^{13} molecules cm^{-2} was found, and 2×10^{14} molecules cm^{-2} on the stepped surface. The adsorption is activated and can be increased, by heating to 550 K, to 2×10^{14} molecules cm^{-2} on the basal plane and 3.5×10^{14} molecules cm^{-2} on the stepped surface. Ordered structures are now seen in LEED. Comparison of these results with previous results from polycrystalline rhenium indicate that the dissociation of β-CO on the latter surface must occur at defects other than steps.

1. Introduction

The adsorption of carbon monoxide by transition metal surfaces has been studied extensively by numerous experimental techniques [1]. The system has proved to be complex and there is as yet much discussion as to the nature of the β-CO species. It is generally agreed that the low temperatures α states, with adsorption heats in the range 20–30 kcal mole^{-1}, correspond to adsorbed molecules of carbon monoxide, but the binding configuration of the higher energy β states, 50–70 kcal mole^{-1} remains controversial. Recent studies in this laboratory [2] on polycrystalline rhenium, showed that CO adsorbs dissociatively in the β states. The results were interpreted in terms of a lateral interaction model, originally proposed by Goymour and King [3]. In an attempt to relate experimentally observed binding states to surface structure, we have studied the adsorption of CO on two single crystal faces of rhenium, the basal (0001) plane and a stepped plane, Re |14 (0001) $\times$ (10$\bar{1}$1)|. We use the notation proposed by Somorjai et al. [4] to describe the stepped surface, which has monatomic steps of orientation (10$\bar{1}$1) and terraces of 14 atom spacings width, of orientation (0001).

In the discussion section, the results obtained on these two surfaces will be compared with those from polycrystalline rhenium ribbons.

2. Experimental

The apparatus will be described in detail elsewhere [5]. Briefly it consists of two separately pumped chambers, joined by a small hole of 3 mm diameter. The main chamber contains a Varian 4 grid LEED optics, which doubles as an Auger spectrometer for the purposes of controlling surface cleanliness, and a crystal manipulator. The sample can be turned in front of the hole, in which position it has direct line-of-sight to the ionisation chamber of a quadrupole mass spectrometer (Riber QS 100) situated in the second chamber. Using this arrangement thermal desorption spectra can be recorded from the front face of the sample without interference by desorption from crystal edges, supports, etc. Vacua of 10^{-10} Torr were regularly attained in the main chamber following bakeout.

The samples were single crystals, cut and mechanically polished to expose the desired face. They were heated by electron bombardment from a filament situated a few millimeters behind them. Cleaning was carried out in situ, by the normal practice of heating in oxygen, followed by high temperature flashing in vacuum. Surfaces prepared in this way gave sharp diffraction patterns, and no surface impurities were detectable with AES.

3. Results

3.1. Adsorption on the basal plane

At 300 K, the maximum coverage of CO is 5.1×10^{14} molecules cm^{-2}. Desorption spectra recorded at increasing coverages are shown in fig. 1. There are two α states, with a maximum population of 4.7×10^{14} molecules cm^{-2} and a relatively small β state, population 4×10^{13} molecules cm^{-2}. From the observation that the temperature maxima of the desorption peaks are independent of coverage, we conclude that the desorption process follows first order kinetics. It is interesting to note that the α and β states populate simultaneously after an exposure of 3×10^{-7} Torr sec. The sticking probability was calculated to be constant at around 0.1 during a large part of the adsorption. No additional ordered structure was seen with LEED, neither after adsorption at 300 K nor after progressive heating of a saturated layer to a maximum of 700 K, at which temperature the β state begins to desorb.

The population of the β state can be increased to a maximum of 2×10^{14} molecules cm^{-2} by adsorption at 550 K. The α states no longer exist at this temperature. On cooling to room temperature, LEED shows an ordered structure, in the form of a six-pointed star, which is shown schematically in fig. 2. This diagram is interpreted [6] as resulting from three domains of rectangular sublattices of adsorbate, of dimensions $2a \times a\sqrt{3}$ R $30°$, oriented $120°$ with respect to one another (fig. 3). Thermal desorption of this layer shows a β state with the same characteristics as that from a layer adsorbed at 300 K.

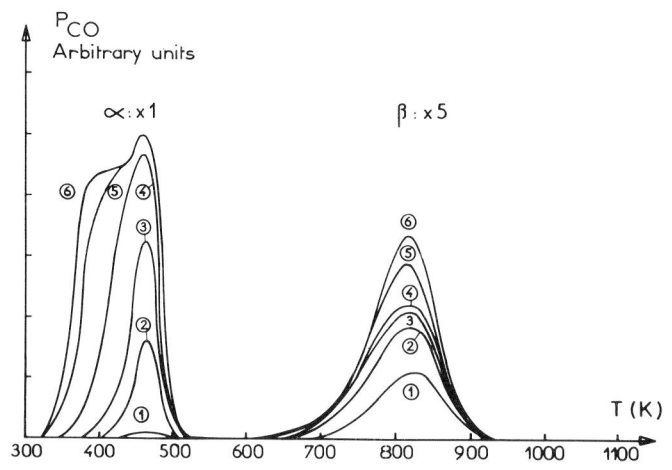

Fig. 1. CO thermal desorption spectra from Re(0001) after the following exposures (in Torr sec): (1) 2×10^{-8}, (2) 4×10^{-7}, (3) 7.5×10^{-7}, (4) 1.5×10^{-6}, (5) 1.2×10^{-5}, (6) $\sim 10^{-4}$. Heating rate: 16 K sec^{-1}. The β state has been enlarged 5 times with respect to the α state. Desorption energies: α states 24 and 27 kcal mole^{-1}; β state 50 kcal mole^{-1}.

3.2. Adsorption on Re $|14\,(0001) \times (10\bar{1}1)$

After adsorption of CO on this surface at 300 K, the maximum population is 6.5×10^{14} molecules cm^{-2}. LEED shows a poorly resolved (2×2) pattern with some disorder along the step direction. The desorption spectra are similar to those

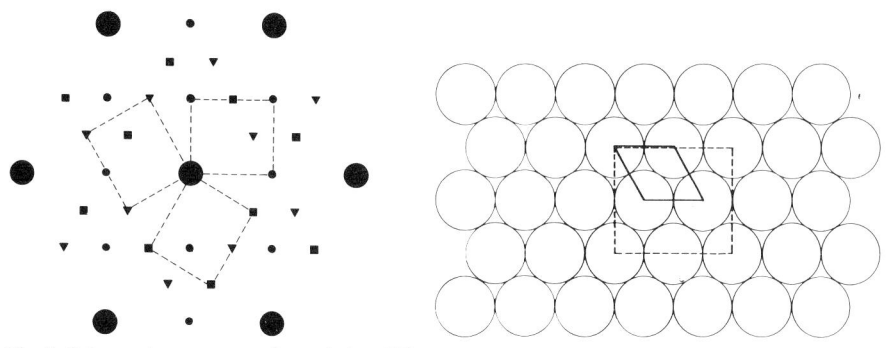

Fig. 2. Schematic representation of the diffraction pattern obtained after increasing the β-CO population on Re(0001) by adsorption at 550 K. The large black circles are the substrate diffraction pattern, the three rectangular domains are outlined with dotted lines.

Fig. 3. Schematic diagram of the surface unit mesh (solid line) and the adsorbate unit mesh (dotted line) which produces the star diffraction pattern, 3 domains of such meshes are imagined oriented at 120° to each other.

for CO adsorbed on the basal plane. There are two α states (population 4.4×10^{14} molecules cm^{-2}) and one β state of considerably higher population than on the basal plane, 2.1×10^{14} molecules cm^{-2}. A major difference is the sequence of filling of the various adsorption states as shown clearly in fig. 4; on this plane the β state fills first and afterwards the α states, once the β is almost filled. As before, the desorption kinetics are first order for all the states, since the temperature maxima are independent of coverage.

Adsorption at 550 K results in an increase of the β state population, to a maximum of 3.4×10^{14} molecules cm^{-2}. A well resolved (2×1) structure appears in LEED, with the double spacing along the step direction.

In fig. 5, we show the desorption spectrum from a saturated CO layer adsorbed at 300 K on the stepped surface after ion bombardment. Before the adsorption was started, the surface was examined with the LEED and no ordered pattern was seen. In the desorption spectrum the α states are no longer resolved as they were on the ordered surfaces. The most interesting feature is however, the appearance of a high temperature shoulder on the β state peak. The β state desorption now begins to resemble that from polycrystalline ribbons, also shown in fig. 5.

Lastly, in order to more precisely characterise the sites for adsorption we studied the coadsorption of CO with oxygen. The stepped surface was first covered by varying amounts of oxygen and then CO was adsorbed on top, to saturation. Desorption

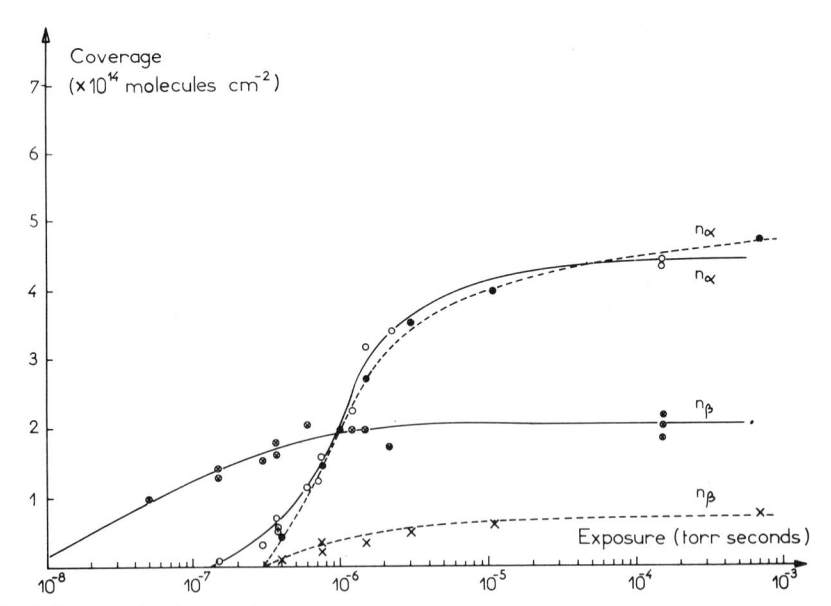

Fig. 4. Coverage in the α and β states as a function of exposure (solid line: stepped surface Re|14 (0001) × (10$\bar{1}$1)|, dashed lined: basal plane Re(0001)), showing the sequence of state-filling on these surfaces.

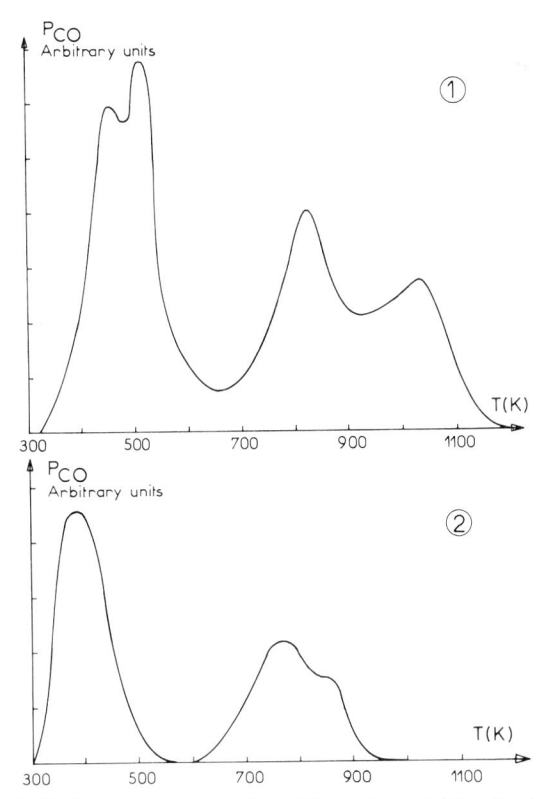

Fig. 5. Comparison of CO desorption spectra from (1) a polycrystalline rhenium ribbon (reproduced from ref. [2] with permission), and (2) the stepped single crystal after ion bombardment.

spectra of CO for varying initial oxygen coverages are shown in fig. 6. The results are surprising, for small initial oxygen coverages the α states are strongly attenuated but the temperature maxima are not changed, and the β state is unaffected. For high oxygen coverages the β state population is decreased and the desorption peak temperature decreases.

4. Discussion

On the two single crystals studied here and on polycrystalline ribbons [2], the sticking coefficient for CO is relatively high, between 0.1 and 0.5 and remains constant during a large part of the adsorption. To account for this behaviour it is normally assumed that adsorption takes place via a physisorbed precursor state which is weakly bound to the surface and hence mobile. The molecules in this precursor "explore" the surface until vacant chemisorption sites are found, where-

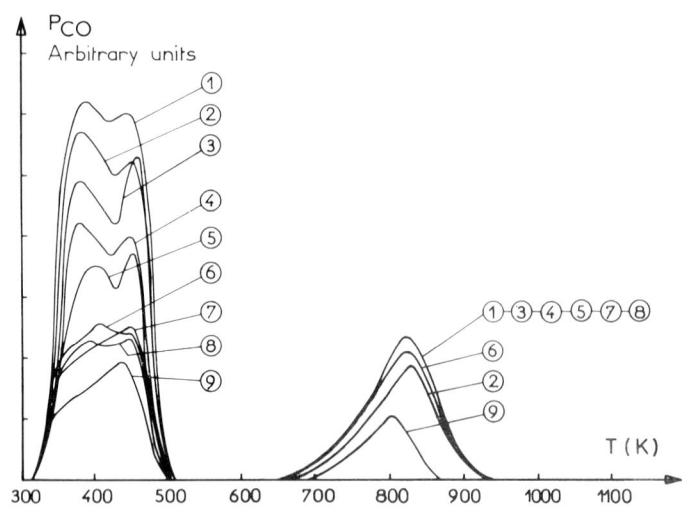

Fig. 6. Effect of increasing amounts of preadsorbed oxygen upon CO adsorption. Oxygen pre-exposures (in Torr sec): (1) clean surface; (2) 3×10^{-7}; (3) 6×10^{-7}; (4) 1.2×10^{-6}; (5) 2.4×10^{-6}; (6), (7), (8) 4.8×10^{-6}; (9) 2.7×10^{-5}. (These results were obtained on a second stepped plane, with a slightly lower step density, hence the smaller $\beta : \alpha$ population ratio.)

upon the higher energy states populate initially, followed by the lower energy states. This is exactly what was seen on the stepped surface, Re $|14(0001) \times (10\bar{1}1)|$ and on polycrystalline rhenium. Work function measurements conducted on the stepped surface revealed the existence of two distinct binding states at room temperature. Similar results were obtained previously on polycrystalline ribbons [2]. However, because of the simultaneous filling of the α and β states on the basal plane and the very low maximum coverage of the β state, it would appear that the β state results from transfer from the α state during thermal desorption, and does not exist on the surface at room temperature. Unfortunately, we have no work function data for this surface to confirm this idea.

The density of defects on Re(0001) is low compared to both the stepped surface and polycrystalline ribbons. If we assume that these defects act as sites for adsorption into the β state, then we can safely assume that the mobility of the CO overlayer is not high enough to extensively populate the β state. Heating the substrate to 550 K, however, increases the mobility and the β state can be populated even on the basal plane.

It is clear that the α states are more or less insensitive to the surface structure: the desorption spectra are similar and yield similar coverages on all three surfaces. This is not true for the β states. If we assume that the population of the β state on the single crystal surfaces is proportional to the number of step atoms, we can calculate, by comparison of the maximum β populations, the number of "step atoms" or defects on the basal plane. This number is found to be one atom in 75, or in other words, the basal plane is composed of monatomic steps with terraces 75 atom spacings wide.

On both single crystal surfaces the adsorption of CO is activated and allows the development of ordered structures probably originating at the defect atoms. These structures form on the terraces and evidently are dependent on the surface structure. The formation of three domains of adsorbate, each at 120° with respect to the others, can be explained by the presence of steps along the three major axes of the surface. On Re $|14(0001) \times (10\bar{1}1)|$, the regularly spaced steps all in one direction allow only the formation of structures with the same directional orientation. Interpretation of the diffraction patterns indicates a density of atoms on the stepped surface higher than that on the basal plane. We have no precise explanation for this, but it is possible that the electronic properties of the surface are modified by the presence of the steps. This modification could take the form of a long-range interaction between the steps through the crystal band structure, similar to that proposed by Grimley [7] for two atoms adsorbed on a metal surface.

The adsorption of oxygen on Re $|14(0001) \times (10\bar{1}1)|$ at 300 K produces a (2×2) structure which must develop on the terraces. It would seem normal, therefore, that the β-CO state which develops essentially along the steps, is not affected by low oxygen coverages. The α-CO states, which we assume populate the terraces, are instantly affected by the preadsorbed oxygen. Certainly on this stepped surface, the α state cannot be regarded as a "gap-filling" phase.

Lastly, as mentioned earlier, the β-CO states on polycrystalline rhenium were interpreted in terms of recombination of dissociated CO molecules. The desorption kinetics revealed in the spectra reported here from the two single crystals are first order and hence reject the notion of complete dissociation of the CO molecule on these surfaces. We realise, however, that the first order kinetics could also arise from desorption of partially dissociated species which are adsorbed in discrete patches or islands. It is clear, nevertheless, that the β-CO is not completely dissociated into C and O atoms and spread at random over the entire surface. We must therefore assume that the dissociation of CO on polycrystalline rhenium ribbons is due to defects other than steps. The exact nature of these defects is, however, far from clear. The experiment, mentioned above, of adsorbing CO on a completely disorded surface, produced by ion bombardment, did not produce very similar desorption spectra. We must thus draw the conclusion that these defects must at least be ordered in some way.

Acknowledgements

Acknowledgement is made to the ATP of the CNRS for provision of research funds.

References

[1] R. Gomer, in: Proc. 2nd Intern. Conf. on Solid Surfaces, 1974, Japan. J. Appl. Phys. Suppl. 2, Pt. 2 (1974) 213;
R. Ford, Advan. Catalysis 21 (1970) 51.
[2] M. Alnot, J.J. Ehrhardt, J. Fusy and A. Cassuto, Surface Sci. 46 (1974) 81.
[3] C.G. Goymour and D.A. King, J.C.S. Faraday I, 69 (1973) 749.
[4] B. Lang, R.W. Joyner and G.A. Somorjai, Surface Sci. 30 (1972) 440.
[5] G. Piquard, J.P. Mihe, R. Ducros and A. Cassuto, to be published.
[6] F. Delamare and G.E. Rhead, Surface Sci. 35 (1973) 185.
[7] T.B. Grimley, J. Phys. Chem. Solids 14 (1960) 227.

A STUDY OF THE INTERACTION OF SULFUR-CONTAINING ALKANES WITH CLEAN NICKEL

C.F. BATTRELL, C.F. SHOEMAKER and J.G. DILLARD

Department of Chemistry, Virginia Polytechnic Institute and State University, Blacksburg, Virginia 24061, USA

The chemical adsorption and reaction of dimethyl disulfide (DMDS) and bis(trifluoromethyl) disulfide (BTFMDS) on clean polycrystalline nickel at 25°C have been investigated using XPS (X-ray photoelectron spectroscopy) and TD-MS (thermal desorption with mass spectrometric analyses). XPS results for DMDS indicated that only one carbon and one sulfur species were present on the nickel surface. From the measured binding energies it appears that the carbon species is a CH_3 or a CH_3S type adsorbed species while the sulfur species is probably sulfide ($S^=$) or the CH_3S species. For BTFMDS adsorption, only sulfur as sulfide was detected on the nickel surface. Thermal desorption spectra have been measured as a function of exposure at 1, 10 and 60 L. These TD measurements revealed that the adsorption process was predominantly dissociative for DMDS and BTFMDS. Adsorption and reaction processes consistent with the XPS and TD-MS results are presented and discussed.

AN AES STUDY OF THE SURFACE COMPOSITION OF COBALT FERRITES

F. GARBASSI, G. PETRINI and L. POZZI

Montedison S.p.A., "G.Donegani" Research Institute, I-28100 Novara, Italy

G. BENEDEK

Gruppo Nazionale di Struttura della Materia del CNR, Via Celoria 16, Milano, Italy

and

G. PARRAVANO

Department of Chemical Engineering, University of Michigan, Ann Arbor, Michigan 48109, USA

A series of cobalt ferrite samples, $Co_{1-y}Fe_{2+y}O_4$, where $-0.1 \leq y \leq 0.1$, was examined by AES. A remarkable Fe surface enrichment was observed near the stoichiometric composition for y both positive and negative, and found to be in quite good agreement with theoretical calculations. The submission of samples to various treatments, at different temperatures and gas atmospheres, produced changes in the surface composition, in a fashion associated with the oxygen concentration both in gas and solid phase.

1. Introduction

The chemical composition, stability ranges and defect structure of cobalt ferrite, $CoFe_2O_4$, have been extensively studied in recent years [1]. These studies have stimulated investigations on the surface reactivity and catalytic properties of cobalt ferrite [2,3]. In the latter studies a decisive factor for obtaining meaningful kinetic correlations was the composition of the reacting surface, which could be inferred only from the bulk composition. With the introduction of Auger electron spectroscopy (AES) into surface studies, the composition of surface layers may be obtained directly. Consequently, we have submitted a series of non-stoichiometric cobalt ferrite samples, $Co_{1-y}Fe_{2+y}O_4$, already used in the earlier experiments, to AES analysis. The results of these experiments, together with observations on the influence of thermal treatments under various gas atmospheres are collected in the present communication and interpreted on the basis of the thermodynamic theory of surface segregation [4].

2. Experimental

Cobalt ferrite samples were prepared by wet milling Fe_2O_3 and $CoCO_3$ and firing in air in a furnace at 1340 K for 12 h [2]. Cooling to 1090 K was carried out in the furnace, followed by an air quench. Bulk compositions were measured by X-ray fluorescence spectrometry, while the homogeneity of the spinel phase was checked by X-ray diffraction. The AES analyses were performed in a commercial system (Physical Electronics Ind., Eden Prairie, Minn.), equipped with a cylindrical mirror analyzer, a 5 keV integral electron gun, a 1 keV sputter-ion gun and an electron bombardment heater. The base pressure was about 1.33×10^{-7} Pa. All the Auger measurements were done on oxide grains of suitable dimensions, at normal incidence and under the following experimental conditions: beam current 50 μA; primary energy 3 keV; modulation voltage 3 V; time constant 0.001 sec. Charge effects caused large shifts in the scan energy. However, due to the relative simplicity of the spectra, they could be adjusted easily to the correct position on the energy scale. Heat treatments on cobalt ferrite samples were carried out in Pt crucibles set in a tubular furnace, with a gas flow of about 5 l/h. Samples were cooled in the same atmosphere.

3. Results

Since the usual contaminants (Cl, C, S) were present in relatively small amounts (up to 5%), a preliminary cleaning of the surfaces was not considered necessary, nor were the former taken into account in the calculation of surface compositions. The low energy $M_{2,3}M_{4,5}M_{4,5}$ peaks of Fe and Co were sometimes unresolved and masked by charge effects, consequently the $L_3M_{2,3}M_{2,3}$ peak of Fe (at 600 eV) and the $L_3M_{4,5}M_{4,5}$ peak to Co (at 775 eV) were chosen for the analysis, together with the $KL_{2,3}L_{2,3}$ peak of O. Surface compositions were determined by the method of elemental sensitivity factors [5]. Values of 1, 0.31 and 0.60 were used for the sensitivity factors of O, Fe and Co respectively. The above values were determined with the aid of AES spectra of sputter-etched samples. Generally, 50 to 90 min at 1 keV and 8 μA of ion current, with an Ar pressure of 6.5×10^{-3} Pa were necessary to achieve a constant value of the metal peaks ratios. A selective enrichment induced by the sputtering can be excluded since the Co and Fe sputter yields are similar [6]. Some examples of depth profiles obtained in this way are presented in fig. 1. The sensitivity factor values obtained from the standard spectra [7] were slightly corrected by making the Co/Fe ratio measured after sputtering to reproduce the respective bulk ratio, and by adjusting the oxygen–cation ratio of the stoichiometric sample, which remains constant during sputtering, to the bulk value 1.33. By comparison with Auger spectra of Fe_3O_4 and Co_3O_4, however, the oxygen surface concentrations can be considered affected by an error not larger than 20%. The overall error in the Co/Fe ratio was estimated to be ±5%, this value including possible sample inhomogeneity.

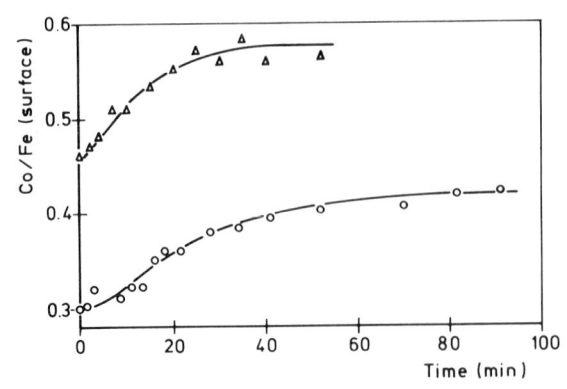

Fig. 1. Composition depth profile of cobalt ferrite, versus sputtering time. Open circles and triangles are referred to Co/Fe bulk values of 0.43 and 0.575 respectively.

Bulk and surface compositions of the untreated samples are reported in table 1 and the corresponding surface versus bulk Co/Fe ratios are shown in fig. 2a. The surface values represent averages of several analyses (up to six) performed on different pellets of the same sample. The experimental surface compositions obtained after sputtering are reported in table 2, together with those measured after various thermal treatments, and plotted in fig. 2b versus bulk Co/Fe ratios. In fig. 2c, the values of the Co/Fe ratio obtained after a thermal treatment in N_2 at 1273 K for 6.5 h are reported. Rather similar results, as shown in fig. 2d, were found using He instead of N_2. Another series of samples was heated at 1073 K for half an hour in the UHV chamber at a residual pressure of 1.33×10^{-7} Pa, and a remarkable increase in the Co surface concentration was found (fig. 2e).

Whenever possible, X-ray diffraction analyses were performed on the samples which underwent the thermal treatments. The results of the X-ray analyses are sum-

Table 1
Surface composition of cobalt ferrite

Surface composition (%)			Co/Fe [a]	
O	Fe	Co	(Surface)	(Bulk)
59.2	31.5	9.3	0.30	0.43
57.3	28.9	13.8	0.48	0.455
57.5	28.3	14.2	0.50	0.495
57.3	28.0	14.7	0.52	0.505
60.5	26.9	12.6	0.47	0.535
59.8	26.2	14.0	0.53	0.575

[a] Atomic ratio.

Table 2
Surface composition of cobalt ferrite: (a) 50–90 min sputter ion etching in 6.5×10^{-3} Pa Ar; (b) 1273 K, 6.5 h in 1 atm N_2; (c) as (b) in He; (d) 1073 K, 0.5 h at 1.33×10^{-7} Pa

Co/Fe (bulk)	(a)		(b)		(c)		(d)	
	Co/Fe	O/(Co+Fe)	Co/Fe	O/(Co+Fe)	Co/Fe	O/(Co+Fe)	Co/Fe	O/(Co+Fe)
0.43	0.42	1.23	0.18	1.51	0.24	1.61	0.57	1.43
0.455	0.44	1.10	0.22	1.41	0.36	1.44	0.52	1.36
0.495	0.48	1.32	0.38	1.45	0.38	1.54	0.58	1.38
0.505	–	–	0.41	1.48	0.44	1.49	0.60	1.27
0.535	0.56	1.17	0.44	1.86	0.52	1.49	0.64	1.44
0.575	0.56	1.25	0.39	1.62	0.43	1.89	0.64	1.36

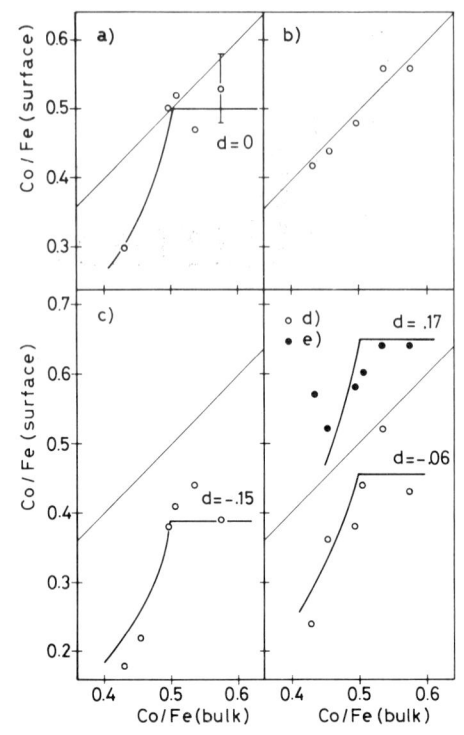

Fig. 2. Co/Fe surface versus bulk ratio in cobalt ferrite samples. (a) Untreated samples. (b) 50–90 min sputter ion etching in 6.5×10^{-3} Pa Ar. (c) 1273 K, 6.5 h in 1 atm N_2. (d) as (c), in He. (e) 1073 K, 0.5 h at 1.33×10^{-7} Pa. Curves represent the results of theoretical calculations, assuming the d values indicated (for the definition of d, see text).

Table 3
X-ray diffraction of cobalt ferrite after thermal treatments; phases outer than $CoFe_2O_4$ are indicated: (a) after heating to 1273 K for 6.5 h in 1 atm N_2; (b) as (a) in He atmosphere; (c) after heating to 1073 K for 0.5 h at 1.33×10^{-7} Pa

Co/Fe (bulk)	(a)	(b)	(c)
0.43	–	–	1% CoO
0.455	–	–	n.d.
0.495	2–3% CoO	1–2% CoO 1% Co_3O_4	n.d.
0.505	–	2% CoO	1% CoO
0.535	1–2% CoO	–	2–3% CoO
0.575	n.d.	3–5% CoO 1% Co_3O_4	4–6% CoO

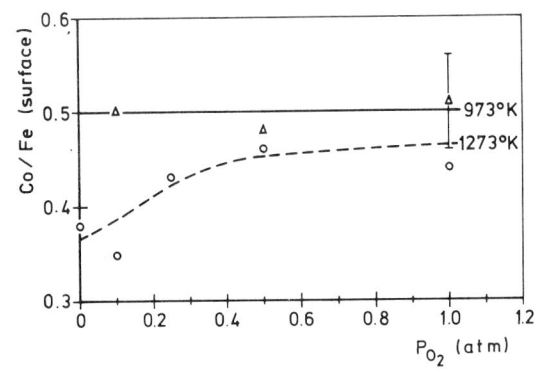

Fig. 3. Co/Fe ratio versus oxygen partial pressure, pO_2, during pretreatment at high temperatures. Co/Fe (bulk) = 0.495; treatment time 6.5 h.

marized in table 3. In most of the samples, CoO as a separate phase was determined in amounts increasing with the bulk Co concentration. Treatment in an O_2 atmosphere was performed on a sample with a bulk Co/Fe ratio of 0.495 with pO_2 varying from 0.1 to 1 atm (He was the carrier gas) at 973 and 1273 K. The results are shown in fig. 3. At 973 K no variation of the Co/Fe ratio was found, while at 1273 K there was a gradual increase of the ratio with increasing pO_2. In this sample, no separate CoO phase was observed by X-ray diffraction.

4. Discussion

A general feature of the untreated samples (fig. 2a) is the surface iron enrichment, with respect to the bulk composition, both for samples having a Co/Fe ratio in defect or in excess as regards the stoichiometric composition. After a prolongated etching, the surface Co/Fe ratios converge to the bulk values (fig. 2b). These results were used to check the reliability of the sensitivity factor, while the depth profile shapes (fig. 1) indicated a gradual composition change.

In the treated samples, where a surface oxygen enrichment was always found (table 2), an increased iron enrichment is observed (figs. 2c and 2d). However, the surface cleaning by heating the samples in vacuo at 1073 K (fig. 2e) yields a remarkable cobalt enrichment of the surface, in spite of the fact that a surface oxygen excess is still found.

The correlation between surface oxygen amount and cation distribution was investigated on the basis of equilibrium thermodynamics.

Oxygen exchange between surface and gas phase takes place as both lattice oxygen loss and chemisorption. In the first process two electrons per oxygen atom are left in the crystal and captured by the trivalent cations. In the second process, molecular oxygen is assumed to be chemisorbed as O_2^-, the additional electron being provided by divalent cations. Hence we write the surface composition as

$$Fe^{3+}(Co^{2+}_{1-y_s}Fe^{2+}_{y_s+d}Fe^{3+}_{1-d})O^{2-}_{4-b}(O^-_2)_c \qquad (y_s > 0),$$

$$Fe^{3+}(Co^{2+}_{1+d}Co^{3+}_{-y_s-d}Fe^{3+}_{1+y_s})O^{2-}_{4-b}(O^-_2)_c \qquad (y_s < 0), \qquad (1)$$

where $d \equiv 2b - c$ is either positive or negative; b and c are functions of gas-phase pressure and temperature. The bulk compoistion is given by (1) as well, with b, $c = 0$ and y instead of y_s.

The equilibrium condition between bulk and surface reads for both $y > 0$ and $y < 0$

$$\frac{1 - |y_s|}{1 - |y|} \frac{y}{y_s + d} = \exp(\Delta^{(i)}/kT), \qquad (2)$$

where $\Delta^{(2)}$ and $\Delta^{(3)}$ are the changes of the standard chemical potential for the bulk-surface exchange of divalent cations ($y > 0$), and trivalent cations ($y < 0$), respectively. From a theoretical calculation [4] we found $\Delta^{(2)} = -0.142$ eV and $\Delta^{(3)} = 0.682$ eV. The calculated curve $\rho_s = (1 - y_s)/(2 + y_s)$ versus $\rho = (1 - y)/(2 + y)$ for $b = c = 0$ compares reasonably well with the untrated sample experimental points (fig. 2a). In treated samples, the experimental average values of the oxygen-to-cation ratio $\frac{1}{3}(4 - b + 2c)$ are 1.56 for N_2 flux treatment and 1.58 for He flux treatment, against 1.33 for the ideal case. Hence we have an oxygen excess at the surface due to chemisorption. A reasonable fit of the two above sets of experimental data is obtained with $d = 0.15$ and $d = -0.06$, respectively (fig. 2c and 2d). Combining these values with those of the oxygen-to-cation ratios we find a small number (few percents) of surface oxygen vacancies, largely compensated by chemisorption: we get $b = 0.13$, $c = 0.40$ for N_2 treated samples and $b = 0.21$, $c = 0.47$ for He treated samples.

For the last set of data, those obtained after a UHV treatment at high temperature, the average oxygen-to-cation ratio reduces to 1.38, and cobalt excess always occurs at the surface. Now, the theoretical fitting is quite poor (fig. 2e). Nevertheless we obtain $b = 0.16$, a value quite close to those found in treated samples, and $c = 0.15$, which is instead quite lower than the values found above, because of the surface cleaning. It is interesting to remark that now $c \simeq b$, which means that at the surface we have as many chemisorbed oxygen molecules as vacant oxygen lattice sites. This could be interpreted as due to the fact that oxygen molecules chemisorbed at vacant oxygen lattice sites are tightly bound, in such a way they survive to UHV heating treatment.

Finally, the X-ray diffraction data given in table 3 are in agreement with the known features of the Fe–Co–O phase diagram [1] where p_0 and T necessary to obtain a single phase with given Co/Fe ratio can be predicted; fig. 3 suggests that there is a connection also between T, p_0 and surface Co/Fe ratio.

References

[1] W. Müller and H. Schmalzried, Ber. Bunsenges. Physik. Chem. 68 (1964) 270;
J.D. Tretjakow and H. Schmalzried, Ber. Bunsenges. Physik. Chem., 69 (1965) 396.
[2] R.G. Squires and G. Parravano, J. Catalysis 2 (1963) 324.
[3] G. Parravano, in: Proc. 4th Intern. Congr. on Catalysis, Moscow, 1968, vol. I (Akademiai Kiado, Budapest, 1971) p. 149.
[4] G. Benedek, F. Garbassi, G. Petrini and G. Parravano, submitted to J. Phys. Chem. Solids.
[5] P.W. Palmberg, Anal. Chem. 45 (1973) 549A.
[6] N. Laegreid and G.K. Wehner, J. Appl. Phys. 32 (1961) 365.
[7] P.W. Palmberg, G.E. Riach, R.E. Weber and N.C. MacDonald, Handbook of Auger Electron Spectroscopy (Edina, 1972).

Surface Science 68 (1977) 294-304
© North-Holland Publishing Company

THE SURFACE COMPOSITION OF THE NICKEL–COPPER ALLOY SYSTEM AS DETERMINED BY AUGER ELECTRON SPECTROSCOPY

F.J. KUIJERS and V. PONEC

Gorlaeus Laboratoria, Rijksuniversiteit Leiden, P.O. Box 75, Leiden, The Netherlands

The Ni-Cu alloys were prepared by evaporation of the specpure metals in UHV onto a quartz substrate. Spectra were obtained from clean as well as from gas covered surfaces. The Auger signal intensity of a monolayer of both metals was determined for the low energy electrons (102–105 eV) and for the high energy electrons (716–920 eV). The overlapping peaks of Cu and Ni in the low energy region (102–105 eV) were evaluated by comparing them with computer simulated alloy spectra. The results of the sintered alloys are interpreted by means of a model by Gallon and Jackson, using the experimentally determined signal intensity of a monolayer. Several surface enrichment data were used to predict the experimentally observed Auger signal intensities. A clear indication of surface enrichment of Cu was obtained; this is in good agreement with previous conclusions based upon hydrogen adsorption and work function measurements. An explanation is suggested why previous work with AES and CO chemisorption did not reveal any surface enrichment.

1. Introduction

The main purpose of the paper is to obtain additional information on the surface composition of Ni–Cu alloys by Auger electron spectroscopy. However, some aspects of this work are more general and the procedure developed may be of more general importance for future work on alloys.

In the past, Ni–Cu alloys have repeatedly attracted attention of many catalytic and surface-chemistry scientists. First, Sachtler et al. [1,2] showed by means of work function measurements that the surface composition of evaporated Ni–Cu alloy films was constant and the surface enriched in Cu, over a wide range of bulk compositions. Recently, the previous experimental results [1,2] were confirmed by new experimental evidence obtained in a broader range of sintering temperatures (420–670 K) [9]. Some points in the explanation of data [1,2] had to be modified but the essential point – surface enrichment in Cu – was fully confirmed. Conclusions with regard to the surface composition were also supported by hydrogen adsorption and benzene hydrogenation measurements on Ni–Cu films [3–5]. The behaviour of alloy powders sintered at $T \sim 600$ K indicated that conclusions made for films actually were applicable to powders as well [6–8].

However, an apparent controversy existed between the data just mentioned and data from two other sources of information: (i) Auger electron spectroscopy in the region 0.8–1.0 keV [10–12] and (ii) volumetric adsorption measurements with CO at room temperature [13,14]. The first controversy was partially removed by Helms et al. [15–17] who have shown that low-energy Auger electrons (80–100 eV) have to be used in order to be able to detect a surface enrichment. However, no attempt has been made by these authors to confirm the conclusions by quantitative analysis. The second controversy has been explained recently [14] by the assumption that CO can bind Ni also from the sub-surface layer and eventually even atract Ni atoms to the surface. Nevertheless, the need of additional data by Auger spectrometry remained.

2. Experimental

The experiments were performed in a standard stainless steel UHV apparatus (RIBER, France) equipped with a CMA from Physical Electronics (USA). The alloy films were prepared by evaporating the pure metals (specpure Ni and Cu, Johnson & Matthey, England) onto a quartz substrate. To protect the internal of the UHV system and the CMA optics from contamination with the evaporated metals a special evaporator was constructed of which the experimental details will be published elsewhere. The pressure during evaporation and sintering (473 K) was below 1×10^{-10} Torr. The gases used (CO and Ar) were obtained from L'Air Liquide, Belgium and had a purity of 99.999%. Auger spectra were obtained using a primary energy of 2000 eV with a beam current of 5 μA. For a more detailed analysis, the spectra could be recorded on a Racal instrumentation recorder (Store 4) which permitted computer treatment (such as integrating) of the spectra.

The analytical procedure applied below relies on the theory of Gallon [18]. It is necessary for this theory to know the signal of the bulk ($I(\infty)$) and of a monolayer ($I(1)$) of the material investigated. To this end, thick (thicker than 250 Å) sintered (500 K) films were prepared and $I(\infty)$ determined. Reproducibility of this determination was better than 95%. These films were then covered by evaporation of the second component (substrate temperature was 300 K). The variation of signal intensities during the evaporation was used to determine the parameter $I(1)$. After this, the composed layers of both metals were equilibrated. The results on surface composition were independent on the sequence of evaporation (Ni onto Cu, Cu onto Ni and simultaneous evaporation). The different preparation modes led to the same state of equilibrium.

2.1. Calibration

Since the low-energy signals of Cu and Ni have a considerable overlap a procedure is necessary to separate the peaks. The integrated experimental spectra showed no changes such as peak broadening upon alloying so that essentially the peak heights of the differentiated spectra were used in the calculations of the surface composition. The procedure was quite simple: the low energy signals of pure Cu and Ni films are recorded and, after corrections for varying background, mixed in a computer to simulate the spectra of a series of Ni-Cu alloys (interval concentration 1%). These simulated spectra are then compared with the measured spectra of the equilibrated alloys after which the contribution of the Cu and Ni signals can be determined. The relative sensitivity factors can be reliably determined from the spectra of the pure metals since the difference in cross-sections and back-scattering coefficients of both metals is negligible (within the experimental error). The relative sensitivity coefficients ($f_L = I_{Cu}/I_{Ni}$ were 1.79 for the low energy electrons and 2.52 for the high energy electrons (f_H).

Results

3.1. Determination of parameters of the Gallon [18] equation and data evaluation

The primary data are the peak-to-peak intensities of the Ni and Cu signals (from Ni, Cu and Ni–Cu films), measured under the same conditions. There are several possible ways to use these data for a quantitative analysis of the surface composition of alloys. Some authors, [21–24] for example, have used an equation which is derived under the assumption of an exponential decay of the signal intensity with the distance, Z, from the surface:

$$I = I(0)\, e^{-Z/\lambda}, \qquad (1)$$

where $I(0)$ is the intensity of the signal at the place of its origin and λ is the mean free escape depth. Another way is to use the theory and formalism of Gallon and Jackson [18–20]. We were particularly attracted by the possibilities which this simple theory offers to the experimentalist and we shall, therefore, describe how this theory can be used for quantitative analysis. The theory calculates the intensity $I(n)$ of an Auger signal originating from n layers for a material for which the parameters $I(\infty)$, the signal of the bulk material, and $I(1)$, the signal of a monolayer, are known:

$$I(n)/I(\infty) = 1 - (1 - I(1)/I(\infty))^n. \qquad (2)$$

It is an advantage of the Ni–Cu system that information can be gained from two signals in the low and high energy region (E_1, E_2). In order to apply Gallon's model, $I(1)$ and $I(\infty)$ must be known. It is possible to determine $I(\infty)$ experimentally and to *calculate* the $I(1)$ parameters from the literature data on $\lambda_1(E_1)$ and $\lambda_2(E_2)$.

The other possibility is to *determine* $I(1)$ experimentally and calculate the $\lambda(E)$ value from the theoretical curve fitting the experimental data. The value $\lambda(E)$ determined in this way can then be compared with the literature data. We used the second procedure because it might – to some extent – be an independent check of the procedure.

The curves calculated according to eq. (2) show that the high energy signal grows

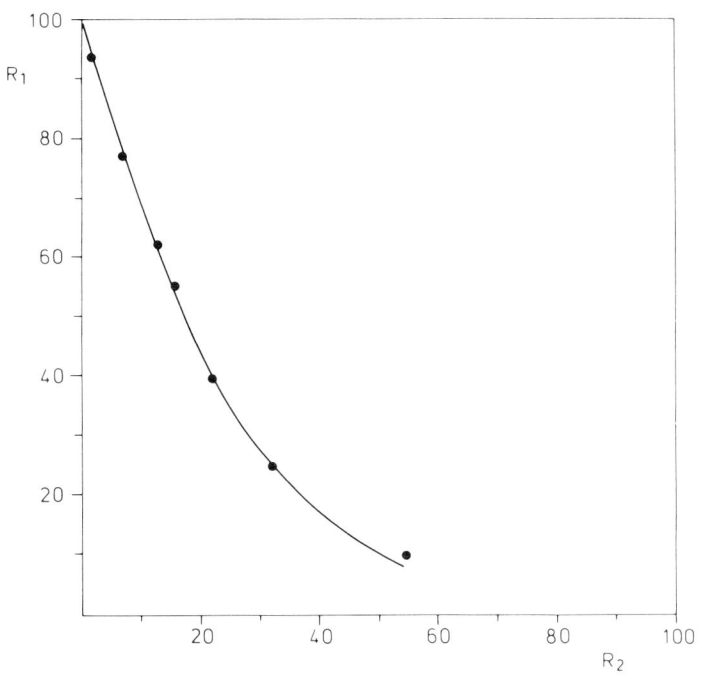

Fig. 1. Decrease of the normalized Ni(102 eV) peak (R_1) as a function of the normalized Cu(920 eV) signal (R_2) (see text for definition of R_1 and R_2, both are in %).

almost linearly with n over 2–3 layers while the low energy signal reveals a clear bend already at the first monolayer (see e.g. fig. 1 in ref. [19]). We monitored the signals upon evaporating Cu over a Ni film and plotted the low energy signal ratios R after various evaporation periods t ($R_1 = I(t)102\mathrm{eV}/I(\infty)102\mathrm{eV}$) versus the high energy level ratios ($R_2 = I(t)920\mathrm{eV}/I(\infty)920\mathrm{eV}$). As can be seen, the curve starts to deviate from linearity at $R_2 \cong 15-20\%$ and $R_1 \cong 50\%$. These values can be taken as trial values for $I(1)(102\ \mathrm{eV})$ and $I(1)(920\ \mathrm{eV})$ whereafter the theoretical curves can be calculated with these trial values. The curve which fits the points in fig. 1 was calculated with $I(1)102\mathrm{eV}/I(\infty)102\mathrm{eV} = 0.5$ and $I(1)920\mathrm{eV}/I(\infty)920\mathrm{eV} = 0.18$.

The experiments just described were performed at room temperature in order to avoid alloying of the substrate with the overlayer. Changing beam voltage and current did not influence the peak intensities which showed that the beam did not cause any change in surface composition.

However, when the opposite procedure was applied and Ni evaporated on Cu, it appeared that alloying already occurred with a beam voltage of 1000 eV. (This is easily understood if one considers the minimum in free energy of alloy formation in the Cu-rich alloys [3], so that Cu-rich alloys are formed very easily.)

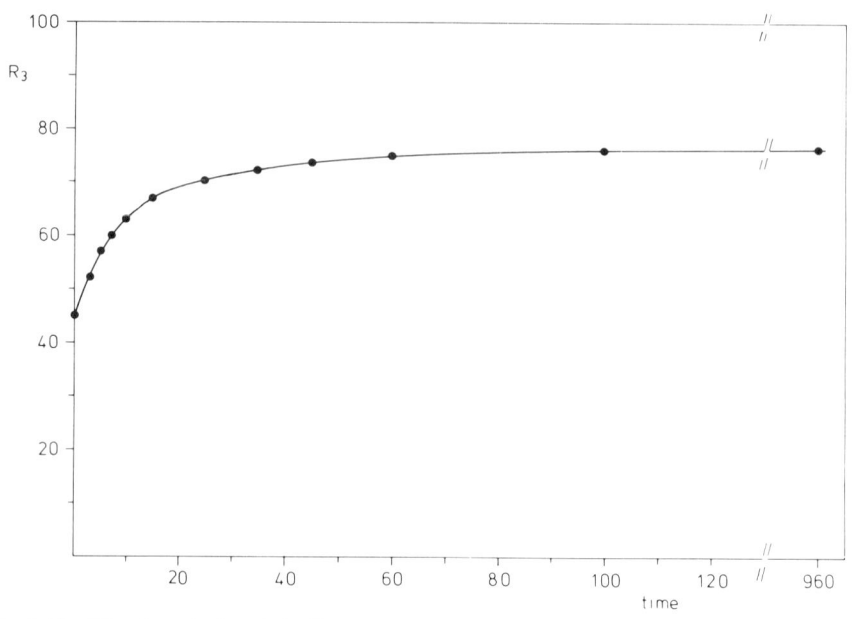

Fig. 2. Equilibration of a "sandwich" film prepared by evaporation of Cu onto Ni. Composition of the surface, characterized by the ratio R_3 as a function of time (in minutes). Here, $R_3 = \{f_L\text{Ni}(102\text{ eV})/[f_L\text{Ni}(102\text{ eV}) + \text{Cu}(105\text{ eV})]\} \times 100\%$. Temperature of annealing is 473 K.

3.2. Equilibration

All alloys were sintered at 473 K during various periods. It can be seen from fig. 2 that equilibrium is reached within two hours. In the first thirty minutes diffusion takes place very rapidly, resulting in a nearly equilibrium state. A check by X-ray diffraction showed that alloying was achieved (see for more details e.g. ref. [9]) under the conditions of the experiment.

3.3. Information on the surface composition of the alloys

The results obtained with equilibrated alloy films are summarized in fig. 3. Instead of plotting the surface sensitive signal Ni(102eV) against the bulk composition we have plotted this signal against the high energy signal. This is a value which is measured with a higher accuracy than the film composition determined by, e.g. X-ray diffraction. The plot as in fig. 3 can be converted into a plot of surface composition versus bulk composition using one of the models described in section 4. Since one of these models must be applied in any case to interpret the results we prefer this presentation instead of introducing an inevitable uncertainty inherent to the determination of the bulk composition of the alloy film.

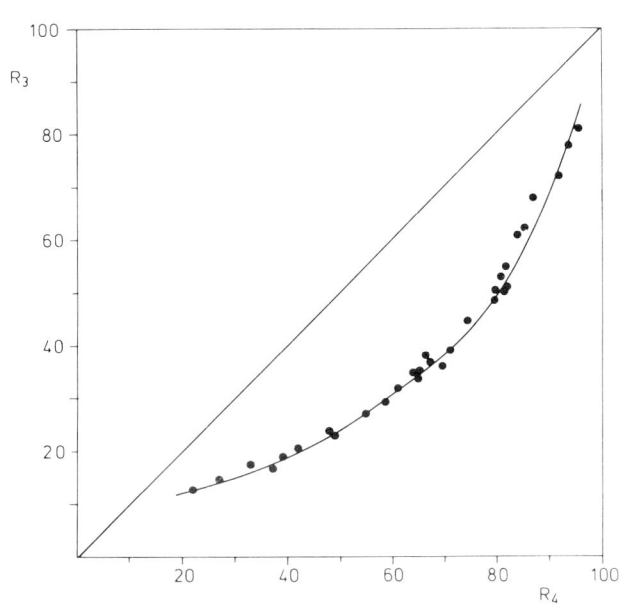

Fig. 3. Normalized peak ratios: R_3 (for definition, see fig. 2) as a function of R_4, where $R_4 = \{f_H \text{Ni}(716 \text{ eV})/[f_H \text{Ni}(716 \text{ eV}) + \text{Cu}(920 \text{ eV})]\} \times 100\%$, for the equilibrated alloys. The surface sensitive ratio R_3 reflects the surface enrichment in a more pronounced manner than the high ratio R_4, so that the curve shown gives already a rough picture of the surface composition.

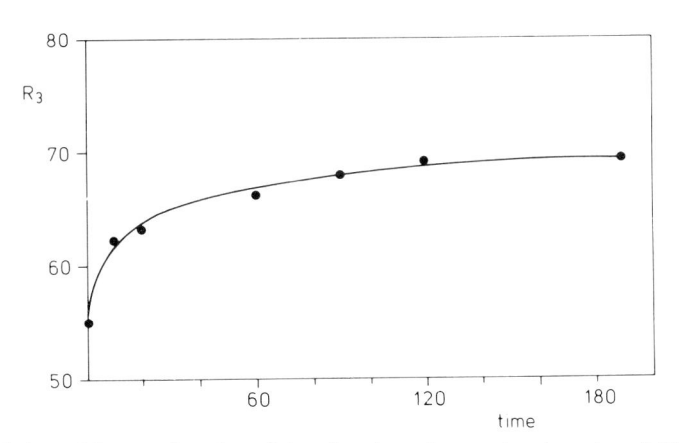

Fig. 4. Variation of R_3 as a function of time (in minutes) upon chemisorption of CO at 473 K. Gas-induced segregation of Ni is clearly observable. Corresponding R_4 ratio is 82% (for definition of R_3 and R_4, see figs. 2 and 3).

3.4. Sputtering

Several samples were sputtered with Ar^+ ions and annealed at 473 K in order to check whether a sputter and annealing procedure can change the surface composition. It appeared that although sputtering results in a Ni enrichment of the surface, after annealing for 2 h the original surface composition was reestablished.

3.5. Carbon monoxide adsorption

CO adsorption was monitored at 293 K and 1×10^{-6} Torr. No significant change in surface composition was observed during an exposure period of two hours. Care was taken to prevent CO dissociation by the electron beam by decreasing beam voltage and beam current as far as possible. When the temperature was increased to 473 K a change in surface composition became clearly visible. In fig. 4 the process of Ni enrichment due to corrosive CO chemisorption is shown.

4. Discussion

There are several points which deserve discussion in more detail. First, the correctness of the procedure of data handling and determination of the necessary constants and second, the information which is finally obtained on the surface composition of equilibrated Ni–Cu alloy films.

The whole analysis as presented in this paper relies very much on the model of Gallon and Jackson [18–20]. The first point which might be questioned in our procedure is the determination of the $I(1)$ parameters. While a thick film of a sintered metal may form a continuous layer [31], it is questionable whether the amount of Cu characterized here by two signals $I(1)$ (105 and 920 eV) is really in the form of a monolayer or forms irregular islands. Metal condensed on metal can, in principle, form regular layers [20,29] and the fast decrease of the surface sensitive signal in fig. 1 indicates that this can be the case. Moreover, the values of $I(1)$ lead to $\lambda(E)$ values in good agreement with the literature data as can be seen in table 1. The

Table 1

E (eV)	λ (measured)	λ (lit.) [28]
102	1–2 atomic layers ~4 Å	~5 Å
920	~5 atomic layers ~12 Å	10–15 Å

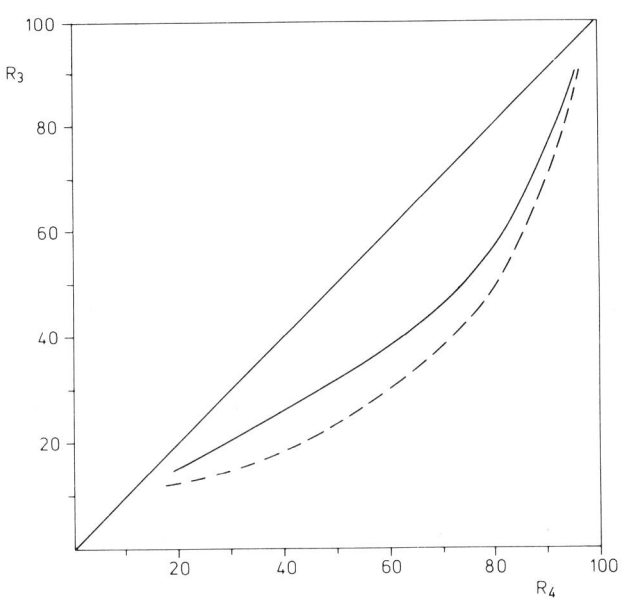

Fig. 5. Experimental data (– – –) compared with a theoretical curve, calculated for the following model: the upmost layer has a composition as determined by hydrogen adsorption, the rest of the film has an average bulk composition (for definition of R_3 and R_4, see figs. 2 and 3).

result indicates that the model of Gallon and Jackson can be applied reliably in this case.

As mentioned in section 3.1, the Gallon–Jackson model has not been applied for the analysis of surfaces since most authors preferred the formulas based upon the exponential decay of the signal with distance from the surface (eq. (1)). We found it therefore interesting to explore the possibilities which the simple eq. (2) offers *.

The second problem to be discussed is that of the surface composition. When Auger electron spectroscopy was added to the arsenal of the surface sensitive methods, the expectations were very high and the surface sensitivity of the method was rather overestimated. However, eq. (2) and, among others, also the data in fig. 1 demonstrate that several layers contribute considerably to the total Auger signal. This means that, in principle, it is possible that a given signal intensity of a metal is obtained by either a certain concentration of that metal in the first layer or by a lower concentration in the first layer combined with a higher concentration in the

---

* During the preparation of this manuscript, Professor T. Yamashina kindly sent us a preprint of his paper [32], where the exponential formulas were applied to the analysis of the same system. The results are very similar to ours, which is not surprising, when considering the mathematical basis of the formulas applied here and in his paper.

second layer. To discern between these possibilities several alternative procedures are possible.

(i) The data obtained by another analytical method (selective chemisorption, ion scattering) are used as an information regarding the first layer composition and it is checked whether the experimental Auger signal can be obtained using the Gallon-Jackson model combined with a reasonable model for the depth profile.

(ii) A theory of surface enrichment is used which predicts both the surface enrichment and the depth profile after which the theoretical and experimental values for the Auger signals are compared.

(iii) A formula for a depth profile is used with one adjustable parameter, whereafter this parameter and the surface composition are calculated from the two signals at different energies.

The last mentioned procedure has been recently applied (see footnote) to the Ni–Cu system. Let us analyse the other possibilities. Fig. 5 shows the signal ratios predicted on the basis of the following model (model 1): the surface has a composition as determined by selective H_2 chemisorption [5], the other layers have an average bulk composition. The calculations were also performed for another model (model 2): the upmost layer has a composition determined by low energy ion scat-

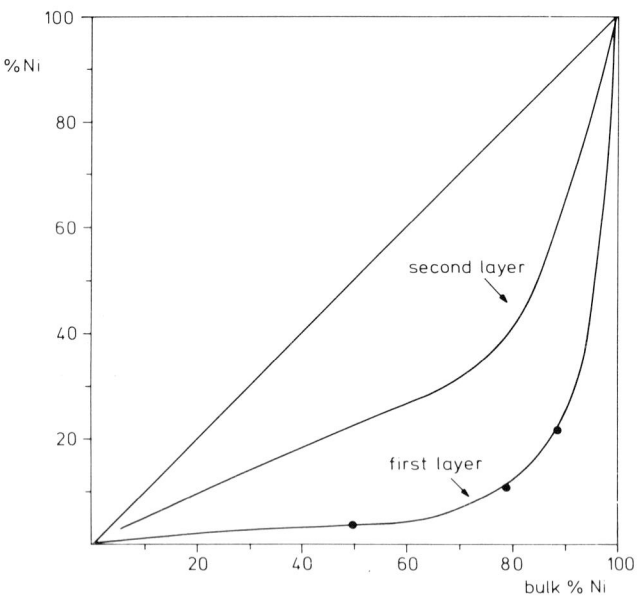

Fig. 6. Concentration of Ni in the first and second layer as a function of the bulk concentration in Ni. With the concentrations as shown in this figure the experimental curve in fig. 3 can be exactly reproduced. The three points shown are the concentrations determined experimentally at 500°C by Brongersma and Buck [25] by ion scattering spectroscopy.

tering [25] the rest has bulk composition. This results in signal ratios that do not differ very much from those for model 1.

Fig. 6 shows which depth profile would fit the experimental data when the surface has a composition given by ion scattering [25]. The experimental curve is almost matched even by the first simple model and it is evident that the determination of the surface composition will be — at the present state of knowledge — always accompanied by some uncertainties. However, it now seems fairly well established that the surface composition is better reflected by H_2 adsorption [3–5] or by work function measurements [1,2,9] and slow ion scattering [25] than by CO adsorption [13,14].

Other work performed in our laboratory using the thermal desorption method [14] and infrared spectroscopy [26] indicated already what could be the reason of this apparent controversy between the H_2 and CO adsorption. It has been shown [14,26] that Ni is attracted to the surface by adsorbed CO (gas-induced segregation has been observed first by Bouwman et al. [27]) even at low temperatures (the temperature of the infrared beam). This phenomenon is again confirmed by results in this paper. The observation leads to another interesting conclusion. When CO can attract Ni atoms to the surface, it must be bound rather strongly by those atoms; in other words, Ni atoms of the second layer are somehow accessible and can bind CO. Because of this effect the surface composition (in Ni) as determined by CO adsorption is higher than the real one.

Acknowledgements

The investigations were supported by the Dutch Foundation for Chemical Research (SON) with financial aid from the Dutch Organization for the Advancement of Pure Research (ZWO). The authors are grateful to Professor Dr. H.H. Brongersma for the possibility to use the ISS results on Ni–Cu prior to their publication. We are indebted to Mr. C. Veefkind for his skilful developing and construction of the evaporator and sample holder.

References

[1] W.M.H. Sachtler and G.J.H. Dorgelo, J. Catalysis 4 (1965) 654.
[2] W.M.H. Sachtler and R. Jongepier, J. Catalysis 4 (1965) 665.
[3] W.M.H. Sachtler and P. van der Plank, Surface Sci. 18 (1969) 62.
[4] P. van der Plank and W.M.H. Sachtler, J. Catalysis 12 (1968) 35.
[5] V. Ponec and W.M.H. Sachtler, J. Catalysis 24 (1972) 250.
[6] J.M. Beelen, V. Ponec and W.M.H. Sachtler, J. Catalysis 28 (1973) 376;
    A. Roberti, V. Ponec and W.M.H. Sachtler, J. Catalysis 28 (1973) 381.
[7] D.A. Cadenhead and N.J. Wagner, J. Phys. Chem. 72 (1968) 2775;
    J.H. Sinfelt, J.L. Carter and D.J.C. Yates, J. Catalysis 24 (1972) 283.

[8] M. Araki and V. Ponec, J. Catalysis 44 (1976) 439;
    W.A.A. van Barneveld and V. Ponec, Rec. Trav. Chim. 93 (1974) 243.
[9] P.E.C. Franken and V. Ponec, J. Catalysis 42 (1976) 398.
[10] G. Ertl and J. Küppers, J. Vacuum Sci. Technol. 9 (1972) 829.
[11] D.T. Quinto, V.S. Sundaram and W.D. Robertson, Surface Sci. 28 (1971) 504.
[12] G. Ertl and J. Küppers, Surface Sci. 24 (1971) 104.
[13] G.D. Lyubarskii, Problemy Kinetika i Kataliz 14 (1970) 129;
    S. Engels, G. Höfer, J. Höfer, J. Radke and M. Wilde, Z. Chem. 15 (1975) 459.
[14] J.C.M. Harberts, A.F. Bourgonje, J.J. Stephan and V. Ponec, J. Catalysis, in press.
[15] C.R. Helms, K.Y. Yu and W.E. Spicer, Surface Sci. 52 (1975) 217.
[16] C.R. Helms, J. Catalysis 36 (1975) 114.
[17] C.R. Helms and K.Y. Yu, J. Vacuum Sci. Technol. 12 (1975) 276.
[18] T.E. Gallon, Surface Sci. 17 (1969) 486.
[19] D.C. Jackson, T.E. Gallon and A. Chambers, Surface Sci. 36 (1973) 381.
[20] A. Chambers and D.C. Jackson, Phil. Mag. 31 (1975) 1357.
[21] R. Bouwman and P. Biloen, Surface Sci. 41 (1974) 348.
[22] R. Bouwman, L.H. Toneman, M.A.M. Boersma and R.A. van Santen, Surface Sci. 59 (1976) 72.
[23] C.C. Chang, Surface Sci. 48 (1975) 9.
[24] J.M. McDavid and S.C. Fain Jr., Surface Sci. 52 (1975) 161.
[25] H.H. Brongersma and T. Buck, private communication.
[26] W.L. van Dijk, J.A. Groenewegen and V. Ponec, J. Catalysis 45 (1976) 277.
[27] R. Bouwman and W.M.H. Sachtler, J. Catalysis 19 (1970) 127.
[28] Data derived from the curve in: J.C. Rivière, Contemp. Phys. 14 (1973) 513.
[29] G.E. Rhead, J. Vacuum Sci. Technol. 13 (1976) 603.
[30] V. Ponec, in: Electronic Structure and Reactivity of Metal Surfaces, Eds. Derouane and Lucas (Plenum, New York, 1976) Part IV, p. 537.
[31] J.V. Sanders, in: Chemisorption and Reactions on Metallic Films, Ed. J.R. Anderson (Academic Press, London, 1971) Vol. 1, p. 1.
[32] K. Watanabe, M. Hashiba and T. Yamashina, Surface Sci. 61 (1976) 483.

FIELD EMISSION STUDY OF AMMONIA ADSORPTION AND CATALYTIC DECOMPOSITION ON INDIVIDUAL MOLYBDENUM PLANES

M. ABON, G. BERGERET and B. TARDY

Institut de Recherches sur la Catalyse, 69626 Villeurbanne Cédex, France

A probe-hole field emission microscope was used to investigate the crystallographic specificity of ammonia adsorption at 200 and 300 K on (110), (100), (211) and (111) molybdenum crystal planes. Chemisorbed NH_3 causes a large work function decrease, especially at 200 K in agreement with an associative adsorption model which can also explain that this decrease is more important on the crystal planes of highest work function (At 200 K, $\Delta\phi = -2.25$ eV on Mo(110) compared to $\Delta\phi = -1.55$ eV on Mo (111). The decomposition of NH_3 was followed by measuring the work function changes for stepwise heating of the Mo tip covered with NH_3 at 200 K. On the four studied planes NH_3 decomposition and H_2 desorption are completed at about 400 K. $\Delta\phi$ changes above 400 K depend on the crystal plane and have been related to two different nitrogen surface states. No inactive plane towards NH_3 adsorption and decomposition has been found but the noted crystallographic anisotropy in this low pressure study is relevant to the structure sensitive character of the NH_3 decomposition and synthesis reactions.

1. Introduction

It is frequently assumed that catalytic synthesis or decomposition of ammonia are structure sensitive reactions. According to Boudart et al. [1], the (111) plane of iron would be particularly active in NH_3 synthesis. McAllister and Hansen [2] have reported that the rate of NH_3 decomposition is higher on the (111) plane than on the (100) and (110) planes of tungsten. It has also been suggested that the (100) plane of W should be the most active [3,4] and the (110) plane the least [5,6] in the adsorption and decomposition of NH_3. A crystal plane specificity in the adsorption of NH_3 may be therefore expected.

In the present work, the chemisorption and subsequent thermal decomposition of NH_3 has been investigated on the (110), (100), (211) and (111) molybdenum planes by "probe-hole" field emission microscopy (FEM). Mo has the same (bcc) structure as Fe and W, and is also a very efficient catalyst for the decomposition and synthesis of NH_3 [7–10].

2. Experimental

The fully bakeable "probe-hole" field emission microscope used in this work has been described previously [11–13]. Changes in the average or local work function of the Mo crystal during NH_3 adsorption or decomposition were determined from measurements of the total or local field emission current–voltage characteristics using the Fowler–Nordheim (FN) equation [14,15]. A base pressure of 10^{-10} Torr was obtained using an ion pump and a titanium sublimator. Anhydrous high purity $^{15}NH_3$ was admitted from a supply bottle via a metal leak valve. The purity of this gas was checked using a quadrupole mass spectrometer mounted in the vicinity of the microscope. Decomposition of $^{15}NH_3$ gave $^{30}N_2$ so that contamination by ^{28}CO could be detected and was found to be low.

3. Ammonia adsorption at 200 and 300 K on (110), (100), (211) and (111) Mo planes

At 200 K and in high vacuum there should not be any physical NH_3 adsorption [16]. The tip was cleaned by heating at 1800 K and cooled in vacuum [17,18] before the introduction of 10^{-6} Torr NH_3 for 2 min. To eliminate NH_3 decomposition on hot filaments, the ionisation gauge and the mass spectrometer were turned off during the experiment. Measurements were made after NH_3 pumping.

3.1. Results

After NH_3 adsorption, the field emission pattern shows a bright emission around the (100) planes as previously described for NH_3 on W by Dawson and Hansen [16].

Results are summarized in table 1. Relative work functions ϕ_0 of clean Mo single-crystal planes have been determined in a previous work [12]. This table shows that NH_3 adsorption results in a large work function decrease especially at 200 K, in good agreement with related works on W [4,5,16,19]. NH_3 adsorption causes also a decrease in the FN pre-exponential term, $\Delta \log A$ being between -1 and -2 for total emission and individual planes.

Table 1
Relative clean Mo work function ϕ_0, atom densities n_0 and work function changes $\Delta\phi$ by NH_3 adsorption at 200 K and 300 K at full coverage

	Total emission	(110)	(100)	(211)	(111)
ϕ_0 (eV)	4.20	5.00	4.45	4.60	4.20
n_0 (10^{14} atoms/cm^2)		14.3	10.1	8.28	5.86
$\Delta\phi$ (eV) – 200 K	-1.80	-2.25	-1.95	-1.90	-1.55
$\Delta\phi$ (eV) – 300 K	-1.0	-1.2	-1.1	-1.0	-0.9
$\Delta\phi$ (200 K)/$\Delta\phi$ (300 K)	1.8	1.9	1.8	1.9	1.7

3.2. Ammonia adsorption model

At 200 K, total dissociation of NH_3 would give H and N adsorbed species which is incompatible with the observed work function decrease [20]. Partial dissociation leading to NH or NH_2 may be considered, but according to Gutman et al. [21], these species on Mo would be electron acceptors and would therefore increase the work function. An associative chemisorption model with bonding by partial transfer of the NH_3 lone-pair electrons into the vacant orbitals of the metal seems the most likely interpretation [4,5,16]. However, the bonding would be mainly covalent, in contrast to ionic potassium adsorption which leads to an increase of log A [22] whereas the observed decrease is usual for covalent type gas chemisorption on metals [16,20]. Additional experimental evidence for the presence even at 300 K of a molecular NH_3 adsorbed complex comes from recent spectroscopic studies [23,24]. However there is a sharp reduction in the work function decrease at 300 K compared to 200 K (table 1) which may be explained by the presence on the surface at 300 K of products coming from the decomposition of a fraction of the NH_3 adsorbed molecules [16,24].

3.3 Crystal plane specificity

Dawson and Hansen [16] demonstrated that NH_3 decomposition occurs before surface diffusion on W. Then NH_3 migration should not explain that no inactive plane towards NH_3 adsorption was found in this study. The nearly constant value of the ratio $\Delta\phi$ (200 K)/$\Delta\phi$ (300 K) (table 1) gives some support to the statement that NH_3 surface migration on Mo is also unsignificant up to 300 K and shows that the same kind of crystal plane specificity exists at 200 and 300 K. Owing to strong repulsion between dipoles and to steric reasons [25], the NH_3 population should not depend very much on crystal plane. The anisotropy in $\Delta\phi$ may be therefore tentatively ascribed mainly to differences in dipole moment μ. High $\Delta\phi$ would correspond to high μ and then to strong bonding. NH_3 would be more strongly held on Mo(110) than on Mo(111).

4. Work function changes with temperature

At 200 K, the tip was first fully covered with NH_3. After 2 h pumping, the tip was stepwise heated for 60 sec at increasing temperatures. NH_3 is very difficult to pump and owing to re-adsorption (P_{NH_3} in the 10^{-9} Torr range), $\Delta\phi$ data (fig. 1) were obtained using the "single point" method [15]. Results are in general agreement with FEM studies of NH_3 decomposition on W by Dawson and Hansen [16] and Wilf and Folman [19]. In these studies, the microscope itself was immersed in liquid helium [16] or liquid nitrogen [19]. It is suggested that NH_3 re-adsorption which occured in the present work where cryogenic pumping was not used, might

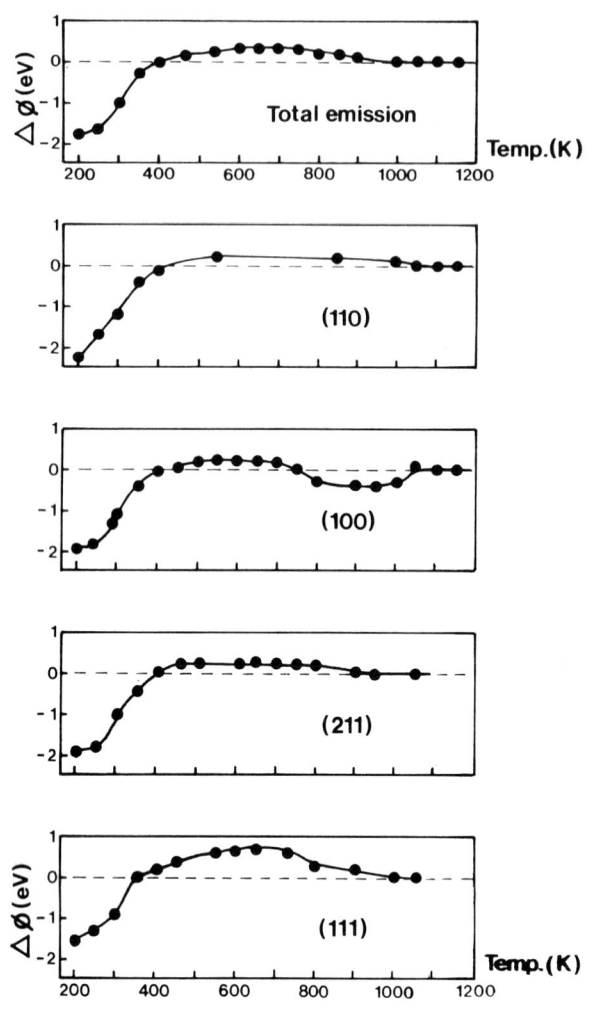

Fig. 1. Work function changes for stepwise heating of the Mo tip covered with NH_3 at 200 K for total emission and for (110), (100), (211) and (111) planes.

account for the differences which concern mainly the total emission and the (100) plane.

4.1. Work function changes in the 200–400 K temperature range

The important rise in $\Delta\phi$ which occurs between 200 and 400 K (fig. 1) indicates NH_3 decomposition followed by H_2 desorption, in agreement with thermal desorption results [26]. A near simultaneous decomposition is observed on all planes studied though decomposition is easier on (111) plane where $\Delta\phi = 0$ at 350 K instead of $\simeq 400$ K on the other planes. NH_3 surface migration is likely to be unsignificant as stated before and then cannot account for the very similar behaviour of the different planes. The observed broad temperature range of NH_3 decomposition might be explained by a large variation in the activation energy of decomposition with coverage. Wilf and Folman [19] observed the same $\Delta\phi$ rise though somewhat sharper on some planes such as the W(211). It is considered that the differences with our results are not great enough to suggest that NH_3 re-adsorption can explain the gradual character of the rise in $\Delta\phi$.

4.2. Work function changes above 400 K

Nitrogen is more electronegative than Mo and its presence explains the positive $\Delta\phi$ maximum observed at 600–700 K on all planes studied. Above 700 K, nitrogen migration followed by desorption at $\simeq 1000$ K might account for the observed evolution of the curves. However, a negative $\Delta\phi$ minimum is observed only on the (100) plane at 950 K. Thermal desorption experiments [26] showed that NH_3 readsorption can provide an interpretation to the FEM results. In addition to the high temperature β-nitrogen peak, a low temperature nitrogen peak with a greater total nitrogen coverage results of NH_3 re-adsorption on a Mo filament first covered with NH_3 at 240 K and then heated up to 750 K. On W, this low temperature nitrogen state has been designated "X-nitrogen" by Matsushita and Hansen [25,27] and "δ-nitrogen" by Peng and Dawson [28]. The (β + X)-nitrogen would be electronegative [25,33] and its gradual formation by NH_3 re-adsorption above 350 K would explain the rise in $\Delta\phi$ and the positive $\Delta\phi$ maximum at 600–700 K on all the planes. Above 700 K, this (β + X)-structure would be progressively destroyed with a partial desorption of nitrogen giving near 950 K a β-nitrogen state. This transition would account for the decrease in $\Delta\phi$ above 700 K and for the negative minimum only observed on the (100) plane at 950 K. It is well known that β-nitrogen decreases the work function of W(100) [30–32]. General agreement is found with the results of Wilf and Folman [19] with the main exception of the W(100) plane where a positive $\Delta\phi$ maximum was not observed. As stated before, NH_3 re-adsorption on W(100) at 78 K was probably very limited and NH_3 decomposition directly leaves above 400 K the electropositive β_2 state which lowers $\Delta\phi$ by $\simeq 0.9$ eV [19] instead of 0.4 eV on Mo(100) (fig. 1). On further exposure to NH_3 and heating to 800 K a less strongly-bonded β_1-state which raises $\Delta\phi$ by 0.15 eV on W(100) has been observed [19]. It is not clear whether one may relate this β_1-state to the X-nitrogen state. Wilf and Folman found also this β_1-state even when molecular nitrogen is adsorbed on W(100) [31], but it has been demonstrated [25,28] that the X-state (or δ-state) cannot result of non-activated molecular nitrogen interaction with W. Thermal desorption experiments [26] supported the formation of the X-nitrogen state in good agreement with related works on Mo [27] and W [25,28].

5. Conclusion

NH_3 chemisorption on Mo leads to a large work function decrease especially at 200 K. $\Delta\phi$ results support molecular NH_3 adsorption at 200 K whereas at 300 K a fraction of the NH_3 layer may be decomposed. The observed crystal face specificity has been tentatively related to dipole moments variations. The dipole associated with the NH_3 adsorbed molecule would be the greatest on high work function planes. This would mean that NH_3 is more strongly adsorbed on the (110) plane than on other planes and especially on (111) plane. It is also worth stressing that no inactive plane towards NH_3 adsorption has been found. NH_3 decomposition followed by H_2 desorption may explain the $\Delta\phi$ rise between 200 and 400 K. Above this temperature, NH_3 interaction with the β-nitrogen deposit left on the surface would give on all the crystal planes an electronegative $(\beta + X)$-structure $(\Delta\phi > 0)$ of a greater nitrogen surface coverage than the β-layer. Above 700 K, partial nitrogen desorption leaves a β-nitrogen layer: $\Delta\phi > 0$ for all the crystal planes with the exception of the (100) where $\Delta\phi = -0.4$ eV. Results have been compared with related works especially with the study of NH_3 on W by Wilf and Folman [19].

This low pressure study on Mo crystal planes of high cleanliness showed that the only species left on the surface above 400 K would be nitrogen and this observation may be related with high pressure kinetic studies which usually favoured adsorption-desorption of nitrogen as rate limiting step in the catalytic synthesis or decomposition of NH_3 [9,36]. The strong crystal face specificity towards N_2 adsorption [30,34] gives the impression that the (100) plane may be the most active in NH_3 synthesis or decomposition reaction. This work shows that NH_3 may be indeed adsorbed and decomposed on the (100) plane as on other planes but NH_3 decomposition would be easier on the (111) plane.

References

[1] J.A. Dumesic, H. Topsøe and M. Boudart, J. Catalysis 37 (1975) 513.
[2] J. McAllister and R.S. Hansen, J. Chem. Phys. 59 (1973) 414.
[3] P.T. Dawson and R.S. Hansen, J. Chem. Phys. 45 (1966) 3148.
[4] P.J. Estrup and J. Anderson, J. Chem. Phys. 49 (1968) 523.
[5] J.W. May, R.J. Szostak and L.H. Germer, Surface Sci. 15 (1969) 37.
[6] R.C.A. Contaminard, R.C. Cosser and F.C. Tompkins, in: Adsorption–Desorption Phenomena, Ed. F. Ricca (Academic Press, London, 1972) p. 291.
[7] S. Kiperman and M.I. Temkin, Acta Physicochim. USSR 2i (1946) 267.
[8] M.R. Hillis, C. Kemball and M.W. Roberts, Trans Faraday Soc. 62 (1966) 3570.
[9] W.G. Frankenburg, in: Catalysis III, Ed. Emmett (Reinhold, New York, 1955), p. 185.
[10] S. Tsuchiya and A. Ozaki, Bull. Chem. Soc. Japan 42 (1969) 344.
[11] G. Bergeret, M. Abon and S.J. Teichner, J. Chim. Phys. 10 (1974) 1299.
[12] G. Bergeret, M. Abon, B. Tardy and S.J. Teichner, J. Vacuum Sci. Technol. 11 (1974) 1193.
[13] G. Bergeret, M. Abon, B. Tardy and S.J. Teichner, Vide 30 (A) (1975) 104.

[14] R. Gomer, Field Emission and Field Ionization (Harvard Univ. Press, Cambridge, MA, 1961).
[15] L. Schmidt and R. Gomer, J. Chem. Phys. 45 (1966) 1605.
[16] P.T. Dawson and R.S. Hansen, J. Chem. Phys. 48 (1968) 623.
[17] K. Matsushita and R.S. Hansen, J. Chem. Phys. 51 (1969) 472.
[18] P.T. Dawson and Y.K. Peng, J. Chem. Phys. 52 (1970) 1014.
[19] M. Wilf and M.Folman, Faraday Trans I, 72 (1976) 1165.
[20] M. Abon and S.J. Teichner, Nuovo Cimento Suppl. 5 (1967) 521.
[21] E.E. Gutman, I.A. Myasnikov and E.V. Bol'shun, Russian J. Phys. Chem. 49 (1975) 24.
[22] L. Schmidt and R. Gomer, J. Chem. Phys. 42 (1965) 3573.
[23] T. Kawai, K. Kunimori, T. Kondow and K. Tamaru, Japan. J. Appl. Phys. Suppl. 2 (1974) 513.
[24] W.F. Egelhoff, J.W. Linnett and D.L. Perry, Faraday Trans. I, Faraday Discuss. Chem. Soc. 60 (1975) 127.
[25] K. Matsushita and R.S. Hansen, J. Chem. Phys. 52 (1970) 4877.
[26] G. Bergeret, B. Tardy and M.Abon, to be published.
[27] K. Matsushita and R.S. Hansen, J. Chem. Phys. 54 (1971) 2278.
[28] Y.K. Peng and P.T. Dawson, J. Chem. Phys. 54 (1971) 950.
[29] A. Ignatiev, F. Jona, D.W. Jepsen and P.M. Marcus, Surface Sci. 49 (1975) 189.
[30] D.L. Adams and L.H. Germer, Surface Sci. 27 (1971) 21.
[31] M. Wilf and M. Folman, Surface Sci. 52 (1975) 10.
[32] T.A. Delchar and G. Ehrlich, J. Chem. Phys. 42 (1965) 2686.
[33] K. Matsushita and R.S. Hansen, J. Chem. Phys. 52 (1970) 3619.
[34] S.P. Singh-Boparai, M. Bowker and D.A. King, Surface Sci. 53 (1975) 55.
[35] P.T. Dawson, J. Catalysis 33 (1974) 47.
[36] G.C. Bond, Catalysis by Metals (Academic Press, New York, 1962) p. 374.

ADSORPTION OF GOLD ON LOW INDEX PLANES OF RHENIUM

S.J.T. COLES and J.P. JONES

University College of North Wales, School of Electronic Engineering Science, Dean Street, Bangor, Gwynedd LL57 1UT, UK

On atomically rough areas of a thermally cleaned rhenium field emitter, adsorbed gold behaves like it does on tungsten. The average work function $\bar{\phi}$ increases at low average gold coverage $\bar{\theta}$ due to formation of gold-rhenium dipoles, and at high coverage a structural transformation in the gold layer leads to a $\bar{\theta}$-independent work function. Broadly similar behaviour is found for gold on the low-index planes of tungsten, but on low-index rhenium planes gold behaves rather differently. When thermally cleaned at >2200 K and annealed below 800 K, the work function, ϕ(clean), of $(10\bar{1}1)$ takes one of two values 5.25 ± 0.04 eV, and 5.36 ± 0.04 eV, which are tentatively attributed to the two possible structures of this plane. Similar behaviour is expected and observed for $(10\bar{1}0)$, but the values taken by ϕ(clean) are not well defined. Both forms of $(10\bar{1}1)$ are thought to undergo reconstruction above 800 K forming a single structure with ϕ(clean) = 5.55 ± 0.03 eV. $(11\bar{2}0)$ and $(11\bar{2}\bar{2})$ each have only one possible structure, and in keeping with this, ϕ(clean) has a single well-defined value for each plane. The flatness of $(10\bar{1}1)$ and $(10\bar{1}0)$ leads to field reduction at their centres which produces an increase in their measured work functions by up to 10%. The initial increase in ϕ produced by gold condensed at 78 K and spread at low equilibration temperatures T_s on $(11\bar{2}\bar{2})$, $(10\bar{1}1)$ and $(11\bar{2}0)$ is attributed to gold–rhenium dipoles, which, on the latter two planes approximate to the Topping model, giving dipoles characterised by $\mu_0(1011) = 0.1 \times 10^{-30}$ C-m with $\alpha = 10$ Å^3 and $\mu_0(11\bar{2}0) = 0.32 \times 10^{-30}$ C-m with $\alpha = 22$ Å^3, where μ_0 is the zero-coverage dipole moment and α its polarizability. Failure of the Topping model on $(11\bar{2}\bar{2})$ is attributed to its atomically rough structure. No dipole effect is seen on $(10\bar{1}0)$. Energy spectroscopy of electrons field emitted at $(20\bar{2}1)$ and $(10\bar{1}1)$ demonstrates the non-free character of electrons in rhenium, while the small effect of adsorbed gold strengthens the belief that gold is bound through a greatly broadened 6s level centred 5.6 eV below the Fermi level and the dipolar nature of the bond supports this model. At higher values of T_s and $\bar{\theta}$ gold appears to form states which are well-characterised by a coverage-independent work function. $(10\bar{1}0)$, $(10\bar{1}1)$ and $(11\bar{2}0)$ each form two such states, one in the range $2 < \bar{\theta} < 4$ (state 1), and the second at $\bar{\theta} > 4$ (state 2). The atomic radii of gold and rhenium are thought to be sufficiently similar to allow the possibility that state 1 is a replication of the Re plane structure by gold. The high work function and thermal stability of state 2, taken together with the observed temperature dependence of the transformation of state 1 to state 2, encourages the belief that state 2 results from atomic rearrangement of state 1 into a close-packed Au(111) structure. State 2 also forms on $(11\bar{2}\bar{2})$ and the absence of state 1 on this plane suggests some surface alloying at coverages below $4\bar{\theta}$.

1. Introduction

Probe-hole field emission microscopy has shown that the dipolar character of the adsorption bond formed by atoms in the first monolayer of gold on tungsten depends upon the work function of the substrate [1]. Second and higher layers of gold, formed at temperatures below 400 K yield a structure of low work function which on heating can be transformed to form a surface of high work function which is unaffected by further additon of gold [2]. In order to gain some insight into the extent to which the substrate affects the behaviour of gold, rhenium was chosen as a substrate because it differs from tungsten both in its crystal structure and its work function and also forms no alloy with gold [3].

2. Experimental technique

The probe-hole field emission microscope has been described previously [4] and is based on the design of Van Oostrom [5]. It was equipped with two gold sources, one for near-head-on evaporation, and the other for side-on depositon, together with an ionization gauge head and molybdenum getter. After processing to achieve a pressure of 5×10^{-9} Torr on a diffusion-pumped vacuum system, the microscope was sealed off by collapsing the glass pumping stem, and gettered to reduce the active gas pressure to less than 1×10^{-11} Torr, as judged by the rate at which a thermally cleaned rhenium tip became contaminated.

The tip was formed by electropolishing a short length of 0.006 inch diameter rhenium wire of 99.95% purity (Goodfellow Metals Ltd.), spot welded to a support loop of 0.010 inch diameter rhenium wire equipped with potential leads of the same material for temperature measurement and control.

Electropolishing was accomplished by anodic dissolution of the wire in 12N sulphuric acid at 4.5 V dc.

Tip temperatures were controlled to within ±2 K by a servocircuit which sensed and controlled the voltage which developed between the potential leads, and using established resistance temperature data [6], quoted temperatures are thought to be accurate to within ±1%.

Work function data and energy spectra were obtained by techniques which have been described earlier [1]. The rhenium surface was considered to be clean when heating for short periods at progressively higher temperatures produced no change in the emission pattern up to the point at which rapid blunting became evident. Judged by this criterion thermal cleaning was complete after a brief heating at 2200 K. It should perhaps be added that tips formed from a sample of rhenium from another source could not be cleaned thermally, and microprobe analysis of this material indicated that iron and nickel were present in significant amounts. The thermally cleaned pattern, fig. 1a, resembles closely that obtained by others for $(11\bar{2}0)$-oriented wire [7–9] and following their practice we adopt the value 4.92 ± 0.03 eV for the average work function of rhenium.

3. Results

3.3. Changes in work function

Changes in the average work function, $\bar{\phi}$, and in that of each $(hklc)$ plane ϕ_{hklc} produced by small increments of condensed gold which had been thermally equilibrated at a temperature T_s were obtained by measuring the $i-V$ characteristics for the total field emitted current i_{tot} and that emitted by the chosen plane i_p. Magnetic trimming of the image was necessary to maintain the centre of the chosen plane over the probe hole as the voltage V was varied. Work functions were acceptable when the least squares error was no greater than ±0.02 eV.

Annealing the tip at T_s was found to increase ϕ at $(10\bar{1}0)$ and $(10\bar{1}1)$ and to minimise this effect, after flash-cleaning at 2200 K, tips were annealed for 2 min at the chosen temperature T_s before dosing with gold.

3.1.1. Changes in $\bar{\phi}$

There is a close resemblance between the $\bar{\phi}-\bar{\theta}$ curves for gold equilibrated on tungsten and rhenium at the same temperature as can be seen in fig. 1b. Because of this close similarity we have adopted the same form of coverage scale as that used for gold on tungsten, thus gold coverage is quoted in units of $\bar{\theta}$ where $1.0\ \bar{\theta}$ is the amount of gold required to achieve $\bar{\theta}_{max}$ on rhenium.

The principal difference between the $\bar{\phi}-\bar{\theta}$ curves for gold on tungsten and rhenium seems to be that the latter lies 0.1 eV above the former giving the impression that this behaviour reflects some properties of gold which are almost independent of the particular substrate. However, gold on the low index planes behaves in a strikingly different manner.

3.1.2. The $10\bar{1}1$ plane

When pre-annealed at $T_s < 800$ K, ϕ adopts one of two values (i) 5.25 ± 0.03 eV and (ii) 5.36 ± 0.03 eV, but formation of these states does not seem to depend in any systematic way on T_s.

The initial rise $\Delta\phi$ in ϕ produced by adsorbed gold, fig. 2a, also depends upon ϕ(clean) fig. 2b, and for state (i) is well-described by the Topping model [10] as shown by the linear relationship [11] between $\bar{\theta}/\Delta\phi$ and $\bar{\theta}^{3/2}$ fig. 2c. This suggests that $\Delta\phi$ results from dipole formation, and if we assume that the gold monolayer density is the same as that of the rhenium surface, 7×10^{18} atoms m^{-2}, the Topping relationship yields for the gold–rhenium bond a zero-coverage dipole moment μ_0 of 0.10×10^{-30} C·m having a polarizability α of 10 Å^3. The initial increase $\Delta\phi$ from state (ii) is not well described by the Topping equation and $\Delta\phi$ reaches its maximum at a lower coverage, which indicates that the number of sites available for dipole formation has become reduced.

For $T_s > 800$ K, the constancy of ϕ clean, the smallness of $\Delta\phi$ (0.03 eV) and

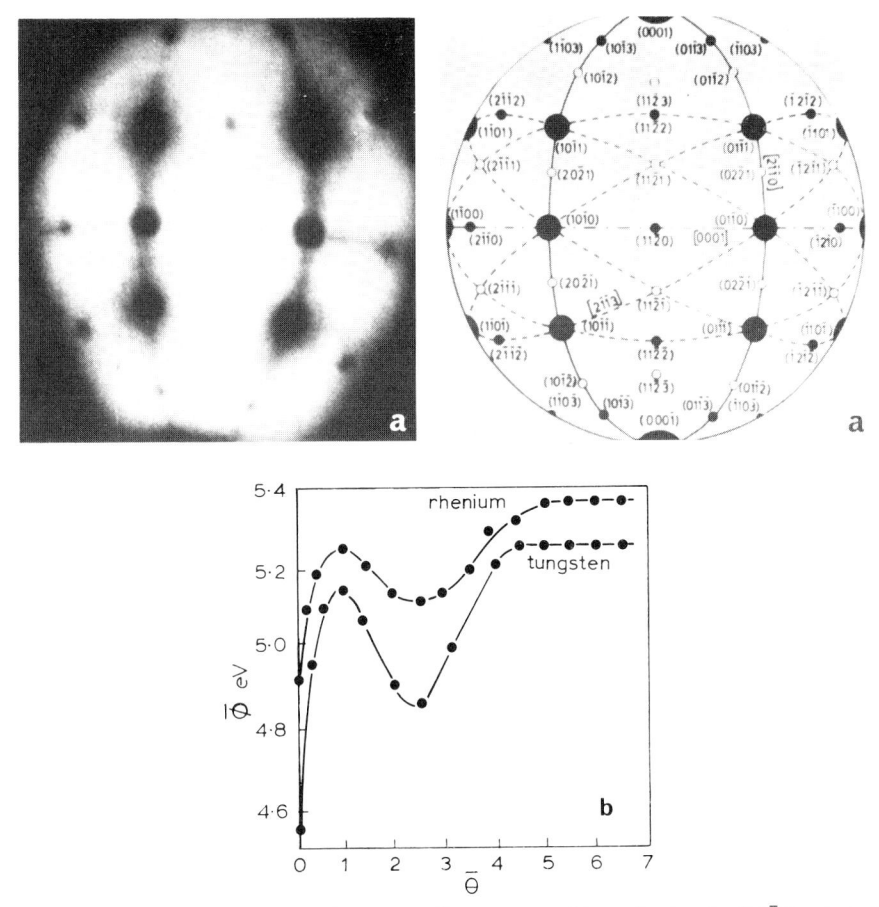

Fig. 1. (a) Field emission image and corresponding stereographic projection for $(11\bar{2}0)$-oriented rhenium. (b) Comparison of changes in average work function $\bar{\phi}$ with average coverage $\bar{\theta}$ for gold spread at 600 K on tungsten and on rhenium.

failure of the Topping model, all point to reconstruction of the clean plane to form a more stable surface structure.

The $10\bar{1}\bar{1}$ plane can exist in two possible forms [12], and the reconstruction which is observed on $(10\bar{1}\bar{1})$ during thermal annealing of a field-evaporated end form has been interpreted [13] as transformation from one possible structure to the other which is more stable thermally. In the present study, however, all surfaces formed on the thermally cleaned end form are quasi-equilibrium structures which are continuously generated during dissolution of the surface as blunting proceeds at high temperature, and then frozen-in at the end of the cleaning stage. We therefore tentatively propose that the two states formed at $T_s < 800$ K are the two alternative forms of $(10\bar{1}\bar{1})$, the apparently random appearance of each being determined

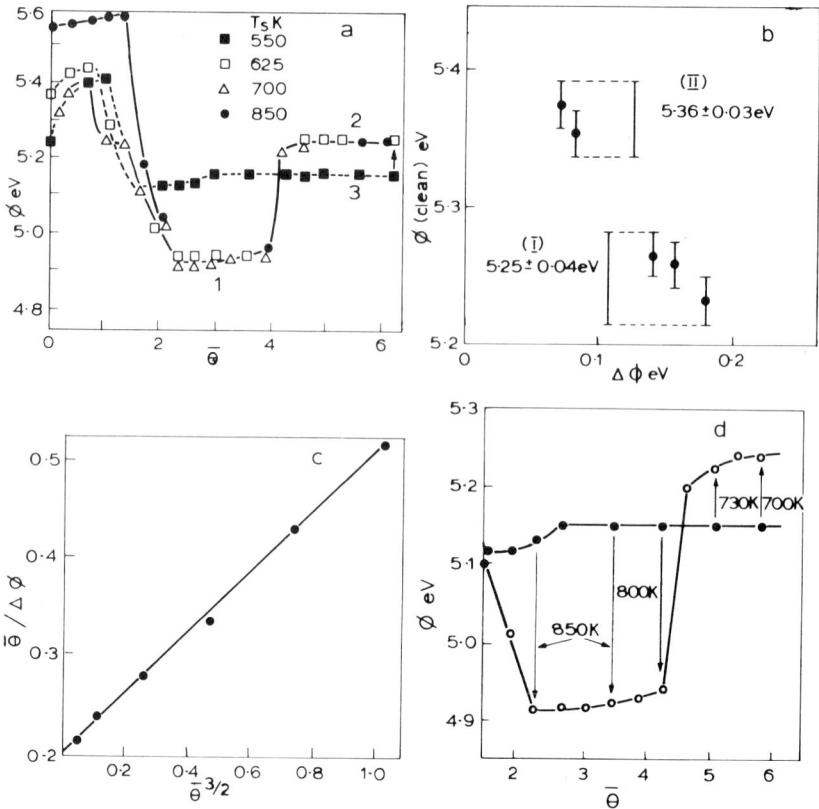

Fig. 2. Gold on Re(10$\bar{1}\bar{1}$). (a) Dependence of the ϕ–$\bar{\theta}$ curve on T_s; (b) annealing at <800 K leads to 2 values of ϕ(clean), and also to 2 values of the initial increase $\Delta\phi$ in ϕ; (c) linearity of the plot of $\bar{\theta}/\Delta\phi$ versus $\bar{\theta}^{3/2}$ demonstrates the applicability of the Topping model, (d) transformation from the low (T_s < 550 K) to the high (T_s > 550 K) temperature characteristic at the indicated temperature.

by the particular form of plane structure which is frozen-in. Subsequent annealing at T_s < 800 K serves only to enlarge the plane area. Of the two possible forms of (10$\bar{1}\bar{1}$), that having the higher work function is most likely to possess the least furrowed surface.

It is clear from field ion microscopy that surface atom movement at 800 K is sufficient to permit rearrangement of the (10$\bar{1}\bar{1}$) surface, it is therefore considered likely that the stable surface found at T_s > 800 K is derived by reconstruction of both forms of (10$\bar{1}\bar{1}$).

The small but reproducible increase in ϕ with gold cover on the high-temperature surface (fig. 2a) indicates that either gold is absent at $\bar{\theta}$ < 1.3 or that adsorbed gold generates no significant dipole moment. At $\bar{\theta}$ > 1.3 gold clearly invades the plane forming at 2.3 $\bar{\theta}$ state 1 which is very well defined up to 4 $\bar{\theta}$ for a wide range of T_s.

The relatively low work function of this state suggests an open structure and the observed reduction in the emitting area, fig. 7, lends some support to this view. The transformation to state 2 which proceeds at $\bar{\theta} > 4$ shows no dependence on T_s unlike that seen on $(10\bar{1}0)$.

For $T_s < 550$ K and $\bar{\theta} > 1$, state 1 is replaced by state 3 which remains stable up to the onset of crystallite growth at 8 $\bar{\theta}$ either because higher layers cannot form on the state 2 surface, or, more likely, because replication of the state 3 structure in higher layers does not alter ϕ. Transformation from state 3 to states 1 or 2 occurs at appropriate coverages, but as fig. 2d shows, the transformations to states 1 and 2 need considerably higher temperatures than are required to form them by incremental addition of gold. This suggests that the transformation from state 3 involves simultaneous reconstruction of more than one layer of gold.

State 2 is thermally stable and lies in the range 5.2 to 5.3 eV, and since on W(110) gold forms the Au(111) [14] surface of work function 5.3 ± 0.03 eV [15], it is tempting to assign to state 2 the Au(111) structure. Such an assignment must be tentative, firstly because the flatness of $(10\bar{1}1)$ leads to an overestimate of ϕ by up to 10% (section 3.2) and secondly because such comparisons depend heavily on the "right" choice of $\bar{\phi}$ for tungsten and rhenium emitters.

3.1.3. The $10\bar{1}0$ plane

Of all the planes which are accessible to study on the $(11\bar{2}0)$-oriented tip this is the most closely packed, giving rise to a high work function which has a mean value of 5.95 ± 0.15 eV. This range of values lies well outside the range of experimental error and is believed to result from the random appearance of the two possible alternative forms of $(10\bar{1}0)$ [12]. However, unlike the case of $(10\bar{1}1)$ no clear-cut grouping of values for ϕ(clean), could be seen within the observed range.

Gold condensed head-on onto the tip and spread at T_s changes ϕ in the manner shown in fig. 3a, and fig. 3b shows the behaviour at low coverages in more detail. $\phi(10\bar{1}0)$ exhibits three regions in the first of which ϕ changes very little from the clean plane value. Fig. 3b shows that gold is present on the plane at all temperatures and the temperature-dependence of the sharp peak, fig. 3b, indicates that this might result from an adsorbate-induced structure incorporating rhenium. It is evident however, that the adsorbate bond possesses very little dipolar character on this plane.

Between 1.5 $\bar{\theta}$ and 2.0 $\bar{\theta}$, ϕ increases sharply to 6.15 eV (state 1) and for $T_s < 650$ K remains constant up to the onset of crystallite growth at 8 $\bar{\theta}$. The sharpness of the increase in ϕ and the low temperature at which it proceeds, indicate that the new state is well-defined, requiring little energy for its formation, which suggests that no drastic changes in structure take place at $(10\bar{1}0)$. We tentatively propose that state 1 is formed by adsorption of gold into the troughs on a Re($10\bar{1}0$) surface which may be expanded by 20% perpendicular to $(21\bar{1}0)$, fig. 3c (i), to form at ~2 $\bar{\theta}$ an Au(110) surface, fig. 3c (ii) the degree of expansion depending upon the effective size of the gold adatom.

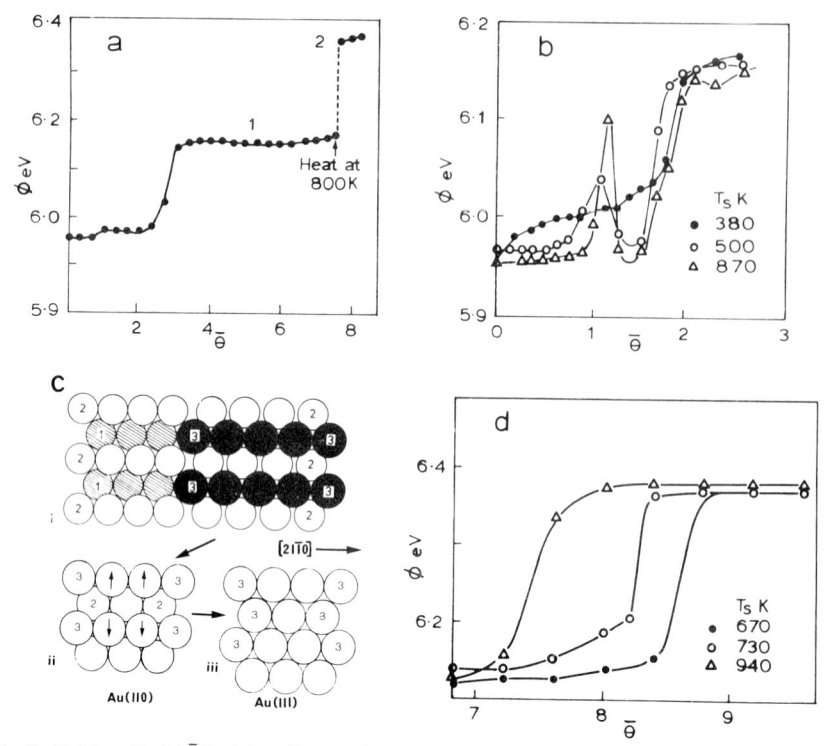

Fig. 3. Gold on Re($10\bar{1}0$). (a) $\phi-\bar{\theta}$ curve for gold spread at 400 K showing conversion of state 1 to state 2 at 800 K. (b) $\phi-\bar{\theta}$ at low coverage showing the peak formed at ~1 $\bar{\theta}$. (c) Schematic of gold on ($10\bar{1}0$) forming an Au(110)-like structure (i) and expansion of the structure to form an Au(111) surface (iii); (1) recessed Re atom, (2) surface Re, (3) gold adatom. (d) Temperature dependence of the high temperature transition.

The fact that ϕ then remains unchanged from 2 $\bar{\theta}$ to 8 $\bar{\theta}$ indicates that either replication of Au(110) leaves ϕ unchanged, or that no new layers can be nucleated on the state 2 surface. The latter is thought to be less likely in view of the coverage-dependence of the transition to state 2, fig. 3d. Both the sharpness and temperature dependence of the transition resemble that of gold on W(211) [1].

When gold is admitted to the plane by surface diffusion from a side deposit state 1 is easily reproduced, but state 2 is not formed at any temperature up to 950 K or coverage up to the onset of crystallite growth. A topographical barrier to diffusion onto the state 2 surface is ruled out because the highest temperatures employed approach those for thermal desorption. Thus state 2 appears to be a phase which can form only in the presence of gold which has been condensed directly onto state 1, presumably because it requires nucleation by condensed gold.

The high work function and thermal stability of state 2 point to formation of Au(111) at this stage, but the value of 5.4 eV observed for ϕ(Au(111)) lies well below that of state 2. Evidence presented in section 3.2 shows that the flatness of $(10\bar{1}0)$ can give rise to field reduction at the plane centre which leads to measured values of ϕ that are higher than the true value by up to 10%; but even if we allow a 10% correction a value of 5.8 eV for state 2 still lies above that expected for Au(111). However, it is hard to conceive of any gold plane structure which is likely to have a higher work function than that of the close-packed Au(111) plane, and surface alloying is not expected to give rise to anomalously high work functions [16]. Accordingly we suggest that state 2 has the Au(111) structure, is formed from state 1 by a process of nucleation, and requires some degree of expansion in the surface lattice, as indicated schematically in fig. 3c (iii). The anomalously high measured work function must then result from field reduction produced by extension of the flat plane by adsorbed gold, and perhaps from failure of the Fowler–Nordheim model along $(10\bar{1}0)$ preventing measurement of the true work function at this plane.

3.1.4. The $11\bar{2}\bar{2}$ plane

Of the planes studied, this is the roughest on an atomic scale, a factor reflected in its relatively low work function. It has only one possible structure [12], and in agreement with this, ϕ(clean) is well-behaved at 4.95 ± 0.02 eV. This low value leads to an expected similarity between its $\phi-\bar{\theta}$ curve, fig. 4a, and $\bar{\phi}-\bar{\theta}$, fig. 1. The 420 K curve is representative of behaviour in the range $370 < T_s < 600$ K and the value of ϕ attained at $\bar{\theta} > 4$ remains constant at 5.22 ± 0.03 eV up to at least $10\,\bar{\theta}$. The initial rise in ϕ does not follow the Topping model, fig. 4c, and the first maximum, ϕ_{max}, is dependent upon T_s as shown in fig. 4b. This, taken together with the variation in the following minimum value of ϕ, strongly suggests some variability of the surface condition perhaps by formation of a surface alloy. Such behaviour is also suggested by the dependence of ϕ_{sat} on T_s which is shown in fig. 4d to settle at 5.35 eV only at $T_s > 800$ K. Clearly on this plane the transition to the final state (state 2) is not well defined.

Fig. 4e is a schematic view of $(11\bar{2}\bar{2})$, and if pairwise interaction governs the bond strength, atoms will first occupy sites A, and at coverages above half a monolayer will occupy B sites to complete the monolayer. In view of the similarity of atom sizes and the rough character of $(11\bar{2}\bar{2})$ it is difficult to see how gold can do other than replicate the structure of the rhenium surface at $T_s < 600$ K, in which case each gold atom dipole will be partly screened from its neighbour by rhenium atoms, and this may explain the failure of the Topping description. Rhenium atoms are known to be mobile on $(11\bar{2}\bar{2})$ at 800 K [13], thus the increase in ϕ_{max} at $T_s > 600$ K, fig. 4b, is thought to reflect some re-ordering of the surface structure by gold-assisted movement of rhenium atoms. The fall in ϕ_{max} at $T_s > 780$ K almost certainly results from more extensive structural changes which might be expected in this temperature range, perhaps leading to formation of a surface alloy. The high work function and thermal stability of state 2 which persists up to the onset of crystallite growth suggests that it may have the Au(111) structure. State 3, fig. 4a,

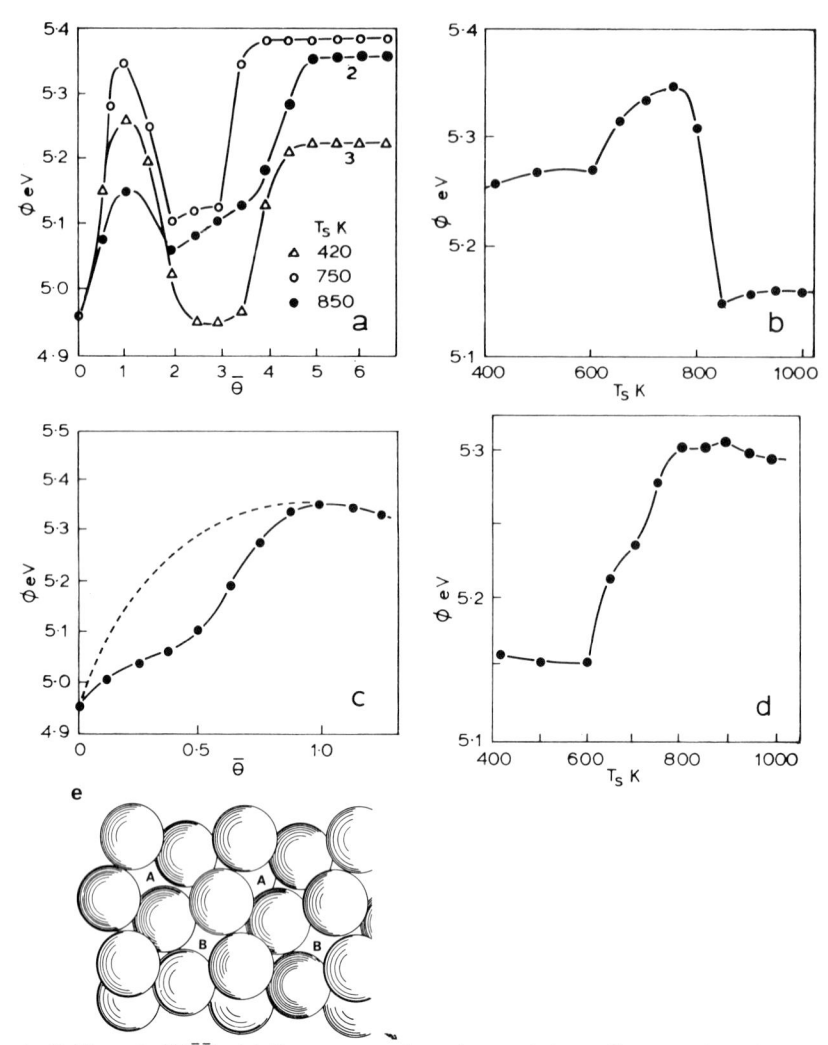

Fig. 4. Gold on Re(11$\bar{2}\bar{2}$). (a) Temperature dependence of the ϕ–$\bar{\theta}$ curve. (b) Variation of ϕ_{max} ($\bar{\theta}$ = 1) with temperature showing the rise at 600 K and sharp fall at 800 K. (c) ϕ–$\bar{\theta}$ at low coverage does not follow the Topping model (– – –). (d) Variation in ϕ with T_s at 5 $\bar{\theta}$ shows the rapid rise in ϕ at T_s > 600 K. (e) Plan view of (11$\bar{2}\bar{2}$) showing sites A and B into which gold can be adsorbed.

forms only below 600 K and is probably a gold layer of unknown but more open structure than that of state 2.

3.1.5. The 11$\bar{2}$0 plane

In keeping with the fact that (11$\bar{2}$0) has only one structure, ϕ(clean) = 5.08

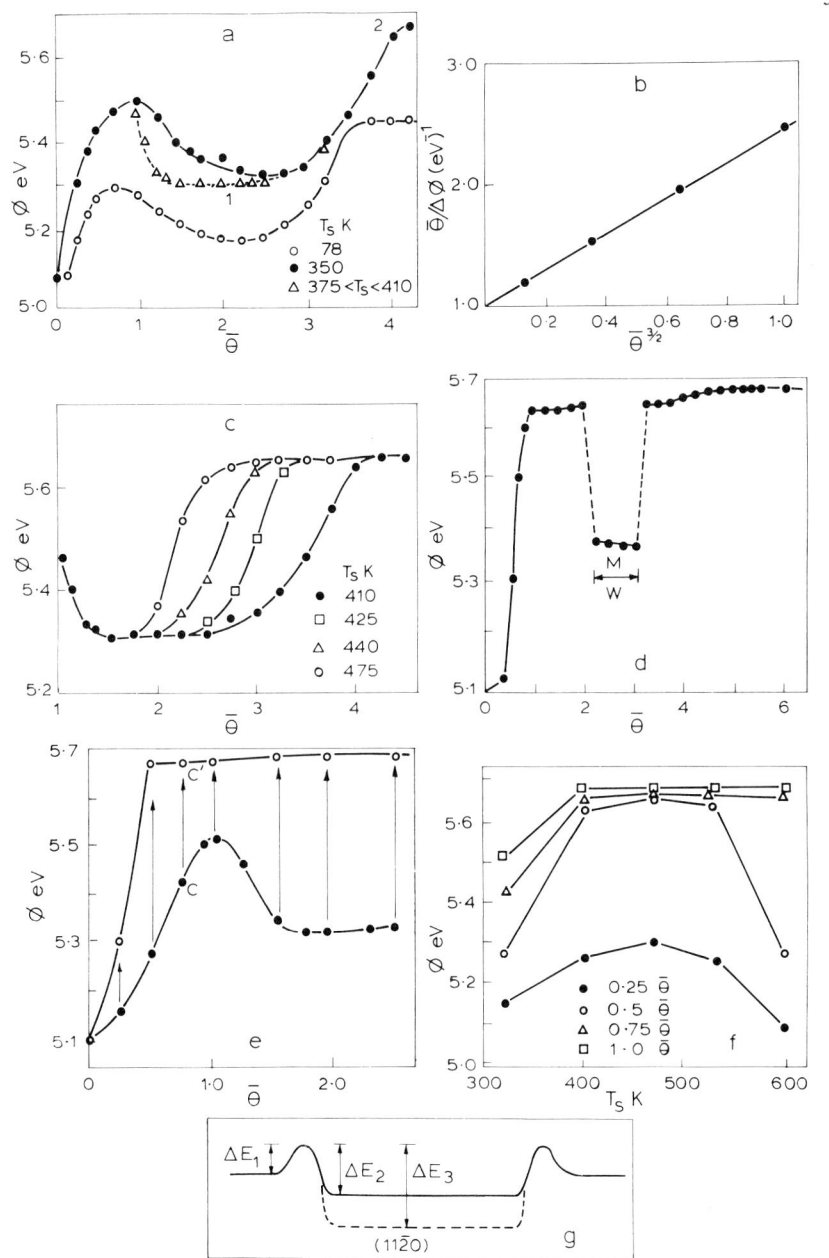

Fig. 5. Gold on Re(11$\bar{2}$0). (a) Temperature dependence of the ϕ–$\bar{\theta}$ curve for $T_s < 410$ K. (b) Linear plot of $\bar{\theta}/\Delta\phi$ versus $\bar{\theta}^{3/2}$ shows good fit to the Topping model. (c) Temperature dependence of the transformation to state 2. (d) ϕ–$\bar{\theta}$ curve for $T_s > 470$ K showing metastable state M. (e) Change from the low to high temperature characteristic (●) doses spread at 320 K, (○) after heating to 500 K. (f) Change in ϕ with increasing T_s for low coverage of gold showing loss of material from the plane only at $\bar{\theta} < 0.75$. (g) Schematic of the energy barriers surrounding (11$\bar{2}$0).

± 0.02 eV. Study of the ϕ–$\bar{\theta}$ relationship on $(10\bar{1}\bar{1})$ and $(11\bar{2}\bar{2})$ at $T_s < 400$ K was limited by the onset of crystallite growth, which produces an apparent continuous decrease in ϕ with increasing $\bar{\theta}$. On the smoother $(11\bar{2}0)$ gold adatom mobility is high at 300 K and these aggregates do not form. Fig. 5a shows the effect of gold condensed from the head-on source at $T_s < 400$ K. The increase in ϕ at $\bar{\theta} > 3$ produced by gold condensed at 78 K evidently occurs even when the condensed layer is relatively disordered. This increase is normally associated with thermally assisted reorganisation of the deposited material [2], thus is seems that on $(11\bar{2}0)$ diffusion is sufficiently easy for some structural ordering of the gold layer. Increasing T_s to 350 K leads, as expected, to an increase in ϕ, presumably due to increased ordering. The initial rise in ϕ is well described by the Topping model, fig. 5b, which leads to a dipole of moment $\mu_0 = 0.32 \times 10^{-30}$ C·m and polarizability of 22 Å. This latter value is unexpectedly high for gold. At $T_s > 375$ K the minimum ϕ_{min} at 5.30 eV extends over the range $1 < \bar{\theta} < 2.5$ to form a well-defined state (state 1) and at $\bar{\theta} > 2.5$, ϕ attains the thermally stable ϕ_{sat} of 5.65 ± 0.02 eV, the transformation depending on T_s as shown in fig. 5c. It is tentatively suggested that this state (state 2) has the Au(111) structure. For all spreading temperatures above 470 K the ϕ–$\bar{\theta}$ curve is of the form shown in fig. 5d, in which the final work function compares well with that obtained at lower temperatures, but at coverages in the range $2 < \bar{\theta} < 4$, ϕ drops sharply to form a "metastable" state M so termed because although it commences in the range 2.0 ± 0.25 $\bar{\theta}$ its width in coverage terms varies from 0.5 $\bar{\theta}$ to 1.7 $\bar{\theta}$. The work function of this state is well-defined at 5.37 ± 0.02 eV which compares with ϕ_{min} at lower temperatures. An attempt was made to investigate the dependence of this state on T_s, and although the reproducibility was not high, the state width W decreases with increasing T_s. Because of lack of reproducibility, measurement of the rate of transformation between the two states proved to be possible only at low coverages as depicted in fig. 5e. The transformation lettered C → C' in fig. 5e proceeds at a measureable rate in the temperature range 445–480 K with an activation energy of 1.27 ± 0.05 eV and pre-exponential 10^{13} sec^{-1}.

When gold is admitted to the plane only by diffusion from a side deposit there is no detectable change in ϕ until 0.5 $\bar{\theta}$ when ϕ rises to the saturation value of 5.66 ± 0.02 eV. This behaviour, and that which is summarised in fig. 5e gave rise to the suspicion that the increase to ϕ_{sat} at $\bar{\theta} < 1$ resulted from the scavenging of gold by $(11\bar{2}0)$ from its surroundings, but no evidence for a decrease in local gold concentration in the environs of $(11\bar{2}0)$ could be found.

The sharp increase in work function to the saturation value at $\bar{\theta} < 1$, fig. 5e, is most simply ascribed to plane invasion over some form of diffusion barrier. The model which we propose tentatively is depicted in fig. 5g. At $T_s < {\sim}410$ K the barrier to diffusion ΔE_1 prevents detectable invasion of the plane which thus contains a gold population obtained by direct deposition which behaves in the "normal" manner depicted in fig. 5a.

At $T_s > 410$ K the plane is invaded by a proportion of the gold which was deposited near the plane. As T_s increases, an increasing proportion of each deposited dose invades the plane, effectively "compressing" the coverage scale as shown in fig. 5c.

At $T_s > 470$ K, figg. 5f shows that ϕ returns to its clean value of $\bar{\theta} < 0.5$ presumably by loss of gold from the plane over ΔE_2, fig. 5g. Above 0.7 $\bar{\theta}$ this cannot occur because coverage of the plane is now sufficiently high to stabilise the structure which gives ϕ_{sat} and which increases to ΔE_3 the barrier to egress of gold, fig. 5g. Examination of the 470 K characteristic, fig. 5d, supports this picture for ϕ_{sat} is attained at 0.8 $\bar{\theta}$ but only beyond 0.5 $\bar{\theta}$ does ϕ rise significantly above the clean plane value.

The random appearance of the metastable state strongly indicates that its formation requires some form of nucleation followed by rapid completion of the structure.

3.2. Effect of plane size on ϕ

The possibility that field reduction at the plane centre is responsible for the high measured work function of $(10\bar{1}0)$ was examined by measuring the current profile across the plane in the manner described by Plummer and Rhodin [17]. The result is shown in fig. 6a, and in fig. 6b the effect of annealing on the current profile of $(10\bar{1}\bar{1})$ is also presented. All data was obtained after annealing for 3 min at the indicated temperature following flash cleaning at 2200 K. The parabolic dependence of current on distance from the plane centre is demonstrated in fig. 6c and strongly indicates a local flattening of the tip profile at these locations as found for (110) tungsten [18].

Widening of the current profile of $(10\bar{1}\bar{1})$ at the higher temperature suggests that the increase of ϕ(clean), fig. 2a, results from extension of the plane, and $(10\bar{1}0)$ is subject to the same effect, because after flash cleaning $\phi(10\bar{1}0) = 5.9$ eV but rises to 6.18 eV after 5 min annealing at 1300 K.

Cranstoun and Pyke [13] in a study of atom movement on rhenium with the field ion microscope have shown that plane enlargement by diffusion over step edges occurs on $(10\bar{1}0)$ at 1030 K while, on $(10\bar{1}\bar{1})$ at $T > 700$ K plane enlargement may be accompanied by rearrangement of the surface structure. It seems clear therefore that $\phi(10\bar{1}0)$ and $\phi(10\bar{1}\bar{1})$ obtained directly using the Fowler–Nordheim equation are likely to be too high due to field reduction which is caused by extension of these planes on annealing. It is estimated that the measured values of ϕ on $(10\bar{1}0)$ and (1011) may be too high by up to 10%.

3.3. Emitting areas

Emitting areas A were calculated according to the method of Van Oostrom [5] and the dependence of A on $\bar{\theta}$ for gold spread at 850 K is presented in fig. 7. There is little significant change in A on any plane except perhaps $(10\bar{1}\bar{1})$ and this contrasts with the larger changes produced by gold on the low-index tungsten planes

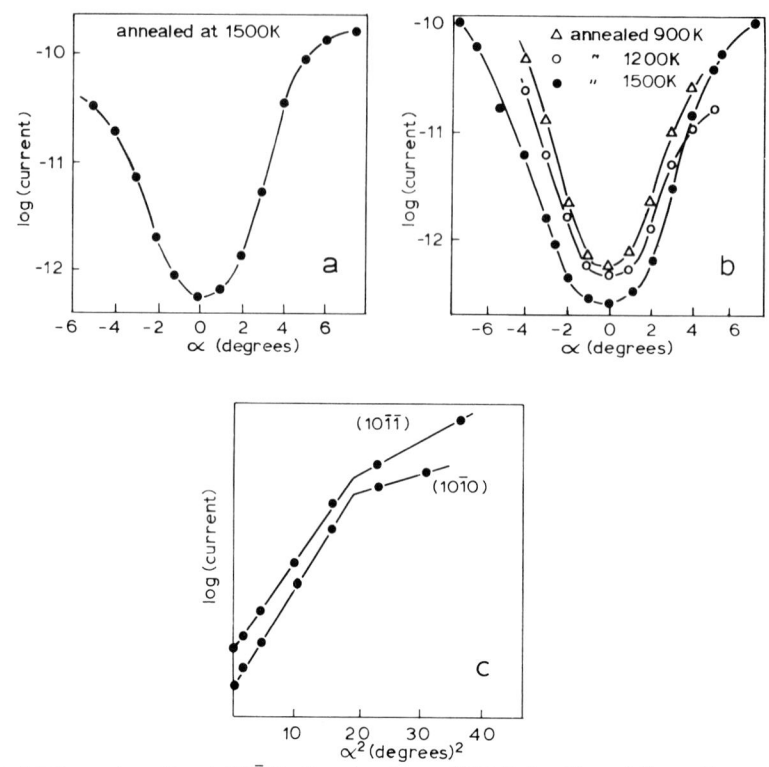

Fig. 6. (a) Current profile at (10$\bar{1}$0) after cleaning at 2200 K for 10 sec followed by annealing at 1500 K for 3 min. (b) Changes in the current profile at (10$\bar{1}$1) after flash cleaning and annealing for 3 min at the indicated temperatures. (c) Parabolic dependence of current on the square of the angle α away from the plane pole.

[1]. Lack of any change in A is significant because it rules out the possibility that at any stage in the high temperature ϕ–$\bar{\theta}$ characteristics, very small nuclei or islands of a particular phase are forming, for the presence of such features would be expected to produce a significant change in A.

3.4. Total energy distribution (TED)

The only recognisable low index planes which were accessible by magnetic deflection without producing unacceptable deterioration in the analyser performance were (20$\bar{2}$1) and (10$\bar{1}$1). Deviation from free-electron behaviour is displayed in fig. 8a by plotting the enhancement factor $R(\epsilon)$ [19], and addition of gold can be seen to have only a slight effect on the TED.

On (20$\bar{2}$1) there is some evidence for a clean-plane resonance 0.4 eV below the

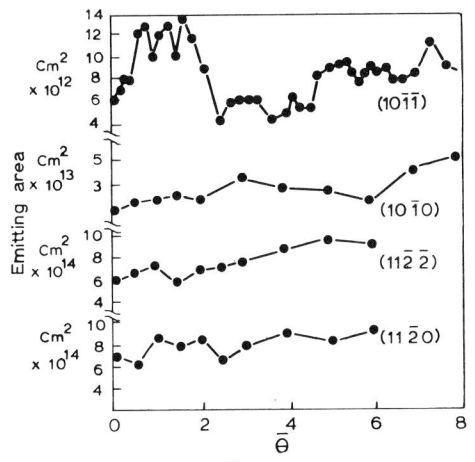

Fig. 7. Changes in emitting area with $\bar{\theta}$ for gold spread at 850 K on each plane.

Fermi energy E_F and addition of gold suppresses this while enhancing a deeper lying resonance.

A resonance on $(10\bar{1}\bar{1})$ centred ~0.55 eV below E_F is broadened and shifted upwards by addition of $3\,\bar{\theta}$ of gold and partly suppressed at $7\,\bar{\theta}$. The fact that gold introduces no strong change in the TED is indicative of an extremely broad resonance which might well be centred at the point suggested by Newns [20] halfway between the ionization I and electron affinity A levels. For gold I lies 9.22 eV and A 2.1 eV below the vacuum level so that the resonance level will lie 5.66 eV below the vacuum level or 0.3 eV below E_F on $(10\bar{1}\bar{1})$ as indicated on fig. 8a. The precise

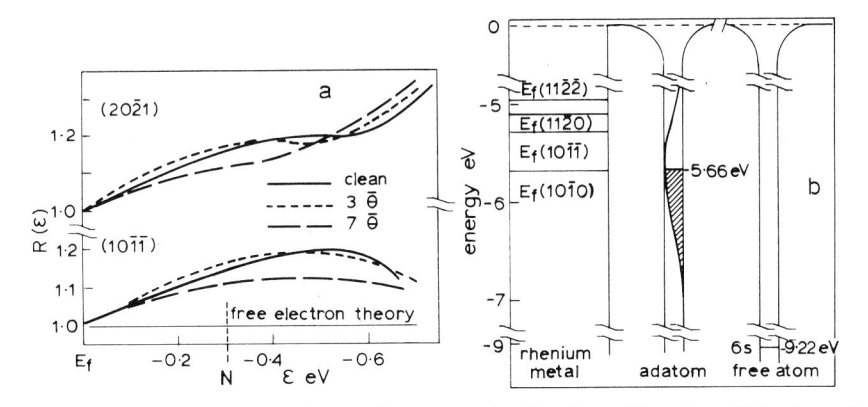

Fig. 8. (a) Plot of enhancement factor $R(\epsilon)$ versus ϵ showing that gold produces little change in electron energy structure. N signifies the position of the Newns level. (b) Proposed model showing bonding to rhenium through a broadened gold 6s level.

position of this resonance will depend upon the adatom-metal separation but we take its mean point to lie 5.6 eV below the vacuum level, and this situation is depicted schematically in fig. 8b. The broadened Au 6s level is half-filled and equilibration of this resonance with the electrons of the substrate will result in net transfer of electronic charge to the gold atom on ($11\bar{2}\bar{2}$), ($11\bar{2}0$) and ($10\bar{1}\bar{1}$) leading to the observed dipole structure in the first monolayer. Figure 2b shows that the expected electron transfer from gold to the substrate does not take place at ($10\bar{1}0$), and we propose that this is due to an overestimate of $\phi(10\bar{1}0)$ for reasons already discussed. Indeed the absence of any dipole effect on ($10\bar{1}0$) indicates that this work function is nearer 5.6 eV, so that practically no electron transfer takes place.

4. Conclusions

The observed variation in ϕ(clean) on ($10\bar{1}\bar{1}$) and ($10\bar{1}0$) is thought to result from the fact that on both planes two structures are possible. In keeping with this, at ($11\bar{2}0$) and ($11\bar{2}\bar{2}$), which have only one structure, ϕ(clean) behaves normally.

At coverage less than 1 $\bar{\theta}$, gold adsorbed at $T_s < 400$ K replicates the rhenium substrate forming a dipole structure on all planes, and on the smoothest of these the Topping model is applicable. The sign and magnitude of the dipole moment is governed by the position of the rhenium Fermi level relative to the highest filled level in the broad gold resonance which is centred about 5.6 eV below the vacuum level.

The effective radius of an adatom clearly depends upon the nature of its bonding orbitals and therefore on its environment [21], moreover, the Goldschmidt radii of Re and Au (1.37 and 1.41 Å respectively) are sufficiently close to make the assumption that adsorbed gold and substrate rhenium have the same atomic size, appear not unreasonable. This assumption makes possible the proposal that state 1, which exists in the range $2 < \bar{\theta} < 4$ on all but ($11\bar{2}\bar{2}$) is formed by replication of the rhenium surface structure by adsorbed gold. On ($11\bar{2}\bar{2}$) which is the roughest plane, the absence of state 1 and temperature-dependence of ϕ suggest that a surface alloy of variable composition is formed.

Based on its high work function and thermal stability, it is tentatively proposed that state 2 has a surface structure which approaches that of Au(111). In support of this, the work function of state 2 on ($10\bar{1}\bar{1}$), ($11\bar{2}0$) and ($11\bar{2}\bar{2}$) agree reasonably well with that measured for gold on W(110) which is known to have the Au(111) surface structure [14]. On ($10\bar{1}0$) it is proposed that our measurement of ϕ does not yield the true work function of state 2.

At low temperatures, states are formed on ($10\bar{1}\bar{1}$) and ($11\bar{2}\bar{2}$) which are not stable, (state 3) transforming into state 2 at temperatures above 600 K.

Further insight into the nature of these gold states will require use of other techniques such as LEED/Auger spectroscopy.

Acknowledgements

The authors are grateful to the Science Research Council for financial support for this work, and for the award of a maintenance grant to one of us (SJTC). We also thank Mr. D. Whitehead for his skilful assistance.

References

[1] J.P. Jones and N.T. Jones, Thin Solid Films 35 (1976) 83.
[2] A. Cetronio and J.P. Jones, Surface Sci. 44 (1974) 109.
[3] A. Hansen, Constitution of Binary Alloys (McGraw-Hill, New York 1958).
[4] E.W. Roberts, Ph.D. thesis, University of Wales (1974).
[5] A.G.J. van Oostrom, Philips Res. Repl. Suppl. 1 (1966).
[6] G.B. Gaines and C.T. Sims, Proc. Am. Soc. Testing Mater. 57 (1957) 759.
[7] J.R. Anderson, J. Appl. Phys. 34 (1963) 2260.
[8] W. Kollmer and D. Stark, Z. Physik 178 (1964) 39.
[9] R. Klein and W. Little, Surface Sci. 6 (1967) 193.
[10] J. Topping, Proc. Roy. Soc. (London) A114 (1927) 67.
[11] R.A. Collins and B.H. Blott, J. Phys. D4 (1971) 102.
[12] A.J. Melmed, Surface Sci. 5 (1966) 359.
[13] G.K.L. Cranstoun and D.R. Pyke, Surface Sci. 46 (1974) 101.
[14] P.D. Augustus and J.P. Jones, Surface Sci. 64 (1977) 713.
[15] J.P. Jones and E.W. Roberts, to be published.
[16] Y. Takasu, H. Konno and T. Yamashina, Surface Sci. 45 (1974) 325.
[17] E.W. Plummer and T.N. Rhodin, J. Chem. Phys. 49 (1968) 3479.
[18] C.J. Todd and T.N. Rhodin, Surface Sci. 36 (1973) 353.
[19] R.D. Young and E.W. Plummer, Phys. Rev. B1 (1970) 2088.
[20] D.M. Newns, Phys. Rev. 178 (1969) 1123.
[21] L. Pauling, The Nature of the Chemical Bond (Oxford Univ. Press, 1960).

A DIGITAL SCANNING AUGER ELECTRON MICROSCOPE

R. BROWNING, M.M. EL GOMATI and M. PRUTTON

Department of Physics, University of York, Heslington, York YO1 5DD, UK

A new instrument, a scanning Auger microscope with digital scanning and a concentric hemispherical analyser, is described together with some preliminary demonstration results. The sample, a titanium ribbon, was hand polished and resistively heated. After a short heating to remove some of the overlay carbon and oxygen, the sample showed some irregular areas of very low SEM contrast which were attributed to calcium from the $L_{2,3}VV$ Auger process at 288 eV. Using this calcium peak a scanning Auger map of the areas showed high contrast and the corresponding titanium Auger maps demonstrate that this indeed was an overlayer of calcium. The probe resolution was 50 nm and a single picture was produced in 3 min.

1. Introduction

The distribution of elements at a surface is of great technological and scientific interest and in order to investigate this area we have constructed a scanning Auger microscope (SAM) to form two-dimensional maps of elemental distributions at a surface. The short mean free paths of electrons with kinetic energies between 10 and 1000 eV are not only responsible for the surface specificity of Auger electron spectroscopy but also are fundamental for microscopy. This is because, besides the generation of Auger electrons by backscattered electrons, the image resolution is expected to depend largely on the size of the electron probe. Thus the resolution of the scanning Auger microscope can be expected to show considerable advantage over the X-ray microprobe in this respect. The present instrument is a stage in the development of the scanning Auger field that was initiated by the work with thermionic cathodes and resoltuions between 100 and 0.5 µm [1–3] but the extension of the field to the truly microscopic (<100 nm) has implied the use of high brightness field emission sources in order to obtain sufficient signal [4–8]. Using such a source spatial resolutions as low as 30 nm have been reported [7].

The signal criteria chosen for the instrument described here are the production of a 128 × 128 point image in ~10^2 sec with a signal-to-noise ratio of 3 : 1.

2. The instrument

Fig. 1 is a schematic of the apparatus. The microscope has been based upon a V.G. Microscope Ltd HB200 UHV scanning electron microscope. The microscope is

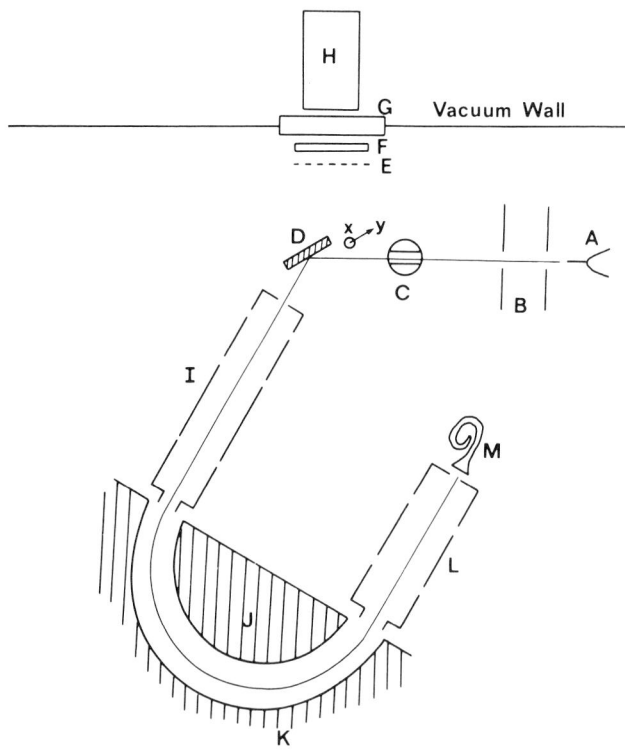

Fig. 1. Arrangement of the FE gun, SEM detector and electron energy analyser. A [310] field emitting tip, B electrostatic lens, C scanning coils, D sample, E retarding grid, F scintillator, G window in vacuum wall, H photomultiplier, I analyser input lens, J inner hemisphere, K outer hemisphere, L analyser output lens, M channeltron detector.

constructed of stainless steel and mu-metal throughout. The normal SEM image can be displayed at TV rates because of the high source brightness. It has a Crewe-type [9] field emitting electron gun with a [310] tungsten tip of radius between 200 and 1,000 Å. This delivers between 10^{-9} and 10^{-8} amps into a probe of approximately 500 Å diameter at 25 kV. The beam current typically decays from 10^{-8} to 3×10^{-9} A in 10^3 sec after cleaning which is done under a reverse field [9]. The reverse field is found to be vital in cleaning and forming the tip for good [310] emission.

3. The analyser

The analyser is based on an instrument previously reported by Bassett [10]. Its major parts are a retarding input lens which has been modified to four element operation, concentric hemispheres, and an output lens which produces a real image

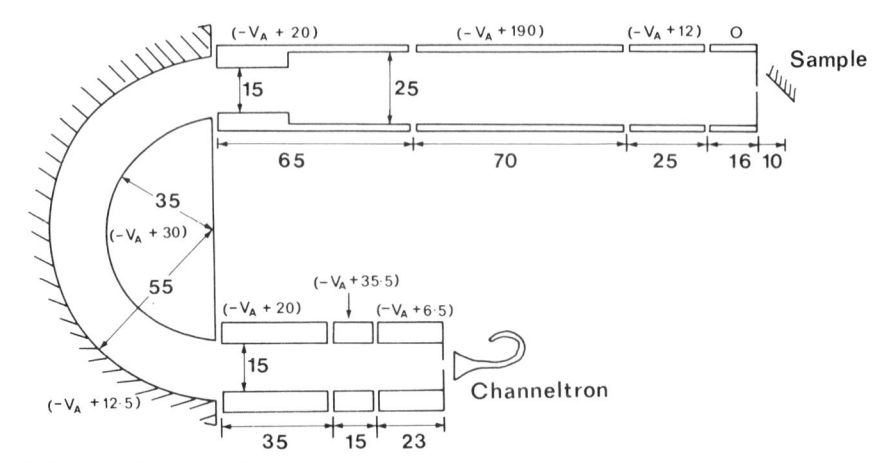

Fig. 2. The CHA and its input and output lenses. The potentials shown indicate the conditions for focussing electrons with kinetic energies eV_A using a constant pass energy of 20 eV and variable retardation by the input lens. All dimensions in mm and potentials in volts. The lens gaps are all one tenth of the lens diameters.

on the detector (fig. 2). The detector has been changed to a Mullard channeltron (B 318 BL) to allow pulse counting up to a maximum count rate of 500 kHz.

The resolution of the analyser is related to the retardation of the input lens and the dispersion of the concentric hemispheres. The resolving power is

$$S_{anl} = (V_1/V_2)S_H ,$$

where S_{anl} = resolving power of the analyser system, S_H = resolving power of the hemispheres, and V_1 and V_2 are the initial and retarded electron energies.

The input lens was designed empirically to maximise the input signal given the geometrical constraints in our present chamber and therefore there was no a priori theoretical estimate of S_H, this depending mainly on the aberrated image size in the dispersion plane of the hemisphere. However experimentally, with a hemisphere pass energy of 20 eV, S_H was found to be approximately 10.

The analyser can be operated in two modes by changing the potentials, one with constant energy window (1), and the second with constant resolving power (2).

(1) The hemispheres are operated at constant pass energy and the input lens retardation is changed when sweeping through the spectrum. Fig. 3. shows that in practice this gives an approximately constant energy window ΔE of 2 eV from 20 eV to 800 eV when the retardations are varying between 1 : 1 and 40 : 1. This is useful for faithful reproduction of $N(E)$ and for maximising the signal from low energy Auger peaks.

(2) The output lens is operated at constant retardation and the pass energy is swept to record a spectrum. This mode gives a constant resolving power $E/\Delta E$. Fig. 3 shows the operation of the system with a fixed input lens retardation of 15 : 1 and

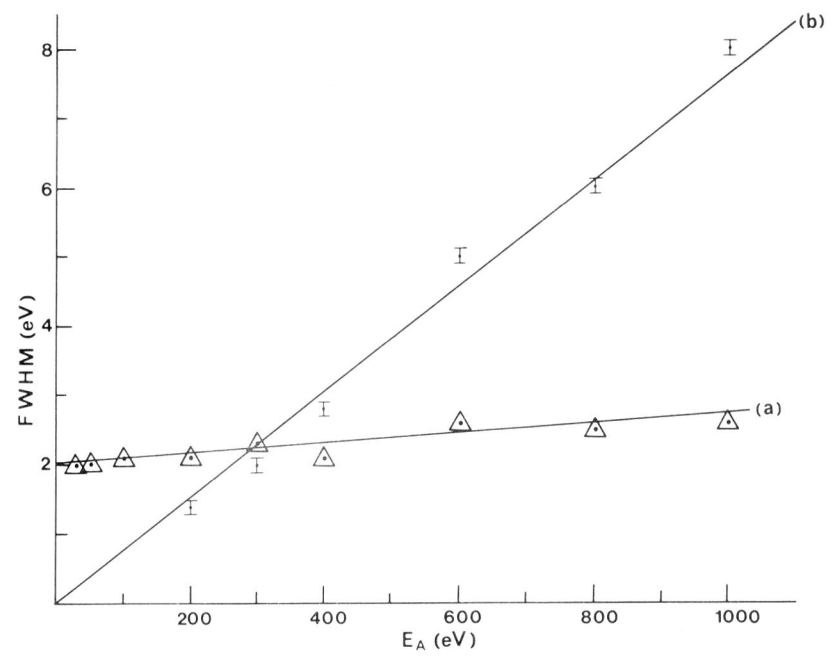

Fig. 3. Full width at half maximum (FWHM) of the elastic peak scattered from a polycrystalline Titanium sample. (a) "Constant" bandpass mode using an analyser pass energy of 20 eV. (b) Constant retardation mode with the input lens focussing at 15 : 1 retardation. $E/\Delta E \simeq 130$.

a constant resolving power of approximately 130 and in addition gives a comparison with the constant window mode of pass energy 20 eV.

The input lens has a useful range of focus from 0.5 to 2 cm achieved by tuning the second element but is normally operated with the sample inclined at 55° to the optic axis and 1 cm from the input lens aperture. With an input slit of 2 mm × 4 mm the arrangement has a collection efficiency of about 1.6% of the electrons emitted from the sample into 2π steradians. This figure compares quite well with the CMA efficiencies that have been used in scanning Auger microscopy (~5.6% from the arrangement described by Venables et al. [7]) with the added advantages of variable working distance and greater flexibility in resolving power and mode of operation. The field of view of the analyser is approximately 0.5 cm diameter.

4. The electronics

The microscope controls, analyser supplies and detection electronics are illustrated in the block diagram fig. 4.

The microscope is controlled from a purpose-built digital microscope control

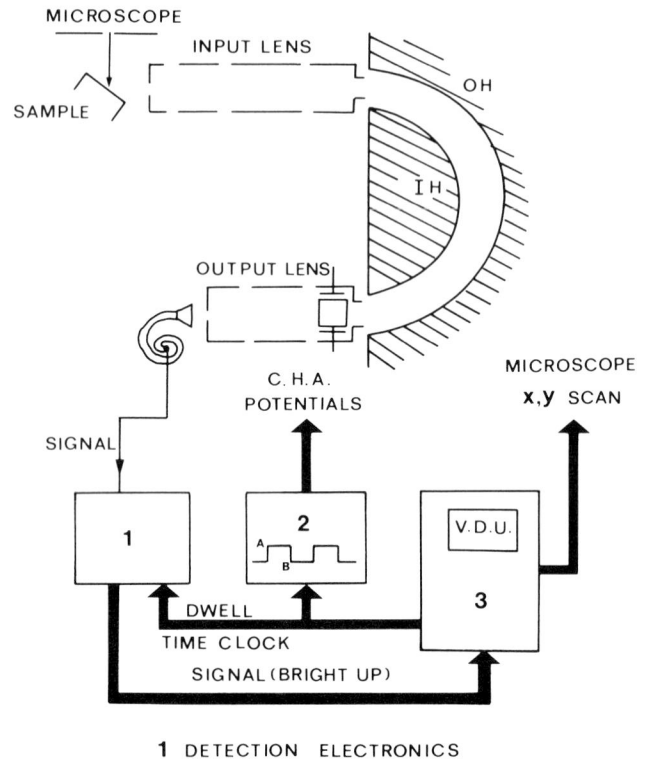

Fig. 4. System electronics block diagram. The detection electronics are a Brookdeal 5Cl photon counter under the control of the DMC.

(DMC) that can be switched in to replace the analogue scans of the HB 200. It has great flexibility in choice of the image matrix, running from 32 to 1024 points per line or lines per frame independently, with dwell times per point between 1 sec and 10 μs. Frame, Line, Single Frame, Single Line, and Spot are pushbutton options, the Line and Spot functions going to the chosen addresses that are set up and displayed on the front panel. The Spot position can also be displayed as a bright point in the frame or line scan. The DMC supplies timing clocks for the rest of the system's electronics and frame and clock control signals that can be tape recorded with the picture information allowing long frame times (~1/2 hr) to be played back quickly for photography and display on a grey tone storage VDU (Tektronix 605).

The analyser potentials are supplied by potentiometer chains from fast slewing operational power supplies, and in order that the potentials have a low enough impedance to drive connecting cables and the spectrometer, high voltage transistors are used as emitter followers so that fast switching of the spectrometer (2×10^5 V/sec) is possible with little current dissipation.

The operational power supplies which determine the analysing energy are driven from an energy controller that gives (1) an analogue scan to generate spectra, (2) allows the setting up of two independent fixed energies and (3) switches between the set energies under the control of the DMC for imaging.

The detection electronics consists of a charge preamplifier with a bandwith of 1 MHz, a Brookdeal Photon counter and an analogue divider. The Brookdeal Photon counter under the control of the DMC produces the Auger signal for each point in a frame and this signal is divided by the average SEM signal over that period. This division is necessary due to the fluctuations in primary beam current that are characteristic of field emission sources.

The Auger signal used in the imaging process is obtained by counting in the un-

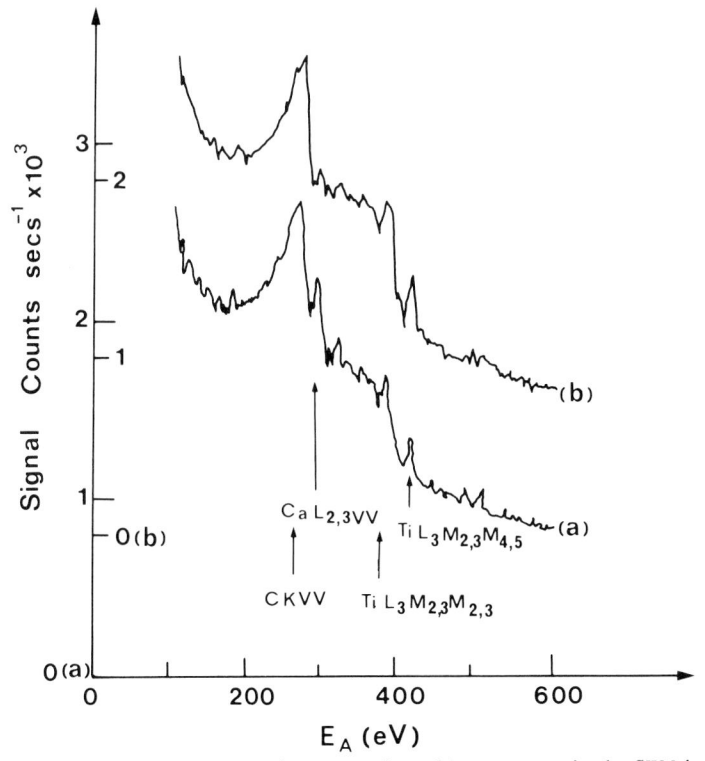

Fig. 5. Frame scans and spectra on and near a region of low contrast in the SEM image. $i_b = 10^{-9}$ A; $\tau = 10^{-1}$ sec; $T_F = 27$ min; $E_A = 288$ eV; $E_B = 296$ eV; $E_P = 25$ keV. (a) Spectrum on the region; $\Delta E = 2$ eV. (b) Spectrum off of the region; $\Delta E = 2$ eV. (c) Auger image; 128×128 points. (d) Auger image on the edge of the bright region in (c); 128×128 points.

c ⌞2 5 µm⌟

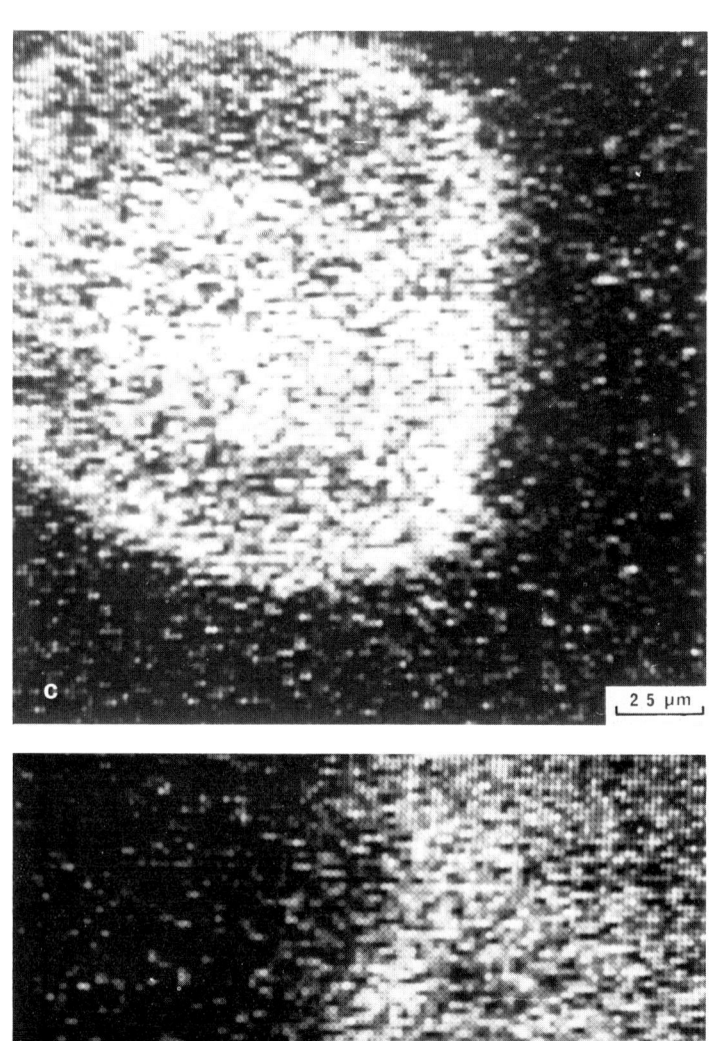

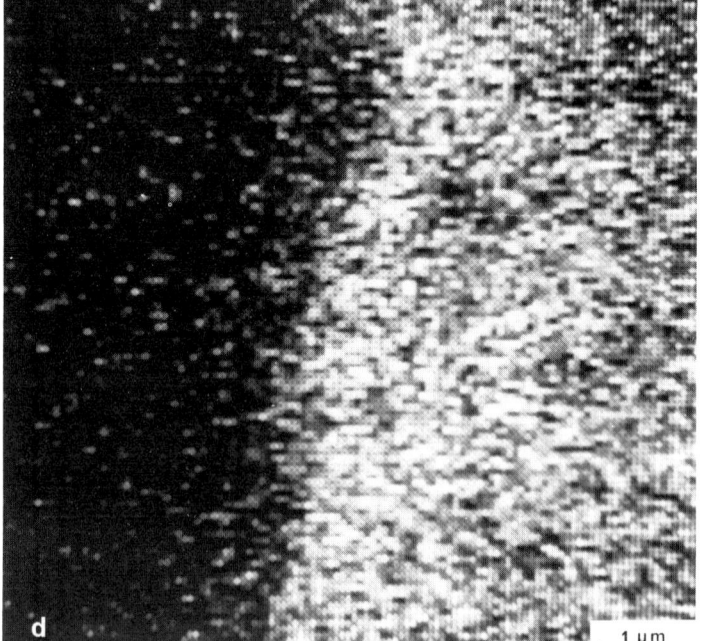

d ⌞1 µm⌟

differentiated $N(E)$ mode. Two potentials are set up by front panel switches on the energy programmer, one potential is at the peak of an Auger signal and in imaging this signal is counted for half an image point dwell time giving a count (A). The spectrometer is then switched by the DMC in the second half of the dwell time to the second potential just above or below the Auger peak to measure the background count (B). The two counts are then subtracted digitally by the Brookdeal Photon Counter to produce the Auger signal. The signal is then divided by the average SEM intensity (count C) over that period, and this result $(A - B)/C$ is used as the image bright-up signal. This has been found to be very satisfactory for a surface with little topographical content, the division in that case merely normalising beam current fluctuations. However it has been found that this is not very accurate for rough surfaces and the division technique of Janssen et al. [11]. will have to be investigated for this instrument. Besides this, however, it is felt that the ease of setting up the spectrometer for maximum signal on a peak is simpler and more obvious with the present digital switching of the spectrometer than with conventional $N'(E)$ methods and also allows one greater freedom for maximising signal on complex peak shapes without losing resolution due to modulation effects (Prutton [12]).

5. Results

The sample was a hand polished polycrystalline titanium ribbon mounted on the input lens of the analyser in a holder that allowed resistive heating of the ribbon in situ. After a short heating to remove part of the overlay carbon and oxygen the sample showed some irregular areas of very low SEM contrast.

Fig. 5a shows the spectrum that was obtained from within the areas, and fig. 5b from outside on the surrounding sample. As can be seen the sample is not clean titanium and still has an overlayer of carbon. The spectrum 5a shows only one peak that is not in the background (fig. 5b) and this was attributed to the calcium $L_{2,3}VV$ transition at 288 eV. Figs. 5c and 5d are scanning Auger images of the sample using this peak and it can be seen that the contrast is very high (~100%). Images generated using the titanium $L_3M_{2,3}M_{4,5}$ signal showed approximately 50% contrast and were deficient in the areas that the calcium images were bright. The carbon KVV peak gave images with little contrast implying that the carbon was indeed an overlay. The calcium signal to background ratio was 0.14 and there were about 2000 counts per second at the peak with a beam current of 10^{-9} A. Frame scans of 128×128 points with a dwell time of 0.1 second per point were used, giving about 27 min for a frame.

After further heatings through several stages the sulphur KVV Auger line at 150 eV grew significantly and the carbon KVV line disappeared, but the sulphur signal showed only diffuse concentration gradients with no marked feature at or around grain boundaries. Though it must be pointed out that the experiments were not exacting, the growth of the sulphur signal was fairly homogeneous [8] [13] whereas the resolution of the microscope was well within the grain size of 10—100 μm and any inhomogeneous concentrations within 50 nm of the grain boundaries would have been expected to show up in an image with a signal to noise ratio of 3 : 1 (10^{-8} A, 128 × 128, 0.01 sec). No such details were observed even with signal to noise ratios of 10 : 1 (10^{-8} A, 128 × 128, 0.1 sec) where they were expected to show up more clearly.

After heating for an integrated time of approximately an hour at red heat the surface became distorted by the growth of the grains and it became clear by comparison with the sulphur, titanium and an unassigned low energy peak near 40 eV that our present technique of divising by the SEM signal was not accurate enough and "false" Auger information was generated. The "false" information in this case was less than 20% of the total signal but is nonetheless misleading and it is evident that care must be taken in the interpretation of Auger images in the case of rough surfaces.

6. Conclusions

A scanning Auger microscope has been constructed using a concentric hemispherical analyser and with digital scan. The analyser is normally operated in a mode that gives an approximately constant band pass window of 2 eV from 20 eV to 1000 eV. With beam currents of 10^{-8} A and a collection efficiency of 1.6% the system's sensitivity is sufficient to generate a 128 × 128 point Auger map in 3 min. This is similar in performance to the instrument described by Venables et al. [7], also a field emitting microscope but equipped with a CMA. The present results, which are of a preliminary nature, illustrate the power and the limitations of the present instrument. The identification of calcium as an unknown and the production of an image has great technological import but the difficulties with topology may well be a limitation in the investigation of a surface.

References

[1] N.C. MacDonald and J.R. Waldrop, Appl. Phys. Letters 19 (1971) 315.
[2] M.P. Sea and C. Lea, Scanning Electron Microscopy; Systems and Applications, Newcastle upon Tyne, 1973 (Institute of Physics Conference Series No. 18).
[3] T. Ishida, M. Uchiyama, Z. Oda and H. Hashimoto, J. Vacuum Sci. Technol. 13 (1976) 711.
[4] H.E. Bishop, Materials Development Division, AERE, Report No. 7899 (1974).
[5] G. Todd, H. Poppa, D. Moorhead and M. Bales, J. Vacuum Sci. Technol. 12 (1975) 953.
[6] B.D. Powell, D.P. Woodruff and B.W. Griffiths, J. Phys. E (Sci. Instr.) 8 (1975) 548.
[7] J.A. Venables, A.P. Janssen, C.J. Harland and B.A. Joyce, Phil. Mag. 34 (1976) 495.

[8] R. Browning, P.J. Bassett, M.M. El Gomati and M. Prutton, Proc. Roy. Soc. (London) to be published.
[9] A.V. Crewe, D.N. Eggenburger, J. Wall and L.M. Welter, Rev. Sci. Instr. 39 (1968) 576.
[10] P.J. Bassett, J. Phys. E7 (1974) 461.
[11] A.P. Janssen, C.J. Harland and J.A. Venables, Surface Sci. 62 (1977) 277.
[12] M. Prutton, J. Electron Spectr., 11 (1977) 197.
[13] H.E. Bishop, J.C. Rivière and J.P. Coad, Surface Sci. 27 (1971) 1.

LOW BEAM CURRENT DENSITY AUGER SPECTROSCOPY AND SURFACE ANALYSIS

C. LE GRESSUS, D. MASSIGNON and R. SOPIZET

CEN-Saclay, B.P. 2, 91190 Gif-sur-Yvette, France

Auger and secondary electron spectroscopy become a more and more routine technique in surface characterization. Even with primary electron beam current density as low as 10^{-2} or 10^{-3} A cm^{-2} beam damage were reported in both Auger and LEED experiments. So we developed and compared counting method, brightness modulation and Harris' modulation techniques in terms of signal to noise ratio. The two first methods offer the advantage of a primary beam current density decreasing about 10^4 times. So various mechanisms of beam damage were identified as thermal, chemical and electrical. The advantage of the method is shown with hydrocarbons adsorption layer; the beam cracking of the organic chain produces a chemical shift of the C_{KLL} maximum Auger line about 5 eV. This progressive shift is observed with current densities of 10^{-5} A cm^{-2} order of magnitude. The reproducibility of this low current density Auger spectroscopy allowed the study of the background and the true secondary electron yield modifications when adsorbed layers are built up.

1. Introduction

Electron spectroscopy (AES, energy loss spectroscopy, ...) should ideally be non-destructive, but surface and bulk modifications have been reported when the incident electron dose D is higher than 10^{18} electrons cm^{-2} [1–3]. As a high spatial resolution is necessary for local analysis of heterogeneous surfaces, the primary beam intensity is to be lowered in order to avoid damage.

We shall compare here three methods of electron spectroscopy, the Harris [4] derivative method (AEM), the primary beam modulation (GBM) [5] and a counting technique (CT) [6], by the level of signal to noise ratio observed. All these methods use as signal the output voltage of the cylindrical mirror analyser (CMA). The arguments presented here do not include specific consideration of the background in the Auger spectrum and so are most relevant when high primary energies are being used.

2. Experimental techniques

2.1. Signal

The electron spectrometer and associated high vacuum scanning electron microscope [7] used here has been adapted [5] to use either of two modulation methods:

(i) Analyser energy modulation (AEM) or Harris derivative method: a sinusoidal perturbing voltage is applied on the CMA, resulting in a 2 E_m peak-to peak modulation of the detected secondary electron energy E. The rms value of the CMA output signal is, for small E_m,

$$S_{AEM}(E) = \frac{1}{\sqrt{2}} \delta E_m \frac{dn}{dE}, \qquad (1)$$

where $\delta = RE$ is the CMA energy window ($R = 0.003$ [7]) and $n(E)$ the spectral energy density of the secondary electrons.

(ii) Gun brightness modulation (GBM), or modulation of the primary electron beam intensity according a sine law [5] instead of a beam blanking [8]. The rms value of the CMA output signal is now:

$$S_{GBM}(E) = (1/\sqrt{2}) \beta \delta n(E), \qquad (2)$$

where β is a function of the amount of modulated primary intensity, introduced here for comparison sake of signal (1) and (2): for a 80% modulated primary beam, $\beta = 0.94\sqrt{2}$.

In counting method (CM) [6], the CMA output signal is:

$$S_{CT}(E) = \delta n(E), \qquad (3)$$

practically the same as (2).

Let us now compare (1), (2) and (3). Assuming an ideal Gaussian shape for an Auger emission line [9,10], the corresponding peak is detected by the minimum of dn/dE (1) or by the maximum of $n(E)$, (2), (3):

$$\min. S_{AEM} = -\frac{1}{2\sqrt{\pi e}} \frac{1}{\sigma} \frac{\delta E_m}{\sigma} i = -0.17 \frac{\delta E_m}{\sigma \sigma} i \qquad (4)$$

$$\max. S_{GBM} = \frac{1}{\sqrt{2}} \frac{1}{\sqrt{2\pi}} \beta \frac{\delta}{\sigma} i = 0.375 \frac{\delta}{\sigma} i \qquad (5)$$

where i is the Auger emission intensity and FWHM = 2.35 σ its full-width at half maximum.

Therefore, the comparison between the three techniques rests on the value chosen in AEM for the energy peak-to-peak modulation $2E_m$. These values ought to be taken between tow limits, $2E_m = 0.54$ FWHM (10% contamination of S_{AEM} in (1) by a negative d^3n/dE^3 term [9] and $2E_m = \delta$ (ensuring an equal energy resolution for the three techniques). A numerical example is given by the KLL Auger line of carbon: $E = 268$ eV, FWHM = 15 eV; for both limit values of $2E_m$ (8 and 0.8 eV) the GBM or CT methods give signals 1.75 and 17.5 times more intense than AEM the method. Another illustration is given by oxidized Li KVV, O KLL and Al LVV lines recorded in AEM and GBM (fig. 1).

2.2. Noise

In electron spectroscopy, it is well known that the noise varies as the square root of the primary beam intensity [11]; but this noise is not a white noise and decreases with increasing frequency.

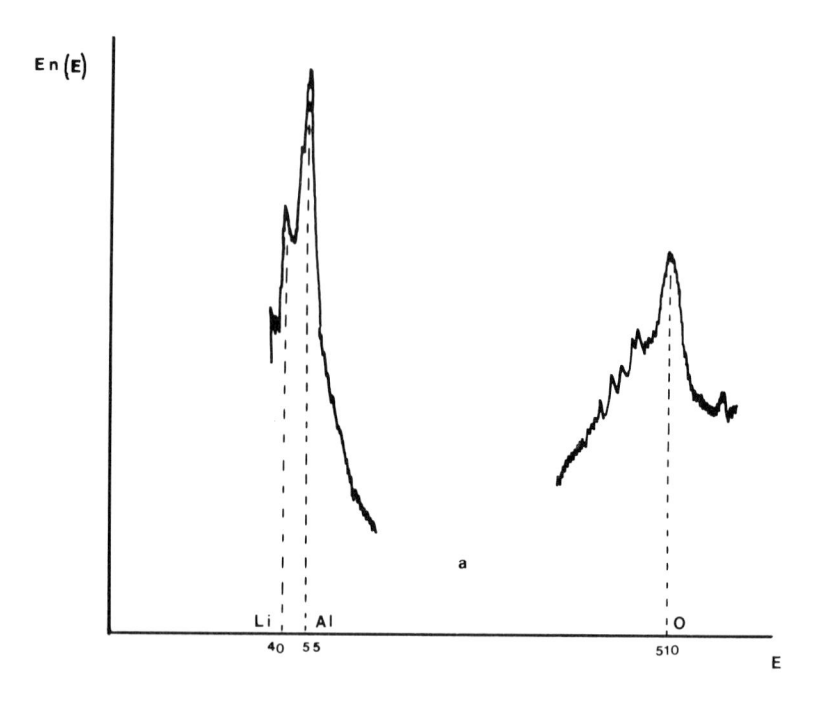

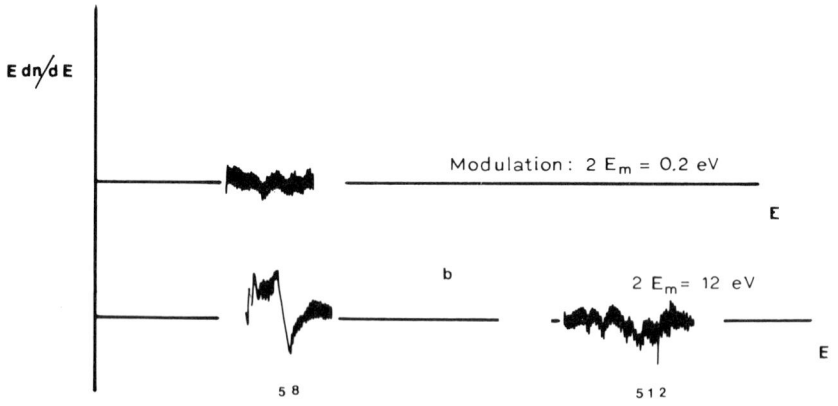

Fig. 1. Signal-to-noise ratio comparison in GBM (a) and AEM (b). $I_p = 10^{-8}$ A. Lock-in: $s = 300$ μV, $\tau = 100$ ms, $\nu = 3$kHz.

With our electron spectrometer, the noise reduction observed between 3 and 30 kHz is better than a factor 5 (figs. 2b, d versus a, c), but this advantage can only be used in GBM method (fig. 2c, d). In the AEM method, a cut-off frequency around 3 kHz, related to the application of a sinusoidal voltage to the CMA outer cylinder, gives a 6 dB/octave signal reduction.

Therefore, if an equal energy resolution is achieved in the spectrum, and for a given Auger line intensity, the signal to noise ratio in AEM at 3 kHz is 100 times smaller than in GBM at 30 kHz. And for a given signal to noise ratio, the primary

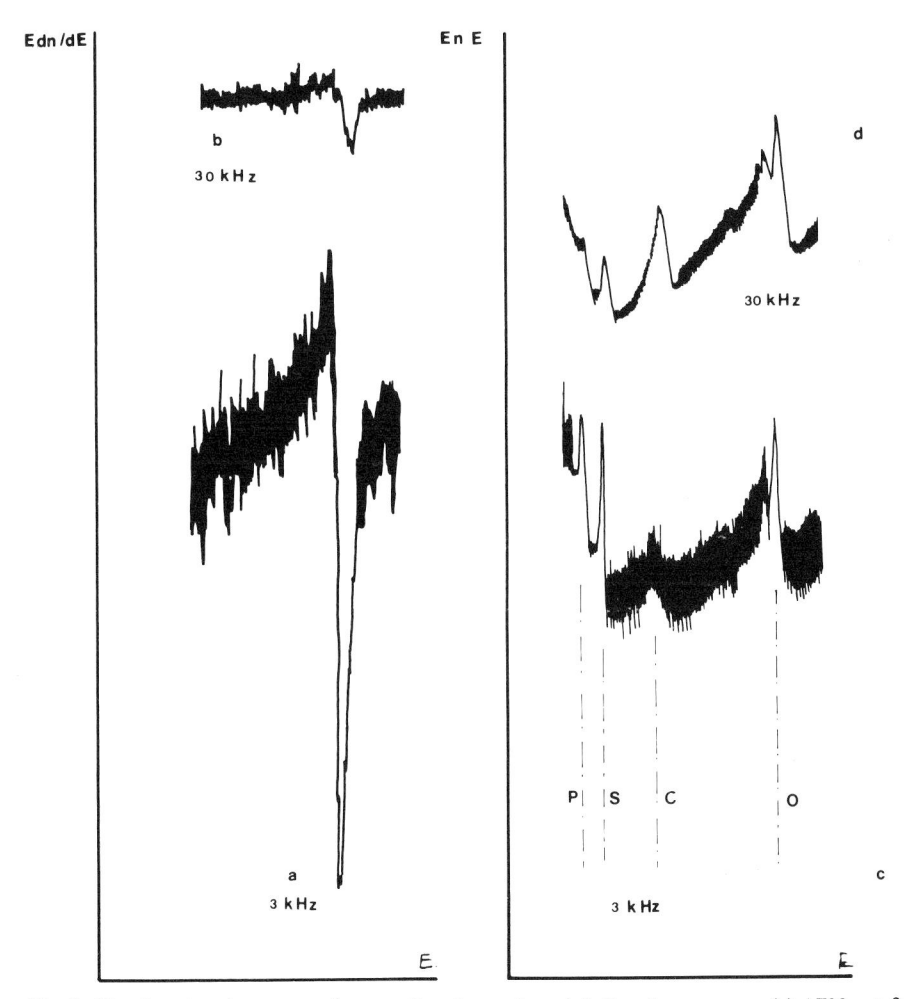

Fig. 2. Signal and noise attenuations as functions of modulation frequency ν: (a) AEM, $\nu = 3$ kHz; (b) AEM, $\nu = 30$ kHz; (c) GBM, $\nu = 3$ kHz; (d) GBM, $\nu = 30$ kHz.

beam intensity may be lowered in GBM by a factor 10^4. The same reduction may also be obtained in CT without losing any information [6].

However, if a low energy resolution is enough for a routine work, it is possible either to increase the energy modulation in AEM ($2E_m$ = 8 eV instead 0.8 eV), or to use a larger CMA aperture in GBM or in CT ($\delta < 0.003\,E$).

3. Electron beam damage

The possibility of using lower primary beam intensity has important applications in local analysis of a surface either in lowering proportionnally the irradiation dose in a given spot area, or in achieving a better spatial resolution with a given irradiation dose, as will show the following examples:

(i) *Thermal damage*. With a 10 kV accelerating voltage, a 10^{-9} A primary beam intensity, a 500 s recording time and a 1000 Å spot diameter, the total irradiation dose D reaches 5000 C cm^{-2} (or 8×10^{22} electrons cm^{-2}), the current density 10 A cm^{-2} and the power to be dissipated 10^5 W cm^{-2}. These values are high enough to produce a sensible temperature rise in the irradiated spot of the sample under the beam. According Baker's temperature calibration method [12], a temperature uprise of several hundred degrees Celcius could thus be obtained. This is also in agreement with the influence of surface temperature uprise under the beam on the oxidation of pure polycrystalline nickel surface reported by Verhoeven [13].

By defocussing the primary beam to observe spots of 1, 10 or 100 μm on the same sample, we register the spectrum difference due to irradiation doses varying from 1 to 10^4. We have found two effects with increasing irradiation dose:

(1) An enrichment in surface impurities, such as S, P, C, O, ..., in various samples (polycrystalline nickel, polycrystalline gold, iron) which follows a diffusion like law in root square of time. We did not measure for each sample the energy of activation of the process, but we attribute the effect to a diffusion of impurities from the bulk. This hypothesis is compatible with the results of studies of C diffusion in polycrystalline nickel [14]. The role of residual gas under a 10^{-9} Torr high vacuum is of a second order of importance.

(2) An attenuation of the recorded signal: the Auger peak/background ratio remains constant but the total signal (Auger peak + background) decreases when the time of irradiation increases. This effect is mainly sensible with insulators and the signal attenuation may reach a factor of 3 (fig. 3). That attenuation can be avoid by cooling the sample. Further investigations of this effect are in progress.

(ii) *Oxide reduction of SiO_2*: in good agreement with Johannessen [15] we find, on a SiO_2 thin layer of 1000 Å on a monocrystalline silicon, that, when the surface is exposed to a flux of 10^{17} e$^-$ cm^{-2} or more, the oxygen reduction may reach about 50% of the initial concentration. This effect is usually attributed to a thermal enhancement of the electron desorption cross section. These critical doses are much lower than the doses of section (i) or than the usual doses in Auger analysis.

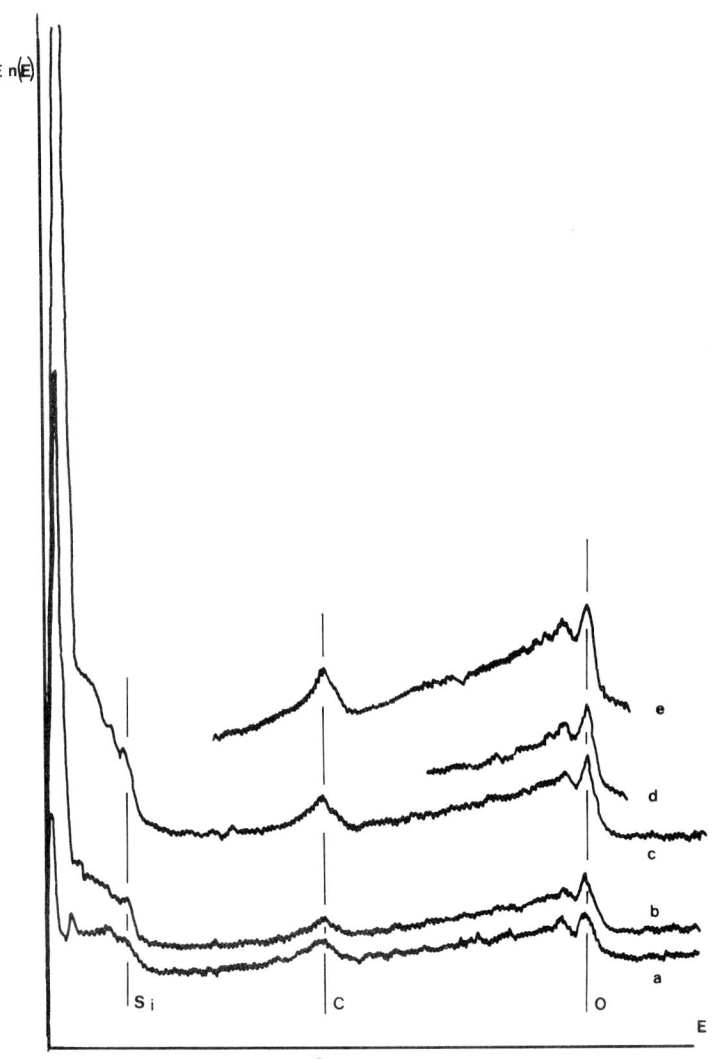

Fig. 3. Sample temperature effect. $I_p = 10^{-8}$ A. Lock-in: time constant $\tau = 100$ ms, sensitivity = 1mV. Ion etching: 2 keV, 10^{-5} A cm^{-2}, argon ions. (a) After 5 mm ion etching, focussed beam (1 μm); (b) after 5 mm ion etching, defocussed beam (30 μm); (c) sample at 20°C; (d) sample cooled at $T = 0°$C; (e) sample cooled at $T = -60°$C.

(iii) *Electrical damage:* when the current density becomes higher than 1 A cm^{-2} the "true secondary" electrons emission reaches a minimum while the surface composition may be modified (see the C_{KLL} peak on fig. 4). Secondary electron image contrast in SEM is, of course, also modified. This may be attributed to a space charge effect induced by the beam which may produce ion migration at room tem-

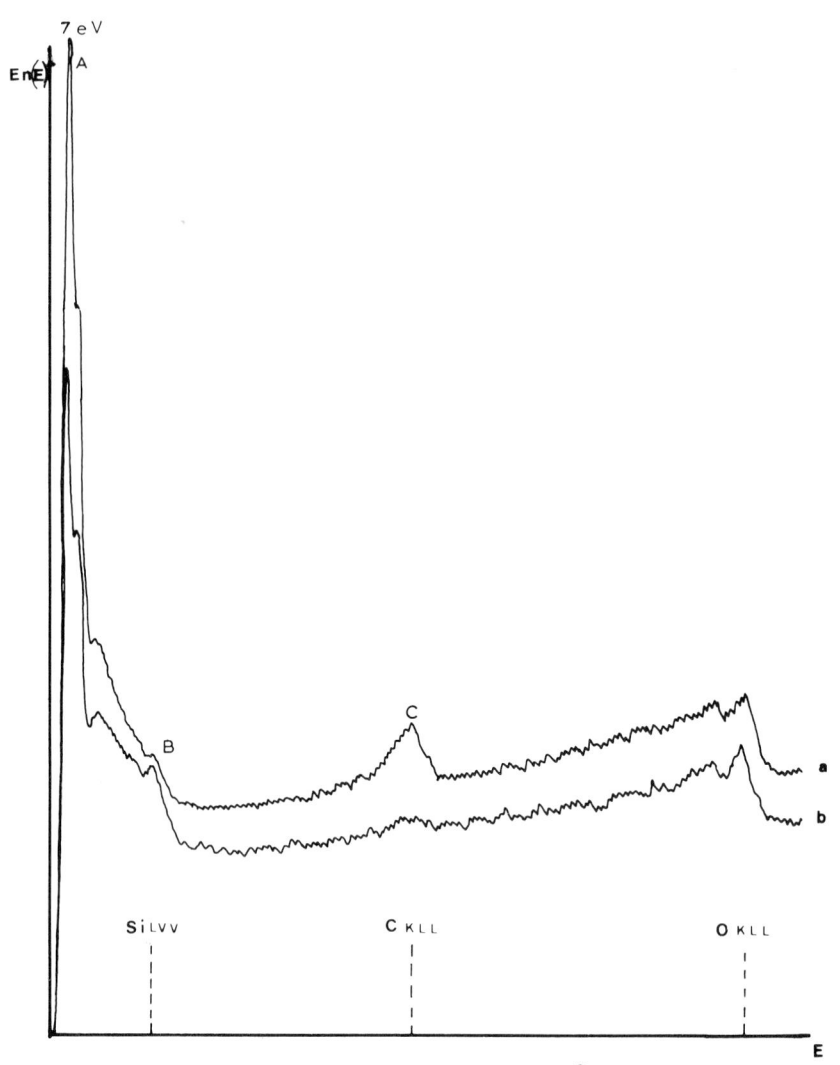

Fig. 4. Carbon contamination of a SiO$_2$ thin layer. $I_p = 5 \times 10^{-9}$ A. (a) Spot diameter 5 μm; (b) spot diameter 5000 Å. When the spot diameter varies from 5 μm (a) to 5000 Å (b), the Si LVV peak increases, the 7 eV "true secondary" peak and the CKLL Auger peak decrease.

perature or higher temperature (in the beam-irradiated spot). Similar effects on insulators have been already reported in AES [16], as well as in ion microprobe [17]. Further investigations should elucidate the mechanism and the correlation with the secondary emission intensity.

(iv) *Organic thin films analysis:* we found that the destructive dose on a 100 Å thick acrylate layer deposited on a steel substrate, or on a monolayer of calcium stearate deposited on aluminum, is about 10^{14} electrons cm^{-2}. A dose is considered destructive when, after AES analysis, the spot can be observed in secondary electron image mode (in SEM). Assuming an average area of 10 Å^2 per molecule, this destructive dose corresponds to 1 electron/10 molecules. The beam cracking may produce graphite or polymerized surface layers, and therefore limits the use of AES in biology. When the sample is damaged, its true secondary electron emission is modified while the C_{KLL} shifts from 263 eV to 268 eV in the case of acrylate layer on steel.

4. Conclusion

The evolution of Auger microscopy towards higher spatial resolution, stimulated by the higher signal to noise ratio in brightness modulation or counting methods, is however still limited by the radiation damage and by the electron multiplier gain. To reach a resolution of 500 Å or less without severe radiation damage would necessitate a primary beam intensity not higher 10^{-12} A: the corresponding dose is 10^{+17} electrons cm^{-2} sec^{-1}.

References

[1] M. Isaacson, J. Chem. Phys. 56 (1972) 1803.
[2] J.P. Coad, M. Gettings and J.C. Rivière, Faraday Discussions Chem. Soc. 60 (1975) 269.
[3] C.C. Chang, Surface Sci. 25 (1971) 53.
[4] L.A. Harris, J. Appl. Phys. 39 (1968) 1419.
[5] C. Le Gressus, D. Massignon and R. Sopizet, Compt. Rend. (Paris) B 280 (1975) 439.
[6] P.W. Palmberg, J. Electron Spectrosc. Related Phenomena 5 (1974) 691 (figs. 6 and 7).
[7] A. Mogami, H. Hotta, H. Hashimoto, C. Le Gressus and D. Massignon, in: Proc. 8th Intern. Congr. on Electron Microscopy, Canberra I, 1974, p. 60.
[8] M.P. Seah, Surface Sci. 32 (1972) 303.
[9] H.E. Bishop and J.C. Rivière, AERE R 5854 (1968).
[10] N.J. Taylor, in: Techniques of Metal Research, Ed. R.F. Bunshah (Interscience, New York, 1971) ch. VII, p. 117.
[11] D. Chatterji, The Theory of Auger Transitions (Academic Press, 1976) ch. VIII, p. 224.
[12] B.G. Baker and B.A. Sexton, Surface Sci. 52 (1975) 353.
[13] J. Verhoeven and J. Los, Surface Sci. 58 (1976) 566.
[14] J.F. Mojica and L.L. Levenson, Surface Sci. 59 (1976) 447.
[15] J.S. Johannessen, Faraday Discussions Chem. Soc. 60 (1975) 313.
[16] C.G. Pantano, D.B. Dove and G.Y. Onada, Jr., J. Vacuum Sci. Technol. 13 (1976) 414.
[17] J.W. Coburn, J. Vacuum Sci. Technol. 13 (1976) 1037.

DESORPTION KINETICS OF CONDENSED TWO-DIMENSIONAL PHASES ON A SINGLE CRYSTAL SUBSTRATE

G. LE LAY, M. MANNEVILLE and R. KERN

Centre des Mécanismes de la Croissance Cristalline CNRS, Centre de Saint-Jérôme, 13397 Marseille Cédex 4, France

We emphasize the fact that the knowledge of the order of the reaction kinetic upon desorption is of major physical interest and permits to characterize twodimensional phases for a given deposit–substrate system.

This order can be obtained directly from isothermal desorption spectroscopy (ITDS) experiments or indirectly from classical thermal desorption spectroscopy (TDS) ones thanks to new developments of the Bauer and Poppa's method of analysis of the spectra we propose.

Assuming that 2D phases on a single crystal substrate form islands which desorb indirectly via a dilute adsorbed gaz we develop a general model from which zero order or half order kinetics can be derived according to the respective influence of surface diffusion.

We have performed ITDS experiments with AES for both already thoroughly investigated Au/Si(111) and Ag/Si(111) systems and TDS experiments for the Ag/Si(111) one, in the submonolayer range.

Confrontation with our model is very encouraging; especially values of physical parameters: cohesive energy of 2D phases and vibrational frequencies deduced are in fairly good agreement with separate determinations for both Au/Si(111) and Ag/Si(111) systems and for the Xe/graphite(0001) too, another well-known system, to which the model is also applied.

AES ANALYSIS OF OXYGEN ADSORBATED ON Si(111) AND ITS STIMULATED OXIDATION BY ELECTRONIC BOMBARDMENT

M.C. MUÑOZ and J.L. SACEDÓN
Instituto de Física de Materiales, Serrano 144, Madrid-6, Spain

The growth of the oxygen KLL peak is compared with the attenuation of the Si $L_{2,3}V_1V_1$ peak, and with the rise up of the SiO_2 75 eV peak.

The final oxygen signal height is calibrated in monolayers by the intensity attenuation of the Si KLL transition. From this coverage, the calculation of the SiO_2 signal growth as a function of oxygen KLL is obtained. This calculation matches satisfactorily with experimental results. For the Si $L_{2,3}V_1V_1$ peak the agreement is only obtained for coverages greater than one monolayer.

At the end of the adsorbtion process without electronic stimulation the oxygen saturation is 0.3 monolayers without SiO_2 formation. During this process one characteristic peak in the Si spectrum is observed, which is attenuated during posterior oxidation.

Finally an interpretation of the SiO_2 Auger spectrum is proposed, based upon the band calculation of Pantelides and Harrison (1976) for SiO_2.

SHAPES AND SHIFTS IN THE OXYGEN AUGER SPECTRA

P. LÉGARÉ *, G. MAIRE *, B. CARRIÈRE ** and J.P. DEVILLE **

Université Louis Pasteur, 4 rue Blaise, Pascal, 67000 Strasbourg, France

The oxidation of silicon and platinum single crystal faces, of polycrystalline supported catalysts and of some alloy surfaces has been studied by AES and as far as possible by LEED. A comparison of the oxygen Auger spectra obtained during the oxidation process with those found on oxides has been made; it shows that the modification of the fine structure of the oxygen Auger peaks gives some information about the binding state of oxygen. Two different structures, which compete one with the other, are described. In one case, a spectrum where three lines dominate is obtained; in the other case, a "quasi-atomic" spectrum characterized by five features is observed: multiplet splitting in the two-hole final state is predominant. Besides these differences in the fine structure of the Auger spectra one can notice shifts of several eV for the main feature. They have been correlated with the various observed LEED patterns. Physisorption, chemisorption, solution of oxygen in the metal lattice, growth of ordered or amorphous oxides are the different possibilities which are discussed.

1. Introduction

Several authors [1–5] have dealt with the interpretation of electron-induced Auger spectra, trying to use them not only for atomic identification but to obtain additional information about the chemical effects during adsorption studies. We have also recently considered [6] the use of Auger electron spectroscopy (AES) in relation to oxidation processes; we showed that the explanation of the energy shifts was not straightforward and had to be further discussed; besides a careful look at the general shape of the oxygen "KVV" spectrum was necessary.

In this paper, we consider the Auger spectra of oxygen: (i) adsorbed on platinum and silicon single crystal faces; (ii) adsorbed on supported catalyst and alloy polycrystalline surfaces; (iii) bound in bulk oxides. We focussed our attention on the detailed shape of the spectrum from 440 to 540 eV and on the energy shifts observed on the main Auger peak in the vicinity of 510 eV. As far as possible the LEED pattern corresponding to a given Auger spectrum was observed and discussed.

At low coverage, during the oxidation of metals and alloys, the oxygen Auger spectrum has been characterized by a fine structure where three major peaks domi-

---

* Laboratoire de Catalyse et Chimie des Surfaces, ERA 385 du CNRS.
** Laboratoire de Minéralogie, ERA 07 du CNRs.

nate; at higher coverage, we found a change in the fine structure which turned into a five-line spectrum. At the same time the main peak was shifted toward lower energies. The oxides generally showed a five-line spectrum but since the shifts are affected by charging effects, the discussion is difficult in that case.

In the case of platinum a correlation between the different LEED patterns (which depend on the oxygen exposures and on the various heat treatments) and the Auger spectra will be discussed.

2. Experimental

The results described in the next section were obtained in two kinds of analyzing systems. The first one is a LEED–Auger four-grid apparatus and the AES curves were recorded using the retarding field method (RFA). In this system we investigated bulk oxides (silica, quartz, alumina and magnesium oxide), the (111) and (110) faces of silicon, the (111) and (100) faces of platinum and a stepped surface of platinum: Pt(S) [6(111 × (100)], i.e. a vicinal of Pt(111). Clean surfaces of oxides were obtained by breaking or cleaving the samples under ultra-high vacuum; for silicon and platinum the standard cleaning procedures have been described previously [7,8]. The second apparatus is fitted with a cylindrical mirror analyzer (CMA) and has LEED facilities. It was mainly used to study alloys (Cu–Ni, 80%–20%) and powdered supported calalysts (Pt, Sn/Al_2O_3: 10% Pt, 5% Sn, 85% alumina; Pt, Ni/Al_2O_3: 5% Pt, 5% Ni, 90% alumina). Some oxides and a (110) silicon surface were also studied with a CMA to control the effects of resolution which are different with a CMA and a RFA.

It is obvious that we have to take into account the differences in work function of the analyzers and the possible charging effects on insulators when discussing the Auger shifts. It is clear also that the observed differences in the fine structure of the oxygen Auger peaks were seen in both types of analyzers and that we discuss the evolution during experiments where the experimental analyzing conditions were kept constant.

3. Results and discussion

3.1. Analysis of the experimental results

Fig. 1a shows as an example an oxygen Auger spectrum obtained on an air-saturated silicon (110) face. The main features of this $dN(E)/dE$ curve (maxima, minima, crossing of the zero level, inflection points) are labelled from 1 to 15. It is clear that features 2, 7, 9 and 12 actually correspond to maxima in the $N(E)$ distribution. This $N(E)$ distribution, shown on fig. 1b has been found by integration of fig. 1a. The intensities of the various Auger peaks are taken as the height

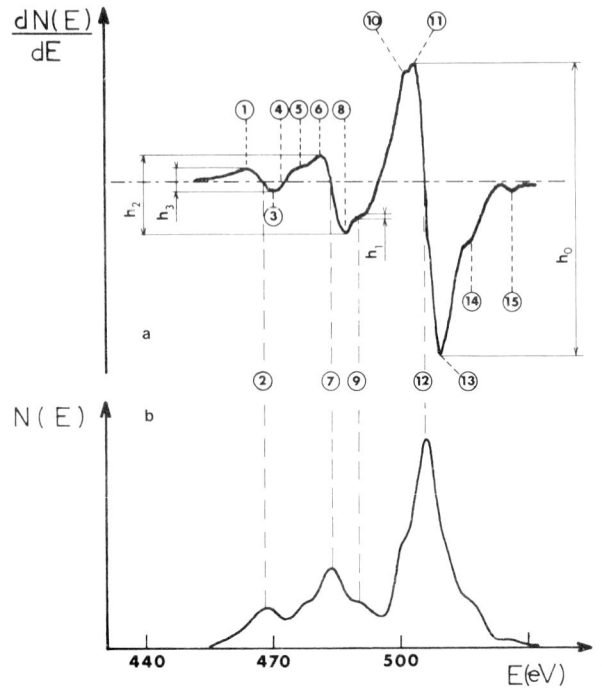

Fig. 1. (a) Oxygen Auger spectrum of an air-saturated silicon (110) surface: $dN(E)/dE$. (b) Integral curve of (a).

between the extrema in the derivative of $N(E)$: they will be thereafter refered as h_0, h_1, h_2 and h_3 as indicated on fig. 1a. For the sake of convenience h_0 will be taken as 100.

Fig. 2 shows a spectrum which is characteristic of an intermediate stage in the silicon (111) face oxidation; one can easily see that some differences occur between this Auger spectrum and the one described in fig. 1a; for example, several features (5, 9, 14 and 15) are missing or hardly visible. Ratios h_2 and h_3 are also quite different than in fig. 1a.

Table 1 summarizes the most important results which were obtained in this study. We have indicated the energies of points 2, 7, 9, 12, 14 and 15, followed by the energy differences between these points and feature 12 which corresponds to the largest peak in the $N(E)$ distribution. These differences will be called:

$\Delta_1 = E(9) - E(12)$, $\quad \Delta_2 = E(7) - E(12)$, $\quad \Delta_3 = E(2) - E(12)$,

$\Delta_4 = E(14) - E(12)$, $\quad \Delta_5 = E(15) - E(12)$.

Moreover information on the type of analyzer has been added.

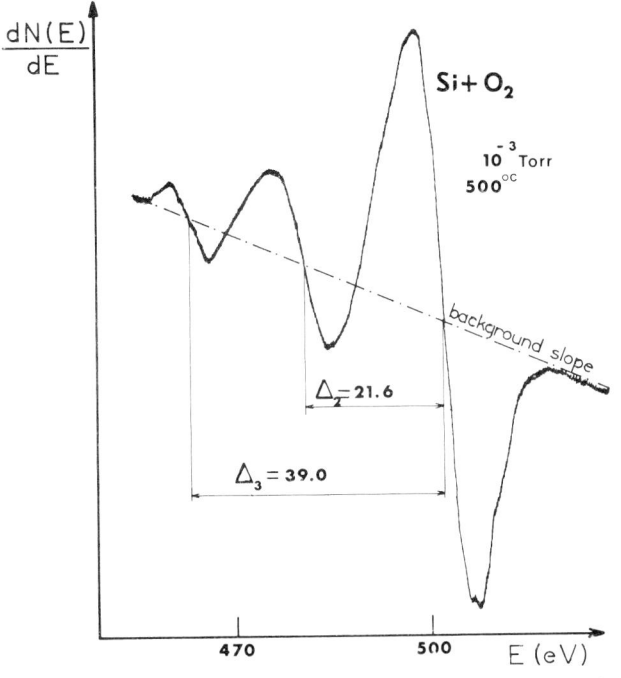

Fig. 2. Oxygen Auger spectrum of an intermediate stage of oxidation of Si(110).

3.2. Oxygen Auger spectra form oxides

Nearly all the oxygen Auger spectra recorded on bulk oxides (silica being until now the exception) have given very similar fine structures. We found four well-characterized peaks (2, 7, 9 and 12) and some less defined features (5 and 10). Sometimes, feature 14 is present but we are not able to really monitor its appearance. In all these spectra the energy differences Δ_1, Δ_2 and Δ_3 are very much alike (respectively, -12.5 eV, -20.5 eV and -33.0 eV). These spectra are closely related to those recorded on MgO by Bassett et al. [9] and by Salmeron and Baró [10], on TiO_2 by Thomas [11]. Agreeing with Bassett's conclusions we think that all these oxides (MgO, Al_2O_3, Fe_2O_3, ...) lead to a "quasi-atomic" spectrum; since the oxygen atoms are doubly ionised their L shell is filled. The L–S coupling theory for KLL transitions is applicable and predicts five lines which are related to the different final states: the energy difference between the initial state ($1s^12s^22p^6$) and each final state (two holes in the L shell) is given to an electron ejected in an Auger process with a characteristic energy. Siegbahn et al. [12] have calculated these energies

Table 1
Energies (in eV) of the various features of the oxygen Auger spectra (values in brackets give the energy difference between a given feature and feature 12)

	$E(2)$ (Δ_3)	$E(7)$ (Δ_2)	$E(9)$ (Δ_1)	$E(12)$	$E(14)$ (Δ_4)	$E(15)$ (Δ_5)	Type of analyzer
Fe_2O_3	471.0 (−33.9)	484.4 (−20.6)	492.3 (−12.6)	504.9			CMA
MgO	460.0 (−35.3)	474.7 (−20.6)	483.2 (−12.1)	495.3	504.1 (8.8)		RFA
Al_2O_3	462.5 (−35.6)	478.1 (−20.5)	486.7 (−11.4)	498.1	507.2 (9.1)	528.1 (20.0)	RFA
SiO_2	491.5 (−44.2)	473.4 (−22.3)		495.7			RFA
Si + O_2 (1 × 1) pattern	466.2 (−36.0)	480.7 (−21.5)		502.2			RFA
Si + O_2	463.5 (−39.0)	481.6 (−20.8)	488.1 (−14.3)	502.4			RFA
Si + O_2 air oxidized	468.4 (−36.4)	482.9 (−21.9)	489.4 (−15.4)	504.8	515.1 (10.3)		CMA
Pt + O_2 [6(111) × (100)]	not meas.	480.3 (−20.3)		500.6	515.5 (14.9)	527.0 (26.4)	RFA
Pt + O_2 [6(111) × (100)]	not meas.	480.5 (−20.6)	491.1 (−10.0)	501.1	513.9 (12.8)	525.0 (23.9)	RFA
Pt + O_2 [6(111) × (100)]	473.4 (−33.6)	487.5 (−19.5)	495.1 (−11.9)	507.0			RFA
Pt(111) + O_2 LEED struct. A	not meas.	483.6 (−20.1)	494.5 (− 9.2)	503.7	514.5 (10.8)		RFA
Cu–Ni 80–20	469.0 (−33.0)	482.0 (−20.0)	490.0 (−12.0)	502.0	513.0 (11.0)		CMA
Pt, Sn/Al_2O_3	466.0 (−37.1)	483.6 (−19.5)	492.6 (−10.5)	503.1	513.5 (10.4)		CMA
Pt, Ni/Al_2O_3	467.0 (−35.0)	482.0 (−20.0)	492.0 (−10.0)	502.0	515.0 (13.0)	522.0 (20.0)	CMA

and found the following values for oxygen:

KL_1L_1, ^1S $(2s^02p^6)$: 474 eV; predicted intensity: 4;

$KL_1L_{2,3}$, ^1P $(2s^12p^5)$: 486 eV; predicted intensity: 12;

$KL_1L_{2,3}$, ^3P $(2s^12p^5)$: 495 eV; predicted intensity: 6;

$KL_{23}L_{23}$, ^1S $(2s^22p^4)$: 504 eV; predicted intensity: 11;

$KL_{23}L_{23}$, ^1D $(2s^22p^4)$: 507 eV; predicted intensity: 100.

We have a very good fit if the peaks 2, 7, 9, 10 and 12 are respectively attributed to the five predicted lines. Since the energy difference between the two $2s^2 2p^4$ states is small, feature 10 will be difficult to resolve. The intensity ratios ($h_1 = 1$; $h_2 = 20$; $h_3 = 7$) between the various transitions also agree fairly well with the predicted values.

We think that feature 5 is due to the bulk plasmon loss on the ^1D transition which is around 23 eV for many oxides. Feature 14, attributed to a double-ionization effect by Bassett et al. [9] will be discussed later.

The case of silica is somewhat different since we found only three lines in the spectrum. Moreover Δ_2 and Δ_3 are larger than for the other oxides. Two explanations may be offered: the first one is that the chemical bond being less ionic for SiO_2, the oxygen atoms do not have a filled L-shell. At that time the approximation of the central-field limit will be more appropriate than the L–S coupling approach to describe the KLL transitions: it predicts indeed a three-line spectrum. The second one is that, due to the insulating properties of silica, there is an important solid-state broadening of the Auger peaks which would smear out the weak features of the "quasi-atomic" spectrum. As a matter of fact the peak to peak widths are rather large in the case of silica (and in the case of oxidized silicon as seen on fig. 2) but experiments done in the same experimental conditions (incident electron beam intensity, modulation on the grids, ...) on MgO and Al_2O_3 do not show the smearing out of the small features which are still easily visible. This would favor the first explanation.

3.3. Oxidation of the silicon (111) and (110) faces

At the beginning of the silicon (111) face oxidation (up to 10^{-3} Torr; up to 500°C; 5 mn; negligible carbon contamination) the Auger spectrum has three dominant lines (fig. 3a): $\Delta_2 = 21.5$ eV, $\Delta_3 = 39$ eV, $h_1 = 33$ and $h_3 = 10$. This spectrum corresponds to a (1 × 1) LEED pattern. After a longer exposure, feature 9 begins to appear and at higher temperatures and higher exposures the LEED pattern is lost and we just observe a diffuse background: at that time the five-line Auger spectrum is found (fig. 3b). The energy differences are smaller and the intensity ratios are weaker ($\Delta_1 = 14$ eV; $\Delta_2 = 21$ eV; $\Delta_3 = 38$ eV; $h_1 = 28$; $h_2 = 2$; $h_3 = 7$). If one leaves this oxygen-covered surface overnight without any more exposure to oxygen and to the electron-beam, the (1 × 1) LEED pattern and an Auger spectrum where feature 9 is hardly visible are restored. The oxidation of the Si(110) face leads to the same AES results.

We never found on the Si(111) or on the Si(110) faces a spectrum identical to the one obtained on silica: Δ_3 is always significantly smaller than 40 eV. Indeed it was found by XPS (13) that our adsorption conditions were not sufficient to build up a SiO_2 overlayer: more drastic conditions are required.

As in the case of silica, two explanations are possible to understand our results. The first one is that at the initial stage of oxidation we have indeed a "delocalized"

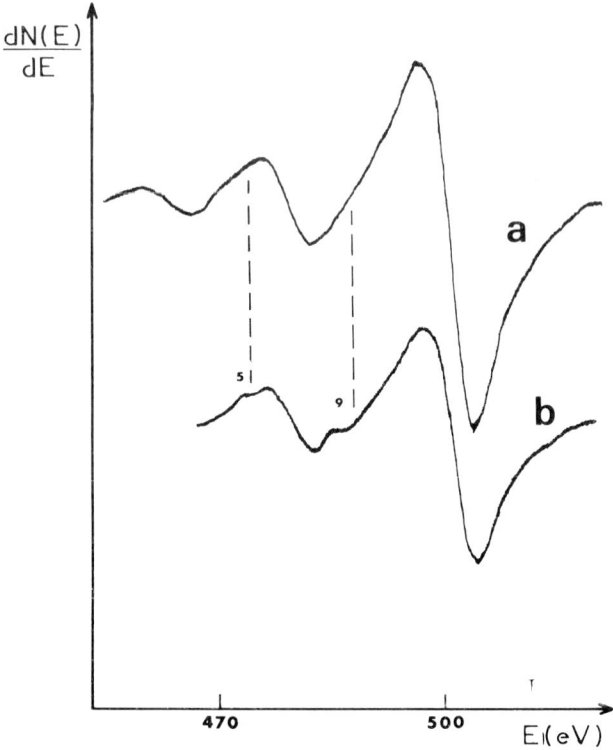

Fig. 3. Oxygen Auger spectra recorded during the oxidation process of Si(111). Most striking differences arise at features 5 and 9.

three-line spectrum which predominates. This could be due to the oxygen atoms which saturate the dangling bonds. After a more complete exposure and probably with the help of the electron beam some O^{2-} ions are formed: they would be responsible of the five-line spectrum which is obviously superimposed to the three-line one. When the surface is left without the electron-beam effect under ultra-high vacuum the O^{2-} ions are either desorbed or dissolved into the bulk of the crystal: we go back to the same LEED pattern and an identical Auger spectrum as in the initial stages of oxygen adsorption. The second possible explanation would be the superimposition of two "quasi-atomic" spectra due to different chemical shifts and extra-electron relaxation. The fact that one of them would be very weak would explain that feature 9 is hardly visible. Considerations about the peak intensities would be more in favor of the first assumption. Anyway, there is an evidence that two different types of oxygen atoms are observed in our experiments.

3.4. Oxidation of platinum single crystal faces

Low-pressure oxygen adsorption on platinum surfaces gave a simple LEED pattern and an Auger spectrum in which three lines dominate. Larger exposures to oxygen lead to a "quasi-atomic" spectrum which is practically the same as the one recorded on ionic oxides. This evolution is not reversible as it is for silicon. The appearance of an Auger spectrum characteristic of the presence of O^{2-} ions agrees well with the modification of the LEED pattern which gives evidence that a PtO_2 overlayer is progressively built up [14].

But there are not only changes in the fine structure of the KLL oxygen Auger spectrum; we found that these peaks were also shifted in energy. These shifts can be discussed much more easily than in the case of silicon since there are no charge problems. Table 2 shows a correlation between the observed LEED structures on three platinum faces and the shifts in energy of feature 13 in the oxygen Auger spectrum. We have chosen this feature for the measurement of the shifts since the precision is better on a minimum of the $dN(E)/dE$ curve. In this table, the results concern three different faces of platinum: Pt(111), Pt(100) and Pt(S) [6(111) × (100)], a stepped surface, for: (1) low oxygen exposures at room temperature; (2) large oxygen exposures at 500°C; (3) heat treatment of the oxygen-covered surfaces at increasing temperatures; (4) extended heating of the surfaces during more than 60 h at temperatures comprised between 600 and 1000°C.

From this table, one can see that low pressure adsorptions give the highest energy for the KL_2L_2 transition and the simplest LEED pattern. When the exposures are increased the LEED pattern becomes more complicated and the main Auger peak is shifted toward lower energies. If the temperature is then increased while keeping the surface under ultra-high vacuum conditions, the PtO_2 overlayer is ordered and the Auger features are shifted toward the lowest energies. A very long annealing procedure of the oxidised surface allows this to reach 504 eV which is a very low energy for the main oxygen Auger peak.

3.5. Discussion of the Pt + O_2 Auger spectra

If the oxygen adsorption is done at very high coverages the LEED data give the evidence of PtO_2 layered oxide formation: at that time a "quasi-atomic" spectrum of oxygen ions is predominant. In this case features 5 and 14 are generally detected. Feature 5 can be attributed to the bulk plasmon loss on the ^1D transition and favors the hypothesis that a definite oxide layer is built up on the surface. Feature 14 which has been attributed to a crossed transition [9] could be in fact the result of the superimposition of two types of Auger spectra: we think that it could be the KL_2L_2 transition of oxygen atoms adsorbed on the oxide layer ("episorbed" atoms). There is indeed a difference of about 6 eV between features 13 and 14: this is the overall shift observed when one goes from the initial stage of oxygen adsorption toward the ultimate stage.

Table 2
LEED structures and oxygen main Auger peak energy for three platinum surfaces; the LEED structures are described: (a) in ref. [14], (b) in ref. [7], (c) in ref. [15]

	LEED structure	Auger peak energy (eV)	LEED structure	Auger peak energy (eV)	LEED structure	Auger peak energy (eV)
Clean surface	Pt(111) (1 × 1)	–	Pt(100) (5 × 1)	–	Pt(S) [6(111) × (100)] (1 × 1)	–
Low pressure adsorption; <100 L; $t = 20°C$	(1 × 1)	510 (a)	(5 × 1) + (1 × 1)	510 (b)	(111) (2 × 2)	507 (c)
High pressure adsorption; >10^7 L; $t = 500°C$	A = D($\sqrt{3} \times \sqrt{3}$) – R30°	512 (a)	(100) (2$\sqrt{2} \times \sqrt{2}$) – R45°	507 (b)	(111) (2 × 2) + D + (111) ($\sqrt{3} \times \sqrt{3}$) – R30°	506 (c)
Heat treatments after exposure UHV; $t = 850°C$	B = D(4$\sqrt{3} \times 4\sqrt{3}$) – R30°	510 (a)	(100) (2$\sqrt{2} \times \sqrt{2}$) – R45°	504 (b)	(111) ($\sqrt{79} \times \sqrt{79}$) – R18° 7'	507 (c)
UHV; $t = 925°C$	C = D(2 × 2)	506 (a)	(100) (2$\sqrt{2} \times \sqrt{2}$) – R45°	504 (b)	(111) (4 × 2$\sqrt{3}$) – R30°	507 (c)
UHV; $t = 1000°C$	D = PtO$_2$(0001) (1 × 1)	504 (a)				
15 to 60 h 600 < t < 1000°C; UHV	D = PtO$_2$(0001) (1 × 1)	504 (a)	PtO$_2$(0001) ($\sqrt{3} \times \sqrt{3}$) – R30°	504 (b)	(111)3(1d)	504 (c)

The observation of supported catalysts where feature 14 appears as a true peak and not only as an inflection point is in good agreement with this explanation: two states of oxygen chemical bonding can be distinguished; the atoms from the bulk oxide (in this case Al_2O_3) give the spectrum at low energy (504 eV) and the "episorbed" atoms lead to the spectrum at high energy (510/512 eV).

4. Conclusion

This work allowed us to show that a careful study of the oxygen Auger spectra can give information about the oxidation process of metals and semiconductors. In the case of platinum and silicon, at the initial stages of oxygen adsorption, a spectrum in which three lines dominate is seen. This spectrum could be compared to the so-called "carbide" one in the case of carbon. This spectrum is generally competing with another one where five lines can be distinguished. In this case the oxygen Auger spectrum can be considered as "quasi-atomic". To this modification of the fine structure of the Auger spectrum correspond shifts of several eV toward the lower energies in the case of platinum oxidation.

Some difficulties in the interpretation of the "three-line" spectrum still arise; it is difficult to ascertain that the small peaks (features 9 and 10) are not smeared out by peak broadening. However, comparisons between alumina and silica, between different stages of oxidation in the case of silicon and platinum do show two competing Auger spectra. This is particularly clear for supported catalysts and this fact offers very interesting outlooks for catalysis if more detailed AES studies are carried out with more sensitive and better resolving analyzers.

References

[1] M.A. Chesters, B.J. Hopkins and P.A. Taylor, J. Phys. C (Solid State Phys.) 9 (1976) L329.
[2] H. Nozoye, Y. Matsumoto, T. Onishi, T. Kondow and K. Tamaru, J. Phys. C (Solid State Phys.) 8 (1975) 4131.
[3] J.T. Grant and M.P. Hooker, Solid State Commun. 19 (1976) 111; Surface Sci. 55 (1976) 741.
[4] J.M. Rojo and A.M. Baró, J. Phys. C (Solid State Phys.) 9 (1976) L543.
[5] K. Kunimori, T. Kawai, T. Kondow, T. Onishi and K. Tamaru, Surface Sci. 54 (1976) 525.
[6] B. Carrière, J.P. Deville, G. Maire and P. Légaré, Surface Sci. 58 (1976) 618.
[7] B. Lang, P. Légaré and G. Maire, Surface Sci. 47 (1975) 89.
[8] C.C. Chang, Surface Sci. 23 (1970) 283.
[9] P.J. Bassett, T.E. Gallon, M. Prutton and J.A.D. Matthew, Surface Sci. 33 (1972) 213.
[10] M. Salmerón, A.M. Baró, and J.M. Rojo, Surface Sci. 53 (1975) 689.
[11] S. Thomas, Surface Sci. 55 (1976) 754.
[12] K. Siegbahn, C. Nordling, A. Fahlman, R. Nordberg, K. Hamrin, J. Hedman, G. Johansson, T. Bergmark, S.E. Karlsson, I. Lindgren and B. Lindgren, ESCA, Atomic, Molecular and

Solid State Structure by means of Electron Spectroscopy. Nova Acta Regiae Soc. Sci., Uppsala Ser. IV, 20. (1967).
[13] B. Carrière, J.P. Deville, D. Brion and J. Escard, J. Electron Spcectrosc. Related Phenomena 10 (1977) 85.
[14] B. Carrière, P. Légaré and G. Maire, J. Chim. Phys. 71 (1974) 355.
[15] G. Lindauer and G. Maire, to be published.

ON THE INFLUENCE OF SIZE AND ROUGHNESS ON THE ELECTRONIC STRUCTURE OF TRANSITION METAL SURFACES

M.B. GORDON, F. CYROT-LACKMANN and M.C. DESJONQUÈRES

Groupe des Transitions de Phases, CNRS, B.P. 166, 38042 Grenoble Cedex, France

The dependence of surface electronic structure on local and farther environment for transition metals is investigated in the tight binding approximation associated with the moment method. Local densities of states (LDS) on fcc cubooctahedral and on icosahedral clusters of increasing size, from 13 up to 2057 atoms, are studied for corner sites and central sites on differently oriented faces. The general LDS features are determined by the nearest neighbour shell but outstanding details, such as the occurrence of a strong central peak, are dependent on each particular cluster site and dimension. Only for particles larger than one thousand atoms the LDS at the center of (100) and (111) faces look like those of the corresponding infinite surfaces. The importance of symmetry is pointed out by the comparison of LDS at equivalent sites on cubo-octahedra (cubic symmetry) and icosahedra (five-fold symmetry). These results might explain the variation of activity with the catalyser dispersion, as the adsorption energy depends on the relative position of the adatom atomic level and the substrate LDS peaks and Fermi level.

1. Introduction

It has recently been shown that the electronic properties of cleaved transition metals are strongly dependent on the density of cleavage planes. The local densities of electronic states (LDS) on dense surfaces are similar to bulk ones, but LDS on non dense surfaces show a resonant peak of surface states in the middle of the band whatever their crystalline structure. These studies were made on semi-infinite crystals or periodic stepped surfaces [9].

A lot of experimental work is done on finite samples which have a great proportion of surface atoms on differently oriented faces. Thus, it is interesting to see how sensitive the electronic properties are to roughness. Small clusters of variable size and shape are suitable to study this effect and to investigate how their properties tend towards those of infinite solids. Moreover, a better knowledge of electronic properties of small particles is important due to their wide field of applications, from catalysis to photographic reactions.

The experiments show that clusters shape and size strongly influence their properties. For example, it was established that catalytic properties of Ni depend drastically on the size of Ni particles [1].

The crystalline structures of very small clusters show either the symmetries of the nuclei of the crystal phases or non crystallographic structure with five fold symmetry such as icosahedra or pentagonal dipyramids [2]. When the size of clusters increases, the five fold symmetry particles transform into the usual fcc structures. One can wonder if these structural phase changes and the rapid variation of catalytic properties and of kinetics of crystal growth, which seem to occur at the same range of sizes, are related phenomena.

There are only very few studies of electronic structure of small transition metal clusters, generally for clusters of less than 60 atoms, using methods of quantum chemistry which requires the diagonalization of the hamiltonian [3]. No systematic comparison between clusters of different sizes or structures has yet been reported.

In this paper, we study the electronic structure of transition metal clusters as a function of their shape and their size by means of the local density of states (LDS). Results are presented for face centred cubic (fcc) and isosahedral clusters from 13 to 2057 atoms using tight binding approximation and neglecting s–d hybridisation.

We have studied LDS for various sites on differently oriented faces and a comparison is made with bulk and cleaved semi-infinite crystals to investigate their dependence with local and farther environment.

2. Geometrical features

There are two possible families of fcc cubo-octahedra. The first ones (hereafter referred to as cubo-octahedra I) have been studied in a previous paper [5]. They have square (100) and hexagonal (111) faces and can be grouped into two sequences -even and odd- of increasing size. Here we study the second ones (fig. 1a) which can be distorted into regular icosahedra (fig. 1b) [4].

The smaller cubooctahedron is built up by surrounding an atom by its first coordination fcc polyhedron. A second layer of atoms close packed over the first one gives rise to the next size cubooctahedron, and a series can be generated by adding up such layers one by one. These cubooctahedra have six square (100) and 8 triangular (111) faces. They have 12 equivalent corners and 24 equivalent edges. The number m of atoms on an edge is equal to the number of shells (we take the central atom as being shell 1) and then it clearly identifies the size of the cluster. In table 1 we give the main structural features of the cubooctahedra studied.

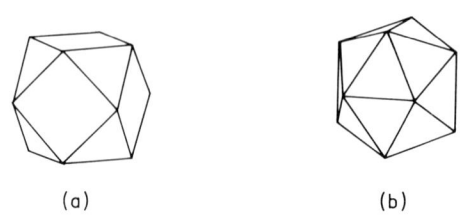

Fig. 1. (a) The cubooctahedron, (b) the icosahedron.

Table 1
Number of atoms of different environment for cubooctahedral and icosahedral clusters

m	N_b	N_s	N_c	N_{100} (Cub. only)	N_e Cub.	N_e Icos.	N_{111} Cub.	N_{111} Icos.	N_T	$\dfrac{N_s}{N_T}$	ϕ (Å)
2	1	12	12	0	0	0	0	0	13	0.92	5
3	13	42	12	6	24	30	0	0	55	0.76	10
4	55	92	12	24	48	60	8	20	147	0.63	15
5	147	162	12	54	72	90	24	60	309	0.52	20
6	309	252	12	96	96	120	48	120	561	0.45	25
7	561	362	12	150	120	150	80	200	923	0.39	30
8	923	492	12	216	144	180	120	300	1415	0.35	35
9	1415	642	12	294	168	210	168	420	2057	0.31	40

m = Number of shells or number of atoms on an edge.
N_b = Number of "bulk" atoms (with their first coordination shell complete).
N_s = Number of surface atoms.
N_c = Number of corner atoms.
N_{100} = Number on (100) square face atoms (they only exist on cubooctahedra).
N_e = Number of edge atoms.
N_{111} = Number on (111) triangular faces atoms.
N_T = Total number of atoms.
ϕ (Å) = Diameter of clusters taking 2.5 Å as interatomic distance.

Table 2
Coordination number for surface atoms

Surface site	Coordination number	
	Cubooctahedron	Icosahedron
Corner	5	6
Edge	7	8
(111) face	9	9
(100) face	8	Become edge or (111) face atoms

As pointed out by Mackay [4], a distortion leads from the cubooctahedron to the icosahedron of equal number of atoms. Each square face splits into two equilateral triangles. The local environment of each atom is slightly modified by the distortion. The icosahedra have 12 corners as the cubooctahedra, but 20 triangular equivalent faces and 30 equivalent edges. The faces have distorted (111) orientation. Each "bulk" atom has 6 neighbours in the same shell, 3 on the shell below and 3 above. The inter-shell nearest neighbour distance is taken equal to the interatomic distance in cubooctahedra. The intra-shell ones are dilated uniformly by 5%, and so tha packing in the icosahedra is lower than the cubic close packing. The main feautes of the icosahedra are also given in table 1.

It is interesting to note that the coordination number for surface atoms on icosahedra is higher than on cubooctahedra. This can be seen in table 2 where we give the coordination number for surface sites in both types of clusters.

3. Methods, results and discussion

The d states of the clusters are described by a tight-binding hamiltonian in a basis of 5 d orbitals per atom. We investigate the LDS at site i:

$$n_i(E) = \frac{1}{5} \sum_{\lambda=1}^{5} n_{i\lambda}(E), \qquad (1)$$

$$n_{i\lambda}(E) = \sum_{n} |\langle i\lambda | n \rangle|^2 \, \delta(E - E_n), \qquad (2)$$

where $|i\lambda\rangle$ is the λ d orbital ($\lambda = 1, \ldots, 5$) at site i; $|n\rangle$ the hamiltonian eigenfunction with eigenvalue E_n. The LDS is the density of states modulated by the square of the wave function and it reflects the "spatial distribution" of the energy levels on the cluster. We calculate (1) and (2) by the moments and continued fraction method (ref. [5] and references therein).

We made the calculations on cubooctahedral clusters with the 3 band parameters of Ni [6]:

$$dd\sigma(R_0) = -0.0417 \text{ Ryd}, \quad dd\pi(R_0) = 0.0188 \text{ Ryd}, \quad dd\delta(R_0) = -0.0023 \text{ Ryd},$$

(3)

where R_0 = Ni nearest neighbour distance. For icosahedra, we took the Ni parameters for the inter-shell hopping integrals. For the intra-shell hopping integrals, the nearest neighbour distance R being 5% longer, we took the law [7]:

$$ddi(R) = ddi(R_0)(R/R_0)^{-5}, \quad i = \sigma, \pi, \delta.$$

(4)

The decrease of compactness and the increase of mean coordination number of the icosahedra with respect to the cubooctahedra have two opposite effects on the bandwidth of LDS for the former. These two effects can be analyzed by means of the first moments μ_p of the LDS. The second moment μ_2 gives the mean bandwidth and the ratio of the fourth to the square of the second moment μ_4/μ_2^2 gives the relative importance of the wings of the LDS. In table 3, we give the ratio of these two quantities for different sites on icosahedral and cubooctahedral structures. One sees that the importance of wings depends on the sites considered but that the mean bandwidths of the icosahedra are always smaller than or equal to those of the cubooctahedra. This is so because μ_2 is proportional to the number of nearest neighbours and to the hopping integrals between them. For the icosahedra, the number of nearest neighbours is in general bigger than for the cubooctahedra (table 2) but the decrease of the hopping integrals (4) for the former overcompensates this effect giving finally a narrower band for the icosahedra than for the fcc clusters.

Fig. 2 shows the LDS on the central atom for the $m = 7$ cubooctahedron and icosahedron. There is a trough at the center of the icosahedral LDS which does not exist in the corresponding fcc cluster. This effect is probably related to their difference in compacity. The LDS of icosahedra will be intermediate between that of a more compact structure and a less compact one, such as the bcc one [6]. Although the mean bandwidth is the same, wings are less important for icosahedral than for cubooctahedral LDS at the central atom, due also to this effect of packing which lowers the ratio μ_4/μ_2^2 (table 3).

Fig. 3 shows the convergence, with increasing size, for the central (100) face

Table 3
Icosahedral to cubooctahedral moments ratio for different sites

	Central atom	Center (111) face atom	Corner atom
μ_2^{icos}/μ_2^{cub}	1	0.74	0.81
$\left(\frac{\mu_4}{\mu_2^2}\right)^{icos}/\left(\frac{\mu_4}{\mu_2^2}\right)^{cub}$	0.85	1.03	0.97

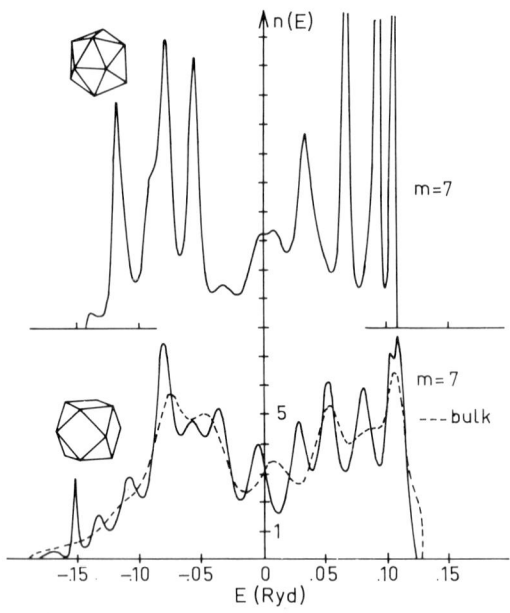

Fig. 2. Central atom LDS for both structures, size $m = 7$. Nickel bulk LDS is shown for comparison [6].

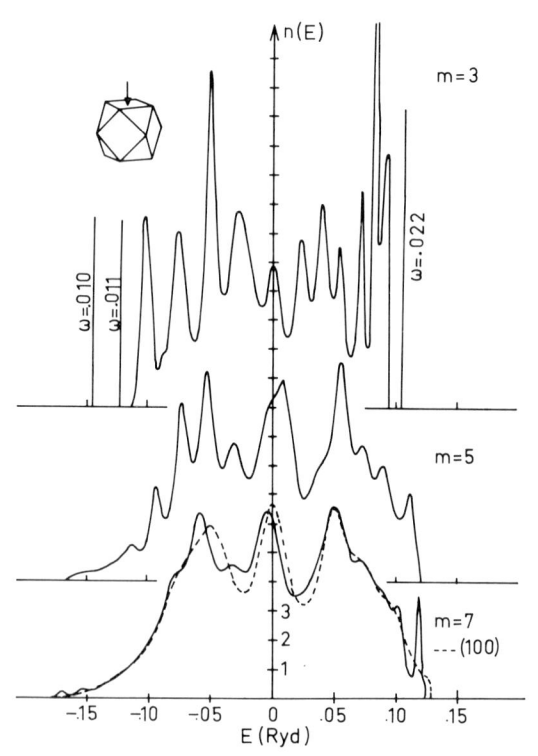

Fig. 3. LDS at the center of (100) square face for 3 cubooctahedra of increasing size. Nickel (100) surface LDS is shown for comparison [6].

atom for cubooctahedra. Even for the largest cluster ($m = 7$), convergence is not entirely achieved because it has only the 7 first moments equal to those of the infinite surface.

For the central atom on the triangular faces, the mean bandwidth is smaller for the icosahedra than for the cubooctahedra but the wings have the same importance on both LDS (table 3). Thus, we find that, as expected, the icosahedral LDS have a central trough as compared to the cubooctahedral ones.

Figs. 4 and 5 show the LDS for the corners. The existence of resonant states near the middle of the band can be easily understood in the same way as for semi-infinite non dense surfaces [8]. Corner atoms are at very "rough" sites. They have a low coordination number and can be considered as almost adsorbed atoms coupled to their nearest neighbour LDS.

We do not find the alternance of peaks and troughs on LDS when increasing cluster size as in cubooctahedra I [5], but a continuous smoothing of curves. Indeed this alternance is related to the existence or not of an atom at the center of the cluster, which is a geometrical peculiarity of cubooctahedra I.

The differences between LDS on various sites of clusters and their change when increasing size may explain the rapid variation of catalytic properties of small parti-

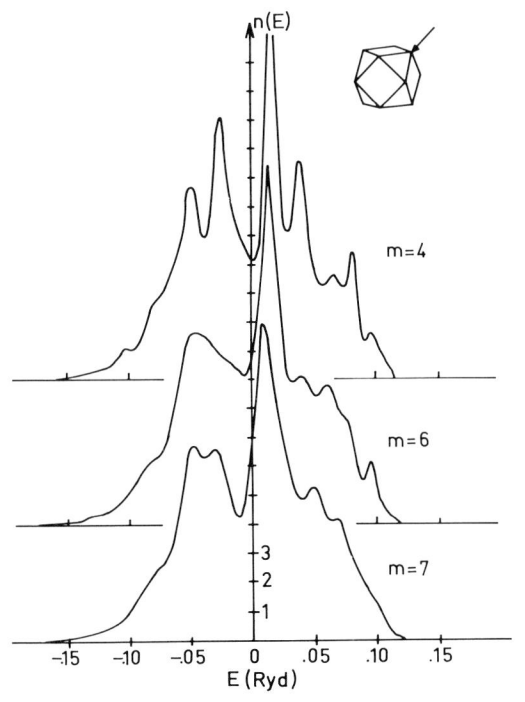

Fig. 4. LDS at corner sites for cubooctahedra.

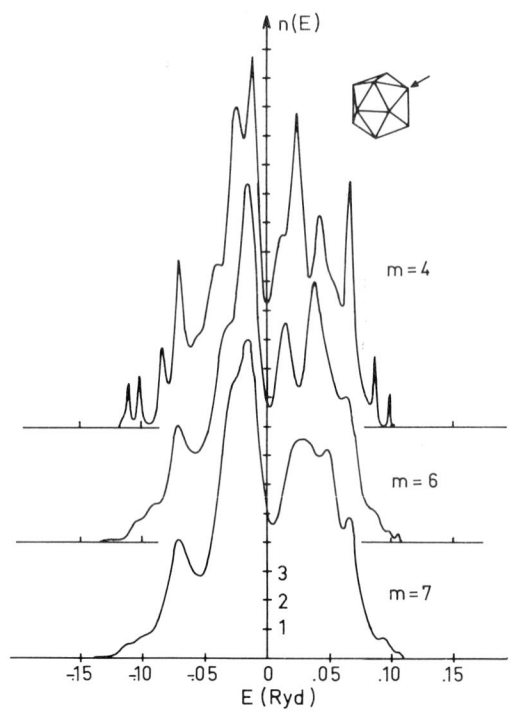

Fig. 5. LDS at corner sites for icosahedra of same size as cubooctahedra of fig. 4.

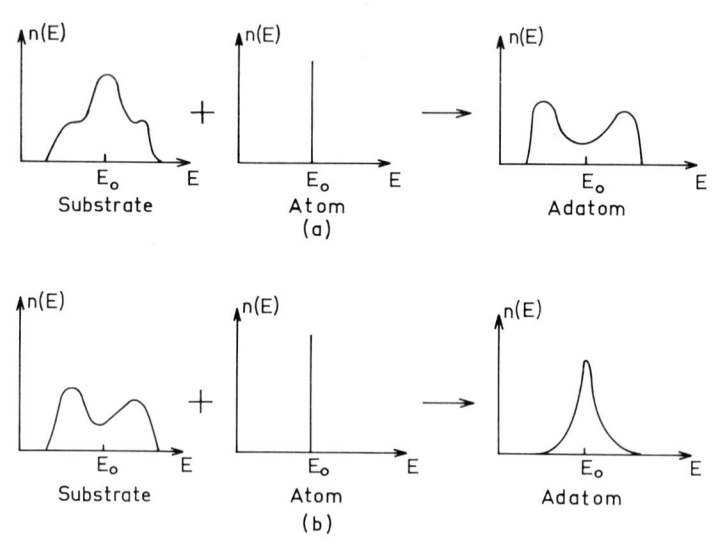

Fig. 6. Schematic adsorption process and its effects on LDS; E_0 is the atomic level. (a) substrate with a central peak on LDS, (b) substrate with a central trough on LDS.

cles with their preparation and dispersion. Roughness, which is greater for smaller particles, will favorize the appearance of strong resonance peaks on LDS. When an atom with atomic level falling in a peak of the substrate LDS is chemisorbed, a surface molecule is formed with strong adsorption energy (fig. 6a). On the other hand, if its atomic level falls on a trough, we expect a weak interaction between adsorbate and substrate (fig. 6b) [10]. These two different regimes will lead to some selection of adsorption sites. These are guiding ideas which remain to be developed as some fundamental problems are not yet solved. One of them is the possible large charge transfer on the clusters between surface, bulk atoms and adsorbates which would require self-consistent calculations.

Nevertheless, our approach to treating the electronic structure of small particles seems to be a powerful way of understanding the difficult problem of adsorption in a catalytic process.

References

[1] R. Van Hardeveld and F. Hartog, Advan. Catalysis 22 (1972) 75.
[2] J.G. Allpress and J.V. Sanders, Surface Sci. 7 (1967) 1.
[3] R.C. Baetzold and R.E. Mack, J. Chem. Phys. 62 (1975) 1513;
    G.B. Blyholder, Surface Sci. 42 (1974) 249;
    P.J. Jennings, G.S. Painter and R.O. Jones, Surface Sci. 60 (1976) 255 and references therein.
[4] A.L. Mackay, Acta Cryst. 15 (1962) 916.
[5] F. Cyrot-Lackmann, M.C. Desjonquères and M.B. Gordon, Intern. Meeting on Small Particles and Inorganic Clusters, Lyon, 1976, J. Phys. (Paris), to be published.
[6] M.C. Desjonquères and F. Cyrot-Lackmann, Surface Sci. 53 (1975) 429.
[7] V. Heine, Phys. Rev. 153 (1967) 673.
[8] J. Friedel, J. Phys. (Paris) 37 (1976) 883.
[9] M.C. Desjonquères and F. Cyrot-Lackmann, Solid State Commun. 18 (1976) 1127.
[10] M.C. Desjonquères, Thesis, Université de Grenoble (1976);
    F. Cyrot-Lackmann and M.C. Desjonquères, to be presented at the 3rd Intern. Conf. on Solid Surfaces, Vienna, 1977.

EXCHANGE CORRECTIONS TO THE DENSITY–DENSITY CORRELATION FUNCTION AT A SURFACE

Greger LINDELL

NORDITA, Blegdamsevej 17, DK-2100 Copenhagen Ø, Denmark

The problem of calculating the density–density correlation function and of including the effects of exchange and correlation are discussed. It is shown in a model calculation how a particularly simple surface model can be easily extended to include exchange corrections. The effects of these corrections on the surface plasmon dispersion are found to be small.

1. Introduction

An important concept in the theory of the homogeneous electron gas is the density–density correlation function

$$\chi(r, r', t - t') = \frac{i}{\hbar} \theta(t - t') \langle [n(r, t), n(r', t')] \rangle, \qquad (1)$$

which contains information about both the static and the dynamical properties of the bulk system. This quantity is also useful for the study of extremely inhomogeneous electron systems such as metal surfaces, and has received attention from many workers [1–5] who have studied the properties of surface plasmons and the surface energy in different surface models. This paper will deal with two aspects of the density–density correlation function at surfaces, how to go beyond the Hartree or mean-field approximation to treat exchange and correlation and how to formulate the Hartree problem such that it can be numerically solved on a computer for any surface model.

In section 2 we study the general problem of introducing local field corrections to the mean-field theory. This is done by a generalization of the method of Singwi et al. [6] to the surface case. In section 3 we look at the shape of the exchange hole around an electron close to a surface. In section 4 we illustrate how the very simplest exchange corrections to the Hartree response function, also in approximate form, leads to an almost analytically solvable expression for the surface plasmon dispersion relation.

We will use the (semi-) Classical Infinite Barrier Model [7] where the response function for the non-interacting electron gas at a surface is approximated by a direct and a reflected part

$$\chi^0(r - r', \omega) = \chi^0_{\text{bulk}}(r - r', \omega) + \chi^0_{\text{bulk}}(r^* - r', \omega). \tag{2}$$

The vector r^* is the mirror image of r in the surface plane.

2. Correlation corrections to the response function

The response of a classical gas of interacting particles to an external potential can be studied by linearizing the equation of motion. The equation thus obtained can also be applied to the electron gas as was pointed out by Singwi et al. [6] (forthwith referred to as SSTL). They were able to reduce the quantum mechanical equation of motion for the density–density response function to the classical form by a truncation of higher-order Green's functions. In this way they were able to calculate corrections to the simplest RPA treatment of the response function in an electron gas. These local field corrections are expressend in terms of the static correlation function $g(r_1, r_2)$.

Before writing down a formula for the induced charge in the SSTL formalism valid at a surface we have to describe our nomenclature briefly. The surface is assumed to be perpendicular to the z-direction and all vectors, both in real and in fourier space, will be divided into a component along the z-axis and a remaining part. Furthermore, this separation will be explicitly shown for the fourier transform of a function.

$$r = R + ze_z, \qquad q = Q + qe_z,$$

$$f(r) = \sum_{Q,q} \exp(iQz)\exp(iqz) f(Q, q). \tag{3}$$

The Fourier transform of $g(r_1, r_2) - 1$ is written as

$$g(r_1, r_2) - 1 = \sum_{Q, k_1, k_2} H(Q, k_1, k_2)$$

$$\times \exp[iQ \cdot (R_1 - R_2)] \exp[i(k_1 z_1 - k_2 z_2)]. \tag{4}$$

Note that two variables are needed for the transformation in the z-direction. The ordinary bulk result is simply retrieved by putting

$$H(Q, k_1, k_2) = H_{\text{bulk}}(Q, k_1) \delta_{k_1 k_2}, \tag{5}$$

whereupon the RPA-assumption that χ^0, the irreducible polarization diagram, connects the effective potential with the induced density

$$\rho(r, t) = \int dr' \, dt \, \chi^0(r', t') V^{\text{eff}}(r', t'), \tag{7}$$

can be used to obtain an integral equation for the induced charge:

$$\rho(Q,z) = \int dz''\, \chi^0(Q,z,z'')\, V^{\text{ext}}(Q,z'')$$

$$+ \int dz'\, \rho(Q,z') \int dz''\, \chi^0(Q,z,z'')\, v(Q,z''-z')$$

$$+ \int dz'\, \rho(Q,z,z'') \sum_q \exp[iqz'']$$

$$\times \sum_{k_2,q'} H(Q-Q', q-q', k_2)\, \frac{q \cdot q'}{q^2}\, v(q') \exp[-i(k_2+q')z'] \,. \tag{8}$$

Or, if we would like, an integral equation for the response function χ that connects the *external* potential V^{ext} with the induced charge ρ. This step was the natural continuation in the bulk case, where an analytical solution could be obtained. In this article we will restrict ourselves to some of the consequences of eq. (8) for the surface plasmon dispersion in a simple model. In order to do this, we will first study the implications of a surface on the Hartree correlation function which will be used as a basis for refinements that include exchange.

3. Calculation of the exchange hole

We have seen how the exchange-correlation hole or $g(r_1, r_2)$ enters the calculation of induced charge as a function of external potential, and also the plasmon dispersion. In this section we will calculate this depletion of charge around a particular electron as due to exchange effects only. This exchange hole will then be included in the formulae of the previous section as a first approximation. In this way we will essentially have made a treatment of the surface response equvalent to what was done in the bulk case by Hubbard [8].

Let us start by writing down the formal relation between the response function, the structure factor, and the probability $P(r_1, r_2)$ of finding one electron at r_1 and another at r_2:

$$S(r_1, r_2) = -\frac{1}{\pi} \int_0^\infty d\omega\, \text{Im}\, \chi(r_1, r_2, \omega)\,,$$

$$S(r_1, r_2) = P(r_1, r_2) - n(r_1)\, n(r_2) + n(r_1)\, \delta(r_2 - r_1)\,. \tag{9}$$

To keep only exchange effects amounts to replace χ in this expression by χ^0.

It turns out to be rather cumbersome to calculate $g(r_1, r_2)$ in this fashion. Only for the case, consistent with the CIBM model, where we neglect all terms except those that depend on $z - z'$ or $z + z'$ is it possible to rearrange the terms in a simple way to give

$$g^{CIBM}(r_1, r_2) = 1 - \frac{9}{2} \frac{j_1^2(k_F|r_1 - r_2|)}{(k_F|r_1 - r_2|)^2} - \frac{9}{2} \frac{j_1^2(k_F|r_1 - r_2^*|)}{(k_F|r_1 - r_2^*|)^2},$$

$$r_2^* = (x_2, y_2, -z_2), \qquad (10)$$

where r_2^* is the mirror image of r_2.

A simpler approach is to use the connection between $P(r_1, r_2)$ and the one-particle density matrix for fermions:

$$\rho(r_1, r_2) = \sum_{|k| < k_F} \psi_k^*(r_1) \psi_k(r_2). \qquad (11)$$

The function $g(r_1, r_2)$ is then given by

$$g(r_1, r_2) = 1 - \frac{1}{2} \frac{|\rho(r_1, r_2)|^2}{\rho(r_1, r_1)\rho(r_2, r_2)}, \qquad (12)$$

where $\rho(r_1, r_1)$ is the density at r_1. The calculaton of $\rho(r_1, r_2)$ immediately reduces to the evaluation of a one-dimensional integral and can easily be performed on a computer. In the regions of flat potential it is in fact possible to isolate an ordinary bulk and a surface contribution:

$$\rho(r_1, r_2) = \int_0^{k_F} dk \, (k_F^2 - k^2)^{1/2} \, \varphi_k(z_1) \varphi_k(z_2) J_1(R(k_F^2 - k^2)^{1/2})$$

$$= \frac{k_F^2}{\pi^2} \frac{j_1(k_F|r_1 - r_2|)}{k_F|r_1 - r_2|}$$

$$+ \frac{1}{\pi^2 R} \int_0^{k_F} dk \, (k_F^2 - k^2)^{1/2} \cos[k(z + z') + 2\delta_k] J_1(R(k_F^2 - k^2)^{1/2}). \qquad (13)$$

It is assumed in this expression that the electronic wavefunctions are plane-wave-like parallel to the surface and that their z-dependent parts are phase-shifted trigonometric functions inside the surface potential barrier. A possible ansatz for this barrier is an infinite potential step, the Infinite Barrier Model (IBM). The exchange hole assumes a particularly simple form in this model

$$g(r_1, r_2) = 1 - \frac{9}{2} \left\{ \frac{j_1(k_F|r_1 - r_2|)}{k_F|r_1 - r_2|} - \frac{j_1(k_F|r_1 - r_2^*|)}{k_F|r_1 - r_2|} \right\}^2. \qquad (14)$$

There are structural similarities between this result and the corresponding expression in the CIBM model as given by eq. (10), where only the cross product between

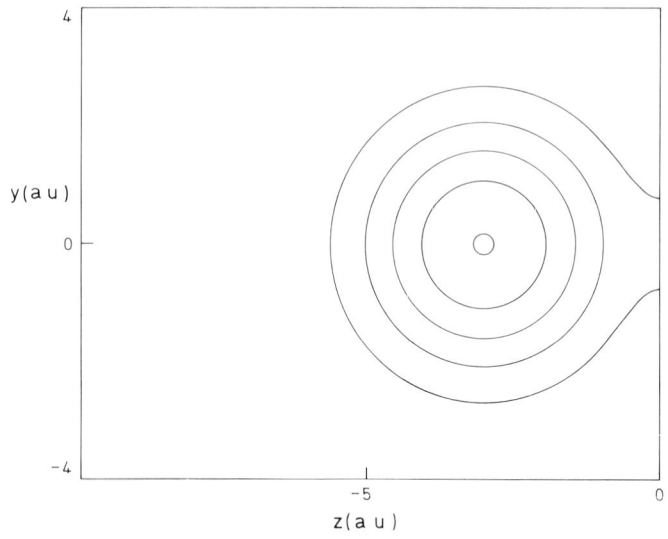

Fig. 1. The exchange hole in the CIBM model for $k_F = 1$ ($r_s = 1.92$). The contour lines are drawn where g assumes the values 0.9, 0.8, etc.

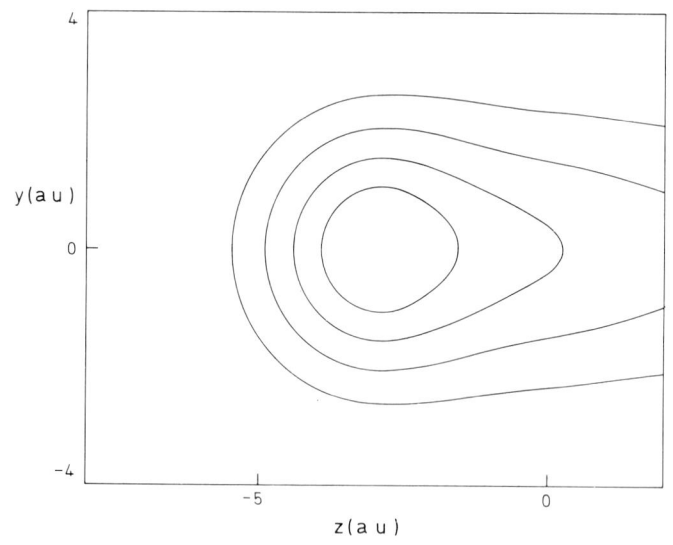

Fig. 2. The exchange hole evaluated with wavefunctions solving a step barrier model with $k_F = 1$ and the step height = 1.5 times the Fermi energy.

the two terms to be squared in eq. (14) is missing. These terms would depend on both z and z' and are physically significant since the exchange hole in the CIBM model can assume values below 0.5 close to the surface. A contour plot of the exchange hole in the CIBM model and a model where the surface is represented by a finite step barrier are shown in figs. 1 and 2. The surface in these plots is located at $z = 0$ and the CIBM hole resembles the overlap between two holes located symmetrically relative to this surface. In the finite barrier case, the hole will be defined for all values of z and will extend into the vacuum region.

We would like to point out that the functions calculated here are almost equivalent to the ones studied by Moore and March [9] by quite another method, except that for our purposes it has been more convenient to concentrate on $g(r_1, r_2)$ rather than on $P(r_1, r_2)$ as they have done.

4. Surface pasmon dispersion in the CIBM model with exchange corrections

In section 2 we saw that the induced charge of the surface plasmons fulfill the integral equation

$$\rho(Q,z) = \int_{-\infty}^{0} dz' \, \rho(Q, z') \int_{-\infty}^{0} dz'' \, \chi^0(Q, z, z'', \omega) \, F(Q, z'', z') , \qquad (15)$$

where $F(Q, z'', z')$ contains both the Hartree potential and the local field corrections.

In this section we will solve this equation with two approximations. Firstly, the function H will contain only the exchange part of the exchange-correlation hole, and secondly this hole will be given only by the first two terms in the CIBM approximation (eq. (10)). This, in fact, leaves us only with *bulk* corrections to the Hartree picture, on the other hand it leads to an integral equation of a particularly simple form. In eq. (16) is shown the explicit form of $F(Q, q, \omega)$ including a further approximation to the analytical form of $G(Q, q)$ made by Hubbard (see Singwi et al.) [8,6].

$$F(Q, q) = v(Q, q) \, [1 - G(Q, q)] = v(q) - 1/2(Q^2 + q^2 + q_F^2) . \qquad (16)$$

The Fourier representation — in the CIBM model — of $\chi^0(Q, z, z'', \omega)$ is

$$\chi^0_{\text{CIBM}}(Q, z, z'', \omega) = \int_{-\infty}^{\infty} dq \, \{\exp[iq(z - z'')] + \exp[iq(z + z'')]\} \, \chi^0_L(q, \omega) , \qquad (17)$$

where χ^0_L is the Lindhard function.

It is tempting to proceed directly by integrating the variable z'' but one must be careful here, since the integration range only extends over a half-space so we do not obtain a simple delta-function of $q - q'$ but rather

$$\int_{-\infty}^{0} dz'' \exp[i(q' - q)z''] = -iP \frac{1}{q - q'} + \pi\delta(q - q') , \qquad (18)$$

and it is the principal parts term that includes the information relevant to our surface plasmon calculation. A few lines of algebra lead to the result

$$\rho(Q, z) = \frac{1}{2\pi} \int_{-\infty}^{0} dz' \, \rho(Q, z') \int_{-\infty}^{\infty} dq \, \{\exp[i(qz + Qz')]$$

$$\times [1 - \epsilon(Q, q, \omega)] + \exp[i(qz + (Q^2 + q_F^2)^{1/2})] \, [1 - \epsilon(Q, q, \omega)] \, G(Q, q)$$

$$+ \{\exp[iq(z - z')] + \exp[iq(z + z')]\} \, [1 - \alpha(Q, q, \omega)]\} , \qquad (19)$$

where $\epsilon(Q, q, \omega)$ and $\alpha(Q, q, \omega)$ are defined as

$$\epsilon(Q, q, \omega) = 1 + v(Q, q) \, \chi_L^0(Q, q, \omega) ,$$

$$\alpha(Q, q, \omega) = 1 + v(Q, q) \, \chi_L^0(Q, q, \omega) \, G(Q, q) . \qquad (20)$$

One now takes the cosine transformation of eq. (19) since this immediately will replace the integral over the first two (bulk) terms to simple products. It has to be a cosine rather than a Fourier transform since our functions only are defined for $z < 0$.

The integral equation finally reduces to

$$\rho(Q, k) = \int_{-\infty}^{\infty} d\kappa \, K(k, \kappa) \, \rho(Q, \kappa) , \qquad (21)$$

where the kernel $K(k, \kappa)$ is of the separable type

$$K(k, \kappa) = \sum_{i=1}^{2} u_i(k) \, v_i(\kappa) ,$$

$$u_1(k) = \frac{\epsilon(Q, q, \omega) - 1}{\alpha(Q, q, \omega)} , \qquad v_1(\kappa) = \frac{Q}{2\pi} \frac{1}{Q^2 + \kappa^2} , \qquad (22)$$

$$u_2(k) = -G(Q, k) \, u_1(k) , \qquad v_2(\kappa) = \frac{(Q^2 + q_F^2)^{1/2}}{Q} v_1(\kappa) , \qquad (22)$$

and a solution exists if Det $(1 - A)$ vanishes since this is a homogenous equation. A is given by

$$a_{ij} = \int_{-\infty}^{\infty} d\kappa \, u_j(\kappa) \, v_i(\kappa) . \qquad (23)$$

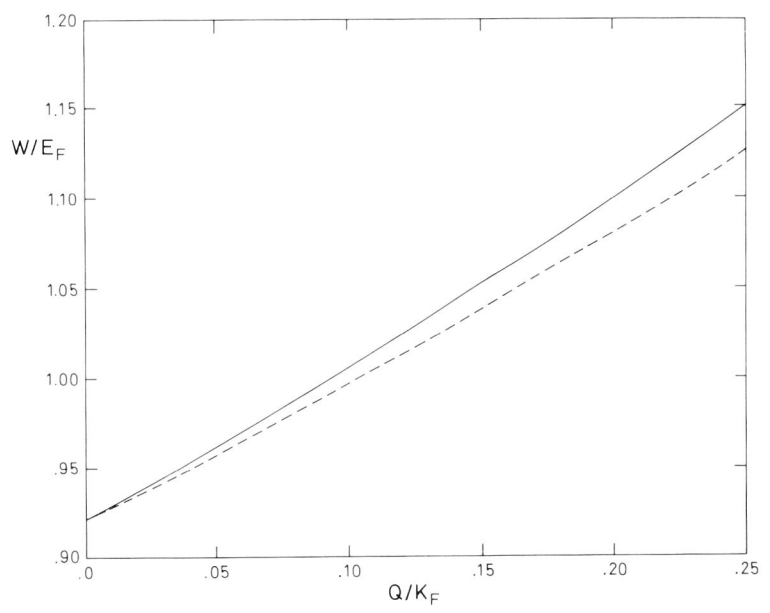

Fig. 3. The surface plasmon dispersion in the Richie–Marusak model (———) and with exchange corrections (------) for $k_F = 1$.

Notice that when $G(Q, q)$ is zero, the criterion reduces to

$$1 - \int_{-\infty}^{\infty} \frac{d\kappa}{2\pi} \frac{\epsilon(Q, \kappa, \omega) - 1}{\epsilon(Q, \kappa, \omega)} \frac{Q}{Q^2 + \kappa^2} = 0, \tag{24}$$

which is exactly the relation found by Richie and Marusak [10].

In fig. 3 we show the surface plasmon dispersion relation including exchange corrections compared with the Ritchie–Marusak result. The inclusion of exchange corrections does lower the dispersion curve although not to a significant degree. We believe that a realistic calculation of this quantity with a better approximation to the irreducible polarization diagram should include exchange corrections at least to the order outlined here. Harris [11] has also treated exchange corrections to surface plasmon dispersion and found corrections of the same magnitude as we have found here.

5. Summary

This article has dealt with some aspects of the calculation of response functions at surfaces. The treatment of realistic potentials and of exchange-correlation effects has been outlined. It has been shown that the exchange effects in a simple surface model can be included without significantly increasing analytical or numerical complexity.

References

[1] D.M. Newns, Phys. Rev. B1 (1970) 3304.
[2] V. Peuckert, Z. Physik 241 (1971) 191.
[3] P.J. Feibelman, Phys. Rev. B9 (1974) 5077.
[4] J.E. Inglesfield and E. Wikborg, Solid State Commun. 14 (1974) 661; 16 (1975) 335.
[5] D.C. Langreth and J.P. Perdew, Solid State Commun. 17 (1975) 1425.
[6] K.S. Singwi, M.P. Tosi, R.H. Land and A. Sjølander, Phys. Rev. 176 (1968) 589.
[7] A. Griffin and J. Harris, Can. J. Phys. 54 (1976) 1396.
[8] J. Hubbard, Proc. Roy. Soc. (London) A243 (1957) 336.
[9] I.D. Moore and N.H. March, Ann. Phys. (NY) 97 (1976) 136.
[10] R.H. Ritchie and A.L. Marusak, Surface Sci. 4 (1966) 214.
[11] J. Harris, J. Phys. C5 (1972) 1757.

Surface Science 68 (1977) 377-384
© North-Holland Publishing Company

CHANGES IN WORK FUNCTION DUE TO CHARGE TRANSFER IN CHEMISORBED LAYERS [†]

J.L. MORAN-LOPEZ [*] and A. TEN BOSCH
Institute for Theoretical Physics, Freie Universität Berlin, 1 Berlin 33, Germany

In a simple tight-binding model well suited to transition and noble metals, we calculate the work function change due to adsorption of alkali metals. A decrease is found to be caused by ionic bonds formed on the surface through charge transfer between the two metals. The local density of states and the number of electrons in the substrate and the chemisorbed layer are evaluated for various alkali metals and coverages. The method is applicable in both the high and low coverage limits in contrast to existing theories.

1. Introduction

It is a well-known but as yet not clearly understood fact that the adsorption of alkali atoms causes a significant lowering of the work function of noble and transition metals [1]. Phenomenological arguments have been suggested which explain the reduction by a transfer of charge between adatoms and the substrate. But, as yet, no theory has been proposed satisfactory at both low and high coverages limits and applicable to transition as well as simple metals.

Such a theory is presented in the following. We explain the electronic mechanism which determines the sign and magnitude of the charge transfer and show how this leads to a non-linear coverage dependence of surface charge and therefore of work function. We demonstrate in the case of noble and simple metals that the amount of charge transferred from the adatoms to the metal is proportional to the contact potential or difference in work functions of the pure bulk metals.

2. Model

In the present paper, we consider only adsorption on noble or simple metals. For such systems, the narrow d-bands if present lie well below the Fermi level and do not play an active role in the charge transfer process. We can thus neglect s–d hybridization and d-electrons. The effect of the d-electrons is important for chemisorp-

[†] Work supported by DFG.
[*] Partially supported by Consejo Nacional de Ciencia y Tecnología (México).

tion on transition metal surfaces and will be studied in a future publication.

The hybridised s–p bands of adatoms and substrate are described within the tight-binding model in order to take changes in the local environment into account. The Hamilton operator of the system of N_A alkali metal atoms A adsorbed on the surface consisting of N simple or noble metal atoms B is given by the standard expression

$$H = \sum_i \epsilon_i n_i + \sum_{j \neq i} t_{ij} c_i^\dagger c_j . \tag{1}$$

Here, $c_i^\dagger$ and c_i are the creation and annihilation operators of Wannier states at site i with n_i electrons. The hopping integral t_{ij} describes electronic transitions from site i to j and causes the adatom levels to widen into bands at high coverage. The single site energies ϵ_i depend on changes in the number of electrons through the Coulomb interactions [2]

$$\epsilon_i = \epsilon_i^0 + U_i \Delta n_i + \sum_{j \neq i} U_{ij} \Delta n_j , \tag{2}$$

where $\Delta n_i = n_i - n_i^0$ is the deviation in the amount of electronic charge assigned to the site i per spin relative to the neutral state with number n_i^0. The repulsive intra-atomic Coulomb interaction is denoted by U_i. The interatomic interaction $U_{ij} = 2e^2/r_{ij}$ acts to stabilize the ionic adsorption. For a single adatom, this term is given by the image potential [3]. As the coverage increases, the repulsive interaction with the other chemisorbed atoms must also be included [4]. For sufficiently high coverage we can assign the same value of charge Δn_i to all atoms i of a given layer parallel to the surface. Because of charge neutrality inside the bulk, charge transfer can take place only between the adatoms and the first layer of the substrate and the charge conservation condition must be fulfilled

$$\theta \Delta n_A + \Delta n_{1,B} = 0 . \tag{3}$$

Here, $\theta = N_A/N$ measures the adatom coverage. We then obtain for the single site energy of the adatom:

$$\epsilon_A = \epsilon_A^0 - U_1 \theta \Delta n_A + U_0 \Delta n_A , \tag{4}$$

and of the first layer of the substrate:

$$\epsilon_{1,B} = \epsilon_B^0 - U_1 \Delta n_{1,B} + U_0 \Delta n_{1,B} , \tag{5}$$

where U_0 is the intra-atomic Coulomb interaction and U_1 is the effective average electrostatic potential of the double layer consisting of charge $2e\Delta n_{1,B}$ on the lattice sites of the substrate surface and charge $2e\theta \Delta n_A$ on the plane of the adatoms [5].

The change on the work function due to the potential drop through the dipole layer of thickness d and surface density for lattice constant a is then

$$\Delta\phi = (4\pi e^2/\epsilon)d(2\theta\Delta n_A)/a^2 , \tag{6}$$

where ϵ is the dielectric constant.

3. Calculations

The charge transfer Δn_i on site i is determined from the local density of states $N_i(E)$:

$$\Delta n_i = \int^{E_F} dE\, N_i(E) - n_i^0 , \tag{7}$$

where E_F is the Fermi energy. We use the method of Green functions to calculate $N_i(E)$ from

$$N_i(E) = -\pi^{-1}\, \text{Im}\, G_{ii}(E) . \tag{8}$$

The single site electronic Green function $G_{ii}(E)$ is given by

$$G_{ii}(E) = (E - \epsilon_i - \Delta_i)^{-1} , \tag{9}$$

where the self-energy Δ_i results from all hopping processes starting and ending on site i. This is described by the expansion

$$\Delta_i = \sum_{j\neq i} \frac{t_{ij}^2}{E - \epsilon_j - \Delta_j^i} + \cdots , \tag{10}$$

and

$$\Delta_j^i = \sum_{l\neq i,j} \frac{t_{ij}^2}{E - \epsilon_l - \Delta_l^{ij}} , \tag{11}$$

where Δ_l^{ij} takes hopping from site j into account avoiding i, etc. As discussed elsewhere, we close the expansion on the surface by using $\Delta_l^{ij} = \Delta_l^i$ [6]. We restrict hopping to nearest neighbours and consider the three uppermost layers in detail. In the first and second layer of the B-type substrate, the number of nearest neighbours within the same layer is denoted by Z_0 and within one of the nearest neighbour planes by Z_1. To study the dependence on adatom coverage along the substrate surface, we assume a random distribution of A-type atoms chemisorbed in the top position of the (100) surface of a simple cubic lattice (fig. 1). Each A-atom has then θZ_0 nearest neighbours within the chemisorbed layer and Z_1 nearest neighbours within the B-metal substrate. The self-energy of the adatom is then given by

$$\Delta_A = \frac{Z_0 \theta t_{AA}^2}{E - \epsilon_A - \Delta_A^A} + \frac{Z_1 t_{AB}^2}{E - \epsilon_{1,B} - \Delta_1^A} , \tag{12}$$

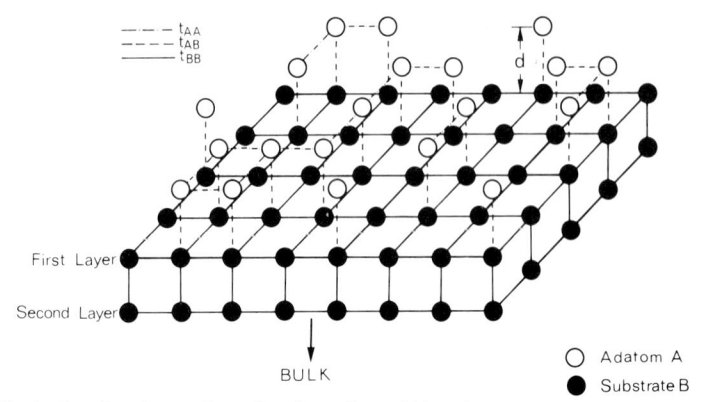

Fig. 1. Illustration for the random chemisorption of N_A adatoms A on top position of N substrate atoms B. The coverage is $\theta = 0.5$. d is the distance between the chemisorbed layer and the first substrate layer, t_{AA}, t_{AB} and t_{BB} are the nearest neighbour hopping integrals between two adatoms, one adatom and one substrate atom and two substrate atoms respectively.

where

$$\Delta_A^A = \frac{(Z_0 - 1)\theta t_{AA}^2}{E - \epsilon_A - \Delta_A^A} + \frac{Z_1 t_{AB}^2}{E - \epsilon_{I,B} - \Delta_I^A}, \quad (13)$$

$$\Delta_I^A = \frac{(Z_1 - 1)\theta t_{AB}^2}{E - \epsilon_A - \Delta_A^I} + \frac{Z_0 t_{BB}^2}{E - \epsilon_{I,B} - \Delta_I^I} + \frac{Z_1 t_{BB}^2}{E - \epsilon_{II,B} - \Delta_{II}^I}. \quad (14)$$

Here $\Delta_j^i = \Delta_A^A$ if i and j are in the chemisorbed layer and similarly for Δ_I^A, Δ_I^I, Δ_{II}^I, etc. The self-energies Δ_I^I, Δ_{II}^I, etc. are determined from similar equations discussed in detail in a previous paper [6]. This set of coupled equations is solved self-consistently in the charge transfer Δn_A and $\Delta n_{I,B}$. Because charge neutrality must be preserved in the bulk, the single site energies of the third and lower layers will be pinned to the Fermi level, so that

$$\epsilon_{II,B} = \epsilon_B^{bulk} = E_F. \quad (15)$$

4. Results

The calculations were performed as a function of coverage for various values of $\epsilon_A^0 - \epsilon_B^0$ and interatomic Coulomb interaction U_1. The half-band width of the bulk metals $W_s = 7$ eV and the intra-atomic Coulomb interaction $U_0 = 3.5$ eV were used as typical for wide s–p bands. We neglect for simplicity differences in the hopping integrals between adatoms and substrate and restrict calculations to a square lattice with nearest neighbours $Z_0 = 4$ and $Z_1 = 1$.

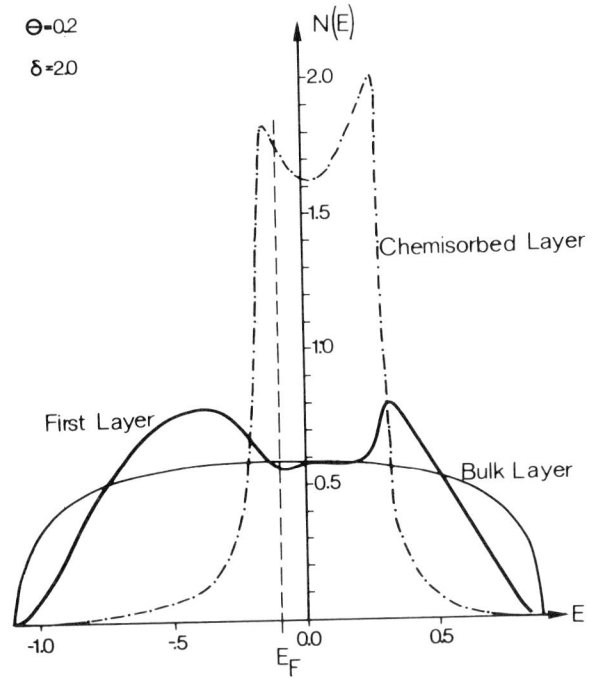

Fig. 2. Local density of states on a site in the chemisorbed layer, in the first substrate layer and in a bulk layer. We used $\theta = 0.2$, $\epsilon_A^0 = 0.1$, $\epsilon_{I,B}^0 =$ abulk $= -0.1$, $U_0 = 3.5$ eV, $U_1 = 7.0$ eV and $t_{AA} = t_{AB} = t_{BB}$. The energy is given in units of half band width $W_s = 7.0$ eV.

Fig. 2 shows the results for the electronic density of states of the adatom, the first layer of the B-metal surface and the B-metal bulk at $\theta = 0.2$. As the coverage increases the adatom band widens and shifts to higher energies. The density of states of the B-metal top layer is shifted down relative to the bulk. In the second layer of the substrate, we obtain essentially the density of states of the bulk. This behaviour is shown in fig. 3 for $\theta = 0.8$.

The shifts in the bands are due to a transfer of electronic charge from adatom onto the metal surface as described by eqs. (4) and (5). The resulting loss in energy is balanced by the electrostatic potential created on the surface which raises the bulk level and the Fermi energy.

The charge $2e\Delta n_A$ on the adatoms is positive and decreases nonlinearily with inreasing coverage. The resulting change in work function is proportional to the total transferred charge $\theta \Delta n_A$ and is shown in fig. 4. A rapid decrease is found at low coverages with an inflection point and change of slope near $\theta = 0.4$. This is especially visible in the case of large $\delta = (\epsilon_A^0 - \epsilon_B^0)/W_s$. The largest value of charge transfer at a given coverage is found for large values of δ. This is demonstrated in fig. 5 for different values of interaction U_1. The role of electronegativity is thus

$\Theta = 0.8$

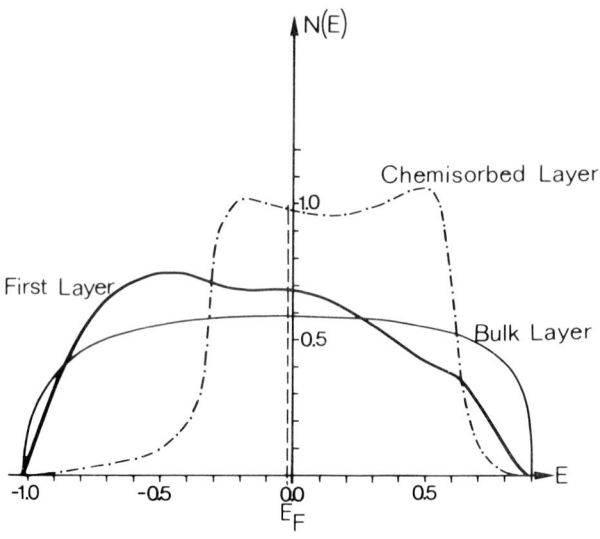

$\delta = 0.2$

Fig. 3. Local density of states on a site in the chemisorbed layer, in the first substrate layer and in a bulk layer for the coverage $\theta = 0.8$. We used the same parameters as in fig. 2.

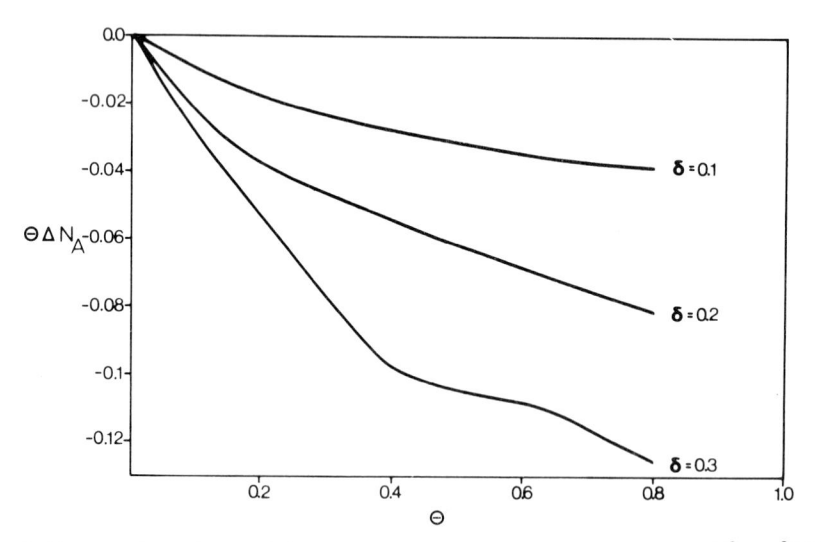

Fig. 4. Coverage dependence of the work function difference for various $\delta = (\epsilon_A^0 - \epsilon_B^0)/W_s$; $U_0 = 3.5$ eV and $U_1 = 7.0$ eV.

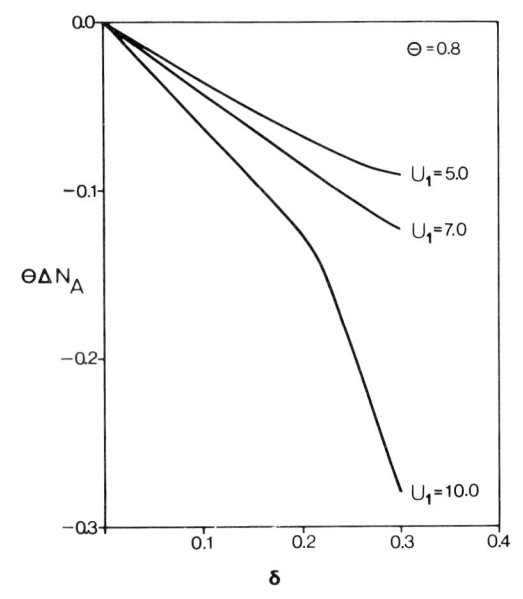

Fig. 5. Dependence of the work function difference on $\delta = (\epsilon_A^0 - \epsilon_B^0)/W_s$ for various interatomic Coulomb interactions U_1. We used $\theta = 0.8$.

assumed here by the difference in the center of energy of the pure bulk metals. For monovalent metals this is equivalent to the difference in the Fermi energies of the pure metals as given by the bulk contact potentials or roughly to the difference of the pure metal work functions. Thus the effects of charge transfer increase along the series Li adsorbed on Ag, Cu and Au or along the series Li, K, Cs adsorbed on Au.

5. Discussion

For higher than monolayer coverage, no further change in the amount of charge transfered from the adatoms to the surface is to be expected [5] and the work function will saturate to the value of bulk A-metal. The density of states of the second adsorbed layer of A-atoms will not deviate substantially from the bulk A-density of states and the Fermi energy lies halfway between ϵ_A^0 and ϵ_B^0. The limiting case of a bulk metallic contact has thus been reached for $\theta > 1$ and no further effects dependent on the number of adsorbed atoms should occur.

We have assumed here a random distribution of adatom positions on the surface. Because of charge transfer and resulting repulsive electrostatic interactions between chemisorbed particles, this will surely correspond to a nonequilibrium configuration. Indeed, the coverage dependence of the work function obtained here is similar to measurements made before thermal equilibrium is established on the surface [7]. In contrast, the work function to coverage curves in the case of equilibrium distribution of adatoms go through a noticeable minimum below $\theta = 1$. This may be connected to an order–disorder transition in the adsorbed phase [8] and it would be of interest to investigate the effect of charge transfer on adatom position.

In summary, we have shown that the chemical bond of adsorbed metal atoms has a polar character which increases with increasing work function difference and decreases with increasing coverage. A partial electron transfer from the adatoms to the substrate occurs if the Fermi energy of the bulk adsorbate is higher than that of the substrate. This is indicated by a shift in the work function due to an effective dipole moment per adsorbed particle which is a nonlinear function of coverage. This is explained by a decrease in attractive electrostatic potential on the electrons on adatom sites with increasing coverage which makes charge transfer unfavourable.

Charge transfer reactions are important in many areas of research in metallurgy, catalysis, electrochemistry [8] and biophysics [9]. The present paper is a step toward the understanding of the physics and chemistry of such systems.

Acknowledgement

We would like to thank Prof. K.H. Bennemann for his assistance and support.

References

[1] N. Lang, Phys. Rev. B4 (1971) 4234.
[2] J. Giner, F. Brouers, F. van der Rest and F. Gautier, J. Phys. F6 (1976) 1281.
[3] A.C. Hewson and D.M. Newns, Proc. 2nd Intern. Conf. on Solid Surfaces, Japan. J. Appl. Phys. Suppl 2 (1974) 121.
[4] J.P. Muscat and D.M. Newns, J. Phys. C7 (1974) 2630.
[5] M. Leynaud and G. Allen, Surface Sci. 53 (1975) 359.
[6] J.L. Morán-López, G. Kerker and K.H. Bennemann, J. Phys. F5 (1975) 1277; Surface Sci. 57 (1976) 540.
[7] L.D. Schmidt and R. Gomer, J. Chem. Phys. 45 (1966) 1605.
[8] D.M. Kolb, M. Przasnyski and H. Gerischer, Electroanal. Chem. Interfacial Electrochem. 54 (1974) 25.
[9] A.P. Nelson and D.A. Mcquarrie, J. Theor. Biol. 55 (1975) 13.

THE ADSORPTION OF OXYGEN AND CARBON MONOXIDE ON CLEAVED POLAR AND NONPOLAR ZnO SURFACES STUDIED BY ELECTRON ENERGY LOSS SPECTROSCOPY

R. DORN * and H. LÜTH

2. Physikalisches Institut der Rheinisch Westfälischen Technischen Hochschule Aachen, D-5100 Aachen, F.R. Germany

Electron energy loss spectroscopy (ELS) with primary energies $E_0 \leq 80$ eV has been performed on ultrahigh vacuum (UHV) cleaved nonpolar ($1\bar{1}00$) and polar zinc (0001) and oxygen ($000\bar{1}$) surfaces of ZnO to study the adsorption of oxygen and carbon monoxide. Except for CO on the nonpolar surface where no spectral changes in ELS are observed a surface transition near 11.5 eV is strongly affected at 300 K on all surfaces by CO and O_2. At 300 K clear evidence for new adsorbate characteristic transitions is found for oxygen adsorbed on the Zn polar surface near 7 and 11 eV. At 100 K on all three surfaces both CO and O_2 adsorb in thick layers and produce loss spectra very similar to the gas phase, thus indicating a physisorbed state.

1. Introduction

Since ZnO belongs to the catalytically active metal oxides the study of gas adsorption is of central interest and has been performed since a long time [1,2]. The effect of oxygen and atomic hydrogen on the surface conductivity, i.e. formation of depletion and accumulation space charge layers is well known [1]. In contrast to the enormous influence of simple gases on the electrical surface properties no dramatic effects could be observed in UV photoemission spectroscopy (UPS) on annealed nonpolar ZnO surfaces upon oxygen, hydrogen and carbon monoxide adsorption [3]. It is therefore interesting to investigate the adsorption process of simple gases by other surface sensitive techniques such as electron energy loss spectroscopy with low primary energies (ELS). It is furthermore necessary to get more information about differences in the adsorption properties of the three crystallographically different surfaces, the nonpolar ($1\bar{1}00$) and the polar Zn(0001) and O($000\bar{1}$) faces. ZnO has the interesting feature that these three types of surfaces can be prepared by cleavage in ultrahigh vacuum (UHV). This is particularly important since the chemical composition of the polar surfaces might be varied by annealing at moderate temperatures [4].

---

* Present address: Standard Elektrik Lorenz AG, Research Centre, D-7000 Stuttgart 40, F.R. Germany.

2. Experimental

The ZnO crystals had been grown from the vapor phase in this laboratory. During the growing process the crystals were slightly copper doped giving rise to a bulk conductivity of about 10^{-3} (ohm cm)$^{-1}$. Etch patterns on the different surfaces were used to determine direction and sign of the crystallographic c-axis [5] prior to mounting the crystal on the holder. The samples used had a length of 13 mm and a thickness of about 3 mm.

Mirrorlike nonpolar ($1\bar{1}00$) surfaces were obtained in UHV by cleavage parallel to the hexagonal c-axis by means of the double wedge technique. Cleaved polar (0001) and (000$\bar{1}$) surfaces were prepared by pressing the free end of the clamped crystal against an edge of the crystal holder. These polar surfaces obtained by cleavage normal to the c-axis are usually slightly convex with a high density of atomic steps [6]. Leads for resistive heating (Au) and for the thermocouple (Au–constantan) were attached to the crystals by pressure bonding. The sample could be cooled down to about 80 K by liquid nitrogen.

The UHV system was a standard stainless steel chamber (operating pressure $<10^{-10}$ Torr) which could be separated from the ion pump (200 ℓ/sec) by a gate valve. By means of a manipulator the sample surface could be positioned in front of a single pass cylindrical mirror energy analyser (CMA) with integral electron gun (Varian) and a quadrupole mass spectrometer (UTI).

Second derivative electron energy loss spectra (ELS) at low primary energies ($E_0 < 100$ eV) were measured by modulating the analyser pass voltage (1 V peak to peak) and recording the collector current of the CMA at the first harmonic. The energetic resolution given by the fullwidth at half maximum (FWHM) of the elastically scattered electrons was about 0.8 eV. The energetic position of spectral structures could be read with an accuracy of ±0.2 eV. The recording time for one spectrum was typically 5 min.

3. Results

Loss spectra with primary energy $E_0 = 40$ eV for all three types of *clean surfaces* are shown in figs. 1, 2, and 3. On the nonpolar (fig. 1) and on the polar Zn surface (fig. 3) the spectra of the clean cleaved surface could be remeasured also after moderate annealing to about 800 K, whereas the spectrum of the clean cleaved O face (fig. 2) changed considerably after annealing to about 600 K. Exact positions of the loss peaks have been taken from a number of original recorder plots: The clean surfaces exhibit losses at: nonpolar ($1\bar{1}00$): 4.3, 7.4, 9.1, 10.9, 12.6, 15.8 eV; polar Zn(0001): 4.3, 7.6, 9.1, 10.7, 15.3 eV; polar O (000$\bar{1}$): 4.2, 7.8, 9.0, 11.2, 12.9, 15.5 eV. The 7.5 eV loss is much more prominent on the nonpolar than the corresponding peaks on the polar surfaces. The loss peak near 11 eV is extremely sensitive to surface treatment. On the polar surfaces this peak vanishes after oxygen and

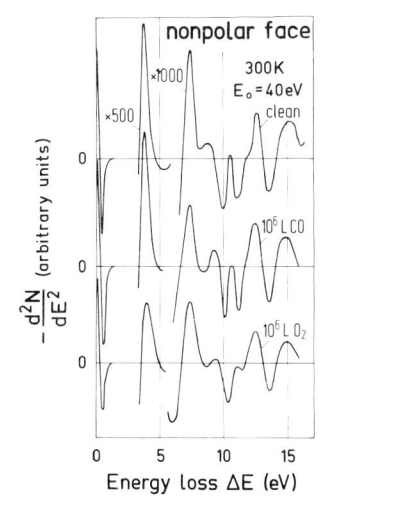

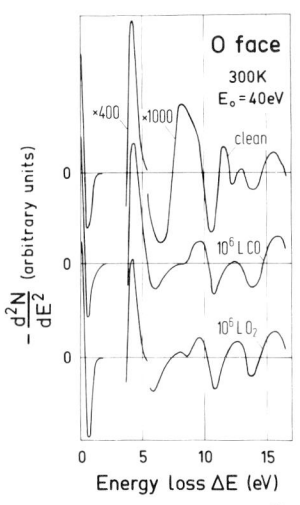

Fig. 1. Second derivative loss spectra of the cleaved clean and gas covered nonpolar ($1\bar{1}00$) surface obtained at room temperature (1 L = 10^{-6} Torr sec).

Fig. 2. Second derivative loss spectra of the polar O($000\bar{1}$) surface at room temperature.

CO adsorption (figs. 2, 3), and on the nonpolar surface it decreases after O_2 adsorption at room temperature (fig. 1).

As is seen from fig. 3 *adsorption* of CO and of O_2 on the *Zn surface* changes the loss spectrum of the clean cleaved surface. Beside the suppression of the 11 eV loss

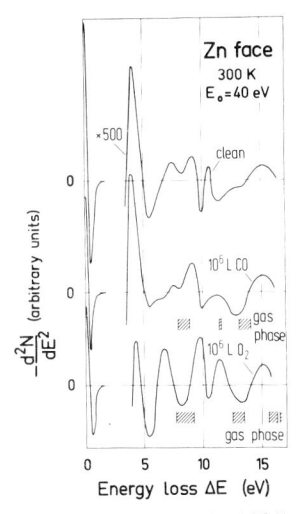

Fig. 3. Second derivative loss spectra of the polar Zn(0001) surface at room temperature. Gas phase data are taken from ref. [9].

oxygen adsorption furthermore induces two loss peaks near 7.0 and 11.7 eV. Slight changes in peak positions upon gas-adsorption are attributed to an increase or decrease, respectively, of neighbouring loss intensities.

On the *O-polar face* the 11.2 eV loss vanishes after oxygen and carbon monoxide adsorption (fig. 2) similar to the 10.7 eV loss on the Zn face. Both gases change the spectrum of the clean surface in a similar way, i.e. losses characteristic for oxygen or carbon monoxide can not be found. The intensity of the 7.8 eV loss decreases considerably by exposure to both gases and instead of the peaks at 11.2 and 12.9 eV a broad structure with its maximum near 12.5 eV appears.

The effect of oxygen and of carbon monoxide on the loss spectrum of the *nonpolar ($1\bar{1}00$) surface* is even less pronounced (fig. 1): CO exposure up to dosages of about 10^6 L does not change the spectrum of the clean surface. Beside a decrease of the 7.4 eV loss peak height oxygen adsorption suppresses the 10.9 eV loss considerably. New bands characteristic for oxygen or carbon monoxide can not be seen after adsorption.

At *lower temperatures (100 K)* thicker layers of oxygen and carbon monoxide appear to be adsorbed as can be seen from the stronger suppression of the ZnO loss

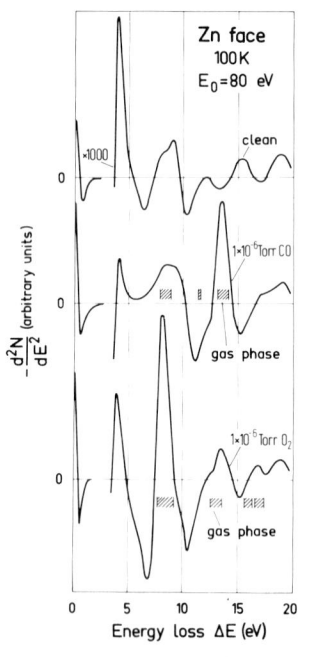

Fig. 4. Second derivative loss spectra of the Zn(0001) face cleaved at 100 K. The spectra of the gas covered surfaces were measured in ambients of 10^{-6} Torr at 100 K. Gas phase data are taken from ref. [9].

peaks in fig. 4. For reasons of experimental convenience the low temperature spectra have been measured with $E_0 = 80$ eV. The spectra in fig. 4 are obtained by exposing a clean cleaved Zn face to an ambient of 1×10^{-6} Torr at 100 K. Similar spectra were obtained on the other surfaces, too. After pumping out the ambient the strong effects of O_2 and CO disappear. The reversible CO adsorption produces two prominent loss bands near 8.3 and 13.6 eV, O_2 exposure induces structures near 8.2, 12.4, 13.6 and 17.1 eV. Obviously the peaks near 4 and 19 eV are remnants from losses of the clean surfaces. Since upon warming up the loss intensity near 13.6 eV starts to decrease at lower temperatures than the 8.2 eV structure (fig. 4) this peak might partially be caused by CO admixtures during the O_2 inlet.

4. Discussion

The energy loss spectra of the clean surfaces have more extensively been discussed in previous publications [7,8] and have been related to theoretical and experimental band structure data [8]. Bulk and surface plasmons on all three types of surfaces have been identified by the observation of single and multiple excitations near energies of 19 and 15.5 eV, respectively. Bulk interband transitions are observed near 4.3, 7.8 and 9.1 eV on all faces, at 12.9 eV on the nonpolar face and at 12.6 eV on the O face. The dependence on primary energy of the loss peak near 11 eV on all faces [8] suggests an interpretation in terms of a surface transition. This interpretation is confirmed by the present observation of its high sensitivity to gas adsorption (figs. 1—3). More involved is the discussion concerning the 7.8 eV loss on the nonpolar face. A careful study of its dependence on primary energy reveales an admixture of a surface transition to the underlying bulk excitation [8]. The fact, that the surface sensitive structure is only observed on the nonpolar face, where both Zn and O atoms form the topmost layer, can be explained by the assumption, that O and Zn "dangling bonds" are initial and final states respectively of the corresponding transition.

At lower temperatures ($T \approx 100$ K) reversible adsorption of CO and O_2 in an ambient could be observed. Thick layers of the adsorbed gases, suggested by the suppression of the losses characteristic for the clean surfaces, gave rise to the same new structures independent of the different types of surfaces (fig. 4). These losses correspond very well to excitations observed in gas phase spectra by other authors [9]. Therefore, it is concluded that considerable amounts of both O_2 and CO are physisorbed at ≈ 100 K. This is supported by the observation that the strong spectral changes vanish upon pumping out the gas from the chamber.

Room temperature adsorption leading to lower coverages strongly affects the surface loss near 11 eV. On all faces except for CO on the nonpolar face this transition is suppressed upon exposure to 10^6 L CO and O_2, respectively (figs. 1—3). Neither CO nor O_2 exposure changes the spectral structure observed on the clean *nonpolar face* drastically. While CO does not show up in the spectrum at all oxygen is adsorbed at least in small quantities as can be seen from the suppression of the surface loss near 11 eV.

On the *O-face* both gases O_2 and CO change the spectral structure in the same way (fig. 2). In preliminary experiments it could be shown, that the spectrum of the clean surface in the region near 12 eV loss energy is very sensitive to annealing at 600 K, which might be explained by a change of the atomic arrangement at the surface. A relation might be given to a negative surface charging that has been observed near temperatures of 600 K in work function measurements [10]. The similar change of the spectra after O_2 and CO adsorption, therefore, might be interpreted tentatively in terms of a reconstruction change initiated by (small?) coverages of adsorbed gases. Further experiments are necessary to give more evidence for this hypothesis.

While CO adsorption changes the loss structure of the *Zn-face* similar to the O-face, characteristic new loss peaks can be observed on the Zn-face after oxygen adsorption (fig. 3). Their energetic positions are 7.0 and 11.7 eV, respectively. In UPS two new bands are found on the Zn-face after O_2 adsorption near 10.1 and 14.5 eV binding energy [8]. The difference of 4.4 eV in binding energy corresponds to the difference of 4.7 eV in loss energy of the oxygen characteristic losses. It is therefore assumed, that the corresponding transitions have the same final state with their initial states in the filled bands observed in UPS. From ELS data alone detailed conclusions about the type of bonding should not be drawn. An oxygen bond continuing the ZnO lattice does not seem unreasonable, since the bulk transition near 7.6 eV has the O-2p derived upper valence band as initial state [11]. From a comparison with the low temperature data (fig. 4) it is obvious that oxygen is chemisorbed at room temperature.

With the present state of knowledge it seems difficult to correlate the presented results with the extensively studied changes in surface conductivity [1] due to oxygen adsorption since both types of measurements might be sensitive to totally different adsorbate amounts. Surface conductivity, i.e. band bending, responds to small coverages in the 10^{11} atoms/cm^2 regime whereas ELS is sensitive to coverages higher that 1% of a monolayer.

5. Conclusions

Similar to previous UPS results [3] the present ELS data show that ZnO surfaces are relatively inert against simple gases like O_2 and CO, i.e. room temperature equilibrium coverages are expected to be far below a monolayer. The most pronounced effect in this study is observed for room temperature adsorption of oxygen on the Zn polar surface where oxygen appears to be chemisorbed in a state different from molecular oxygen. The surface specificity to the gases studied shows that further investigations of adsorption and reaction mechanisms must take into account a different behaviour of the three types of crystallographically different surfaces. This is of particular interest for the investigation of the catalytic activity of ZnO powders where all types of surfaces are expected to contribute.

Acknowledgement

The authors would like to thank Professor G. Heiland for critical reading of the manuscript. The work was financially supported by the Deutsche Forschungsgemeinschaft in the Sonderforschungsbereich 56, Aachen.

References

[1] G. Heiland, E. Mollwo and F. Stöckmann, in: Solid State Physics, Vol. 8, Eds. F. Seitz and D. Turnbull (Academic Press, New York, 1959) p. 191.
[2] H. Lüth, G.W. Rubloff and W.D. Grobman, Solid State Commun. 18 (1976) 1427; G.W. Rubloff, H. Lüth and W.D. Grobman, Chem. Phys. Letters 39 (1976) 493.
[3] H. Lüth, G.W. Rubloff and W.D. Grobman, unpublished.
[4] D. Kohl, M. Henzler and G. Heiland, Surface Sci. 41 (1974) 403.
[5] A. Klein, Z. Physik 188 (1965) 352.
[6] M. Henzler, Surface Sci. 36 (1973) 109.
[7] R. Dorn and H. Lüth, J. Appl. Phys. 47 (1976) 5097.
[8] R. Dorn, H. Lüth and M. Büchel, Phys. Rev. B (1977), to be published.
[9] E.N. Lassettre, A. Skerbele, M.A. Dillon and K.J. Ross, J. Chem. Phys. 48 (1968) 5066; E.N. Lassettre, Can. J. Chem. 47 (1969) 1733; H.J. Hinz, in: Proc. 4th Intern. Vacuum UV Conf. Hamburg, Germany 1974, p. 33.
[10] H. Moormann, private communication.
[11] S. Bloom and J. Ortenburger, Phys. Status Solidi (b) 58 (1973) 561.

OXIDATION PROPERTIES OF InSb(110) SURFACES

E.W. KREUTZ, E. RICKUS and N. SOTNIK

Institut für Angewandte Physik der Technischen Hochschule Darmstadt, Schlossgartenstr. 7, D-6100 Darmstadt, Fed. Rep. Germany

The structural and electronic properties of InSb(110) surfaces, which were cleaned by argon bombardment annealing technique, have been investigated by LEED and surface conductivity measurements during oxygen adsorption. The diffraction patterns before and during exposure exhibit only diffraction spots which are compatible with the bulk periodicity. The exposures resulted in a gradual decrease of all beams. The surface conductivity increases during exposures. The magnitude of the adsorption induced changes is determined by the coverage and by the density of surface defects. In view of these results the oxidation process on the InSb(110) surface is discussed.

1. Introduction

Recent studies indicate the importance of steps on the electronic and chemical properties of clean semiconductors surfaces as shown experimentally [1–6] and as confirmed theoretically [7,8]. The surface properties are influenced or even dominated by steps as expected from the determinant role of electronic interactions in surface phenomena, which can follow a peculiar behaviour of electrons near steps.

Defects in the surface especially steps have received little detailed experimental attention [1,5] for III–V semiconducting compounds. According, the present paper is dealing with the influence of surface roughness on the adsorption of oxygen on InSb(110) surfaces cleaned by ion bombardment annealing technique. The roughness of the surface was altered by the preparation methods and by the cleaning conditions. During oxygen adsorption the structural and electrical properties of the surfaces have been followed by LEED and by measurements of surface conductivity.

2. Experimental

The samples were cut from M.C.P. Ltd., Alpterton Wembley, Middlesex InSb single crystals (excess hole concentration about 5×10^{14} cm^{-3}) grown in [211] direction with a Bridgman technique. The large sample faces were oriented to within 1° of (110) planes by the Laue X-ray backscattering method. The specimens investigated were rectangular parallelepipeds with approximate dimensions of 9 × 5 × 0.5 mm^3. The InSb specimens were lapped with SiC powder in distilled water. The initial polish was followed by etch polishing (3 to 6 sec) in CP4A [9]. The etching was stopped by flooding with distilled water.

In order to produce clean surfaces we have resorted to ion bombardment and annealing techniques. Ar$^+$ ion bombardment of the etched surfaces was performed at room temperature with current densities of 0.6 μA cm^{-2} for periods of several hours at an ion energy of 140 eV. The ion bombardment was followed by heating at 350°C (type I surfaces) and at 450°C (type II surfaces), respectively.

The roughness of the InSb(110) surfaces was studied by transmission electron microscopy (Siemens Elmiskop 1). Determinations were made on sections covered with tylose (C 10, Kalle, Wiesbaden) foil in aqueous solution on which amorphous carbon and platinum films are deposited. Photographs were taken of these replicas mounted on a copper mesh. During replication it was assumed that the surface roughness is not severely changed.

Electrical contacts for current measurements and as potential probes were made by thermo compression of Ag wires or by pressing stainless steal wires to the specimens. Further details especially specimen mounting and manipulator, cooling and heating facilities have been described [10]. The experimental arrangement for LEED measurements, for registration of LEED intensities, and for measurement of surface conductivity have been reported previously [10–12].

The electronic structure of the cleaned surfaces was altered by exposure to oxygen. Throughout each exposure the gases were continuously admitted through a leak valve from glass flasks. During gas exposure the pressure was increased stepwise up to 10^{-4} Torr holding the samples for 5 min at a certain pressure. For exposures up to 10^{-1} Torr min the pressure maintained at 10^{-4} Torr and the exposure time was varied. After reevacuation the measurements have been performed at $T = 95$ K. The oxygen exposures were made with the ion pump non-operating. The total pressure was measured by a millitorr gauge encased in a bent tube attached to the chamber.

The oxygen partial pressure was controlled by a Vacuum Generators QM7 quadrapole mass spectrometer. In addition to oxygen hydrogen, argon, nitrogen, and carbon monoxide were also found in the mass spectra. The millitorr gauge reading has been corrected for the oxygen partial pressure. It was assumed that the influence of carbon monoxide on the results is negligible since the CO partial pressure never exceeded 1%.

3. Results

As to be seen from typical replicas of transmission electron micrographs the etched surfaces show irregular terraces which probably are to some extent of atomic dimensions. During type II surface treatment hillocks are formed on the flat portions of the terraces present in the surfaces whereas no changes of the surface morphology could be detected in the electron micrographs. The influence of the Ar^+ ion energy on the surface roughness as investigated by RHEED measurements have been discussed in more detail elsewhere [13].

The development of the LEED pattern as a function of the cleaning procedure has been reported in an earlier paper [10]. The diffraction patterns of the InSb-(110) surface cleaned by ion bombardment annealing exhibit independent of surface treatment (type I and II) only spots which are compatible with the bulk periodicity.

The relative change of surface conductivity as a measure for the coverage of the InSb(110) surface with adsorbed oxygen is plotted in fig. 1 versus exposure. For type II surfaces (section 2) no changes of the surface conductivity could be observed up to 10^{-8} Torr min. Starting from the surface conductivity, measured immediately after ion bombardment annealing, the surface conductivity then increases with oxygen coverage (fig. 1) in agreement with earlier observations [10,11]. After exposures of more than 10^{-3} Torr min the surface conductivity undergoes no remarkable changes. For type I surfaces (section 2) the surface conductivity, however, is not affected by exposures up to 0.2 Torr min. This result does not imply that the surface conductivity of type I surfaces might not be

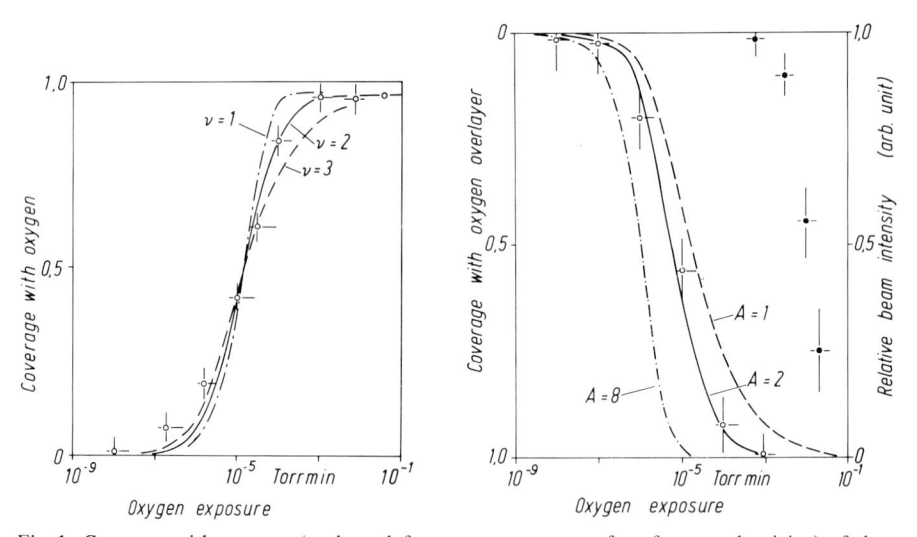

Fig. 1. Coverage with oxygen (evaluated from measurements of surface conductivity) of the InSb(110) surface (type II) versus oxygen exposure at $T = 95$ K and comparison with model calculations (section 4).

Fig. 2. Coverage with oxygen overlayer (evaluated from measurements of relative LEED intensity) of the InSb(110) surface [(○) type II, (●) type I] versus oxygen exposure at $T = 95$ K and comparison with model calculations (section 4).

changed at higher exposures which cannot be measured because of experimental limitations.

The LEED measurements during oxygen exposure give no indication for ordered adsorption with unit cell dimensions differing from the bulk [10]. The relative intensity (defined as the ratio of total beam intensity to background intensity) as a measure for the coverage with an oxygen overlayer is shown in fig. 2 as a function of oxygen exposure. Up to an exposure of about 10^{-8} Torr min of oxygen no changes in relative intensity have been observed for the type II surfaces (section 2). At 10^{-5} Torr min the relative intensity has decreased appreciably (fig. 2). After 10^{-3} Torr min the diffraction pattern has been obliterated. For type I surfaces (section 2) the relative diffraction intensity qualitatively shows a similar behaviour but quantitatively the relative intensity changes at higher exposures. Between 10^{-3} and 10^{-2} Torr min the relative intensity decreases with oxygen exposure (fig. 1). After exposures of about 4×10^{-2} Torr min only a diffuse background could be measured.

4. Discussion

The III–V compound semiconductor InSb crystallizes in the zincblende structure. A number of the low index faces have been examined by LEED reflecting for the (110) surface ideal (1×1) unit cells and for the other low index surfaces enlarged unit cells in agreement with the results presented (section 3). Nevertheless, recent analyses of LEED intensities have revealed that the (110) surface of III–V compound semiconductors is reconstructed.

Following the procedure of other authors [14] we have performed model calculations to get information on the number of adsorption sites and to decide whether the oxygen is adsorbed in a molecular or an atomic state. The saturation of the change in surface conductivity (fig. 1) which is attributed to an adsorbate monolayer as discussed in more detail elsewhere [10] must be due to the decreasing sticking coefficient s with coverage θ. Under the assumption that adsorption is possible only on empty sites, that ν sites are required per adsorption event, and that the adsorbate has free mobility on the surface s is given by

$$s = s_0 (1 - \theta)^\nu, \tag{1}$$

where s_0 is the sticking coefficient on the clean surface and $\theta = \nu N/N_0$ is the ratio of adsorbed particles N to the density of adsorption sites N_0. Depending on ν fomulas for θ [14] are evaluated as a function of exposure. Using the exposure for $\theta = 0.5$ the coverage is calculated for $\nu = 1, 2,$ and 3. As to be seen from fig. 1 the model calculations for $\nu = 2$ mainly fit the experimental findings, i.e. two adsorption sites are required for one adsorption event. To decide whether the adsorbate is mobile or not mobile model calculations have been performed using modifications of eq. (1). Within the experimental errors the calculations for mobile adsorbate coincide with those for not mobile adsorbate that an unequivocal decision experimentally is impossible. For further calculations and conclusions a mobile adsorbate was assumed.

Different concepts [15–18] have been reported for the site of oxygen adsorption on the (110) surface of III–V compound semiconductors. The weakening and disappearing of the diffraction pattern, the missing of fractional order spots during adsorption ([10], section 3) as well as RHEED measurements [13] at first stages of oxygen adsorption imply that the oxygen possibly adsorbs in an amorphous layer or a polycrystalline arrangement in agreement with the observations of other authors [17] for the InSb(100) and InSb(111) surfaces. Following the experimental observations ([10], section 3) we suggest a tentative model for the site of oxygen adsorption. Oxygen might be adsorbed in a molecular or an atomic state at regular lattice sites (In or Sb atoms) or in bridging positions (between In atoms, between Sb atoms, or between In and Sb atoms). The adsorption sites may be deteriorated by surface defects. The mixing of different adsorption sites and the perturbation of the chemical environment may originate in an oxygen adsorption layer showing no superstructure with respect to the underlying semiconductor lattice (section 3). If excited oxygen is present additionally, bonds between the surface atoms and the rest of the crystal might be broken leading to true oxides of In and Sb as supported by RHEED measurements on heavily oxidized surfaces [13].

The coverage α with oxygen overlayer increases with exposure as demonstrated by the disappearing of the ordinary spots (fig. 2). For a mathematical description we have resorted to a model developed by other authors [14]. Using eq. (1) and the assumption that only adsorbed molecules make structural conversion α is given by

$$\alpha = (1 - \theta)^{A/\nu}, \tag{2}$$

where A is the number of adsorption sites in a patch of the surface without any superstructure (1 × 1, section 3) converted by the adsorbed particles. With the parameters determined by the calculations for eq. (1) the coverage α was computed for different values of A. As to be seen from fig. 2 the model calculations for $A = 2$ reveal the experimental curve, i.e. the number of adsorption sites necessary for an elementary adsorption process equals the number of adsorption sites converted to an oxygen overlayer. This is also obviously seen from fig.s 1 and 2 that for type II surfaces structural and electrical properties vary simultaneously with exposure indicating this correlation of crystallographic and electronic structure of the ion bombarded and annealed InSb(110) surface.

As evaluated from ac field effect measurements on real surfaces [19] the surface state density on InSb surfaces consists within the energy gap out of a continuous distribution and increases within the energy bands. Following experimental results [20] and theoretical calculations [21] the surface state density of the InSb(110) surface consists of two bands near the band gap. An empty surface state band above the conduction band minimum, which is primarily In-derived, and a filled surface state band below the valence band maximum which is primarily Sb-derived. The high surface state densities [19] within the bulk energy bands assumingly correspond to these In- and Sb-derived surface state bands. In addition, for the clean GaAs(110) surface a band of surface states with its upper edge near midgap has been attributed to defect induced surfaces states [1] because of the cleavage dependent and the oxygen sensitive properties of these states. Under the assumption that these results [1] hold also for the InSb(110) surface the density of surface states in the gap is higher for a surface with a larger number of surface defects (type II).

If we assume that the InSb(110) surface shows hole conduction in the space charge layer [22], then the bands are bent upwards. During oxygen adsorption the bands generally are bent more upward for type II surfaces whereas the band bending remains unchanged for type I ones. Apparently the higher surface state density of the type II surfaces is responsible for the different behaviour during oxygen adsorption. The surface states present in high density at the Fermi level for type II surfaces are influenced (probably occupied) during adsorption [1] resulting in a stronger band bending at the surface, i.e. positive changes of the surface conductivity (fig. 1). Apparently the density of surface states at the Fermi level is too low for type I surfaces to get a reasonable change of the band bending, i.e. no conductivity changes during oxygen adsorption (section 3). It is presumed, that the oxygen picks up electrons during adsorption resulting in the observed changes of the charge distribution within the surface. This would be in agreement with EPR measurements [17] on clean survaces of III–V compound semiconductors favouring negatively charged oxygen molecules on the sites of group III surface atoms after adsorption. On the other hand, oxygen adsorption might modify the reconstruction of the (110) surface and lead to electronic rearrangement as supported by calculations [23] including an oxygen layer. As suggested by other authors [1] the structural defects in the surfaces may do the same thing following the observations on clean Si(111) surfaces [24].

The different density of structural defects would also explain the exposure dependent decrease of the relative LEED intensities (fig. 2). If sites at the structural defects act as localized adsorption centers these sites would be occupied at low exposures and the absolute amount on step sites should be proportional to the number of structural defects (type II surfaces). The adsorption on regular surface sites should be observed at higher exposures (type I surfaces).

5. Conclusion

LEED measurements and measurements of surface conductivity on InSb(110) surfaces cleaned by ion bombardment annealing show the important part played by surface defects for the structural and electronic properties of p-InSb(110) surfaces. The (1 × 1) structure of (110) surfaces with a low number of surface defects is converted to an oxygen overlayer at higher oxygen exposures than surfaces with a large number. The surface conductivity of (110) surfaces with a large number of surface defects increases during oxygen exposure whereas the surface conductivity remains unchanged for surfaces with a low number.

Acknowledgement

The authors should like to thank Prof. Dr. W. Waidelich for much support. The authors are much obliged to Dipl.-Phys. P. Daab and to Mr. G. Jourdan for the investigations of the surface morphology. The authors are most grateful to the Deutsche Forschungsgemeinschaft for financial support within the framework Kr 516/2. They are very grateful to Mrs. H. Salow for cooperation in typing the manuscript.

References

[1] G.M. Guichar, C.A. Sébenne and G.A. Garry, Phys. Rev. Letters 37 (1976) 1158.
[2] M. Henzler, Surface Sci. 19 (1970) 159; 22 (1970) 12; 36 (1973) 109.
[3] H. Ibach, K. Horn, R. Dorn and H. Lüth, Surface Sci. 38 (1973) 433.
[4] J.E. Rowe, S.B. Christman and H. Ibach, Phys. Rev. Letters 34 (1975) 874.
[5] A. Huijser and J. Van Laar, Surface Sci. 52 (1975) 202.
[6] M. Henzler, Appl. Phys. 9 (1976) 11.
[7] V.T. Rajan and L.M. Falicov, J. Phys. C9 (1976) 2533.
[8] M. Schlüter, K.M. Ho and M.L. Cohen, Phys. Rev. B14 (1976) 550.
[9] A.F. Bogenschütz, Ätzpraxis für Halbleiter (Hanser, München, 1967).
[10] E.W. Kreutz, E. Rickus and N. Sotnik, Thin Solid Films, in press.
[11] E.W. Kreutz, E. Rickus and N. Sotnik, Phys. Letters 47A (1974) 363.
[12] N. Sotnik, Thesis, Technische Hochschule Darmstadt.
[13] E.W. Kreutz, E. Rickus and N. Sotnik, Verh. Deut. Physik. Ges. 1 (1977) 355.
[14] M. Henzler and J. Töpler, Surface Sci. 40 (1973) 388.
[15] P.E. Gregory and W.E. Spicer, Phys. Rev. B13 (1976) 725.
[16] R. Ludeke and A. Koma, J. Vacuum Sci. Technol. 13 (1976) 241.
[17] D. Haneman, Phys. Rev. 121 (1961) 1093.
[18] W. Ranke and K. Jacobi, Surface Sci. 47 (1975) 525.
[19] E.W. Kreutz and P. Schroll, Surface Sci. 37 (1973) 410.
[20] D.E. Eastman and J.L. Freeouf, Phys. Rev. Letters 35 (1975) 1624.
[21] C. Calandra and G. Santoro, J. Phys. C9 (1976) L51.
[22] G.W. Gobeli and F.G. Allen, Phys. Rev. 137 (1965) 245.
[23] W.A. Harrison, Surface Sci. 55 (1976) 1.
[24] J.E. Rowe, S.B. Christman and H. Ibach, Phys. Rev. Letters 34 (1975) 87.

Surface Science 68 (1977) 399–407
© North-Holland Publishing Company

DIFFRACTION INTENSITIES IN HELIUM SCATTERING; TOPOGRAPHIC CURVES

N. GARCIA

Sección de Física Fundamental, Instituto de Física del Estado Sólido (CSIC), Universidad Autónoma de Madrid, Cantoblanco (Madrid), Spain

and

Departamento de Matemáticas, Universidad Nacional de Educaión a Distancia Ciudad Universitaria (Madrid), Spain

G. ARMAND and J. LAPUJOULADE

Centre d'Etudes Nucléaires de Saclay, Service de Physics Atomique, Section d'Etudes des Interactions Gaz–Solides, BP 2, 91190 Gif-sur-Yvette, France

Diffracted beam intensities have been calculated using the hard corrugated wall potential and the so-called GR method, for numerous values of the corrugation parameter and incident angle. Curves of equal intensity are drawn as a function of these two parameters (topographic curves). They give a clear description of diffraction phenomena. Comparing to experimental data, the potential corrugation is determined for alkali halide, for dense metallic surfaces and tentatively for tungsten (112) face. So the microscopic roughness seems to be the parameter which can explain the quite different behaviour of these surfaces regarding the scattering of helium. The potential corrugation of the repulsive part of the potential is discussed in terms of localization of delocalization of electrons in the vicinity of atom cores.

1. The model

In the scattering of helium atoms by alkali halide surfaces, the diffraction phenomenon has been observed in different experimental situations [1–4]. In contrast, no diffraction peaks have been detected when helium atoms are scattered by clean metallic surfaces [5–9] if one excludes the (112) face of tungsten. In this case, in-plane first-order diffraction peaks have been observed when the incident plane contains the [1$\bar{1}$0] direction [10,11]. In these two experiments, out-of-plane measurements were not possible.

In order to explain this strikingly different behavior, two kinds of arguments, which are not exclusive, have been prpoposed: (a) in the experimental condition, the Debye–Waller factor is always very much smaller for metallic surfaces than for alkali halide surfaces [12], and (b) the surface corrugation is lower on metallic surfaces, that is to say the Fourier components of the potential are strongly weaker than those of alkali halide surface [13].

The latter explanation is based on the results of diffraction peak intensities calculation, in which the interaction potential is represented by an infinitely hard corrugated surface (HCS model) [13–16]. If the $\vec{OZ}$ direction is normal to the surface plane and R (XY) denotes a position vector belonging to the surface, the potential is taken to be:

$$V(R, z_1(R)) = \infty \quad z \leq z_1(R),$$
$$= 0 \quad \text{otherwise}, \quad (1)$$

when for a square unit cell of side a

$$z_1(R) = \tfrac{1}{2} a \beta_0 [\cos(2\pi X/a) + \cos(2\pi Y/a)]. \quad (2)$$

Assuming that only elastic events occur, the time-independent Schroedinger equation is solved on the condition that the total wave vanishes for $z = z_1(R)$, and one obtaines the following set of linear equations [13,14] *:

$$\sum_G A_G \exp\{(k_{Gz} + k_z) z_1(R) + G \cdot R]\} = -1, \quad (3)$$

in which A_G is the scattered amplitude in the channel given by the two-dimensional reciprocal lattice vector G, and k_z and k_{Gz} are the normal wave vector components respectively for the incident beam and for the diffracted beam in channel G. The conservation of energy gives:

$$k_{Gz}^2 = k_i^2 - (K_i + G)^2,$$

where K_i is the incident particle wave vector component parallel to the surface. Any good solution of (3) must give results which satisfy the unitarity condition, i.e.:

$$\sum_F P_F = 1, \quad P_F = (k_{Fz}/k_z) |A_F|^2, \quad (4)$$

F being a G vector for which $k_{Fz}^2 = k_{Gz}^2 > 0$.

The A_G values have been first calculated in the framework of eikonal approximation [13]. The oscillating behavior of diffraction peak intensities, known as "rainbow pattern" and experimentally observed [4], is very well depicted. But in order to remove this approximation and to take multiple scattering into account, two methods have been used: (a) the Fourier transform methods [14,15], and (b) the so-called GR method, developed by García et al. [14,16]. A set of eq. (3) is solved by taking a finite number N of non-equivalent points, regarding translational symmetry, in the unit cell (N R values) and N corresponding points in the reciprocal space (N G vectors). The number N is chosen large enough so that the uni-

---

* Eq. (3) corresponds to the Raleigh hypothesis.

tarity condition is satisfied. The calculated results [16] are in good agreement with the experimental ones of Boato et al. [4].

Thus, provided that resonances with bound states do not influence the diffracted beam intensities (as experimentally occurs for He thermal beams [4]), the potential model and the calculation method seem to give a good description of the experimental results.

It is important to notice that although the calculations this paper are done with the Rayleigh hypothesis [17] (eq. (3)), the results presented ($\beta_0 \leqslant 0.12$; $\beta_0/2 \simeq 0.06 < 0.072$) are the exact ones, as has been shown recently with the exact solution by Cabrera and García [18]. That is to say, for the corrugation regime presented in this paper, the Rayleigh hypothesis leads to the exact solutions of the scattering of a wave by an HCS model.

Therefore, we have calculated these quantities for different values of the corrugation parameter β_0 and incident angle θ_i. The results describe the overall evolution of the phenomena in function of these parameters. Thus, with the available experimental data, we can determine the β_0 value for different materials. Conversely, if for a given material one can estimate the β_0 value, the expected diffraction peak intensities can be deduced and the most favorable incident angle chosen.

With these possibilities in mind, results are presented and discussed in the following section.

2. Conditions of calculation; results

We consider an helium incident beam scattered by a surface, the unit cell of which is a square of side a. The particle indicent wave vector and angle are respectively denoted by k_i and θ_i (the direction $\theta_i = 0$ coincides with the normal to the surface). The incident plane contains the [110] direction. k_i is chosen so that $k_i a = 31.2$, giving respectively in the cases of LiF(001), NaCl(001) and Cu(001) surfaces, a k_i value equal to 11, 7.8 and 12.2 Å^{-1}.

These conditions correspond to the experiments of refs. [3], [4] and [9] for LiF and copper. In the case of NaCl [3], the incident wave vector is slightly different being approximately equal to 11 Å^{-1}.

For all cases, the number N is equal to 196. This is exactly the number of channels included in the calculation. With $k_i a = 31.2$, the number of open channels is of the order of 70 for all the values of θ_i.

The diffraction peak intensities have been calculated under these conditions for different values of the parameters β_0 and θ_i. For β_0 less than 0.12 and for all θ_i values the unitarity condition is fulfilled as the sum (4) is lying between 0.9999 and 1.

Thus, the set of equal intensity curves in function of β_0 and θ_i can be drawn for different diffracted beams. The resulting diagram consists of these curves which are called topographic. Such topographic curves are presented for in-plane beams 00,

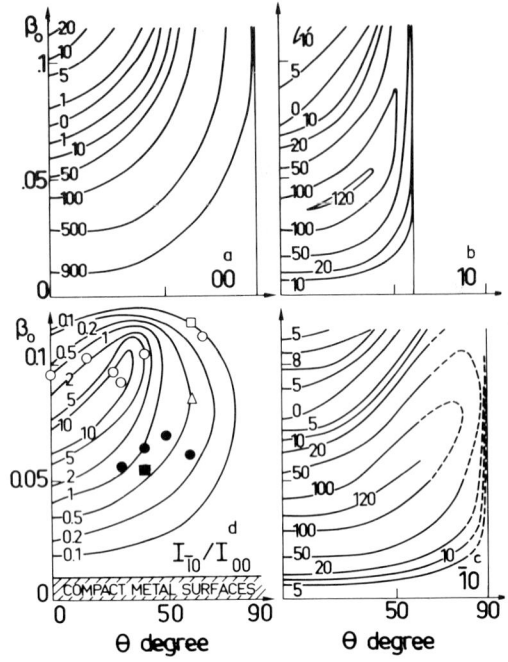

Fig. 1. (a)–(c) Curves of equal intensity of topographic curves in function of corrugation parameter β_0 and incident angle θ_i, for (00), (10) and ($\bar{1}$0) beams. (d) Topographic curves for the ratio $I_{\bar{1}0}/I_{00}$. The values ○, □, △, ● and ■ are obtained respectively from refs. [4], [3], [3], [10] and [11]. The shaded region correponds to dense or compact metallic surfaces.

10, $\bar{1}0$ respectively in figs 1a, b and c, and for out-of-plane beams 11, $\bar{1}\bar{1}$, $2\bar{2}$, 22 respectively in figs. 2a, b, c and d. In these two figures, the incident beam is scaled 1000.

Complementarily, fig. 1d depicts the topographic curves for the intensity ratio $I_{\bar{1}0}/I_{00}$, which will be used later to determine the corrugation of different materials.

3. Discussion

3.1. General behavior

The specular 00 beam is very much greater than the others for very small β_0 values and for all β_0 when approaching grazing incidence. A large "valley" appears for the three in-plane beams when complete extinction of these beams occurs. These intensities vanish in approximately the same β_0, θ_i region. As it could be expected from the unitarity condition, the out-of-plane beams then reach their maxima.

It is now interesting, in order to get a clear insight into diffraction phenomena,

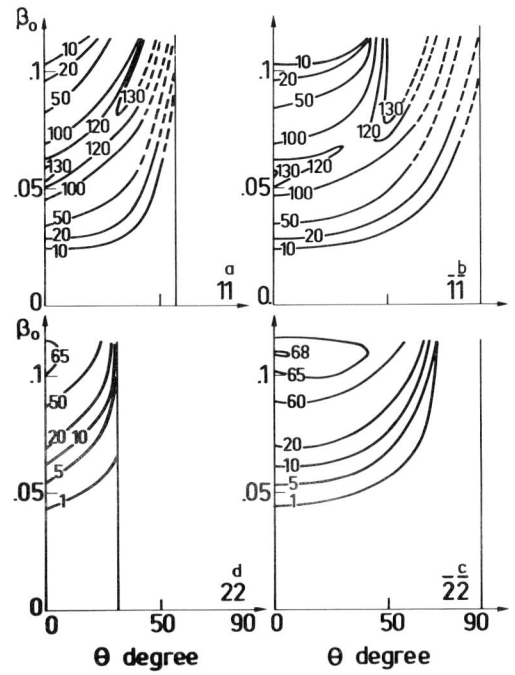

Fig. 2. (a)–(d) Topographic curves for the (11), ($\overline{11}$), (22) and ($\overline{22}$) out-of-plane diffracted beams.

to examine the evolution of intensities as θ_i varies given a value of the corrugation parameter β_0. Doing so, one can distinguish three β_0 regions as far as the evolution of the in-plane peaks are concerned: (a) For low β_0 value, approximately for β_0 less than 0.03, the specular beam increases with θ_i, whereas the 10 and $\overline{1}0$ decrease. (b) For β_0 lying between 0.03 and 0.09, a region where a maximum appears for 10 and $\overline{1}0$ beams (scaled 120), these beams first increase with θ_i, reach a maxima and decrease for large incident angle. The specular peak intensity increases monotonically, but its intensity at normal incidence is drastically lowered in such a way that it can completely vanish. (c) For high β_0 value, but obviously less than 0.12, the 00 peak intensity decreases first, goes to extinction and increases again. The 10 beam has a very low intensity in this region as well as the $\overline{1}0$ for θ_i less than 45–50°. Beyond this value it increases slightly, going after that to very low value for grazing incidence.

The out-of-plane beams have a different behavior. The 11 and $\overline{11}$ decrease monotonically as θ_i increases for β_0 approximately less than 0.045. For β_0 greater than this value, they have a maxima for θ_i value around 40–50° (scaled 130). On the other hand, the 22 and $\overline{22}$ decrease monotonically. All these out-of-plane beams have very low intensities for low β_0 value as it could be expected: the particles are scattered into in-plane beams mainly in the specular.

These results outline the necessary in an experiment to measure the different intensities for different incident angles, not only for the in-plane beams but also the out-of-plane, especially if the former are not very intense. For illustration of this point, suppose a material with a β_0 value about 0.08–0.09. The 10 and $\bar{1}0$ beams remain practically undetectable for all θ_i values, as we have to multiply the intensities given in the figures by a Debye–Waller factor. On the contrary, the 11 and $\bar{1}\bar{1}$ peaks reach a maxima which is as large as the specular for θ_i equal to 40–60°. On the other hand, the 22 and $\bar{2}\bar{2}$ peaks are also greater than the specular in the neighborhood of normal incidence. From only in-plane measurements in this case, one could conclude that the surface potential is very smooth.

3.2. Corrugation of different materials

In order to obtain the corrugation potential of a material, we have to compare the experimental data with our calculated results.

For lithium fluoride (001) face, we consider first the extended results of Boato et al. [4], the Debye–Waller factor being unknown. Nevertheless, it can be shown that this factor has practically the same value for the specular and $\bar{1}0$ beams if the incident angle is less than 45°, because the incident wave vector modulus k_i is large compared to the modulus of the smallest reciprocal vector G ($G/k_i \simeq 0.2$) and the crystal temperature is low ($T_s = 80$ K). With this in mind, the points corresponding to the measured ratio $I_{\bar{1}0}/I_{00}$ for $\theta_i = 0°$, 15°, 27°, 30° and 40° are put in fig. 1d (○). They fall on practically the same line, giving a mean β_0 value equal to 0.10. The point at $\theta_i = 65°$ lies on a higher β_0 value. However, in this case the Debye–Waller factor ratio for these two beams must certainly be less than one and consequently, the ratio $I_{\bar{1}0}/I_{00}$ for $T_s = 0$, that is to say the calculated one, should be greater than the experimental ratio. This correction would lead to a translation of the representative point to lower β_0 values. The same correction should be applied to the point (□) deduced from the experimental data of Bledsoe and Fischer [3]. It could be more important in this case as the crystal temperature is equal to 300 K.

Concluding this analysis, one can say that the corrugation parameter for (001) face of LiF is equal to 0.10 ± 0.01, leading to a full corrugation potential ($\beta_0 a$) of 0.28–0.31 Å. A more precise calculation introducing a new Fourier component in the potencial has been made [16] giving a better fitting of the data [4].

As far as the NaCl(001) face is concerned, we have just one point (△) [3]. It is difficult in this case to estimate the Debye–Waller correction. Compared to the LiF case (□), the surface temperature and incident angle are the same, the G/k_i ratio (0.145) and the bulk Debye temperatue are lower to reduce and increase respectively the difference between the Debye–Waller factor of $\bar{1}0$ and 00 beams. Thus, one can tantatively say that the β_0 value for this surface has probably around 0.085

giving a corrugation potential of 0.34 Å. New experimental results are needed in order to obtain a more precise value.

The case of metals is quite different as no diffraction peak has been observed except on the (112) face of tungsten. In a recent experiment on the (100) face of copper [9], a very narrow specular peak has been detected (0.5°) which is composed of elastically scattered particles, as shown by velocity analysis. Its intensity has been measured for different incident angles and surface temperature. Taking into account the particle acceleration in the stationary attractive part of the potential (Beeby [12]), the usual Debye–Waller plot gives the value of the "apparent" surface Debye temperature interpreted in terms of simultaneous interaction of the incident helium aton with the four copper atoms belonging to a surface unit cell. This plot gives also the intensity of the 00 beam, supposint the crystal at rest, equal to 1000 ± 100. Fig. 1a leads to a β_0 value less than 0.01 for this surface. Therefore, the $I_{\bar{1}0}/I_{00}$ ratio would be lower than 0.02 for a crystal at rest, and in the above experimental condition, the $\bar{1}0$ beam intensity must be less than 0.004 of the incident beam intensity. As the experimental sensitivity is 0.005, the diffracted beam, if it exists, could not be observed.

A preliminary calculation with a rectangular unit cell has shown that this difference introduces a not too significant variation of the calculated intensities. There-, fore we have included in fig. 1d the points relative to the (112) face of tungsten, the incident plane containing the [1$\bar{1}$0] direction following the data of Stickney [10] (●) and Merrill [11] (■). The points fall in the same β_0 range. The correction for the Debye–Waller factor could be important here as the crystal temperature is high (1200 K). Following the preceding discussion, this correction tends to increase the $I_{\bar{1}0}/I_{00}$ ratio and the points would be translated a little to higher β_0 value. However, one can tentatively adscribe to this face for the [1$\bar{1}$0] direction a corrugation parameter of the order of 0.06. The corresponding full corrugation potential would be of the order of 0.27 Å.

When the incident plane contains the [11$\bar{1}$] direction, no diffraction peaks have been detected. As the rugosity perpendicaulr to this direction is great, and according to these calculated results, one may think that out-of-plane beams must be detected. Unfortunately, the experimental device did not allow such as measurement.

4. Conclusions

The topographic curves calculated with hard corrugated wall potential and presented above, give a clear description of diffraction phenomena. They allow one to predict the evolution of in-plane or out-of-plane diffracted beam intensities if one can a priori determine the corrugation potential of the studied surface.

Unfortunately, such a determination is quantitatively impossible as we are not able, at the moment, to calculate the potential from first principles. Thus, only comparison with experimental results could give the potential corrugation value $\beta_0 a$ or the first Fourier components of the potential. In this way we have found for:

LiF(001): $\quad \beta_0 a = 0.30$ Å, $\quad \beta_0 \simeq 0.11$;

NaCl(001): $\quad \beta_0 a \simeq 0.34$ Å, $\quad \beta_0 \simeq 0.085$;

Cu(001): $\quad \beta_0 a \leqslant 0.025 - 0.03$ Å, $\quad \beta_0 \simeq 0.01$;

(or dense faces of metals)

W(112) [1$\bar{1}$0]: $\quad \beta_0 a \simeq 0.27$ Å, $\quad \beta_0 \simeq 0.06$.

During an experiment with dense faces of metals, the crystal temperature is often equal or greater than room temperature in order to avoid surface contamination. The corrugation is low. Therefore, only the in-plane 10 and $\bar{1}$0 beams can be detected, but their intensities seem to be below the experimental sensitivity. So, to date, no diffracted peaks have been observed in a scattering experiment from these clean faces.

On the contrary, the hard wall corrugation of W(112) face in the [1$\bar{1}$0] direction seems to be $\beta_0 \simeq 0.06$ and for LiF(001) $\beta_0 \simeq 0.12$. Consequently, diffracted beams are observed with this metallic surface even at high temperature.

These important differences between materials could be understood qualitatively if one considers that the surface potential is due to the interaction of incident atom electrons with the surface electronic structure. Particularly, the short range repulsive part is, roughly speaking, dependent on the distance between the helium electron shell and the surface electronic charges. The electrons are probably strongly localized in the vicinity of ion cores on alkali halide surfaces and partly delocalized on metallic surfaces. So, for surfaces of the same density such as (001) faces of lithium fluoride and Copper, the potential corrugation is certainly more important in the former than in the latter.

On a tungsten surface, the electron delocalization can also occur. In the [1$\bar{1}$0] direction of a (112) face, however, this delocalization is certainly lower than on the copper (100) face, as the distance between nearest neighbor atoms is larger (W 4.5 Å – Cu 2.55 Å). This greater distance produces a larger potential corrugation. Along the [11$\bar{1}$] direction, the nearest neigbor distance (2.7 Å) is equivalent to that of copper atom on a (100) face. In this direction, the potential corrugation is certainly low, reducing the possibility of multiple scattering effects which are, on the contrary, not hindered on the LiF surface. Therefore, from the point of view of diffraction, the behavior of W(112) face must be intermediate between the behavior of dense metallic and alkali halide surfaces. On the latter, particularly with LiF, it is well known that numerous in-plane and out-of-plane beams have been observed. Assuming a better experimental sensitivity, in-plane 10 and $\bar{1}$0 beams can be detected on dense metallic surfaces. On W(112) face, we expect that not only these in-plane beams, but also out-of-plane 11 and $\bar{1}\bar{1}$ beams can be observed especially when the incident plane contains the [11$\bar{1}$] direction.

Acknowledgements

We are indebted to Professors N. Cabrera and J. Solana for many discussions. We would also like to thank Drs. D.A. Degras and C. Manus for discussions and critical reading of the manuscript.

References

[1] D.R. O'Keefe, J.N. Smith Jr., R.L. Palmer and H. Saltsburg, J. Chem. Phys. 52 (1970) 4447; Surface Sci. 20 (1970) 27.
[2] B.R. Williams, J. Chem. Phys. 55 (1971) 3220.
[3] J.R. Bledsoe and S.S. Fisher, Surface Sci. 46 (1974) 129.
[4] G. Boato, P. Cantini and L. Mattera, Surface Sci. 55 (1976) 141.
[5] R.B. Subbarao and D.R. Miller, J. Chem. Phys. 10 (1969) 4679.
[6] J.N. Smith, Jr., H. Saltsburg and R.L. Palmer, in: 6th Rarefield Gas Dynamics (Academic Press, New York, (1969) p. 1141.
[7] S. Yamamoto and R.E. Stickney, J. Chem. Phys. 53 (1970) 1594.
[8] A.G. Stoll, D.L. Smith and R.P. Merrill, J. Chem. Phys. 54 (1971) 163.
[9] G. Armand, J. Lapujoulade and Y. Lejay, J. Phys. Lettres 37 (1976) L187; Surface Sci. 63 (1977) 143.
[10] D.V. Tendulkar and R.E. Stickney, Surface Sci. 27 (1971) 516.
[11] A.G. Stoll and R.P. Merril, Surface Sci. 40 (1973) 405.
[12] J.L. Beeby, J. Phys. C4 (1971) L359.
[13] V. Garibaldi, A.C. Levi, R. Spadacini and G.E. Tommei, Surface Sci. 48 (1975) 649.
[14] N. García, J. Ibáñez, J. Solana and N. Cabrera, Solid State Commun. 20 (1976) 1159; Surface Sci. 60 (1976) 385.
[15] H. Chow and E.D. Thompson, Surface Sci. 54 (1976) 269; 59 (1976) 225.
[16] N. Carcía, Phys. Rev. Letters 37 (1976) 912; J. Chem. Phys. (15 July 1977).
[17] J.W. Strutt (Lord Rayleigh), Proc. Roy. Soc. (London) A79 (1907) 399.
[18] N. Cabrera and N. Carciá, Topical Invited Paper to the 7th Intern. Vacuum Congr. and 3rd Intern. Conf. on Solid Surfaces, Vienna, 1977, to be published.

THRESHOLD AND LENNARD-JONES RESONANCES AND ELASTIC LIFETIMES IN THE SCATTERING OF ATOMS FROM A CORRUGATED WALL AND AN ATTRACTIVE WELL SURFACE MODEL

M. GARCIA

Departamento Fisica Fundamental C-III, Universidad Autonoma de Madrid, Canto Blanco, Madrid, Spain

In this communication we apply the GR method for solving the scattering equations of atoms from a hard corrugated surface, on accelerated particles above a hard corrugated surface and a hard corrugated surface with an attractive well models. The solutions are given for the Rayleigh hypothesis that under the range of corrugation presented in this paper lead to the exact ones. Threshold resonances are studied observing that the appearance and disappearance of beams must be for a general theory with vertical tangent. The structure of the Lennard-Jones resonances are given for the model mentioned above. For the first time it is stressed that Lennard-Jones resonances are not observed in metal surfaces in general, and, of course, they are unobserved in compact metallic surfaces. This is correlated to the fact that diffraction has not been observed. Both facts are due to the very weak gasmetal interaction potential, in terms of relative corrugation $\beta_0 \leq 0.01$. According to our results, the Lennard-Jones resonances in metals present greater difficulties to be observed experimentally. We also note that the absence of diffraction in compact metal surfaces is because they are almost plane $\beta_0 \leq 0.01$ and not because of the Debye–Waller effect. Finally, we have calculated the lifetimes of the atoms at the crystal surfaces. These are larger, the smaller the incident energy and the larger the corrugation β_0. But the lifetimes are particularly large at resonance condition (10^{-11} s). This implies diffusion of the particles at the surface and large interchange of energy, thereby increasing the chemical reaction rates.

GRAIN-BOUNDARY GROOVE EVOLUTION IN THE PRESENCE OF AN EVAPORATION

Vu Thien BINH, Y. MOULIN, R. UZAN and M. DRECHSLER

Départment de Physique des Matériaux, associé au CNRS, Université Claude Bernard Lyon 1. 43 Boulevard du 11 Novembre 1918, 69621 Villeurbanne Cédex, France

Former theories (Mullins, Robertson, Vu Thien Binh et al.) predict that grain-boundary groove evolution by surface diffusion only is a steady-state profile evolution, i.e. the linear geometrical parameters as the groove depth or the groove width increase with the heating time following the $t^{1/4}$ law, while the groove shape remains the same. On the contrary, in the presence of a free evaporation:

(1) The profile evolution is not any more a steady-state evolution, the groove shape is time depending.

(2) The groove depth and the groove width as a function of time are not any more a $t^{1/4}$ law.

(3) The groove evolved until a limit groove profile which remains constant in spite of continuous surface diffusion and evaporation.

INFLUENCE OF SOME ADDITIONAL ELEMENTS ON THE SURFACE TENSION OF COPPER AT INTERMEDIATE AND HIGH TEMPERATURES

M.F. FELSEN and P. REGNIER

Section de Recherches de Métallurgie Physique, Centre d'Etudes Nucleaires de Saclay, B.P. 2, 91190 Gif-sur-Yvette, France

The influence of some additional elements on the surface tension of copper at intermediate and high temperatures was measured. In the intermediate temperature range ($\sim 0.4\ T_m$) the voids annealing technique was used with thoroughly outgassed copper samples, implanted with He, O_2, S and Kr before irradiation with 500 keV Cu^+ ions, in the swelling peak temperature region. In the high temperature range ($\sim 0.9\ T_m$) the zero creep rate technique was used to check the influence of environmental He and Si on the surface tension of bamboo-structure wires. These experiments were made with a versatile and original device which allows the strain variations to be followed through those of the electrical resistance at constant volume.

1. Introduction

The need for information about surface tension has been emphasized in recent years because this parameter plays a central role in several solid state phenomena such as sintering, diffusion creep, scratch smoothing and fracture. To contribute towards this goal, we have chosen to study the influence of some additional elements on the surface tension of copper at intermadiate and high temperature. This choice was motivated by the fact that copper is the most well known metal from the theoretical point of view and by the opportunity that its surface is one of the most easily cleaned. The impurities we have selected were oxygen, sulphur, silicon, helium and krypton. Sulphur and oxygen because they are the usual impurities of copper, silicon because most of the experiments on copper are done inside a silica vessel, helium and krypton because both of them are rare gases but stimulate in a very different way the nucleation of voids in irradiated materials [1]. Sulphur and oxygen are also efficient enhancers of voids nucleation. From this last point of view the most appropriate technique to deduce surface tension is to follow the shrinkage kinetic of voids. Unfortunaly this technique can only be used conveniently in a short range of temperatures, let us say from 400 to 550°C. So, to obtain some results at higher temperatures, in order to study precisely the effect of temperature, we have begun a set of zero creep rate experiments on copper wires.

2. Shrinkage of voids

Voids produced in metals by irradiation or by precipitation of quenched in lattice vacancies are metastable and tend to shrink by diffusion to minimize the surface energy of the sample. When the phenomenon is controlled by bulk diffusion of vacancies, the evolution of the radius r of an isolated spherical void located in an infinite medium is given by the relation [2]

$$r^3 = r_0^3 - 6(D\gamma_s \Omega/kT)t, \qquad \text{for } r \gg kT/2\Omega\gamma_s, \tag{1}$$

where D is the self-diffusion coefficient, γ_s the surface tension and r_0 the initial radius of the void. Thus from the annealing kinetics of voids, the surface tension can be deduced if the self-diffusion coefficient is already known from other experiments. This recently developed technique which has only been used to study pure copper [3] and aluminium [4] seems to us very appropriate to check the influence of small amounts of additional elements on the surface tension of solids. So, we decide to apply it for the first time to a metal having a small specific impurity content. For that, we must assume that the self-diffusion coefficient and the annealing kinetics are not affected by these impurities.

2.1. Experimental procedure

The voids were produced by 500 keV Cu^+ ions bombardement of thin copper foils. Some of these foils were thoroughly outgassed and ion implanted with various doses of additional elements (helium, krypton, oxygen or sulphur) before irradiation and the others were not outgassed and not implanted.

Both kinds of samples were copper discs of 3 mm diameter and 0.1 mm thickness (electropolished with the usual double jet technique). This geometry assures good thermal contact with the heating holder of the microscope and minimizes extra heating by the electron beam, two crucial points for the in situ annealing experiments we have done.

In order, to fulfill the condition of isolated voids, we have selected voids widely separated from each other and from the surface. The anealing of such voids was continuously followed, typically from a radius of 2000 Å to 200 Å, and their size were measured on photographs at a magnification of 100,000, with an accuracy of ±20 Å. The error due to annealing duration is negligible. Temperature was measured to ±5°C by the thermocouple adjacent to the sample, but in fact we compared results of experiments done with the same thermocouple and having the same geometry, so the relative error in temperature measurements is less than ±1°C.

2.2. Results

During an anneal, the initial polyhedric shape of the voids gradually smoothes and soon turns into a sphere, as can be seen in fig. 1 which shows the shrinkage of

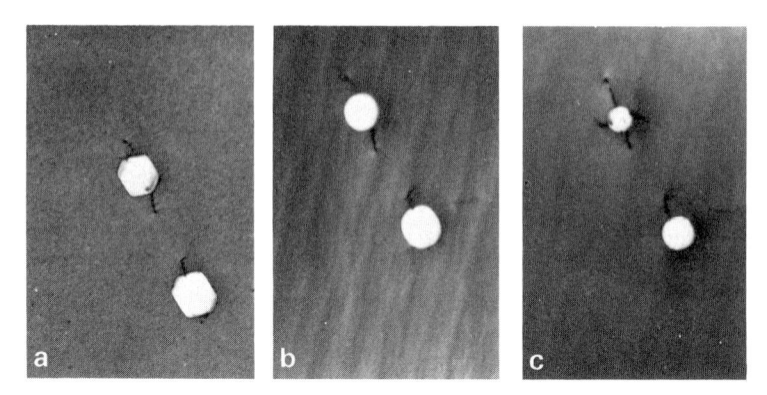

Fig. 1. Sequence of void annealing in a not ougassed and not implanted sample: (a) at room temperature; (b) at 509°C, $t = 0$ mn; (c) at 509°C, $t = 60$ mn.

some voids in a not outgassed and not implanted sample. For every sample, the annealing of about ten voids was followed and the kinetics were analysed on r^2, r^3 and r^4 versus t plots [5]. The only linear curves were those that plot r^3 versus t, which indicates that the annealing kinetics of voids is controlled by bulk diffusion of vacancies. A linear relationship r^2 versus t would correspond to interface control, and a r^4 versus t to pipe diffusion control.

For the samples having their natural gas content, our results are in good agreement with those reported by Bowden and Baluffi [6] at the same temperature

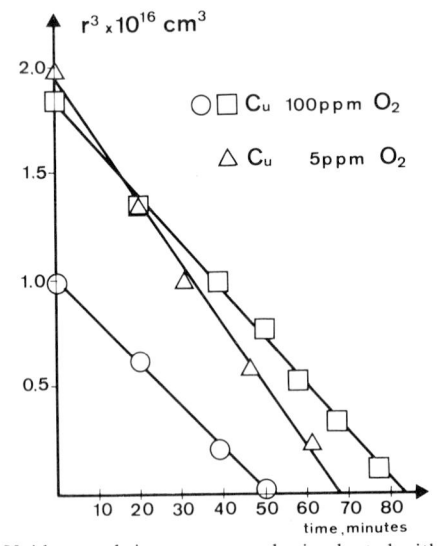

Fig. 2. Voids anneals in copper samples implanted with oxygen.

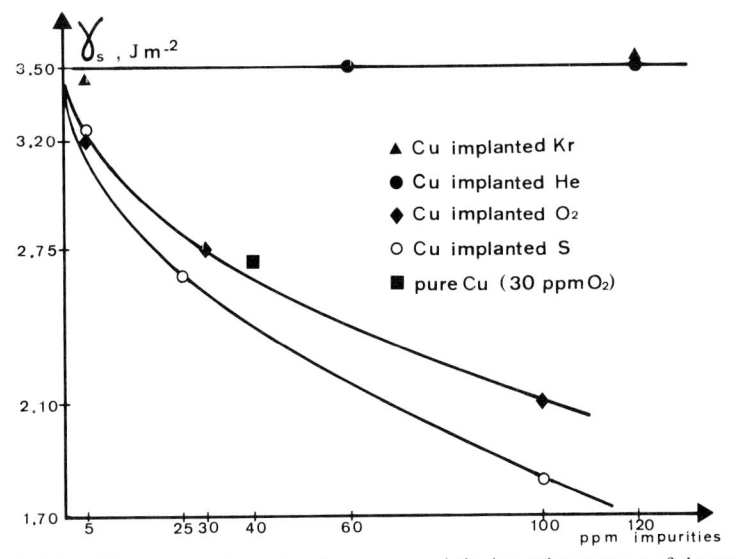

Fig. 3. Relationship between the surface tension γ_s and the impurity content of the samples.

$(509 \pm 5°C)$:

$$D\gamma_{s(\text{Balluffi})} = 0.17 \times 10^{-17} \text{ J/s}, \qquad D\gamma_{s(\text{this study})} = 0.25 \times 10^{-17} \text{ J/s}.$$

Assuming then that the self-diffusion coefficient of copper is not affected by very small amount of impurities, and picking up its value through the extrapolation of the high temperature diffusion data of Kuper et al. [7], we have deduced the surface tension γ_s, from the product $D\gamma_s$. This procedure has given us the results plotted in fig. 3, which clearly indicates that the rare gases helium and krypton do not affect the surface tension of copper while oxygen and sulphur drastically lower it, in agreement with the well established fact that rare gases do not interact with

Table 1
Derivative of the surface tension with respect to concentration for vanishing concentration

System	Technique	Temperature	$(d\gamma_s/dX)$ (erg cm^{-2}/at% X)	Ref.
AgO	Zero creep	912–943	580	[12]
CuO	Shrinkage of voids	519	5×10^5	Present work
CuS	Radiotracer	560	1.7×10^6	[13]
CuS	Shrinkage of voids	475	1.5×10^6	Present work

X is the bulk concentration.

the copper surface [8] while oxygen and sulphur are on the contrary very strongly absorbed [9,11]. The strong influence of oxygen and sulphur on the surface tension of copper is particularly evident in table 1, which gives the derivative of the surface tension with respect to concentration for vanishing concentration. It is interesting to note for sulphur the agreement between the present data and those of Oudar [13], who used a radio tracer technique.

Moreover, though the variation of γ_s with temperature cannot be predicted for impure metals, it is interesting to note that at higher temperature Hondros [14] has found an identical lowering of the surface tension of copper foils surrounded by a low oxygen pressure, with the zero creep rate technique.

So the results of our experiments are consistent and in agreement with those of the literature. It appears that annealing of voids is a convenient method to determine the influence of additional elements on the surface tension of copper.

3. The zero creep rate experiments

At temperatures near its melting point a very thin wire tends to shorten under the combined effects of surface and intergranular tension. The wire, however, will elongate by creep if it is placed under sufficient tension. Therefore, it exists a critical load which gives a zero creep rate, and from the determination of which the surface tension can be deduced if a relation between surface and intergranular tension is known. The expression of the critical load as a function of surface and intergranular tension can be derived either by a virtual work argument [15–17] or by a stress analysis [15,18–20]. Both procedures reach the same result and give for a bamboo structure wire the formula

$$\gamma_s = \frac{r\sigma_c}{1 - (\gamma_g/\gamma_s)(r/l)} \tag{2}$$

where r is the wire radius, σ_c the critical stress, l the grain length and γ_g the intergranular tension.

3.1. Experimental procedure

All the experiments were done using a versatile and original device which allows the strain variations of the samples to be followed by measuring changes in their electrical resistance at constant volume. This device, which will be described in another paper [21], allows the determination of the strain rate to 1% for strains as small as one part in 10^4.

The wires used were of 99.999% copper and of 120 μm diameter. To produce the desired bamboo structure, the differently loaded specimens were held for two days at 1070°C in the chosen atmosphere, which was hydrogen or helium purified by passing through a palladium or a titanium oxyde cell. This recrystallization stage

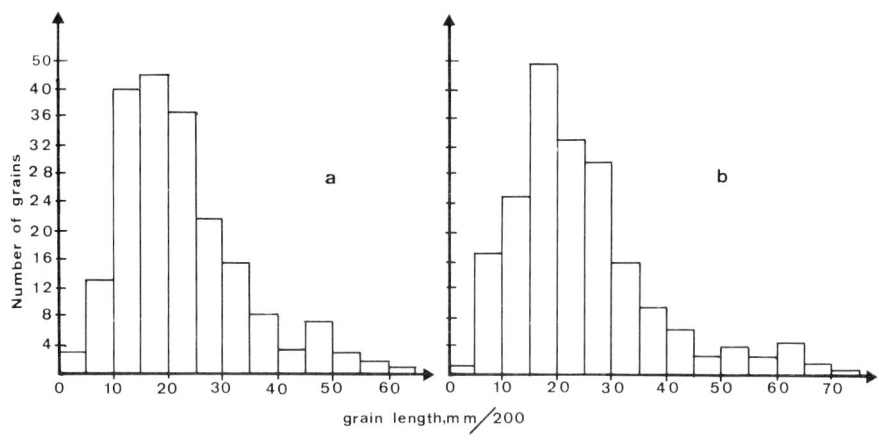

Fig. 4. Histograms of grain size: (a) after the recrystallization stage; (b) after the creep stage.

gave grains which occupied the whole cross section of the wire and whose lengths were of the order of the wire diameter. As shown by the two histograms of fig. 4, the grain size so obtained was stable during the creep study, provide the rest of the experiment was done at a lower temperature than the recrystallization stage.

3.2. Results

A set of experiments was done in a hydrogen atmosphere to eliminate the usual impurities of copper surfaces (carbon, oxygen, sulphur and chlorine), which come from the bulk or from the environment. These impurities were detected by Auger analysis on the surface and in the grain boundaries (after a thick layer was removed by sputtering in the Auger spectroscope) after the first tests in a helium atmosphere without any prior cleaning in a hydrogen atmosphere.

With our devide, we have recorded very linear creep curves such as those shown in fig. 5. From these curves we have deduced the creep rate and plotted it versus the applied stress. Following the usual procedure the critical stress was interpolated from this last plot; surface tension was then derived through the formula (2), putting $\gamma_g/\gamma_s = 1/3$ [22].

Thus we obtain $\gamma_s = 1.46$ J/m^2 at 1020°C. This value was quite reproducible and similar to those already published (cf. table 2).

Unfortunately microprobe X-ray analysis has shown silicon contamination on the surface of the wire which appears like hillocks typical of a condensation phenomenon. So to avoid as much as possible silica reduction while keeping clean surfaces the following procedure was adopted in a new set of experiments: the wire surface was cleaned by a four days anneal at 700°C in a hydrogen atmosphere, then hydrogen was evacuated and the device was filled with purified helium. With this procedure, the $\dot{\epsilon}$ versus t plot (fig. 6) was fairly good and the γ_s values were repro-

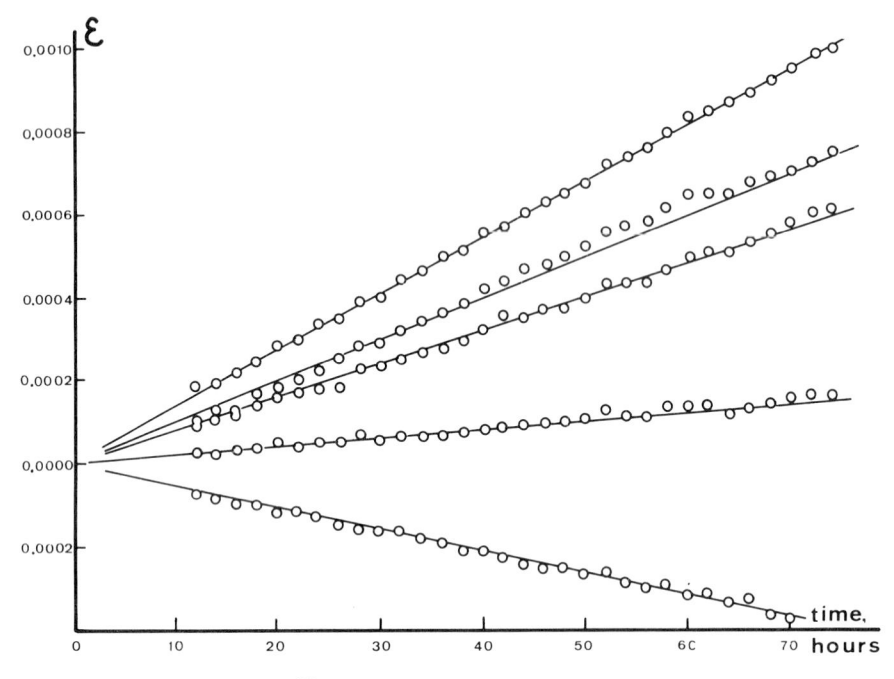

Fig. 5. Results of a typical creep test.

ducible and identical to the previous ones, but a slight silicon background was still detected on the surface by analysis after the experiment. We conclude that the measured surface tension does not change between the two sets of experiments either because: (a) saturation coverage of the surface by silicon was achieved or (b) at high temperature in such conditions silicon has no noticeable effect on the surface tension of copper. Confidence in the latter thesis is increased by the fact that the value of γ_s measured here is very near to that of Hondros whose experiment

Table 2
Surface tension of copper at high temperature

Ref.	Temperature (°C)	Atmosphere	γ_s (J/m^2)
Udin [16]	1010	Vacuum	1.425
Hoage [23]	900	Vacuum	1.755
Inman et al. [24]	950	He	1.58–1.91
Pranatis et al. [25]	1002	H_2 or He	1.8
Hondros et al. [14]	920	H_2	1.42
Present work	1020	H_2	1.46
Present work	1020	He	1.46

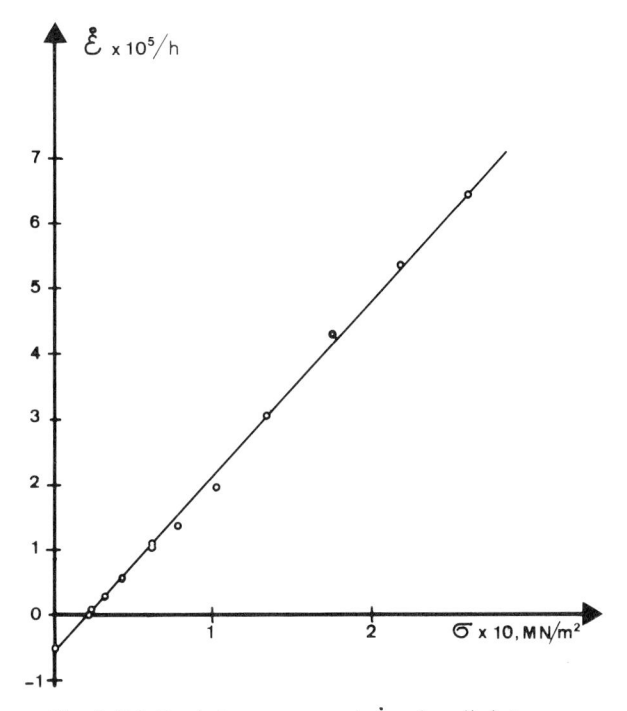

Fig. 6. Relation between creep rate $\dot\epsilon$ and applied stress σ.

employed an alumina vessel. This views is further supported by an experiment done by Cousty [26] in our laboratory. After a cleaning stage of two months in hydrogen in a silica vessel, a thick specimen of copper was placed in a Auger spectroscope and cleaned by sputtering before an anneal of 12 h at 400°C. An analysis detected no silicon, but after, the sample was cooled to room temperature and reanalysed, silicon was found to be present. It is believed therefore that silicon included in the bulk during the cleaning stage of the last set of experiments was no longer present on the surface during the creep stage and that $\gamma_s = 1.46$ J/m^2 is the surface tension of a clean surface.

4. Conclusion

Thus at high temperature near the copper melting point we have used the classical zero creep rate technique to measure the surface tension of copper while at intermediate temperature we have used the annealing of voids technique.

With the latter technique, we have shown in agreement with what is usually claimed that helium and krypton do not affect the surface tension of copper when oxygen and sulphur considerably lower it.

The two experiments give a value of surface entropy of 3.8×10^{-3} J m^{-2} K^{-1} which is somewhat higher than the values reported in the litterature and obtained with the zero creep rate technique [27]. But it is well known that surface entropy measurements is a very difficult task with the zero creep rate technique and that such measurements are not indeed very reliable.

Acknowledgements

We would like to thank C. Le Gressus (CEN, Saclay) for performing the Auger analysis.

References

[1] L.D. Glowinski and C. Fiche, J. Nucl. Mater. 61 (1976) 29.
[2] C.P. Flynn, Phys. Rev. 134 (1964) A241.
[3] A. Johnston, P.S. Dobson and R.E. Smallman, Crystal Lattice Defects 1 (1969) 47.
[4] K.H. Westmacott, R.E. Smallman and P.S. Dobson, Metals Sci. J. 2 (1968) 177.
[5] M.F. Felsen and P. Regnier, Scripta Met. (Feb. 1977) 133.
[6] H.G. Bowden and R.W. Balluffi, Phil. Mag. 19 (1969) 1001.
[7] A. Kuper, H. Letaw, L. Slifkin, E. Sonder and C.T. Tomizuka, Phys. Rev. 96 (1959) 1224.
[8] R. Sau and R.P. Merrill, Surface Sci. 34 (1973) 268.
[9] F. Gronlund and P.E. Hojlund Nielsen, Surface Sci. 30 (1972) 388.
[10] A. Oustry, L. Lafourcade and A. Escaut, Surface Sci. 40 (1973) 545.
[11] J.L. Domange and J. Oudar, Surface Sci. 11 (1968) 124.
[12] F.H. Buttner, E.R. Junk and H. Udin, J. Phys. Chem. 56 (1952) 657.
[13] J. Oudar, Thesis, Paris (1960).
[14] E.D. Hondros and M. McLean, in: Proc. CNRS Conf. on La Structure et les Propriétés des Surfaces Solides, Paris, 1969.
[15] H. Udin, A.J. Shaler and J. Wulf, Trans. AIME 185 (1949) 186.
[16] H. Udin, Trans. AIME 189 (1961) 63.
[17] D. McLean, Grain Boundary in Metals (Oxford Univ. Press, 1957) p. 58.
[18] J. Fisher and C.G. Dunn, in: Imperfections in Nearly Perfect Crystals (Wiley, New York) p. 317.
[19] H. Udin, in: Metal Interface (Am. Soc. Metals, Cleveland, Ohio, 1952) p. 114.
[20] E.D. Hondros, Proc. Roy. Soc. (London) A286 (1965) 475.
[21] M.F. Felsen, N. Sacovy and P. Regnier, J. Phys. E, submitted.
[22] M. McLean, J. Mater. Sci. 8 (1973) 571.
[23] J.H. Hoage, US Atomic Commission Report H.W. 78132 (1963).
[24] M.C. Inman, D. McLean and H.R. Tipler, Proc. Roy. Soc. (London) A273 (1963) 538.
[25] A.L. Pranatis and G.M. Pound, Trans. AIME 203 (1955) 664.
[26] J. Cousty, private communication.
[27] H. Jones and G.M. Leak, Metals Sci. J. 1 (1967) 211.

THE OXIDATION OF A Cu(100) SINGLE CRYSTAL STUDIED BY ANGLE-RESOLVED PHOTOEMISSION USING A RANGE OF PHOTON ENERGIES

D.R. LLOYD, C.M. QUINN and N.V. RICHARDSON

Chemistry Department, University of Birmingham, P.O. Box 363, Edgbaston, Birmingham B15 2TT, England

The changes which occur in the angle-dispersive photoemission spectra of a Cu(100) single crystal with increasing exposure to oxygen are described. They are interpreted in terms of an initial chemisorption stage followed by a gradual incorporation of oxygen and reconstruction of the surface region to a "Cu_2O" type layer. The use of different photon energies and angle-resolving methods facilitates an identification of the various stages by emphasising the appropriate spectral features.

1. Introduction

The oxidation of copper has been extensively studied by a variety of techniques [1–15]. Apart from some discrepancies in the quoted oxygen exposures required to reach the various oxidation stages, it seems to be quite clear that an initial stage of chemisorption is followed by incorporation of the oxygen into the metal surface and the formation of a Cu_2O structure at the surface. Very high exposures can produce a Cu(II) surface oxide which reverts to a Cu(I) oxide on heating because of the greater thermodynamic stability [11,14]. The changes in the He I (21.2 eV) and He II (40.8 eV) photoemission spectra from polycrystalline copper samples are not very marked until the relatively high exposures needed for "Cu_2O" formation are reached [13,14]. This insensitivity is due, in part, to the width of the Cu 3d band (from 2 to 5 eV below the Fermi level) in such spectra and in part to the relatively low cross-section for ionisation from states of oxygen 2p character. The use of Ne I (16.8 eV) and Ar I (11.7 eV) photons and angle-resolved emission from a single crystal reduces the structure in the Cu 3d emission region and allows the low oxygen exposure regime to be monitored more closely. At the higher oxygen exposures, when Cu_2O is thought to be present, there are features in the spectra which show an angular dependence from which information about initial state symmetries might be obtained.

2. Experimental

The experiments were performed in a spectrometer with a base pressure better than 5×10^{-11} Torr and which allows independent setting of polar photon incidence angle and polar and azimuthal electron emission angles. Full details of this instrument are provided elsewhere [16]. The photon beam is produced by a resonance lamp and is necessarily unpolarised.

The copper crystal was cut by spark machining and oriented to the (100) plane within 1°. It was cleaned by cycles of Ar^+ bombardment (800 V, 8 μA, 5 min) and annealing at 700 K. In earlier experiments with this crystal, it was exposed to oxygen several times which removes carbonaceous material from the sample surface. The oxygen (Matheson CP grade) was admitted via a needle valve and a gas-handling line. For exposures to 10^6 L the oxygen was allowed to flow through the spectrometer at pressures to 10^{-4} Torr. For higher exposures, the chamber was closed to the pumping lines.

3. Results

3.1. Clean metal

The angle-dispersed He I and Ne I excited photoemission spectra from a Cu(100) single crystal have been reported elsewhere [13,17,18]. The anisotropy of the emission is marked, and in certain directions very sharp features (FWHM ca. 120 meV) have been observed [13]. The Ar I induced, normal emission spectrum for a clean Cu(100) face is dominated by two symmetric bands at 0.5 and 2.2 eV * whose relative intensity is critically dependent on photon incidence angle [19]. Fig. 1 shows that the Ar I excited emission is also strongly anisotropic.

3.2. Low oxygen exposures

New features (at 3.1, 4.0 and 4.8 eV) appear in the Ar I excited, normal emission spectra at exposures of only 10 L and grow in intensity with increasing exposure to ca. 500 L. Fig. 2 shows the spectrum for exposures of 10 and 100 L compared with that from the clean metal. The intensity of the low energy feature (0.5 eV) decreases over this exposure range and the band at 2.2 eV broadens. Off-normal the effects are not so marked because of the greater number of bands in the clean metal spectrum. However, in appropriate directions, a shoulder can be seen to develop on the low energy side of the Cu 3d features. Similar changes are noted in the Ne I excited emission. Fig. 3 compares a clean metal Ne I spectrum with that obtained after 200 L exposure to oxygen.

---

* All energies are referred to the Fermi level (E_f).

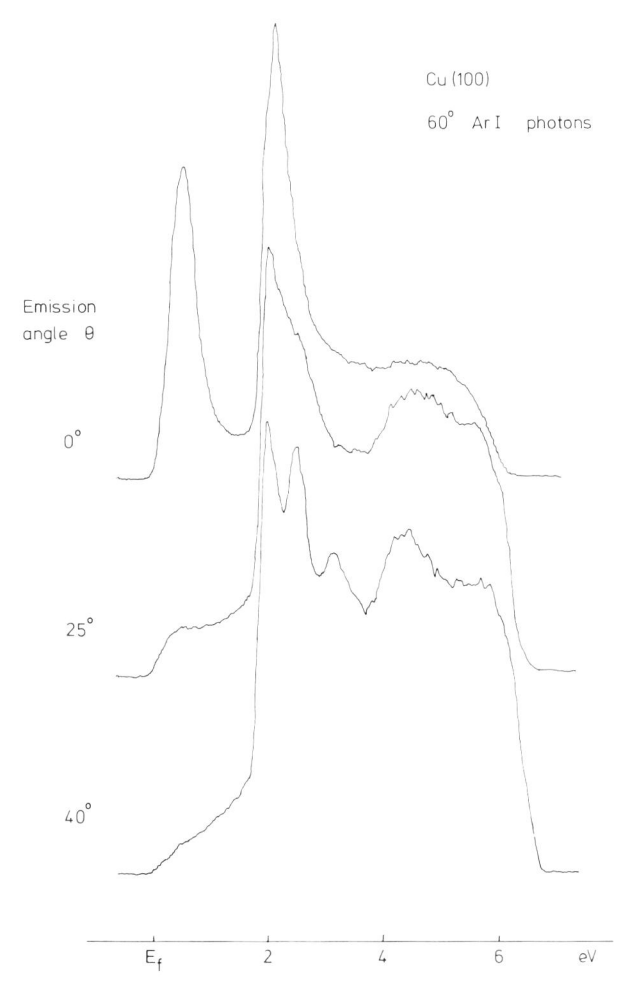

Fig. 1. The photoemission from a Cu(100) single crystal surface excited by Ar I (11.7 eV) photons incident at 60° to the surface normal. The emission is in the ΓXKX plane of the Brillouin zone.

The clean Cu(100), normal emission spectrum excited by He I is very weak in the region below the Fermi level to 2 eV. This enables the shoulder on the low ionisation energy side of the Cu 3d band to be observed at low exposures ($<10^3$ L). The other changes in the spectrum are very small. Some increase in emission can be detected at ca. 5–6 eV. It is important to note that, at exposures less than 10^3 L, the features in the Cu 3d band remain essentially unchanged.

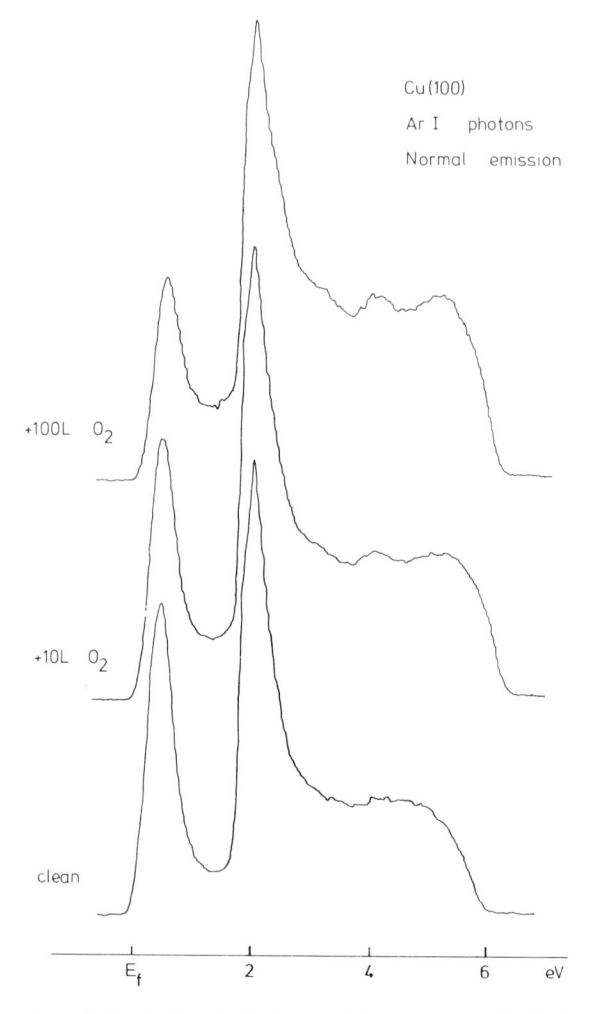

Fig. 2. A comparison of the Ar I excited photoemission spectrum obtained normal to a clean Cu(100) face with the corresponding spectra after 10 and 100 L exposure to oxygen.

3.3. Oxygen exposures 10^3-10^6 L at 10^{-4} Torr

This regime is characterised by a gradual disappearance of the fine structure in the Cu 3d energy region (fig. 4). There is a growth of the shoulder at ca. 1.6 eV and in the emission between 5 and 6 eV. Under Ar I excitation, this second feature overwhelms the relatively sharp peak at 4.8 eV in the normal emission spectrum. By 10^6 L the feature at 0.5 eV in the Ar I spectrum has disappeared entirely. For oxygen pressures below ca. 10^{-3} Torr, changes in the spectra are very gradual even to very high exposures.

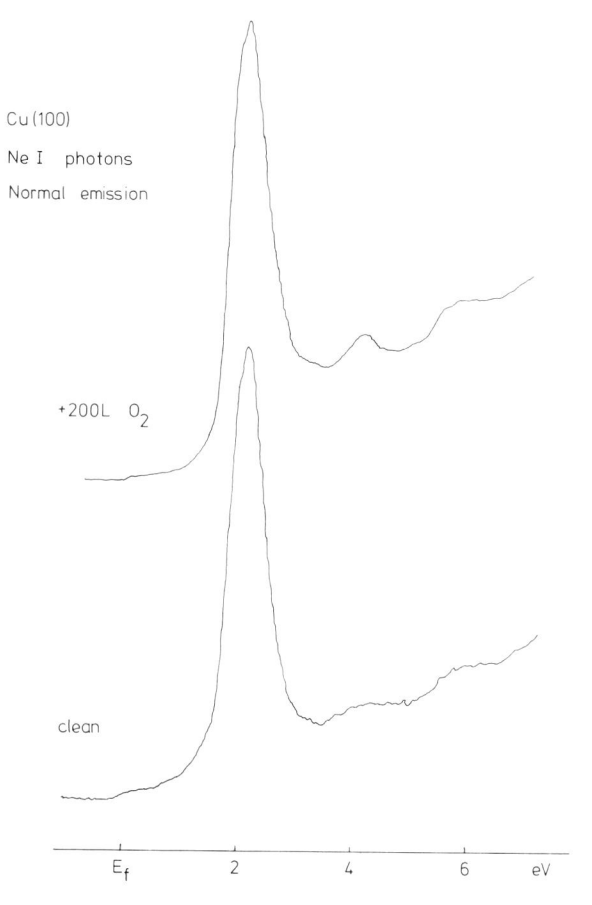

Fig. 3. A comparison of the normal emission, Ne I (16.8 eV), excited photoemission data for a clean Cu(100) face and the same face exposed to 200 L of oxygen.

3.4. Oxygen exposure at higher pressures

If the Cu(100) crystal face is exposed to oxygen at 0.5 Torr for 20 sec (10^7 L), or if the crystal is heated (680 K), after exposure to one atmosphere of oxygen, for 30 min, then distinctive spectra are obtained. They are similar to those obtained by other workers and ascribed to a Cu_2O layer [12–15]. The copper 3d emission now shows an almost symmetric peak with little angular variation. There are two spectral regions of particular interest. Overlapping features are observed at 5.0 and 6.1 eV. The relative intensity of these two features shows a marked variation with polar emission angle. Fig. 5 shows the He I excited spectra at 0° and 70° emission for photons incident at 60°. There is no detectable variation with photon incidence

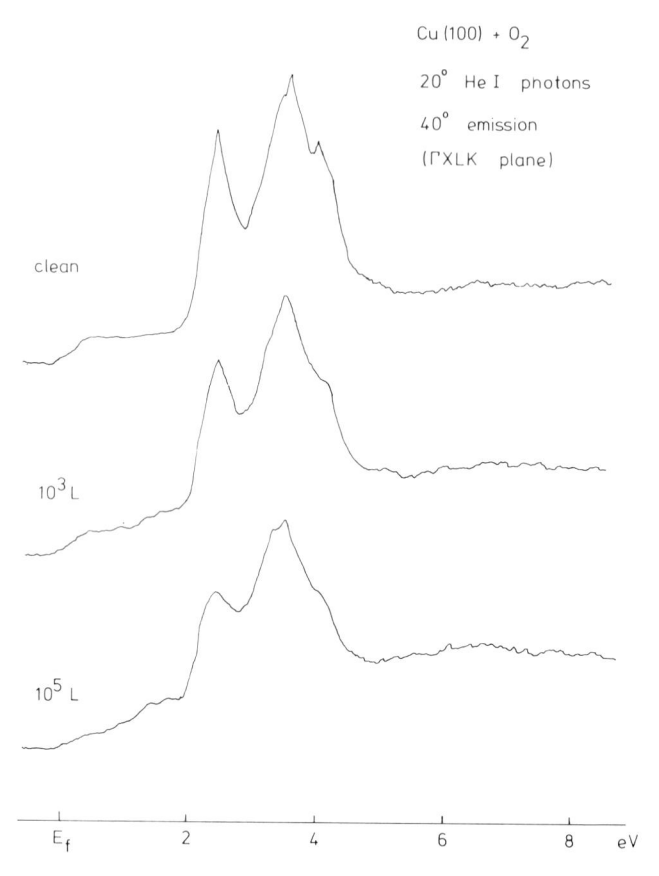

Fig. 4. A comparison of the photoemission data for a clean and oxygen exposed Cu(100) face using He I (21.2 eV) photons and an emission angle of 40° in the ΓXLK plane.

angle or azimuthal emission angle. Similar results are obtained for He II and Ne I excitation, but this region is difficult to characterise with Ar I photons because of the proximity of the zero energy cut-off. The second region of interest is that immediately below the Fermi level. The general increase in emission with oxygen exposure has been noted by other workers [12–15]. However, the spectra in fig. 5 show that, like the emission in the 5–6 eV region, there are two components whose relative intensities vary as a function of polar emission angle. The feature closest to the Fermi level decreases in intensity with increasing emission angle (fig. 5).

There is also a weak band at 8 eV in the He I and He II spectra which was noticed for oxidised polycrystalline copper by Evans et al. [12–14]. This feature is more noticeable at high emission angles.

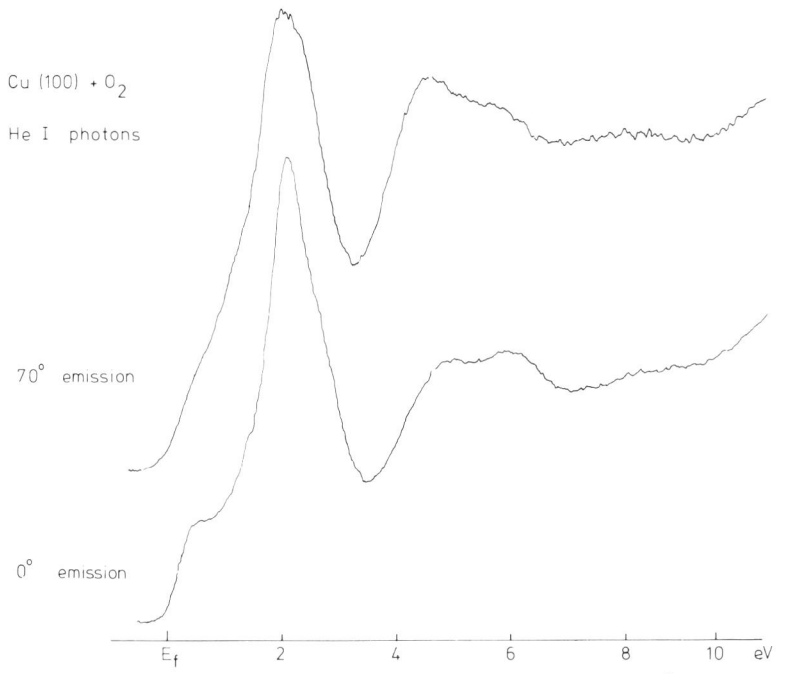

Fig. 5. The He I excited spectrum from a Cu(100) face after exposure to 10^7 L of oxygen at 0.5 Torr.

3.5. Exposure to oxygen at atmospheric pressure

The He I spectrum of a Cu(100) face exposed to one atmosphere of oxygen for thirty minutes shows only two bands centred at 2.9 and 5.5 eV. They are both broad and show no angular dependence.

4. Discussion

It has been proposed, on the basis of LEED and XPS measurements, that oxygen exposures to ca. 2×10^3 L at room temperature result only in chemisorption. Although there is no definite change in the UPS results at this exposure, below this level there is little observable change in the Cu 3d emission. However, in the angle-resolved, normal emission Ar I spectrum, new features have been observed. The peaks at 3.1 and 4.0 eV are at unlikely energies for interpretation as oxygen 2p derived structure *, and although the band at 4.8 eV is in a more acceptable energy

---

* Photoelectron bands in the 5–6 eV region are generally assigned to ionisation from O 2p derived levels [23,27,35,36].

range, its narrowness, appearance at such low exposures (<10 L) and absence from the He I spectrum all suggest that it may be related to the peaks at 3.1 eV and 4.0 eV. We propose that these three features arise from initial states of the copper 3d bands in the surface region because of changes in the photoemission mechanism. The prominent emission features in the angle-resolved spectra from the clean Cu(100) surfaces excited by Ar I, Ne I and He I radiation have been interpreted in terms of direct transitions in the bulk band structure [17–19]. The absence of prominent features corresponding to initial states lying below 3 eV in the Ar I normal emission spectrum arises because of the band gap in the relevant final state structure in the ΓX direction [20]. If the adsorption of oxygen gives rise to virtual levels in this region, then metal to oxygen transfer transitions might be available to enhance the emission from states below 3 eV. This argument gains some support from studies of secondary emission from a clean and oxygen covered single crystal of tungsten. In this work, new peaks were observed at 1–3 eV above the vacuum level when the W(110) crystal was exposed to oxygen. These features appear in a tungsten band gap.

The very weak enhancement in the emission at ca. 5.5 eV in the He I spectra at low exposures (fig. 4) and the shoulder on the low IE side of the Cu 3d band are assigned to ionisations from states of some oxygen 2p character. They are weak for all photon energies and do not become very significant until the next stage of oxidation. Certainly, it is not possible to detect the angular variations of the kind which have been predicted for chemisorbed oxygen [22,23] and which have been shown for oxygen on Ni(100) [23].

Between 2×10^3 L and 10^6 L, the LEED pattern changes to $(\sqrt{2} \times 2\sqrt{2})R45°$ [11]. The oxygen 1s signal moves to higher binding energy and broadens by 0.2 eV [11]. This has been attributed to the onset of surface reconstruction with formation of a Cu(I) oxide layer. In the UV angle-resolved photoemission, this oxygen exposure is characterised by a steady broadening of peaks and disappearance of the strong anisotropy of emission. This also suggests a rearrangement of copper atoms. The spectra can no longer be related to direct transitions in the metallic Cu lattice. Indeed, the band close to the Fermi level in the ΓX direction excited by Ar I had disappeared at an exposure ca. 5×10^4 L. As well as the Cu 3d emission changes, the low energy shoulder and the emission at 5–6 eV continue to increase in intensity and begin to show structure and an angular dependence. The peaks in the low oxygen exposure Ar I spectrum become part of the broad Cu 3d electron emission or, in the case of the 4.8 eV peak, disappear because of the increased intensity of the oxygen derived band at ca. 5.0 eV.

A more highly ordered "Cu_2O" type surface region is produced by the higher oxygen pressures (0.5 Torr) [11] or by heating a heavily oxidised surface [11,14]. The spectrum is then rather simple. The Cu 3d band under He I excitation is almost symmetrical but broadens to higher emission angles. The decrease in the intensity of the 6.1 eV feature with increasing emission angle is taken to indicate that it arises from states of predominantly oxygen $2p_z$ character (where z identifies the surface normal), and the reverse behaviour of the 5.0 eV peak indicates oxygen $2p_x$, $2p_y$ character. The feature on the low energy side of the Cu 3d band has been interpreted as the antibonding combination of the Cu 3d and O 2p levels. The bonding combination of these levels is split into two components which might be described as having σ character (6.1 eV) and π character (5.0 eV) depending on their symmetry with respect to the surface plane. It is to be expected that the antibonding combination should also contain σ and π components. The angular dependence of the emission above the Cu 3d band suggests that the σ components lies closer to the Fermi level than the π component, as expected from the ordering of the bonding combination.

The band at 8.0 eV in the He I spectrum at this oxidation stage is of some interest. It seems to be rather too high in ionisation energy for a band of predominantly oxygen character. There seem to be three other possibilities. It might be a contaminant peak. This seems unlikely since it has been observed in XPS experiments, also, by several workers using different instruments [12,14,24–26] and always at the same relative intensity. A similar peak is present at 9.5 eV in the spectrum of nickel surfaces exposed to oxygen in quantities sufficient to produce a thin oxide film. Secondly, the structure might arise from a multi-electron process [24,25]. X-ray induced ionisation of core-levels shows such a satellite, and a feature at 8 eV in the valence region of transition metal oxides has been assigned to multi-electron excitation [28]. However, in the oxides, this peak is 2–3 eV wide and decreases rapidly in intensity with decreasing photon energy over the range (100–20 eV). It would be unlikely that it could be observed at He I energy. The third possible explanation is that this feature arises from the high density of initial states, at the centre of the Cu_2O Brillouin zone, which are predominantly copper 4s in character [29]. For the clean metal, they are predicted [20] to lie at about 8.5 eV but are not observed, presumably because of the low cross-section for ionisation of s bands with He I or He II photons. An admixture of oxygen 2p character would increase the ionisation cross-section.

The two peak spectra produced by very high oxygen exposures which show little angular variation are thought to arise from a Cu(II) oxide surface layer. The X-ray PE spectra of copper treated in this way reveal satellite structure on the Cu 2p peaks [11,14] which is usually taken to indicate the presence of Cu(II) species [30–34].

Conclusions

The usefulness of angle-resolved photoemission studies with different photon energies for the investigation of the oxidation of a copper single crystal has been demonstrated. Three interesting new observations have been made. In the initial stages of exposure to oxygen, enhanced emission from certain initial states of the clean metal band structure is attributed to the occurrence of new, oxygen derived, final states in a one-dimensional band gap. In the later stages of oxidation when Cu_2O is thought to be present, the angular dependence of the new emission (5–6 eV) allows an assignment to initial states of different symmetries. The related emission from the corresponding antibonding combination of levels shows a similar angular dependent behaviour and this has been used to assign initial state symmetries to these states also. Finally, the possibility that the band at 8 eV in the "Cu_2O" surface can be assigned to the bottom of the upper 4s band has been explored.

References

[1] R.N. Lee and H.E. Farnsworth, Surface Sci. 3 (1965) 461.
[2] G. Ertl, Surface Sci. 6 (1967) 208.
[3] G.W. Simmons, D.F. Mitchell and K.R. Lawless, Surface Sci. 8 (1967) 130.
[4] A. Oustry, L. Lafourcade and A. Escaut, Compt. Rend. (Paris) B274 (1972) 1402.
[5] A. Oustry, L. Lafourcade and A. Escaut, Surface Sci. 40 (1973) 545.
[6] A. Oustry, L. Lafourcade and A. Escaut, Compt Rend. (Paris) B278 (1974) 189.
[7] K. Okada, T. Matsushika, H. Tomita, S. Matoo and N. Takahashi, Shinkū 13 (1970) 371.
[8] L. McDonnell and D.P. Woodruff, Surface Sci. 46 (1974) 505.
[9] L. McDonnell, D.P. Woodruff and K.A.R. Mitchell, Surface Sci. 45 (1974) 1.
[10] C.M. Quinn and M.W. Roberts, Trans Faraday Soc. 60 (1964) 899.
[11] M.J. Braithwaite, R.W. Joyner and M.W. Roberts, Discussions Faraday Soc. 60 (1975) 89.
[12] S. Evans, D.E. Parry and J.M. Thomas, J.C.S. Faraday Discussions 60 (1975) 103.
[13] Discussion remarks, J.C.S. Faraday Discussion 60 (1975) 137.
[14] S. Evans, J.C.S. Faraday II, 71 (1975) 1044.
[15] L.F. Wagner and W.E. Spicer, Surface Sci. 46 (1974) 301.
[16] D.R. Lloyd, C.M. Quinn, N.V. Richardson, D. Latham, A. Jones and P.M. Williams, submitted for publication.
[17] D.R. Lloyd, C.M. Quinn and N.V. Richardson, J. Phys. C8 (1975) L371.
[18] L. Ilver and P.O. Nilsson, Solid State Commun. 18 (1976) 667.
[19] D.R. Lloyd, C.M. Quinn and N.V. Richardson, Solid State Commun. 22 (1977) 721.
[20] G.A. Burdick, Phys. Rev. 129 (1963) 138.
[21] R.F. Willis, J.C.S. Faraday Discussion 60 (1975) 245.
[22] A. Liebsch, Phys. Rev. B13 (1976) 544.
[23] K. Jacobi, M. Scheffler, K. Kambe and F. Forstmann, Solid State Commun. 22 (1977) 17.
[24] G.K. Wertheim and S. Hüfner, Phys. Rev. Letters 28 (1972) 1028.
[25] S. Hüfner and G.K. Wertheim, Phys. Rev. B8 (1973) 4857.
[26] R.F. Roberts, J. Electron Spectr. 4 (1974) 273.
[27] D.E. Eastman and J.K. Cashion, Phys. Rev. Letters 27 (1971) 1520.
[28] D.E. Eastman and J.L. Freeouf, Phys. Rev. Letters 34 (1975) 395.
[29] L.F. Mattheiss, Phys. Rev. B5 (1972) 290.
[30] D.C. Frost, A. Ishitani and C.A. McDowell, Mol. Phys. 24 (1972) 861.
[31] T. Robert, M. Bastel and G. Offergeld, Surface Sci. 33 (1972) 123.
[32] T. Novakov and R. Prins, Solid State Commun. 9 (1971) 1975.
[33] G. Schon, Surface Sci. 35 (1973) 96.
[34] K.S. Kim, J. Electron Spectr. 3 (1974) 217.

PLASMA RESONANCE ABSORPTION IN INTERFACIAL PHOTOEMISSION FROM VERY THIN SILVER FILMS ON Cu(111)

J.K. SASS, S. STUCKI and H.J. LEWERENZ

Fritz-Haber-Institut der Max-Planck-Gesellschaft, Faradayweg 4-6, D-1000 Berlin 33, Germany

Very thin silver films deposited on a Cu(111) substrate were studied with the photoemission-into-electrolyte technique. The optical resonance absorption at the silver bulk plasma frequency was observed for films thicker than two monolayers. The correlation between optical absorption and photoemission intensities depends in a complicated fashion on the photon energy as well as the film thickness.

1. Introduction

Although there have been several experimental observations of the long-wavelength optical plasma resonance in thin metal films [1], a minimum critical film thickness for the onset of this effect has as yet not been determined. Obviously, such a determination requires the preparation of parallel-sided films in the monolayer range and accurate control of their thickness. In the present investigation, these requirements were to a large extent fulfilled by depositing thin silver films electrochemically [2] on a very smooth Cu(111) substrate.

It has been demonstrated that the enhanced optical absorption in the vicinity of the bulk plasma frequency manifests itself also in photoemission experiments [3–5]. By nature of its surface sensitivity, photoemission appears to be particularly suited for the investigation of very thin films. In the case of silver the plasma resonance ($\hbar\omega_p \sim 3.8$ eV) is located energetically below the vacuum work function. By means of the photoemission-into-electrolyte technique [6], however, the effect may still be studied because the threshold can be considerably lowered at a metal–electrolyte contact. This special feature of interfacial photoemission [7] has previously been exploited in a study of thick Ag(111) films [8].

The optical excitation of the plasma resonance in thin metal films occurs only with p-polarized light. The absorption of p-polarized light A_p in the film may exceed that of s-polarized light A_s by more than an order of magnitude. The ratio A_p/A_s is particularly high for very thin films and approaches the rather low values of bulk material with increasing thickness. This strong film-thickness dependence provides a unique opportunity for the study of the correlation between the absorption ratio A_p/A_s and the corresponding vector ratio Y_p/Y_s of the photoemission yields in the same metal.

2. Experimental

Smooth Cu(111) substrates were prepared [9] by evaporation onto air-cleaved mica at temperatures of 250–300°C in a standard high vacuum system ($p \sim 1 \times 10^{-6}$ Torr, deposition rate $\sim$50 Å/s). In the electrochemical cell silver films are deposited onto these substrates by dissolving a separate "generator" electrode electrochemically in controlled amounts. A special arrangement for both electrodes [2] ensures that all the resulting silver ions are homogeneously deposited onto the copper and the coverage can then be determined by recording the charge associated with the electrochemical deposition. In vacuo, the ionization method for monitoring alkali metal atoms is based on the same principle. The thicknesses are quoted in terms of monolayers (ML) in the following section on the assumption that the silver deposits grow in a layer-by-layer fashion (Frank–Van der Merwe mechanism) and that they exhibit a (111)-orientation [10]. The single crystal character was confirmed by taking electron micrographs of the deposits. Experimental details relating to the photoemission yield measurements in an electrochemical cell have recently been given elsewhere [9]. The adjustable emission threshold [7] was kept at $\sim$3.0 eV in all measurements.

3. Results and discussion

The photoemission properties of thick Cu(111) films at low photon energies have been described in a recent paper [9]. Fig. 1a shows the spectral dependence of the total yields obtained at oblique incidence of the plane-polarized light. The yields Y_p and Y_s differ appreciably and the vector ratio Y_p/Y_s exhibits a very characteristic spectral structure which has been interpreted in terms of the copper bulk band-structure. In the present investigation the emission from the copper represents the interfering background which, fortunately, is quite low in the energy range below 4.1 eV [9].

Besides the emission from the blank substrate fig. 1b also shows the two yield spectra which one obtains when three monolayers of silver are deposited onto the copper. The feature of interest in the present study is the shoulder in the yield Y_p appearing in the vicinity of the silver bulk plasma frequency of 3.8 eV which is not observed in the yield spectrum of s-polarized light. We attribute this effect to the optical excitation of a long-wavelength plasma resonance in this very thin silver film.

Compared to fig. 1a this resonance manifests itself more clearly in the vector ratio Y_p/Y_s. In fig. 1b spectra of Y_p/Y_s are shown for silver coverages between two and four monolayers. The effect appears now as a pronounced peak and increases rapidly as the film becomes thicker. When the coverage is below approximately two monolayers the vector ratio more or less resembles the structure of the blank copper although changes in the yields Y_p and Y_s of the order of 20% are observed

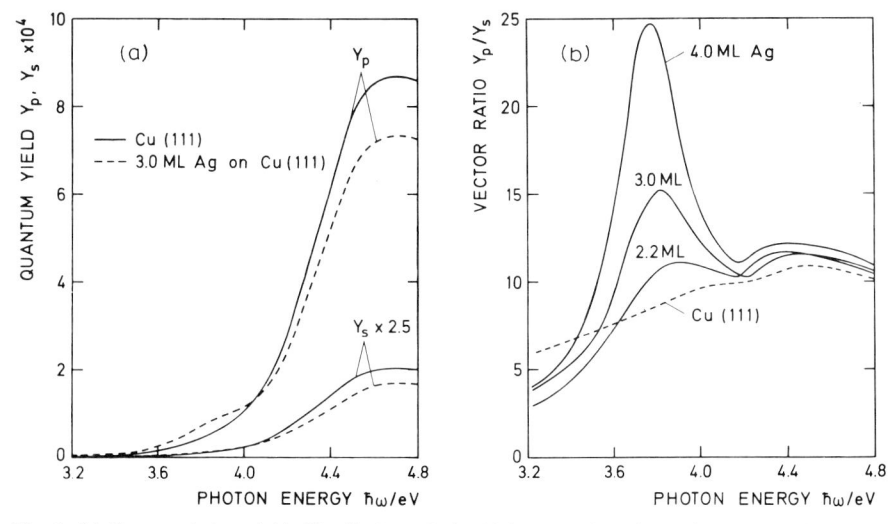

Fig. 1. (a) Photoemission yields Y_p, Y_s from Cu(111) for p- and s-polarized light and effect of a thin, electrochemically deposited silver film; angle of incidence $\alpha = 67.5°$, threshold ~3 eV. (b) Appearance of the plasma resonance absorption in the vector ratio Y_p/Y_s for very thin silver films, $\alpha = 67.5°$.

[11]. Thus, we may conclude that the minimum critical film thickness for the occurrence of the resonance is two monolayers (~5 Å).

When still more silver is deposited onto the copper the vector ratios continue to increase and at 15 monolayers, for example, a spectrum of Y_p/Y_s as shown in fig. 2a is obtained (full line). The resonance maximum is now located energetically at 3.72 eV, approximately 0.05 eV below the value of the bulk plasma frequency which is usually quoted [1]. A possible origin of this low value can be found when one assumes that the electrodeposited films slightly deteriorate at larger thicknesses. Such a correlation * of the film quality and the energetic position of the resonance maximum has been observed by Schulz and Zurheide [12]. To verify this hypothesis 15 monolayers of silver were evaporated onto the copper substrate in the vacuum chamber. The spectrum obtained from such a film is also shown in fig. 2a and the maximum now occurs at about 3.78 eV.

The peak in the photoemission spectrum is clearly the result of an enhanced optical absorption in the thin silver film. In a simple approximation this absorption may be calculated by use of the bulk optical constants of silver [13]. Such an approach obviously neglects the contribution from longitudinal fields to the optical absorption [14]. In addition, when comparing this absorption to the photoemission

---

* A somewhat poorer film quality leads to a reduced electron life time, i.e., an increased internal damping which shifts the resonance frequency to lower energies.

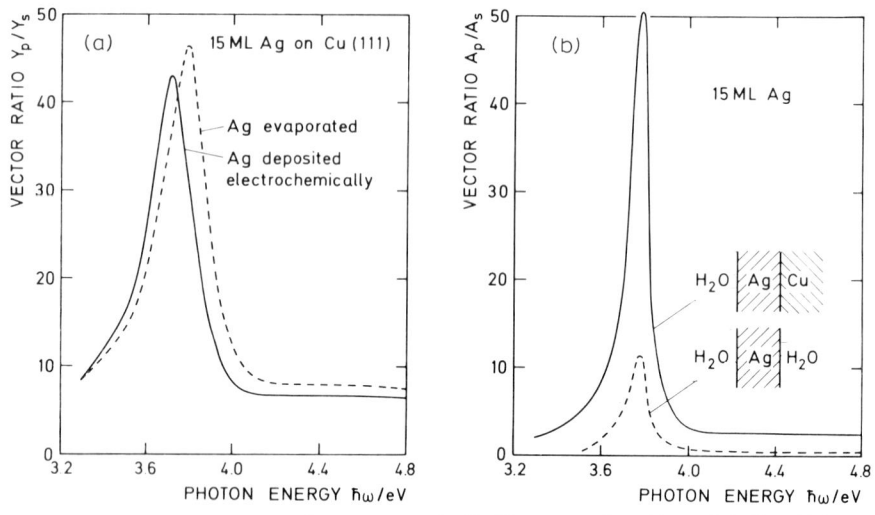

Fig. 2. (a) Spectral dependence of the vector ratio Y_p/Y_s in two silver films of similar thickness (15 monolayers); $\alpha = 67.5°$. The peak shift to lower energies in the electrodeposited film is attributed to an increase of the internal damping due to structural imperfections. (b) Calculated [15] absorption ratios A_p/A_s in a silver film of 15 monolayers bounded by water and copper (———) and water on both sides (- - - - -) as indicated in the figure; $\alpha = 67.5°$.

yield the anisotropic character of the photoelectron excitation process has to be taken into account.

Keeping these limitations in mind, we have calculated the "classical" absorption of p-polarized and s-polarized light in a silver film of 15 monolayers. In fig. 2b the resulting absorption vector ratios A_p/A_s are shown for two different cases as indicated in the figure. The interesting result is that in a thin film, bounded on both sides by water, the resonance in A_p/A_s is not as pronounced as on a copper substrate. This is largely due to a substantially smaller A_s in the latter case.

The overall impression when comparing figs. 2a and 2b is one of general similarity of both the measured photoemission vector ratios and the calculated absorption ratios. Two decades ago, Thomas and Mayer [4] obtained similar results in an investigation of thin potassium films and concluded that the photoemission yield is always strictly proportional to the volumetric [15] absorption in a region near the surface which is determined by the electron escape length. In the very thin films considered here the volumetric absorption varies essentially linearly across the film and in the simplest photoemission theory [4,8,15] the vector ratio Y_p/Y_s is then expected to be approximately equal to A_p/A_s. In fig. 3 the angle-of-incidence dependence of the measured A_p and the calculated A_p are compared for the evaporated film of fig. 2a and the overall spectral agreement seems again satisfactory.

However, a detailed analysis shows that the optical absorption cannot provide the whole photoemission story. Comparing the half-widths of the peaks in fig. 2

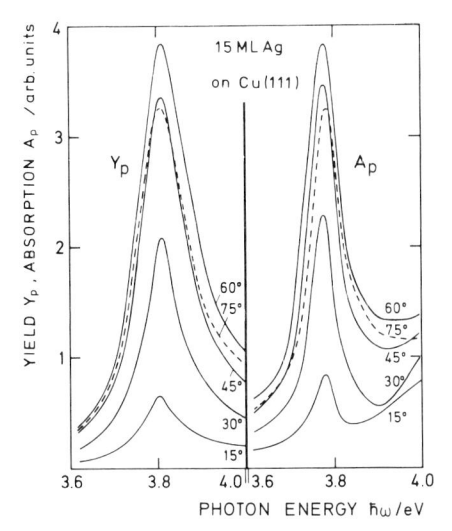

Fig. 3. Comparison of the yield Y_p in a silver film of 15 monolayers with the absorption A_p at various angles of incidence.

large values of Y_p/Y_s are seen to occur at higher energies than the A_p/A_s curve would suggest. This fact is most clearly demonstrated by evaluating the expression $(Y_p/Y_s)/(A_p/A_s)$ which, according to the simple picture, should equal unity. Fig. 4a reveals that this quantity is highly structured and ranges between one and four. For comparison the same quantity obtained for a thick silver film [8] is also shown in fig. 4a and the deviations from unity are even larger in this case. Thus fig. 4a indicates that the correlation of Y_p/Y_s and A_p/A_s depends also on the thickness in a complicated fashion.

Supporting evidence for this conclusion is also found in fig. 4b where the dependence of Y_p/Y_s and A_p/A_s on film thickness is shown for two photon energies near the resonance maximum. In accordance with the conclusions of ref. [12] that, regardless of the film quality, the shape of the transmission curve remains essentially the same (see also fig. 2a) we have shifted the energy scale in the A_p/A_s calculation by 0.05 eV for the electrodeposited films. As pointed out earlier in connection with fig. 1b, the vector ratio Y_p/Y_s in fig. 4b is seen to rise steeply at about two monolayers where the onset of plasma resonance absorption occurs. The slope of this rise seems also to indicate that the emission from the Cu(111) substrate is more rapidly suppressed than would be expected from the usual mean-free-path arguments because the electron wave-matching conditions may impose serious restrictions on the number of photoelectrons crossing the interface from the copper to the silver.

Even when accounting for these aspects the correspondence between Y_p/Y_s and A_p/A_s is very poor. The fact that the photoemission intensities do not resemble the

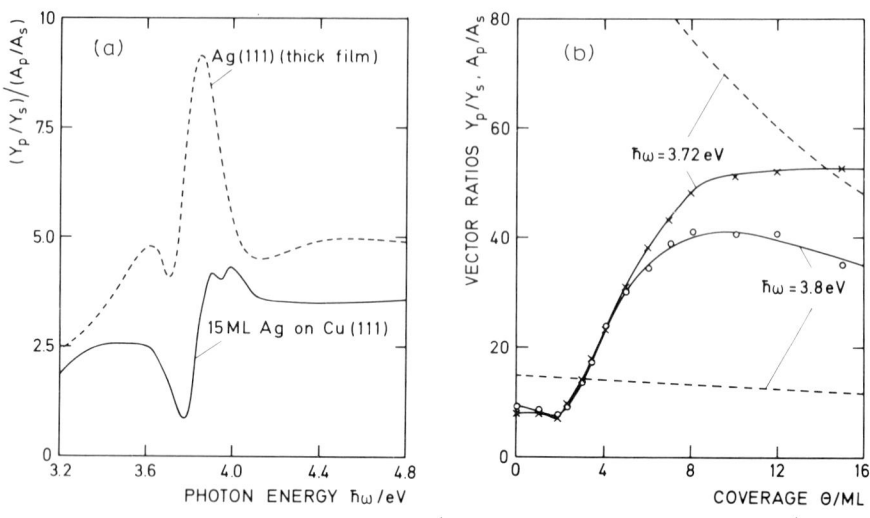

Fig. 4. (a) Spectral dependence of the vector Y_p/Y_s referred to the absorption ratio A_p/A_s for a silver film of 15 monolayers and a thick Ag(111) film; $\alpha = 67.5°$. (b) Film thickness dependence of the vector ratios Y_p/Y_s (———) and the calculated absorption ratios A_p/A_s (-----) for two photon energies; $\alpha = 67.5°$ (see text).

optical absorption reflects the anisotropic character of the photoelectron excitation mechanism [16] in a metal. In recent analysis of vector ratios in thick films it has been shown [8,9] that both the optical response of the metal and the details of the metal band structure require consideration. Particularly in the case of silver it is expected that nonlocal effects again play an important part in the thin films [14,17]. Further analysis of the results presented in this paper must therefore await a correct treatment of the tranverse and longitudinal electromagnetic fields in thin silver films. In addition, the photoelectron excitation mechanism leading to band gap emission from Cu(111) and Ag(111) at low photon energies [9] is not fully resolved.

Acknowledgement

The authors would like to thank Prof. Dr. H. Gerischer for his support and many stimulating discussions. The expert technical assistance of E. Piltz is gratefully acknowledged.

References

[1] See for example: W. Steinmann, Phys. Status Solidi 28 (1968) 437;
    H. Raether, in: Physics of Thin Films, Vol. 7 (Academic Press, 1977) p. 145.

[2] J.K. Sass and S. Stucki, to be published.
[3] H.E. Ives and H.B. Briggs, J. Opt. Soc. Am. 28 (1938) 330.
[4] H. Mayer and H. Thomas, Z. Physik 147 (1957) 419.
[5] M. Skibowski, B. Feuerbacher, W. Steinmann and R.P. Godwin, Z. Physik 211 (1968) 342.
[6] G.C. Barker, A.W. Gardner and D.C. Sammon, J. Electrochem. Soc. 113 (1966) 1182.
[7] J.K. Sass and H. Gerischer, Interfacial Photoemission, ch. 16 in: Photoemission from Surfaces, Eds. B. Feuerbacher et al. (Wiley–Interscience, 1977).
[8] J.K. Sass, H. Laucht and K.L. Kliewer, Phys. Rev. Letters 35 (1975) 1461.
[9] H. Laucht, J.K. Sass, H.J. Lewerenz and K.L. Kliewer, Surface Sci. 62 (1977) 106.
[10] E. Bauer and H. Poppa, Thin Solid Films 12 (1972) 167;
K.A.R. Mitchell, D.P. Woodruff and G.W. Vernon, Surface Sci. 46 (1974) 418;
M.J. Gibson and P.J. Dobson, J. Phys. F: Metal Phys. 5 (1975) 1828;
H. Neddermayer, Habilitationsschrift, Universität München (1976).
[11] J.K. Sass, H. Laucht and S. Stucki, in: Proc. Intern. Symp. on Photoemission, Noordwijk, The Netherlands, Eds. R.F. Willis et al. (European Space Agency, 1976).
[12] D. Schulz and M. Zurheide, Z. Physik 211 (1968) 165.
[13] G.B. Irani, T. Huen and F. Wooten, J. Opt. Soc. Am. 61 (1971) 128.
[14] W.E. Jones, K.L. Kliewer and R. Fuchs, Phys. Rev. 178 (1969) 1201;
A.R. Melnyk and M.J. Harrison, Phys. Rev. B2 (1970) 835.
[15] S.V. Pepper, J. Opt. Soc. Am. 60 (1970) 805.
[16] See for example: B. Feuerbacher and R.F. Willis, J. Phys. C9 (1976) 169; and ref. [7].
[17] I. Lindau and P.O. Nilsson, Phys. Scripta 3 (1971) 87;
M. Anderegg, B. Feuerbacher and B. Fitton, Phys. Rev. Letters 27 (1971) 1565.

CALCULATIONS OF RUMPLING IN THE (100) SURFACES OF DIVALENT METAL OXIDES

M.R. WELTON-COOK and M. PRUTTON

Department of Physics, University of York, Heslington, York, England

Differential relaxation of the anion and the cation has been calculated using a simple shell model for the (100) surface of the divalent metal oxides MgO, NiO, CaO and SrO. Only the surface layer of atoms is allowed to relax. Pairwise interaction energies are used. The polarisation of the surface ions is related to the force constants and shell charges of the shell model in a new way. The results give a surface polarisability which is different from that in the bulk. The calculations will be described and results for MgO(100) and NiO(100) will be compared with the results obtained from LEED theory/experiment comparisons.

FURTHER DYNAMICAL AND EXPERIMENTAL LEED RESULTS FOR A CLEAN W{001}-(1 × 1) SURFACE STRUCTURE DETERMINATION

M.K. DEBE, David A. KING and F.S. MARSH

The Donnan Laboratories, The University of Liverpool, P.O. Box 147, Liverpool L69 3BX, England

Full dynamical layer-doubling calculations have been made for comparison with precision LEED spectra for the clean W{001}-(1 × 1) surface at approximately 470 K. Using 45 beams and 10 phase shifts, multi-layer spacing calculated spectra are critically compared with 12 experimental curves involving 5 different beams and 5 incidence angles. Both visual judgements and a semi-quantitative peak deviation/penalty evaluation yield the same result. A surface-bulk layer spacing of 1.51 ± 0.05 Å is concluded, a 4.4% contraction, in contrast to the most recent other determination of 1.40 ± 0.03 Å. This analysis re-emphasizes the need for a reliable and objective criterion for comparing observed and calculated LEED spectra, and corrects a potentially important input to the analysis of more complex systems. For example, at low temperatures (<370 K) the W{001} clean surface rearranges to a c(2 × 2) structure.

1. Introduction

Low-energy electron diffraction (LEED) dynamical calculations are proving to be more and more successful at uniquely determining the surface structures of both clean surfaces and overlayer/substrate structures [1]. Adsorption on W{001} has been extensively investigated by a variety of techniques, and it is important to characterise the clean W{001} surface structure precisely, both because of its widespread use as a substrate and because of the recently established c(2 × 2) clean surface rearrangement that occurs near room temperature [2].

Relatively little surface crystallography has been done on the body-centred cubic materials. On W{001}, the original LEED experimental work was done by Wei [3] and theoretical calculations for the purpose of structural analysis were first carried out by Van Hove and Tong [4]. A surface layer spacing contracted 0.1 ± 0.1 Å from the bulk value was concluded, though on the basis of only 3 beams and a limited energy range. A more recent investigation, using different experimental data and based on a comparison of 4 beams over a more extensive energy range [5], gave a top layer spacing reduced by 0.18 ± 0.03 Å. Although both were done by some of the same investigators and the differences ascribed to the degree of reliability of the different data bases, the current work shows that the ambiguity was not correctly resolved.

We have used full dynamical LEED calculations in comparison with a larger amount of precise data, consisting of 12 spectra involving 5 different beams and 5 incidence angles, to show that the W{001}-(1×1) clean surface is contracted 0.07 ± 0.05 Å, in contrast to the most recent other determination [5] of 0.18 ± 0.03 Å.

2. Experimental

Of the many experimental and calculated spectra obtained, the relevant ones used in the analysis were: $\theta = 0°$; $0\bar{1}$, $\bar{1}\bar{1}$, $\bar{1}\bar{2}$, $\bar{2}0$: $\theta = -5°$; $0\bar{1}$, $1\bar{1}$, $0\bar{2}$: $\theta = +9°$; $0\bar{1}$, 00: $\theta = 14°$; $0\bar{1}$, 00, and $\theta = +18°$; 00. Fig. 1 identifies the beams, the axis of θ-rotation, and the azimuth ϕ (equal to 90° for all beams).

The measurements were performed in an all-metal, ion plus Ti-sublimator pumped UHV chamber, containing a Varian 4-grid LEED diffractometer with moveable Faraday cup (FC). The ion pump was well shielded magnetically and Helmholtz coils were used to null out stray magnetic fields. The 0.5 ×0.5 ×0.15 cm crystal was supported by spring-loaded 0.05 cm diameter W wires through two parallel, X-ray oriented holes, spark-eroded through its centre along the [010] direction. A W–W75Re25% thermocouple junction was inserted well into a spark-burned hole in the top of the crystal.

The primary beam current was regulated above ~46 eV to 0.51 μA, and the data below 46 eV was normalized when the data was digitized. The electron beam focus and FC suppressor voltage were optimized to yield the maximum current consistent with the minimum angular full-width-at-half-maximum (FWHM) of a diffracted beam. The result was a FC to filament bias of 0.0 V and an angular FWHM of 2.8 ± 0.1° for the specular beam at ~85 eV. The contact potential difference was measured to be 0.0 ± 0.1 eV.

By obtaining identical non-specular spectra for symmetrically equivalent beams at normal incidence and identical (00) spectra for incidence angles equivalent by reciprocity, up to ±18°, the conditions for true normal incidence were established

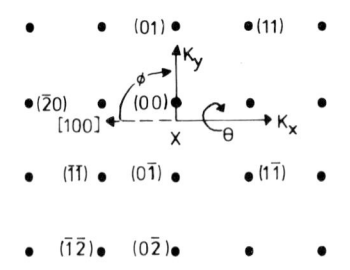

Fig. 1. Diffraction pattern from W{001} defining the beam nomenclature and the axis of rotation. The × marks the position of the electron gun for an angle of incidence of +θ°, with ϕ defined as the angle between the plane of incidence and the [100] direction in the crystal.

and complete stray magnetic field trimming was verified [6]. Hence the absolute calibration of the incidence angles is believed accurate to within $\pm\frac{1}{2}^\circ$ for the $|\theta| \leq 5^\circ$ spectra $\pm 1^\circ$ for the $9^\circ \leq |\theta| \leq 14^\circ$ spectra and $\pm 1\frac{1}{2}^\circ$ for $|\theta| = 18^\circ$.

The crystal had previously been cleaned of bulk carbon by prolonged heating in oxygen in a glass system. In the metal system it was found sufficient to clean the crystal by heating it (by electron bombardment) to ~1900 K in front of a molecular beam of oxygen followed by flashes to over 2500 K. Ultimate Auger C(270 eV) to W(350 eV) peak-to-peak height ratios of $\leq 1/10$ after a cleaning flash implied [7] a coverage of $\leq 1.5\%$ monolayer of residual carbon.

The surface exhibited two types of spot-splitting defects at various locations: twist dislocation boundaries and steps. The twist dislocation was also observed in bulk Laue photographs, which gave a 2° relative azimuthal rotation of the two crystallite domains. (The crystal mounting and alignment holes were burned at an azimuth midway between the [010] direction of each domain, producing an automatic 1° error in ϕ.) The steps were observed on fewer areas of the crystal, and the terrace width increased from one side to the other. However, places of high order on both twist dislocation domains were easily found and the IV spectra for nonspecular beams at normal incidence on the two domains were identical in every detail. Also, the calculated (00) beam spectra at $\theta = 18^\circ$ for $\phi = 87.5^\circ$ and $\phi = 90^\circ$ were visually identical.

The spectra were all taken between 4 min and 24 min after a regenerative cleaning flash to >2500 K, for which a mean crystal temperature of ~470 K was determined. These times were determined as the interval within which the maximum change in any relative peak height due to cooling or H_2 adsorption was less than 10%. The specular beam spectra could be taken in ~2–3 min, but the mode of nonspecular beam measurement, sweeping the FC back and forth across the beam position while energy sweeping at 0.25 eV/sec, required up to 10–20 min. A total ionisation gauge pressure (uncorrected for X-rays) of ~3×10^{-11} Torr was standard without liquid-nitrogen cooling the Ti-sublimator shroud and half that or less with the shroud chilled. LEED spectra taken over one year apart, for the same incidence conditions, were identical in minute detail despite a very large number of heat treatments and annealings over this period.

3. Theoretical calculations

The calculated spectra were computed using Pendry's [8] basic layer-doubling routine, adapted for multi-distance and multi-overlayer calculations, temperature dependent phase shifts, and inclusion of more beams and phase shifts [9]. The phase shifts used [10] were those by Van Hove and Tong [4] calculated from a Mattheis potential. The crystal potential was the usual muffin-tin, with the no-reflection matching condition used to describe the surface–vacuum boundary. Ten phase shifts were used throughout, with 37 beams included up to 127 eV and 45 beams for higher energies. The difference between normal incidence spectra calculated with 37 beams and 45 beams over the range 127 to 139 eV was very small, indicating that 45 beams were sufficient up to 180 eV.

The "inner potential", V_0, was taken as energy independent and equal to -12 eV, that being the value yielding the deepest minimum in a plot of the beam-averaged (using 12 beams) magnitude of the deviation between calculated and experimental peak positions versus surface layer spacing (see section 4.2). Three values of the imaginary part of the bulk potential, β, the damping parameter, were tried for single distance calculations at normal incidence, viz. -2.7, -4.1 and -5.4 eV. In the final multi-distance calculations the value of -5.4 eV was used. Additional calculations using -2.6 eV at normal incidence up to 127 eV indicate an energy-dependent damping parameter would perhaps be better.

Separate Debye temperatures were used to describe the bulk and surface atom vibrations and two different values for each were tried; 380 and 450 K for θ_D^{bulk} and 268 and 318 K for θ_D^{surf}, where in each case $\theta_D^{bulk} = \sqrt{2}\, \theta_D^{surf}$. The 450 K value was determined from experimental log I versus T plots, and since the different values had little effect on the single distance test spectra, the values chosen for the complete multi-distance calculations were 450 and 318 K respectively for θ_D^{bulk} and θ_D^{surf}. Finally, the spectra were calculated for a crystal temperature of 470 K, the average temperature when cooling to room temperature for a spectrum started 4 min after a flash to 2500 K.

The differences of the above non-structural parameter choices from those used by Van Hove and Tong, viz. $V_0 = -10$ eV, $\beta = -5$ eV, $\theta_D^{bulk} = \theta_D^{surf} = 320$ K and $T = 550$ and 300 K, for references [4] and [5] respectively, can probably completely account for the negligible visual differences between their published spectra and ours. The accuracy of the present routines was verified by comparing the calculated values for nickel {001} with the corresponding test values supplied by Pendry [8] for both RFS and layer doubling schemes, applied to both substrate and overlayer programs.

Spectra were calculated for nine spacings about the bulk spacing of 1.58 Å for normal incidence, and for five spacings centered on 1.53 Å for off-normal incidence.

4. Results and layer spacing determination

To determine the surface-bulk layer spacing, d_z, the experimental and calculated spectra were compared in two ways; a visual evaluation and a semi-quantitative analysis of the difference in peak positions of the experimental and theoretical curves, with penalties assigned for "extra" and "missing" peaks.

4.1. Visual evaluation

This criterion is the standard means for assessing the agreement between a calculated and experimental spectrum, and has the advantage that several spectral characteristics are accounted for, e.g. general curve shapes, relative intensities and the energy positions of peaks, shoulders and minima. The eye can be fooled however, and the advantages and disadvantages of visual judgement have been reviewed by Zanazzi and Jona in the development of a more general and quantitative reliability factor [11].

The results of our visual evaluation are shown in fig. 2, which displays for each of the 12 beams, the experimental spectrum bracketed by the three calculated spectra judged to compare most favourably. Three independent visual comparisons produced an estimated d-spacing from each beam. These estimates were then averaged over the 12 beams for each observer, to yield three estimates for d_z, viz 1.50, 1.51 and 1.52 Å. We conclude then, that on the basis of the visual evaluation d_z^{vis} = 1.51 ± 0.05 Å.

4.2. $\langle|\overline{\Delta E}|\rangle_{hk}$ evaluation

An alternative criterion for comparing experimental and calculated spectra that has been used in the recent past, although with some variations [12,13], is based almost solely on a comparison of peak positions, without regard for relative intensities. The advantage is that a crude quantitative measure of the fit for each d-spacing or model is obtained. The disadvantages of this criterion have been clearly pointed out [11]. Peak positions may still be the most important characteristic of a spectrum, however, at least for some materials, and particularly when only the d-spacing, not the registry, is in question. Changes in damping and to a much lesser extent, θ_D, can change calculated relative intensities and peak shapes, but not generally the positions of features. For certain beams and incidence angles, the calculated peak shapes and relative intensities are very insensitive to changes in layer spacing, whereas peak positions do change, e.g. the (01) beam at $\theta = 14°$, the (02) at 0° and the (11) at 0°. On the other hand, experimental peak shapes and heights can be very sensitive to small changes (<2°) in incidence angles, and thus very susceptible to uncertainties in these parameters. So unless non-structural parameterization is truly optimized and the reliability of the data parameters improved, putting more emphasis on relative intensities does not seem warranted, at least for tungsten.

The approach used here is similar to that used by Marcus et al. [12]. In general, for $\beta = -5.4$ eV, the features (peaks, shoulders, etc.) in the multidistance calculated spectra for all beams could be put into 4 categories according to how they evolved with changing d-spacing: (a) features which could obviously be associated with similar experimental peaks at all d-spacings, (b) features present in the calculated spectra at all spacings or which died out as d changed, but which were absent from the experimental curve, (c) features present in the experimental curve which were absent at all or some d spacings and (d) envelopes of a peak and a shoulder, clearly corresponding to only a single experimental peak, in which the intensity shifted completely from one side to the other as d changed. For features in category (a) the deviation in peak position was obvious. For the "extra" peaks of category (b), a

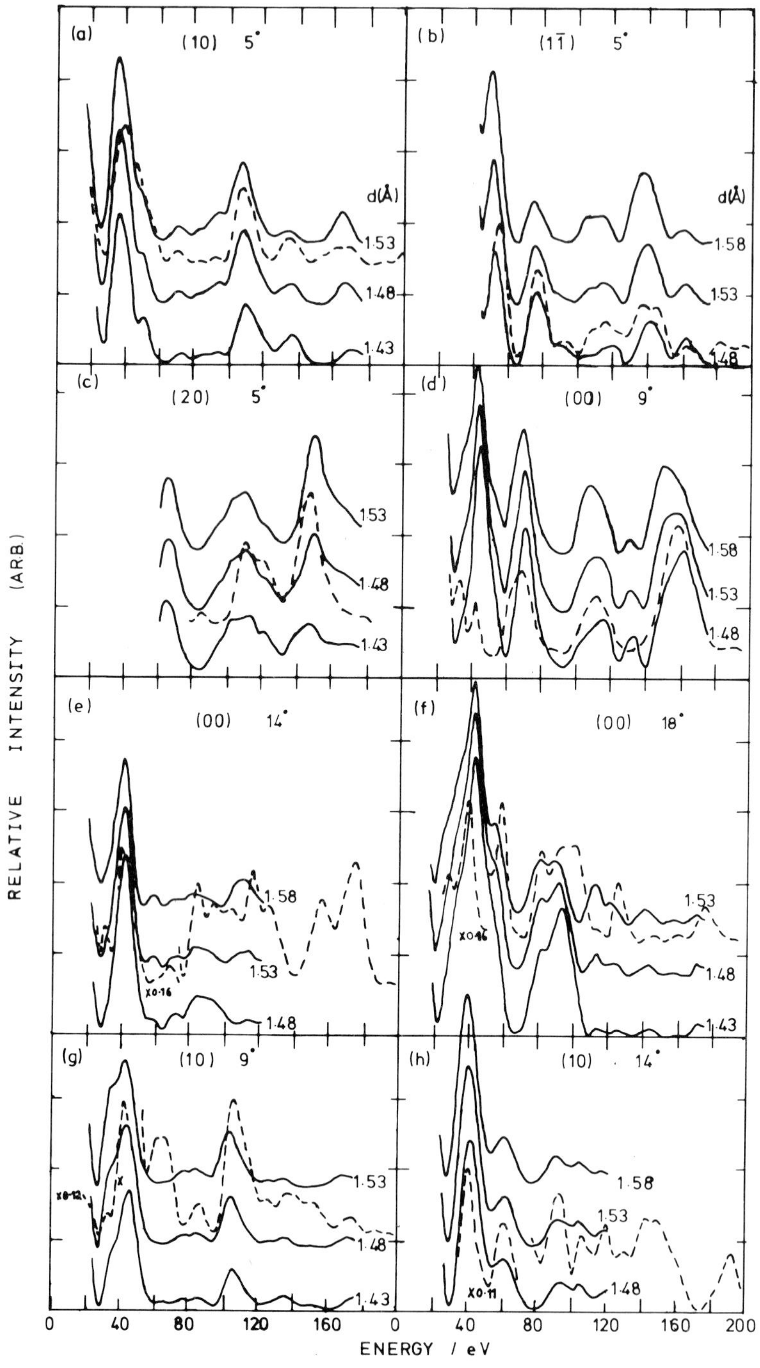
(a)
(10) 5°
d(Å)
1.53
1.48
1.43
(b)
(1$\bar{1}$) 5°
d(Å)
1.58
1.53
1.48
(c)
(20) 5°
1.53
1.48
1.43
(d)
(00) 9°
1.58
1.53
1.48
(e)
(00) 14°
1.58
1.53
x0.16
1.48
(f)
(00) 18°
1.53
x0.16
1.48
1.43
(g)
(10) 9°
x0.12
1.53
1.48
1.43
(h)
(10) 14°
1.58
1.53
x0.11
1.48
RELATIVE INTENSITY (ARB.)
ENERGY / eV
0 40 80 120 160 0 40 80 120 160 200

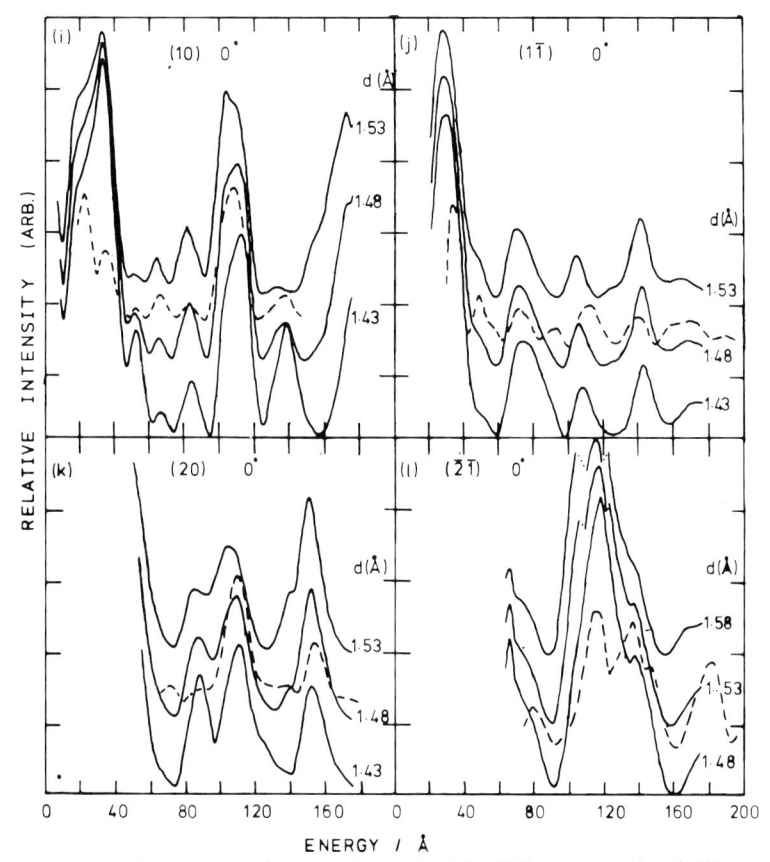

Fig. 2. Calculated and experimental spectra for each of the 12 beams considered. Nine spectra, for nine spacings centered on the bulk value of 1.58 Å, at normal incidence and five spectra centered on 1.53 Å for off normal incidence were calculated. Shown for each observed spectrum, are the computed curves judged on the basis of the visual evaluation to agree closest to the experimental curve. In this way the visually estimated d-spacing was determined for each beam by three independent judges, with the result $d_z^{vis} = 1.51 \pm 0.05$ Å.

penalty of 3, 6 or 9 eV was applied according to whether it was judged as weak, moderate or prominent respectively. If it died out as d progressed, then penalties were assigned in proportion to the maximum penalty charged determined by whether its maximum size was weak, moderate or prominent. For the "missing" features of type (c), a similar penalty was applied according to the experimental peak size, or if the missing feature grew in, the penalty was weighted by the degree to which it had come back to the same relative size as the experimental peak. Finally, for (d) it was usually possible to estimate the peak position from a stacked series of spectra by drawing a straight line between the maxima of the "envelopes" for the d-spacings where the intensity had shifted completely to one side or the other giving a single peak. The peak positions for the intermediate spacings were then taken as the point where the straight line cut the tops of the envelopes, with the spectra displayed vertically to make the envelope maxima approximately equally spaced.

Fig. 3 shows a stacked series of calculated spectra for the (00) beam at $\theta = 9°$ which shows features belonging to each of the four categories just described. For each beam then the mean magnitude of the deviation was found for each of the calculated spectra according to $|\overline{\Delta E}| \equiv (1/N)\Sigma_i |\Delta E_i|$, where either $\Delta E_i = E_i^{\text{exp}} - E_i^{\text{calc}}$ or ΔE_i is the penalty assigned to the ith feature, and N is the number of features considered in the experimental curve.

Plots of $|\overline{\Delta E}|$ versus d-spacing for each of the 12 beams considered, for different

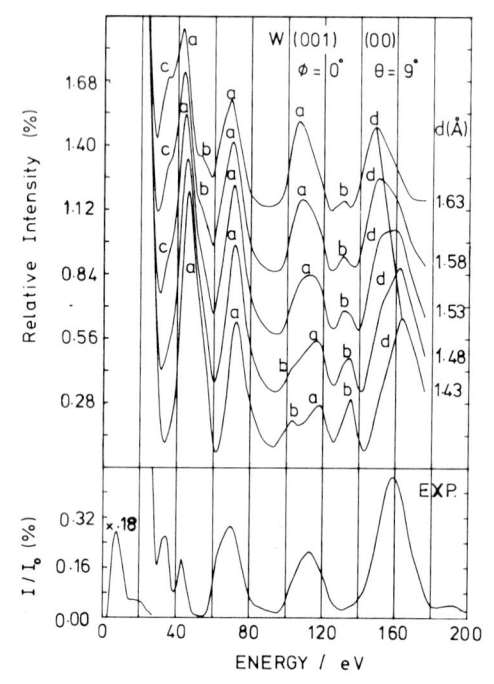

Fig. 3. Calculated spectra for the (00) beam at $\theta = 9°$ for different d-spacings. Each of the four types of feature categories for determining peak deviations and penalties are illustrated: (a) the feature type which remains identifiable with a single experimental peak at all spacings; (b) features which are not identifiable with an experimental peak, which may evolve and are penalized accordingly; (c) features which are identifiable with an experimental peak but which evolve away as d changes and are penalized accordingly; and (d) features which appear to consist of a peak and a shoulder with the intensity transferring from one side to the other as d changes. Intersection of the straight line with the evenly spaced maxima gives a single peak position without assigning a penalty. The (00) $\theta = 9°$ calculated spectra are unusual in that they display all four types of features.

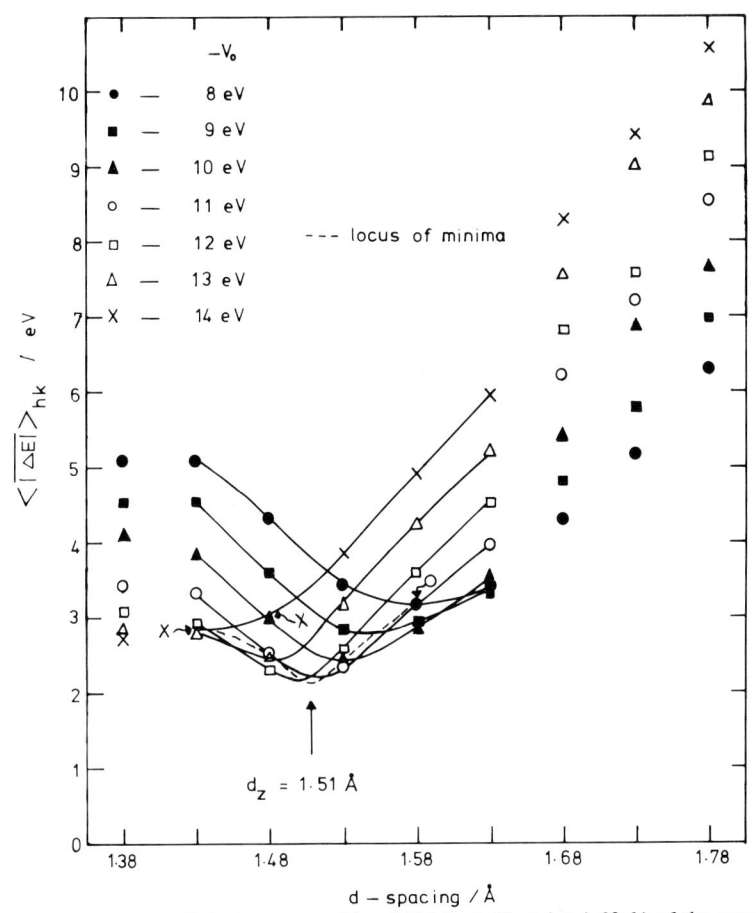

Fig. 4. The average over all the beams considered (12 for $1.43 \leq d \leq 1.63$ Å) of the mean peak-deviation/penalty magnitudes versus d-spacing, for seven different choices of inner potential. The trend for the curve minima to shift to lower d-spacing values and go through a minimum themselves as $|V_0|$ increases, suggests an optimum value for V_0 and the d-spacing. The $d = 1.38$ Å and $d \geq 1.68$ Å points are averages over only 4 beams and were not used to determine the curve minima. The dashed curve is a plot of the minima against the value of d-spacing at which they occur. The optimum value of d-spacing and the "best" inner potential are chosen on the basis of this curve to be $d_z^{\Delta E} = 1.51$ Å and $V_0 = -11$ to -12 eV.

values of inner potential, show the general trend that as $|V_0|$ increases, the minima shift to lower d-spacing and the deepest minima occur for $V_0 = -12$ eV. This trend is brought out clearly in fig. 4 where $\langle|\overline{\Delta E}|\rangle_{hk}$, the magnitude of the mean deviation averaged over all 12 of the beams (only 4 beams for $d = 1.38$ Å and $d > 1.68$ Å) is plotted versus d-spacing with V_0 as a parameter. The deepest minima occur for $V_0 = -11$ and -12 eV. This shifting and deepening with increasing $|V_0|$ of the minima in the $\langle|\overline{\Delta E}|\rangle_{hk}$ versus d-spacing curves is similar to the more dramatic changes in the curves of reliability factor versus d_z^s for different inner potentials reported by Zanazzi and Jona [11], and suggests that the optimum d-spacing and the complementary "best" inner potential are those corresponding to the minimum in a plot of those minima versus the value of d-spacing for which they occur. Such a plot is shown by the dashed curve in fig. 4. On this basis we conclude that the peak deviation and penalty evaluation yields an optimum choice for the surface layer spacing of $d_z^{\Delta F} = 1.51$ Å, with $V_0^{best} = -11$ eV to -12 eV.

5. Conclusions

The beam-averaged peak deviation/penalty magnitude is a crude reliability factor as it only takes account of two of the most important characteristic features of a spectrum, i.e. peak position and "extra" or "missing" features. Two characteristics not accounted for, are the positions and depths of minima and the relative intensities of successive portions of the spectra. However, because of the sometimes strong effects on relative intensities of small errors in incidence angles and imperfect theoretical modeling, we believe an evaluation using peak positions and the degree of "extra" or "missing" features is more reliable than a visual evaluation of 12 beams, and possibly more precise as well because of its semi-quantitative nature. Since the peak deviation/penalty evaluation result agrees with the visual evaluation, there is no conflict in taking the $\langle|\overline{\Delta E}|\rangle_{hk}$ determination to be just as definitive. We conclude that the W{001}-(1 × 1) clean surface—bulk layer spacing is 1.51 ± 0.05 Å.

A close inspection of the figures published by Lee et al. [5] reveals several reasons for the discrepancy between their result for the surface layer spacing and ours, i.e. 1.40 ± 0.03 Å and 1.51 ± 0.05 Å respectively. The most important is that their determination of d_z appears, *on the basis of their published results*, to depend heavily on the existence of one major high energy peak in the (10) spectrum at ~265 eV. In choosing the calculated spectrum which reproduces this peak most closely, the positions of 6 lower energy peaks and their minima are misrepresented, the latter agreeing much better with $d_z = 1.48$ Å. Even the relative intensity of some of those peaks agree better with the 1.48 Å calculated spectrum. Of the other three spectra, the (20) and (11) in our visual judgement also support a spacing closer to 1.48 Å while the (21) is very insensitive to such small changes in d-spacing. Comparing with their published spectra, our data for these beams agree very well below 200 eV, while the calculated spectra are essentially identical. In our analysis, the visual or otherwise, these 4 beams alone imply a spacing of 1.49 Å. The differences in results are thus attributable to their decision to emphasize the energy range above 200 eV. However, comparison of the three different sets of data, Wei, Lee et al. and the present work, shows good agreement below ~150 to 200 eV and much poorer above, for all four beams. Since we have found the data to be very sensitive to small changes in θ, these differences above 200 eV may be attributable to differences in incidence angle as small as 2°. The high energy range is therefore particularly sensitive to experimental conditions, so unless the incidence parameters can be absolutely verified, the high energy range (>150 eV) should be viewed as less reliable and certainly not weighted more.

The above re-emphasizes that for reliable structure determinations using LEED, a procedure for objectively comparing all the characteristics of calculated and experimental spectra is necessary, and a large number of beams (10–20) is needed [11].

Acknowledgements

The awards of an equipment grant, a postdoctoral fellowship to M.K.D. and a studentship to F.S.M. by the Science Research Council are gratefully acknowledged. We also acknowledge the University Computer Laboratory for making the time and core available.

References

[1] For a current review, see M.A. Van Hove, to be published.
[2] M.K. Debe and D.A. King, J. Phys. C10 (1977) L303.
 T.E. Felter, R.A. Barker and P.J. Estrup, Phys. Rev. Letters 38 (1977) 1138.
[3] P.S.P. Wei, J. Chem. Phys. 53 (1970) 2939.
[4] M.A. Van Hove and S.Y. Tong, Surface Sci. 54 (1976) 91.
[5] B.W. Lee, A. Ignatiev, S.Y. Tong and M.A. Van Hove, J. Vacuum Sci. Technol. 14 (1977) 291.
[6] D.P. Woodruff and L. McDonnell, Surface Sci. 40 (1973) 206.
[7] M. Housley and D.A. King, Surface Sci. 62 (1977) 81.
[8] J.B. Pendry, Low Energy Electron Diffraction (Academic Press, London, 1974).
[9] F.S. Marsh, Ph.D. Thesis, Univ. of Liverpool (1977).
[10] We are indebted to S.Y. Tong for sending us the tungsten phase shifts.
[11] E. Zanazzi and F. Jona, Surface Sci. 62 (1977) 61.
[12] P.M. Marcus, J.E. Demuth and D.W. Jepsen, Surface Sci. 53 (1975) 501.
[13] M.A. Van Hove, S.Y. Tong and N. Stoner, Surface Sci. 54 (1976) 259;
 M.A. Van Hove and S.Y. Tong, Phys. Rev. Letters 35 (1975) 1092.

Surface Science 68 (1977) 448-456
© North-Holland Publishing Company

LEED STUDY OF THE EPITAXIAL GROWTH OF THE THIN FILM Au(111)/Ag(111) SYSTEM

F. SORIA and J.L. SACEDON

C.I.F., L. Torres Quevedo, Serrano 144, Madrid-6, Spain

and

P.M. ECHENIQUE and D. TITTERINGTON

Cavendish Laboratory, Cambridge CB3 0HE, U.K.

An analysis of LEED data from the Ag(111) surface at room temperature and $5° \leq \Theta \leq 16°$, $\phi = 12°$ has been carried out in order to test three different model potentials for the exchange and correlation part of the one-electron LEED potential. Clean Au(111) surfaces have been grown on Ag(111) at room temperature at a deposition rate of 0.15 Å s^{-1}. Similar method of calculation and potentials have been employed for the Au overlayer on Ag(111). After the deposition of $\simeq 2.5$ monolayers of Au/Ag(111) the growth of Au can proceed in two different ways. One of them matches satisfactorily with the theoretical calculation for the Au(111) overlayer on Ag(111) following the fcc sequence. The other seems to be concerned with the diffusion of Ag during the Au growth. Similar curves have been obtained during the diffusion of Ag through 350 Å of Au(111).

1. Introduction

One of the possible applications of LEED is the study of metal/metal monocrystalline growth. The good agreement between the theoretical calculations and experimental results for some metallic clean surfaces, and the possibility of $(I-V)$ curve calculations for metal/metal adsorbate structures allows the collection of basic information about metal/metal growth by direct comparison between theoretical calculations and experimental results. At the same time, it is possible to gain some light about basic aspects of the LEED interaction with bimetallic systems.

In this work the growth of Au(111) on to Ag(111) is studied by means of a comparison between experimental $(I-V)_{00}$ curves and theoretical calculations. Because both metals have the same lattice parameter from the point of view of LEED, and a difference of 4 eV in the inner potential, the experimental results will show whether or not there is a discontinuity through the interface in such a quantity in overlayer structures and its agreement with calculations for the same monolayer structure in order to test this type of growth.

Eventually, some attempts (not included in this work) of relating the atom per cent compositions determined by AES, with $(I-V)_{00}$ curves obtained during the surface segregation of Ag through a thin film of Au(111) wil show the high sensitivity of $(I-V)$ curves to this segregation. From those experiments we should be able to extract information about the possible diffusion of small quantities of the metal substrate through the interface during the growth.

2. Experimental methods and results

Following the technique developed by Pashley [1] and Gibson and Dobson [2] we could easily obtain films of Ag(111)/mica of thicknesses greater than 1 μm, but these films presented double-positioned LEED diagrams. To obtain single-positioned Ag(111) films we had to change the growth conditions. These were: residual pressure 1×10^{-5} Torr, substrate temperature 400–420°C and a deposition rate of 10 Å s^{-1}. The films were prepared in a conventional HV system. The thin film samples obtained are placed in the holder of a LEED system. Once UHV conditions had been attained ($p \simeq 10^{-10}$ Torr) the films were annealed at 400°C for 6 to 12 h. After the LEED observations, if the LEED pattern was dim, 500Å of clean Ag was deposited by means of a evaporator included in the system until finally a bright LEED diagram was obtained. The Au was evaporated on the Ag(111) surface without breaking the vacuum conditions and whilst monitoring with a quartz crystal. This deposition is interrupted at different thicknesses to obtain the corresponding LEED diagrams. These growth conditions are: deposition rate of 0.15 Å s^{-1}, residual pressure during the growth of 1×10^{-8} Torr. In those conditions single-positioned LEED patterns were always obtained for all thicknesses.

During all the deposition satellite spots do not appear in the diffraction pattern, the same $(I-V)_{00}$ curve being obtained in one kind of growth for a given coverage. From these observations it seems that the gold atoms are arranged following the lattice periodicity of the substrate, the growth proceeding by means of platelets.

From these surfaces we have recorded $(I-V)$ curves for the (00) spot with a spot-photometer for angles of incidence $5° \leq \Theta \leq 16°$, and azimuthal angle $\phi = 12°$. The spot-photometer detects an angular interval of 2° seen by the crystal, covering the whole diffraction spot. In fig. 1, we present the experimental results, $\Theta = 6°$, $\phi = 12°$, for a clean Ag(111) surface and the theoretical calculations for different model potentials. In this figure, as in the others, the intensity is given in arbitrary units, and the energy in electron volts. Our experimental results for $\Theta = 5°$, $\phi = 12°$, and $\Theta = 8°$, $\phi = 12°$, are similar to those obtained by Rovida et al. [3] for $\Theta = 5°$, $\phi = 0°$, and Berndt and Forstmann [4] for $\Theta = 8°$, $\phi = 0°$, showing that for this surface the azimuthal angle (ϕ) does not influence very much the $(I-V)$ curves for the (00) beam.

After de deposition of 2.5 monolayers the growth can proceed in two different ways giving different peak positions, which lead to two different structures, named for convenience Au 1 and Au 2 (fig. 2). So far we have not defined different growth

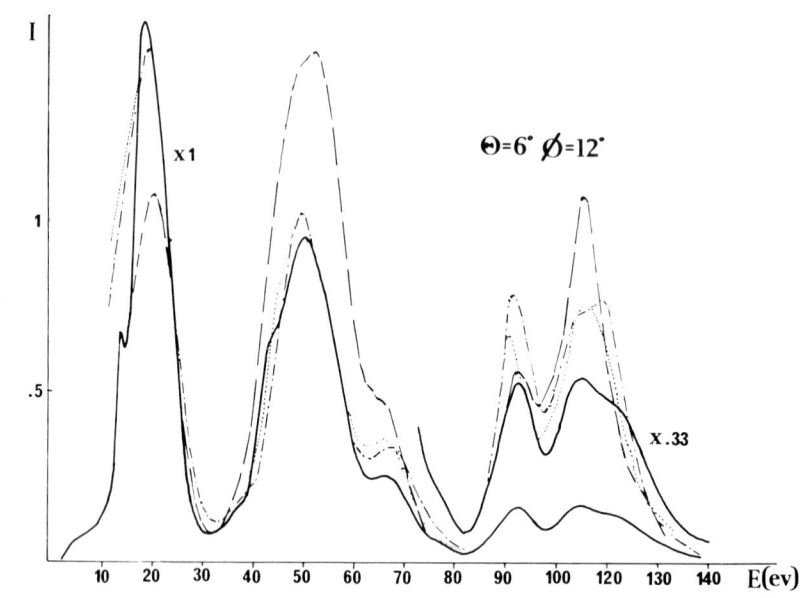

Fig. 1. Comparison between the experimental $(I-V)_{00}$ curve ($\Theta = 6°$, $\phi = 12°$) and the theoretical calculation for a clean Ag(111) surface: (———) experimental curve; (– – –) calculation with the Slater potential; (· · · · ·) calculation with the KSG potential; (– · – · –) calculation with the HF+HL potential.

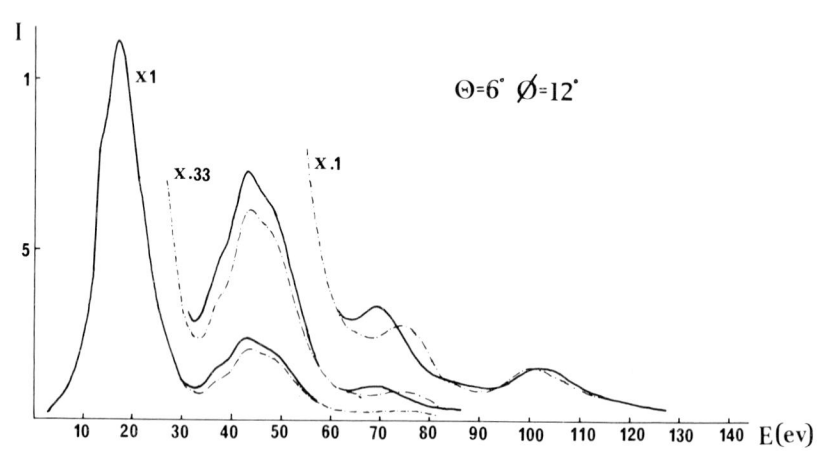

Fig. 2. Experimental $(I-V)_{00}$ ($\Theta = 6°$, $\phi = 12°$) for: (———) clean Au 1(111) surface; (– · – · –) clean Au 2(111) surface.

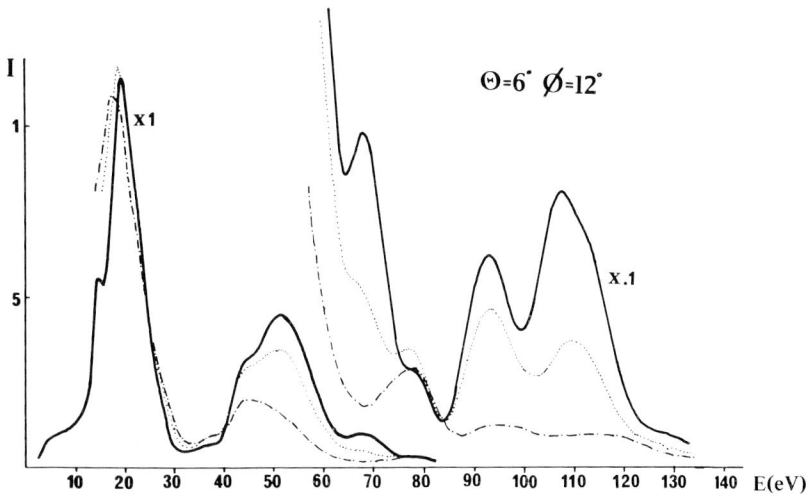

Fig. 3. Experimental $(I-V)_{00}$ ($\Theta = 6°$, $\phi = 12°$) for: (———) clean Ag(111) surface; (· · · · · ·) after deposition of 0.9 monolayers; (– · – · –) after deposition of 3.5 monolayers.

conditions for these structures but rather one or the other appears for apparently the same growth conditions. From $5° \leq \Theta \leq 11°$, $\phi = 12°$, these differences are in the energy position of the Bragg peak at high energy ($\simeq 100$ eV) and in the position of the secondary peaks, but in the range $12° \leq \Theta \leq 16°$, $\phi = 12°$, these differences influence all Bragg's and secondary peaks given $(I-V)_{00}$ curves with notorious morphological differences.

After de deposition of six monolayers there are no more changes in the $(I-V)_{00}$ curves, i.e., both growths present continuously their differences.

In fig. 3 it is shown the experimental $(I-V)_{00}$ curve for $\Theta = 6°$, $\phi = 12°$, of a clean Ag(111) surface, and after deposition of 0.9 monolayer and 3.5 monolayers. The 3.5 monolayer curve correspond to Au 2 characterized by a secondary peak at 75 eV and a Bragg peak at 99.5 eV, while in Au 1 these peaks appear at 69.5 eV and 101.5 respectively (fig. 2).

3. Calculations and comparison with experiments

For calculations of the Ag(111) surface three different model potentials were tried. The first and the second were constructed according to the Mattheis prescription [5] which assumes overlapping atomic charge. The first uses the Slater treatment of exchange [6] and the second uses that proposed by Kohn and Sham [7,8]. Their final result is equivalent to the use of a Slater "two-thirds" coefficient. The third potential uses the more sophisticated Hedin and Lundqvist scheme [9]. This potential, named Hartree–Fock+Hedin–Lundqvist, has been successfully employed by Echenique [10] and by Anderson, Pendry and Echenique [11] in their calculations for Na(110).

We assume thermal vibrations to be isotropic, temperature effects being included by means of complex temperature phase shifts. We have used eight of them to describe the scattering by the vibrating atoms. In both Ag and Au we used the bulk Debye temperatures [12] 226.2 K for Ag and 162.4 K for Au. We know that the effects of Debye temperature for Ag [13], particularly at low energies, will be less energy dependent for the LEED case. This will probably also be the case in Au but since we are mainly interested in the peak positions and in the relative value of the peak intensities and not in the absolute intensities, which are quite unreliable at the present state of the technique, we feel that these Debye temperatures are good enough for our purposes.

To calculate the LEED reflected intensities we use Pendry's formulation of the layer method [14] and the RFS perturbation scheme [15] to calculate the scattering properties of the crystal. This method converges very effectively over the entire range of energies ($15 \leqslant E \leqslant 120$ eV). We have used 19 beams up to 3.5 Hartrees and 31 up to 5.5 Hartrees in order to minimize (<1%) the errors in the calculations.

A constant value for the imaginary part of the self-energy, $V_{0i} = -4$ eV has been used for both Ag and Au. The "inner potential", the difference between the energy zeo of the calculation and of the measurements, is determined by shifting the intensity curves until the structures agree.

In fig. 1 is shown $(I-V)_{00}$, $\Theta = 6°$, $\phi = 12°$, for a clean Ag(111) surface, and its comparison with the calculation using the Slater potential, the Kohn–Sham–Gaspar potential, and the Hartree–Fock–Hedin–Lundqvist potential, for a clean Ag(111) surface without surface dilatation. All three give the peak positions correctly but the KSG and HF+HL gives better agreement in the relative intensities.

From this agreement we found a value of (-8.5 ± 0.5) eV for the inner potential, without evidence for any energy dependence in the 0–120 eV range of incidence energies. The energies of the Bragg peaks are measured within an error of ± 0.5 eV in all our experiments.

In fig. 4 is shown the experimental $(I-V)_{00}$ curve for $\Theta = 6°$, $\phi = 12°$, after the deposition of 0.9 monolayer of Au onto Ag(111), with the calculation for a monolayer of Au(111) on the Ag(111) surface using the KSG potential for Ag and Au, and the HF+HL potential for Ag and KSG for Au, following the fcc sequence, without surface dilatation. The theoretical calculation has been shifted -8 eV to match, as much as possible, with the experimental result. Also, the result is shown for the calculation using the KSG potential for Ag and Au following the hcp sequence without surface dilatation.

Fig. 5 shows the experimental $(I-V)_{00}$ for $\Theta = 12°$, $\phi = 12°$, after the deposition of 0.9 monolayers of Au onto Ag(111), with the calculation using the same potentials as in fig. 3; following the fcc sequence, the KSG potential for Au and Ag, and the HF+HL potential for Ag and KSG for Au, and following the hcp sequence with

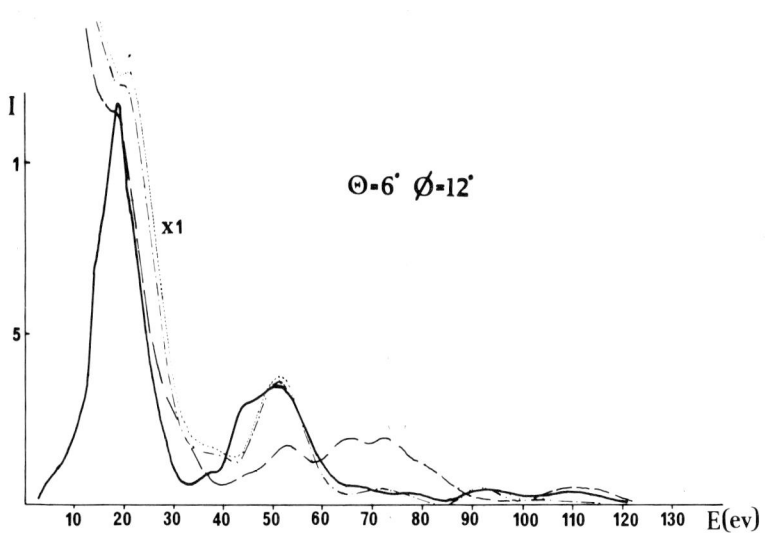

Fig. 4. Comparison between the experimental $(I-V)_{00}$ ($\Theta = 6°$, $\phi = 12°$) and the theoretical calculation for a monolayer of Au(111) onto Ag(111): (———) experimental curve; (·····) calculation with the KSG potential for Au and Ag following the fcc sequence; (– · – · –) calculation with the KSG potential for Au and HF+HL for Ag following the fcc sequence; (– – –) calculation with the KSG potential for Au and Ag following the hcp sequence.

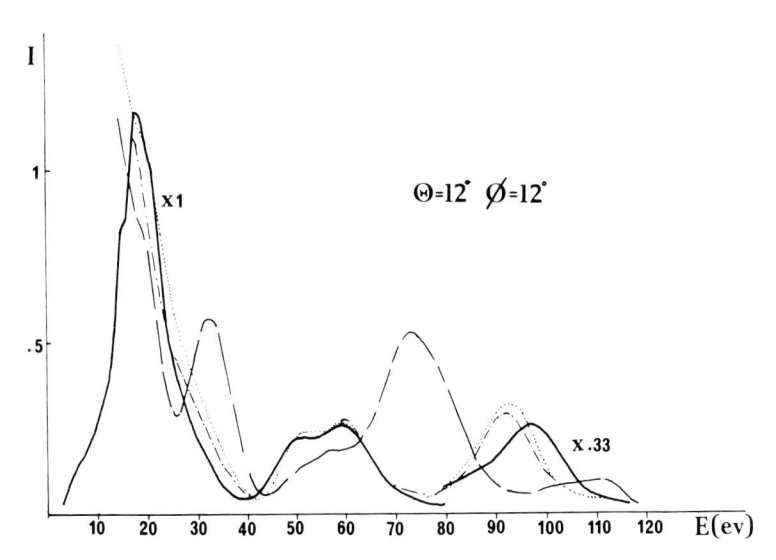

Fig. 5. Comparison between the experimental $(I-V)_{00}$ curve ($\Theta = 12°$, $\phi = 12°$) and the theoretical calculation for a monolayer of Au(111) onto Ag(111). The curves have the same meaning as in fig. 4.

the KSG potential for Ag and Au. The theoretical calculation has been shifted -10 eV to match with the experimental result.

It is observed that the comparison for $\Theta = 12°, \phi = 12°$, is much better than for $\Theta = 6°, \phi = 12°$. For $\Theta = 6°, \phi = 12°$ the Bragg peak at 19 eV needs a shift of 11 eV to match the theoretical peak. close to the value found for a clean Au(111) surface (-12.5 eV) while the peaks at 51.5 eV and 93 eV need a shift of 8 eV to match the theoretical value, close to the value for a clean Ag(111) surface (-8.5 eV). For $\Theta = 12°, \phi = 12°$, the shift for the 18 eV is 12 eV and for the 52 and 59 eV this shift is 10 eV. This difference is 6.5 eV only for the peak at 96.5 eV.

From these observations it can be inferred that for the case of the deposition of a monolayer of Au(111) on Ag(111), the peaks shifts do not have a constant value, but are a function of the angle of incidence and energy of the incident electrons. It looks as if the peaks of higher energies are influenced by the substrate's inner potential, this influence being diminishes for incidence at $\Theta = 12°$ reflecting more the characteristics of the Au overlayer.

Fig. 6 shows the experimental $(I-V)_{00}$ curve for $\Theta = 6°, \phi = 12°$, of a clean Au 1 (111) surface, and its comparison with the theoretical calculation for a clean Au(111) surface without dilatation using the Slater potential and the KSG potential. The agreement is good with a inner potential of (-12.5 ± 0.5) eV with no energy dependence in the 0–120 eV range. Nevertheless, the relative intensities of peaks in the range 43–48 eV appear altered. The Slater potential gives a double

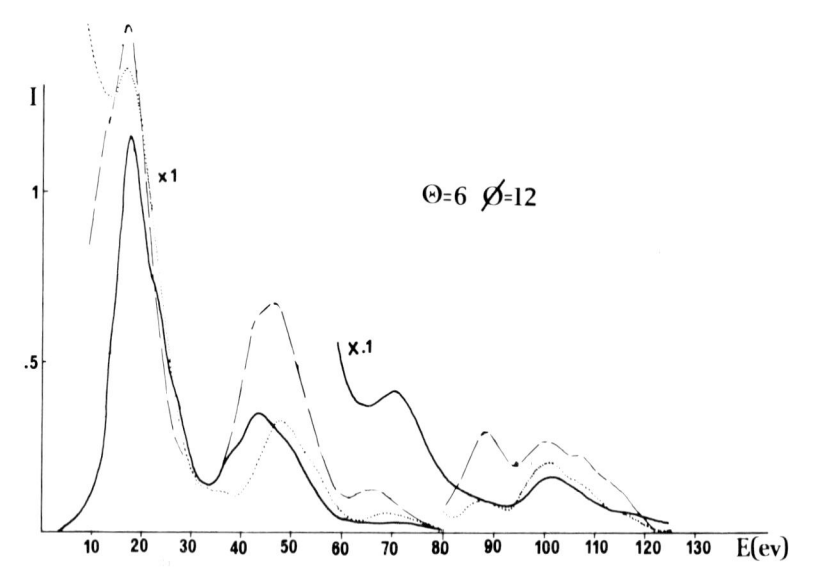

Fig. 6. Comparison between the experimental $(I-V)_{00}$ curve ($\Theta = 6°, \phi = 12°$) for a clean Au 1(111) surface and the theoretical calculation for a clean Au(111) surface: (———) experimental curve; (– – –) calculation with the Slater potential; ($\cdots\cdots$) calculation with the KSG potential.

peak at 43 and 46 eV in agreement with the experiment, while the KSG does not. On the other hand, the KSG potential gives better agreement in relative intensities.

The comparison between the calculations and the Au 2(111) surface is not satisfactory. In experiments of surface segregation of Ag up to the Au(111) surface carried out in our laboratory [16], we have detected the Au(111) 2 structure for a small segregation of Ag.

4. Discussions

The comparison of the experimental results with the theoretical calculation for a clean Ag(111) surface without dilatation is excellent. The Kohn–Sham–Gaspar and Hartree–Fock+Hedin–Lundqvist potentials seem to be more appropriate than the Slater potential for the Ag(111) surface.

In the case of Au grown epitaxially onto Ag(111) substrates, this agreement is obtained by shifting the theoretical curves for $\Theta = 6°$, $\phi = 12°$, and $\Theta = 12°$, $\phi = 12°$, 8 eV and 10 eV respectively. The agreement between the $(I-V)$ experimental curves for a monolayer of Au and the theoretical calculation is satisfactory for a structure following the fcc sequence. Therefore this growth seems to proceed monolayer to monolayer; by means of platelets of monoatomic height, but only introducing in the theory a variation of inner potential through the interface, an exact matching in position peaks could be achieved and an exact confirmation of monolayer growth be obtained. This growths begins always following a unique structure up to 2.5 monolayers. Afterwards, the growth can follow two different structures. The $(I-V)$ curves of one of them are very similar to the theoretical curves obtained for a clean surface of gold (111), the other gives similar $(I-V)$ curves to the experimental ones obtained during the segregation of Ag through a thin film of Au (111).

Acknowledgments

We would like to express our thanks to Professor M. Blackman of Imperial College, London, for all the facilities and helpful discussions and specially to Dr. P.J. Dobson for his guidance throughout all the experimental work. Finally, we are gratefully indebted to the Royal Society for the provision of a grant for one of us (F.S.).

References

[1] D.W. Pashley, Phil. Mag. 4 (1959) 316.
[2] M.J. Gibson and P.J. Dobson, J. Phys. F5 (1975) 864.

[3] G. Rovida, F. Pratesi, M. Maglietta and E. Ferroni, Surface Sci. 43 (1974) 230.
[4] F. Forstmann, Japan. J. Appl. Phys. Suppl. 2 Pt 2 (1974) 657.
[5] L.F. Mattheis, Phys. Rev. 134 (1964) 970.
[6] J.C. Slater, Phys. Rev. 81 (1951) 385.
[7] L.J. Sham and W. Kohn, Phys. Rev. 145 (1966) 561.
[8] R. Gaspar, Acta Phys. Hung. 3 (1954) 263.
[9] L. Hedin and B.I. Lundqvist, J. Phys. C4 (1971) 2064.
[10] P.M. Echenique, J. Phys. C9 (1976) 3193.
[11] S. Andersson, J.B. Pendry and P.M. Echenique, Surface Sci. 65 (1977) 539.
[12] C. Kittel, Introduction to Solid State Physics, 3rd ed. (Wiley, New York, 1966).
[13] J.T. McKinney, E.R. Jones and M.B. Webb, Phys. Rev. 160 (1967) 523.
[14] J.B. Pendry, J. Phys. C4 (1971) 2501.
[15] J.B. Pendry, J. Phys. C4 (1971) 3095.
[16] F. Soria and J.L. Sacedón, to be published.

A LEED STUDY OF THE Si(100) (1 × 1)H SURFACE STRUCTURE

S.J. WHITE [*], D.P. WOODRUFF, B.W. HOLLAND and R.S. ZIMMER [**]

Physics Department, University of Warwick, Coventry CV4 7AL, England

Adsorption of hydrogen in the presence of an ion gauge (believed to be associated with atomic hydrogen) on a clean Si (100) (2 × 1) surface leads to a (1 × 1)H structure. Detailed LEED experiments have been performed on this surface and an extensive range of intensity data collected. Clearly the most obvious explanation for the structure is that of bulk structure Si atom layers with hydrogen atoms adsorbed on to the resulting dangling bonds. As the hydrogen atoms should be weak scatterers, their role in influencing the LEED intensities should be minimal. Dynamical theory calculations based on an ideal un-reconstructed Si(100) surface give satisfactory agreement. This indicates that model parameters used in the calculations also appear to be satisfactory and should therefore be suitable for calculations aimed at a structural interpretation of the clean surface (2 × 1) structure.

---

[*] Now at Coordinated Sciences Laboratory, University of Illinois at Urbana-Champaign, Urbana, Illinois 61801, USA.
[**] Now at Institute of Educational Technology, The Open University, Milton Keynes, Buckinghamshire, England.

XPS, UPS AND XAES STUDIES OF OXYGEN ADSORPTION ON POLYCRYSTALLINE Mg AT ~100 AND ~300 K *

J.C. FUGGLE

Institut für Festkörperphysik, Technische Universität München, D-8046 Garching, W. Germany

Our results show that at both ~300 and ~100 K the adsorption of oxygen on evaporated Mg slows dramatically when oxide layers ~7 Å thick have formed on the surface. At 100 K the initial sticking coefficient is approximately 0.1. The oxidation goes through two phases at both temperatures; initial adsorption of up to approximately one monolayer, followed by formation of a layer with many characteristics similar to those of bulk MgO. During initial adsorption at 300 K the bulk of the oxygen adsorbed must sit below the metal surface, at 100 K its position is unclear. At both 100 and 300 K the photoelectron spectra from the surface layer produced by initial adsorption of oxygen are unlike those from either MgO or Mg. Even at "saturation" coverage, XAES provides evidence for a layer unlike MgO, or Mg, at the MgO/Mg interface. It is found that shifts of 0.5 eV in the photoelectron peaks from the oxide layer can be induced by only ~5 × 10^{13} oxygen atoms cm^{-2}.

---

* Details of this work will be published in Surface Science.

CORE AND VALENCE LEVEL PHOTOEMISSION STUDIES OF IRON OXIDE SURFACES AND THE OXIDATION OF IRON

C.R. BRUNDLE, T.J. CHUANG and K. WANDELT

IBM Research Laboratory, San Jose, California 95193, USA

The core and valence level XPS spectra of Fe_xO ($x \sim 0.90-0.95$); Fe_2O_3 (α and γ); Fe_3O_4; and FeOOH have been studied under a variety of sample surface conditions. The oxides may be characterized by a combination of valence level differences and core-level effects (chemical shifts, multiplet splittings, and shake-up structure). Fe^{II} and Fe^{III} states are distinguishable, but octahedral and tetrahedral sites are not. The O 1s BE cannot be used to distinghuish between the oxides since it has a nearly constant value. Fe 3d valence level structure spreads some 10 eV below E_F, much broader than suggested by previous UPS and photoelectron-spin-polarization (ESP) measurements for Fe_xO and Fe_3O_4. Fe surfaces (films, foils, (100) face) yield predominantly Fe^{III} species when exposed to high exposures of oxygen or air, though there is evidence for some Fe^{II} also. At low exposures the Fe^{II}/Fe^{III} ratio increases.

1. Introduction

The variety of structures possible in the bulk oxides is considerable [1,2]. Fe_xO (wustite) has a cubic lattice with Fe^{II} ions octahedrally coordinated. α-Fe_2O_3 (haematite) has the rhombohedral Al_2O_3 structure ($\alpha = 55°$) with the Fe^{III} ions coordinated by a distorted oxygen octahedron. Fe_3O_4 (magnetite) is a spinel structure with the Fe^{II} ions in octahedral sites, and the Fe^{III} ions half in octahedral and half in tetrahedral sites. The γ-Fe_2O_3 structure is the same as that of Fe_3O_4 with vacancies and Fe^{III} ions replacing the Fe^{II} ions in octahedral sites. We take the view that the best way to be sure of the interpretation of any one oxide spectrum is to examine, in detail, the spectra of all the oxides under a variety of surface conditions.

We then use the characterized oxide spectra to aid the interpretation of the Fe/O_2 and air reactions.

2. Experimental

Both commercially obtained oxide powders and single crystals were studied. All were bulk characterized by X-ray diffraction. The XPS data on the oxides was taken using an HP 5090 spectrometer (base pressure $\sim 1 \times 10^{-9}$ Torr) with monochromated Al Kα radiation. All powder samples [3] (α-Fe_2O_3, γ-Fe_2O_3, Fe_3O_4, FeOOH) and a single crystal of Fe_xO were first studied in the as received, air-exposed form. They were then subjected to cycles of argon ion bombardment cleaning (1 kV, etch rate a few Å per minute), spectra being recorded after each cycle. The single crystal samples (Fe_xO, α-Fe_2O_3, Fe_3O_4) were studied by crushing and loading under argon. They were then argon ion bombarded and re-studied, and finally air-exposed and re-studied. All BE calibrations were made against Au $4f_{7/2}$ at 83.95 eV by mixing samples with Au powder.

The Fe oxidation studies were carried out in three spectrometers. The oxidation of Fe foils, cleaned in situ by argon ion bombardment, was followed in an HP spectrometer. Initial oxygen chemisorption on iron foils was studied in a Physical Electronics XPS/Auger system. Preliminary XPS/UPS work on Fe(100) was performed in a V.G. Scientific system equipped with XPS, UPS, LEED, Auger, and SIMS.

3. Results

Fe $2p_{3/2}$, $2p_{1/2}$, 3p, 3s, O 1s, O 2s, and valence band spectra were recorded. Lack of space prevents complete presentation here, so we report only selected Fe $2p_{3/2}$, O 1s, and valence band spectra.

The Fe $2p_{3/2}$ spectrum of a clean Fe foil is shown in fig. 1a for reference (BE = 707.0 eV). That of freshly crushed, as-received α-Fe_2O_3 powder is shown in fig. 1b. The spectrum of single crystal α-Fe_2O_3, crushed and inserted under argon is practically identical. The Fe $2p_{3/2}$ spectrum of freshly crushed, as-received Fe_3O_4 powder was also very similar to fig. 1b. The spectrum of the Fe_xO crystal, lightly abraided prior to insertion, was similar to fig. 1b, but the Fe 2p peak showed a slight swelling on the low BE side. The spectrum of the freshly crushed, as-received γ-Fe_2O_3 powder is shown in fig. 1d. The Fe $2p_{3/2}$ peak is slightly narrower than in α-Fe_2O_3, but the main features are the same. The O 1s spectrum of FeOOH (fig. 1c), exhibits two O 1s peaks of equal intensity, one in the oxide postion and one about 1 eV to higher BE. The spectra of the single crystal samples of Fe_3O_4 and Fe_xO crushed and inserted under argon are shown in figs. 1e and 1f. The Fe 2p regions are quite different from the equivalent air-exposed spectra, and from each other. The O 1s spectra are, however, very similar to the other spectra, particularly with respect to the BE of the main peak.

A controlled minimum sputtering of the Fe_3O_4 powder sample was able to reproduce the shape of the spectrum in fig. 1e, though the low BE Fe $2p_{3/2}$ shoulder was not clearly resolved. Controlled sputtering of the air-exposed Fe_xO sample reproduced the spectrum of fig. 1f, but controlled further sputtering produced the spectrum of fig. 1g.

For all the oxide surfaces extensive ion bombardment results in intensity in the Fe $2p_{3/2}$ peak being transferred from higher to lower BE as a function of sputter-

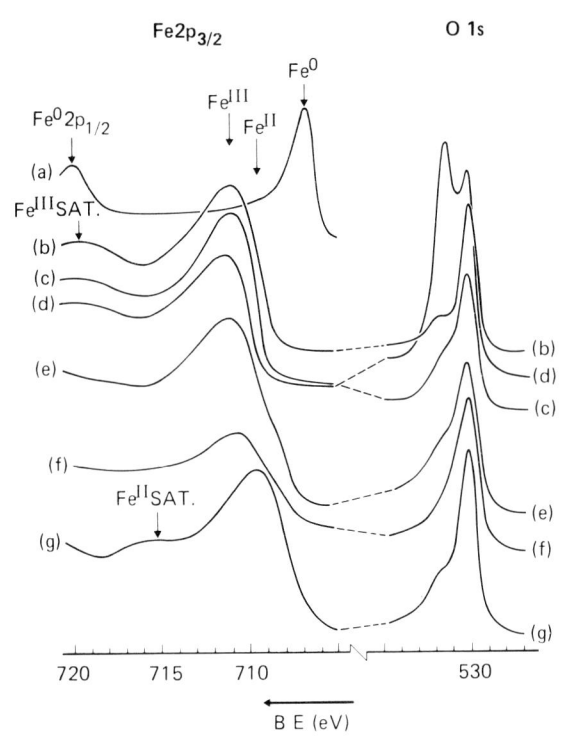

Fig. 1. Fe $2p_{3/2}$ and O 1s spectra from iron oxide surfaces. Relative intensity scales are arbitrary. (a) Ar$^+$ cleaned Fe foil. (b), (c), (d) Freshly crushed α-Fe$_2$O$_3$, γ-Fe$_2$O$_3$, FeOOH powders respectively. (e) Single crystal Fe$_3$O$_4$, crushed and inserted under argon. (f) Single crystal Fe$_x$O crushed and inserted under argon. (g) Sample (f) briefly sputtered.

ing time, and the characteristic Fe metal feature appears quite rapidly. The effect on the O 1s spectra is to first decrease the intensity of the high BE peak or shoulder, and then reduce the main O 1s/Fe 2p intensity ratio. The effect on FeOOH is dramatic with the 1 : 1 ratio of low BE, high BE O 1s peaks changing to 5 : 1 within 50 sec of bombardment.

From the behavior of the Fe 2p spectra of the air exposed surfaces, crushed crystal surfaces, and ion bombarded surfaces, plus similar data in the Fe 3s and 3p regions not discussed here, we are satisfied that the Fe $2p_{3/2}$ spectra of figs. 1b, c, e, and g are characteristic of the genuine surfaces of α-Fe$_2$O$_3$, FeOOH, Fe$_3$O$_4$, and Fe$_x$O, respectively. The reported spectrum of γ-Fe$_2$O$_3$ is too similar to α for us to be sure it does not represent an α-Fe$_2$O$_3$ covered surface.

In fig. 2 we show the XPS valence band spectra corresponding to the core-level spectra for the above oxides.

The XPS core-level spectra for the oxygen adsorption and oxidation of Fe polycrystalline surfaces are shown in fig. 3.

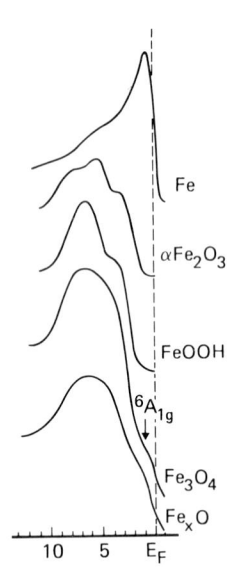

Fig. 2. XPS valence band spectra of Fe and the oxide surfaces.

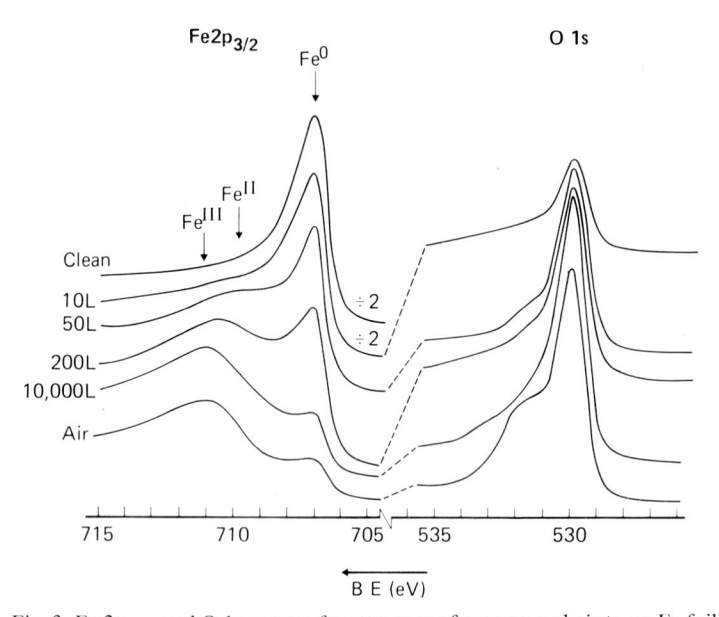

Fig. 3. Fe $2p_{3/2}$ and O 1s spectra for exposure of oxygen and air to an Fe foil.

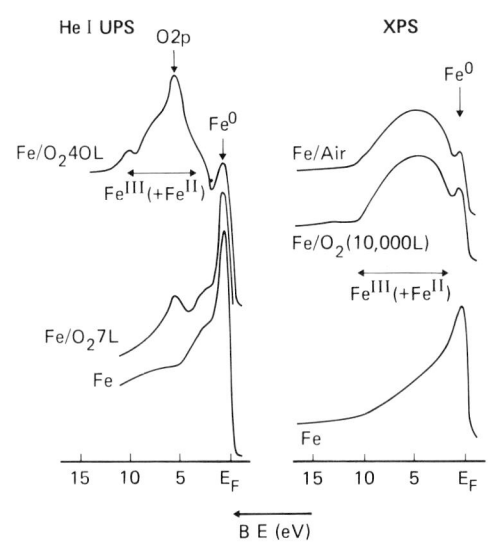

Fig. 4. (a) HeI spectra for Fe(film)/O₂ [22]. (b) XPS valence band spectra for Fe(foil)/O₂.

Fig. 4 shows some of the equivalent XPS and UPS valence band spectra for the oxidation sequences.

4. Discussion

4.1. The oxide spectra

The Fe $2p_{3/2}$ data lead us to the following conclusions concerning their use for characterizing the oxide surfaces.
(1) An Fe^{III} species has a BE of ~711.2 eV and a peak width of ca. 4.5 eV, irrespective of the particular oxide (i.e., Fe_2O_3, Fe_3O_4, FeOOH).
(2) Fe^{III} octahedrally coordinated is not easily distinguishable from tetrahedrally coordinated by using the Fe $2p_{3/2}$ core-level spectra.
(3) Fe^{II} has a BE of ~709.7 eV and a peak width of ~4.5 eV in Fe_xO. Deconvolution of the Fe_3O_4 spectrum into two-thirds Fe^{III} and one-third Fe^{II} leads to a value of ~709.5 eV for the latter.
(4) The broad satellite centered at ~719.8 eV (8.5 eV above the Fe^{III} $2p_{3/2}$ position) is characteristic of an Fe^{III} species, independent of the oxide and the coordination number. The broad satellite at ~715 eV (6 eV above the Fe^{II} $2p_{3/2}$ position) is characteristic of an Fe^{II} species.
(5) The combined use of the main $2p_{3/2}$ line and its satellite, plus similar use of the $2p_{1/2}$, 3s, and 3p lines provide a reliable identification of Fe^{III} and Fe^{II} species, even in mixtures, or in mixed oxides (e.g., Fe_3O_4).

Of the several lower resolution previous core-level studies of iron oxide surfaces, those of Allen et al. [4] and Asami et al. [5] seem to be the most reliable. The shapes of the Fe $2p_{3/2}$ spectra reported by Allen et al. for Fe_xO (powder form, ion-bombarded) and α-Fe_2O_3 (powder form) are compatible with our spectra, but the absolute BE for Fe_xO is given as 710.3 eV, 0.6 eV higher than our value, which would make distinction of Fe^{II} and Fe^{III} species more tenuous than we believe is true from the present data. Their spectra of FeOOH differs from ours in the lack of any obvious satellite structure. They do not show an Fe_3O_4 spectrum, but report that though a line broadening could be observed on heating to 900 K, no distinction states could be made, in contrast to our results.

Asami et al. report the spectra of α-Fe_2O_3, grown as a high temperature oxide on iron, and a mixture of $FeO/Fe_2O_3/Fe_3O_4$ produced by decomposition of iron oxylate. They rely on deconvoluting the Fe^{III} peak of α-Fe_2O_3 from the broader $FeO/Fe_2O_3/Fe_3O_4$ spectrum to obtain an Fe^{II} spectrum. The BE's they obtain for Fe^{II} and Fe^{III} are 709.2 eV and 711.0 eV, corrected to our reference calibration, in fair agreement with our data. The peak width they obtain for Fe^{II} by the deconvolution is much too narrow however (~2 eV). They do not discuss satellite structure.

The striking feature of the O 1s spectra is the invariance of the BE for the various oxides. This has been noted before for FE [4] and other metals, as has the only small difference in BE between the oxides of one transition metal and another (e.g., Ni [6], Co [7], Fe), reinforcing our opinion that the O 1s BE can not be related to the oxidation state of the cation present [7] (despite claims to the contrary [8]). The complete story of the higher BE O 1s still eludes us. For the Fe oxides Allen et al. believe it is entirely due to hydroxyl species contaminating the surfaces. Likewise, Haber et al. [8] believe O 1s features in this BE region produced during adsorption studies on the metals represent hydroxyl formation. While agreeing that OH O 1s certainly comes in the 531 eV region (witness the resolved peak in FeOOH), we do not believe that it is the *only* oxygen containing species capable of this BE, since our surfaces produced by crushing under argon still showed its presence, and at least for some metals, it can be generated during oxygen exposure under fully UHV conditions [6]. We therefore tend to believe that it can also represent non-stoichiometric surface oxygen atoms.

The interpretation of the core-level spectra in terms of the electronic structures of the oxides is complex, but general points can be made. Both Fe^{II} and Fe^{III} species are high spin states, and it is therefore expected that the Fe 2p lines will be broad with unresolved multiplet splitting [9]. Secondly it has been empirically observed, though not well understood theoretically, that high spin states of transition metal ions have strong shake-up core-level satellites; whereas low spin states have weak or zero satellite structure [10]. We thus expect, and observe, satellite structure for both Fe^{II} and Fe^{III}. This contrasts with the Co^{II} and Co^{III} results where Co^{III}, which is low spin, has only very weak satellite structure [7]. Finally the relative BE's observed for the $2p_{3/2}$ and $2p_{1/2}$ lines follow the intuitively expected order $Fe^{III} > Fe^{II} > Fe^\circ$, again in contrast to cobalt where $Co^{II} > Co^{III}$ [7].

The valence level XPS spectra of the oxide surfaces (fig. 4) are interesting both for further means of distinguishing Fe^{II} from Fe^{III} and from an electronic structure point of view. The main distinguishing feature is the presence of structure at ~1 eV below E_F when Fe^{II} is present. It is generally recognized that for the photoemission process final state interactions can play a dominant role in the experimental valence band spectrum and that when such interactions are large, the experimental spectrum cannot usually be simply related to an initial density of states [11]. In the present case, in the fully ionic approximation the final state ions for Fe 3d photoemission are Fe^{3+} $3d^5$ from Fe^{II} and Fe^{4+} $3d^4$ from Fe^{III}. The effect of the crystal yield of the surrounding O^{2-} ligands results in a splitting into seven states [12] for the octahedrally coordinated Fe^{3+} ($t_{2g}^3 e^2$ and $t^4 e^1$ configurations). The $^6A_{1g}$ state, which results from photoemission of the minority spin electron of Fe^{II}, is expected to lie closest to E_F and has been assigned in UPS studies [13] to the ~1 eV feature. Both UPS [13] and spin polarized photoemission (ESP) [11] measurements for Fe_xO and Fe_3O_4 supposedly show that the total Fe 3d final state structure is confined between 0–5 eV below E_F. This is clearly qualitatively incompatible with our XPS results, where a spread to about 10 eV below E_F is indicated for Fe_xO, Fe_2O_3 and Fe_3O_4. We believe the interpretation of the UPS/ESP data to be in error owing to incorrect subtraction of O 2p and background from the total experimental spectrum. At Al Kα photon energies we know that the O 2p contribution is low (from relative Fe 3d/O 2p cross-sections [14]) and can also confidently assign a *maximum* background subtraction. The valence band structure of Fe_xO is discussed fully elsewhere together with an ab initio crystal field calculation of the final state structure [15].

4.2. Chemical effects of argon ion bombardment on iron oxide surfaces

The standard technique for depth-profiling the atomic composition through an interface is to sputter away the surface by argon ion bombardment while simultaneously monitoring Auger intensities [16]. An extension is to analyze the chemical composition by using XPS or Auger chemical shifts during this process. Many such studies involve metal or semiconductor surfaces oxidized either in dry or aqueous media [17]. Our results for the bulk iron oxides indicate that sputter profiling of such passivated surfaces can give highly misleading results. For instance, starting with bulk α-Fe_2O_3 we had to sputter only 50 sec before observing a significant Fe^{II} concentration produced by reduction from Fe^{III} by the Ar^+ beam, even though the beam sputters at a rate of only a few Å a minute. A further 370 sec sputtering results in further depletion of Fe^{III} from the escape depth leaving a mixture of Fe^{III}, Fe^{II}, and Fe metal. The differential sputtering effect of Argon ions has been previously documented in XPS studies [18], and has also been pointed out in Auger studies [19], but it does not seem to be appreciated how serious a problem it can be for profiling measurements.

4.3. Oxygen interaction with Fe surfaces

From the assignment of Fe^{III} and Fe^{II} $2p_{3/2}$ positions at 711.2 eV and 709.7 eV, the observed shape of the $2p_{3/2}$ level for Fe_3O_4, and the knowledge that FeOOH has two resolved equal intensity O 1s peaks, it becomes possible to interpret the XPS data for Fe/O_2 and Fe/air, given in fig. 3. It is clear that high oxygen exposures of films or foils result in an Fe^{III} species, with some Fe^{II} present. Though a weak high BE O 1s shoulder is present, the major peak is at 530.1 eV, ruling out FeOOH as the product and leaving Fe_2O_3 or Fe_3O_4 as possibilities. We tentatively favor the latter assignment, in agreement with RHEED studies on Fe(100) [20]. Air exposure produces very similar Fe 2p spectra, but the high BE O 1s peak is now quite intense, indicating some additional hydroxyl formation. At the low exposures we have either a mixture of Fe^{II} and Fe^{III} species, with the Fe^{II}/Fe^{III} ratio much larger than at high exposures, or some single species which is intermediate in electronic character. The very low exposure shown in fig. 3 shows so little chemically shifted Fe 2p intensity that one cannot be certain about an assignment. This problem should be alleviated by working at grazing angle electron ejection to decrease the effective escape depth and hence accentuate the surface component contribution to the spectrum [21].

It is instructive to examine the valence band data (UPS [22] and XPS) for Fe/O_2 in the light of the information obtained from the core-levels, and from the valence band spectra of the bulk oxides. Changes in the UPS spectrum show up at much lower exposures than in the XPS because of the differences in escape depths (fig. 4). The position of the O 2p structure (5.5 eV) is very clear in the UPS data owing to the high O 2p/Fe 3d ionization cross-sections ratio at HeI energies. The low O 2p/Fe 3d cross-sections ratio for Al Kα energies[14] means that Fe 3d derived structure dominates the XPS spectrum. In the UPS a 40 L exposure clearly results in transfer of Fe^0 3d intensity from ~0.6 eV to higher BE's. If Fe^{II} were the dominant product, a strong peak at ~1 eV should be observed (the $^6A_{1g}$ final state), which is not the case. The main new Fe 3d structure lies between ~3 and 10 eV, i.e., compatible with an Fe^{III} species. The presence of the residual 0.6 eV Fe^0 substrate signal partly obscures the $^6A_{1g}$ state region, however, so that we cannot rule out Fe^{II} as a minor species. The UPS data are thus compatible with the XPS core-level data interpretation that an Fe^{II}/Fe^{III} mixture is obtained at intermediate exposures. The XPS valence band spectra shows little at the lower exposures, because of the larger escape depths, and at the high exposures we know from the core-level data that Fe^{III} is dominant. The 10,000 L exposure XPS spectrum of fig. 4 then merely serves as confirmation that the band width of an Fe^{III} oxide is nearly 10 eV (cf. the bulk oxide spectra). Preliminary results on an Fe(100) surface indicate similar results to the films and foils.

4. Conclusions

We have obtained the characteristic XPS spectra of the surfaces of Fe_xO, $\alpha\text{-}Fe_2O_3$, Fe_3O_4, and FeOOH. Fe^{II} and Fe^{III} species are readily distinguished but octahedral and tetrahedral sites are not. The oxide O 1s BE is nearly invariant at $\sim$530.1 eV. The hydroxyl O 1s is $\sim$1 eV higher. The XPS valence band Fe 3d structures of the oxides spreads nearly 10 eV below E_F. These results are considered more reliable than previous USP and ESP Fe 3d valence band estimates where O 2p and scattered electron backgrounds are of large, but uncertain, magnitude.

Fe_3O_4 and Fe_xO samples were covered with a surface layer of Fe_2O_3. Ar^+ bombardment can remove the overlayer, but rapid reduction also takes place to eventually produce metal.

Oxygen or air exposure of Fe surfaces generates dominant Fe^{III} species at large exposures, but there is also evidence for Fe^{II}. Fe_3O_4 is tentatively considered to the product. At low oxygen exposures ($<$2 monolayer oxide product) the Fe^{II}/Fe^{III} ratio is greater, but definitive assignment will require more detailed studies at grazing emission angles to increase surface sensitivity.

Acknowledgments

Dr. J.L. Freeouf is thanked for the single crystal of Fe_xO, which was from the same batch as used in ref. [13]. Dr. R. Remeika is thanked for the $\alpha\text{-}Fe_2O_3$ crystal and Professor H. Siegmann for the Fe_3O_4 crystal. We are grateful for stimulating discussions with Dr. P.S. Bagus, Professor H. Siegmann, and Dr. W.Y. Lee concerning aspects of this work.

References

[1] A. Taylor and B.J. Kagle, Crystallographic Data on Metal and Alloy Structures (Dover, New York, 1963).
[2] J. Smith and H.P.J. Wijn, Ferrites (Wiley, New York, 1959).
[3] $\alpha\text{-}Fe_2O_3$ and Fe_3O_4 samples supplied by Fisher Scientific Co.; $\gamma\text{-}Fe_2O_3$ and FeOOH supplied by Mapico.
[4] G C. Allen, M.T. Curtis, A.J. Hooper and P.M. Tucker, J. Chem. Soc. Dalton Trans. (1974) 1525.
[5] K. Asami, K. Hashimoto and S. Shimodaira, Corrosion Sci. 16 (1976) 35.
[6] C.R. Brundle and A.F. Carley, Chem. Phys. Letters 31 (1975) 423.
[7] T.J. Chuang, C.R. Brundle and D.W. Rice, Surface Sci. 59 (1976) 413.
[8] J. Haber, J. Stoch, and L. Ungier, J. Electron Spectr. 9 (1976) 459.
[9] P.P. Gupta and S.K. Sen, Phys. Rev. B10 (1974) 71; B12 (1975) 15.
[10] D.C. Frost, C.A. McDowell and J.S. Woolsey, Mol. Phys. 27 (1974) 1473.
[11] S.F. Alvarado, M. Erbudak and P. Munz, Phys. Rev. B14 (1976) 2740, and references therein.
[12] P.S. Bagus, J.L. Freeouf and D.E. Eastman, Phys. Rev. Letters, to be published.

[13] D.E. Eastman and J.L. Freeouf, Phys. Rev. Letters 34 (1976) 395.
[14] J.H. Scofield, J. Electron Spectr. 8 (1976) 129.
[15] P.S. Bagus, C.R. Brundle, T.J. Chuang and K. Wandelt, to be published.
[16] P.W. Palmberg, in: Electron Spectroscopy, Ed. D.A. Shirley (North-Holland, Amsterdam, 1972) p. 835.
[17] M. Seo, J.B. Lumsden and R.W. Staehle, Surface Sci. 50 (1975) 541.
[18] K.S. Kim, W.E. Baitinger, J.W. Amy and N. Winograd, J. Electron Spectr. 5 (1976) 351.
[19] S. Thomas, Surface Sci. 55 (1976) 754.
[20] P.B. Sewell, D.F. Mitchell and M. Cohen, Surface Sci. 33 (1972) 535.
[21] R.J. Baird, C.S. Fadley, K.S. Kamamoto, S.M. Mehta, R. Alvares and J.A. Silva, Anal. Chem. 48 (1976) 61.
[22] C.R. Brundle, Surface Science, to be published.

HIGH RESOLUTION LMM AUGER ELECTRON SPECTRA OF SOME FIRST ROW TRANSITION ELEMENTS

G.C. ALLEN, P.M. TUCKER and R.K. WILD

Central Electricity Generating Board, Berkeley Nuclear Laboratories, Berkeley, Glos., England

Auger spectra have been recorded from elements of the first transition series using a hemispherical analyser. Highly resolved LMM spectra were obtained showing for the first time the composite nature of these peaks for many of the elements studied. The recorded spectra show a general similarity for the elements Sc → Zn but interesting differences emerge. At the beginning of the transition period the L_3 based transitions have the relative intensities $L_3M_{2,3}, M_{2,3} > L_3M_{2,3}M_{4,5} > L_3M_{4,5}M_{4,5}$ whereas towards the end of the series the order $L_3M_{4,5}M_{4,5} > L_3M_{2,3}M_{4,5} > L_3M_{2,3}M_{2,3}$ is observed. Pronounced chemical shifts have been observed upon oxidation. The spectra are interpreted in terms of an L–S coupling scheme and the fine structure discussed in terms of effects produced by multiplet splitting.

1. Introduction

The Auger process [1] has developed through a number of instrumental and electronic improvements to its present sophisticated state [2–4]. Recent Cylindrical Mirror Analysers (CMA) with $\Delta E/E \sim 5\%$ provide perfectly adequate resolution for Auger transitions of energy less than 100 eV, but show a progressive degradation as the energy of the Auger electron increases above this value so that the advantage of highlighting fine structure by differentiation is not always evident. Concurrent with developments in Auger spectroscopy has been the introduction of X-ray photoelectron spectroscopy (XPS) [6,7]. In order to obtain the maximum benefit from the use of XPS to derive information on the electronic structure of solids, instruments have been produced providing constant high resolution over a wide energy range. The collected signal is rarely differentiated though, so the recorded spectrum often contains Auger transitions superimposed upon an undulating background.

In this paper we illustrate the advantages of combining both methods of measurement. Examples are given which show the fine structure associated with the LMM Auger transitions for elements in the first transition series, and the changes that occur on oxidation. The LMM transitions themselves are also discussed in relation to the d shell configuration.

2. Experimental

Spectra were recorded using the modified ES 300B electron spectrometer shown schematically in fig. 1. This instrument was constructed with two sample preparation chambers (A and B), one on each side of the analysing chamber (C). The sample mounted horizontally on the probe was cleaned in the preparation chamber by a combination of 5 kV argon ion bombardment and heating to 600 K in a vacuum of $\sim 5 \times 10^{-8}$ Nm^{-2}. In the analysing chamber examination at $<2 \times 10^{-9}$ Nm^{-2} by AlK$_\alpha$ or MgK$_\alpha$ radiation, by monochromatic AlK$_\alpha$ radiation or by a focussed beam of electrons of up to 5 keV energy was possible. In practice though, a combination of excitation sources were used to ionise the L shell of the transition metals. Photon bombardment was used where the Auger peaks were well separated from the photoelectron peaks, but when interference occurred 4 kV electron excitation was used and X-rays employed only to confirm that beam effects from the electron gun could safely be discounted. When further cleaning was required the sample was heated and subjected to low energy argon ion bombardment in the analysing chamber.

The samples used were obtained from Goodfellow Metals Limited in the form of metal foils 0.125 mm in thickness. Oxidation was carried out at 600 K by leaking research grade oxygen (British Oxygen Company Limited) into the reaction chamber at a partial pressure of $\sim 10^{-6}$ Nm^{-2} for measured periods.

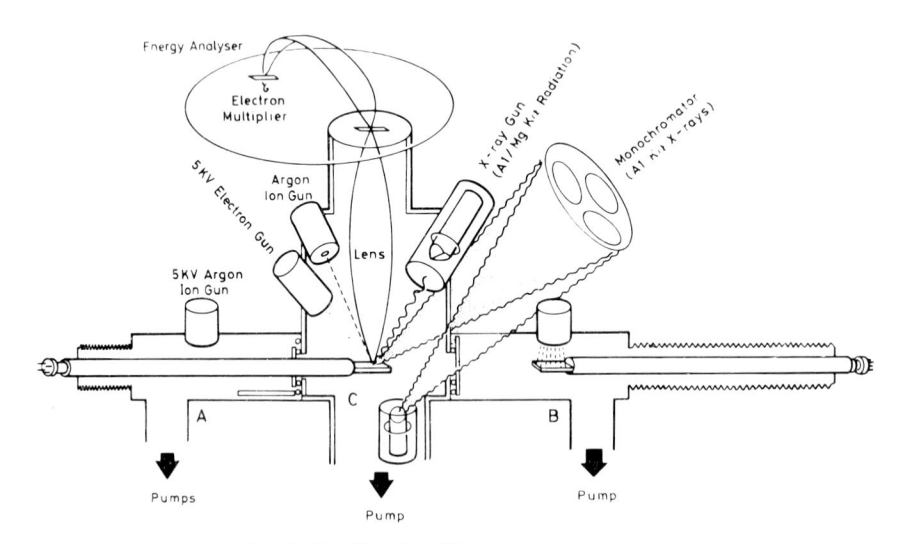

Fig. 1. Modified ES 200B electron spectrometer.

2.1. Area ratio determination

For each element the undifferentiated $N(E)$ Auger spectrum was recorded in the best detail possible on a 250 eV scan. The $L_{23}M_{23}M_{23}$, $L_{23}M_{23}M_{45}$ and $L_{23}M_{45}M_{45}$ transition groups (vide infra) with associated energy-loss contributions lay on a curving background due to constantly altering spectrometer transmission (the spectrometer was operated at constant resolution; 65 eV Fixed Analyser Transmission). Peak areas were obtained from each LMM envelope by first drawing the "best fit" line through the base of the contributing peaks reflecting as well as possible the constantly varying background. Having determined the overall peak shape and background variation, the gross area, including energy loss and satellite peaks of each LMM transition envelope was determined using a planimeter. These values were corrected to constant sensitivity using the spectrometer transmission function.

3. Results and discussion

The LMM Auger transitions recorded from the surface of some first row transition elements are shown in figs. 2a and 2b. Both the $N(E)$ and $N'(E)$ spectra are reproduced. Despite variations in intensity the main features of these spectra show a marked similarity and, rather conveniently, can be divided into three major groups belonging to Auger transitions of the following types: $L_3M_{23}M_{23}$, $L_3M_{23}M_{45}$ and $L_3M_{45}M_{45}$ [8,9]. To the high binding energy side of each group the considerably less intense set of $L_2M_{23}M_{23}$, $L_2M_{23}M_{45}$ and $L_2M_{45}M_{45}$ bands is also visible for elements towards the end of the transition series.

In recent years high resolution measurements have been reported for the metals Ni, Cu and Zn [10–13] and the spectra in fig. 2b do not show a significant improvement over previously published spectra for Cu and Zn. For the lighter elements though, high resolution data is less plentiful and the spectra shown in fig. 2a do represent an improvement over earlier results. Pronounced splitting is present in the $L_3M_{23}M_{45}$ groups as well as a number of fine splittings in the other lines.

Several authors [8,14] have remarked upon the intensity changes which occur across the series. Here both the $N(E)$ and $N'(E)$ spectra illustrate this phenomenon rather well. For elements at the beginning of the transition series where the 3d shell is sparsely populated (e.g. Ti) the L_3MM Auger transitions have relative intensities in the sequence $L_3M_{23}M_{23} > L_3M_{23}M_{45} > L_3M_{45}M_{45}$. For iron though, towards the middle of the series these transition groups have very similar intensities (see fig. 2a), but reference to fig. 2b shows that towards the end of the series where the 3d level is nearly filled the initial order is reversed.

These observations indicate that the electron population in the 3d shell has a direct bearing on the events which follow photoionisation when the vacancy so created is filled by an electron from an outer shell. Although the energy released in the second process may appear as a photon it is far more likely in the range of

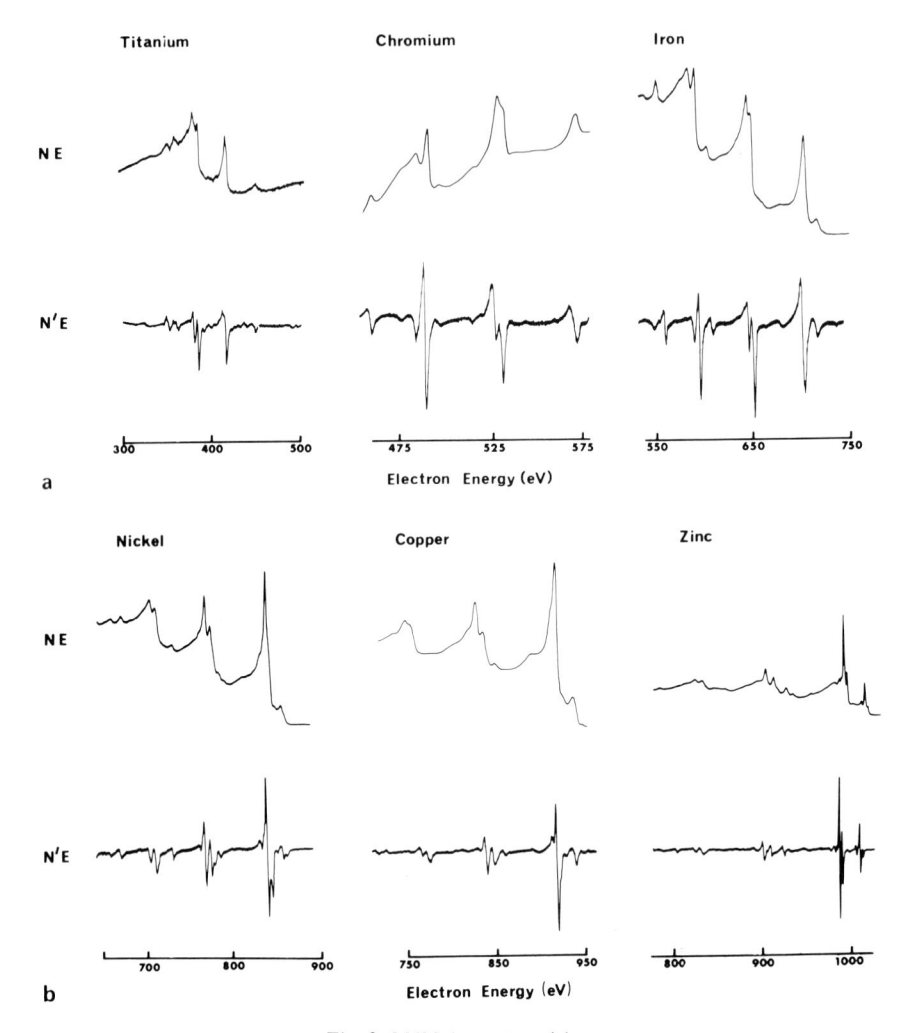

Fig. 2. LMM Auger transitions.

energies below 2 keV that a further electron will be released as an Auger electron [15].

Quantum mechanics can be used to explain the spectral lines observed when atoms decay from excited states on the basis of three selection rules; $\Delta n \geqslant 1$, $\Delta l = \pm 1$ and $\Delta j = 0, \pm 1$. For certain LMM transitions the sequence of events which follows electron ejection from the L_{23} shell contravenes the Δl and Δj rules because the Auger effect may be regarded as a radiationless transition due to the direct interaction of the two electrons concerned in the decay process.

It is necessary therefore if the observed intensities are to be rationalised, to consider all the nine possible transition groups which can occur following a primarily L_{23} hole state, namely, $L_{23}M_1M_1$, $L_{23}M_1M_{23}$, $L_{23}M_{23}M_1$, $L_{23}M_1M_{45}$, $L_{23}M_{45}M_1$, $L_{23}M_{23}M_{23}$, $L_{23}M_{23}M_{45}$, $L_{23}M_{45}M_{23}$ and $L_{23}M_{45}M_{45}$. Atoms in solids when undergoing Auger transitions can behave as though they are "quasifree". Yin et al. [10] noted that the $L_3M_{45}M_{45}$ Auger spectra of Cu and Zn are essentially free of solid state band structure. Thus it has been assumed here that it is equally probable that an electron will be ejected from an s, p or d shell and that an s, p or d electron will fall back to the L_{23} level during the process of de-excitation. Support for this notion in terms of p and d electrons may be taken from the fact that the $L_{23}M_{23}M_{23}$, $L_{23}M_{23}M_{45}$ and $L_{23}M_{45}M_{45}$ transitions are essentially equally intense in the spectrum recorded from Fe with the outer shell electronic configuration $3p^63d^64s^2$.

The intensity of the possible transitions may be written:

$$f_{L_{23}M_1M_1} = \left\{\frac{s^n}{(s^n+p^n+d^n)}\right\}\left\{\frac{(s^n-1)}{(s^n+p^n+d^n-1)}\right\}, \tag{1}$$

$$f_{L_{23}M_1M_{23}} \equiv f_{L_{23}M_{23}M_1} = \left\{\frac{s^n}{(s^n+p^n+d^n)}\right\}\left\{\frac{p^n}{(s^n+p^n+d^n-1)}\right\}, \tag{2}$$

$$f_{L_{23}M_1M_{45}} \equiv f_{L_{23}M_{45}M_1} = \left\{\frac{s^n}{(s^n+p^n+d^n)}\right\}\left\{\frac{d^n}{(s^n+p^n+d^n-1)}\right\}, \tag{3}$$

$$f_{L_{23}M_{23}M_{23}} = \left\{\frac{p^n}{(s^n+p^n+d^n)}\right\}\left\{\frac{(p^n-1)}{(s^n+p^n+d^n-1)}\right\}, \tag{4}$$

$$f_{L_{23}M_{23}M_{45}} \equiv f_{L_{23}M_{45}M_{23}} = \left\{\frac{p^n}{(s^n+p^n+d^n)}\right\}\left\{\frac{d^n}{(s^n+p^n+d^n-1)}\right\}, \tag{5}$$

$$f_{L_{23}M_{45}M_{45}} = \left\{\frac{d^n}{(s^n+p^n+d^n)}\right\}\left\{\frac{(d^n-1)}{(s^n+p^n+d^n-1)}\right\}, \tag{6}$$

where $f_{L_xM_yM_z}$ = probability of transition $L_xM_yM_z$ and s^n, p^n, d^n = number of s, p of d electrons in the shell of the neutral atom; and for the transitions of direct interest the following intensity ratios can be derived,

$$\frac{I(L_{23}M_{23}M_{45})}{I(L_{23}M_{23}M_{23})} \propto \frac{2 f_{L_{23}M_{23}M_{45}}}{f_{L_{23}M_{23}M_{23}}} = \frac{2\,d^n}{(p^n-1)}, \tag{7}$$

$$\frac{I(L_{23}M_{45}M_{45})}{I(L_{23}M_{23}M_{23})} \propto \frac{f_{L_{23}M_{45}M_{45}}}{f_{L_{23}M_{23}M_{23}}} = \frac{d^n(d^n-1)}{p^n(p^n-1)}, \tag{8}$$

$$\frac{I(L_{23}M_{45}M_{45})}{I(L_{23}M_{23}M_{45})} \propto \frac{f_{L_{23}M_{45}M_{45}}}{2 f_{L_{23}M_{23}M_{45}}} = \frac{(d^n-1)}{2\,p^n}; \tag{9}$$

for $Z = 21-30$ $p^n = 6$ giving

$$\frac{I(L_{23}M_{23}M_{45})}{I(L_{23}M_{23}M_{23})} \propto d^n, \qquad (10)$$

$$\frac{I(L_{23}M_{45}M_{45})}{I(L_{23}M_{23}M_{23})} \propto d^n(d^n - 1), \qquad (11)$$

$$\frac{I(L_{23}M_{45}M_{45})}{I(L_{23}M_{23}M_{45})} \propto (d^n - 1). \qquad (12)$$

That these simple relationships appear to hold is evident from the plots of d^n against $I(L_{23}M_{23}M_{45})/I(L_{23}M_{23}M_{23})$ and $d^n(d^n - 1)$ against $I(L_{23}M_{45}M_{45})/I(L_{23}M_{23}M_{23})$ shown in figs. 3a and 3b respectively; lending weight to the notion that Auger transitions do not obey simple selection rules.

3.1. Splitting of the LMM peaks

All of the LMM Auger spectra show some fine structure. The observed $L_3M_{23}M_{23}$ splitting shows a gradual increase across the transition series from ~5 eV in the Ti spectrum to ~9 eV for Cu (the Zn value of ~8.0 eV is less reliable in view of the low intensity of the $L_3M_{23}M_{23}$ transition for this element).

Coad [16] has shown that the $L_3M_{23}M_{23}$ transition results in the two hole final state configuration $3p^4$ $3d^n$. The dominant energy term, he argues, is the spin exchange splitting between the singlet ($S = 0$) and triplet ($S = 1$) states.

Splitting of the same order is present in the transitions designated $L_3M_{23}M_{45}$ where the two hole final state configuration may be written $3p^5$ $3d^{n-1}$. For titanium the splitting is not obvious in fig. 2, but careful examination of expanded spectra indicates that a small splitting of the order of 1–2 eV is present. On crossing the transition series to Cr though, a splitting of 4.0 eV is observed and as in the case of the $L_3M_{23}M_{23}$ transitions this separation continues to increase more or less uniformly until a value of 8.5 eV is attained in the Cu and Zn spectra. Previous measurements have indicated splittings between 6–9 eV for Ni, Cu and Zn [17] and smaller splittings for Fe and Co [18].

To account for the origin of this structure the transitions may be represented in terms of the entire "two-hole" outer electronic configuration in both instances, i.e. $3p^4$ $3d^n$ $4s^m$ or $3p^5$ $3d^{n-1}$ $4s^m$ as the case may be. The observed separations may then be related to exchange interaction between the unpaired electrons in the "two-hole" state in a manner somewhat analogous to that observed in X-ray photoelectron spectra of open shell compounds of these elements.

3.2. N(E), N'(E) spectra and chemical shifts

The advantage of displaying spectra in two forms is exemplified by the Zn spectra shown in fig. 4. The $N(E)$ curve shows the $L_3M_{45}M_{45}$ transition as a large peak

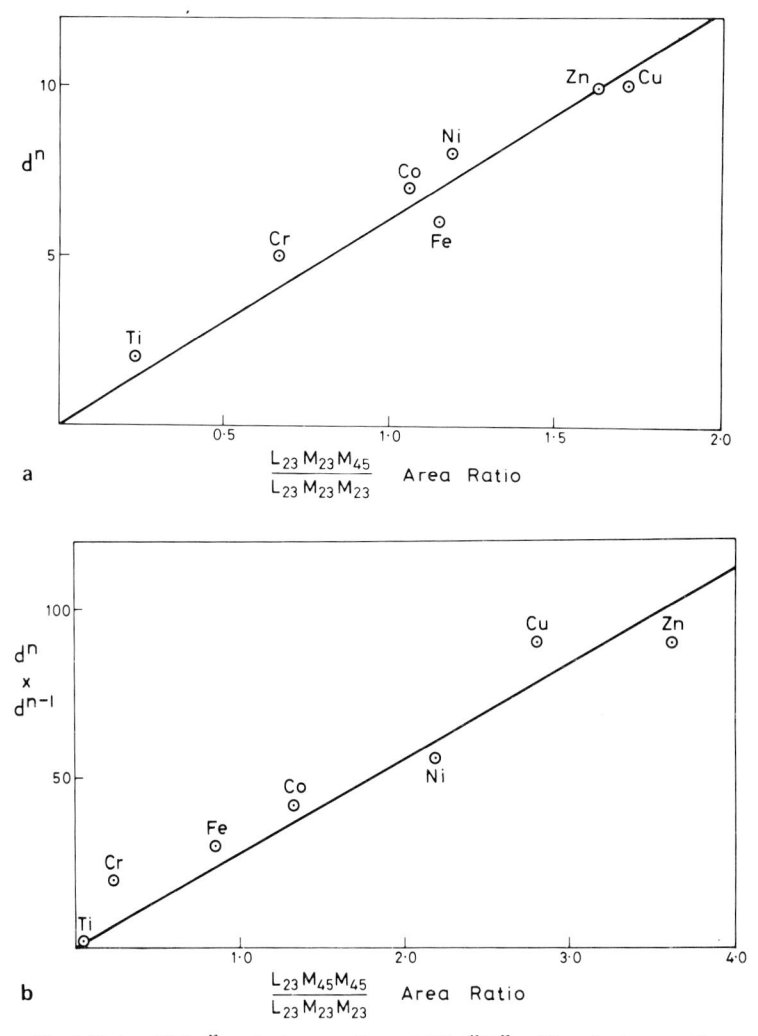

Fig. 3 Plots of (a) d^n against area ratio and (b) $d^n(d^n - 1)$ against area ratio.

with a shoulder to the high energy side. Although less intense, the corresponding $L_2M_{45}M_{45}$ transition has a similar form. As indicated in the figure, shoulders on these peaks can be identified more readily in the $N'(E)$ spectrum where the position of the transition responsible for the shoulder can be obtained from the minimum of the differential curve. However, the $N'(E)$ spectrum is less useful for intensity measurements and in fig. 4 it can be seen that this representation does not give a true impression of the relative peak heights or the low energy tail to one side of the main peak.

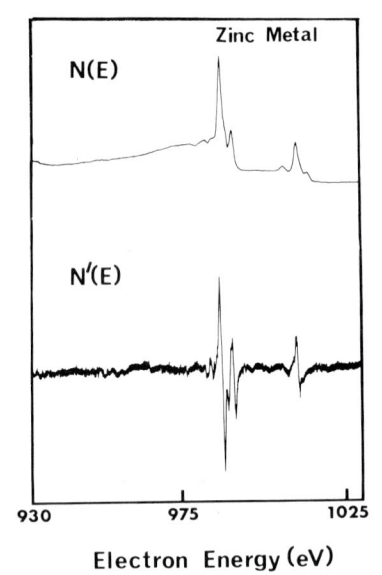

Fig. 4. $N(E)$ curves for zinc.

It is interesting to reflect that if the spectrum shown in fig. 4 had been recorded using a CMA with $\Delta E/E = 0.5\%$, the resolution at 1000 eV would be 5 eV and the $L_3M_{45}M_{45}$ transition envelope would be visible as a single peak rather than the five components shown here. Moreover, increased resolution aids the interpretation of

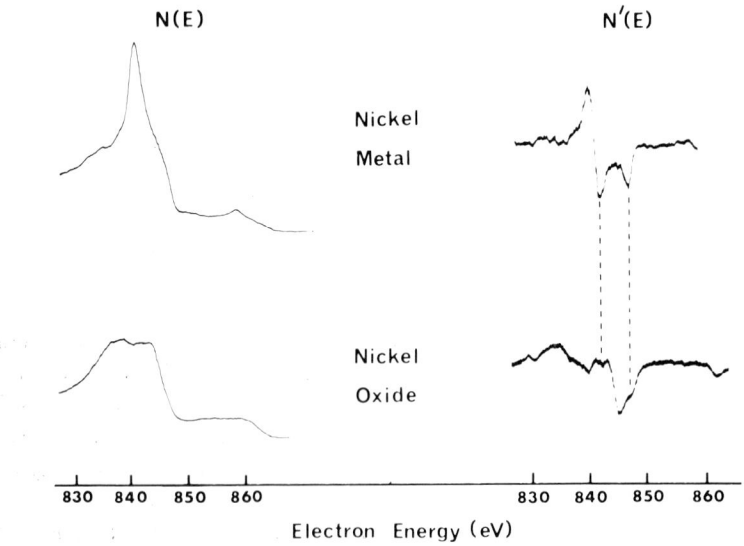

Fig. 5. $N(E)$ and $N'(E)$ curves for nickel and nickel oxide.

changes which occur in Auger spectra recorded following oxidation of the surface under investigation. An example of this is shown in fig. 5. In fig. 5a the Ni $L_3M_{45}M_{45}$ transition at 848 eV has been recorded from a clean nickel surface and reproduced in the $N(E)$ and $N'(E)$ form. After an exposure of 14.4×10^3 L oxygen at 600 K the spectrum shown in fig. 5b was obtained. The $L_3M_{45}M_{45}$ transition from the clean surface is a doublet which the $N'(E)$ spectrum portrays as two peaks of almost equal intensity. The $N(E)$ spectrum though, consists of a large peak with a weaker peak at slightly higher kinetic energy. On oxidation these spectra show considerable change. The two components in the metal $N(E)$ curve remain, but the high energy shoulder shows a considerable increase in amplitude. In the $N'(E)$ spectrum though the high energy peak appears to have moved to lower energy and is now the dominant component although the metal peaks are just visible through the oxide.

Fiermans, Hoogewijs and Vennik [19] interpreted their $L_3M_{45}M_{45}$ spectra from NiO by identifying the high energy shoulder with the presence of a shake-up satellite. The present results indicate that the two-component spectrum in the metal $N(E)$ curve remains in the two-component form on oxidation, but the shoulder increases in amplitude. In this respect the present results for Ni resemble those of Fiermans et al. [19] for Co and Cu and we conclude therefore, that shake-up satellites are absent in the Auger spectra of these elements.

Acknowledgements

This paper is published by permission of the Central Electricity Generating Board.

References

[1] J.J. Lander, Phys. Rev. 91 (1953) 1382.
[2] L.A. Harris, J. Appl. Phys. 39 (1968) 1419.
[3] R.E. Weber and W.T. Peria, J. Appl. Phys. 38 (1967) 4355.
[4] P.W. Palmberg and T.N. Rhodin, J. Appl. Phys. 39 (1968) 2425.
[5] P.W. Palmberg, G.K. Bohn and J.C. Tracy, Appl. Phys. Letters 15 (1969) 254.
[6] K. Siegbahn, C. Nordling, A. Fahlman, R. Nordberg, K. Hamrim, J. Hedman, G. Johansson, T. Bergmark, S-E. Karlsson, I. Lindgren and B. Lindberg, ESCA-Atomic Molecular and Solid State Structure Studied by Means of Electron Spectroscopy, Nova Acta Regia Soc. Sci. Upsaliensis Ser. IV, 20 (1967).
[7] K. Sieghbahn, J. Electron Spectrosc. 5 (1974) 3.
[8] M.F. Chung and L.H. Jenkins, Surface Sci. 28 (1971) 637.
[9] L. Yin, T. Tsang, I. Adler and E. Yellin, J. Appl. Phys. 43 (1972) 3464.
[10] L. Yin, I. Adler, T. Tsang, M.H. Chen, D.A. Ringers and B. Crasemann, Phys. Rev. A9 (1974) 1070.
[11] P. Weightman, J.F. McGilp and C.E. Johnson, J. Phys. C9 (1976) L585.

[12] E.D. Roberts, P. Weightman and C.E. Johnson, J. Phys. C8 (1975) L301.
[13] K.S. Kim, S.W. Gaarenstroom and N. Winograd, Chem. Phys. Letters 41 (1976) 503.
[14] L. Yin, I. Adler, M.H. Chen and B. Crasemann, Phys. Rev. A7 (1973) 897.
[15] H.E. Bishop and J.C. Rivière, J. Appl. Phys. 40 (1969) 1740.
[16] J.P. Coad, Phys. Letters 37A (1971) 437.
[17] J.P. Coad and J.C. Rivière, Z. Physik 244 (1971) 19.
[18] S. Aksela, M. Pessa and M. Karrass, Z. Physik 237 (1970) 381.
[19] L. Fiermans, R. Hoogewijs and J. Vennik, Surface Sci. 47 (1975) 1.

PHOTOELECTRON SPECTROSCOPY OF LOCALIZED LEVELS NEAR SURFACES: SCATTERING EFFECTS AND RELAXATION SHIFTS

M. ŠUNJIĆ *,** and Ž. CRLJEN *

Institute of Theoretical Physics, Chalmers University of Technology, Fack, S-402 20 Göteborg 5, Sweden

and

D. ŠOKČEVIĆ

Institute R. Bošković, Zagreb, Croatia, Yugoslavia

We calculate the spectrum of photoelectrons excited from localized levels in solids far into the continuum (XPS situation), in a model that takes into account exactly their interaction with a boson-like spatially non-uniform field. The general results are applied to discuss the spectra of photoelectrons in metals, where the dominant interaction is the coupling to bulk and surface plasmons. In particular, we study the modification of spectral sum rules connecting the strength of inelastic processes with the energy shifts in the spectrum, showing that the electron ("extrinsic") scattering changes the average position of the spectrum with respect to the pure "intrinsic" result in the "sudden approximation", which leads also to its dependence on the excitation energy and position of localized level. We estimate these deviations in a simple model which nevertheless reproduces correctly the dynamical aspects of the problem.

1. Introduction

Inelastic effects are notoriously difficult to take into account in the theory of photoemission from solids, though they are known to be important for any quantitative calculation of the photoelectron spectrum. A systematic approach has been developed by Chang and Langreth [1] who used the Green function perturbation method to calculate the effects of excited electron and hole interaction with bulk and surface plasmon fields in metals, which seems to be the dominant inelastic scattering mechanism in X-ray photoemission. Several authors [2,3,5] used a different, physically equivalent, nonperturbative approach, based on the semiclassical calculation of the correlation function, describing the response of the plasmon field to the creation of the electron-hole pair. This approach gave useful quantitative predictions for the strengths of multiple scattering processes, complementing the results

* Permanent address: Institute R. Bošković, Zagreb, Croatia, Yugoslavia.
** Also at NORDITA, Copenhagen, Denmark.

of perturbative calculations [1]. It became obvious that the hole ("intrinsic") and electron ("extrinsic") scattering, as well as their interference, depend in a nontrivial way on the parameters of the experiment, in particular on the excitation energy, distance of the localized level from the surface and the electron exit angle. Theoretical calculations and analysis of experimental data [4–6] confirmed the importance of "extrinsic" processes in solid state photoemission. This is in contrast to similar gas phase experiments, where the high energy electron is usually supposed to leave the atom or molecule without scattering ("sudden approximation"), so that its spectrum is given as a replica of the hole spectral function $A_h(E)$. This latter assumption leads to a set of simple sum rules [7–10], e.g. relating the screening ("hole relaxation") energy shift ΔE_h, the (adiabatic) threshold peak E_r and the intensity of the inelastic scattering (in our case plasmon satellites), i.e. the shape of the spectrum [11]:

Spectral sum rule: $\quad \int A_h(E - \omega) E \, dE = E_{fo} = E_r - \Delta E_h$. (1)

Here ω is the photon energy and E_{fo} is the "frozen orbital" energy, corresponding to the "unrelaxed" core hole state. The scattering effects in the final state redistribute the spectral weight but not the average or Koopmans' theorem energy of the photoelectrons. In other words, the energy gets redistributed between the hole and (plasmon) excitations, but the average energy of the system is unchanged.

However, if "extrinsic" effects become important, photoelectron spectrum is no longer given by the hole spectral function, but has to be derived taking inelastic effects into account, e.g. from the correlation function in the quadratic response [12,13].

In section 2 we therefore calculate the spectrum, assuming a localized initial state and a high energy final electron, interacting with a general boson like field. In section 3 we analyze the general features of this spectrum, and find the modification of the spectral sum rule (1). In section 4 we apply the results to the case of XPS from metals, where the charges interact with bulk and surface plasmons. We see that, e.g. for adsorbed core levels, the outgoing electron can deexcite the system and partially suppress the hole scattering, and thus modify the sum rule (1). If the "extrinsic" effects dominate, and this seems to be the case in bulk XPS, the "shifts", or better, the average weight of the photoelectron spectrum actually increase from their "sudden approximation" values and lose any direct relation to the "hole relaxation shifts" ΔE_h.

2. Calculation of the photoelectron spectrum

The system contains electrons in localized initial states λ at lattice sites R_j and extended (wave-packet like) final states with average kinetic energy $E_p = p^2/2m$:

$$H_{el} = \sum_{\lambda,j} E_\lambda c_\lambda^\dagger c_\lambda \int |\phi_\lambda(r - R_j)|^2 \, dr$$

$$+ \sum_p c_p^\dagger c_p \int\int \phi_p^*(r - r_e) \left(E_p - \frac{i}{m} p \cdot \nabla\right) \phi_p(r - r_e) \, dr \, dr_e \, . \tag{2}$$

r_e is the final electron position operator, and in H_{el} we have neglected the terms in kinetic energy corresponding to the quadratic deviation from the mean electron velocity p. This particular form of the hamiltonian is necessary because we want to take into account explicit spatial dependence of the interaction matrix elements $V_i(x)$ [14].

The electrons are coupled to the boson field:

$$H_b = \sum_i \omega_i a_i^\dagger a_i \, , \tag{3}$$

by the interaction:

$$H_{int} = \sum_i \int dx \, V_i(x) \, a_i \{\rho_0(x) + \rho_f(x)\} + \text{h.c.} \, , \tag{4}$$

where

$$\rho_0 = \sum_{\lambda,j} c_\lambda^\dagger c_\lambda \, |\phi_\lambda(x - R_j)|^2 \, , \qquad \rho_f = \sum_p c_p^\dagger c_p \int |\phi_p(x - r_e)|^2 \, dr_e \, , \tag{5}$$

are the localized and extended electron densities, respectively.

We consider photoexcitation from the single localized orbital E_λ at the site R_j into the continuum state with asymptotic energy $E = E_p - V_0$ ($V_0 =$ inner potential). Photoelectrons are excited by the coupling

$$h = -\frac{1}{c} \int dr \, j(r) \cdot A(r, t) \, e^{\eta t} \tag{6}$$

of the electron current $j(r)$ to the external monochromatic radiation field of frequency ω:

$A(r, t) = A(r) \, e^{i\omega t} + \text{h.c.}$

The steady current at R is given by the expression [12,13]:

$$j_\mu(R) = \frac{e^2}{c^2} \sum_{\sigma,\nu} \int\int dr_1 \, dr_2 \, A_\nu(r_1) A_\sigma^*(r_2) \int\int_{-\infty}^{0} dt_1 \, dt_2$$

$$\times \exp[i\omega(t_1 - t_2) + \eta(t_1 + t_2)] \, R_{\mu\nu\sigma}(r_1, r_2, R, t_1, t_2) \, , \tag{7}$$

where

$$R_{\mu\nu\sigma} = \langle \Psi | j_\nu(r_1, t_1) j_\mu(R, 0) j_\sigma(r_2, t_2) | \Psi \rangle \, . \tag{8}$$

Here $|\Psi\rangle = |E_\lambda, [n_i]\rangle$ is the eigenstate of the full hamiltonian

$$H = H_0 + H_{\text{int}}, \qquad H_0 = H_{\text{el}} + H_{\text{b}},$$

containing a localized state E_λ coupled to the boson field, which is described by its occupation numbers $[n_i]$. The current operators also depend on H:

$$j(r, t) = e^{iHt} j(r) e^{-iHt}.$$

Transforming the correlation function (8) into the interaction representation and assuming the system to be decoupled adiabatically ($H_{\text{int}} \to 0$) at time $t \to -\infty$, we find:

$$|\Psi\rangle = U(0, -\infty)|\Psi_{\text{I}}(-\infty)\rangle, \qquad |\Psi_{\text{I}}(-\infty)\rangle = |E_\lambda\rangle|[0]\rangle,$$

where we have taken the boson system to be initially in the ground state $|[0]\rangle$. The time evolution operator is

$$U(t, t_0) = T \exp\!\left(-i \int_{t_0}^{t} \widetilde{H}_{\text{int}}(\tau)\, d\tau\right),$$

$$\widetilde{H}_{\text{int}}(\tau) = \exp(iH_0\tau)\, H_{\text{int}}\, \exp(-iH_0\tau). \tag{9}$$

For the particular form of the H_{int} and H_{el} it is possible to calculate $\widetilde{H}_{\text{int}}$ and $U(t, t_0)$ exactly, neglecting the distortion of the localized orbital and the wave packet due to emission/absorption of the boson quata [15]. The result is:

$$\widetilde{H}_{\text{int}} = \sum_i \int d\mathbf{x}\, V_i(\mathbf{x})\, a_i \exp(-i\omega_i\tau) \left[\sum_\lambda c_\lambda^\dagger c_\lambda |\phi_\lambda(\mathbf{x} - \mathbf{R}_j)|^2 \right.$$

$$\left. + \sum_p c_p^\dagger c_p \int d\mathbf{r}_e \left|\phi_p\!\left(\mathbf{x} - \mathbf{r}_e - \frac{\mathbf{p}}{m}\tau\right)\right|^2 \right] + \text{h.c.} \tag{10}$$

We notice that

$$H|\Psi\rangle = (E_\lambda + \Delta E_\lambda)|\Psi\rangle,$$

where *"intrinsic" energy shift* is

$$\Delta E_\lambda = \sum_i |Q_i^{\text{h}}|^2 \omega_i$$

$$Q_i^{\text{h}} = \int_{-\infty}^{0} d\tau \int d\mathbf{x}\, V_i(\mathbf{x})\, |\phi_\lambda(\mathbf{x} - \mathbf{R}_j)|^2 \exp(-i\omega_i\tau + \eta\tau). \tag{12}$$

The photoelectron spectrum is given by

$$\frac{dj}{d\Omega} = \lim_{R \to \infty} R^2\, \hat{R} \cdot \mathbf{j}(\mathbf{R}) = \int \frac{d^2 j}{d\Omega\, dE}\, dE \tag{13}$$

In order to find the asymptotic limit $R \to \infty$ we can express the current operator $j(R, 0)$ in (8) in terms of asymptotic states [16], insert a complete set of final (interacting) states $|p,[n_i']\rangle = U(0, \infty)|p\rangle|[n_i']\rangle$ and perform the time integrations. Exactly as in the noninteracting electron case, this procedure would yield the energy conserving δ-functions for a specific choice of final states. In our case we prefer to obtain a nonperturbative solution including all possible final states by removing $1 = \Sigma |[n_i']\rangle\langle[n_i']|$ from the correlation function (8) and calculating the (boson) ground state average. After some algebra involving exponential boson operators we obtain a result that looks rather familiar [16,23]:

$$\frac{d^2 j}{d\Omega\, dE} = e \frac{2\pi m}{\hbar^2} \left(\frac{2mE}{\hbar^2}\right)^{1/2} |\langle E_\lambda|h|p\rangle|^2 A(E + V_0 - \omega) , \tag{14}$$

where the matrix element is

$$\langle E_\lambda|h|p\rangle = -\frac{i\hbar}{2m} \int dr\, \phi_p^*(r)\, A(r) \cdot \nabla \phi_\lambda(r) , \tag{15}$$

and the final energy $E = E_p - V_0$, is measured in vacuum.

The difference from the noninteracting electron case is contained in the spectral function $A(E)$, which now differs from the sharp δ-function shape, and was the main object of our calculation:

$$A(E) = \int \frac{dt}{2\pi} \exp\{i(E - E_\lambda - \Delta E_\lambda)t + B(t)\} \tag{16}$$

$$B(t) = \sum_i |Q_i^h + Q_i^e|^2 [\exp(-i\omega_i t) - 1] . \tag{17}$$

$B(t)$ is the "satellite generator" [8] of the photoelectron spectrum. Q_i^h is given by (12), and similarly:

$$Q_i^e = \int_0^\infty d\tau \int dx\, V_i(x)\, |\phi_p(x - R_j - \frac{p}{m}\tau)|^2 \exp(-i\omega_i \tau - \eta\tau) . \tag{18}$$

We note that both Q's depend in principle on R_j through the matrix element $V_i(x)$, Q_i^e also depends on the final momentum p/m. The factorized form of the spectrum (14) is possible due to the fact that the interaction (10) describes essentially forward scattering processes, so the final electron current is conserved after the excitation in each direction separately, as we shall see in the next section.

3. General properties of the spectral function $A(E)$

From (16)–(18) we can see the following general features of the photoelectron spectrum $A(E)$:

(i) The spectrum is normalized to one electron per absorbed photon:

$$\int A(E) \, dE = 1 \tag{19}$$

(ii) The spectrum (at $T = 0$) begins at $E_r = E_\lambda + \Delta E_\lambda$ (the position of the main or "no-loss" line), which is the energy of the electron state λ in the fully interacting system ("relaxed hole") [18].

(iii) Satellite peaks in the spectrum occur at multiples of excitation frequencies ω_i away from the relaxed level E_r:

$$E + V_0 - \omega = E_r - \sum_i n_i \omega_i, \tag{20}$$

corresponding to the eigenstates of the interacting hole–boson system.

(iv) The average (first moment) of the spectrum is not given by the spectral sum rule (1), but instead

$$E_{av} = \int E \, A(E) \, dE = E_\lambda + \Delta E_\lambda - \sum_i |Q_i^h + Q_i^e|^2 \, \omega_i. \tag{21}$$

Here E_λ corresponds to the "frozen orbital" energy E_{fo} in (1), and ΔE_λ is given by (11). The sum rule (1) is recovered when the electron–boson scattering term Q_e^i vanishes, which is in general not the case. Therefore we can write:

$$E_{av} = E_{fo} - \Delta E_e - \Delta E_i; \tag{22}$$

"Extrinsic" shift $\quad \Delta E_e = \sum_i |Q_i^e|^2 \, \omega_i;$

Interference term $\quad \Delta E_i = 2 \, \text{Re} \sum_i (Q_i^h)^* (Q_i^e) \, \omega_i < 0.$

Depending on the relative magnitude of these "extrinsic" and interference terms E_{av} can be higher or lower than E_{fo}.

4. X-ray photoemission from metals: interaction with bulk and surface plasmons

We now apply these results to discuss the influence of electron and hole interaction with bulk and surface plasmons in XPS from metals. The strengths of plasmon satellites in the spectrum have been discussed previously [1–4,19], so we shall concentrate on the discussion of energy shifts and sum rules when "extrinsic" scattering becomes important.

In the following we shall assume a perfectly localized initial state at R_j (see refs. [5] and [20] for a discussion of a more general case), and a narrow final wave packet, approximating the densities in (12) and (18) by δ-functions. We use the dispersionless plasmon model [21] which has the advantage that the results can be ob-

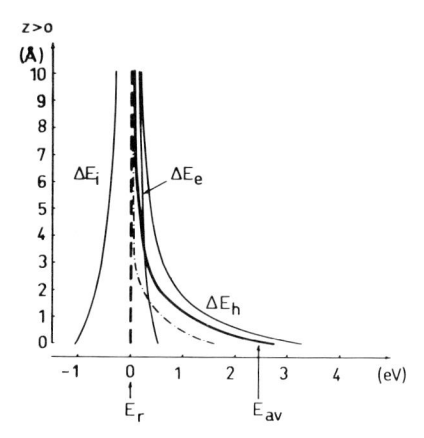

Fig. 1. Spatial variation of energy shifts for electrons excited from core levels of adsorbed atoms at distance z outside the solid. (Parameters: $E = 1176$ eV, surface plasmon energy $\hbar\omega_s = 10.75$ eV, exit angle $\theta = 0$.) ΔE_h is the upward shift of the adiabatic threshold line. E_{av} is here plotted with respect to the fixed energy $E_r(z \to \infty)$, which otherwise varies with z. Dash-dotted line denotes the position of E_{av} in the case of (unphysically) low kinetic energy ($E = 50$ eV), and is shown here to illustrate the extreme cancellation effects.

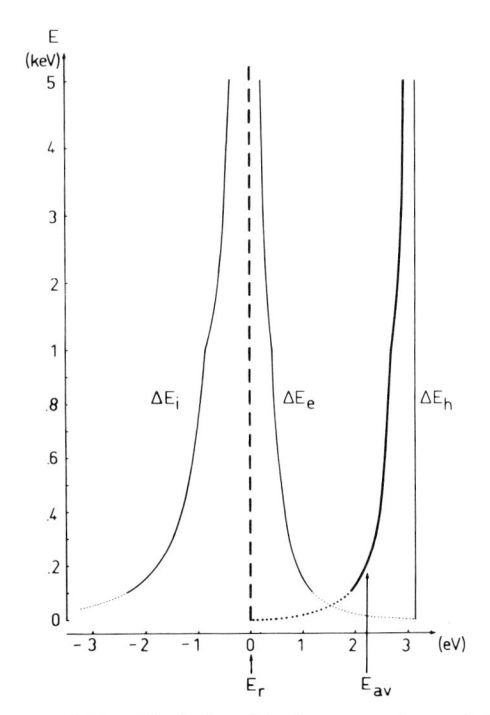

Fig. 2. Variation of energy shifts with electron kinetic energy, for excitation from the surface ($z = 0$). (Parameters same as in fig. 1.) Note change of scale at $E = 1$ keV.

tained analytically, though the treatment could be in principle extended to a more realistic plasmon hamiltonian. We shall omit somewhat lengthy analytic expressions, but instead discuss the qualitative features of the results presented in figs. 1–3.

Let us first consider excitation from the localized level outside the solid (figs. 1 and 2), e.g. from the core levels of adsorbed atoms [22]. The interference terms ΔE_i are always larger than the "extrinsic" terms ΔE_e, which leads to an overall reduction of the shift ΔE. Due to the fact that interaction with surface plasmons is localized near the surface, this situation in a way corresponds to the atomic case, where one also expects a reduction of the strength of inelastic structures in the photoelectron spectrum for lower electron kinetic energies [23]. It must be stressed

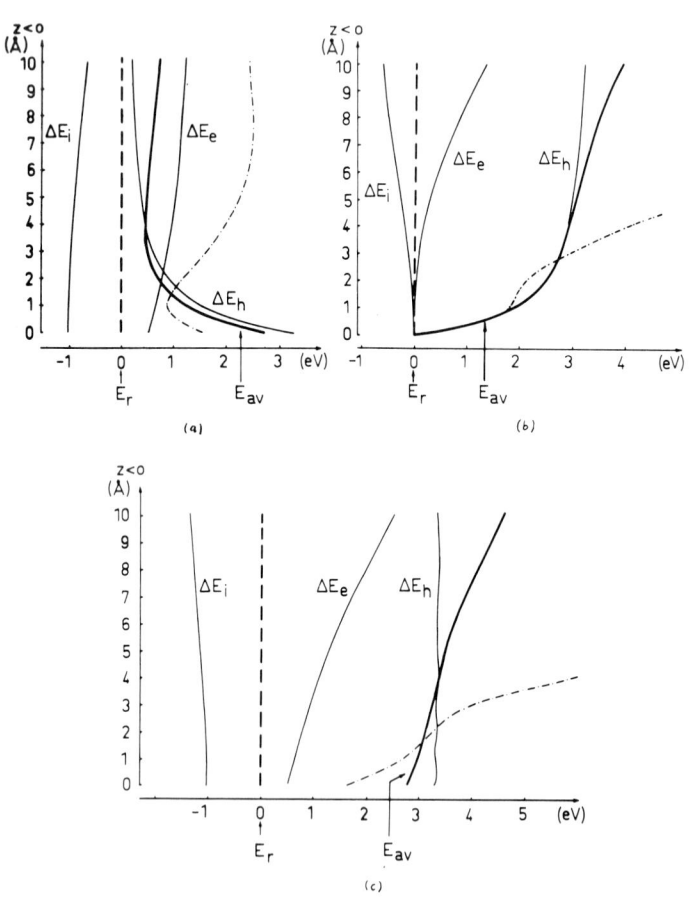

Fig. 3. Spatial variation of energy shifts for electrons originating from distance z inside the solid, due to (a) surface and (b) volume plasmon excitation, and (c) the total shifts. (Parameters same as in fig. 1, $\hbar\omega_p = 15.2$ eV.)

that the interaction (surface plasmon excitation) differs from that in atoms, having a very long range compared to the excitation mechanisms in atoms. This enhances the interference terms, because the correlation between hole and electron scattering persists for larger separations between them, so that electrons partially screen the hole. The importance of this cancellation could be estimated by comparing the average time τ the electron with velocity v spends in the region of the potential, which has a characteristic range d: $\tau = d/v$ with the characteristic time for the response of the boson system $1/\omega_i$. The importance of electron coupling should decrease for $\tau \ll \omega_i^{-1}$, i.e. with increasing v, as is visible in fig. 2.

The situation is very much different when the electrons originate from inside the solid (figs. 3a, b, c), where they can couple to the volume plasmons. The interaction of outgoing electrons now occurs over an extended region of the solid, so that the extrinsic coupling (ΔE_e) can dominate over the interference terms ΔE_i (fig. 3c). In our specific calculation this occurs at $z \sim 4$ Å inside the solid for relatively fast electrons ($E = 1176$ eV), and this distance decreases at lower energies. As a consequence the total shift is *increased* with respect to the "sudden approximation" result ΔE_h, which is not recovered even for relatively large electron kinetic energies. We can say that at this point the influence of the localized (surface plasmon) potential scattering (fig. 3a) is overcome by the effect of the extended (bulk plasmon) potential (fig. 3b), so the situation changes from the atomic-like to the solid-state transport-like regime.

A word of caution should be added here: The inclusion of plasmon dispersion in general would reduce the hole screening, especially close to the surface region, so that the "extrinsic" terms ΔE_e may actually dominate even closer to the surface.

5. Conclusions

Apart from the discussion of the photoelectron spectrum given in section 3 and the analysis of XPS from metals in section 4 we have to add a few general comments.

Calculations of atomic photoionization spectra have already shown that the concepts of relaxation shifts and spectral sum rules become difficult to apply when the final state scattering processes are taken into account systematically [23]. Nevertheless, they often persist (especially in solid state physics) and even prove intuitively useful. Therefore we tried here to show on a model problem (sections 2 and 3) how final state scattering leads to the breakdown of the "sudden approximation" and the new set of spectral sum rules.

Furthermore, when applied to XPS from metals, our model clarifies the problem of the "natural energy scale" for this process [24], and confirms the theoretical [5] and experimental [24] conclusions that take the "relaxed" energy E_r as the relevant threshold binding energy.

The plasmon scattering studied in section 4 represents an extreme case – the long range interaction which consequently emphasizes the role of interference terms in the matrix elements. Other short range scattering processes in the solid photoemission (secondary electron excitation, ionization losses), as well as all intraatomic relaxation mechanisms, probably do not require such an elaborate treatment. In these cases we would expect photoemission process to be adequately described by the step-model, which treats "intrinsic" and "extrinsic" scattering separately.

Acknowledgements

It is a pleasure to thank Professors Stig Lundqvist, Alf Sjölander and Göran Wendin for useful discussions and hospitality at the Institute of Theoretical Physics, Göteborg. Financial support from NORDITA, Copenhagen is gratefully acknowledged.

References

[1] J.-J. Chang and D.C. Langreth, Phys. Rev. B8 (1973) 4638.
[2] G.D. Mahan, Phys. Status Solidi (b) 55 (1973) 703.
[3] M. Šunjić and D. Šokčević, Solid State Commun. 15 (1974) 165.
[4] W.J. Pardee, G.D. Mahan, D.E. Eastman, R.A. Pollak, L. Ley, F.R. McFeely, J.P. Kowalczyk and D.A. Shirley, Phys. Rev. B11 (1975) 3614.
[5] A.M. Bradshaw, W. Domcke and L.S. Cederbaum, Phys. Rev. B15 (1977).
[6] S.A. Flodström, R.Z. Bachrach, R.S. Bauer, J.C. McMenamin and S.B.M. Hagström, J. Vacuum Sci. Technol. 14 (1977) 303.
[7] B.I. Lundqvist, Physik Kondens. Materie 9 (1969) 236.
[8] S.Doniach, in: Computational Methods in Band Theory, Eds. P.M. Marcus, J.F. Janak and A.R. Williams (Plenum, New York, 1971) p. 500.
[9] R. Manne and T. Åberg, Chem. Phys. Letters 7 (1970) 282.
[10] See also J.W. Gadzuk, in: Photoemission from Surfaces, Eds. B. Feuerbacher, B. Fitton and R.F. Willis (Wiley, to be published) ch. 7, for a recent review of many-body effects in photoemission.
[11] For a particularly clear discussion of this sum rule, see D.C. Langreth, in: Collective Properties of Physical Systems, Eds. B.I. Lundqvist and S. Lundqvist (Academic Press, New York, 1974) p. 210.
[12] W.L. Schaich and N.W. Ashcroft, Phys. Rev. B3 (1971) 2452.
[13] H. Hermeking, Z. Physik 253 (1972) 379.
[14] The set of wave packet states p is not orthogonal to the set of localized states λ. (We thank Professor T.B. Grimley for drawing our attention to this point.) However, it can be shown (see M.L. Goldberger and K.M. Watson, Collision Theory (Wiley, New York, 1964) Appendix C) that the wave functions for these two scattering channels are indeed orthogonal when projected into the subspace of asymptotic scattering states, as we shall do when calculating the photoemission current.
[15] See e.g. E.P. Gross, in: Mathematical Methods in Solid State and Superfluid Theory (Oliver and Boyd, Edinburgh, 1967) p. 46.
[16] H. Hermeking and R.P. Wehrum, J. Phys. C8 (1975) 3468.

[17] G.D. Mahan, Phys. Rev. B2 (1970) 4334.
[18] Much of the discussion in this section was anticipated in ref. [5] on the basis of a semiclassical (trajectory) model.
[19] M. Šunjić, D. Šokčević and A. Lucas, J. Electron Spectrosc. 5 (1974) 964.
[20] J.W. Gadzuk, Phys. Rev. B14 (1976) 2267.
[21] M. Šunjić and A. Lucas, Phys. Rev. B3 (1971) 719.
[22] This case has also been studied, e.g., by: J. Harris, Solid State Commun. 16 (1975) 671;
M. Šunjić and D. Šokčević, Solid State Commun. 18 (1976) 373;
A. Datta and D.M. Newns, Phys. Letters 59A (1976) 326;
and refs. [5] and [19].
[23] See G. Wendin, in: Photoionization and Other Probes of Many-Electron Interactions, Ed. F. Wuilleumier (NATO ASI Series, Plenum 1976), and references therein.
[24] Y. Baer, Solid State Commun. 19 (1976) 669.

THEORY OF ANGLE-RESOLVED PHOTOEMISSION FROM LOCALISED ORBITALS AT SOLID SURFACES

B.W. HOLLAND

Physics Department, University of Warwick, Coventry CV4 7AL, England

The theory of angle resolved photoemission from localised orbitals is reviewed and is cast in a form requiring the calculation of the purely outgoing wave emanating from an emitting atom, that describes the final state of the photoelectron, rather than using the more usual approach based on time reversed scattering states. An explicit expression is written down for the superposition of partial waves that results from emission from an atomic orbital, and it is pointed out that emission from more complex initial states such as localised bonds or Bloch and surface states, can be described by coherently combining such sets of partial waves. The effect of the crystal surface environment in damping and scattering the waves is described briefly. Model calculations are performed to investigate the major influences on the angular distribution of the photoelectrons. The profound effect of varying the polarisation direction of the incident light relative to the surface is discussed, with examples from the literature, showing how it can be used to determine the type of initial orbital. Emission from directed orbitals is studied and it is shown that scattering by the emitter potential can be an important effect, so that the radial wave function of the outgoing electron, which determines the amplitudes and phases of the outgoing waves, must be calculated with care. Different choices of these quantities lead in the model calculations to very different angular profiles, that sometimes bear little relation to the shape of the initial orbital. The consideration of emission from localised bonds shows that provided that bonds are not too strongly polarised, interference between waves from different centres is always significant at higher energies, and can also be important at energies of a few eV relative to the vacuum. Scattering by the ion cores of the surface region can strongly distort the angular distribution, or may have little effect. But it is generally difficult to decide a priori which influences are dominant for a particular case, so that the interpretation of angular profiles must be based on careful calculations including all these effects. The optimum energy range for the interpretation of experimental data is that from about 30 to 100 eV.

1. Introduction

In this paper we review the basic physics of photoelectron emission from localised orbitals at solid surfaces, with emphasis on those factors that are believed most significant in determining the angular distribution of the emitted electrons. The influence of these factors is illustrated by means of model calculations. The results lead us to some recommendations on the collection of angle resolved data, and some conclusions on its interpretation.

It is clear that the angular distribution of the emitted electrons will depend on the initial state, and that the experiment measures the sum of contributions from all degenerate initial states; this summation will generally result in a weakening of the angular structure that would occur for a single state. This averaging effect is most marked for atoms in the gas phase, where the angular profile from an atomic orbital with given l and m values is smoothed out by summing over all m values [1]. But at a crystal surface the atomic valence orbitals are hybridised and the energy levels split. The presence of a crystal field [2] must mean that more of the angular structure characteristic of individual initial states should be apparent in the measured angular distribution. It is therefore of some interest to study the angular profiles associated with emission from bonding orbitals [3], in particular directed orbitals. We are then immediately confronted with the existence of powerful selection rules and the associated dramatic dependence of the angular distribution from a single orbital on the orientation of the polarisation vector of the incident light. This effect can of course be seen using a conventional light source, by varying the angle of incidence [4]. But the polarised nature of synchrotron radiation makes it the natural tool for the full exploitation of polarisation effects [5].

The hybridised valence orbitals of the surface atom form bonds by overlapping with orbitals centred on neighbouring atoms. The emission of an electron from such a bond can be described in terms of outgoing waves, each centred on one of the atoms involved in the bond. In contrast to the case of emission from core states, where a localised hole is left behind, it is in principle impossible to say which atom emitted the electron, so that all the waves must be treated as coherent [6]. Thus interference between waves from different atoms may give rise to angular structure. We expect such structure to be strong when the wavelength is of the order or less than the atomic spacing, i.e. for higher energies, provided the waves emitted from different centres have comparable amplitudes. Such initial state interference effects have been especially emphasised by Gadzuk [3]. While we confine our attention here to localised bonds, there is no new physical principle involved in the problem of unlocalised states, such as Bloch and surface states. We then merely have a superposition of waves of similar amplitude from a very large number of centres, giving rise to extremely sharp structure in the angular distribution. Since however, such states are associated with a continuum of energy levels, a sum over many initial states must always be performed, with consequent smoothing of the angular structure.

The emitted waves are of course scattered by the neighbouring atoms with a resulting distortion of the angular distribution. While it is frequently assumed in interpreting data that such scattering is not very significant, model calculations [7, 8,9] show that in general the opposite is true. Indeed Liebsch, in his work [10] on the azimuthal spectra of TaS_2 in which he obtains a beautiful fit to the data of Smith et al. [11], demonstrates conclusively the importance of final state scattering by showing that its omission destroys the agreement between theory and experiment. This is not to say that scattering effects are always important. For example, if it were possible to find materials where the mean free path for inelastic scattering of the emitted electrons is of the order of a lattice spacing, at energies low enough to make the wavelength greater than a lattice spacing, then the effect of scattering on the angular distribution should be weak. The scattered waves emanating from neighbours would then have small amplitudes relative to the emitted waves and the relative phases of the waves from different centres would vary only slowly with direction. But even in this situation one has to take seriously the scattering of the emitted waves as they propagate out through the emitter potential. Interference between the different coherently emitted partial waves influences the angular distribution, and the phase relations between these waves are themselves determined by the phase shifts induced by the emitter potential. Such effects are of course completely absent in the popular but highly suspect plane wave approximation to the final state. We shall see for example that the prediction of this approximation that the intensity vanishes in directions perpendicular to the polarisation vector, can be quite misleading.

We illustrate all these effects by model calculations, and consider only these physical processes since they are probably the most important influences on the angular distribution. Thus we work within the one-electron approximation, simulating the effect of electron—electron interactions only through the use of a complex optical potential, whose primary effect is to induce damping of the electron wave propagation. It is of course this damping that makes photoemission a surface sensitive technique. Many-body effects such as relaxation in the neighbourhood of the hole, excitation of plasmons, electron—hole pairs etc. in the emission process, and hole-lifetime effects are undoubtedly significant (particularly for emission from core states), in determining the energy distribution of the emitted electrons, but we regard them as of secondary importance in discussing the angular distribution. Similarly, we ignore effects arising from the spatial variation of the vector potential in the surface region. Such screening effects are likely to be very significant for photon frequencies less than or of the order of the plasmon frequency, but not otherwise. We therefore begin by outlining the general theory within the independent electron model, and then present our model calculations illustrating the dominant influences on the angular distribution of photoemitted electrons.

2. General theory of photoemission — an outline

2.1. The Golden Rule formulation

The theory is almost universally presented in a form based on the use of the Golden Rule, expressing the differential cross-section in terms of the square of the matrix element $\langle f|H'|i\rangle$, where H' is the interaction of the electron with the electromagnetic field, and the initial and final states $|i\rangle$ and $|f\rangle$ are eigenstates of the unperturbed Hamiltonian. There is no difficulty about the initial state, but the nature of the boundary conditions to be satisfied by the final state is a more subtle problem which has been discussed by Breit and Bethe [12], and was put in definitive form by Gell-Mann and Goldberger [13]. We require a solution that tends to a free wave pocket with wave vector k in the future, and this is obtained by time reversal from the solution of the scattering problem in which, in the distant past, a free wave pocket with wave vector $-k$ was incident on the target. Thus for photoemission from an atom our final state contains a set of ingoing partial waves incident on the emitter as well as a plane wave [1]; for a crystal it is a time reversed LEED state [14,6,4]. This is hardly a very intuitive way to approach the problem, since we know that the photoelectron wave function is a purely outgoing wave centred on the emitter. We therefore now show how to formulate the problem in a manner that is equivalent to the Golden Rule approach, but is more direct and intuitive, and that requires no more computational effort in making comparisons with experimental data.

2.2. The photoelectron wave function

For simplicity we consider emission from an atomic orbital with quantum numbers n, l, m. We shall see later how to generalise to the case of emission from a crystal surface.

We start from the Lippmann–Schwinger equation for an atom in an electromagnetic field.

$$|i+\rangle = |i\rangle + \frac{1}{E - H^0 + i\eta} H'|i+\rangle,$$

where H^0 is the unperturbed Hamiltonian; the interpretation of the states is that $|i+\rangle$ coincides with $|i\rangle$ in the remote past, but as time passes a contribution from the interaction is added. Thus $|i\rangle$ is the initial state and is of the form

$$|i\rangle = |p\rangle|nlm\rangle,$$

$|nlm\rangle$ being the initial electron state and $|p\rangle$ the initial state of the electromagnetic field.

Then, to first order in the interaction

$$|i+\rangle = |i\rangle + \sum_r \frac{1}{E - H^0 + i\eta} |r\rangle \langle r|H'|i\rangle = |i\rangle + \sum_r \frac{\langle r|H'|i\rangle |r\rangle}{E - E_r + i\eta}, \quad (1)$$

where

$$H^0|r\rangle = E_r|r\rangle,$$

$$H' = -\frac{e}{mc} A \cdot p = -\frac{e}{mc} p \cdot \sum_{q,\alpha} \left(\frac{hc^2}{\omega\Omega}\right)^{1/2} \exp(i q \cdot r)\, \epsilon_{q\alpha}(a_{q\alpha} + a^\dagger_{-q\alpha}),$$

where p is the electron momentum operator, Ω is a volume determining normalisation, $\omega/2\pi$ is the photon frequency, α a polarization index, q a photon wave vector, $\boldsymbol{\epsilon}$ a polarisation vector and a and $a^\dagger$ are photon destruction and creation operators. If there is one photon of wave vector $\boldsymbol{k}$ present initially, the only terms that survive in the sum of (1), on taking the scalar product with the vacuum state $|0\rangle$, are those of the form

$$|r\rangle = |0\rangle |\epsilon l'm'\rangle ,$$

where $|\epsilon l'm'\rangle$ is a solution of the atomic Hamiltonian with energy ϵ. Thus, taking the scalar product of (1) with the vacuum state of the field we find for the final electron state, assuming a long wavelength photon initially:

$$|nlm+\rangle = \gamma \sum_{l'm'} \int d\epsilon \, \rho_{l'm'}(\epsilon) \frac{\langle \epsilon l'm'|\boldsymbol{\epsilon} \cdot \boldsymbol{p}|nlm\rangle}{\epsilon_n + \hbar\omega - \epsilon + i\eta} |\epsilon l'm'\rangle ,$$

where $\rho_{l'm'}(\epsilon)$ is the density of states $l'm'$ at energy ϵ, ϵ_n is the initial electron energy, and $\gamma = -(e/mc)(hc^2/\omega\Omega)^{1/2}$.

In the coordinate representation we have

$$\psi(r) = \gamma \sum_{l'm'} \int d\epsilon \, (2mk/\hbar^2\pi) \frac{\langle \epsilon l'm'|\boldsymbol{\epsilon} \cdot \boldsymbol{p}|nlm\rangle}{\epsilon_n + \hbar\omega - \epsilon + i\eta} Y_{l'm'}(r) [\exp(i\delta_{l'})h_{l'}^+(kr)$$

$$+ \exp(-i\delta_{l'})h_{l'}^-(kr)] ,$$

where r refers to a point outside the range of the atomic potential so that the radial function can be written as a sum of phase shifted outgoing and incoming spherical Hankel functions, the phase shifts being determined by the atomic potential, and normalisation adjusted to give continuity of the wave function.

Remembering that $h_l^+(kr)$ is a polynomial in $(kr)^{-1}$ multiplied by a factor e^{ikr}, while $h_l^-(kr)$ has a factor e^{-ikr}, where $\hbar^2 k^2/2m = \epsilon$, and noting that the pole of the integrand is in the upper half plane, we see that the terms in h_l^- vanish on performing the integral over ϵ. Hence

$$\psi(r) = i\pi\gamma \sum_{l'm'} \exp(i\delta_{l'}) \rho \langle \epsilon_n + \hbar\omega, l'm'|\boldsymbol{\epsilon} \cdot \boldsymbol{p}|nlm\rangle Y_{l'm'}(r) h_{l'}^+(\kappa r) \qquad (2)$$

where $\hbar^2 \kappa^2/2m = \epsilon_n + \hbar\omega$. We see that the final electron wave function is a sum of purely outgoing partial waves as anticipated. $\rho = 2m\kappa/\pi\hbar^2$, and on taking the asymptotic form of the spherical Hankel functions and calculating the radial flux as a function of direction, it is a straightforward matter to show that the differential cross-section calculated from (2) is exactly the same as is found in the Golden Rule formulation [1], as it must be.

2.3. Calculation of matrix elements – selection rules

We need to calculate the matrix elements $\langle \epsilon l'm'|\boldsymbol{\epsilon} \cdot \boldsymbol{p}|nlm\rangle$, where ϵ is the energy of the emitted electron. Put

$$\langle r|\epsilon l'm'\rangle = R_{l'}(r) Y_{l'm'}(\theta, \phi) , \qquad \langle r|nlm\rangle = U_{nl}(r) Y_{lm}(\theta, \phi) .$$

Then

$$\langle \epsilon l'm'|\boldsymbol{\varepsilon}\cdot \boldsymbol{p}|nlm\rangle = -i\hbar \begin{pmatrix} \sin\theta'\cos\phi' \\ \sin\theta'\sin\phi' \\ \cos\theta' \end{pmatrix} \sigma_{nll'} \int Y^*_{l'm'} \begin{pmatrix} \sin\theta\cos\phi \\ \sin\theta\sin\phi \\ \cos\theta \end{pmatrix} Y_{lm} d\Omega , \qquad (3)$$

where θ' and ϕ' define the direction of polarisation and

$$\sigma_{nll'} = \int_0^\infty U_{nl}(r)\, r \left(\frac{d}{dr} \pm \frac{2l+1\pm 1}{2r} \right) R_{l'}(r)\, r\, dr , \qquad l' = l \pm 1 .$$

The angular integral of (3) gives the well known selection rules of atomic spectra, namely

$$l' = l \pm 1, \qquad m' = m, m \pm 1 .$$

We can of course replace our so-called velocity form of the interaction Hamiltonian by two other equivalent forms. Thus using the identity

$$\boldsymbol{p} = m\dot{\boldsymbol{r}} = \frac{m}{i\hbar}[\boldsymbol{r}, H] ,$$

where H is the electron Hamiltonian, and the square bracket is a commutator, we find

$$\langle \epsilon l'm'|\boldsymbol{\varepsilon}\cdot \boldsymbol{p}|nlm\rangle = im\omega\, \langle \epsilon l'm'|\boldsymbol{\varepsilon}\cdot \boldsymbol{r}|nlm\rangle .$$

(3) then remains valid provided we make the replacement

$$\sigma_{nll'} \to -\frac{m\omega}{\hbar} \int_0^\infty r^3 R_{l'}(r) U_{nl}(r)\, dr .$$

This is the length form. Another form, the acceleration form is obtained from the identity

$$(\epsilon_n - \epsilon)\langle \epsilon l'm'|\boldsymbol{p}|nlm\rangle = \langle \epsilon l'm'|[\boldsymbol{p}, H]|nlm\rangle$$

$$= \langle \epsilon l'm'|[\boldsymbol{p}, V]|nlm\rangle = -i\hbar\langle \epsilon l'm'|\nabla V|nlm\rangle$$

where V is the potential seen by the electron. Eq. (3) then remains valid using the replacement

$$\sigma_{nll'} \to \frac{1}{\hbar\omega} \int_0^\infty R_{l'}(r) \left(\frac{dV}{dr}\right) U_{nl}(r)\, r^2\, dr .$$

Since the wave functions used will necessarily be approximate, the three equivalent forms may give different results. The length form emphasises contributions from regions distant from the nucleus, while the acceleration form weights heavily contributions from near the nucleus. Since the wave functions are usually most accurately known at intermediate distances, the velocity form is usually the most accurate [15]. The acceleration form is convenient when using the muffin-tin model for the ion core potentials in a crystal, since the potential outside the miffin tins is constant, so that the only contributions to the matrix element come from within the muffin tin.

2.4. Explicit form of photoelectron wave function

Evaluating the angular integrals of (3), and substituting into (2), we can now write down explicitly the form of the wave function describing the electron emitted from an atomic orbital $|nlm\rangle$. It is

$$\psi(r) = \frac{\pi \hbar \gamma \rho \sigma_{nl,l+1} \exp(i\delta_{l+1}) h^+_{l+1}(kr)}{[(2l+1)(2l+3)]^{1/2}} \begin{cases} -\frac{1}{2}[(l+1+m)(l+2+m)]^{1/2} e^{-i\phi'} \sin\theta' \, Y_{l+1,m+1} \\ +\frac{1}{2}[(l+1-m)(l+2-m)]^{1/2} e^{i\phi'} \sin\theta' \, Y_{l+1,m-1} \\ +[(l+1)^2 - m^2]^{1/2} \cos\theta' \, Y_{l+1,m} \end{cases}$$

$$+ \frac{\pi \hbar \gamma \rho \sigma_{nl,l-1} \exp(i\delta_{l-1}) h^+_{l-1}(kr)}{[(2l+1)(2l-1)]^{1/2}} \begin{cases} +\frac{1}{2}[(l-1-m)(l-m)]^{1/2} e^{-i\phi'} \sin\theta' \, Y_{l-1,m+1} \\ -\frac{1}{2}[(l-1+m)(l+m)]^{1/2} e^{i\phi} \sin\theta' \, Y_{l-1,m-1} \\ +(l^2 - m^2)^{1/2} \cos\theta' \, Y_{l-1,m} \end{cases} \quad (4)$$

The definition of the spherical Hankel functions and spherical harmonics are those of Messiah [16]. Since we are interested in emission from solid surfaces we are not free to choose the z axis to coincide with the polarisation direction; in fact we shall choose it to correspond to the inward surface normal. So in general we have a coherent superposition of six outgoing partial waves describing the photoelectron. An inspection of (4) shows the profound influence of varying the direction of polarisation. For example for polarisation along the z axis, only two of the partial waves have finite amplitudes. We shall discuss this effect in more detail later.

2.5. Generalisation to emission from a crystal

Emission from a crystal surface brings a number of additional complications to the problem. The initial state will not normally be a simple atomic orbital, but it can be expressed as a coherent superposition of atomic orbitals, on one, a few or many centres, according to whether the states are core states, localised bonds or unlocalised states such as Bloch states. Thus, from each centre involved in the initial state there will emerge a superposition of sets of waves of the form given in eq. (4) with amplitudes and phases depending on the initial state.

As the photoelectron propagates out it undergoes strong interactions with the valence electrons. This has two major effects; a shift in energy of the outgoing electron due to polarisation of the valence electron gas, and inelastic scattering. Thus the outgoing waves are attenuated as they propagate through the valence electron gas. It has been established in the context of LEED that these effects can be adequately simulated by introducing a complex optical potential V_0 into the electron Hamiltonian. V_0 is usually taken as uniform within the crystal, and this simple model apparently does not lead to significant disagreement with experimental data. Re V_0 and Im V_0 may be treated as adjustable parameters, determined by fitting LEED data for example, or they may be calculated in some approximation [17], or determined by a variety of empirical procedures [18–20]. The photoelectron propagation constant κ, is then determined by

$$\hbar^2 \kappa^2 / 2m = \epsilon_n + \hbar\omega - V_0,$$

and is of course complex. Mean free paths (half the wave damping length) typically range from 2 to 5 Å in the energy range 30–200 eV, so that photoemission must be regarded as a surface process. At lower energies however, theory indicates a much longer mean free path, though reliable values are difficult to find.

The emitted electron is also scattered by the neighbouring ion-cores (represented by muffin tin potentials), and this scattering is strong; typically the total cross-section is of the order of a square ångström. This means that not only is the distortion of the intrinsic angular distribution produced by scattering strong, but that simple approximations for the calculation of this effect are not a priori justified. One has to consider not only single, but double, triple and indeed multiple scattering to quite high order. Fortunately these formidable calculations can be performed relatively easily by adapting LEED programmes to the task. We shall not go into details here but merely indicate in outline how the scattered wave is genereated from the emitted waves.

If ψ_e^0 is the unscattered wave emitted from a centre at e, the final scattered wave ψ_e can formally be written as

$$\psi_e = (1 + GT) \psi_e^0,$$

where

$$G = (E - H_0)^{-1},$$

T is the transition operator and H_0 the electron Hamiltonian, in the absence of scatterers. We expand T in multiple scattering series; then

$$\psi_e = (1 + G \sum_s t_s + G \sum_{\substack{s,s' \\ (s' \neq s)}} t_s G t_{s'} + \ldots) \psi_e^0$$

$$= (1 + G \sum_{s(\neq e)} t_s + G \sum_{\substack{s,s' \\ (s \neq s', \\ s' \neq e)}} t_s G t_{s'} + \ldots)(1 + G t_e) \psi_e^0, \tag{5}$$

where t_s is the transition operator for the ion core at s. The entity $(1 + Gt_e)\psi_e^0$ is the wave emerging from the emitter, including the effect of scattering by the emitter potential, and for emission from a single atomic orbital would have the form (4) outside the range of the emitter potential. For emission from a linear combination of atomic orbitals on one centre it would of course be a coherent superposition of sets of waves of the form (4). If several or many centres emit coherently the scattered amplitudes must be combined before finding the final angular distribution. The manipulation of the series (5) to facilitate practical calculations of the scattered distribution has been treated in detail from several points of view [21,22,14]; in particular Pendry [23] has given a convenient formalism for the calculation of both scattering and initial state effects for valence band studies.

Finally the photoelectron must pass through the surface barrier where the optical potential V_0 vanishes. The surface barrier should vary smoothly and hence be a weak scatterer, but in angular studies it is important to take account of refraction at the surface, which is easily done by resolving the photoelectron wave function into plane waves at the surface, and using the conditions of conservation of energy, and of momentum parallel to the surface [21,23]. Total internal reflection is an important effect at low energies.

3. The model calculations

3.1. Specific details of the model

All calculations were made for a Cu(111) surface with the top layer displaced to lie immediately over the second layer in order to simulate an ordered adlayer with 1-fold coordination. The scattering properties of the copper atoms were treated realistically using the method described by Loucks [24] to construct the muffin tin potential and find the phase shifts. The scattering calculations were in fact carried out in the single scattering approximation since it appears [25,26] that the angular profiles in this approximation agree qualitatively quite well with those obtained from the full multiple scattering calculation. The extra computational effort required by the multiple scattering calculations, while necessary in comparing with experimental data, was not deemed worthwhile in these model calculations, where details are not of interes.

The optical potential was chosen to agree with that used by Pendry [27] in his successful fit of LEED data, modified for the fact that a single scattering approximation was employed, by adding the forward scattering t matrix element of an ion-core [28]. This gave Re V_0 = 15 eV and a mean free path of 2.7 Å, both parameters being nearly independent of energy over a range up to about 100 eV.

The radial integrals of (4) were not evaluated, but the relations between them were chosen in the following way. Approximating the radial factor $R_{l'}(r)$ by the spherical Bessel function $j_l(kr)$ and using the length form of the matrix element, one finds

$$\sigma_{nl,l+1} = -\sigma_{nl,l-1} . \tag{6}$$

We therefore take all the radial integrals equal in magnitude but with relative phases determined by (6). Explicitly, since we use s, p and d initial states;

$$\sigma_{01} = \sigma_{12} = -\sigma_{10} = \sigma_{23} = -\sigma_{21} . \tag{7}$$

Of course, no real system with the properties of our model exists; but provided cate is taken not to confuse results specific to the model with general conclusions, the model calculations should provide some insight into the influence of the different physical processes on the angular distribution of the photoelectrons.

3.2. Polarisation effects

We have already remarked on the importance of the direction of polarisation of the incident light for photoemission. Eq. (4) shows how strongly the relative amplitudes of the different partial waves arising from emission from an atomic orbital depend on the polarisation direction.

A good example of the fruitful exploitation of polarisation direction is found in the work of Schlüter et al. [5] in their study of the adsorption of chlorine on the (111) surfaces of silicon and germanium. The problem was to determine whether the chlorine atom sits in a 1-fold coordination site on these surfaces. The local density of states was calculated for the two adsorbate geometries. For the 1-fold structure, the characteristic feature is that a sharp peak is associated with the chlorine p_z orbital which is split off from the p_x and p_y orbitals and shifted to lower energy since it is involved in a directed σ bond together with the silicon p_z. In the 3-fold case, no σ bond is formed involving the chlorine p_z, which forms a single peak in the density of states together with the other p orbitals. The photoelectron energy distribution curves were measured using s and p polarised light. For chlorine on silicon, the measured EDC showed a sharp peak for the p polarised case, that was strongly diminished on switching to s polarisation. An examination of eq. (4) shows that for polarisation perpendicular to the z axis or surface normal, for emission from a p_z state, s wave emission is completely suppressed, two d waves with $m = \pm 1$ being emitted, these having small amplitudes in directions away from the surface plane. For a polarisation normal to the surface however, an s wave and a d wave with a lobe protruding out of the surface are emitted. Therefore we expect the intensity emitted along or near the surface normal to be much larger for p than for s polarisation. This is illustrated in fig. 1, where the angular distribution arising from emission from a p_z orbital for the two different polarisations is shown. The crystal optical potential and scattering effects have been taken into account in these results. The polarisation effect is so large that it is readily apparent even for angle integrated measurements. Thus Schluter et al. concluded that on silicon chlorine is adsorbed in the 1-fold geometry. No such sharp peak in the EDC chlorine on

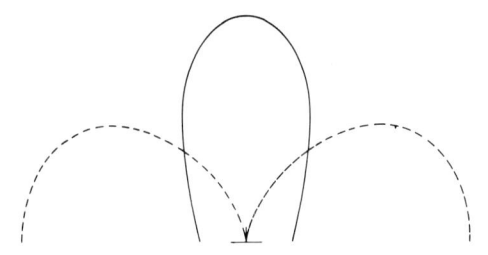

Fig. 1. Polar diagram of the emitted intensity from a p_z orbital; surface normal (z axis) in the vertical direction. Polarisation along surface normal (full line); polarisation in surface plane parallel to the page (dashed line). Scales used are different; if the maximum for the full line is 39.9 units, that of the dashed line is 7.3.

germanium was found and the polarisation effect was weak; hence the conclusion is that on germanium chlorine has the 3-fold coordination.

The same idea was used in the angle resolved work of Jacobi et al. [4] for oxygen on Ni(001). Chemisorption calculations indicate that there should be a level derived from the oxygen $2p_z$ orbital, and a doubly degenerate level derived from the $2p_x$, $2p_y$ obitals, but cannot give the energies reliably. Using unpolarised light at normal incidence, the level at 8 eV below the Fermi surface shows negligible emission normal to the surface, as expected from eq. (4), if it is a p_z state. On the other hand, the level 6 eV below the Fermi surface shows a maximum normal to the surface, as we expect for p_x and p_y orbitals.

As a final example we cite the work of Smith et al. [29] on CO adsorbed on a Ni(001) surface. The data is compared with calculations of Davenport [30] for an oriented CO molecule, and it is concluded that the CO molecule stands on end on the Ni surface, with the carbon atom nearest the surface. A key factor in reaching this conclusion was the observation that emission from the 4σ CO bond shows zero emission normal to the surface for normally incident light, i.e. with zero polarisation component along the bond, but strong emission in this direction for light incident at 47°.

3.3. Emission from directed orbitals

We consider the angular distribution from two different directed orbitals centred on an atom in the second layer of the Cu(111) surface. One of these orbitals is an sp_z hybrid, the other an $sp_z d_z$ hybrid, both pointing directly along the outward surface normal at an atom in the "adlayer" in the 1-fold site. For simplicity the atomic orbitals were taken to have the asymptotic form of hydrogen atom $n = 3$ orbitals, so that the radial factors were identical except for phase. Excluding the common radial function the hybrid orbital wave functions were then of the form (in our coordinate system the z axis is the inward surface normal)

$$\psi_{sp} = [Y_0 - Y_{10}]/\sqrt{2}, \qquad \psi_{spd} = [Y_{00}/\sqrt{6} - Y_{10}/\sqrt{2} + Y_{20}/\sqrt{3}].$$

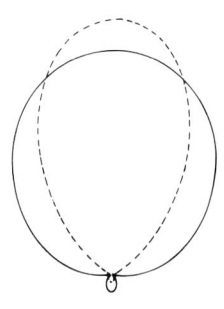

Fig. 2. Polar distribution of electron density for the sp_z hybrid orbital (full line) and the $sp_z d_z$ hybrid orbital (dashed line).

The angular distribution of the electron density is shown in the polar plots of fig. 2 for each of these orbitals.

For polarisation along the surface normal the angular distribution of emitted electrons is easily calculated from eq. (4) using the asymptotic form of the spherical Hankel function, and the results are shown in fig. 3 for the sp_z hybrid. The full line corresponds to zero phase shifts and the choice of phases for the various radial integrals was that given by (7). Though the orbital itself is strongly directed along the outward surface normal, the photoelectron distribution is symmetrical about the surface plane. This result is dependent on our specific choice of phases. If instead of this choice we assume that all radial integrals are identical we obtain the dash-dot

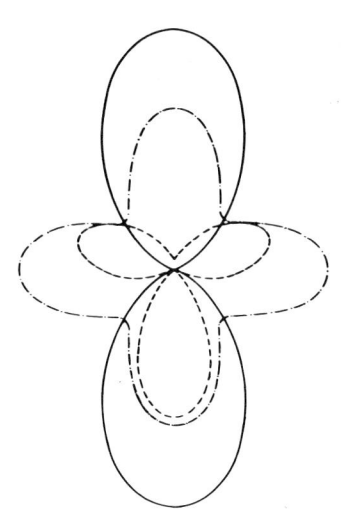

Fig. 3. Emitted intensity from sp_z orbital, z polarisation. Phase shifts zero and σ values given by eq. (6) (full line); phase shifts zero and σ values all equal (dash-dot line); σ values of eq. (6) with $\delta_0 = \pi/2$, $\delta_1 = 0$, $\delta_2 = -\pi/2$ (dashed line).

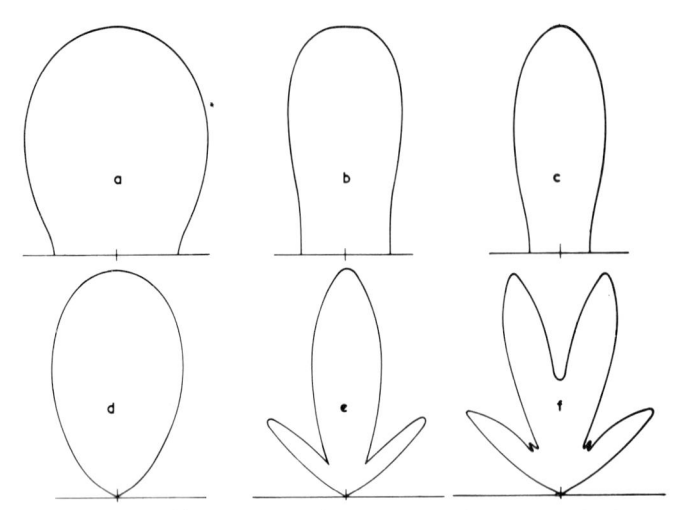

Fig. 4. Emission from sp_z orbital, z polarisation. Unscattered (a); scattering by neighbours (b); inclusion of scattering by emitter potential also (c); all at 12 eV in the vacuum. (d), (e) and (f) are corresponding results for an energy of 94 eV in the vacuum.

line of fig. 3. This bears even less direct relation to the shape of the sp_z orbital electron distribution. Notice also that it has maxima in directions perpendicular to the polarisation vector, a result in direct contradiction to the plane wave approximation to the final state, which fails to take proper account of phase relations and resulting interference effects between the different emitted partial waves [2]. The results for a further choice of phases is shown by the dashed line where we have used (7), together with phase shifts $\delta_0 = \pi/2$, $\delta_1 = 0$, and $\delta_2 = -\pi/2$. The resulting three lobed profile is totally different from the other two cases. It is clear from this that the emitter potential can have a profound effect on the angular distribution emerging from the emitter, and it is therefore very important to calculate the outgoing radial wave functions and phase shifts carefully. The plane wave approximation to the final state can be extremely misleading.

Thus far we have taken no account of the fact that the initial orbitals are situated in a crystal near the surface. The result of including these effects for emission from the sp hybrid is shown in fig. 4. The top tow of polar plots corresponds to an energy of 12 eV in the vacuum, the bottom row to 94 eV. The effect of total internal reflection is evident in the low energy profiles, which have large intensities for grazing emission. The left hand panels show the distributions in the absence of ion core scattering, only including damping and energy shifts due to the valence electron gas.

The middle panels include the effect of scattering by the ion cores around the emitter but do not include phase shifts induced by the emitter potential. At the low energy the scattering effect is not very significant; the mean free path (2.7 Å) lattice parameter (2.5 Å) and wavelength (2.4 Å) are comparable. Thus the relative phases of contributions from scatterers close enough to the emitter to be significant, vary only slowly with direction. At the higher energy this is no longer true; the wavelength is now short enough for scattering to induce marked structure in the angular profile.

The right hand panels show the effect of including the phase shifts arising from the emitter potential, as well as scattering off non-emitting atoms. Again this has no dramatic result at low energy, but at the higher energy has a considerable effect, inducing a deep minimum at normal exit.

It should be pointed out that the value of the mean free path of 2.7 Å, used here for all calculations, is likely to be far too small for most materials at an energy of 12 eV relative to vacuum. Calculations using a longer value of 6 Å show that the profile of the middle panel of figure 4 at this energy acquires distinct shoulders. It is difficult to carry out consistent calculations for much longer values of the mean free path, but it is to be expected that values much greater than the nearest neighbour distance and the wavelength, will induce appreciable angular structure arising from scattering. It is therefore not safe to assume that scattering is insignificant at very low energies without reliable evidence that the mean free path is sufficiently short.

3.4. Emission from localised bonds

As mentioned earlier, the sp_z and $sp_z d_z$ orbitals are centred on atoms in the second layer (top layer of the substrate) and directed towards atoms in the top layer or adlayer. We now introduce s orbitals on the atoms of the adlayer, forming localised bonds by overlap with the directed hybrids. We calculate the angular distribution for electrons emitted from these bonds, treating the emitted waves from the two atoms coherently. The polarisation vector is again normal to the surface.

The polar profiles for the sp_z:s bond are shown in fig. 5. The low energy emission with no scattering included is the upper curve; interference between waves emitted from the two atoms of the bond has not had a very dramatic effect, though there is a small dip at normal emergence. The left hand bottom curve is the higher energy profile with no scattering. Interference between waves from the two centres now produces marked structure in the angular profile; particularly noticeable is the deep minimum for emission along the surface normal. The effect of scattering in this case is profound as is seen in the right hand curve. In particular, the deep minimum at normal exit is changed into a strong peak, and the wings of the unscattered distribution are chopped off.

The results for the $sp_z d_z$:s bond shown in fig. 6 are very different. Now the low energy profile shows a dramatic dip at normal exit due to interference between waves from the two centres. The effect is very similar to that induced by switching the polarisation from normal to parallel to the surface shown in fig. 1. Thus initial

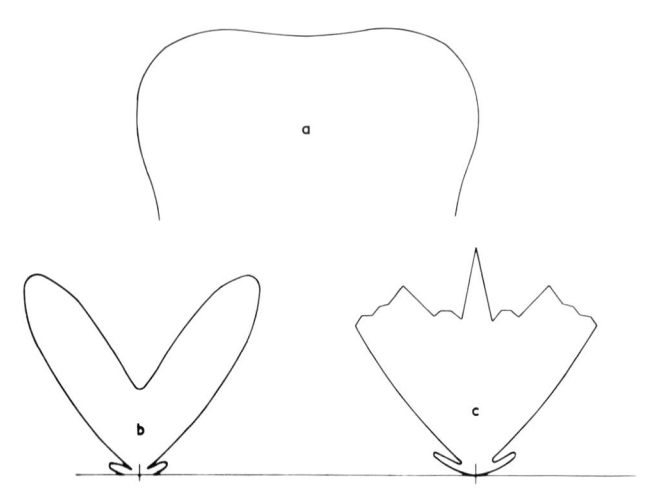

Fig. 5. Emission from an sp_z:s bond, z polarisation. 12 eV in vacuum, no scattering (a); 94 eV, no scattering (b); 94 eV including scattering (c).

state interference effects can be large even at low energies, and even when only two centres are involved. One must recognise that a profile like that of the upper curve of fig. 6 could arise as here through an interference effect, or say by emission from a p_z orbital with polarisation parallel to the surface.

Again the interference effects can induce much more structure at the higher energy, as seen in the left hand panel, but scattering in this case (right hand curve) makes only rather weak modifications to the angular profile.

It seems therefore difficult to make useful generalisations. Initial state inter-

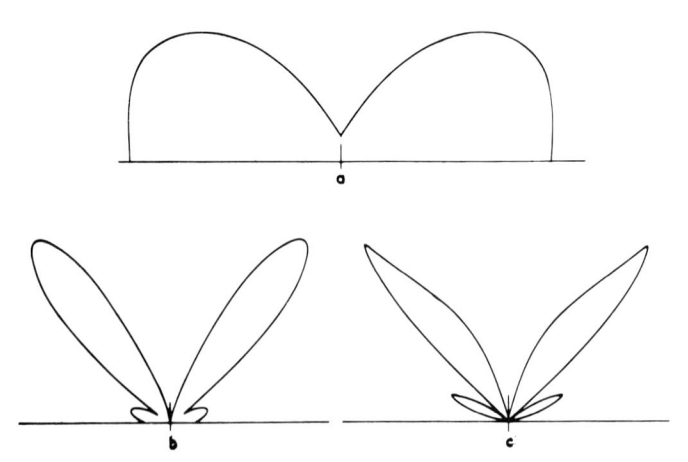

Fig. 6. As for fig. 5 except that the emission is now from an sp_zd_z:s bond.

ference effects can be important at low energies as well as high, and either initial or final state effects can dominate the angular distribution.

4. Conclusion

The basic object in the thoery is the superposition of outgoing partial waves (4), describing emission from an atomic orbital. Emission from more complex initial states can be described by a coherent combination of such sets of waves. The calculation of the appropriate coefficients for this combination, in the most interesting case of adatom-surface bond initial states, is not a problem of photoemission theory as such but of chemisorption theory [31] and may be extremely formidable.

The model calculations show that in evaluating the amplitude (4), it is important to take proper account of scattering by the emitter potential, calculating the phase shifts and the radial factor carefully so that good estimates of the amplitudes (coefficients $\sigma_{nl,l\pm1}$) and phases of the outgoing waves can be made.

While it is often true that with polarisation along the axis of an orbital or bond, emission along this axis is strong and emission perpendicular to it weak, this is by no means universal. As seen in the model calculations of sections 3.3 and 3.4, interference of waves from a single centre, determined largely by the emitter potential, or from two or more different centres can give rise to distributions having no such obvious correlation with the bond or orbital axis.

It is clear that the direction of polarisation of the incident light plays a profoundly important role in determining the angular distribution. Studies in which the polarisation vector is varied relative to the surface have already allowed firm conclusions to be drawn about the initial state symmetries, and orientation of molecules relative to the surface, and we may expect synchrotron radiation users to exploit this powerful weapon fully in the future.

Initial state interference effects and final state scattering effects can both be strong and even dominating influences on the angular distribution, and it seems fom the model calculations of section 3.4 particularly, that it is very difficult to decide without detailed calculation what effect each of these two factors will have. Indeed the model calculations force one to the conclusion that while one can hope to interpret certain features of the data intuitively, this approach needs to be backed by careful calculations so as to ensure that all the factors are being given due weight. The model calculations on the face of it indicate one possible simplification, namely that at sufficiently low energies scattering effects might be ignored. But unfortunately this conclusion is suspect because the mean free path at low energy tends to become very long, and therefore scattering is probably much more significant than our calculations, employing a short mean free path, suggest.

Rather than try to avoid the scattering problem by the dubious approach of working at very low energies, it would be better to work in a range that facilitates the accurate calculation of scattering. This means the range say 40 to 100 eV typically, where the mean free path is usually at its shortest, and where large numbers of partial waves are not needed for an accurate description of the scattering. The short mean free path also ensures maximum surface sensitivity.

References

[1] J. Cooper and R.N. Zare, Lectures in Theoretical Physics XIC (Gordon and Breach, New York, 1969).
[2] J.W. Gadzuk, Phys. Rev. B12 (1975) 5608.
[3] J.W. Gadzuk, Phys. Rev. B10 (1974) 5030.
[4] K. Jacobi, M. Scheffler, K. Kaube and F. Forstmann, Solid State Commun., to be published.
[5] M. Schluter, J.E. Rowe, G. Morgaritando, K.M. Ho and M.L. Cohen, Phys. Rev. Letters 37 (1976) 1632.
[6] J.W. Gadzuk, in: Nato Advanced Study Institute: Electronic Structure and Reactivity of Metal Surfaces, Eds. E.G. Derouane and A.A. Lucas (Plenum, New York, 1976).
[7] L. McDonnell, D.P. Woodruff and B.W. Holland, Surface Sci. 51 (1975) 249.
[8] D.P. Woodruff, Surface Sci. 53 (1975) 538.
[9] S.Y. Tong and M.A. Van Hove, Solid State Commun. 19 (1976) 543.
[10] A. Liebsch, Solid State Commun. 19 (1976) 1193.
[11] B.V. Smith, M.M. Traum and F.J. Di Salvo, Solid State Commun. 15 (1974) 211.
[12] G. Breit and H.A. Bethe, Phys. Rev. 93 (1954) 888.
[13] M. Gell-Mann and M.L. Goldberger, Phys. Rev. 91 (1953) 398.
[14] A. Liebsch, Phys. Rev. B13 (1976) 544.
[15] H.A. Bethe and E.E. Salpeter, Quantum Mechanics of One and Two-Electron Atoms (Academic Press, New York, 1957).
[16] A. Messiah, Quantum Mechanics (North-Holland, Amsterdam, 1965).
[17] B.I. Lundquist, Physik Kondens. Mater. 7 (1967) 117.
[18] D. Norman and D.P. Woodruff, Surface Sci. 68 (1977) 000.
[19] C.J. Powell, Surface Sci. 44 (1974) 29.
[20] I. Lindau and W.E. Spicer, J. Electron Spectrosc. 11 (1974) 212.
[21] J.B. Pendry, J. Phys. C8 (1975) 2413.
[22] B.W. Holland, J. Phys. C8 (1975) 2679.
[23] J.B. Pendry, Surface Sci. 57 (1976) 679.
[24] T.L. Loucks, Augmented Plane Wave Method (Benjamin, New York, 1967).
[25] R. Lindsay and C.G. Kinniburgh, Surface Sci. 63 (1977) 162.
[26] B.W. Holland, unpublished.
[27] J.B. Pendry, J. Phys. C4 (1971) 2514.
[28] P.W. Anderson and W.L. McMillan, in: Scuola Internazionale di Fisica, Varenna, Italy, Ed. W. Marshall (Academic Press, New York, 1967).
[29] R.J. Smith, J. Anderson and G.J. Lapeyre, Phys. Rev. Letters 37 (1976) 1081.
[30] J.W. Davenport, Phys. Rev. Letters 36 (1976) 945.
[31] T.B. Grimley, in: Nato Advanced Study Institute: Electronic Structure and Reactivity of Metal Surfaces, Eds. E.G. Derouane and A.A. Lucas (Plenum, New York, 1976).

ANGLE-RESOLVED UPS MEASUREMENTS IN A MODIFIED LEED SYSTEM

G.L. PRICE * and B.G. BAKER

School of Physical Sciences, The Flinders University of South Australia, Bedford Park, South Australia 5042

A LEED chamber has been modified to include a differentially pumped discharge lamp (He or Ne) and an additional retarding grid electron energy analyser for UPS. This small analyser is located at right angles to the LEED analyser and does not interfere with normal LEED and Auger operations. The UPS signal is amplified by a channel plate multiplier and accelerated onto a phosphor-coated screen. Directional information is obtained by scanning this screen with a collimated photomultiplier detector. A phase-lock amplifier is used to differentiate the signal from the photomultipler. Alternatively the phosphor screen can be used as a collector to measure a total spectrum. The acceptance angle of the UPS analyser is 90°. In the angular resolving mode it is possible to observe emission from a (100) fcc crystal in the ⟨100⟩, ⟨110⟩ and ⟨111⟩ directions with a fixed incident photon angle in the range 20–40° to the normal. The acceptance angle of the detector was usually ~7° but this can be varied by changing the collimating tube on the photomultiplier. The direction dependent features of the d-band spectrum of clean nickel with a (100) surface have been examined. Characteristic features were observed for each of the ⟨100⟩, ⟨111⟩ and ⟨110⟩ directions. These are compared with those reported for crystals with the corresponding surface orientations. The effects resulting from the chemisorption of nitric oxide on this nickel crystal have also been investigated.

1. Introduction

Ultra violet photoelectron spectroscopy (UPS) has proved a useful method of investigating the band structures of solids and surface states [1], and also the interaction of adsorbed gases with adsorbents [2]. Recent work has focussed on angle-resolved photoemission with the aim of providing a more detailed understanding of bulk solid and surface band structure [3,5] and the crystallography of adsorbates [6].

The apparatus described in this paper is an angle-resolving photoelectron spectrometer incorporated into a conventional LEED chamber. This facility enables total UPS and angle-resolved spectra to be taken in conjunction with LEED, Auger spectroscopy and thermal desorption studies of single crystal surfaces.

The results of experiments on the (100) face of nickel are presented. Directional photoemission from this surface is of particular interest for the following reasons:

---
* Present address: School of Mathematical and Physical Sciences, Murdoch University, Murdoch, Western Australia 6153.

(i) The UPS spectrum from the (100), (110) and (111) faces of nickel taken along the principal directions normal to the faces have been published [7,8]. Comparison of these spectra with those measured in the corresponding directions but from one face should show the relative roles of surface and bulk properties on the photoemission.
(ii) The theoretical density of states for nickel has been difficult to fit to the experimental data of both UPS [9,10] and XPS [11].
(iii) The theoretical surface density of states, recently calculated for nickel [12] could be useful in determining the effect of surface states on the spectra.

2. Experimental

The analyser is shown in fig. 1. Light from a glass, water-cooled discharge lamp passes through a differential pumping system consisting of two capillaries (100 × 1 mm) separating a roughing line, a diffusion pumped chamber and the UHV system. One Torr of helium in the lamp raised the pressure in the ion pumped chamber from a background of 5×10^{-11} Torr to $\sim 3 \times 10^{-9}$ Torr of helium. Between the chamber and the lamp system is a bellows for alignment and a straight-through valve for isolation. The light passes through these and then through a 3 mm hole drilled centrally through the various components of the analyser. The whole lamp-analyser assembly is mounted on a 150 mm flange.

The photoelectrons enter a four grid retarding potential spectrometer shielded

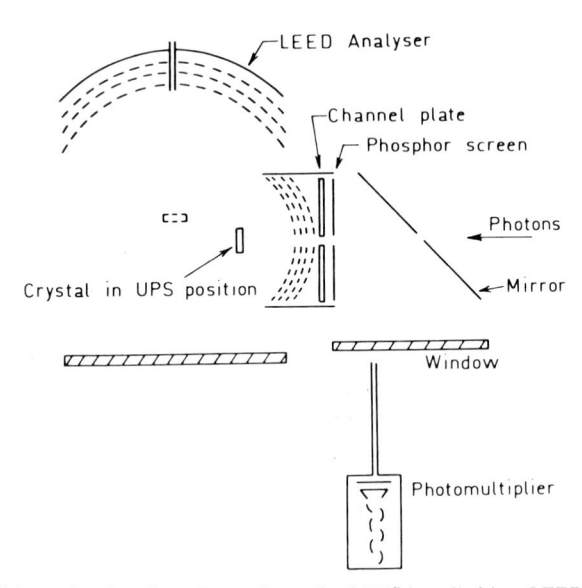

Fig. 1. Schematic of analyser for angle-resolved UPS installed in a LEED system.

by a mu-metal can. The first and fourth grids are at earth, the second and third are ramped and modulated. Electrons passing through the fourth grid are accelerated to 1000 eV and strike a channel plate (Mullard G50) which has 1500 V across it. The gain of the channel plate is about 1000. The electrons leaving have an energy of 2500 V above ground and they are further accelerated to 6000 eV before striking a conducting phosphor screen. The image on this screen is reflected through a 60 mm window which is next to that of the LEED chamber. The analyser grids have a mean radius of 33 mm with an acceptance angle of 90°.

The analyser was operated in two modes: angular and total. In the total mode, the photoelectrons were collected at the phosphor screen and their modulation due to the ac component on the retarding grids was fed via a decoupling capacitor into a phase-lock amplifier harmonic of 270 Hz. In the angular mode, a photomultiplier tube collimated by a 50 mm × 3 mm diameter tube was pointed at specific points of the imaged phosphor screen. Again the modulated output was fed to the phase-lock amplifier though the frequency was dropped to 34 Hz to allow for the slow response of the phosphor.

The ultimate analyser resolution was found to be better than 0.2 eV. Sonewhat lower resolution was obtained when large modulation voltages (0.2 to 0.4 V p-p) were needed to achieve reasonable scan times. The total spectra were recorded with a time constant of 1 s and the angular spectra required time constants of 3–10 s. The angular spectra were generally recorded three times and an average was taken. The apparatus was originally designed in the hope that some symmetry might be seen by eye in the photoelectron display. Nothing of the kind has been observed even though the nickel d-band peak alone caused an appreciable glow on the phosphor screen.

The angular resolution can be computed approximately by assuming that the electrons move in a straight line parallel to the centre line of the spectrometer after leaving the fourth grid. It ranges from 5° at the centre to 8° at the edge of the phosphor screen. Since there were large variations in gain as a function of position over the channel-plate–phosphor combination, no comparison of the intensity as a function of angle were made: interest centred on peak shape and on the ratio of peak heights within a single spectrum.

The reflected light from the crystal was found to have little effect. At normal incidence it was reflected back the way it came; at greater than 25°, it was reflected outside the grids and at angles less than this it was observed as a bright spot on the phosphor, since the u.v. light stimulated the channel plate. Spectra could be taken in the neighbourhood of the spot.

Retarding grid analysers have been used before to analyse UPS spectra on solid surfaces [13] including angle-resolved spectra [14]. The novel features of this device are that it is small enough to be fitted in most LEED chambers and that the angle of photon incidence can be varied from 0° to about 50°. As seen in fig. 1, LEED patterns can still be photographed and Auger spectra taken. In the space of a few minutes one can observe LEED, UPS, AES and thermal desorption spectra.

The nickel crystal was spark out, mechanically polished, electropolished and then mounted, in the manner described previously [15]. A good LEED pattern was obtained after 30 min sputtering and one anneal.

Results

3.1. Clean nickel (100)

The spectra observed in the vicinity of the ⟨100⟩, ⟨110⟩ and ⟨111⟩ directions for photon incidence angle of 20°, are shown in fig. 2. Only the d-bands are shown, as the lower energy parts of the spectra are similar. The three spectra are easily distinguished. The ⟨100⟩ has a fairly plat topped peak with the part near E_F slightly depressed and gives the impression of being composed of two narrow peaks. The peak falls away at $E_F - 1$ eV with substantial positive curvature. The ⟨110⟩ is much broader with a second peak at $E_F - 1.5$ eV. The ⟨111⟩ can be distinguished from the ⟨100⟩ by the sharper top and the linear, verging on negative, curvature of the back of the peak: there is a greater contribution to the peak below $E_F - 1$ eV than for the ⟨100⟩.

In scanning over the range of directions it was found that the spectra changed so as to lose one characteristic shape and assume another. In order to represent the overall picture, qualitative judgements were made to classify each spectrum as having the essential characteristic of a, b, or c as in fig. 2. The results for 40° and 20° incident photon angle are shown in figs. 3a and b respectively. Each point on these diagrams corresponds to a point on the phosphor screen: the distance between the centre and the edge spans 45°. If the spectra at any point had the

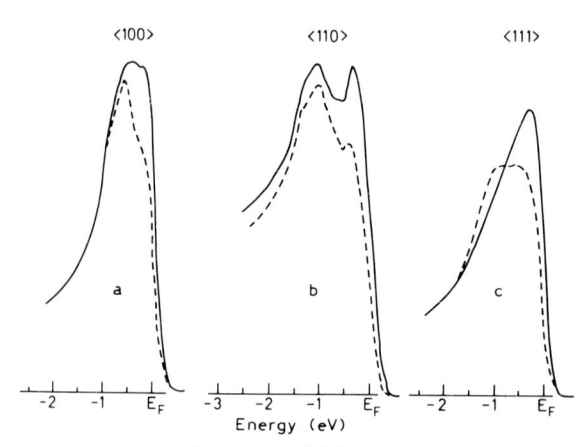

Fig. 2. Angle-resolved spectra from clean nickel (100) taken in three crystallographic directions (full lines); surfaces with adsorbed nitric oxide (broken lines).

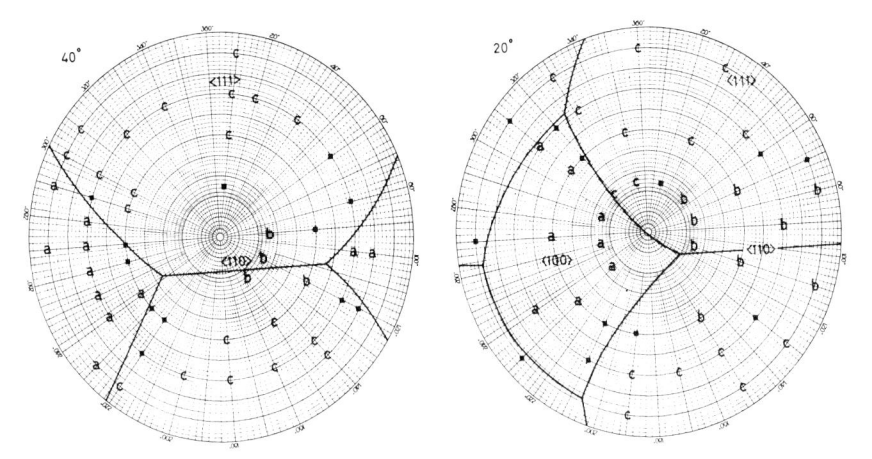

Fig. 3. Identification of angle-resolved UPS spectra from (100) nickel for incident photon angles of (a) 40° and (b) 20°. Spectra as in fig. 2 designated by a, b or c; mixed characteristics by ■. Crystallographic directions shown as e.g. ⟨100⟩; projections of Brillouin zone boundaries by full lines (see text).

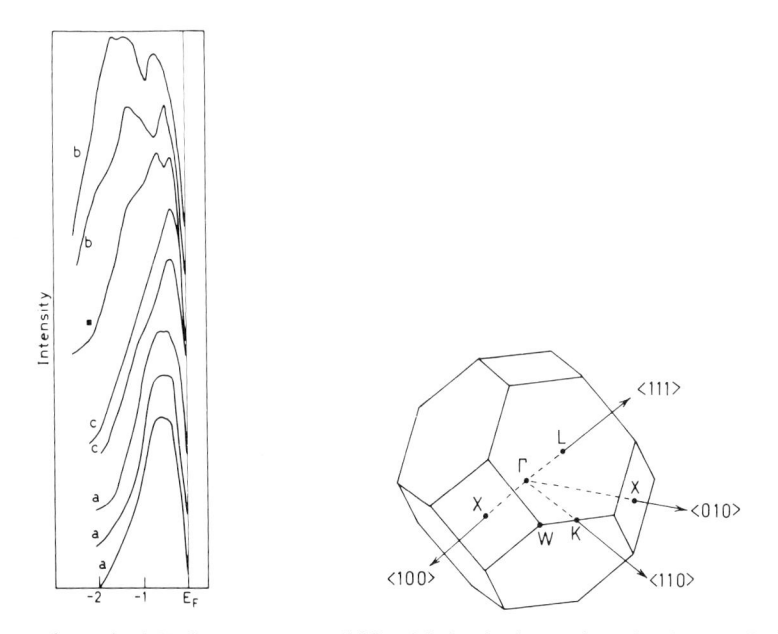

Fig. 4. Angle-resolved UPS spectra from (100) nickel. The inner ring of points of fig. 3b.

Fig. 5. Brillouin of fcc lattice showing principal directions.

essential characteristics of one of the three in fig. 2, then the point was labelled a, b or c. If no one characteristic shape predominated, it was labelled ■. Spectra and their labelling are shown in fig. 4 for the inner ring of points in fig. 3b from 60° to 230° on the polar plot. The projection of the Brillouin zone boundaries (see fig. 5) are also shown in fig. 3. Their positions were calculated from the crystal orientation known from the LEED pattern. An asymmetry in the specimen holder caused the normal to the crystal to dip down as the photon incidence angle was increased.

There is general agreement between the borders of the three spectra and the Brillouin zone lines. Similar results were taken for 10° and 0° incidence angle. The zero incidence results consisted of mainly a with mixtures of a and c spectra at the edges and borders were not well defined. The 10° results were a further rotation of the 20° plot.

Spot checks were carried out at 20° and 40° using 16 eV photons. The spectra were similar and followed the same pattern as the above results.

3.2. Adsorbed nitric oxide

The effect of a monolayer of adsorbed NO on the d-band structure is shown in fig. 2. The monolayer was adsorbed at room temperature and flashed to 100°C to remove molecular NO [16,17]. The spectra show that the front of the peak at the Fermi edge is removed by the chemisorbed gas. In the case of the ⟨111⟩ direction this effect is accompanied by a strengthening of the signal between −1 and −1.5 eV below E_F.

4. Discussion

In this simplified presentation of the angle-resolved spectra (fig. 3) the projection of the Brillouin zones reasonably described the distribution at various photon incidence angles. The ⟨100⟩ spectrum showed constant characteristic features for incidence angles ranging from normal to 45°.

The emitted electrons undergo refraction at the metal—vacuum interface. The crystallographic directions labelled in fig. 3 therefore, do not correspond accurately with the electron emission directions within the metal. An approximate calculation based on the assumptions that the component of the wave vector parallel to the surface is conserved and that the emitted electron is in a nearly plane wave state inside the metal shows that refraction amounts to ~4° at the boundary between ⟨100⟩ and ⟨111⟩ regions. This is within the angular resolution of the spectrometer.

The spectra from (100) nickel observed in the vicinity of the ⟨100⟩ direction are in general agreement with previous measurements [7,8]. The spectra from this same surface measured in the vicinity of the ⟨111⟩ direction have the essential characteristics of spectra reported for a (111) nickel surface [8,18]. A small peak at E_F − 1.2 eV is slightly developed in only some of our spectra. This feature has been

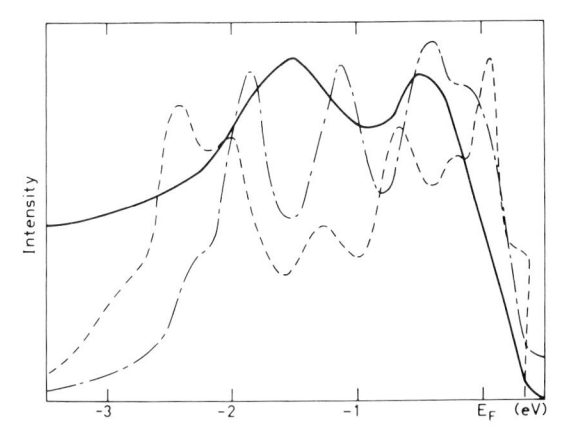

Fig. 6. Comparison of the measured ⟨110⟩ spectrum from a nickel (100) surface (full line) with the calculated surface (− · − · −) and bulk (− − −) densities of states [12].

shown to be strongly angle dependent [18]. In our case the emitted electrons from the true ⟨111⟩ within the crystal are, after refraction, out of range of the screen.

The trend observed in moving from ⟨111⟩ towards ⟨110⟩ is in general agreement with ref. [18]. However, our ⟨110⟩ is completely different from that reported in

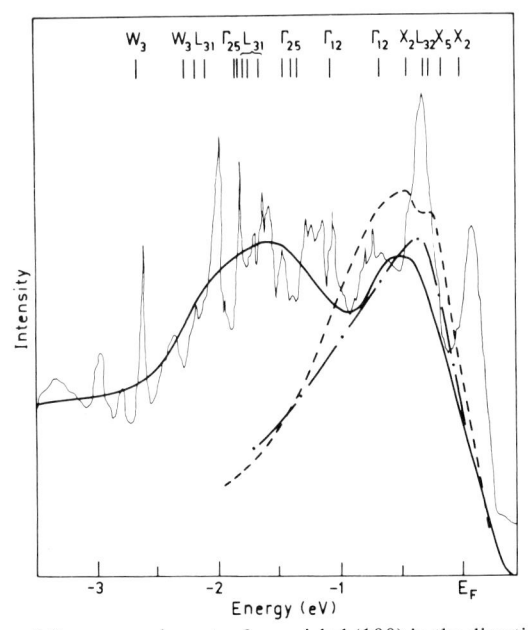

Fig. 7. Comparison of the measured spectra from nickel (100) in the directions. ⟨110⟩ (full line, ⟨100⟩ (− − −), and ⟨111⟩ (− · − · −) with calculated bulk density of states [19].

ref. [8] for the (110) surface. As shown below, ⟨110⟩ fits the calculated bulk density of states rather well and so this supports the suggestion that the spectrum from a (110) surface may be due to a surface state.

Fig. 6 shows a comparison of our results with recent theoretical calculation of the surface and bulk density of states [12]. Only the ⟨110⟩ spectrum is shown as the other two are encompassed by the front peak which would be merely increased in height by their addition. The bulk states calculation gives two broad, well-separated peaks and the surface states calculation gives three peaks. There is poor agreement with either calculation or any mixture of the two.

Fig. 7 shows an earlier bulk density of states calculation [19] compared with our three spectra. From the nickel band structures X_2 and X_5 would be expected to contribute to the ⟨100⟩ peak, in agreement with the results. From fig. 2 it is the X_5 contribution which is attenuated on absorption of NO. L_{32} contributes to the ⟨111⟩ peak with the back of the peak being filled out by Γ_{12} and Γ_{25}. The ⟨110⟩ second peak is due to the high density of states along Γ_{KX} at the energy around Γ_{25}.

The effect of chemisorbed nitric oxide on the spectra in fig. 2 would be explained by the attenuation of the X_5 contribution. It should be noted, however, that a qualitatively similar effects has been reported for adsorption on the (111) nickel surface [20].

5. Conslusions

(i) The UPS spectrometer described here enables the main features of angle-resolved spectra to be determined.

(ii) The angle-resolved spectra of nickel (100) conform to the photoemission expected for bulk nickel and do not show evidence of the existence of surface states.

Acknowledgements

We wish to thank Mr Bruce Gilbert for skilled technical assistance in the construction of the spectrometer.

This work was supported by the Australian Research Grants Committee.

References

[1] B. Feuerbacher and R.F. Willis, J. Phys. C9 (1976) 169.
[2] C.R. Brundle, Surface Sci. 48 (1975) 99.
[3] J. Anderson and G.J. Lapeyre, Phys. Rev. Letters 36 (1976) 376.
[4] P.M. Williams, D. Latham and J. Wood, J. Electron Spectr. 7 (1975) 281.
[5] H. Becker, E. Dietz, U. Gerhardt and H. Angmüller, Phys. Rev. B12 (1975) 2084.

[6] J.W. Gadzuk, Surface Sci. 53 (1975) 132.
[7] P.J. Page and P.M. Williams, Faraday Discussions Chem. Soc. 58 (1974) 80.
[8] P. Heimann and H. Neddermeyer, J. Phys. F (Metal Phys.) 6 (1976) L257.
[9] D.E. Eastman, J. Phys. (Paris) 32 (1971) C1-293.
[10] M.M. Traum and N.V. Smith, Phys. Rev. B9 (1974) 1353.
[11] S. Hufner, G.K. Wertheim, N.V. Smith and M.M. Traum, Solid State Commun. 11 (1972) 323.
[12] M.C. Desjonquères and F. Cyrot-Lackmann, Surface Sci. 53 (1975) 429.
[13] D.E. Eastman and J.K. Cashion, Phys. Rev. Letters 27 (1971) 1520.
[14] B.J. Waclawski, T.V. Vorburger and R.J. Stein, J. Vacuum Sci. Technol. 12 (1975) 301.
[15] G.L. Price, B.A. Sexton and B.G. Baker, Surface Sci. 60 (1976) 506.
[16] G.L. Price and B.G. Baker, to be published.
[17] I.P. Batra and C.R. Brundle, Surface Sci. 57 (1976) 12.
[18] P.M. Williams, P. Butcher, J. Wood and K. Jacobi, Phys. Rev. B14 (1976) 3215.
[19] E.I. Zornberg, Phys. Rev. B1 (1970) 244.
[20] H. Conrad, G. Ertl, J. Küppers and E.E. Latta, Surface Sci. 50 (1975) 296.

INFRARED VIBRATION SPECTROSCOPY OF MOLECULAR ADSORBATES ON TUNGSTEN USING REFLECTION INELASTIC ELECTRON SCATTERING

C. BACKX [*], R.F. WILLIS [**], B. FEUERBACHER and B. FITTON

Surface Physics Group, Astronomy Division, European Space Research and Technology Centre, Noordwijk, The Netherlands

The observation of adsorbate vibrational energies in the range, $30 \lesssim h\nu_{vib} \lesssim 1000$ meV, by electron-energy-loss spectroscopy, provides detailed information on the geometry of atomic and molecular complexes. The "surface normal dipole selection rule", is discussed and illustrated with results obtained for CO and C_2H_2 adsorption on the principal low-index faces of tungsten, viz.: W(100), W(110) and W(111) using a high-resolution electron reflection spectrometer. Specifically, the behaviour of chemisorbed *diatomic* carbon monoxide and *polyatomic* acetylene is compared as a function of coverage and surface crystallography. Comparison is made with the spectral information obtained by reflection infrared spectroscopy and recent ultraviolet photoelectron spectroscopy studies of the chemisorption binding energies. The energy loss spectra are discussed in terms of current adsorbate models and the possible formation of "distorted rehybridized surface molecular complexes" based on molecular orbital theories of organometallic compounds.

1. Introduction

The vibrational states of atoms and molecules adsorbed on metal surfaces at coverages of less than one monolayer was first reported using electron-energy-loss spectroscopy (ELS) just under 10 years ago [1]. It is only very recently, however, that the high-resolution energy analysis techniques required have been developed to a level where it is now possible to begin to identify the nature of atomic [2–4] and molecular [5–8] chemisorption species.

In this paper, we discuss the ELS spectra of CO and C_2H_2 adsorption on the principal low index faces of tungsten, viz.: W(100), W(110) and W(111), as a function of coverage and surface crystallography. These systems have been studied extensively by a variety of methods, i.e., thermal desorption, low-energy-electron diffraction (LEED), ultra-violet and X-ray photoelectron spectroscopy etc. [9] so that we are now able to compare results with the various proposed adsorbate models. In the case of hydrogen chemisorption, previous ELS results [2,4] are representative

---

[*] Present address: Koninklijke Shell Laboratorium (Shell Research B.V.), Amsterdam, The Netherlands.
[**] Address correspondence to this author.

of *atomic* adsorption at all surface coverages on all three faces at ambient temperature. In contrast, CO dissociates at low coverages, but produces *diatomic* molecular species at higher gas exposures. Similarly, acetylene chemisorbs dissociatively at low coverage and produces a *polyatomic* "surface molecular complex" at high coverage. More importantly, the spectral fine structure suggests that only those vibrational modes that are associated with net electric dipole variations perpendicular to the surface are excited strongly. This fact allows us to interpret the spectra in terms of the geometry of specific atomic adsorbate sites [4]; confirms that molecular CO adsorbs in the "upright" configuration [7]; and, in the case of C_2H_2, the measurements indicate that considerable rehybridization of the free molecule occurs on chemisorption [5]. This latter observation is compatible with reflection infrared spectroscopic measurements of olefinic species adsorbed on a variety of transition and noble metal substrates [10]. Recent photoemission studies of C_2H_4 and C_2H_2 adsorption on W(100) surfaces [11] suggest that a "distorted olefinic" species is formed via a Dewar–Chatt-type model of π–d bonding, as in organometallic compounds [12]. It is clear that the distribution of the surface local density of d-states will determine the actual form of the adsorbate complex in these circumstances, bond elongation and distortion occurring due to back-donation of charge into the molecule from the filled d-band metal states.

The aim of the present paper is to examine the ELS results in the light of these recent developments and proposed adsorbate models.

2. The "surface normal dipole selection rule"

The operator which causes excitations in inelastic electron scattering is $e^{i q \cdot r}$, where r is the position of the scattered electron and q the transferred momentum [13]. In the case of a *localized* initial state, this operator may be expanded: $e^{i q \cdot r} = 1 + i q \cdot r + \frac{1}{2}(i q \cdot r)^2 + ...$ The linear term dominates for q sufficiently small and causes dipole transitions; the quadratic term, which gives rise to monopole, quadrupole and higher order cross terms in the matrix element, becomes important for larger q. Thus, when q is small, we observe the same transitions as are seen in optical absorption studies; "forbidden" transitions may be observed at larger q.

In the case of electron reflection from an ideal conducting metal surface, the conduction electrons respond to the incident radiation producing a dipole "image charge" potential normal to the vacuum–metal interface [14]. The time response of this *long range* Coulombic interaction between the incoming charged particle and the surface (plasmon) electric field, which extends out into the vacuum, is shown schematically in fig. 1a, together with the Fourier transform $F(\nu)$ if this time response, fig. 1b [15]. This rather crude representation of a complex scattering process serves to illustrate the point that the electron beam acts as a source of wideband radiation. The optical absorption properties of an adsorbate will determine at which frequencies ν "photons" are adsorbed from this radiation source, correspond-

TIME
(a)

FREQUENCY
(b)

IMAGE DIPOLE CHANGE
(c)

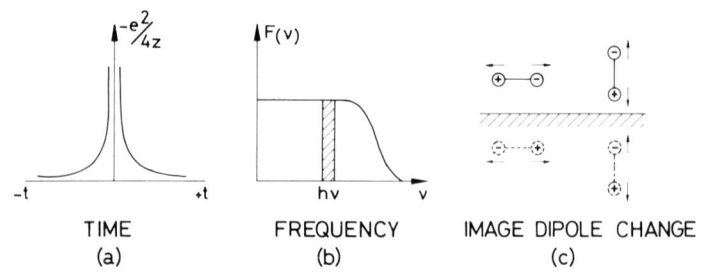

Fig. 1. Illustrating (a) the time dependence of the induced-dipole image potential during electron reflection from a metal surface, (b) the Fourier transform $F(\nu)$ of this time response, and (c) the "image" dipole change during the vibration of a molecular dipole, oriented parallel and perpendicular, to the surface.

ing to an energy loss $h\nu$ of an electron scattered in the specular direction, i.e., relatively small net momentum transfer q associated with the loss.

In addition to this optical dipole selection rule, the electric field normal to the surface, associated with the induced image charge, provides a second selection rule which is important in the interpretation of energy-loss and reflection-infrared spectra. In this case of atomic and molecular adsorbate vibrational excitations at infrared frequencies, only vibrations which give dipole changes perpendicular to the surface will absorb radiation strongly. This *"surface normal dipole selection rule"* is equally true whether the incident radiation is an electron or an infrared photon [16]. It can therefore be readily seen (fig. 1c) that, for example, a bond-stretching vibration, which would give rise to a dipole change parallel to the surface, will produce an equal and opposite change in the induced "image" dipole in the metal substrate; the net dipole change is effectively zero. On the other hand, the oscillating dipole perpendicular to the surface will be reinforced by the oscillating image dipole. Another important point is that the surface carries a permanent normal dipole field (which is the origin of the work function [17]) that effectively polarizes otherwise infrared-inactive species when adsorbed. For this reason, the normal vibration modes of chemisorbed hydrogen produce inelastic scattering of the incident electrons [2,4].

3. Experimental results

Electron energy loss spectroscopy of adsorbate vibrational states occurring in the range 30 to 1000 meV (240 to 8000 cm^{-1}) has been performed with an energy resolution in the range 5 to 50 meV. The condition that the primary beam energy, $h\nu_0 \gg h\nu_{loss}$, ensures that the small q momentum transfer condition can be achieved about the reflected beam direction. For low energy electrons, $h\nu \simeq 5$ eV, various electrostatic deflection systems have been found to be suitable for monochromating and energy analyzing the incident and reflected electrons [1–8]. The measurements presented here for W(110) and W(111) surfaces were performed with a system consisting of a hemispherical monochromator and analyzer both oriented at a fixed angle of 45° to the crystal surface such that the crystal could be rotated in the azimuthal plane. At an incident energy of 4 eV, measurements at 4° out of the plane of incidence/reflection diminish the high elastic background intensity by a factor of 10 relative to the specular beam, while the loss intensities decrease only by a factor of 3. Further details of the spectrometer and specimen cleaning procedure have been published [4,5].

3.1. Diatomic adsorption; CO on tungsten

Thermal desorption spectra of CO adsorbed on tungsten surfaces show exceedingly complex desorption kinetics suggesting the presence of several binding states, the relative population of which is both coverage and temperature dependent [9]. Three main binding states — the so-called α-, β- and "virgin"-CO — have been identified. The simplest interpretation consistent with the data to date on several planes is that the virgin and α-CO phases are molecular while the β states consist of dissociatively adsorbed C- and O-atoms.

A series of ELS spectra obtained for different gas exposures at 300 K for W(110), W(111) and W(100) are compared in fig. 2. The very high resolution spectra obtained by Froitzheim et al. [7] for W(100) show that for small exposures to CO (<1 L), the two loss peaks at 68 meV and 78 meV are due to vibrations of isolated chemisorbed C- and O-atoms respectively, adsorbed in fourfold sites each, confirming dissociative adsorption of the "β-CO state" [9]. The separation of these peaks is not resolved in the loss at 70 meV in W(110) and W(111) spectra presented here due to

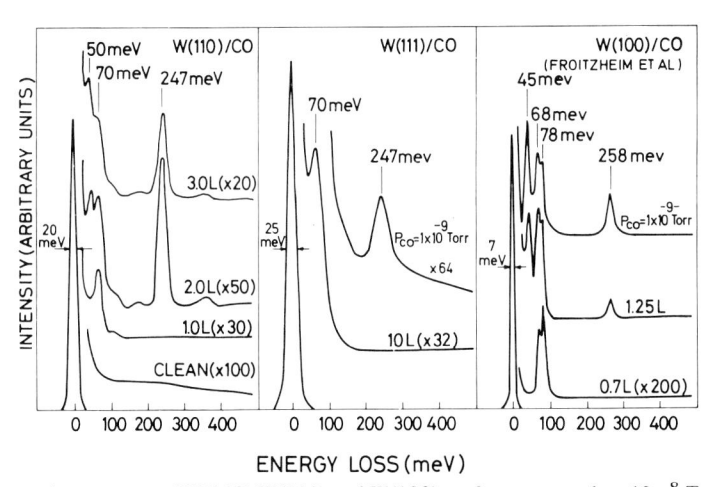

Fig. 2. Energy loss spectra of W(110) W(111) and W(100) surfaces exposed to 10^{-8} Torr partial pressure of carbon monoxide for various periods. The W(100) spectra were obtained by Froitzheim et al. [7].

the coarser resolution of the electron spectrometer which was employed [4,5]. With further exposure to CO, two additional losses at 45 meV and 258 meV are observed to evolve in the (100) spectrum corresponding to similar features resolved at 50 meV and 247 meV on the W(110) surface. Only one loss at 247 meV has been sofar resolved on W(111) and then only at much higher gas exposure and pressure.

Since *two* losses are detected associated with dipole changes normal to the surface, the two peaks around 40 and 50 meV and 240 and 260 meV relate to the characteristic vibrations of the undissociated α-CO state corresponding to a molecule adsorbed in an upright position. For W(100) [7], the 45 meV and 258 meV energy losses are close to the W–C and C–O stretching frequencies in tungsten carbonyl, i.e., $W(CO)_6$, at 46 and 263 meV respectively [18]; the C–O stretching vibration of the free molecule occurs at 269 meV. The appearance of only one frequency (45 meV) for the vibration of the whole CO molecule against the W surface and the correspondance with $W(CO)_6$ has lead to the suggestion [7] that chemisorption occurs at positions directly "on-top" of individual W atoms on the W(100) surface. This is supported by ELS measurements on Ni(100) [8] and Pt(111) [19] which show additional losses appearing simultaneously at lower frequencies for adsorption in higher coordination sites corresponding to bonding between the C-atom and more than one substrate atom. This is consistent with the conventional interpretation of reflection IR-spectroscopy [20] where lower CO stretching vibrations are assigned to sites of higher coordination.

The situation is complicated however by the fact that *two* metal–carbon excitations can arise in a bridge-bonded situation due to a low frequency *symmetric* mode and a higher frequency *asymmetric* mode, the latter being less intense since it effectively represents a displacement of the CO molecule parallel to the surface. Andersson [8] has recently observed a strong loss at 44 meV and a weaker loss around 81 meV for CO adsorption on Ni(100) at 173 K which he associates with these bridge-site modes. Furthermore, for the arguments presented earlier for hy-

Table 1
Energy loss spectroscopy of CO on d-band metal surfaces

Metal	C–O stretch	Metal–CO	Site	Reference
W(100)	258	45	On-top	[7]
W(110)	247	50	Disordered	Present work
W(111)	247	?		
Pt(111)	261	58	On-top	[19]
>0.2 L	232	45	4 Fold	
Ni(111)	256	60	On-top	[8]
	240	45	Out-of-registry bridge	
	249	54	Disordered	

drogen chemisorption [4], while it is likely that the 247 meV loss observed on W(111) also corresponds to the "on-top" position, there is the problem of accounting for a 10 meV shift with respect to the W(100) face (cf. fig. 2). The complications inherent in any simple interpretation of the CO frequencies purely in terms of the simple site geometry arguments used for atomic adsorption [2–4] are illustrated in the results shown in table 1. Not only is the metal–CO stretch frequency sensitive to site geometry and coordination, as is to be expected, but so too is the C–O stretch frequency. This would suggest that the molecular orbitals of the adsorbate mix with the metal wavefunctions, particularly the d-states, the amount of "back-bonding" into the empty anti-bonding $2\pi^*$ orbitals of the molecule being particularly sensitive to the local coordination geometry [21,22]. A similar mechanism will be true for adsorbed olefinic complexes [11,12].

3.2. Polyatomic adsorption; C_2H_2 on tungsten

The adsorption of acetylene on tungsten surfaces is similar to CO in that complete fragmentation of the molecule occurs at low coverages at 300 K [5]. With

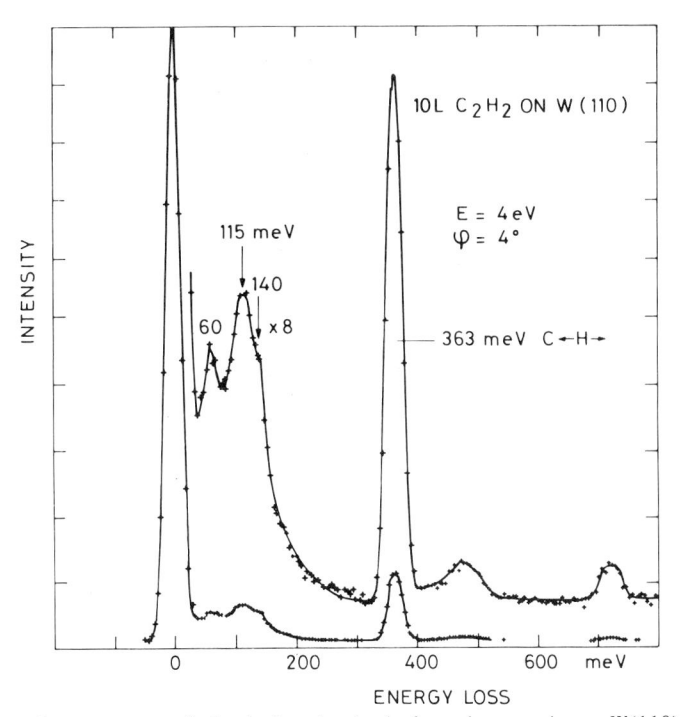

Fig. 3. Energy loss spectrum of adsorbed molecular hydrocarbon species on W(110) following exposure to 10 L C_2H_2. The electron impact energy was 4 eV and the azimuthal collection angle was 4° out of the specular plane at a polar reflection angle of 45°.

increasing exposure, energy loss peaks evolve that are characteristic of hydrocarbon molecular species. This occurs for gas exposures $\gtrsim 2$ L for W(111) [5] compared with $\gtrsim 0.5$ L for the more close-packed W(110) surface, indicating that the very open (111) lattice plane is the more reactive. Measurements of the overall work function change with increasing exposure indicate the behaviour of the W(100) surface to be similar [23].

In fig. 3, a typical energy loss spectrum is shown for the case of a W(110) surface exposed to 10 L C_2H_2. Four intense losses occur at 60, 115, 140 and 363 meV with weaker ones around 480 and 720 meV. These latter have been interpreted as combination frequencies and an overtone frequency respectively of the more intense vibrations at lower energies, 60 to 363 meV [5]. Their appearance in the spectrum leads to the important conclusion that the 60, 115, 140 and 363 meV losses arise due to normal modes of *the same molecular species*. The intense loss at 363 meV is observed in reflection infrared spectroscopy of adsorbed C_2H_4 and C_2H_2 on silica supported Ni, Pd, Pt and Rh particles and is due to a C–H stretching mode [10]. Frequencies observed in the range 2800 to 3000 cm^{-1} ($\sim$345 to 370 meV) "leave no doubt that the hydrocarbon species itself has sp^3 hybridized carbon atoms" [10]. However, the absence of peak connected to the –CH_2 scissor vibration ($\sim$180 meV) in both the infrared and the energy loss spectra (cf. fig. 3) excludes more than one H-atom being attached to a C-atom. A further difficulty encountered earlier in the interpretation of the infrared data was that, assuming a planar M–C–C–M skeleton (M = metal), σ-diadsorbed rehybridized "olefinic" species were expected to have *two* fundamental CH stretch modes, A_1 and B_1, in the range 2800 to 3000 cm^{-1} conforming to an overall C_{2v} symmetry. The reason why this is not so is illustrated in fig. 4.

Recent photoemission measurements [11] have shown that the adsorption of C_2H_2 on W(100) surfaces at 295 K produces molecular energy level shifts which are indicative that rehybridization causes the adsorbed molecule to distort from its linear structure such that the C–C bonds are "stretched" and the C–H bonds take up a position at an angle of approximately 60° to the C–C axis. This distortion occurs despite the fact that some remnant of π-bonding remains. This being the

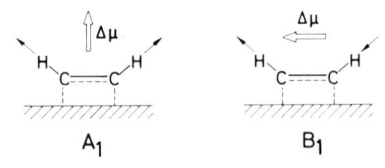

Fig. 4. Illustrating the form of the C–H stretching normal modes of vibration of a planar M–CH–CH–M "olefinic" surface molecular complex of symmetry C_{2v}, which can result from the chemisorption of acetylene on a transition metal surface (M = surface metal atom). Only the mode of symmetry A_1 gives a dipole change normal to the surface although both modes are infrared active for C_{2v} overall symmetry.

case, since only the C–H stretching modes of symmetry A_1 give a dipole change in a direction normal to the surface, only this vibration mode absorbs radiation strongly, the B_1 mode giving a net dipole displacement parallel to the surface (fig. 4). It is also interesting to note that this A_1 mode is symmetry related to an infrared forbidden (Raman active) C–H stretch mode of the parent acetylene molecule, i.e., neither the IR-reflection or energy loss spectrum has any similarity with the IR-absorption spectrum of the free molecule due to rehybridization and the operation of the "surface normal dipole selection rule"; gaseous C_2H_2 has a single strong peak at 410 meV [24].

The ELS spectrum, fig. 3, is compatible with the configuration shown in fig. 4 if, in addition to the above C–H stretch vibration at 363 meV, we assume the following to be true: 3 additional normal modes should occur associated with (a) the metal–carbon vibration of the whole molecule against the surface: (b) a C–C stretch vibration; and (c) and C–H bending vibration normal to the surface. These give rise to the loss peaks at 60, 115 and 140 meV. The question of *why* we should see a C–C stretching mode *parallel to the surface* is a complex one. We envisage that a change in the C–C bond distance causes a net change in the normal dipole moment, brought about by a charge redistribution in the tungsten-to-molecule bond during the vibration cycle. This would explain the observed 4 peaks. It would also explain the relatively strong intensities of the observed losses. Any scission of the C–C bond would not produce such a spectrum. Some evidence for this line of argument comes from the fact that a spectrum of the "diatomic" surface complex MC–H consisting only of *two* loss peaks at about 360 meV (C–H stretch) and 70 meV (C–W vibration) has been observed when the W(110) crystal was exposed to only a very low exposure (~0.5 L) of C_2H_2. Also, photoemission measurements of ethylene adsorption on W(110) endorse the view that the C–C bond remains intact at saturation coverage at 300 K [25]; fragmentation of the C_2H_2 molecule to form adsorbate M–CH, MCH_2 or MC_2H radicals produces very different ionization energy levels to those observed [26]. The 115 meV loss is close to that observed for sp^3 hybridized C–C bonds (112 meV) in the gas phase. On the other hand, the C–H bending mode, occurring at 125 meV in the free sp^3-hybridized hydrocarbons, is shifted up in frequency (140 meV) on the surface. However, a similar shift has been reported in the case of benzene adsorbed on Ni(100) [6], attributable to a steric hindrance of this H-atom motion with the substrate.

4. Surface bonding considerations

One of the problems in assigning particular energy loss features with specific gas phase molecular vibrations is that rehybridization and dipole interactions between the adsorbed species, particularly at high surface coverage, produce shifts in the free molecule modes. This is particularly true in the case of CO and unsaturated hydrocarbon adsorption on transition metal surfaces. Recent photoelectron spectroscopy studies [25,26,27] have revealed that significant molecular distortion occurs during chemisorption at room temperature. Various molecular orbital bonding schemes have been proposed in terms of σ and π-orbital overlap with the crystal-field split metal d-states distribution at the surface, fig. 5 [22,26].

Chemisorbed CO shows a large bonding shift of the 5σ orbital with metal valence orbitals (fig. 5a), in addition to which, a significant amount of back-donation of charge into the unoccupied $2\pi^*$ anti-bonding orbitals is thought to take place [21,22]. This weakens the C–O bond which, in turn, increases the separation of the carbon and oxygen atoms by an amount depending on the degree of back bonding from the metal. This could explain the sensitivity of *both* the C–O stretch and metal–C vibrations to adsorption site (table 1). Such a model will favour sites where donation of the carbon lone-pair 5σ electrons into empty d-orbitals of ϵ_g symmetry can be stabilized by back-bonding via filled t_{2g} orbitals and the CO $2\pi^*$ orbitals. Recent calculations of the orbital symmetries of the surface states on W(100) [28] and Pt(111) [29] confirm that just such a situation occurs for the "on-top" sites. However, if this is the case, the ELS spectra would indicate C–O stretch frequencies (table 1) which "soften" more due to increased back-donation occurring for the higher coordination sites.

The Dewar–Chatt model [12] for the π-bonding of olefins to singly coordinated transition metal ions is similar. The π-orbital of the hydrocarbon forms a bond by donating charge to an unfilled d_z^2 orbital of the metal ion accompanied by back-donation from the d_{yz} (or d_{xz}) orbitals into anti-bonding π^* orbital of the molecule. Vorburger et al. [11] have recently proposed such a scheme for C_2H_2 adsorption centred directly above the metal atom on a W(100) surface; the degree of back bonding from the metal d-states determines the amount of C–C stretch and C–H bond bending, as illustrated schematically in fig. 5b. Such a model is consistent with the relative shifts which they observe for both the σ_{CC} and σ_{CH} binding energies in photoemission spectra. On the other hand, the 363 meV loss peak (fig. 4) is typical of a sp^3 C–H stretch vibration, indicating that a di-σ bonding mechanism might be more applicable in the W(110) case, in spite of the fact that a π-orbital

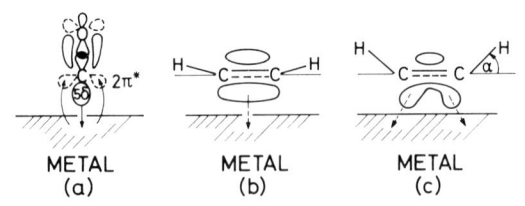

Fig. 5. Illustrating the molecular orbital geometries and charge densities for (a) chemisorbed CO, (b) π–d bonded C_2H_2 and (c) an "olefinic" chemisorbed di-σ-like complex produced by rehybridization of the molecular orbitals with the metal d-states; charge donation occurs to the metal via the molecular π-orbitals with back-donation into the anti-bonding orbitals from the metal.

energy level is identifiable in the photoemission spectrum [25]. Demuth [26] has recently proposed the di-σ-type olefinic species shown, fig. 5c, for similar "surface olefinic species" observed on Pd(111) and Pt(111) surfaces. Such a model is consistent with the classical di-σ adsorbed species discussed in the literature [20], but at the same time retains a remnant of rehybridized π-bonding with the surface not necessarily restricted to a single surface atom site. The important point here is that the formation of such a distorted rehybridized olefinic molecular complex will produce C–C and C–H bond frequencies that are different to those in the original free molecules. Also, any such mixing of the molecular orbitals with the metal d-wavefunctions will vary for different faces of different d-band metal substrates, depending on site coordination. For this reason, unambiguous identification of the C–C stretch and the C–H bend frequencies with either the 115 or 140 meV losses, fig. 2, is difficult [30]. What is certain, however, is that rehybridization and distortion of the original linear C_2H_2 molecule has taken place.

5. Concluding remarks

In spite of a number of remaining uncertainties, results obtained to date have shown that electron energy loss spectra of the vibration excitation of adsorbed species provide considerable insight into the structure and geometry of the surface molecular complexes formed on transition metal surfaces. Ultraviolet photoelectron spectroscopy studies [11,25–27] would indicate, however, that some caution should be exercised in identifying molecular species, in particular, purely in terms of gas phase fundamental vibrational modes and shifts due only to local site geometry.

As pointed out many years ago by Blyholder [21] with regard to the interpretation of infrared spectra of chemisorbed CO, subtle shifts in the C–O stretch frequency are not necessarily due to different adsorption sites on different metals. The degree of back-bonding, and hence a "softening" of the C–O bond, will depend on the local spectral density of d-states at the surface available for providing charge [28,29]. The ELS spectra indicate that the higher the coordination site, the more the C–O stretch vibration softens via the back-donation mechanism. The bcc metals appear to be strongly back-bonding causing the chemisorbed CO molecule to stretch and dissociate [34]. Similar considerations would appear to apply in the case of the chemisorbed olefins. Here the situation appears to be more complex due to a strong mixing of the molecular orbitals and the metal d-wavefunctions leading to strong rehybridization and distortion of the chemisorbed species. Again the extent to which this will occur, however, will be dependent on the nature of the metal surface d-states. The ELS data (fig. 3) are not able to distinguish unequivocally the actual chemical bonding. However, the surface sensitivity of the technique together with the detailed information which is provided on the actual geometry of the adsorbed species will provide a strong test of future developments in this field.

Acknowledgements

We wish to thank M.R. Barnes for his enthusiastic technical assistance and to express our gratitude to the authors (refs. [2–8, 25–27,30]) for sending us pre-print copies of their hitherto unpublished results.

References

[1] F.M. Propst and T.C. Piper, J. Vacuum Sci. Technol. 4 (1967) 53.
[2] H. Froitzheim, H. Ibach and S. Lehwald, Phys. Rev. B14 (1976) 1362; Phys. Rev. Letters 36 (1976) 1549.
[3] S. Andersson, Solid State Commun. 20 (1976) 229.
[4] C. Backx, B. Feuerbacher, B. Fitton and R.F. Willis, Phys. Letters 60A (1977) 145.
[5] C. Backx, B. Feuerbacher, B. Fitton and R.F. Willis, Surface Sci. 63 (1977) 193; in: Proc. Intern. Symp. on Photoemission, Eds. R.F. Willis et al. (European Space Agency Special Publication No. SP118, rev. 1, 1976), p. 291.
[6] G. Dalmai-Imelik, J.C. Bertolini and J. Rousseau, Surface Sci. 63 (1977) 67; in: Proc. Intern. Symp. on Photoemission, Eds. R.F. Willis et al. (European Space Agency Special Publication No. SP118, rev. 1, 1976), p. 285.
[7] H. Froitzheim, H. Ibach and S. Lehwald, Surface Sci. 63 (1977) 56.
[8] S. Andersson, Solid State Commun. 21 (1977) 75.
[9] For a review:
E.W. Plummer, B.J. Waclawski, T.V. Vorburger and C.E. Kuyatt, Progr. Surface Sci. 7 (1976) 149;
E.W. Plummer in: Topics in Applied Physics, Vol. 4, Ed. R. Gomer (Springer, Berlin, 1975) p. 143.
[10] H.A. Pearce and N. Sheppard, Surface Sci. 59 (1976) 205.
[11] T.V. Vorburger, B.J. Waclawski and E.W. Plummer, Chem. Phys. Letters, in press.
[12] See, for example, A.C. Blizzard and D.P. Santry, J. Am. Chem. Soc. 90 (1968) 5749.
[13] See, for example, P.M. Platzman and P.W. Wolff, Waves and Interactions in Solid State Plasmas (Academic Press, New York, 1973).
[14] A.A. Lucas and M. Sunjić, Phys. Rev. Letters 26 (1971) 229;
D.L. Mills, Surface Sci. 48 (1975) 59.
[15] The point at which the electron is reflected from the surface produces a singularity in the Fourier response $F(\nu)$ due to the fact that a *long range* dipole Coulombic scattering law is no longer applicable. This *impact scattering* involves a short range interaction with the lattice and is discussed, for example, by Mills [14].
[16] S.A. Francis and A.H. Ellison, J. Opt. Soc. Am. 49 (1959) 131.
[17] For a review see: N.D. Lang, in: Solid State Physics, Vol. 30, Eds. H. Ehrenreich et al. (Academic, New York, 1973) p. 225.
[18] L.H. Jones, Spectrochim. Acta 19 (1963) 329.
[19] H. Froitzheim, H. Ibach and S. Lehwald, in: Proc. Intern. Symp. on Photoemission, Eds. R.F. Willis et al. (European Space Agency Special Publication No. SP118, rev. 1, 1976) p. 277.
[20] See, for example, L.H. Little, Infrared Spectra of Adsorbed Species (Academic Press, New York, 1966).
[21] G. Blyholder, J. Phys. Chem. 68 (1964) 2772.
[22] G. Doyen and G. Ertl, Surface Sci. 43 (1974) 197.
[23] R. Nathan and B.J. Hopkins, J. Phys. E7 (1974) 851.

[24] G. Herzberg, Infrared and Raman Spectra of Polyatomic Molecules, 11th printing, (Van Nostrand, New York, 1964).
[25] E.W. Plummer, B.J. Waclawski and T.V. Vorburger, Chem. Phys. Letters 28 (1974) 510.
[26] J.E. Demuth, Chem. Phys. Letters, in press.
[27] G. Brodén, T.N. Rhodin and W. Capehart, Surface Sci. 61 (1976) 143.
[28] M.C. Desjonquères and F. Cyrot-Lackmann, J. Phys. F6 (1976) 567.
[29] G. Apai, P.S. Wehner, R.S. Williams, J. Stöhr and D.A. Shirley, Phys. Rev. Letters 37 (1976) 1497.
[30] Recent ELS measurements of C_2H_2 and C_2D_2 chemisorption on Pt(111) (H. Ibach, H. Hopster and B. Sexton, to be published) would suggest that the *higher* energy loss (140 meV on W(110)) is due to the C–C stretch vibration.

INFRARED SPECTRA FOR CO ISOTOPES CHEMISORBED ON Pt{111}: EVIDENCE FOR STRONG ABSORBATE COUPLING INTERACTIONS

Alison CROSSLEY and David A. KING

The Donnan Laboratories, The University of Liverpool, P.O. Box 147, Liverpool L69 3BX, England

Previous results for $^{12}C^{16}O$ chemisorbed on a Pt{111} recrystallised ribbon revealed that the infrared absorption band due to the CO stretch appears at low coverages at 2063 cm^{-1} and shifts to ~2100 cm^{-1} at saturation coverage at 300 K. The cause of this shift is studied in the present work, by investigating the vibrational spectra from a variety of mixtures of $^{13}C^{16}O$ and $^{12}C^{16}O$. The results show that there is a strong dipole–dipole coupling interaction between adsorbate molecules in the overlayer, and provide conclusive evidence that the 35 cm^{-1} frequency shift observed with increasing coverage for $^{12}C^{16}O$ is attributable to coupling.

1. Introduction

Vibrational spectra have been obtained from carbon monoxide chemisorbed on a wide variety of high-area supported metal catalysts [1]. More recently, using reflection absorption infrared spectroscopy [2] and high resolution reflection electron scattering spectroscopy [3], these studies have been extended to well-defined metal surfaces under ultrahigh vacuum conditions. Some 20 years ago, Eischens and Pliskin [4] reported in a study of CO on SiO_2-supported platinum that the dominant high frequency absorption band due to the CO stretch, attributed to a linear structure, shifted continuously towards higher frequencies as the coverage was increased, over the range 2040 to 2067 cm^{-1}. A similar shift was recently reported by Shigeishi and King for CO on Pt{111} [5]. This effect was attributed by Blyholder [6] to increase competition for metal d-electrons back-donated into the CO $2\pi^*$ orbital. According to this model, as back-donation is reduced, the band shifts back towards the gas phase frequency (2143 cm^{-1}). However, Hammaker, Francis and Eischens [7] demonstrated, using isotopes of carbon monoxide, that the coverage-induced frequency shift is consistent with dipole-dipole coupling between chemisorbed species. Although their isotope results are uniquely interpreted in terms of coupling interactions, it was not quantitatively demonstrated that the observed frequency shift is attributable to this effect; subsequent workers have favoured the Blyholder model. Recently, for example, Primet and co-workers [8] concluded from co-adsorption studies with electron donor and acceptor molecules that competition for $d\pi^*$ back-donation was the cause of the frequency shift.

A resolution of this problem is critically important, for two reasons. Firstly, while through-bond lateral interactions between chemisorbed species have been placed on a sound theoretical footing [9] there is as yet little more than a intuitive basis for the concept of increased competition for $d\pi^*$ backbonding. Second, if strong dipole coupling exists between molecules in an overlayer, it is erroneous to use experimentally observed absorption frequencies (except in the limit where the coverage tends to zero) for calculating force constants, without accounting for the coupling force constant. In the present work an experiment is described which conclusively demonstrates strong coupling within a carbon monoxide overlayer, accounting quantitatively for the observed coverage-induced shift in the absorption frequency.

2. Experimental

The ultrahigh vacuum reflection–absorption infrared cell and monochromator used in the present work have been described in detail [10,5]. In order to optimise the signal-to-noise ratio, infrared radiation was used at near-grazing incidence to the metal surface (84° to the surface normal, with an angular spread of ±4°). The spectrum was scanned at 1.8 cm^{-1} sec^{-1}, with a spectral resolution of either ~5 cm^{-1} (0.6 mV) or ~15 cm^{-1} (1.8 mV). To reduce atmospheric water vapour bands in background spectra the monochromator (Grubb Parsons GS-2) was flushed with dry air. The absorbent was a platinum foil cleaned and recrystallised by heating to 1325 K in 10^{-5} Torr O$_2$ for 24 h followed by a vacuum anneal at 1525 K for a further 24 h. In agreement with previous X-ray and LEED observations of recrystallized platinum foils [11], X-ray analyses of the ribbon revealed that it was {111}-oriented. Grain boundaries and stepped crystallites are likely to be exposed at the surface, but the general agreement with CO sticking probabilities, surface coverages and desorption spectra [11,12] and infrared spectra [13] obtained on bulk Pt{111} crystals indicates that such defects have a negligible effect on the results. Gas exposure was achieved by dynamical flow through capillaries and isotopic mixtures were analysed with a small magnetic-sector mass spectrometer attached directly to the infrared cell.

3. Vibrational spectra for CO on Pt{111}, and overlayer structures

Some hysteresis was observed in the intensity and frequency of the absorption band due to CO on Pt{111} during adsorption and desorption sequences at 300 K [5]; the absorption bands obtained after equilibration of the adlayer at temperatures between 400 and 500 K over the coverage range 0.6 to 7.2 × 10^{14} molecules cm^{-2} are shown in fig. 1. A single band was observed, first appearing at 2063 cm^{-1}, shifting strongly with coverage in the range 1.5 to 4.5 × 10^{14} molecules cm^{-2}, and

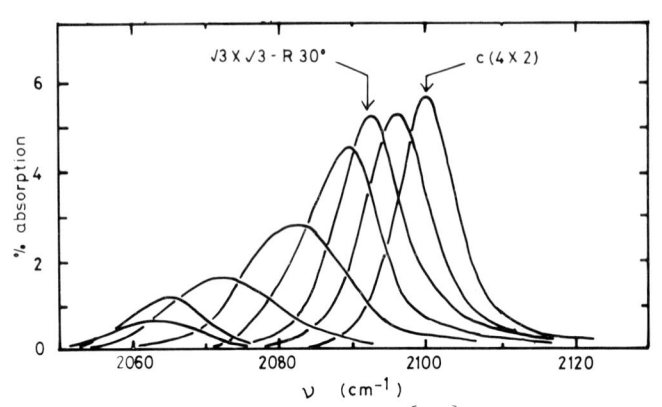

Fig. 1. Development of the infrared band for CO on Pt{111} as a function of increasing coverage at 300 K, after equilibration of the adlayer at 400 to 500 K. Bands corresponding to the $\sqrt{3} \times \sqrt{3}$-R30° and c(4 × 2) structures observed by LEED [12] and characterised here by absolute coverage measurements are indicated with arrows.

finally reaching a value of 2100 cm^{-1} at the room temperature "saturation" coverage of 7.2×10^{14} molecules cm^{-2}. The cause of this frequency shift is the subject of this paper.

The intensity of the observed absorption band reaches a plateau at a coverage of $\sim 4.5 \times 10^{14}$ molecules cm^{-2}, although the band continues to shift to higher frequencies beyond this coverage. An explanation for this result is suggested by a recent electron energy loss study of CO on Pt{111}, in which Froitzheim, Ibach and Lehwald [14] observed a vibrational band at $\sim$258 meV (2081 cm^{-1}) at relatively low coverages, corresponding to the band shown in fig. 1, but, in addition, at high coverages a band was found at 232 meV (1871 cm^{-1}). This band was less intense and broader than the high frequency band, which may explain why it was not observed in the infrared study [5]. It would appear that it has a natural line width of $\sim$80 cm^{-1}. (In the present work a further careful examination of the spectra in the region of 1850 cm^{-1} was made, at the relatively low resolution of 15 cm^{-1}, but again no band was observed; this might indicate the operation of different selection rules for infrared and electron energy loss spectroscopies, although convincing arguments have been presented to show that the "normal-dipole-selection rule" operates in both cases [15,3].) One further discrepancy between the infrared and electron energy loss studies remains, and is puzzling. The natural linewidth of the high frequency band is $\sim$6 cm^{-1} (from fig. 1, allowing for instrumental broadening) from the infrared study, but $\sim$40 cm^{-1} from the electron energy loss study [14].

By comparison with LEED data recently obtained for this system by Ertl, Neumann and Streit [12], Froitzheim et al. [14] propose that at a coverage of 5×10^{14} molecules cm^{-2}, where a $\sqrt{3} \times \sqrt{3}$-R30° structure is observed, all CO molecules are in linear (single-bonded) configurations, while at 7.5×10^{14} mole-

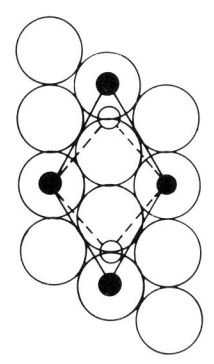

Fig. 2. Proposed $\sqrt{3} \times \sqrt{3}$-R30° (unit mesh indicated by full line) and c(4 × 2) (dashed line) structures for CO on Pt{111}.

cules cm^{-2}, where the LEED pattern shows a c(4 × 2) structure, half the CO molecules are in linear and half in bridged positions. The structures are summarised in fig. 2. While this model is attractive it does not embody a full description of the experimental results without a further refinement. Thus, the c(4 × 2) structure contains *less* molecules in the linear position (3.75 × 10^{14} molecules cm^{-2}) than in the $\sqrt{3} \times \sqrt{3}$-R30° structure (5 × 10^{14} molecules cm^{-2}), but in neither the present study nor the electron energy loss study [14] was a decrease in the intensity of the high frequency band observed during the corresponding coverage change. An explanation for this lies in the work function change measurements for this system [13,12] and in the LEED data [12] which show a strong tendency for disordering at temperatures above 80 K. Thus at 300 K it is likely that some occupation of bridged positions occurs even at or below a coverage of 5 × 10^{14} molecules cm^{-2}. We conclude that only on formation of the c(4 × 2) structure is the overlayer "locked" into a well-defined configuration, with a 1 : 1 ratio of bridged and linear CO molecules.

4. Theory of dipole—dipole coupling interactions

Models for dipole—dipole coupling have been developed by Decius [16], with particular applications to coupling in solid state aragonite-type lattices, and by Hammaker et al. [7] in relation to two-dimensional chemisorbed arrays. We present here only the essential features of the model so as to define the basis of the experiment described in section 5 below.

For a set of N dipoles aligned parallel and perpendicular to a surface, the secular equation has been solved exactly to yield the expression

$$\lambda = \lambda' + \left(\frac{\partial \mu}{\partial r}\right)^2 \frac{1}{M_r \epsilon_0} \sum_{j=2}^{N} \frac{1}{R_{ij}^3} \qquad (1)$$

for the only infrared-active mode of the coupled system, where $\lambda = 4\pi^2 C^2 \nu^2$ (ν is the band frequency in cm^{-1}); λ' refers to the isolated chemisorbed species, or the "singleton" frequency; $\partial\mu/\partial r$ is the dynamic dipole moment; M_r is the reduced mass of the oscillator; ϵ_0 is the permittivity; and

$$\sum_{j=2}^{N} (1/R_{ij}^3)$$

is the "lattice sum" corresponding to a particular configuration for the overlayer, where R_{ij} is the distance between the centres of two dipoles i and j in this configuration. As the coverage tends to zero, provided that island formation, or clustering, does not occur, the lattice sum clearly becomes zero; the coverage-induced frequency shift is thus given by the second term in eq. (1). The lattice sum was empirically calculated for the c(4×2) CO structure on Pt{111}, counting only molecules in linear positions, by summing over circles of increasing radii up to 26.3 Å, where convergence was obtained, giving a value of 0.059 Å^{-3}. Inserting this value into eq. (1), and using the experimentally observed values of the singleton frequency (2063 cm^{-1}) and the c(4×2) frequency (2100 cm^{-1}) (fig. 1), we obtain:

$\partial\mu/\partial r = 9.8 \times 10^{-20}$ C = 2.9 D/Å.

Hammaker et al. [7] have provided the following approximate solutions to the secular equation for the infrared-active modes when one isotopic species, given the symbol A, is completely surrounded by $N-1$ molecules of a second isotope B:

$$\lambda^+ = \lambda'_A + \left(\frac{\partial\mu}{\partial r}\right)^4 \frac{1}{(\epsilon_0 M_r)^2} \sum_{j=2}^{N} \frac{1}{R_{ij}^6} (\lambda'_A - \lambda'_B)^{-1}, \tag{4}$$

$$\lambda^- = \lambda'_B - \left(\frac{\partial\mu}{\partial r}\right)^4 \frac{1}{(\epsilon_0 M_r)^2} \sum_{j=2}^{N} \frac{1}{R_{ij}^6} (\lambda'_A - \lambda'_B)^{-1}, \tag{5}$$

in terms of the singleton values λ'_A and λ'_B for the two isotopes, where $\lambda'_A > \lambda'_B$. Eq. (4) corresponds to the in-phase mode and eq. (5) to the out-of-phase mode. These expressions thus provide a means of calculating the frequencies of the infrared active modes for mixtures of A and B in which the amount of the components A or B tend to zero. For example, for a mixture of ^{12}CO–^{13}CO, eq. (4) provides a means of calculating the position of the high frequency band in the limit where the ^{12}CO mole fraction tends to zero. The inverse sixth power lattice sum (which converges more rapidly than the inverse third power sum) was again evaluated empirically for the c(4×2) structure on Pt{111}, accounting only for molecules in linear positions, and found to be 2.7×10^{-4} Å^{-6}. Using this value, and the value for $\partial\mu/\partial r$ (9.8×10^{-20} C) obtained above from the frequency shift, the second term in eq. (4) was found to correspond to 2 cm^{-1}. Thus, since the singleton frequency for ^{12}CO is 2063 cm^{-1} (fig. 1), the high frequency band is predicted to be at 2065 cm^{-1} for the c(4 ×2) structure in the limit where the ^{12}CO component in the mixture approaches zero. Clearly, in the limit where the ^{13}CO component approaches zero the high frequency band for the c(4 ×2) structure is at the experimentally observed frequency for ^{12}CO, i.e. 2100 cm^{-1}. This suggests a critical experiment to test the self-consistency of the dipole interaction model. Absorption bands obtained at the coverage corresponding to the c(4 ×2) structure obtained for a variety of mixtures of ^{12}CO and ^{13}CO should show a shift in the high frequency band which matches, within 2 cm^{-1}, the coverage-induced frequency shift obtained with a pure isotope.

Theoretical spectra from isotope mixtures of *any* composition may be approximated using the in-phase ^{12}CO–^{12}CO, ^{12}CO–^{13}CO and ^{13}CO–^{13}CO modes, and the out-of-phase ^{12}CO–^{13}CO modes, as described by eqs. (1), (4) and (5). The four modes divide into a high frequency set and a low frequency set, and we here write simple expressions for each set in which each mode is weighted according to its mole fraction *and* its relative intensity:

$$\nu_h = \frac{\theta_{12}^2 \nu_{12-12}^+ + I_{12-13}^+ \theta_{12}\theta_{13}\nu_{12-13}^+}{\theta_{12}^2 + I_{12-13}^+ \theta_{12}\theta_{13}}. \quad (6)$$

$$\nu_l = \frac{\theta_{13}^2 \nu_{13-13}^+ + I_{12-13}^- \theta_{12}\theta_{13}\nu_{12-13}^-}{\theta_{13}^2 + I_{12-13}^- \theta_{12}\theta_{13}}, \quad (7)$$

where θ_{12} and θ_{13} are fractional coverages in ^{12}CO and ^{13}CO, respectively; and the relative intensities of the various modes are given by

$$I_k = (B_k + 1)^2/(B_k^2 + 1), \quad (8)$$

where $B_k = +1$ for the in-phase mode and -1 for the out-of-phase mode when reference molecule and environment are the same isotopic species, but when they are different species [7]:

$$B_k = (\lambda_2' - \lambda_k)/(-1)\left(\frac{\partial \mu}{\partial r}\right)^2 \frac{1}{\epsilon_0 M_r} \sum_{j=2}^{N} (1/R_{ij}^6)^{1/2}. \quad (9)$$

The intensity ratio I_{ratio}, i.e. the ratio of the intensities of the high frequency to the low frequency bands, is given by:

$$I_{\text{ratio}} = \frac{\theta_{12}^2 + \theta_{12}\theta_{13}I_{12-13}^+}{\theta_{13}^2 + \theta_{12}\theta_{13}I_{12-13}^-} \quad (10)$$

We note that with $\partial\mu/\partial r = 9.8 \times 10^{-20}$ C and using the lattice sum for the c(4 ×2) (linear species only) structure, I_{12-13}^-, from eqs. (8) and (9), is 0.11, whereas I_{12-13}^+ is 1.89. There is relatively little intensity in the 12–13 out-of-phase mode. Thus, from eqs. (6) and (7), we anticipate that in an experiment in which the coverage is maintained constant, corresponding to the c(4 × 2) structure, and the isotopic composition varied, two bands should be observed; the high frequency band should vary with composition over the full range 2065 to 2100 cm^{-1}, while the relatively less intense low frequency band should be almost invariant with composition over the range 0 to ~50% ^{12}CO.

5. Vibrational spectra from ^{12}CO–^{13}CO isotopic mixtures

Reflection-absorption infrared spectra obtained from isotopic mixtures of ^{12}CO–^{13}CO in the range 8% ^{12}CO to 100% ^{12}CO, at a spectral resolution of ~5 cm^{-1}, are shown in fig. 3. Isotopic compositions were determined by mass spectrometric analysis of the gas phase during adsorption, and checks were also made on the desorption spectra. Qualitatively, the spectra are in accord with the above predictions. However, in order to improve the intensity of the observed spectra, a second series of experiments were conducted at a spectral resolution of ~15 cm^{-1}. The results are shown in fig. 4. In the mid-range of isotopic mixtures two bands are clearly resolved, and we note in particular the shift of the high frequency band

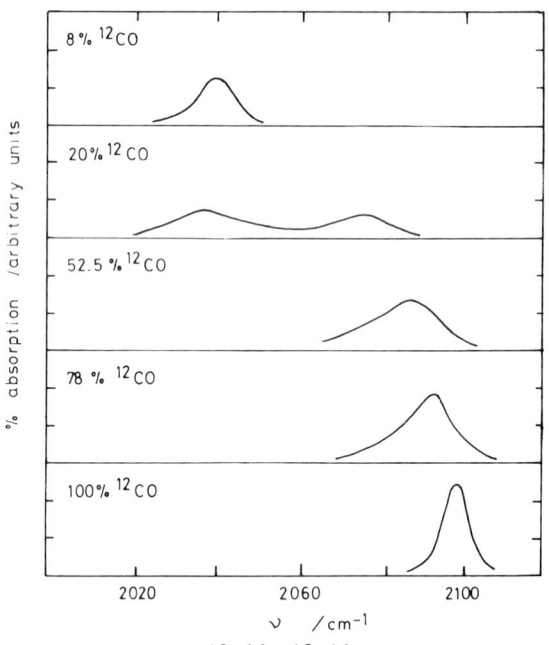

Fig. 3. Infrared absorption bands for ^{12}C^{16}O/^{13}C^{16}O isotopic mixtures adsorbed into the c(4 × 2) configuration (corresponding to saturation at 300 K) on Pt {111}, obtained at a spectral resolution of ~5 cm^{-1}.

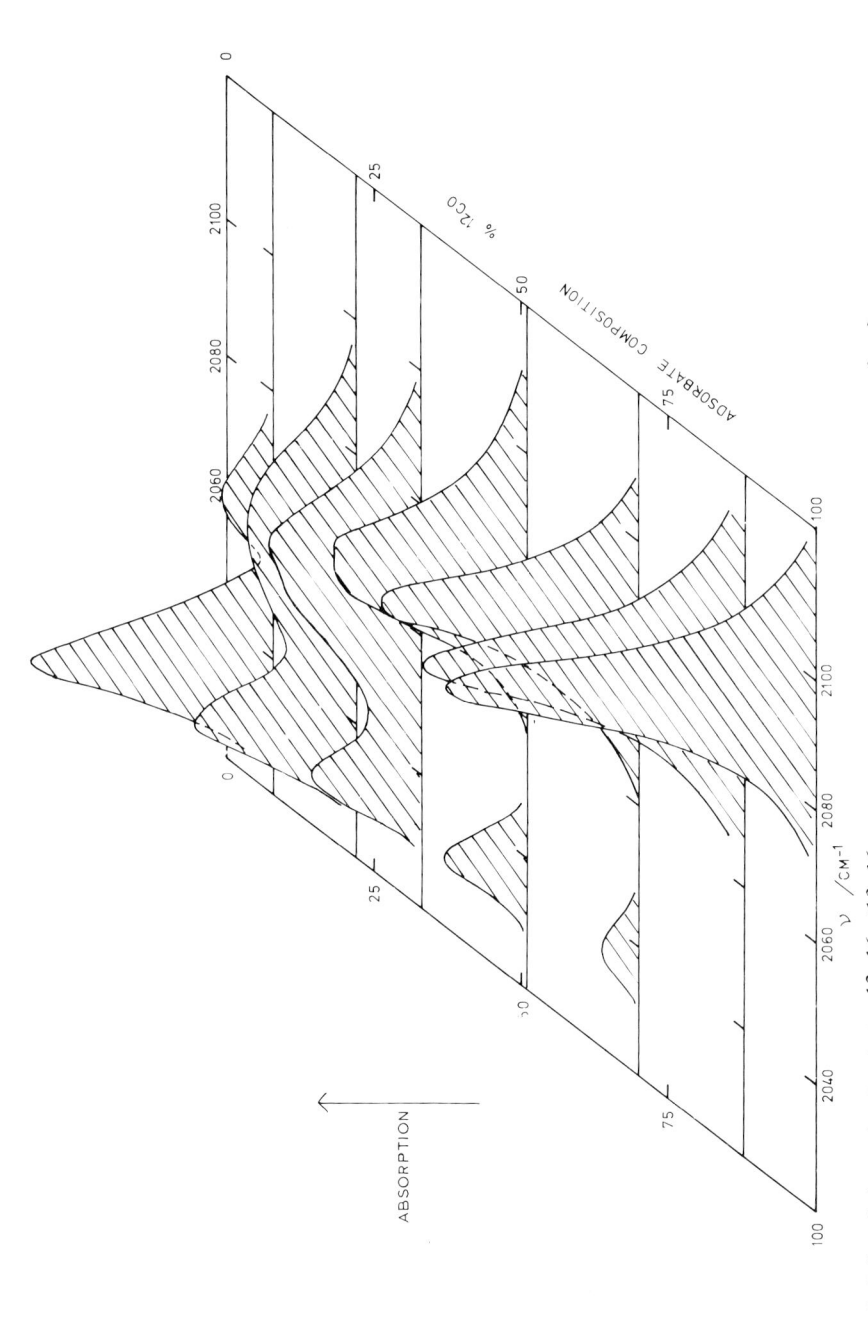

Fig. 4. Infrared absorption bands for $^{12}C^{16}O/^{13}C^{16}O$ mixtures adsorbed to saturation at 300 K on Pt{111}, obtained at a spectral resolution of ~15 cm^{-1}.

from 2100 to 2065 cm^{-1} as the composition varies from 100% ^{12}CO to 8% ^{12}CO, while the low frequency band position is relatively invariant, at 2040 cm^{-1}, over the composition range for which it could be observed.

Fig. 5a shows the experimentally observed variation of both the high and the low frequency band peak positions as a function of adsorbate composition. The variations predicted from eqs. (6) and (7) are shown as full lines on this figure, using the value of $\partial\mu/\partial r$ obtained from the frequency shift with coverage (fig. 1), the lattice sums from the c(4 × 2) structure (linear molecules only) and singleton frequencies 2063 and 2003 cm^{-1}, respectively, for ^{12}CO and ^{13}CO. The agreement between theory and experiment shows quantitatively that the full coverage-induced frequency shift is attributable to dipole interactions. The experimental and theoretical variations of I_{ratio} with isotopic composition are shown in fig. 5b; here the agreement is only qualitative, probably due to the greater experimental difficulty in measuring peak intensities compared with peak positions.

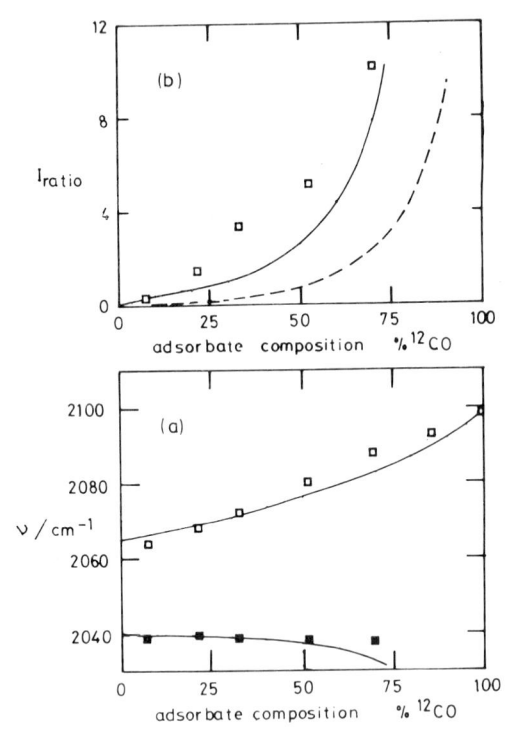

Fig. 5. (a) Variation of frequency at band maximum with adsorbate composition for ^{12}CO/^{13}CO mixtures, for both the high frequency and low frequency bands. Full lines are theoretical curves from eqs. (6) and (7). (b) Variation of I_{ratio} with adsorbate composition. Full line is a theoretical curve from eq. (8), and the dashed line the theoretical curve assuming no coupling in the overlayer.

From the expression [7] $k' = (\partial\mu/\partial r)^2/\epsilon_0 L^3$, where L is the nearest neighbour distance in the c(4 × 2) (linear molecules only) structure and k' is the dipole–dipole coupling force constant, k' was calculated as 0.10 mdyn/Å. Decius et al. [16] give values between 0.01 and 0.15 mdyn/Å for carbonate- and nitrate-containing aragonite structures, while Hammaker et al. [7] found k' to be in the range 0.04 to 0.08 mdyn/Å for CO on silica-supported platinum.

6. Conclusion

Using the dipole interaction models developed by Decius et al. [16] and by Hammaker et al. [7], it is shown that an experiment can be performed in which the coverage-induced frequency shift observed during chemisorption can be simulated at *constant* coverage by varying the isotopic composition of ^{12}CO–^{13}CO mixtures. Under these conditions, any frequency shift due to possible increased competition for metal d-electrons contributing to π^* backbonding [7] is separated from a coupling-induced shift. Since there is complete experimental agreement between the coverage-induced and the isotopic-composition-induced frequency shifts (the latter for the high frequency band), it is concluded that the full 35 cm^{-1} frequency shift observed with increasing coverage for $^{12}C^{16}O$ on Pt{111} is attributable to coupling.

This result has wide implications for the evaluation of vibrational spectra from chemisorbed overlayers. Where, for example, shifts are not observed for relatively intense bands. (i.e. for which $\partial\mu/\partial r$ is large) this would imply that either adsorption proceeds by island formation (so that the molecular environment is virtually independent of coverage) or that there is a coverage-dependent, bonding-induced shift towards *lower* frequencies which counteracts the dipole coupling shift towards higher frequencies. Such an effect has been postulated by Van Hardeveld and Van Montfoort [17] from transmission infrared data for nitrogen on supported nickel catalysts. Finally, a proper evaluation of force constants from vibrational spectra requires that account be taken of the dipole coupling force constant, or the use of true singleton band frequencies, as discussed by Lin et al. [18].

Acknowledgements

The authors express their gratitude to Professor Robert Hammaker for several stimulating and informative discussions. The award of an S.R.C. studentship to A.C. is gratefully acknowledged.

References

[1] Reviewed by:
L.H. Little, Infrared Spectra of Adsorbed Species (Academic Press, New York, 1966);
M.L. Hair, Infrared Spectroscopy in Surface Chemistry (Dekker, New York, 1967).

[2] Recently reviewed by:
J. Pritchard and T. Catterick, in: Experimental Methods in Catalytic Research, Vol III, Eds. R.B. Anderson and P.T. Dawson (Academic Press, New York, 1976).
[3] See, for example, the papers by:
C. Backx, R.F. Willis, B. Feuerbacher and B. Fitton, Surface Sci. 68 (1977) 516;
J.C. Bertolini, G. Dalmai-Imelik and J. Rousseau, Surface Sci. 68 (1977) 539.
[4] R.P. Eischens and W.A. Pliskin, Advan. Catalysis 10 (1958) 1.
[5] R.A. Shigeishi and D.A. King, Surface Sci. 58 (1976) 379.
[6] G. Blyholder, J. Phys. Chem. 68 (1964) 2772.
[7] R.A. Hammaker, S.A. Francis and R.P. Eischens, Spectrochim. Acta 21 (1965) 1295.
[8] M. Primet, J.M. Basset, M.V. Mathieu and M. Prettre, J. Catalysis 29 (1973) 213.
[9] T.B. Grimley, Proc. Phys. Soc. (London) 90 (1967) 751; 92 (1967) 776;
T.L. Einstein and J.R. Schrieffer, Phys. Rev. B7 (1973) 3629.
[10] J.T. Yates, Jr. and D.A. King, Surface Sci. 30 (1972) 479;
J.T. Yates, Jr., R.G. Greenler, I. Ratajczykowa and D.A. King, Surface Sci. 36 (1973) 739.
[11] R.L. Lambert and C.M. Comrie, Surface Sci. 38 (1973) 197.
[12] G. Ertl, M. Neumann and K.M. Streit, Surface Sci. 64 (1977) 393.
[13] K. Horn and J. Pritchard, personal communication.
[14] H. Froitzheim, H. Ibach and S. Lehwald, Surface Sci. 63 (1977) 56.
[15] R.G. Greenler, J. Chem. Phys. 44 (1966) 310.
[16] J.C. Decius, O.G. Malan and H.W. Thompson, Proc. Roy. Soc (London) A275 (1963) 295.
[17] R. van Hardeveld and A. van Montfoort, Surface Sci. 17 (1969) 90.
[18] K.C. Lin, J.D. Witt and R.A. Hammaker, J. Chem. Phys. 55 (1971) 1148.

CO STRETCHING VIBRATION OF CARBON MONOXIDE ADSORBED ON NICKEL(111) STUDIED BY HIGH RESOLUTION ELECTRON LOSS SPECTROSCOPY

J.C. BERTOLINI, G. DALMAI-IMELIK and J. ROUSSEAU

Institut de Recherches sur la Catalyse, 79, boulevard du 11 Novembre 1918, 69626 Villeurbanne Cédex, France

The frequency of the ν-CO stretching vibration measured by HRELS has been followed as a function of the CO coverage and in the presence of coadsorbed hydrocarbons on the Ni(111) face. The ν-CO frequency shifts continuously from 225 meV (1814 cm^{-1}) to 237 meV (1911 cm^{-1}) when the CO coverage increases from 0 to 0.41. Coadsorption of electron donor molecules, such as ethylene and benzene, generates a significant lowering of the ν-CO frequency. Results are discussed in terms of the back donation of metallic electrons into the $2\pi^*$ antibonding orbitals of CO, the dipole–dipole coupling and the coordination number of the CO adsorbed molecules. The back donation is found to play the major role in the range of coverage explored but we cannot exclude some contribution of a dipole–dipole coupling effect.

1. Introduction

The frequency of the ν-CO stretching vibration of adsorbed CO on metals has been found sensitive to the crystallite size [1,2], the presence of other adsorbates [3] and the coverage [4]. Furthermore, the determination of the ν-CO frequency has been widely used to characterize the net flow of electrons from hydrocarbons to the metallic substrate, CO being in these studies considered as a probe molecule [5]. Some conclusions have been drawn in terms of the coordination number of adsorbed CO, of the exposed oriented faces and of the coordination number of the adsorption site. However, the reported results seem ambiguous since they were obtained mainly with powder surfaces. In order to gain more relevant information, well defined surfaces have to be used.

Only a few results have been obtained about the ν-CO frequency upon carbon monoxide adsorption on single crystal surfaces by means of infrared spectroscopy (IR) [6,7] and by High Resolution Electron Energy Loss Spectroscopy (HRELS) [8,9].

The following results concern the study, by HRELS, of the frequency of the ν-CO stretching vibration in function of the CO coverage and in presence of coadsorbates such as hydrocarbons, on the nickel (111) face.

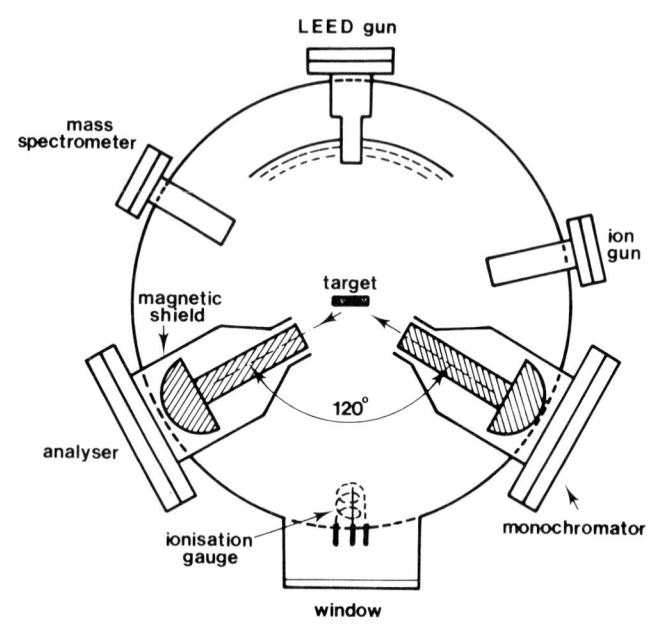

Fig. 1. Schematic representation of the significant parts of the experimental arrangement (top view).

2. Apparatus and materials

The experiments were performed within a UHV system combining LEED, $\Delta\phi$, AES and HRELS techniques: the significant parts of the apparatus are shown in fig. 1. Further details concerning this equipment will be described elsewhere [10].

The Ni(111) sample was "spark cut" from a 3/8 inch diameter cylindrical single crystal (purity 99.95%) issued from "Cristal-Tec" and was first mechanically then chemically polished [11]. The orientation so obtained is better than ±0.5° from the desired one.

The gaseous reactants were "spectroscopically pure" and mass spectrometric examination showed no impurity of importance. The benzene was dried with sodium wires and used after removal of dissolved gases by pumping under vacuum while chilled.

3. Experimental procedure

The clean surface is obtained after combined argon ion bombardment and chemical cleaning as described elsewhere [12]. The sample cleaned this way exhibits a sharp LEED pattern typical of the (1 × 1) structure and AES measurements show no peaks other than those associated with nickel.

The operational conditions for HRELS measurements are as follows:
- the resolution power is about 20–25 meV;
- the incident electron beam intensity is about 10^{-8} A;
- the incident and analysed reflected directions define angles, with respect to the normal to the surface of the sample, equal to $\pm \pi/3$;
- the incident energy is fixed at 3 eV;
- the analyser acceptance angle is equal to 5×10^{-2} rd.

The reactants are passed over the single crystal through a leak valve and gas pressure is continuously monitored by a Bayard–Alpert gauge. Gas analysis are performed with a quadrupole mass-spectrometer "Riber QS 200". Work function change measurements are monitored by the diode method using the LEED gun.

The experiments are done in two different steps. First the LEED patterns and work function changes are followed in terms of the CO exposure. Secondly HRELS spectra are registred for different CO exposures. Several runs are performed in order to control the accuracy and reproducibility of the results. The experiments are carried out on samples maintained at room temperature. HRELS spectra are recorded after pumping off the gas phase and consequently only the irreversible part of CO adsorbed has to be considered.

4. Results

CO adsorption, under 5×10^{-9} Torr on Ni(111) maintained at room temperature, generates an increasing of the work function with $\Delta\phi_{max} = 1.05$ eV.

By pumping off the gas phase only a slight decrease of $\Delta\phi$, to 1 eV, is observed. The sample then exhibits the LEED pattern corresponding to the c(4 × 2) structure. Christmann, Schober and Ertl [13] have demonstrated that $\Delta\phi$ is directly proportional to the coverage θ_{CO} up to $\Delta\phi = 1.1$ eV, according to the relationship: $\theta_{CO} = 0.41 \Delta\phi$, which is therefore used, in this work, to determine θ_{CO} from the $\Delta\phi$ measurements. Fig. 2 shows a plot of $\Delta\phi$ (and θ_{CO}) versus CO exposure at room temperature.

During HRELS measurements, we note a large decrease of the reflectivity of the surface when the CO coverage increases: the intensity of the elastic peak decreases by a factor of roughly 1/1000 at saturated CO coverage. Then, because of the very low intensity of the whole spectrum, the data acquisition requires a great amount of time. The energy loss spectra reported on the fig. 3 (normalized with respect to the elastic peak) are obtained for different CO exposures. The ratio of the respective intensities of the loss peak to the elastic one increases linearly with the CO coverage while the energy related to this loss peak (ν-CO) shifts continuously from 225 to 237 ±2 meV (fig. 4) up to $\theta_{CO} = 0.41$ which corresponds to the irreversible part of adsorbed CO.

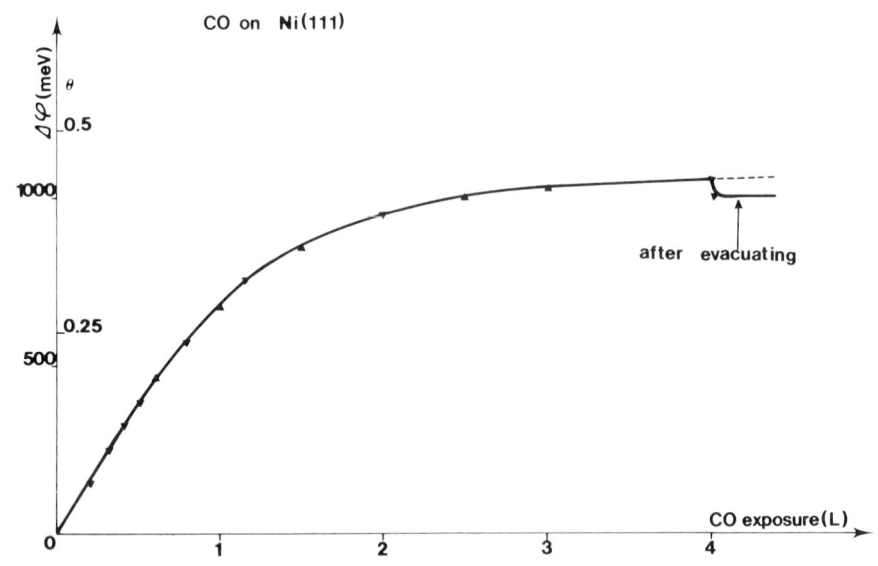

Fig. 2. Work function changes $\Delta\phi$ (and CO coverage) as a function of CO exposure at room temperature and at pressure 5×10^{-9} Torr (uncorrected gauge readings).

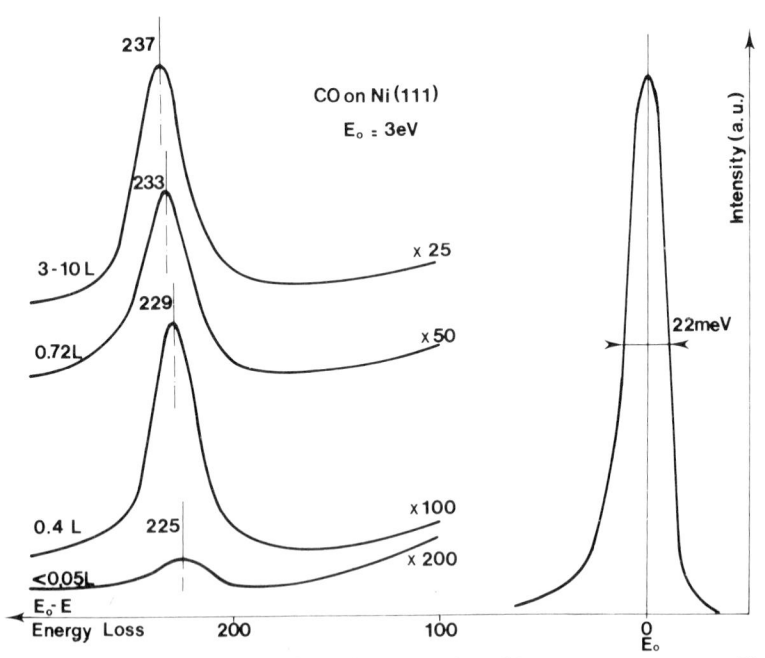

Fig. 3. Electron energy loss spectra of Ni(111) exposed to CO at room temperature. The exposures are the uncorrected gauge readings.

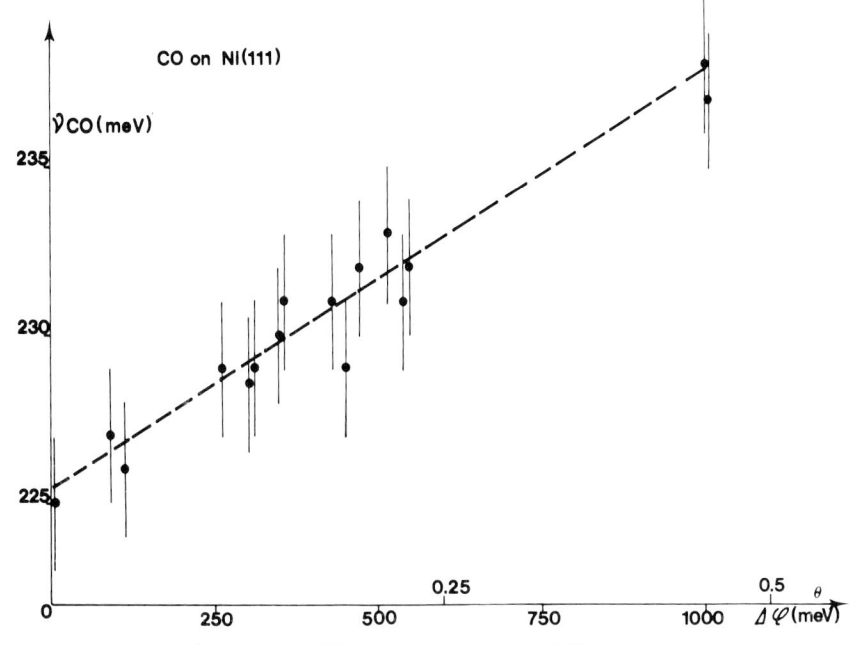

Fig. 4. Measured ν-CO energy versus $\Delta\phi$ (and CO coverage).

The presence of coadsorbed species generates some significant shifts of the ν-CO frequency. The adsorption of C_2H_4 and C_6H_6, at full coverage, on a surface containing very small amounts of adsorbed CO, lowers the ν-CO vibration from 225 meV to respectively 222 and 212.5 meV. In a general way, the lowering of the ν-CO frequency seems to be the largest when the adsorbate electropositivity is the highest ($\Delta\phi$ equal to −0.25 eV for C_2H_4 and −1.1 eV for C_6H_6) [14].

5. Discussion

The CO sticking coefficient on the Ni(111) face is found to be close to unity, up to a coverage $\theta_{CO} \sim 0.2$. At higher coverages, it decreases as shown by the change of the slope of the curve ($\theta, \Delta\phi$) versus CO exposure (fig. 2). At room temperature, the irreversible part of the adsorbed CO may be estimated to correspond to a coverage nearby 0.41 associated with a c(4 × 2) LEED pattern. These results are in good agreement with the data obtained by Christmann et al. [13].

The prominent loss peak, evident in the spectra of the fig. 3, is due to excitation of the C–O stretching vibration of the adsorbed CO molecules. Such a vibration must have a dynamic dipole moment normal to the surface to couple efficiently to the impinging electrons via dipole scattering [15,16]. Although we can not absolutely exclude that some adsorbed molecules may have their C–O bond somewhat parallel to the surface (since they would not be detected in our HRELS experiments), UPS measurements [17] support the assumption of a C–O bond normal to the surface, for CO adsorbed on Ni(111).

Different effects may be invoked to explain the behaviour of the ν-CO stretching vibration in function of the coverage:
– electron transfer from the 5σ orbital of CO to the metal and the back donation of metallic electrons into the $2\pi^*$ antibonding orbital of CO;
– dipole–dipole coupling,
– adsorption site and coordination number of the adsorbed molecules.

We will now discuss the results in terms of these three effects.

CO chemisorption on d metals involves metal–carbon bonds via the CO 5σ orbital and metallic d orbitals, with a "back-donation bonding" via d orbitals and $2\pi^*$ antibonding orbitals. Electron transfer into the $2\pi^*$ orbitals weakens the C–O bond and thus lowers the frequency of the ν-CO vibration. It has been calculated [18,19] that the adsorption energy E_{ad} is mainly determined by the back-donating of the metallic electrons. Therefore the ν-CO frequency is expected to increase when E_{ad} is lowered. Christmann et al. [13] have reported a decrease from 30 to 23.5 kcal/mole of the isosteric heat of adsorption (E_{ad}) of CO on Ni(111), when θ_{CO} varies from 0 to 0.5. According to the previous predictions, these results are in good agreement with the ν-CO energy variations reported in this work (from 225 to 237 meV) with coverage.

The dipole–dipole coupling induces also some variations of the ν-CO frequency with coverage. Its influence has been clearly shown on Pt(111) [7,20] by using mixtures of CO isotopes [20]. On this surface, a continuous increase of the ν-CO frequency versus θ_{CO} has been observed even when the adsorption energy (and consequently the $2\pi^*$ filling) remains constant. On Ni(111), the measured decrease of E_{ad} versus θ-CO is not found to be monotonous [13]. Two plateaus are observed: $E_{ad} = 26.5$ kcal/mole for $0.1 < \theta < 0.3$ and $E_{ad} = 23.5$ kcal/mole for $0.35 < \theta < 0.45$. According to the uncertainty in our ν-CO experimental measurements, we cannot conclude either that there is a continuous increase of the ν-CO frequency versus θ-CO (fig. 4) or the presence of plateaus. Consequently we cannot exclude some contribution to the ν-CO shift, when θ_{CO} increases, from dipole– dipole coupling.

A strong argument in favour of a major role played by the back donation of metallic electrons on the $2\pi^*$ antibonding orbital is evidenced by the coadsorption of CO with electron donor molecules. Ethylene ($\Delta\phi = -0.25$ eV) and benzene ($\Delta\phi = -1.1$ eV) acting as donors promote an electron filling of the $2\pi^*$ orbital and thus induce a lowering of the ν-CO frequency which is largest when the adsorbate electropositivity is the highest.

An elementary relation between the $2\pi^*$ filling and the adsorption energy E_{ad} is not enough for a convincing explanation of the ν-CO frequency values, if we compare the behaviour of the nickel (111) and (100) faces. On the Ni(100) face, during the progressive ordering of the c(2 × 2)CO structure, Tracy [21] has reported a constancy of E_{ad} (30 kcal/mole), whereas Andersson [8] observed two ν-CO vibrations at 239.5 and 256 meV, the lowest tending to disappear when θ_{CO} increases. Preliminary results we have obtained on this face show the presence of only one loss peak at 245 ± 2.5 meV whose intensity increases but whose energy remains constant, when θ_{CO} increases. This is in good agreement with the constancy of E_{ad}. However, as the 245 loss peak is broadened, when the c(2 × 2)CO structure takes place, it is possible that indeed the sum of the two overlapping peaks as reported by Andersson [8] is observed. In all cases, since E_{ad} is higher on Ni(100) than on Ni(111) a lower ν-CO frequency on Ni(100) than on Ni(111) is expected in contrast to the observed values. One may try to explain this discrepancy by the difference in the coordination number of the adsorbed CO or (and) in the $2\pi^*$ filling dependance upon the considered face. A coordination number effect can explain the higher ν-CO frequencies obtained on the (100) than on the (111) face, since the smallest coordination number of the chemisorbed CO (CO linearly bonded to Ni atoms) is effectively observed on the (100) face [8]. Nevertheless, one has to consider carefully the energetic positions of the highest filled chemisorption level with respect to the highest occupied d state in the metal, which determine E_{ad} (i.e. the whole CO chemisorption) according to Doyen and Ertl [18]. In fact, the $2\pi^*$ antibonding orbital is broadened into a resonance state which would be located close to the Fermi level ϵ_F (−0.63 eV below ϵ_F [18]). Presumably, according to the differences in the density of states and the work function values for the (100) and (111) faces, the $2\pi^*$ filling should be somewhat modified generating significant differences, for the behaviour of the two faces, upon CO chemisorption.

6. Conclusion

On the nickel (111) surface the energy shift of the ν-CO frequency versus the CO coverage can be explained in terms of an increase of the competition for the metallic electrons so that there is less charge available to put into the $2\pi^*$ orbital of CO, which induces an increase of the ν-CO frequency, with the CO coverage. Moreover some dipole–dipole coupling may provide a contribution to the frequency shift. Some "site and face effect" namely the coordination number of nickel atoms, the different bond number of adsorbed CO molecules, the relative position of the Fermi level and the $2\pi^*$ orbital, and the relative density of states in the valence band with respect to the face, must be invoked to explain the specific behaviour of different faces.

References

[1] R. van Hardevelt and F. Hartog, Advan. Catalysis 22 (1972) 75.
[2] G. Blyholder, J. Phys. Chem. 68 (1964) 2772.

[3] M. Primet, J.M. Basset, M.V. Mathieu and M. Prettre, J. Catalysis 29 (1973) 213.
[4] L.H. Little, Infrared Spectra of Adsorbed Species (Academic Press, London, 1966) ch. 3.
[5] M. Primet and M.V. Mathieu, J. Chim. Phys. (Paris) 72 (1975) 30.
[6] J. Pritchard, T. Catterick and R.K. Gupta, Surface Sci. 53 (1975) 1.
[7] R.A. Shigeishi and D.A. King, Surface Sci. 58 (1976) 379.
[8] S. Andersson, Solid State Commun. 21 (1977) 75.
[9] H. Froitzheim, H. Ibach and S. Lehwald, in: Proc. Intern. Symp. on Photoemission, Noordwijk, The Netherlands, 1976 (E.S.A. publ.) p. 277.
[10] J.C. Bertolini, G. Dalmai-Imelik and J. Rousseau, J. Microscopie et de Spectroscopie Electroniques, to be published.
[11] W.J. McG. Tegart, The Electrolytic and Chemical Polishing of Metals (Pergamon, London, 1956).
[12] G. Dalmai-Imelik, J.C. Bertolini and J. Rousseau, Surface Sci. 63 (1977) 67.
[13] K. Christmann, O. Schober and G. Ertl, J. Chem. Phys. 60 (1974) 4719.
[14] G. Dalmai-Imelik and J.C. Bertolini, Japan. J. Appl. Phys. Suppl. 2, Part 2 (1974) 205.
[15] A.A. Lucas and M. Sunjic, in: Progress in Surface Science, Vol. 2 (Pergamon, Oxford, 1972) p. 75.
[16] D.L. Mills, Surface Sci. 48 (1975) 59.
[17] P.M. Williams, P. Butcher, J. Wood and K. Jacobi, Phys. Rev. B14 (1976) 3215.
[18] G. Doyen and G. Ertl, Surface Sci. 43 (1974) 197.
[19] T.B. Grimley in: Molecular Processes at Solid Surfaces, Eds. W. Drauglis and R. Gretz (McGraw-Hill, New York, 1969) p. 299.
[20] A. Crossley and D.A. King, Surface Sci. 68 (1977) 528.
[21] J.C. Tracy, J. Chem. Phys. 56 (1972) 2736.

EXPERIMENTAL ASPECTS OF ANGLE-RESOLVED PHOTOEMISSION FROM CLEAN AND ADSORBATE-COVERED METAL SURFACES

D.R. LLOYD, C.M. QUINN and N.V. RICHARDSON

Chemistry Department, University of Birmingham, Birmingham B15 2TT, England

Instrumental techniques for the observation of angle-resolved photoelectron spectra from surfaces are reviewed. Results from the work of various groups on clean metals, particularly tungsten, copper, nickel and palladium, are described. Spectra induced by adsorption of hydrogen, oxygen, carbon monoxide and benzene on some of these metals are discussed.

1. Introduction

I was originally asked to describe some of the work which we are carrying out with angle-resolved UPS at Birmingham. It seems appropriate to try to put our work into a context of that going on elsewhere, and a very good indication of the intensity of effort in this field can be gained by perusal of the excellent collection of papers given recently at the Noordwijk Symposium on photoemission [1]. The basic reason for an angle-resolved experiment is that a theoretical analysis, particularly a correlation with a band structure, is itself in angle-resolved form, and so an angle-resolved experiment can be directly compared with theory. Until the surprisingly recent advent of angle-resolved work it was normal to integrate theoretical results, usually over the entire Brillouin zone. Since Dr. Holland has discussed theoretical considerations [2], and theoretical reviews are available [3,4], I have tried to collect together something of the detail of the actual techniques used in agle-resolved work, with a look at some selected results.

2. Techniques

The photoemission technique itself, using photons in the uv-vacuum uv region, has of course a very distinguished history. Very early work used variable photon energy, concentrating on the electrons of maximum kinetic energy. The natural development from the stopping potential used to examine kinetic energy is the retarding potential, usually applied between grids, which when scanned gives a transmitted current which is the integral of the familiar $N(E)$ versus E curves which we call photoelectron spectra. With a reasonably small irradiated area of the sample, at the centre of spherical sector grids, the spectrum with such a device is obtained with electrons which are well defined in angle, but the spectrum is angle-integrated over the acceptance angle of the grids. Work of this type has been carried out by Spicer and others for many years [5]. Until relatively recently UHV compatible monochromators for the vacuum uv have not been available, and even now are relatively rare and expensive, so such work has usually been carried out only up to the cutoff of LiF windows, ca. 11 eV.

Two developments which had been explored by the gas-phase photoelectron spectroscopists were introduced simultaneously which extended the technique considerably. Extension to higher photon energies was achieved with the rare gas resonance lines produced in differentially pumped discharge tubes; however, although these provide sharp and intense sources of photons, they do so only at widely separated energies. The other improvement was the introduction of small electrostatic deflection analysers for k.e. analysis, which provide better resolution and signal/noise. Pioneer work here was carried out by Eastman and colleagues [6]. Unlike the retarding grids, the deflection analysers are sharply selective in angle, but are not necessarily angle-resolved. Indeed, since the majority of commercial instruments have been designed to achieve maximum sensitivity, they often accept over rather complex angles.

One of the commonest deflection analysers is the spherical sector device. The most usual form is either the hemisphere, with deflection through 180° (e.g. the AEI ES200), or a slightly truncated hemisphere; the VG ESCA 3 deflects through 150°. Such analysers accept approximate rectangular pyramids, or sectors of hollow cones, of solid angle.

An alternative form of the hemispherical analyser, indeed the original form [7], is the one with smaller deflection angle, often 90°, but occupying the full 360° of azimuth around the source–detector axis. This is the form used in the Varian instrument and it has also been employed by Leckey and colleagues [8]. It accepts a complete hollow cone of electrons, usually of mean semi-angle 45°. The other common analyser is the cylindrical mirror (CMA) which accepts a similar complete hollow cone, usually of semi-angle 42.3° to achieve double focussing. Because of the double focussing, the range of angles from the mean can be as high as ±6°.

The complex acceptance shapes of these analysers do not present particular problems of integration for theoretical comparisons so long as the actual integration angles are specified. However, the actual acceptance angle can change through a spectrum. Most analysers operate with a constant deflection potential, and the spectrum is scanned by changing an accelerating or retarding voltage between the deflector mean potential and the sample potential. This has the effect of changing the effective acceptance angle by an amount which depends on the position in the spectrum and on the analyser pass energy, so any discussion of the angular properties of spectra taken on such instruments needs to take this into account.

All of these deflection analysers and the grid retarder are readily converted to fully angle-resolved operation merely by stopping down the aperture. The double

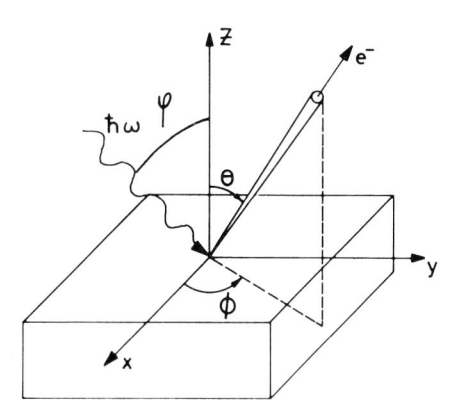

Fig. 1. The variables in angle-resolved photoelectron spectroscopy (from ref. [3]).

focussing quality is not particularly desirable when good angular resolution is required, so simpler analysers such as the cylindrical deflector, in either the 90° or 127° form [3,9], or the parallel plate analyser [10], can be used. In angle-resolved work there are a number of significant variables to consider (fig. 1). These are the polar and azimuth angles θ and ϕ for the emitted electron, the incidence angle ψ of the photons, the photon energy, and, if polarised light is available, the orientation of the polarisation vector A and the azimuth angle of incidence. An ideal instrument would have continuous variation of all these quantities over wide ranges, with high resolution in energy and angle, and high sensitivity. With the advent of synchrotron radiation, particularly from storage rings, it becomes possible to approach closely to this ideal. However, the majority of published work has been limited in one way or another.

The simplest devices for angle-resolved work have analyser and photon source fixed with respect to each other. A very important experiment which can be carried out with such equipment is the study of symmetry oriented spectra in which electrons emitted normally from low index crystal faces are examined [11]. By rotating the crystal about the axis perpendicular to the analyser—photon source plane, it is possible to vary the electron polar emission angle θ; such experiments were carried out by Barber et al. [12], and the group of the late Professor Linnett [13]. However, when θ is varied, ψ inevitably also changes, and disentangling the two effects is not easy.

The introduction of a moving analyser allows a much fuller study of the dispersion of energy spectra with angle — the term angle-dispersive electron spectroscopy (ADES) has been coined by an instrument firm. The first such instruments used miniature plane retarders in front of a channeltron electron multiplier; such an analyser is very compact and light and can easily be mounted on standard sample manipulators. An instrument of this type was described by Koyama and Hughey in 1972 which had a synchrotron light source [14], another with a monochromator and LiF

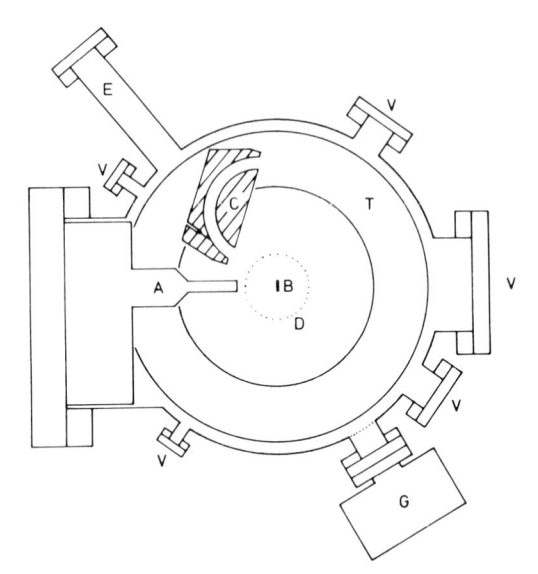

Fig. 2. Plan view of the Birmingham instrument for angle dispersive electron spectroscopy. A: discharge photon source; B: sample, surrounded by grounded screen D; C: hemispherical electron analyser mounted on turntable T; E: electron gun; G: argon ion gun; V: viewports.

window was used by Smith, Traum and DiSalvo in their well-known experiments on the layer dichalcogenides [15].

Better resolution is available with deflection analysers, but these are usually heavier and clumsier than the miniature retarders, so motion is restricted to a plane. The solution in our instrument * is shown in fig. 2: a 5 cm radius hemispherical $150°$ deflection analyser is mounted on a turntable. The photon source is a discharge lamp, so we are currently restricted to unpolarised photons at the fixed energies of the resonance lines. However, θ, ϕ and ψ can be independently varied. A very similar instrument has been described by Hagström and Lindau in 1971 [16] and another has been constructed in the group of Nilsson [17]. An alternative geometry in which the analyser moves in the plane perpendicular to that containing the incident photon beam and the sample normal has been employed by Feuerbacher and Willis [18]; this has the advantage that the reflected photon beam does not interfere with the electron analyser at specular reflection angles, but does not allow ψ variation.

An alternative to moving the bulky analyser is to use one of the $360°$ azimuth analysers with a movable aperture stop which accepts only those electrons in a particular cone of emission. Angular variation is obtained by moving the stop. Experiments of this type were reported by Leckey et al. [8], who used a spherical analyser. With a CMA, the moving aperture stop forms the basis of the powerful instru-

---

* Modified version available from V.G. Scientific Ltd. as ADES 400.

ment of Lapeyre and co-workers, who use the Wisconsin storage ring as a photon source [19]. By appropriate combinations of motions, the variations of emission into most of the 180° solid angle around a sample can be studied.

A number of new experiments become possible if synchrotron radiation, with monochromators, is available. The most obvious is that the photon energy can be continuously varied over a very wide range. The polarisation of the synchrotron radiation allows the use of s-polarised (electric vector lying in the surface plane) or p-polarised light (electric vector in the plane containing the surface normal and the photon propagation direction). The continuously variable nature of the photon source allows an auxiliary technique, constant initial state spectroscopy, to be used in angle-resolved form. By scanning the photon monochromator in step with the analyser energy, with a fixed energy difference E, all the final states which are accessible from a state E below the vacuum level, at the chosen angle, can be scanned [20]. On the Stanford SPEAR storage ring a new type of angle dispersive photoelectron spectrometer has been set up, which uses the very short pulse duration of the photon emission from the ring as the basis of energy analysis by electron flight time. The only part which has to move is then the electron detector, since the intervening space is the analyser. However, energy resolution falls off rapidly with increasing electron energy [21].

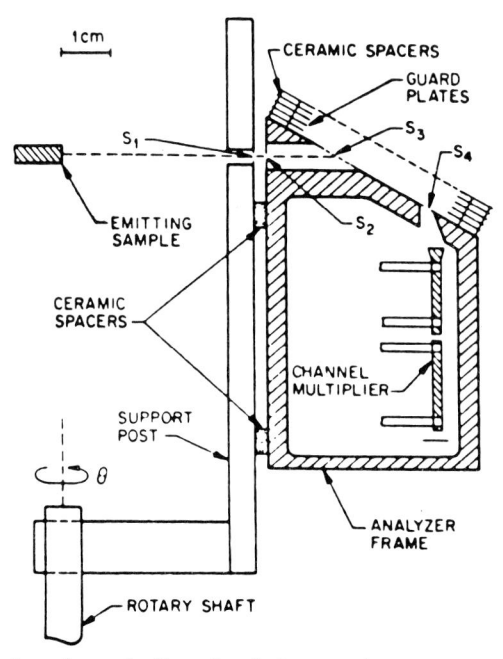

Fig. 3. Miniature analyser for angle dispersive electron spectroscopy mounted on a rotary manipulator [10].

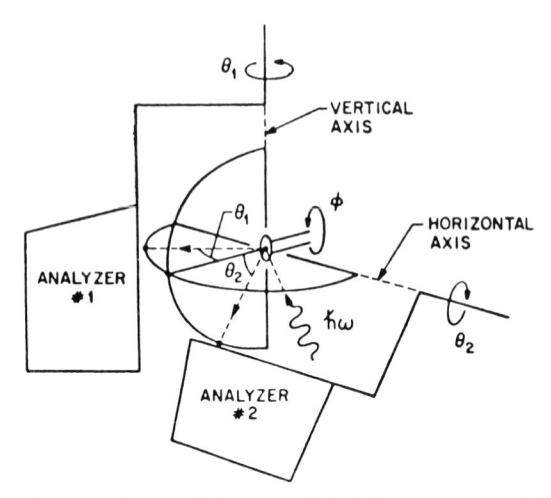

Fig. 4. Arrangement of two analysers of the type of fig. 3 for measurement in two perpendicular circles [10].

Other types of analyser have been used. Smith, Larson, Traum and Chiang describe a miniature parallel plate analyser which can be mounted on a standard rotary manipulator (fig. 3) [10]. It is small enough for two such analysers to be mounted on two perpendicular circles about the sample (fig. 4), and it has been suggested that banks of these analysers could be mounted so as to make better use of machine time by accumulating several spectra simultaneously. The single analyser is probably the simplest angle-dispersive analyser to construct which has adequate resolution. Finally, some instrument designs have returned to the spherical grid retarder system. Plummer et al. use a channeltron detector which can be moved around outside the grids [22]. Waclawski and co-workers have used a channel plate multiplier and phosphor combination so that electrons over a wide range of angles (0.6π sr) are all detected as light intensity at the phosphor [23]. The light output from a selected spot is detected by a photometer. This system has the advantage of no moving parts inside the vacuum except for the sample mount. A similar instrument has been described in these proceedings by Baker and Price [24]. Such instruments have potential for data accumulation at a number of angles simultaneously; Gerhardt and Dietz have described experiments with an array of fixed detectors [25] in a retarding configuration. With moving detectors and fixed grids it is as easy to move out of the plane containing the photon source and sample normal as within it, whereas most of the deflection analysers move in only one plane; on the other hand, energy resolution and signal-noise are superior with the deflection analysers.

3. Experimental results

3.1. Clean metals

The collection of hardware described above represents a substantial investment, and it is apposite to enquire whether all the possible variations described are experimentally significant. High resolution in electron energy is certainly desirable in our experience, at least for copper crystals. Fig. 5 shows a portion of the d band region of the normal emission spectrum from Cu(111). The desirability of high resolution may perhaps depend on one's point of view; under medium resolution only two bands are found in this region, with a hint of a third. This appears to provide a satisfactory correspondence with the three possible direct transitions in the ΓL section of the emitter bulk band structure, shown on the figure. However, under the optimum resolution there are at least six components. Similarly, on Cu(100) at off-normal exit some bands which are apparently single show splittings at high resolution (fig. 6): the half widths of the components are about 130 meV. In some Ar I spectra we have observed additional splittings due to the doublet nature of the light source. The spectrometer resolution for these spectra was about 40 meV. However, resolution of this order is not always observed; thus spectra from nickel [26] show much broader features. This has been explained by Pendry and Titterington [27] as a lifetime effect due to the differing densities of states at the Fermi level. Our spectra for palladium show more distinct features than those from nickel, but none with half-width much smaller than 400 meV. High energy resolution is clearly useful to have on occasion, but it can often be traded for sensitivity.

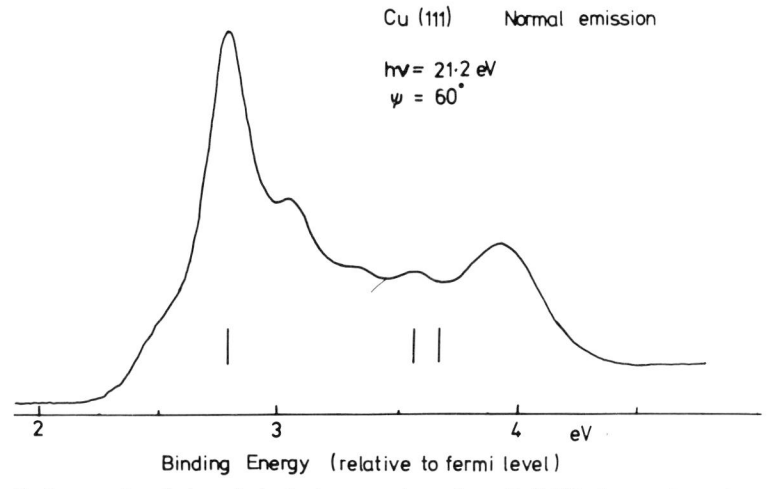

Fig. 5. He I normal emission photoelectron spectrum from Cu(111) at a spectrometer resolution of 40 meV. The dashes mark the positions of direct transitions for the bulk band structure.

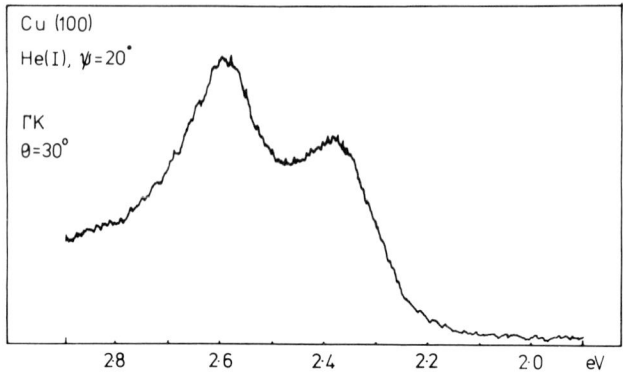

Fig. 6. He I photoelectron spectrum from Cu(001) at a spectrometer resolution of 40 meV.

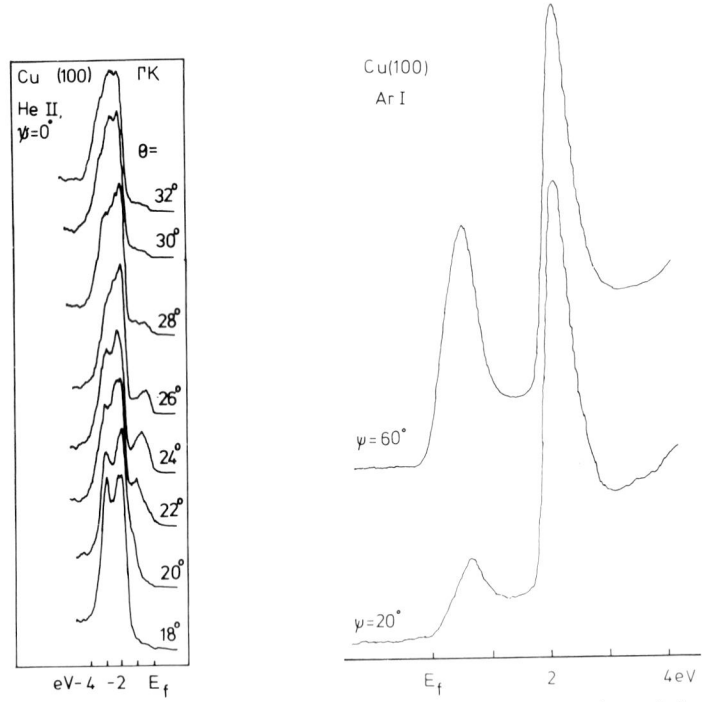

Fig. 7. He II photoelectron spectrum from Cu(001) as a function of polar emission angle θ, with normally incident photons.

Fig. 8. The variation with photon incidence angle ψ of the Ar I normal emission photoelectron spectrum from Cu(001) [30].

The capability for good angular definition is probably even more important. A normal emission spectrum, for instance, is ideally one in which $k_\parallel$ is zero for all electrons entering the analyser. In practice, of course, there is a range of $k_\parallel$ which is sampled due to the angular width of the spectrometer acceptance cone. The actual range of $k_\parallel$ depends on the emitted electron energy, and in the X-ray region, energies for valence electrons are such that a 7° acceptance cone samples more than 1/2 of the Brillouin zone [28]. In the uv region the maximum $k_\parallel$ is smaller by an order of magnitude, and further improvement is obtained by decreasing the acceptance cone to about 1°–3°. Fig. 7 demonstrates that sharp variation in Cu(100) spectra can be obtained over angular intervals of 2° in polar angle θ, at the relatively high energy of the He II line. Changing azimuth angle ϕ allows arcs to be traversed along differing sections of the zone; it also allows a check on surface symmetry. We and others have found 120° repeat patterns for (111) faces and 90° repeat patterns for (100) faces [26,29].

Substantial variations with photon incidence angle ψ are also found. Fig. 8 shows the variation of intensity of the Ar I excited normal emission spectrum from Cu(100). The sharp drop in the intensity of the feature close to E_F can be interpreted quantitatively in terms of bulk dipole selection rules [30]. In the spectrum emitted normally from Cu(111) there is a peak very close to E_F which we showed

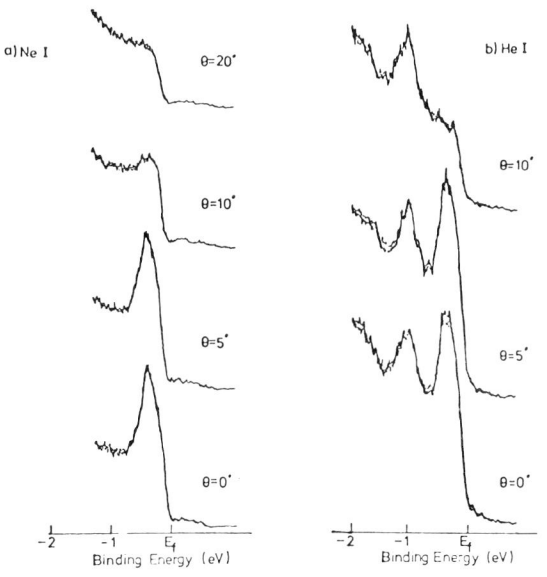

Fig. 9. The variation with θ, around the ΓL direction, of the photoemission close to E_F from Cu(111). The angle-invariant peak at an apparent energy of −1 eV in the He I spectrum is a satellite emission from the main d-band excited by the He I β line at 23.1 eV.

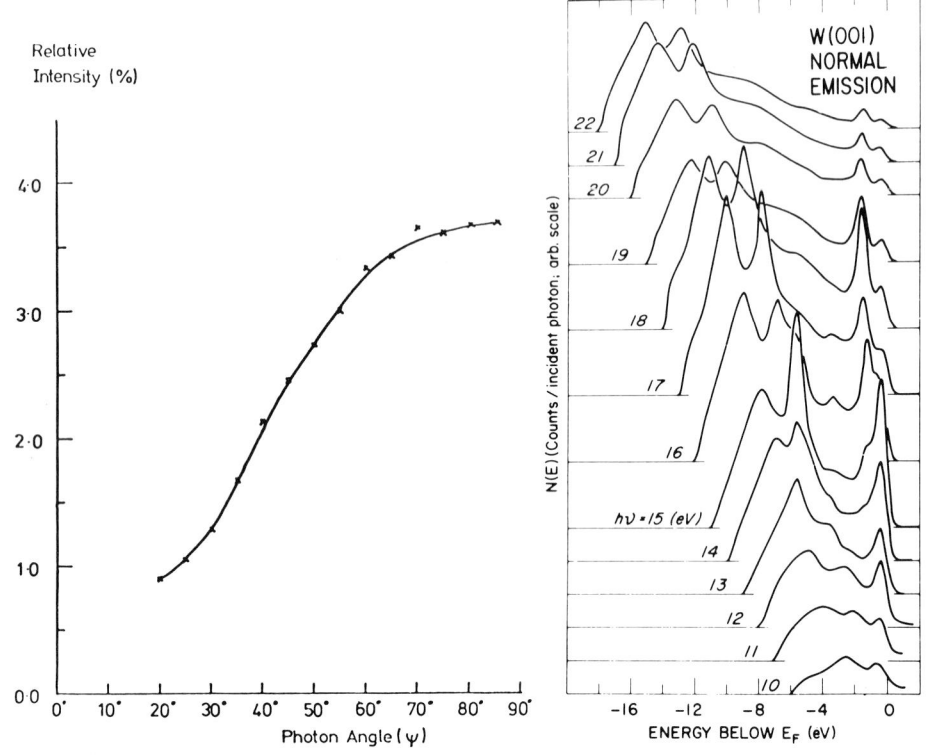

Fig. 10. The variation with ψ of the intensity of the feature closest to E_F in fig. 9, relative to that of the main d-band.

Fig. 11. The photoelectron spectra from W(001) at various photon energies, from 10 to 22 eV, excited by s-polarised light (from ref. [36].

some time ago was very sensitive to adsorbed species [31] and which can be detected only for a very restricted range of $k_\parallel$ (fig. 9). It has been assigned by us and others as emission from a surface state lying in the L gap of the zone [31–33]. The emission intensity from such a state should fall off rapidly as the electric vector becomes parallel with the surface, and this is observed (fig. 10). A similar structure is observed on Ag(111) and Au(111) [33] and at least for Au(111) a similar $k_\parallel$ restriction is observed [34]; for all three metals the intensity of this feature increases with decreasing photon energy.

In the normally emitted photoelectron spectrum from W(100) obtained with unpolarised light, there is a feature with kinetic energy such that it appears just below E_F in the spectrum [35]. This feature is the dominant one in most spectra; its exact nature is the subject of debate but it is undoubtedly surface related. However if s-polarised light is used, as in the experiment of Lapeyre et al. [36], this surface feature is suppressed, and the spectrum close to the Fermi level (fig. 11) can now be interpreted in terms of bulk transitions. The left panel of fig. 12 shows the connection between the photoelectron spectrum and the angle resolved constant initial state spectra. The central panel shows the band structure to which the transitions are related.

Variation of photon energy at closely spaced intervals is a powerful technique for following direct transitions, and this is a substantial limitation on the work with line sources. The available energies are widely separated, so it can be difficult to connect spectra with each other since the spectra can change so drastically; fig. 13 shows an example. Variations in spectra continue out to 150 eV for Cu spectra [37]. In contrast to the rapid variations in band position and band intensity with $h\nu$ and with emission angle which are a characteristic of direct transitions in the bulk, surface photoemission should show relatively slow variations. The surface state on the (111) faces of the Cu group metals varies monotonically from a very high intensity at energies of 6–7 eV to being barely detectable at 21 eV. It has been claimed

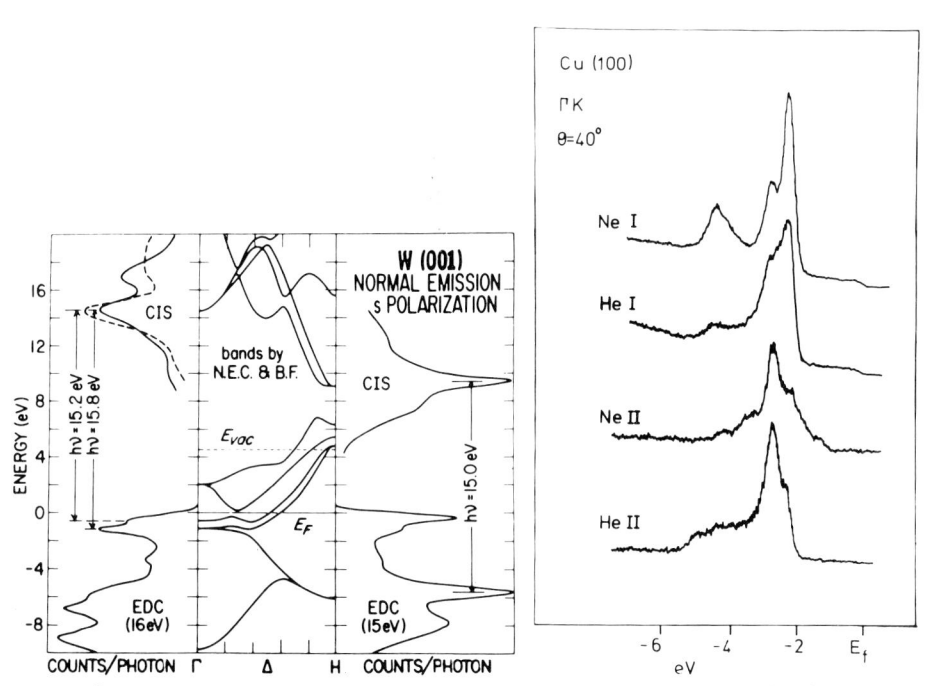

Fig. 12. Photoelectron spectra (EDC) from fig. 11 plotted against the W band structure, showing the connection of initial and final states in angle-resolved constant initial state (CIS) spectra [36].

Fig. 13. Variations in the photoelectron spectra from Cu(100) by using the resonance lines Ne I (16.7 eV), He I (21.2 eV), Ne II (26.7 eV), and He II (40.8 eV).

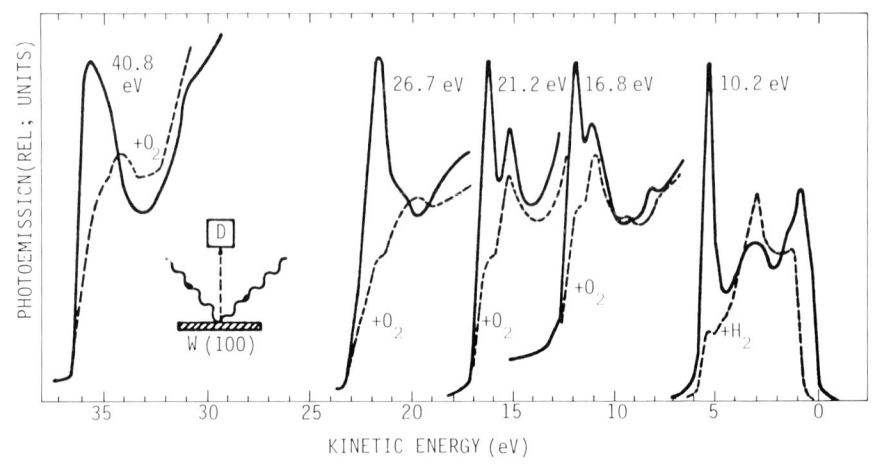

Fig. 14. Variations in the normally emitted photoelectron spectra from W(001) using unpolarised resonance sources, from a clean surface and from adsorbate-covered surfaces (from [38], cf. fig. 11).

that the surface related feature on W(100) shows a similar drop in intensity, but recent work by Feuerbacher and Willis [38] shows that this is not so; the feature remains strong out to 41 eV, but broadens (fig. 14). It has been suggested that the broadening, asymmetry and slight shift in energy are all consistent with ionisation from a surface state whose initial energy is not at E_F but is appreciably lower, and that the broadening is a consequence of the relaxation following ionisation of this state. Such differential relaxation effects are well known in gas phase photoelectron spectra and are generally referred to as a "breakdown in Koopmans' theorem". The surface sensitivity of the festure is also illustrated in fig. 14; the spectra with adsorbed O or H are similar to those in fig. 11 where the surface emission has been removed by the use of s polarisation. However, the interpretation as a surface state is not universally accepted. Adsorption of Hg does not remove the feature from the photoelectron spectrum [39], and the related feature in the field emission spectrum from W(100) is not removed even by a full monolayer of Au atoms [40].

In addition to surface states and surface resonances, surface emission from bulk states may also occur. Such effects are certainly present; thus at all photon energies the copper group metals show weak continuous emission between E_F and the start of the d band emissions, though there are few if any direct transitions available out of band 6 (the s band). Certain features in the spectra from W faces at normal emission have been assigned to surface emission on account of their insensitivity to photon energy [35]. Neddermeyer and co-workers [41] have argued that since the normally emitted spectra from (110) faces of Cu, Ag and Au do not change greatly on changing the photon energy from 21 eV (He I) to 17 eV (Ne I), the emission is therefore mainly surface in nature. At least for Cu(110) this is difficult to reconcile with the angular variation observations of Nilsson and Ilver [42], who find that a change in polar angle to 10° away from normal produces very different spectra at these photon energies, and that a direct transition analysis can be applied to the spectra. The majority of features in most angle-resolved spectra have been generally assigned in terms of a direct transition analysis, modified to some extent by the presence of the surface. However, it is likely that there will be matrix element modulations of individual transition intensities. Shirley and co-workers [43] have assigned variations in high photon energy electron spectra as variations between t_{2g} and e_g type initial states. While this may be valid at high electron kinetic energies which emerge from a region where the effective symmetry is that of the bulk, with lower photon energies it may be more appropriate to look for correlations using the surface symmetry, since the escape depth is small. We have calculated one-dimensional densities of states for the individual bands of Pd along the major symmetry directions and have compared these with the normally emitted spectra at 21 and 17 eV. Fig. 15 shows the spectra for Pd(001); the narrowing of the band in the Ne I spectrum can be most easily associated with a loss of intensity for d_{xz}, d_{yz} initial states. Fig. 16 shows the Pd(111) spectra, and a very similar interpretation can be fitted to the data so long as the d_{xz}, d_{yz} labels are

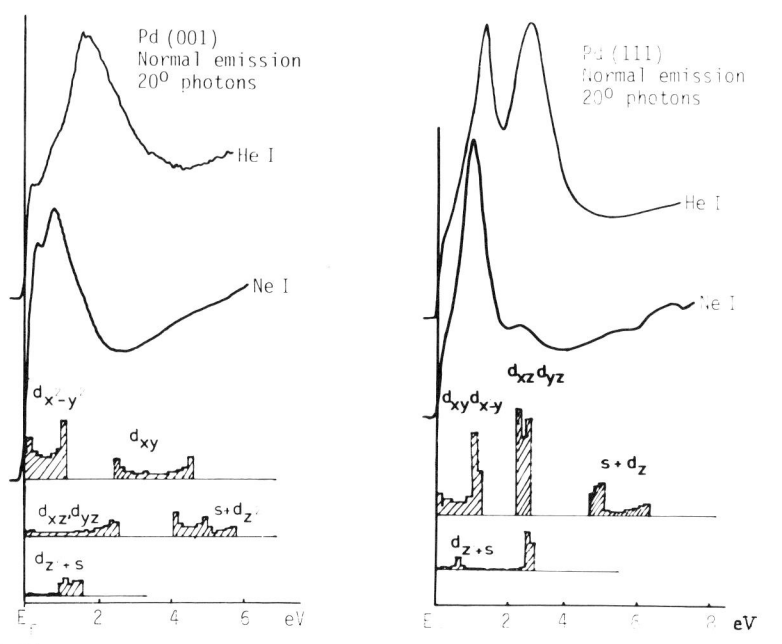

Fig. 15. The normally emitted photoelectron spectra from Pd(001) at He I and Ne I energies compared with individual densities of states along the ΓX direction [44].

Fig. 16. The normally emitted photoelectron spectrum from Pd(111) at He I and Ne I energies compared with individual densities of states along the ΓL direction [44].

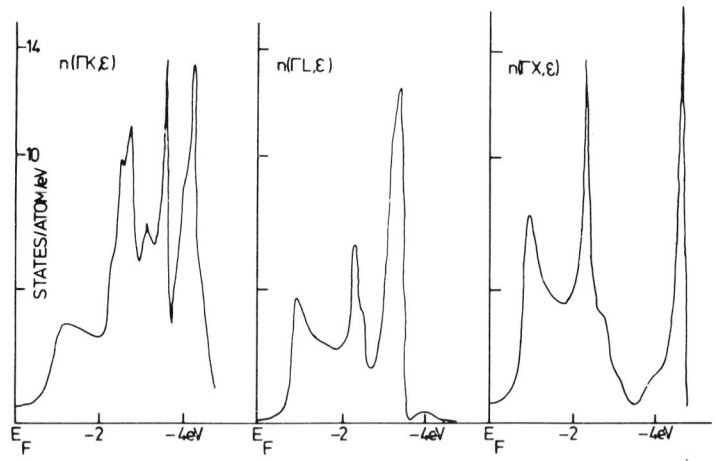

Fig. 17. Calculated densities of states for a 23% Ni–77% Cu random alloy [46]. The feature closest to E_F for ΓX is mainly Ni in character.

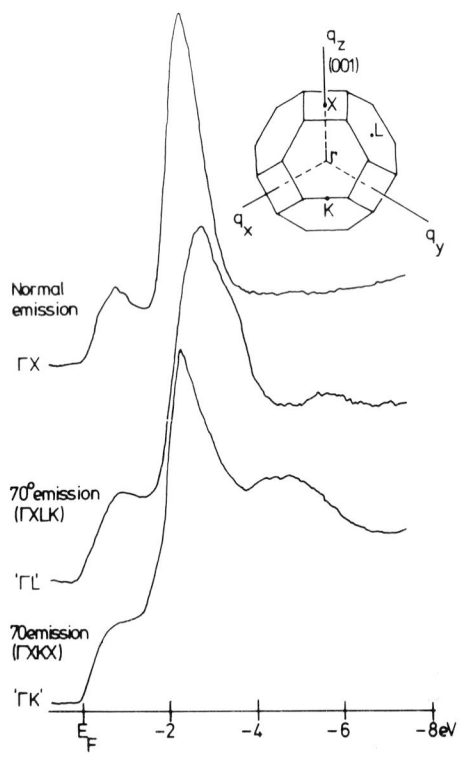

Fig. 18. He I photoelectron spectra from a 23% Ni–77% Cu(001) alloy single crystal [46].

those for z normal to the surface, not those for the bulk [44]. The behaviour of emission from Ag(111) and Ag(100) is very similar [45], and can be interpreted in a similar fashion. On the Ne I spectra from Pd there are also some weak features in the 4–6 eV region which may be emission from band 1, which normally is not observed in He I spectra.

The adsorption of various species on to surfaces generally does not greatly disturb d-band emissions, but usually there is some loss of definition. It is rather surprising that introduction of a gross perturbation into the bulk solid by alloying also does not greatly disturb the d emissions in the one example we have studied so far, a $Cu_{77}Ni_{23}$ alloy crystal with (001) orientation. The main Cu d region is very similar to that in the pure Cu(001) crystal in its variation with angle, though as with adsorbate covered surfaces there is some loss of definition. However, introduction of the nickel does produce an additional group of states close to the Fermi edge. Calculations by Stocks, Gyorffy and Temmerman indicate that these Ni states should be concentrated near the X point (fig. 17) and there seems to be some support for this from the photoelectron spectra (fig. 18) [46].

3.2. Adsorbates

Much of the work carried out for clean metals can be interpreted in terms of band-band transitions. In contrast most work on adsorbates has been described in terms of surface molecules or adatoms with only localised interactions between substrate and adsorbate. There is an important exception to this, the W(001)–H system, where there is evidence for surface bands of H character. The work of Lapeyre, Anderson and Smith [47] on the polarisation dependence provides evidence that these H bands have at least the mirror symmetry of the substrate (fig. 19). For this experiment the analyser is positioned in one of the (100) planes, and the polarisation direction A is set parallel or perpendicular to this plane. The polarisation effects can be associated with the even or odd character of the symmetry species generated by the direct product of the symmetries of the initial and final states, since the dipole operator has even symmetry for $A_\parallel$, odd for $A_\perp$: this assists in the assignment of the transitions. There is also a very strong polarisation effect on the H induced peak at -6.5 eV which demonstrates that the H induced initial state also has symmetry with respect to the $\langle 100 \rangle$ plane of the substrate. By identifying the upper state as a W band with even symmetry, this H induced band is shown to have even symmetry also. At lower H coverages, it has also been shown that a direct transition of the bulk W band structure can be induced to appear in new directions by addition of a surface G vector of the ordered overlayer [19].

There have been very many studies of CO adsorbed on metal surfaces; these studies show two bands * at about 8 eV and 11 eV below E_F. Some tima ago it was

---

* Work by Brundle reported in a post-deadline paper at this meetings shows a complex behaviour for CO on copper, with four bands present.

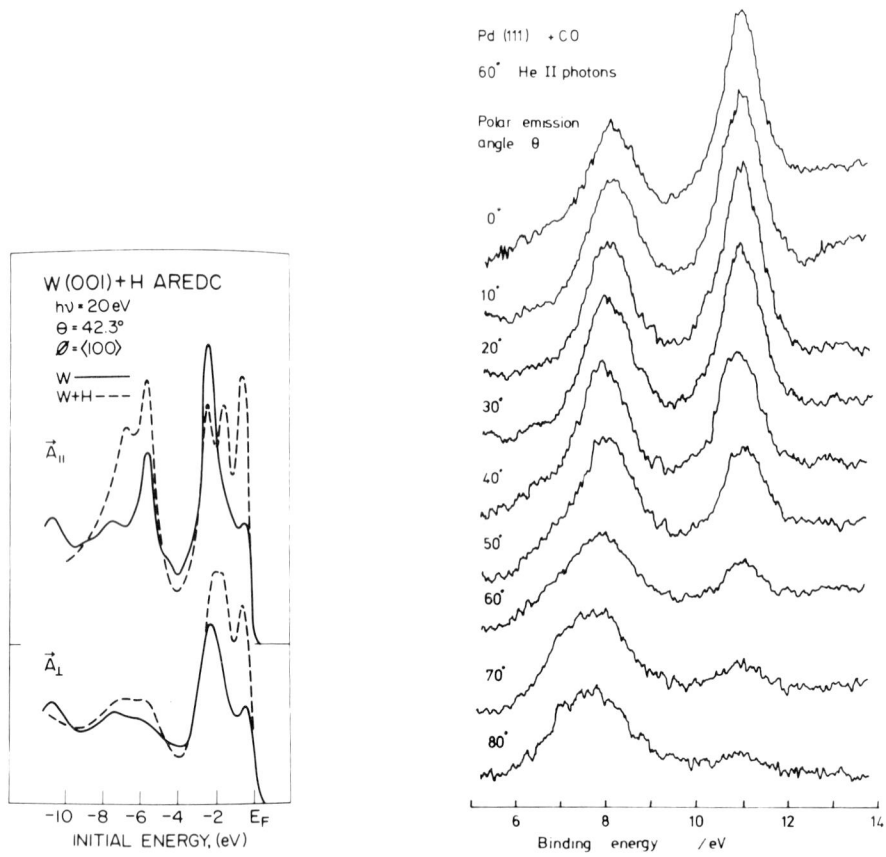

Fig. 19. Photoelectron spectra (AREDC) from clean and hydrogen-covered (W(001)) surfaces using polarised light. The detector lies in a ⟨100⟩ plane of the surface, and the polarisation vector is set parallel ($A_\parallel$) or perpendicular to this plane [47].

Fig. 20. Variation with θ of the He II photoelectron spectrum of the adsorbate-induced bands of CO on Pd(111) [50].

suggested [48] that these bands are most logically explained in the same way as the two major bands observed in the gas-phase spectra of metal carbonyls, i.e. the 11 eV peak as 4σ, the O lone pair, and the 8 eV peak as 5σ (C lone pair, M–C bonding) superimposed on 1π. Confirmation of this has been provided by angle-integrated synchrotron radiation studies of intensity measurements as a function of energy [49]. Since in free CO, 4σ has a resonance in a similar position to one observed in normally emitted electrons in the 11 eV peak, there is experimental evidence that the bonding is in fact through C [47]. Most chemists would be very uneasy about any other bonding geometry for CO to Ni or earlier transition metals, but it is very

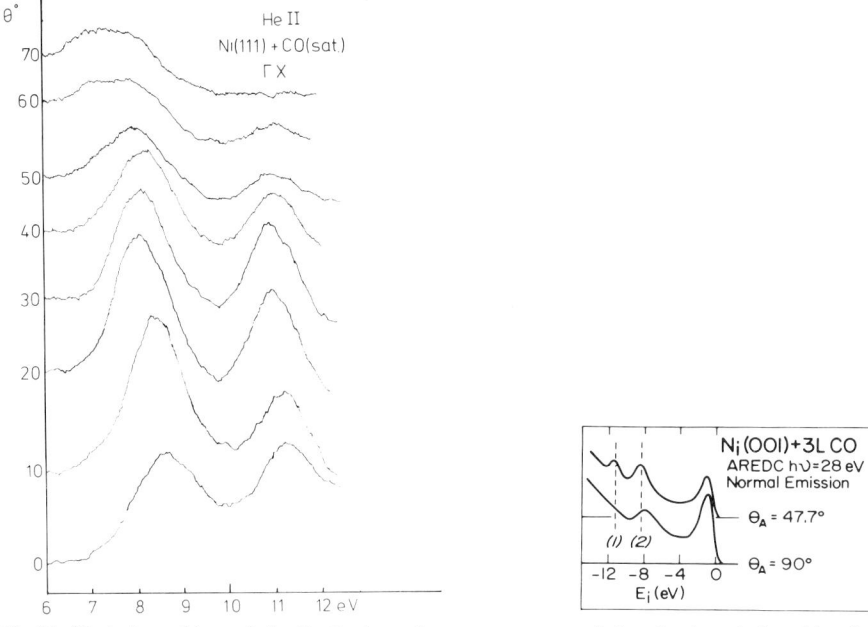

Fig. 21. Variation with θ of the He II photoelectron spectrum of the adsorbate-induced bands of CO on Ni(111) [26]. Note that the direction of increasing θ is opposite to that of fig. 20).

Fig. 22. Variation of the normally emitted photoelectron spectrum from CO adsorbed on Ni(001) as the polarisation of the incident light changes from p (upper curve) to s (lower curve).

useful to have experimental confirmation. In fig. 20 the angular variation of these two peaks for CO on Pd(111) is shown [50] and in fig. 21 the very similar results of Williams, Jacobi and co-workers for CO on Ni(111) (note that the curves show θ increasing downwards in fig. 20, upwards in fig. 21 [26]. Both of these show that the 4σ peak (11 eV) disappears towards grazing exit but that the 8 eV peak is composite, with an apparent shift to lower binding energy with increasing θ. This is best assigned as 1π being the orbital with lowest ionisation energy, which is the residual emission close to grazing exit, with the major component of the 8 eV band at small θ values being due to 5σ. The difference between Ni and Pd is then that 5σ is slightly more stable on Ni than on Pd. A greater stabilisation on Ni is reasonable on chemical grounds, since CO complexes of Pd are unstable at room temperature. Fig. 22 shows the effect of s and p polarised light on the normally emitted electrons [47]: the 4σ peak is absent for s polarised light, and the apparent shift of the 8 eV peak suggests that the 5σ is also absent.

Finally, we have some preliminary data on benzene; fig. 23 shows the angular variations on Pd(111). By changing angles, evidence for all the peaks of the gas-phase molecule is obtained. Each orbital appears to have its own variation with

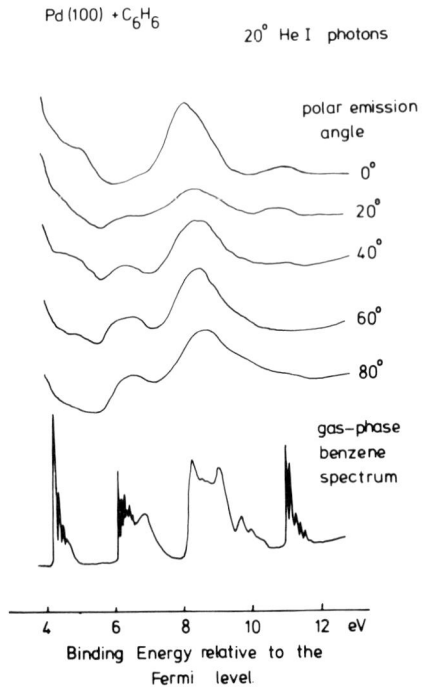

Fig. 23. Variation with θ of the He I spectrum from benzene adsorbed on Pd(111) compared with the gas-phase spectrum from free benzene.

angle: the only simple comment is that all emissions are minimised close to 20° [51]. Experiments with polarised light, and detailed calculations, would seem to be worthwhile.

Acknowledgement

We thank the S.R.C. of Great Britain for support of our work.

References

[1] Photoemission; Proc. Intern. Symp. Noordwijk, Netherlands, September 1976, Eds. R.F. Willis, B. Feuerbacher, B. Fitton and C. Backx (ESTEC, Noordwijk).
[2] B.W. Holland, Surface Sci. 68 (1977) 490.
[3] B. Feuerbacher and R.F. Willis, J. Phys. C. (Solid State Phys.) 9 (1976) 169.
[4] D.R. Lloyd, C.M. Quinn and N.V. Richardson, Chem. Soc. Specialist Periodical Reports, Surface and Defect Properties of Solids 6 (1977) (in press).
[5] W.E. Spicer, K.Y. Yu, I. Lindau, P. Pianetta and D.M. Collins, Surface and Defect Properties of Solids 5 (1976) 103.

[6] D.E. Eastman, in: Electron Spectroscopy, Ed. D.A. Shirley (North-Holland, Amsterdam, 1972) p. 487.
[7] E.M. Purcell, Phys. Rev. 54 (1938) 818.
[8] R.T. Poole, J. Liesegang, J.G. Jenkin and R.C.G. Leckey, Vacuum 22 (1972) 499.
[9] W.F. Egelhoff, D.L. Perry and J.W. Linnett, Phys. Rev. Letters 34 (1975) 93.
[10] N.V. Smith, P.K. Larsen, M.M. Traum and S. Chiang, in ref. [1], p. 119.
[11] B. Feuerbacher and N.E. Christensen, Phys. Rev. B10 (1974) 2373.
[12] R.H. Williams, J.M. Thomas, M. Barber and N. Alford, Chem. Phys. Letters 117 (1972) 142.
[13] W.T. Bordass and J.W. Linnett, Nature 222 (1969) 660.
[14] R.Y. Koyama and L.R. Hughey, Phys. Rev. Letters 34 (1972) 93.
[15] N.V. Smith, M.M. Traum and F.J. DiSalvo, Phys. Rev. Letters 32 (1974) 1241.
[16] S.B.M. Hagström and I. Lindau, J. Phys. E, Sci. Instr. 4 (1971) 936.
[17] P.O. Nilsson and L. Ilver, Solid State Commun. 17 (1975) 677.
[18] R.F. Willis, B. Feuerbacher and B. Fitton, Solid State Commun. 18 (1976) 1315.
[19] J. Anderson and G.J. Lapeyre, Phys. Rev. Letters 36 (1976) 376.
[20] G.J. Lapeyre, J. Anderson, P.L. Gobby and J.A. Knapp, Phys. Rev. Letters 33 (1974) 1290.
[21] R.Z. Bachrach, M. Skibowski and F.C. Brown, Phys. Rev. Letters 37 (1976) 40.
[22] S.P. Weeks and E.W. Plummer, Solid State Commun. 21 (1977) 695.
[23] B.J. Waclawski, T.V. Vorburger and J.R. Stein, J. Vacuum Sci. Technol. 12 (1975) 301.
[24] G.L. Price and B.G. Baker, Surface Sci. 68 (1977) 507.
[25] E. Dietz, H. Becker and U. Gerhardt, Phys. Rev. B12 (1975) 2084.
[26] P.M. Williams, P. Butcher, S. Woods and K. Jacobi, Phys. Rev. B, in press.
[27] J.B. Pendry and D. Titterington, Commun. Phys. (1977) (in press).
[28] L.F. Wagner, Z. Hussain, C.S. Fadley and R.J. Baird, Solid State Commun. 21 (1977) 453.
[29] L. Ilver and P.O. Nilsson, Solid State Commun. 18 (1976) 677.
[30] D.R. Lloyd, C.M. Quinn and N.V. Richardson, Solid State Commun., in press.
[31] D.R. Lloyd, C.M. Quinn and N.V. Richardson, J. Chem. Soc. Faraday II (1976) 1036.
[32] P.O. Gartland, S. Berge and B.J. Slagsvold, Phys. Rev. Letters 30 (1973) 916.
[33] H.F. Roloff and H. Neddermeyer, ref. [1], p. 9.
[34] G.V. Hansson, S.A. Flodstrom and S.B.M. Hagström, ref. [1], p. 29.
[35] B. Feuerbacher and B. Fitton, Solid State Commun. 15 (1974) 301.
[36] R.J. Smith, J. Anderson, J. Hermanson and G.J. Lapeyre, Solid State Commun. 18 (1976) 975.
[37] R.F. McFeeley, J. Stöhr, G. Apai, P.S. Wehner and D.A. Shirley, Phys. Rev. B 14 (1976) 3273.
[38] B. Feuerbacher and R.W. Willis, Phys. Rev. Letters 37 (1976) 446.
[39] W.F. Egelhoff, J.W. Linnett and D.L. Perry, Surface Sci. 54 (1976) 670.
[40] L. Richter and R. Gomer, Phys. Rev. Letters 37 (1976) 763.
[41] P. Heimann, H. Neddermeyer and H.F. Roloff, Phys. Rev. Letters 37 (1976) 775; ref. [1], p. 41.
[42] P.O. Nilsson and L. Ilver, in: Proc. Intern. Symp. on Electron Spectroscopy, Kiev, 1975.
[43] P.S. Wehner, J. Stöhr, G. Apai, F.R. McFeeley and D.A. Shirley, Phys. Rev. Letters 38 (1977) 169.
[44] D.R. Lloyd, C.M. Quinn and N.V. Richardson, Surface Sci. 63 (1977) 174.
[45] H.F. Roloff and H. Neddermeyer, Solid State Commun. 21 (1977) 561.
[46] G.M. Stocks, B.L. Gyorffy, W.L. Temmerman, D.R. Lloyd, C.M. Quinn and N.V. Richardson, unpublished.
[47] G.J. Lapeyre, J. Anderson and R.J. Smith, ref. [1], p. 249.
[48] D.R. Lloyd, Faraday Discussions 58 (1975) 136.
[49] T. Gustafsson, E.W. Plummer, D.E. Eastman and J.L. Freeouf, Solid State Commun. 17 (1975) 391.
[50] D.R. Lloyd, C.M. Quinn and N.V. Richardson, Solid State Commun. 20 (1976) 409.
[51] D.R. Lloyd, C.M. Quinn and N.V. Richardson, Solid State Commun., in press.

DIRECTIONAL UV PHOTOEMISSION FROM CLEAN AND SULPHUR SATURATED (100), (110) AND (111) NICKEL SURFACES

T.T. Anh NGUYEN and R.C. CINTI

Groupe des Transitions de Phases, CNRS, B.P. 166, 38042 Grenoble Cedex, France

Angle resolved photoemission energy distribution curves (EDC's) were obtained on clean and sulphur saturated (100), (110) and (111) nickel surfaces for excitation energies equal to 10.2, 13.5, 16.8 and 21.2 eV. The EDC's of clean surfaces are weakly structured at $\theta = 0°$ and become more rich in features for oblique angles. In the explored energy range, adsorption of sulphur produces two extra-structures at initial energies depending on surface orientation. One of which situated at about -4.5 eV below the Fermi level is in good agreement with Hagstrum ion neutralization spectroscopy results, the other at around -1.8 eV has never been observed before. A remarkable similarity of adsorption effects on the three surfaces is found. These results are compared with experimental data obtained previously on equivalent sulphur saturated surfaces by INS and discussed in relation to recent theoretical calculations on chalcogen adsorption on nickel.

1. Introduction

Angle resolved photoemission has been increasingly used to investigate the electronic structures of clean and adsorbate-covered transition metal surfaces. Encouraging results were obtained for clean surfaces and in many cases electron energy distribution curves (EDC's) were correctly related to bulk bandstructures [1–3]. For adsorption, on the other hand, many works were done by integrated photoemission and presence of adsorption induced states was clearly demonstrated following controlled adsorption of simple atoms or molecules. Some of these states were associated with a good probability, to atomic or molecular states before interaction [4]. Recent theoretical calculations [5–8] and experimental data [9–11] showed that angle resolved photoemission is able to give information on the spatial location of adsorbed species and more precise description of adsorbate–substrate electronic interaction. The present investigation was undertaken in this latter purpose. Sulphur on nickel was chosen for two principal reasons: (a) it has been well studied by conventional surface techniques (LEED, AES and INS) [12–14]; (b) it has similarity to the oxygen/nickel system for which theoretical works have been done.

In previous letters [15,16], preliminary results concerning clean Ni(100) and S on Ni(100) and (110) have been reported. in this paper more detailed results will be presented for clean and S saturated simple Ni faces.

2. Experimental

The apparatus used for this study has been described previously [1]. The Ni samples were cut along the (100), (110) and (111) crystallographic planes to within 0.5°. After mechanical polishings and electropolishings, they were cleaned in situ by successive ion bombardments (500 V, 10 μA, 30 min) and annealings (700°C, 15 min). About 100 of these cycles were necessary to obtain S free surfaces. During experiments, a pressure of about 1×10^{-9} Torr was maintained. 20 min elapsed before CO contamination was detectable by AES. A clean surface was restored by flash heating to 450°C which desorbed CO without segregation of S from the bulk. Two minutes scan times between two heat pulses were used to record spectra. The crystal was then cooling in the temperature range between 450 and 360°C where Ni is paramagnetic. Sulphurated surfaces were obtained either by dissociative H_2S adsorption [12] or by segregation of S from the bulk when crystals were heated to 850°C. No difference was observed in the results obtained with these two methods. Sulphuration was extended to have a saturation concentration on surfaces detected by a constant ratio between the S 152 V and Ni 61 V Auger peaks. This ratio about 0.95 was approximately the same for the three surface orientations. During measurements on these sulphurated surfaces, the temperature was maintained in the same range as for the clean ones. The full acceptance angle of the electron analyzer was 10°. The collection angle θ relative to the surface normal, was varied by rotating the sample around an axis in the plane of the surface, arcs θ projecting onto simple azimuthal directions, [001] for Ni(100) and (110) and [11$\bar{2}$] for Ni(111). No correction for the variation of the analyzer resolution and transmission was applied to the presented spectra.

3. Results and discussion

3.1. Clean surfaces

The EDC's obtained at the normal to the surface from clean (100), (110) and (111) faces of Ni are shown in fig. 1. The bottom scale refers to the initial energy states of electrons before excitation relative to the Fermi level E_F.

A common feature seen on these weakly structured spectra is a stationary structure, labelled a, at 0.2 eV below E_F which appears as predominant peak in all curves except for those of Ni (100) at $\hbar\omega$ = 21.2 eV and Ni(111) at $\hbar\omega$ = 16.8 and 21.2 eV. In these latter EDC's this feature is overlapped by a peak localised at around −0.5 eV. For $\hbar\omega \geqslant 13.5$ eV one or two other shoulders appear depending on excitation energy and crystal face. The width of the EDC's decreases with $\hbar\omega$ and for 21.2 eV spectra the total width is approximately equal to 2.5 eV.

At oblique collection angles the EDC's have more structures. Except for the curves obtained with $\hbar\omega$ = 10.2 eV on Ni(100) and (111), which present a single

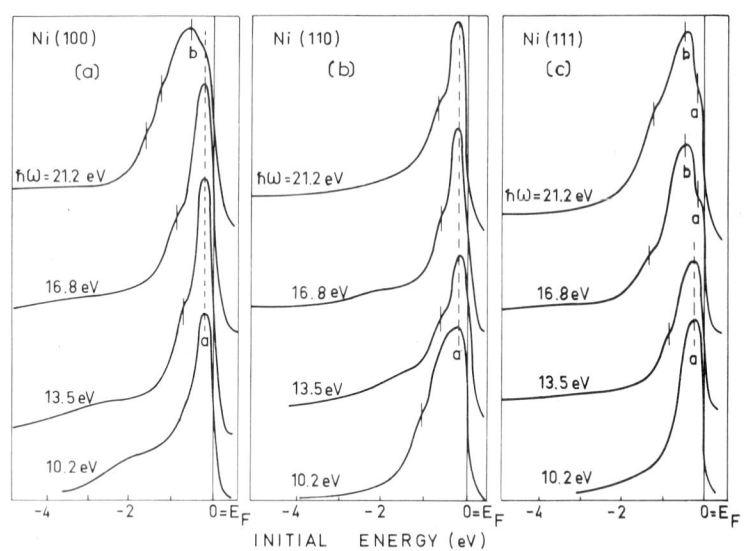

Fig. 1. EDC's from clean Ni along the surface normal for $\hbar\omega$ = 10.2, 13.5, 16.8 and 21.2 eV. (a) Ni(100), (b), Ni(110), (c) Ni(111).

peak, in all other curves clearly appear two or three structures. Generally the weak shoulders observed at $\theta = 0°$ become well defined peaks which move in initial energy when θ is varied. Such a behaviour is illustrated in fig. 2 for Ni(100) with $\hbar\omega$ = 13.5 and 21.2 eV.

Very recently, clean ferromagnetic nickel single surfaces have been investigated by directional photoemission [3,11,17,18]. At $\theta = 0°$ (normal to the surface) the most similar experimental conditions are those used by Heimann and Neddermayer (HN) [3]. These authors measured the EDC's from Ni(100), (110) and (111) along the surface normal using excitation energies $\hbar\omega$ = 21.2, 16.8 and 11.8 eV. For Ni(100) and (111) they found strong correlation between the EDC's and bulk bandstructures. Direct interband transitions and surface emission process [2] were used to relate structures observed on spectra to bandstructures. For Ni(100) they attributed important surface effects to the stationary peak at −0.15 eV.

As in our experiments we observed any detectable difference in the spectra measured at sample temperature below the Curie temperature of Ni, in the range of 250° to 300°C (the same observation has been reported by Pierce and Spicer [19]), we think it is possible to compare our results with HN ones. On this point, for the two common excitation energies (16.8 and 21.2 eV) the results agree fairly well, with minor differences on the weak structures. This discrepancy could be attributed to the difference in the light incidence angles, 48° here and 80° in HN case, and to the difficulty encountered to resolve these weak structures. Our EDC's can then be associated to direct transitions and surface emission in the same manner as HN with however the following remarks:

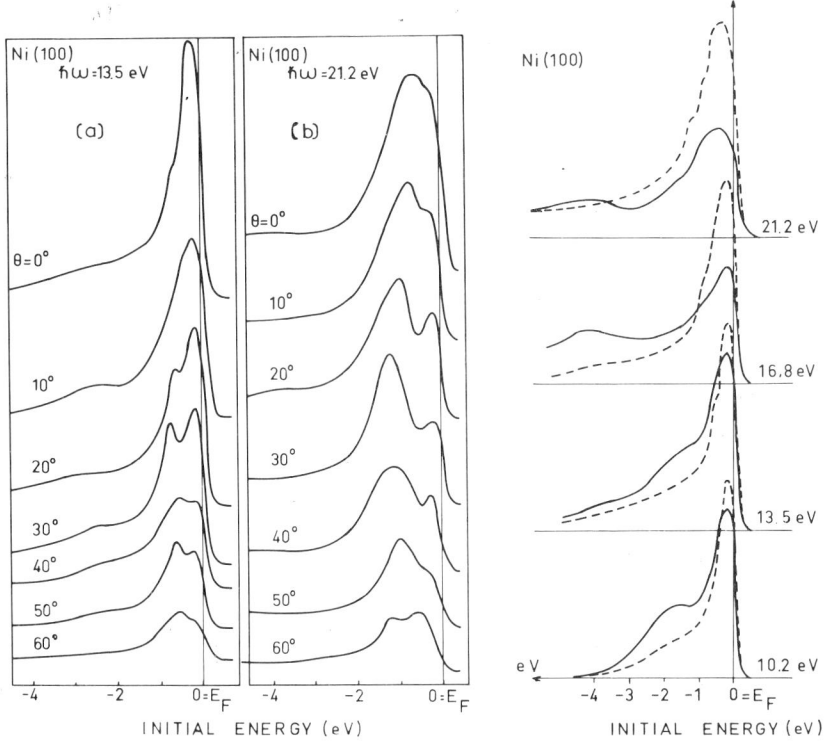

Fig. 2. Angular profiles of the EDC's from clean Ni(100) for collection angles θ varying from 0° (normal to the surface) to 60°, arcs θ projecting into [001] azimuthal direction. (a) $\hbar\omega$ = 13.5 eV, (b) $\hbar\omega$ = 21.2 eV.

Fig. 3. EDC's from Ni(100) surface along the surface normal. Full lines: surface saturated with adsorbed S, dotted lines: clean surface.

(a) The structure near E_F and labelled a in our Ni(100) and (111) EDC's cannot be attributed to direct transitions because there are not enough corresponding empty states for all the four $\hbar\omega$ used. A surface emission process would be more likely.

(b) For the stationary peak of the Ni(110) spectra, if its origin can be assignated to surface effects, we did not find any evidence for surface states because in our adsorption experiments, the intensity decreasing of this feature upon S and O adsorption is of the same order as the similar structure (−0.2 eV) of the two other faces.

At oblique collection angles the EDC's show a strong dependence on the parallel component $k_\parallel$ of the wave vector. The Ni "d band" is then larger than that observed along the surface normal; about 4.5 eV instead of 2.5 eV in concordance with the recent observation of Smith et al [18] by angle resolved photoemission using synchroton radiation. For Ni(111) and $\hbar\omega = 21.2$ eV our θ dependent EDC's are also in agreement with those of Williams et al. [11] for the ΓLKL plane.

3.2. Sulphur adsorption

Sulphur adsorption on Ni(100) has been well studied [12–14]. At saturation a c(2 × 2) LEED structure has been observed and interpreted by an ordered S overlayer having a concentration 1/2, S atoms sitting in fourfold sites in the unreconstructed Ni network.

Fig. 3 shows spectra measured normally to this surface for the four excitation energies used. Corresponding clean surface spectra are also drawn with dotted lines for comparison. One can note the relative weakening of the principal structure originating from the Ni d band when adsorbate is present on the surface. Moreover, a well defined extra peak occurs for $\hbar\omega = 10.2$ eV around 1.8 eV below E_F. Its amplitude decreases for $\hbar\omega = 13.5$ eV and it is no longer clearly apparent for the highest excitation energies used. Conversely for $\hbar\omega = 16.8$ and 21.2 eV, there appears a new structure located at about 4.25 eV below the Fermi level, not detectable at lower $\hbar\omega$, perhaps because of the cut-off function of our analyzer system.

This drastic variation of the ionization cross-section of the highest S induced structure with uv excitation energy is the first interesting observation of this work. It shows the necessity, for adsorption investigations by UPS technique, of the use of a large range of excitation energies. For example, in a previous work on the same

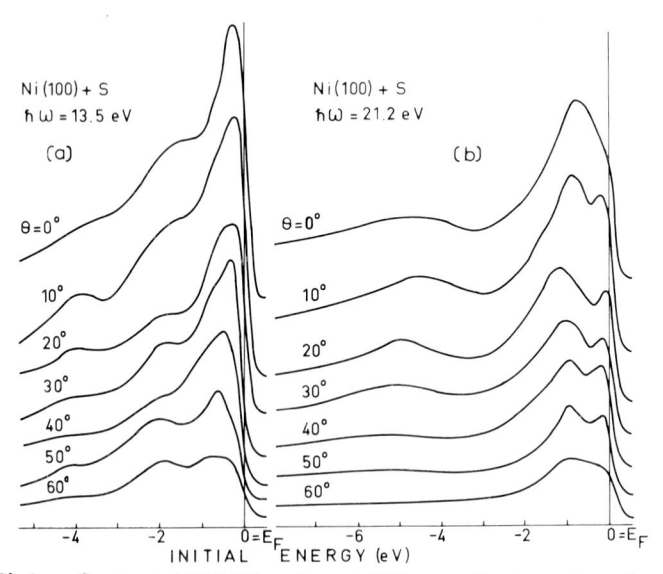

Fig. 4. EDC's from S saturated Ni(100) surface at different collection angles θ from 0° to 60°. (a) $\hbar\omega = 13.5$ eV, (b) $\hbar\omega = 21.2$ eV.

system done by Hagstrum [20], only the second structure was observed because of the high energy used (He I line).

The second interesting observation is the variation of spectra when collection solid angle θ is ruled out from the surface normal. Fig. 4 shows spectra obtained for $\hbar\omega$ = 13.5 eV and 21.2 eV when θ changes from 0° to 60°.

As for the normal emission, the Ni d band contribution is smoothed with regard to the clean sample feature for the two energies. On the 21.2 eV set of curves, the low energy S induced structure decreases in intensity when θ is increased and its position moves toward lower initial energies by more than 1 eV. On the second set of spectra, for which $\hbar\omega$ = 13.5 eV, the first structure due to adsorbate appears clearly up to θ = 60° and its position on the initial energy scale is approximately stable between 1.8 and 2 eV below E_F.

In our previous paper [16], only normal emission measurements (fig. 3) have been done. These partial results were compared with those of Becker and Hagstrum obtained on the same system by means of INS technique. For the c(2 × 2) structure, they also found the diminution of the Ni d band peak and the extra structure at 4.2 eV below E_F. On the other hand, they did not observe the high energy level 1.8 eV below E_F occurring in our lowest excitation energy spectra. Nevertheless, they found two deep initial energy structures located at −8 and −9.5 eV (this latter being probably ascribable to contamination of the covered surface). In the discussion, we underlined the good agreement between the three extra-structures complementary observed by UPS and INS and the theoretical results of Kasowski LCMTO calculations [21] which predict three extra-levels induced on (100)Ni surface by chalcogen adsorption: two of them originating in the atomic p level split by the crystal field, the third being a Ni surface state induced in the s–d gap by the presence of adsorbate atoms.

Our new results showing a polar dependence of the initial energy location for one of the S induced levels suggests a non-negligible interaction between adatoms which delocalizes their electronic states in the surface plane and gives a dispersion relation $E(k_\parallel)$ between their energies and their crystal momentum. Assuming $k_\parallel$ conservation during photoemission process, it is then possible to plot an experimental dispersion curve $E(k_\parallel)$ from our measurements. Fig. 5 shows this curve, a being the surface unit cell parameter of the c(2 × 2)S structure. One can note a relatively good coherency between points originating from different measured spectra for the four excitation energies used. This dispersion property cannot be described by the results of the α-X method [22] which only predict the behaviour of non-interacting adsorbate levels, each Ni_xS cluster being considered as an isolated molecule. Neither can it be described by the Kasowski's results [21] developed only at the Γ point of the O/Ni system. Conversely it is the central point of the recent theoretical work done by Liebsch [8] on oxygen adsorption on Ni. This latter presents calculated energy distributions for uv photoemission from valence states of an ordered monolayer adsorbed on Ni(100) surface. The spectra show the split oxygen 2p resonances below the Ni band and a strong dependence of the position of these levels with $k_\parallel$.

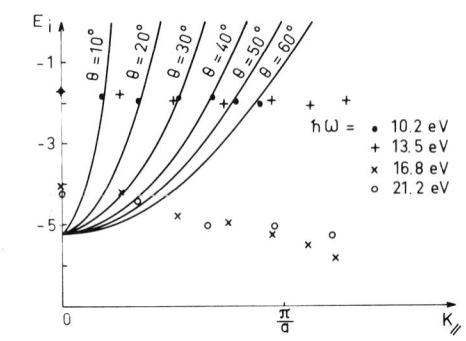

Fig. 5. Variation of the S induced levels on Ni(100) versus $k_\parallel$.

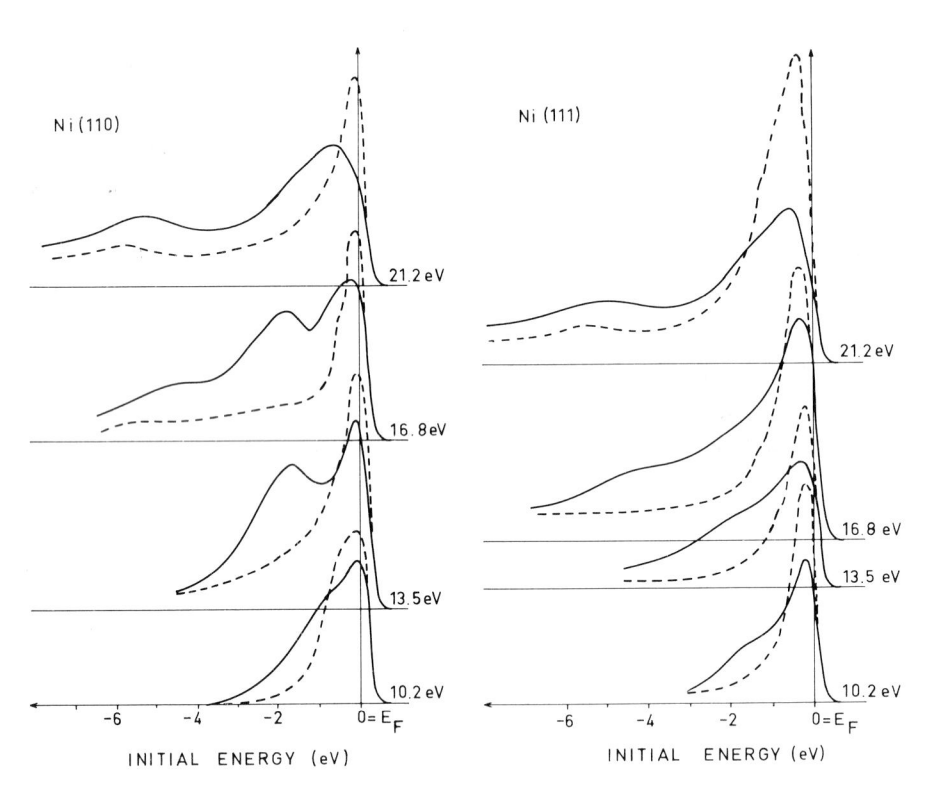

Fig. 6. EDC's from Ni(110) along the surface normal. Full lines: surface saturated with adsorbed S; dashed lines: clean surface.

Fig. 7. EDC's from Ni(111) along the surface normal. Full lines: surface saturated with adsorbed S; dashed lines: clean surface.

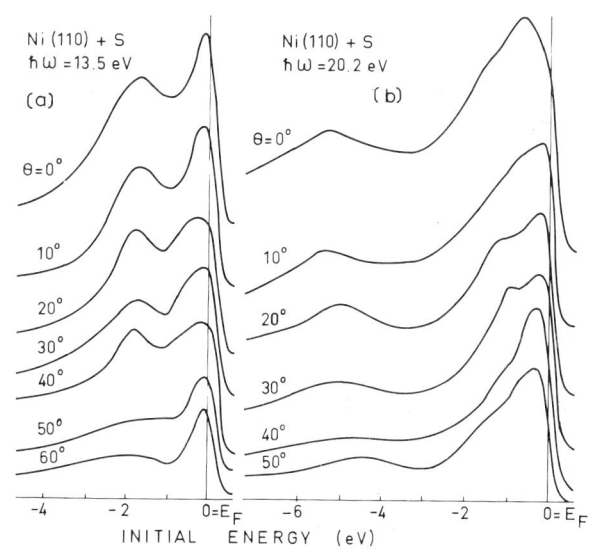

Fig. 8. EDC's from S saturated Ni(110) surface at different collection angles θ from 0° to 60°. (a) $\hbar\omega$ = 13.5 eV, (b) $\hbar\omega$ = 21.2 eV.

For a large part, this dependence may be understood by the dispersion of the three p bands of a single oxygen monolayer. Nevertheless, this development is relative to an (1 × 1) ordered layer and then is not directly applicable in our case. We think that the calculations planned by this author for the more realistic c(2 × 2) structure [8] will allow us to improve the interpretation of our angular profile data.

As one can see from the lack of convergence of various LEED observations [12–14], sulphur adsorption on (110) and (111) surfaces is more complex than on the (100) one. This complexity is perhaps ascribable to surface reconstructions as suggested by Perdereau and Oudar [12].

Energy distributions of normal photoemission from respectively (110) and (111) surfaces are presented on figs. 6 and 7. The two sets of curves show large similarities to the (100) results discussed above. For the two surfaces the deep broad S induced level is detected for $\hbar\omega$ = 16.8 and 21.2 eV at about −5 eV below E_F, nearly 1 eV deeper than on the (100) surface. The second structure also appears at around −1.9 eV, slightly lower than precedently. Its tendency to vanish toward high excitation energies is not as clear as before, especially on the (110) surface where it seems to broaden toward the Fermi level. Fig. 8 gives the polar variations of the (110) spectra for two excitation energies. For $\hbar\omega$ = 13.5 eV, the location of the −1.9 eV structure is nearly stationary on the initial energy scale. Conversely for $\hbar\omega$ = 21.2 eV, this levels which appears as a broad shoulder for $\theta = 0°$ tends to become a more defined structure moving between −1 and −1.5 eV below E_F when θ increases. A dispersion property is also observed for the deepest S induced level on this set of curves.

As underlined above, S adsorption on these two surfaces is more complex than on the (100) one and consequently no corresponding theoretical work has been published. This fact limits our discussion on these results. We can however emphasize the relative similarity of S induced structures observed on the three Ni samples which suggests a close relationship between the S adsorption mechanisms on these surfaces.

4. Conclusion

Angle resolved UPS measurements have been made on the three simple Ni surfaces, clean and saturated with adsorbed S.

On clean surfaces, the EDS's are in fairly good agreement with recent published results obtained in similar experimental conditions. A Ni d band contribution approximately equal to 2.5 eV and 4.5 eV is observed respectively at the surface normal and at oblique analysis angles.

S adsorption induces two extra-levels at comparable initial energies on the three surfaces. The first, located in the middle of the Ni d band has its ionization cross-section rapidly varying with the excitation energy in the explored range (10.2–21.2 eV). The second appears in the lower part of the d band and presents variations with the analysis direction which suggests a dispersion relation $E(k_\parallel)$ in the surface plane. For (100) surface, assuming that I.N.S. observations of Becker and Hagstrum effectively reflect the ground states of S induced levels, the two structures we observed can be reasonably compared with Kasowski's theoretical results. In this optics the lowest one is ascribable to one of the split Sp levels, the second to the "induced surface state" predicted by this author.

Acknowledgments

The authors thank Dr. H. Neddermeyer for communicating conduction bands of nickel calculated by E. Marschall.

References

[1] R.C. Cinti, E. Al Khoury, B.K. Chakraverty and N.E. Christensen, Phys. Rev. B14 (1976) 3296.
[2] N.E. Christensen and B. Feuerbacher, Phys. Rev. B10 (1974) 2349, 2373.
[3] P. Heimann and H. Neddermeyer, J. Phys. F6 (1976) L 257.
[4] See for example:
   J.E. Demuth and D.E. Eastman, Phys. Rev. B13 (1976) 1523;
   D.E. Eastman and J.K. Cashion, Phys. Rev. Letters 27 (1971) 1520;
   T. Gustafsson, E.W. Plummer, D.E. Eastman and J.L. Freeouf, Solid State Commun. 17 (1975) 391.

[5] J.W. Gadzuk, Phys. Rev. B10 (1974) 5030.
[6] S.Y. Tong and M.A. Van Hove, Solid State Commun. 19 (1976) 543.
[7] A. Liebsch, Phys. Rev. B13 (1976) 544.
[8] A. Liebsch, Phys. Rev. Letters 38 (1977) 248.
[9] B. Feuerbacher and R.F. Willis, Phys. Rev. Letters 36 (1976) 1339.
[10] R.J. Smith, J. Anderson and G.J. Lapeyre, Phys. Rev. Letters 37 (1976) 1081.
[11] P.M. Williams, P. Bütcher and J. Wood, Phys. Rev. B14 (1976) 3215.
[12] M. Perdereau and J. Oudar, Surface Sci. 20 (1970) 80.
[13] J.E. Demuth and T.N. Rhodin, Surface Sci. 45 (1974) 249.
[14] G.E. Becker and H.D. Hagstrum, Surface Sci. 30 (1972) 505.
[15] T.T.A. Nguyen, R. Cinti and S.S. Choi, J. Physique Lettres 37 (1976) L 111.
[16] R.C. Cinti and T.T.A. Nguyen, J. Physique Lettres 38 (1977) L 29.
[17] G.P. Williams and C. Norris, in: Proc. Intern. Symp. on Photoemission, Noordwijk, Sept. 1976, p. 55.
[18] R.J. Smith, J. Anderson, J. Hermanson and G.J. Lapeyre, Solid State Commun. 21 (1977) 459.
[19] D.T. Pierce and W.E. Spicer, Phys. Rev. B6 (1972) 1787.
[20] H.G. Hagstrum and G.E. Becker, Proc. Roy. Soc. (London) A331 (1972) 395.
[21] R.V. Kasowski, Phys. Rev. Letters 33 (1974) 1147.
[22] S.J. Niemczyk, J. Vacuum Sci. Technol. 12 (1975) 246.

ANGLE RESOLVED PHOTOEMISSION STUDIES OF THE BAND STRUCTURES OF SEMICONDUCTORS: PbI_2

C. WEBB and P.M. WILLIAMS

VG Scientific Ltd., East Grinstead, Sussex, UK

Angle resolved photoemission data is presented for single crystals of the layered semiconductor PbI_2. The geometric form of the valence electron energy bands has been mapped using a simple approximation and the possible electronic origins of these bands are discussed. Such band dispersion effects in angle resolved measurements are contrasted with HeII excited data for the localised Pb $5d_{5/2}$ core level component, where "flat" band behaviour and only weak residual final state scattering effects may be discerned. Finally we comment on the role of the incoming photon in such experiments.

1. Introduction

Photoemission studies of the surfaces of solids may in principle yield information on the binding energies of electronic states characteristic of both the bulk and of the surface itself. Angle resolved measurements in which the direction of emission of the photoelectron relative to the surface into the energy analyser is carefully specified, may further permit the simultaneous evaluation of the momentum of electron. The symmetries of both surface and bulk states may therefore be demonstrated in such measurements, as originally pointed out by Smith and coworkers [1]. Applications to adsorbate – surface problems thus constitute an important area of effort in this field, as reviewed elsewhere in the present proceedings by Lloyd [2].

Smith [1] and others [3–6] have, however, demonstrated that at the surfaces of certain layer-structured semiconductors and metals, bulk states may dominate the energy distribution curve for the photoelectrons (EDC), even though photoemission has classically been thought of as a "surface" process. Such layered materials may be conveniently prepared in an ideal form for photoemission by cleaving in vacuum to reveal atomically flat, clean surfaces and therefore provide ideal test subjects for angle resolved photoelectron studies. Their predominantly two dimensional character may, in some respects, be regarded as a prototype of the saturated adsorbate – surface system, and there is also considerable interest in the physics and chemistry of these anisotropic solids in their own right.

Lead iodide is such a layered material whose bulk electronic energy band structure has been the subject of some controversey, thereby providing an interesting

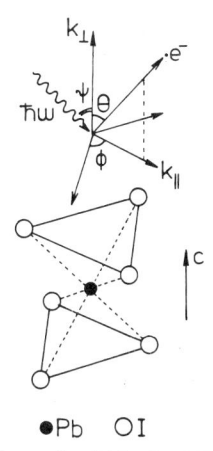

Fig. 1. Local octahedral coordination unit within the CdI_2 structure lattice of the layered semiconductor PbI_2 with (inset) the angles specified in the photoemission experiments.

problem for angle resolved measurement. Crystallising in the CdI_2 structure (fig. 1), the valence electrons comprise iodine 5p and 5s states, admixed with lead 6s electrons. It is the relative ordering of the iodine and lead based bands which has been the subject of debate since suggestions of "cationic" excitonic transitions between an uppermost Pb 6s valence band and a lowest Pb 6p conduction band inferred from optical measurements [7]. Early photoemission measurements [8,9] show considerable structure within the valence band based on iodine 5p states but the data of Mitsukawa and Ishii [10] indicates that the lead 6s band is at 8 eV below this iodine 5p band edge, in contrast with the optical results [7]. More recently, Azonlay et al. [11] have produced evidence to support Mitsukawa whereas Margaritondo et al. [12] find structure in synchrotron excited EDC's which suggests agreement with the optical assignments. It was the purpose of the present investigation, therefore, directly to determine the form of the angular depencence in uv excited EDC's ($\hbar\omega$ = 16.8 to 40.8 eV) from PbI_2 in order to attempt an assignment of the atomic character of the valence states observed previously in uv and X-ray excited photoemission [8].

2. Experimental

Measurements were carried out in an ultra high vacuum system (Model ADES 400) in which sample manipulator and hemispherical energy analyser were both rotatable about a vertical axis permitting independent selection of incident photon polar angle (ψ) and photoelectron exit polar angle (θ) (inset fig. 1). Azimuthal setting of the sample was controlled by a second rotation of the manipulator about the surface normal. Single crystals of lead iodide were cleaved in situ at pressures around 10^{-10} Torr and EDC's recorded for incident photons at 16.8 (Ne I) 21.2 (He I) 26.9 (Ne II) and 40.8 (He II) eV from windowless neon or helium discharges. Typical scan times per EDC were 30–100 sec for all but He II where 300 sec per scan was necessary.

3. Results and discussion

A wealth of data has been so recorded for PbI_2; reasons of space prevent presentation here of all but one typical series of angle resolved EDC's for He I excitation at 30° incidence and for wave vectors contained within the ΓKHAΓ plane (fig. 2). From the kinetic energies E_k of features within these EDC's (determined from the measured values of binding energies relative to E_f, the photon energy and the work

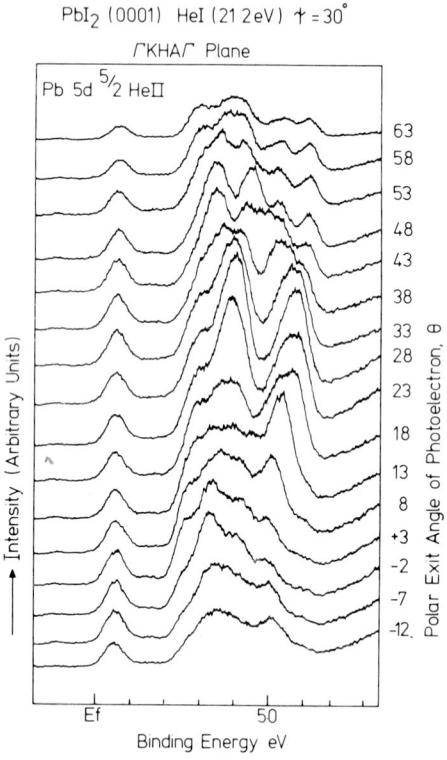

Fig. 2. Series of angle resolved EDC's for wave vectors contained within the ΓKHAΓ plane of PbI_2, using He I photons at 30° incidence (c.f. fig. 5). The level "near" E_f is in fact the overlapping Pb $5d_{5/2}$ core level excited by He II photons. Note the independence in position of this level on θ, contrasted with the dispersion behaviour in the valence band.

function) the component of electron momentum of the free photoelectron parallel to the surface, $K_{||}$, may be determined at each polar angle, since $K_{||} = \sqrt{(2mE_k/h^2)} \cdot \sin \theta$. Previous work [1–6] suggests that this momentum is conserved both during optical excitation and coupling between the excited final state within the crystal and the plane wave state outside observed as the "photoelectron" in these measurements. A graphical presentation of the relationship between binding energies E and momentum components $K_{||}$ of features in such EDC series may therefore be regarded as a projection of energy bands of the solid along the crystal normal, in this case parallel to the hexagonal c^* axis. Such a plot is shown in fig. 3 for the data of fig. 2 and, superimposed, for Ne II (26.9 eV) and He II (40.8 eV) photon energies.

With reference to fig. 2, the feature close to E_f derives in fact from He II excitation of the lead $5d_{5/2}$ core level which overlaps the He I excited valence band when using unfiltered radiation and can be suppressed if the lamp is operated at higher pressure. The weaker spin orbit split $5d_{3/2}$ component is obscured by the main valence band but has little or no effect on the principal features of the latter as simple graphical deconvolution may demonstrate. Happily, the occurrence of this core level provides an ideal opportunity to determine the relative importance of final state scattering in the photoemission process as is discussed further below. Its intensity is therefore plotted radially against polar angle in fig. 4. Finally, fig. 5 demonstrates the role of the incoming photon in these measurements; for the same sample azimuth as fig. 2, both sample and analyser are now moved together to maintain constant exit angle θ whilst varying the incidence angle of the photon.

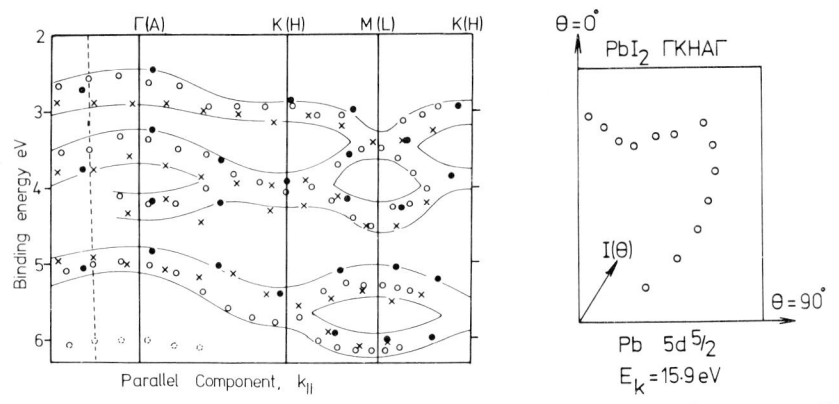

Fig. 3. Plot of binding energy (relative to E_f), E, versus parallel component of momentum, $K_{||}$, for the data of fig. 2 (circles) with, superimposed, data for Ne II at 26.9 eV (dots) and He II at 40.8 eV (crosses). Note the reproducibility in band dispersion for different photon energies.

Fig. 4. Intensity versus θ for the Pb $5d_{5/2}$ component of fig. 2, showing lobe like structure peaking near 35° which may be associated with final state scattering effects.

3.1. The valence states

The dispersion of features in the valence electron region of fig. 2 with θ (and hence $K_\parallel$) is as expected for an initial-state dominated photoemission process. The closely similar dispersion trends in the $E \rightarrow K_\parallel$ representation for three different photon energies (fig. 3) can only satisfactorily be explained if final states of appropriate symmetry are readily available over a broad range of energies as has been previously demonstrated for graphite [5]. That is not to suggest that the matrix element for the process is of no importance; on the contrary, the strength of the transition is largely determined by this factor (see below) but we are here concerned with the thesis that angle resolved photoemission can, usefully, predict the *geometric* form of the dispersion of the energy bands in solids.

Such appears to be the case in fig. 3, although it should be noted that an uncertainty of approximately 0.5 eV must be admitted within the resulting energy bands for different photon energies, as noted also by Margaritondo et al. [12]. Whether this reflects the selection of bands of different momentum perpendicular to the crystal surface as we vary the photon energy is uncertain, although measurements by Smith et al. [13] suggest this may be the case. We further note that the role of the final state in determining the form of the EDC becomes stronger at lower photon energy. Ne I (16.8 eV) measurements (to be fully reported elsewhere) show evidence of strong direct transition modulation of the form of the energy bands so that only at high photon energy do such drastic approximations appear useful.

We contend for the present, however, that the bands of fig. 3 do represent the one dimensional projection along c^* of the initial state valence electron structure. How, then, may those bands be interpreted in the light of previous studies of PbI_2? Firstly, there is little or nothing in the dispersion behaviour of the uppermost band to suggest Pb 6s character although Margaritondo et al. [12] somewhat inexplicably relate the "new" uppermost structure in their non-angle-resolved EDC's with this level. On the contrary, the four bands observed at Γ may be satisfactorily explained in terms of the four expected states based on the iodine 5p electrons: Γ_2^- (P_z antibonding) Γ_3^- ($P_{x,y}$ antibonding) Γ_3^+ ($P_{x,y}$ bonding) Γ_1^+ (P_z bonding), where "bonding" and "antibonding" refer to the relative overlaps between the P orbitals on the two iodine atoms in the unit cell. We tend to agree, therefore, with the assignment of Mitsukawa that the 6s states of lead are below the main valence band. No comparison has been attempted here with calculated energy band schemes, although we once more note that as has been observed previously [3,5] the large values of $K_\parallel$ attainable at these photon energies facilitate such comparisons over the whole of the Brillouin zone. This will be discussed more fully elsewhere.

3.2. Final state scattering: Pb 5d emission

Many theories of photoemission, particularly for adsorbates, place considerable emphasis on scattering in the final state as a determining factor in both the symmetries and intensities of angle resolved EDC's. Here we have an ideal opportunity to test the importance of such effects. The He II excited photoelectrons from the localised Pb $5d_{5/2}$ state are closely adjacent in final state energy to the He I excited valence levels (fig. 2). Scattering effects should therefore affect both to a similar extent. As can be seen, however, intensity modulation of the Pb 5d emission (fig. 4) is relatively weak, although showing the lobe-like structure expected on certain models [14]. Indeed the *total* valence electron intensity follows a similar trend but the intensity variations at a specific binding energy within the valence band are dominated by the initial state dispersion discussed above; having set an "energy window" within the band, the measured intensity as a function of angle will be almost entirely determined by the rate of energy band dispersion across that window, not by final state scattering. We conclude, therefore, that a useful working interpretation of such valence band EDC's may readily discount scattering as of only secondary importance. We note finally the independence of the position of the Pb $5d_{5/2}$ emission on θ (i.e. $K_{||}$), as expected for a localised d state.

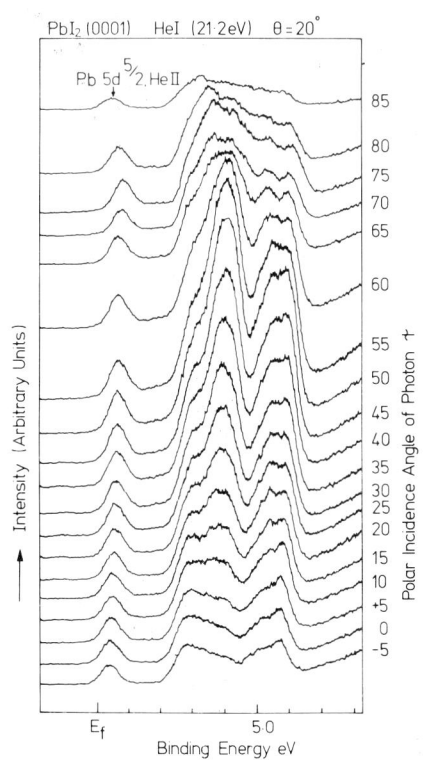

Fig. 5. Photon incidence angle dependence for states near K (fig. 3) and for the same azimuth as fig. 2. The strong level near 4 eV is presumed to have a strong iodine $5P_z$ admixture.

3.3. Photon incidence angle dependence

Of potentially greater importance in determining the form of angle resolved EDC is the role of the incoming photon. It should be noted, for example, that in fig. 2, features in the EDC's at $-12°$ and $+13°$, approximately equivalent in $K_{\parallel}$ exhibit considerable changes in relative intensity via $(A \cdot P)$ terms in the optical matrix elements, although not in their energy positions. Fig. 5 further amplifies this effect. Here, for $\theta = 20°$, within the ΓKHAΓ plane the states lie, approximately, at K (or along KH). By comparison with the polarised synchrotron data of Margaritondo et al. [12], $\psi = 0$ here corresponds to their, s polarised light, $\psi \neq 0$ to S + P components together (the light source is unpolarised). The growth of the strong band near 4 eV with ψ suggests, therefore, a strong P_z component within this band ($A \cdot P \to 0$ for pure P_z based states at $\psi = 0$). Some P_z admixture into a shoulder on the lowest band near 5 eV might also be discerned. The total intensity in the valence band (and in the 5d core level) also exhibits an overall dependence on ψ, peaking near 55° incidence angle. Interestingly, data for the ΓMLAΓ plane (not reproduced here) exhibits this overall dependence but comparatively little relative dependence (on ψ), between features. From fig. 3, however, it can be seen that the bands at K in fact disperse and branch towards M so that their atomic character at M may be heavily mixed. Full analysis of these effects must await detailed comparisons with an energy band scheme, although preliminary comparisons with the results of psendopotential calculations [15] haveproved unfruitful.

Acknowledgements

We thank F. Levy (EPF Lausanne) for the gift of PbI_2 crystals.

References

[1] N.V. Smith and M.M. Traum, Phys. Rev. B11 (1975) 2087.
[2] D.R. Lloyd, C.M. Quinn and N.V. Richardson, Surface Sci. 68 (1977) 547.
[3] D.R. Lloyd, C.M. Quinn, N.V. Richardson and P.M. Williams, Commun. Phys. 1 (1976) 11.
[4] P.M. Williams, D. Latham and J. Wood, J. Electron Spectros. 7 (1975) 281.
[5] P.M. Williams, Nuovo Cimento 38B (1977) 216.
[6] R.H. Williams, Nuovo Cimento 38B (1977) 241.
[7] G. Harbeke and E. Tosatti, Phys. Rev. Letters 28 (1972) 1567.
[8] F.R. Shepherd and P.M. Williams, in: Vacuum Ultraviolet Radiation Physics, Eds. E.E. Koch, R. Haensel and C. Kunz (Pergamon/Vieweg, Braunschweig, 1974) p. 508.
[9] P.M. Williams, in: Optical and Electrical Properties of Materials with Layered Structures, Ed. P.A. Lee (Reidel, Dordrecht, Netherlands, 1976) p. 273.
[10] T. Matsukawa and T. Ishii, J. Phys. Soc. Japan 41 (1976) 1285.
[11] J. Azoulay and L. Ley, to be published.
[12] G. Margaritondo, J.E. Rowe, M. Schluter, G.K. Werlheim, E. Levy and E. Mooser, to be published.
[13] P.K.Larsen, M. Schluter and N.V. Smith, to be published.
[14] B.W. Holland, Surface Sci. 68 (1977) 490.
[15] I. Ch. Schluter and M. Schluter, Phys. Rev. B9 (1974) 1652.

THE USE OF X-RAY PHOTOELECTRON SPECTROSCOPY IN CORROSION SCIENCE

J.E. CASTLE

Department of Metallurgy and Materials Technology, University of Surrey, Guildford, Surrey, GU2 5XH, UK

Most metals in commercial use are protected from reaction with an environment at ambient temperature by an oxide layer of ~5 nm in thickness. This layer may increase in thickness, by solid state diffusion, at high (>300°C) temperature or may dissolve and be replaced by soluble species by electrochemical reactions in aqueous environments. These are the twin processes of corrosion and the problem of preventing such destructive transformations had led to a long-standing interest by corrosion scientists in the thin film: and indeed in sub-monolayers. Prior to the introduction of the techniques of surface derived ion and electron spectroscopy the chemical analyses of films, which had known physical and electronic characteristics, were unavailable. This review attempts to specify the requirements of an analytical technique for thin films and examines the present role of XPS in this context. The areas covered will include elemental analysis, evaluation of valence state, non-destructive information, elemental distribution and finally the use of ion-etch depth profiling. Examples are drawn from corrosion problems in both aqueous and gaseous environments.

1. Introduction

It is now nearly twenty years since, in one of his first papers on the topic, Siegbahn and his co-workers [1] showed the possibility of distinguishing the XPS spectrum on an oxide from that of its parent metal. This is the cornerstone of the application to corrosion science of this technique, which came into regular use in the early 1970's after its years of development by Siegbahn's group [2]. Basic descriptions of XPS may now be found in several recent texts [3–5] and these may be consulted for details of this form of spectrometry. Suffice to say that the spectrum which is examined is that of the photoelectrons derived from a surface by the action of soft X-ray irradiation; that this spectrum is, in the main, a characteristic of the elements and their ions in the outermost 1–2 nm of the material, and that little or no spatial resolution is possible within the overall area of a few millimetres square which contributes to the signal. It is thus complementary to the twin technique of AES [6].

To those working in corrosion science during the 1960's, the gradual emergence of the twin techniques of XPS and AES from the physicists' workshops seemed like the answer to a prayer. The techniques would for the first time permit the analysis of surface films and protective layers which were performing a useful function, rather than those which had failed. Thus the role of alloying elements and inhibitors, and of destructive ionic species might be investigated without the laborious technique of chemical stripping of the film from the metal [7]. In the seven years since intruments became commercially available, XPS has found wide acceptance in many disciplines and technologies, so that to-day corrosion science would not be the first field associated with the technique in the minds of many surface scientists. In this paper I hope to show how, on one hand, corrosion science has benefited from XPS, and on the other, how corrosion research has stimulated the development of XPS. Before going further, however, I am sure those many workers in the field of corrosion who now find XPS invaluable would want me to express our gratitude to Professor Siegbahn and his school for their work in developing the technique.

2. Areas of corrosion research

A list of possible topics, table 1, would start with sub-monolayer coverage and extend to interfaces within thick films (>10 μm). However, in oxidising gases at room temperature the oxide layer on newly exposed metals and alloys will grow rapidly to a limiting thickness of <5 nm. This is almost entirely encompassed by the sampling depth of XPS and workers in the field of passivation are now able to

Table 1
Areas in corrosion research receiving benefit from surface science

1. Passivity — monolayer adsorption — theories relating to d band filling
2. Very thin films ($<$ the tunnelling thickness)
3. The very thin film in aqueous conditions
4. The breakdown of passivity in the aqueous phase
5. The thin film — interface and depth analysis
6. Anodic and chemical conversion layers
7. Thick film — spalling and exfoliation interfaces
8. Coatings — adhesion promotion and fracture route
9. Ancillary uses — monitoring of surface preparation
10. Dynamic situations — depletion effects, selective oxidation phenomena, assessment of mass transport, direction of change, environmental interaction

Spectrometric requirements of corrosion research

1. Differentiation of surface compounds from parent elements, i.e., valence state recognition
2. Interpretation of peak heights as composition, i.e., provision of sensitivity factors
3. Assay of structure, i.e., electron escape depths or mean free paths
4. Ability to locate and analyse key interfaces
5. Stability under X-ray in vacuum
6. Assessment of damage under ion bombardment
7. Elimination of or correction for electrostatic charging

obtain a *mean* analysis of a particular protective film with relative ease. As will be shown below, however, this analysis is heavily weighted to the surface composition and depends greatly on sensitivity factors which may be inaccurate (examples of the value of such work may be found in recent papers by Asami et al. [8] and by Farrell [9]).

XPS is used, of course, to study the growth of the protective film from submonolayer coverage [10–12]. However, because of the initial rapid growth there is generally less interest in monolayer films, which are of such importance in, say, catalysis. Thus, although in the opinion of some authorities [13] the passivity of metals in otherwise aggressive media is related to the interaction of adsorbed molecules with unfilled bands in the alloy, this area of study will be omitted from the present review.

The passive film formed in aqueous conditions is of similar thickness to that formed in gaseous atmospheres but is considered separately because of the special problems associated with its removal from the liquid phase and examination in a vacuum spectrometer. Of key interest here is the mechanism by which passivity might be lost, usually in the presence of certain anions such as chloride [14]; and of the manner by which alloying addition may inhibit corrosion, e.g. as molybdenum in stainless steel [15,16]. Water soluble additives of course also have an important role. Roberts [17] has used XPS to study bonding in the benzotriazole/copper system and others have studied the phosphate/steel interaction [18]. Breakdown, once initiated, usually gives rise to localised pitting which requires spatial resolution and is better suited, with stress corrosion cracking, to study by AES. This is not included within the review.

The protective layer on metals in oxidizing gases increases in thickness with increase of temperature. XPS is not suited to the determination of the small deviations in stoichiometry which account for solid state mobility but is of value in studying surface diffusion, solid state reactions [19–21], transport direction and depletion effects in alloys [22,23], selective oxidation, and catalysis of unwanted side reactions, e.g. carbon deposition from hydrocarbon gases [21,25].

In the aqueous phase at ambient temperature the very thin film may further thicken under the action of electrochemical process such as anodic oxidation, or by chemical conversion of the surface. It will be shown that correlation of XPS with electrochemical data [26,27] has provided excellent substantiation of the quantitative role of XPS. Interfaces within the thick film in the aqueous phase frequently have an ion exchange interaction [28] with the adjacent solution and analysis of these can provide the evidence for pH or ionic gradients within a permeating aqueous phase. Such evidence is illustrated below. This is an example of the use of XPS in an active sense since the concentration gradient implies mass transport. The ability to analyse very thin layers means that the response to very small perturbations of a system can be monitored [29]. In this way it could one day supplant the traditional accelerated test in a predictive ability. In an electrochemical system it is particularly valuable in its ability to reveal the nature and valence state of ions at the electrode surface [27,30] and thus provide evidence for Pourbaix Diagrams.

The nature of the interface between organic or other coatings and a metal or metal oxide is of importance because of the role of paints in the field of protection. It may be possible to distinguish chemical from mechanical bonding [31] and has been used to study adhesion promoters [32].

Finally the important role of XPS as a monitor of surface preparation should not be ignored. It would be true to say that XPS can transform the attitude of a research school in respect of the interaction with polishing media, impure solvents, buffers and heating up procedures.

3. Checklist of requirements for quantitative application

There are certain requirements for application of XPS and the reader is referred to the texts [2–5] already mentioned for a detailed discussion of this form of spectroscopy. However, table 1 provides a check list of the more necessary requirements which should be useful in the context of corrosion science. This list is amplified in the following section.

3.1. Separation of compounds from elements

It has already been indicated that the surface compound [5] must be distinguishable from the parent elements in the substrate. For many oxides this is now possible by use of chemical shift [33–40], satellite production [41] or multiplet splitting [42]. Furthermore there is some hope of distinguishing chlorides [43,44] and hydroxides [45,46]. The chemical shift of oxygen is helpful in distinguishing oxide, hydroxide and bound water forms [46,47]. Similarly the carbon peak may reveal the formation of carbide [48]. However, the more difficult problem is the necessity of making such identification in the presence of electrostatic charging of the surface. In this case the fact that the photoexcited auger lines and the photoelectron lines have differing chemical shifts, so that the difference (the auger parameter [49]) is independent of charging [33,50], may be of help. The possibility of deducing valence state information after use of argon ion bombardment must be ruled out unless each oxide has been checked for stability. Many authors [52, 53] have pointed out the rapid decomposition of higher, and sometimes even lower, valence states on use of this technique. All possible information on valence state should be obtained, if necessary by use of long counting times, before removal of surface "contamination". Decomposition in vacuum under X-radiation is known [54] but is much less severe than ion beam damage.

3.2. Concentration and distribution of material

Following separation of the spectra quantitative estimation of the concentrations represented by the peak areas, and of their distribution is required. The relationship between chemical and physical distribution of the material may be expressed as [55]

$$I[Z] = J \sin \phi [Z] \alpha \int_0^\infty \exp(-x/\lambda \sin \theta) \, dx \exp(-d/\lambda \sin \theta) K \, . \tag{1}$$

I is the number of electrons received from a given quantum level in a given species, J is the photon flux and ϕ is angle to the specimen, Z is the concentration of a given element in a defined valence state, α is the photoemission cross section for the appropriate element and quantum number which will depend on photon energy and on the incident angle, relative to the electron collection angle. The integral term represents the space integrated escape probability of the electrons, the "sampling depth" and within this term is λ, the characteristic length related to the electron mean free path, which is a function of electron kinetic energy and the crystal and electronic structure of the compound. The final exponential term relates to the attenuation of the electron by a contaminant layer of thickness d, and because of the character of λ this will also be a function of electron kinetic energy. K is the instrumental collection efficiency which will be a function of the pass efficiency of the hemispheres, the slit size, the solid collection angle (a function of kinetic energy) and the channeltron or detector characteristic and setting. J and K may be considered so variable as to preclude normal use of absolute count rates and it is usual to use experimental sensitivity factors derived from peak areas of given elements relative to that for fluorine in compounds of known stoichiometry. Some useful sets of data have been published [56–58], but these should be used as a guide only. In quantitative investigations sensitivity factors appropriate to the compounds under investigation should be determined. It should also be pointed out that K and its dependence on electron energy varies between instruments of different designs and particularly so between two classes of instruments, i.e., those which scan the spectrum using a retarding field and those which scan the analyser or input lens potentials. Thus sensitivity factors for one class of instruments cannot be used directly for the other class.

Escape depths [59,60] are difficult to determine and are even more poorly known than sensitivity factors. This will give rise to inaccuracies in analysis, both because of the importance of λ in the space integrated escape probability and, also, because of the overwhelming effect of surface contamination. The term $\exp(-d/\lambda \sin \theta)$ will distort relative intensities in a given spectrum to such an extent that quantitative analysis is impossible unless it is corrected for or the contamination removed.

For the purposes of this review it is assumed that compilations of experimental sensitivity factors, and escape depths, modified as necessary to suit individual systems of study have been obtained and permit the use of eq. (1) in a variety of simplified forms [55]. Certain combinations are particularly useful. Use is made of the fact that I_∞, the intensity from an infinitely thick, clean, substrate is, except at low angles, independent of collection angle. Then, if d_1 is the thickness of an outer absorbing layer and d_2 the thickness of an inner emitting layer, there are three basic cases:

$$I_{\infty d_1 \theta}/I_\infty = \exp(-d_1/\lambda \sin \theta) , \qquad (1)$$

representing the relative signal from a contaminated substrate

$$I_{d_2 \theta}/I_\infty = 1 - \exp(-d_2/\lambda \sin \theta) , \qquad (2)$$

representing the relative signal from an overlying (oxide) layer in the absence of contamination, and

$$I_{d_2 d_1 \theta}/I_\infty = \exp(-d_1/\lambda \sin \theta) - \exp[-(d_1 + d_2)/\lambda \sin \theta] , \qquad (3)$$

giving the relative signal from the overlying layer when covered with contaminent. This will be the usual case when examining the oxide layer formed in aqueous solutions.

When the collection angle is varied eq. (1) becomes

$$I_{\theta_1}/I_{\theta_2} = \exp(-d_1/\lambda) \, | \cosec \theta_1 - \cosec \theta_2 | , \qquad (4)$$

which describes the angular variation of the substrate signal in the presence of an overlayer. Eq. (2) has a similar solution but the equivalent solution for case (3) is

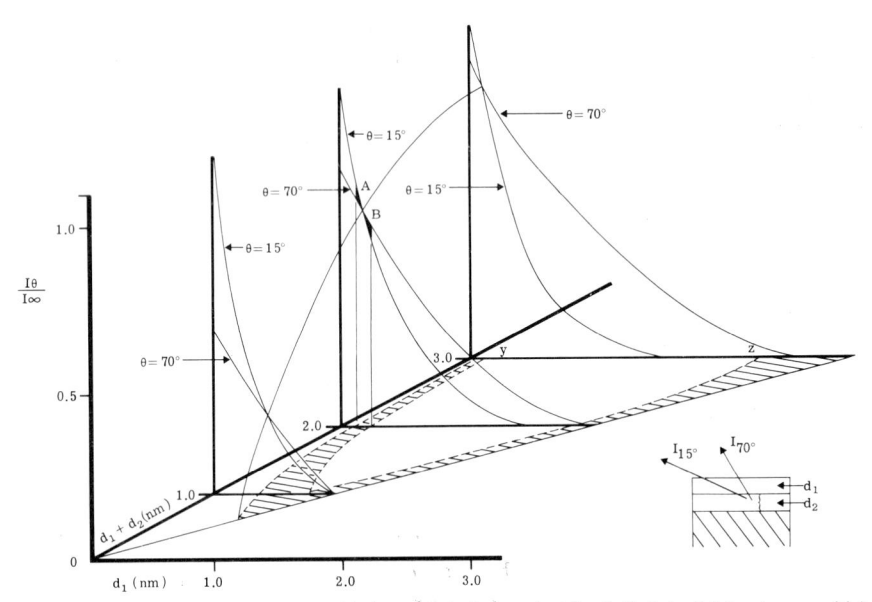

Fig. 1. The signal strength from an oxide layer (relative to that for infinitely thick, clean, oxide) covered with a contaminent. Note that low collection angles only enhance the oxide signal on clean surfaces. In the hatched field the oxide signal is invariant with angle.

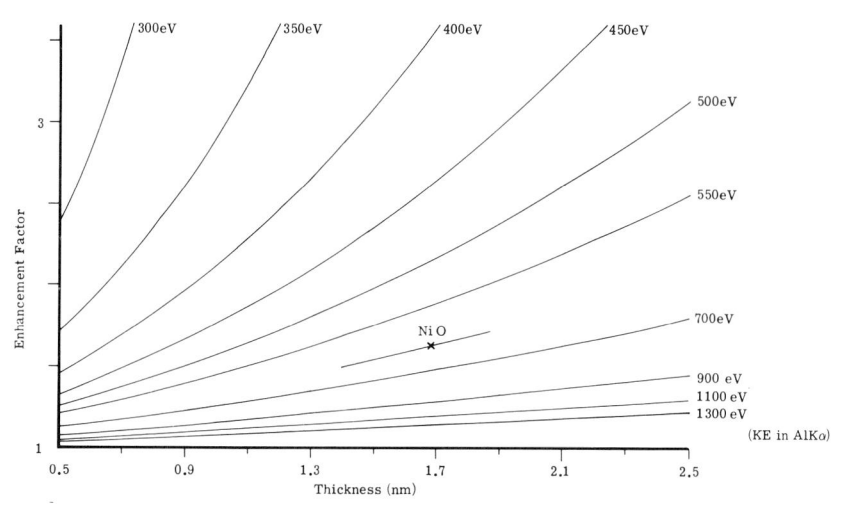

Fig. 2. The enhancement of an oxide signal in low energy photons: curves giving the enhancement of the oxide: metal intensity factor in MgKα relative to that in AlKα. Note, kinetic energies are quoted for AlKα radiation; λ assumed $= 0.5 \sqrt{KE}$; point marked NiO is measured value from figs. 3c and 3d.

described by a surface, as illustrated in fig. 1 (see Castle and Clayton [61]). Thus the angular variation from a contaminated corrosion sample may be relatively complex and the oxide signal is not necessarily a maximum at low collection angles.

A form which has been used by several authors occurs when the signal from the substrate (1) is attenuated by a layer of its own oxide, (2); d_1 then equals d_2 and

$$I_{\infty d\theta}/I_{d\theta} = \exp(-d/\lambda \sin\theta)/1 - \exp(-d/\lambda \sin\theta), \qquad (5)$$

or, as stated by Dickinson et al. [27],

$$\log_e(K + I_m/I_{ox})/(I_m/I_{ox}) = d/\lambda. \qquad (6)$$

This equation may also be used with appropriate sensitvity factors when the overlayer is a deposit of an element different from that in the substrate [75].

It is frequently useful to vary λ, either by comparing electrons of differing kinetic energy derived from the same element [29] or by changing the exciting radiation. This is easily achieved with a twin anode and curves are given (assuming $\lambda \propto E^{1/2}$) in fig. 2 for the AlKα/MgKα combination. Here the enhancement factor F is given by the ratio of eq. (5) written for each exciting radiation, i.e.

$$F = (I_m/I_{ox})_{\lambda_1}/(I_m/I_{ox})_{\lambda_2} = \exp[(d/\lambda \sin\theta)(E_1/E_2)^{1/2} - 1]. \qquad (7)$$

Eq. (7) is particularly useful for confirming the order of a series of layers, and indeed for checking that they are layers and not islands. Its interpretation is simpler than the angular variation for complex layers.

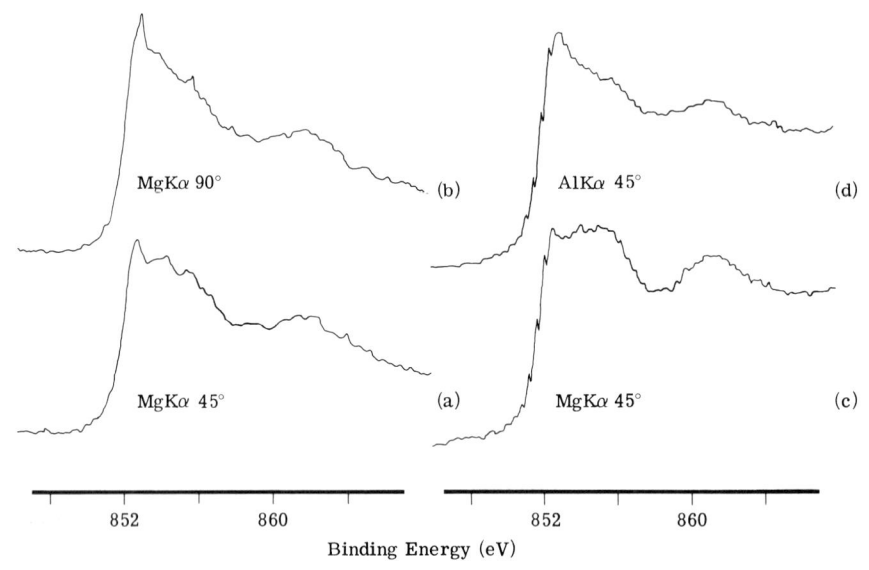

Fig. 3. The enhancement of an oxide signal in low energy photons: comparison with angular variation (a) and (b) from Evans et al. [62] for oxide on single crystal Ni, (c) and (d) data obtained using polycrystalline nickel. The agreement of the enhancement factor with the curves of fig. 2 shows the oxide to be laminar.

The operation of eqs. (4)–(7) is best illustrated by a series of examples. Fig. 3 (a, b) taken from the work of Evans et al. [62] shows the influence of collection angle on the spectra from nickel oxide and the nickel substrate: the oxide here had a mean thickness of $\simeq 1.6$ nm. In fig. 3 (c, d) a similar pair of spectra are shown. The lower is again in MgKα at 45° but in this case the upper curve is also at 45° but excited by AlKα. The close relationship between changing collection angle and photon energy is clearly seen from the shapes of these envelopes. The value of F derived from figs. 3c and 3d points to a laminar oxide of $\simeq 1.7$ nm in thickness when interpreted according to the curves of fig. 2. This is in accord with the close similarity of conditions and spectra of figs. 3a and 3c.

The correlation between thickness and enhancement factor is one way of obtaining evidence that the distribution of oxide is laminar. This is assumed in all the above derivations but is not always found in practise. As an example Briggs [63] has deduced the existance of oxide islands during the oxidation of zinc.

The distortion of relative peak heights by overlying contamination is discussed, by way of case histories, below. However we should note that more fundamental distribution factors will cause sensitivity factors to vary with crystal structure, as pointed out by Hercules and Ng [64] and, as Swingle [65] has shown, with the availability of scattering modes (he illustrates graphite vis a vis polyethylene). Moreover Wagner [56] obtained a range of sensitivity factors for Na ls which was well outside the experimental scatter on values for other elements and would seem to be associated with sodium in complex minerals: a dependence of sensitivity on the size of the unit cell of the material would be quite in accord with the distribution equations and must be noted by the corrosion worker, who will frequently be dealing with such compounds.

4. Case histories

The following sections will follow roughly the outline used in describing the "areas of corrosion science" and will use case histories to illustrate points remaining on the check list in table 1.

4.1. Passive films in gaseous environments

As has been pointed out the mean concentration (weighted according to distribution as required by eq. (1)) of a passive film is relatively easily obtained. Of

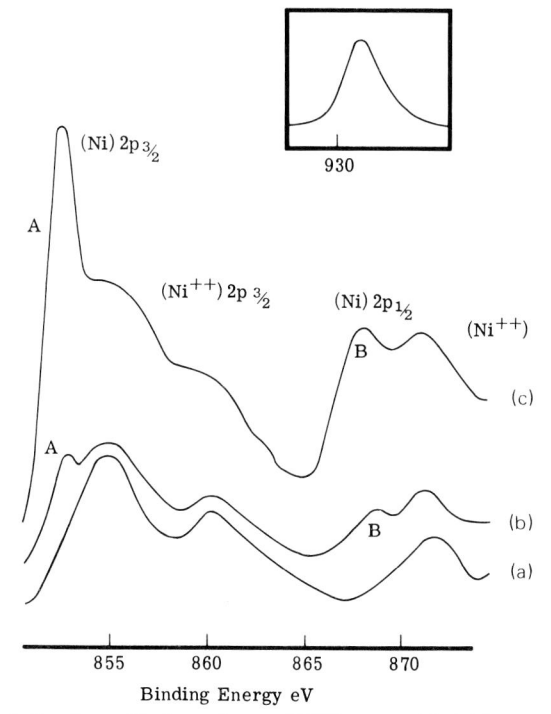

Fig. 4. Substrate depletion by oxide formation on 70/30 cupro-nickel (a) at room temperature. The nickel metal peak at A and B on 50/50 cupro-nickel (b) and pure nickel (c) is replaced by copper (inset) in the 70/30 spectrum.

greater importance and interest is that it is frequently possible, simultaneously to obtain some information from the surface layer of the underlying metal substrate. Thus it is possible to study the phenomenon of alloy depletion. Fig. 4 illustrates this using the nickel 2p spectra from nickel and two copper-nickel alloys. Examination of the respective copper spectra (inset figure) shows the air formed film on all three materials to be exclusively NiO [41]. However whilst the oxide layers are each of similar thickness only the lower alloy shows no evidence of underlying nickel metal. There is (inset) ample evidence of underlying copper and the surface layers of the alloy are thus depleted in nickel. This phenomenon is most important to questions relating to repassivation following scratch damage. This nondestructive use of XPS is more readily applied than the soft X-ray analysis described by Pickering and Holiday [66]. Asami et al. [67] have quantisized the method using a derivation of eq. (1) and applied it to the study of films on Fe/Cr alloys so that they routinely quote oxide composition and interface metal composition in their important series of papers on specimen preparation.

4.2. Solid state reaction

The interaction between substrate and oxide may be followed, dynamically, by measurement of the changes occurring during vacuum heating. Inner layers of low oxygen activity are then able to grow at the expense of outer layers of higher oxygen potential. This use is nicely illustrated by vacuum annealing of copper–nickel alloys carrying the characteristic duplex film formed at temperatures above 260°C. This has an outer layer of cuprous oxide overlying nickel oxide: the reaction, $Ni + Cu_2O = 2 Cu + NiO$, permits this inner layer to grow in the absence of molecular oxygen. Fig. 5 shows data obtained from peak height conversion using sensitivity factors and the chemical shift of the CuLMM Auger peak. Both sets agree in showing the reaction to be complete after 6 h. However the following points of more scientific interest also emerged from this work [20]:
(1) The diffusion of nickel to the surface is driven by the free energy of the reaction, rather than by changes in surface energy.
(2) The nickel oxide layer is porous; it permits the diffusion of reduced copper back into the alloy.
(3) The process has the kinetic form of a diffusion controlled reaction: a consequence of the depleted layer noted above?

Similar effects are shown during vacuum annealing of alloy steels by Castle and Durbin [19], Olefjord [22] and Asami [67]. In each case chromium is enhanced in the surface during the heating up period preceeding oxidation rate measurements. Castle and Durbin showed that this did not occur if the airformed "iron oxide" was first removed by ion etching. Fig. 6 shows another frequently observed effect: manganese diffusion to the surface. Peak height analysis shows that this also is a consequence of reactions with iron oxide and that the reaction goes to completion [68].

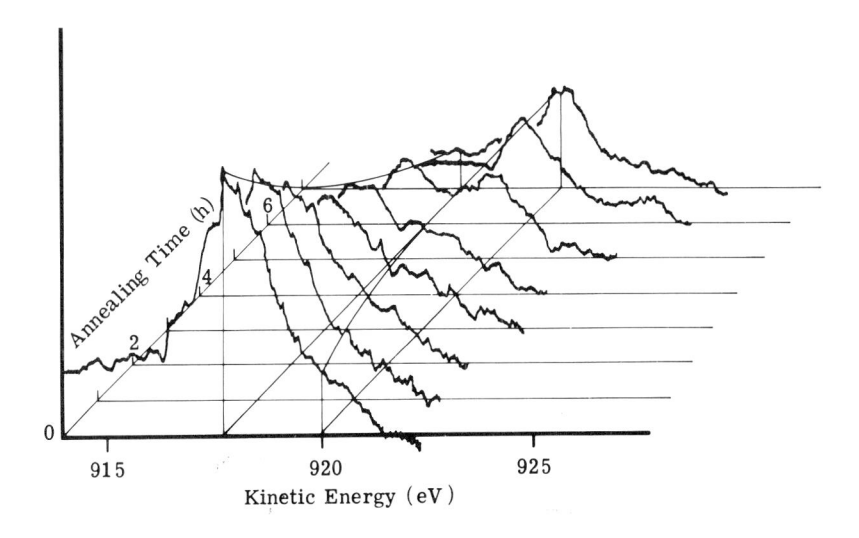

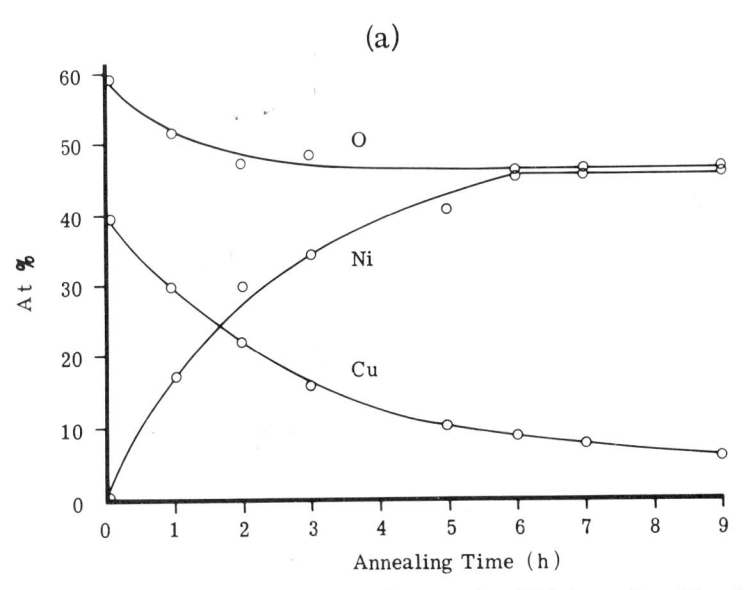

Fig. 5. The progress of a solid-state surface reaction [20] is monitored by the shift in the CuLMM Auger peak (a) and by the at% composition derived from PE peak heights: the agreement is good.

4.3. Electrochemistry and aqueous corrosion

In this branch of the Science we might expect the obvious difficulties of translating the specimen from the aqueous phase to the spectrometer vacuum to give

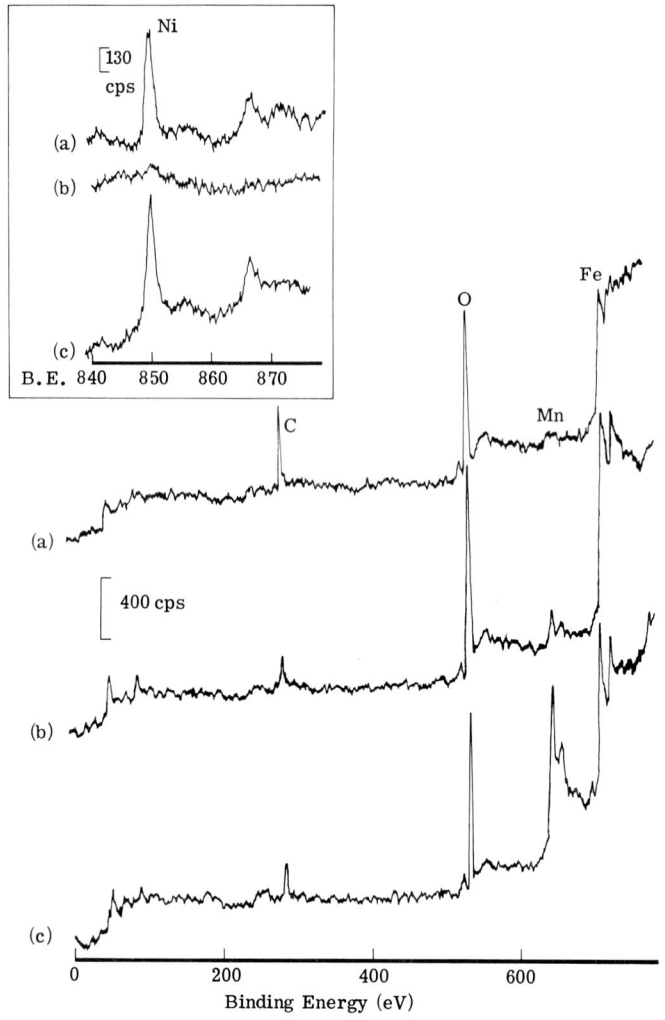

Fig. 6. Manganese enriches in the surface of a 9% Ni-steel by reduction of iron-oxide [19] during heat treatment. This permits diffusion of metallic nickel to the surface: (a) prior to oxidation, (b) after 90 min at 623 K in CO_2 at 133 Pa, (c) after 10 min in vacuo at 773 K.

insuperable difficulties. These have been outlined by Castle [29], Olefjord [23], Sherwood [30] and Asami [67], three points will be considered:
(1) The inevitability of a strongly adsorbed and sometimes multiple layer of water means that peak heights will be distorted in relation to their individual extinction coefficients.
(2) Labile species e.g. hydroxyl groups, amines, etc., may be lost.

(3) Compounds characteristic of the aqueous phase may be transformed to new states if exposed to air.

In fact, some of the best demonstrations of the quantitative use of XPS have come from the electrochemical field. Brinon [26] (fig. 7a) compared the quantity of lead deposited on an electrode by peak area and galvanostatic charge transfer and found a linear correlation. Sherwood [30] (fig. 7b) used eq. (6) to correlate the layer thickness of oxide on gold with the charge transfer with excellent results. However where a large number of elements are involved the impact of surface contamination cannot be ignored. Table 2 shows data from Castle and Epler [29] giving the enhancement of peak heights after 15 sec ion etching. The values are the mean of some 12 specimens described in their paper. We note that carbon and oxygen (as water) are removed by ion etching but that the remaining peaks are intensified in inverse relation to their kinetic energy. Meaningful correlations between species were naturally only obtained *after* this brief ion etch. Fig. 8 gives their correlation of the magnesium and chloride concentrations as a function of electropotential. This illustrates the development of cathodic basicity and anodic acidity after only 0.5 A s cm^{-2} of charge had passed. These data also serve to illustrate a further point: the differing behaviour of the two magesnium peaks, arising from their differing kinetic energies may be used to estimate the thickness of the layer removed — in this case 0.25 nm. The equation is given by Castle and Epler [29], and in a different context by Stoddart et al. [69].

With regard to labile compounds, we may take encouragement from the fact that so many authors, dealing with electrodes or aqueous corrosion, have recorded the presence of bound water: Asami concludes the layer on Fe/Cr alloy to be hydrated

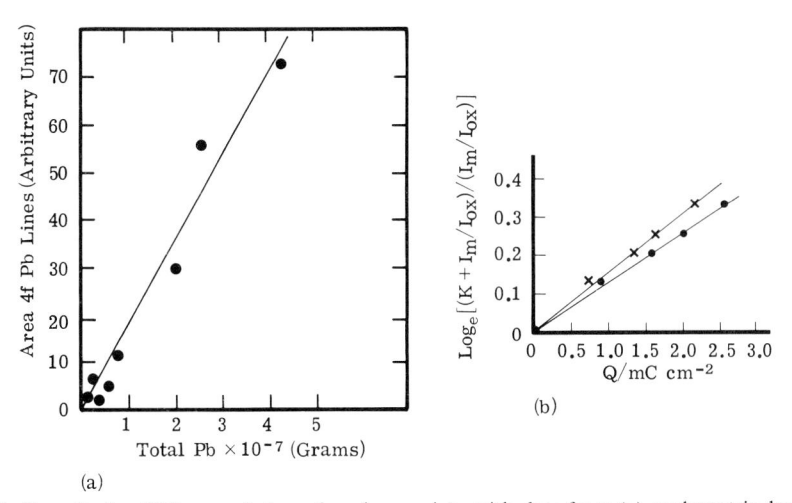

Fig. 7. Quantitative XPS: correlation of peak area data with data from (a) coulometric deposition of lead [26] and (b) reduction of platinum oxide [27].

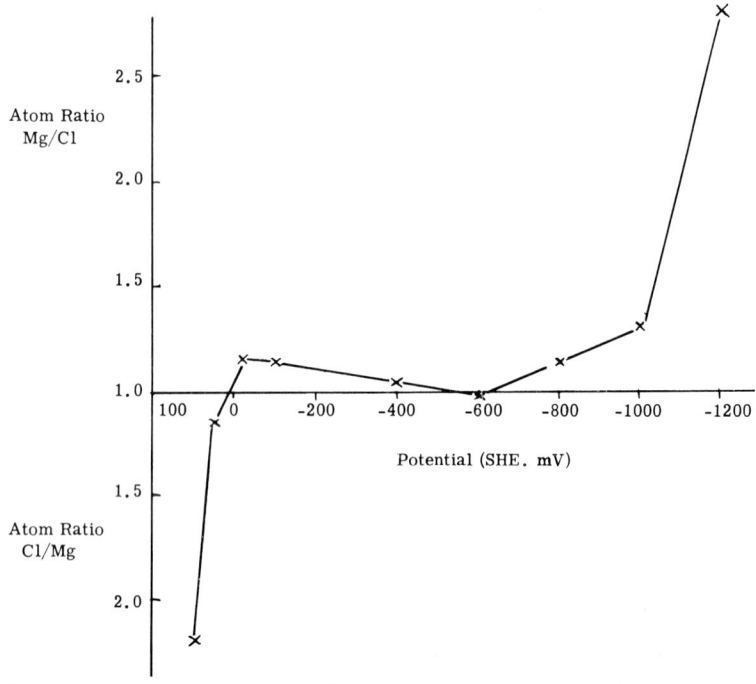

Fig. 8. Correlation of magnesium and chloride pick-up on the surface with electropotential [29].

chromium oxyhydroxide; Olefjord concludes it to be hydrated chromium hydroxide while Castle and Clayton [61] consider the bound water to be stabilised by organic impurities. Dickenson et al. comment on the presence of water on the gold electrode. In each case the water is recognised by the presence of an O_{1s} peak at close to 533 (cf. O_{1s} in oxide $\simeq$ 529–530). The case for the ability of hydroxides to resist vacuum dehydration is even better and Castle, Epler and Peplow [70], examining condenser tube surfaces, have found quantitative agreement of peak heights with mixed hydroxides known to be present from X-ray diffraction. McIntyre et al. also, have described hydroxides [47].

Finally, transfer between environments seems, in many cases, to pose no prob-

Table 2
Enhancement of peaks on removal of over layer; the Mg_{1s}/Mg_{KLL} ratio indicates that 0.25 nm was removed; this reduced the carbon peak to 0.7 original value

Peak	Cl_{2p}	Mg(KLL)	Zn(LMM)	O_{1s}	Cu_{2p}	O(KLL)	Zn_{2p}	Na_{1s}	Mg_{1s}
KE, (eV)	1286	1180	990	954	554	510	466	416	181
Enhancement	1.1	1.6	1.75	0.92	1.45	1.55	2.2	2.6	2.6

lems of reoxidation. Feischmeister and Olefjord, and Castle and Clayton, have shown very significant differences between the films formed in air and those formed in water. In both studies the surface of the aqueous formed material was much richer in chromium. There were differing interpretations of this, Olefjord, working with static aerated water concluded that iron was lost by selective dissolution of the hydroxide whilst Castle and Clayton working with refreshed deaerated cells, could find no influence on the surface concentration of refreshment rate.

The scientific conclusions are obviously an interesting basis for discussion but there is no doubt from this work that films characteristic of formation in the aqueous phase are found after transfer to the spectrometer and they must be presumed sufficiently stable for transfer. Confirmation of this has come from the work of Asami et al. who have shown, using a series of Fe/Cr alloys that the concentration of Fe^{2+} ion in the passive layer formed in 1M sulphuric acid increases at a critical concentration of 12% Cr frequently found to be a critical value in corrosion resistance using other techniques (fig. 9). Any tendency for the valence state of iron to alter, either on air exposure or under the influence of X-ray in vacuum would have masked this effect. Asami and also Dickenson, Sherwood and co-workers have reported several instances in which particular oxidation states, associated with known electropotentials, are subsequently found after transfer to the spectrometer.

The occurrence, on a metal surface, of a cathodic reaction such as the reduction

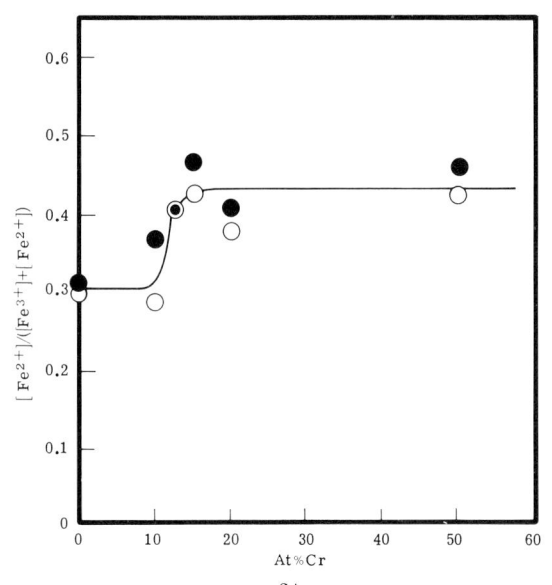

Fig. 9. Aqueous corrosion: correlation of Fe^{2+} content of passive layer in 1 M H_2SO_4 with chromium content of alloy [34].

of oxygen,

$$O_2 + 2 H_2O + 4e = 4 OH^-,$$

leads to a deviation of the surface pH value from that of the equilibrium solution pH. This deviation depends on the rate of the reaction and is increased if the boundary layer thickness is increased by the presence of porous deposits. XPS is well suited to deciphering the changes which occur within thick deposits as a result of such gradients in pH. Thus although the deposits are thick enough for more conventional analysis such methods will average gradients of the various ions which could otherwise act as "tracers" of electrode activity [70]. Work in this area has been so fruitful as to engender a new anode, Si Kα, for the determination of aluminium in the presence of copper [71].

4.4. Ion beam etching

The major problems in the application of XPS to corrosion science comes, not with the passive layers of less than 5 nm nor with thick layers of several micrometers in thickness which can be separated mechanically for interface analysis, but with layers of a few tens of nanometers. This thickness is such that virtually the only way to examine the structure is by ion beam etching. At least one other paper in this meeting will deal with the problems arising from this technique [72] which include, charging, differential etching and, most seriously ion beam decomposition [51,52]. Part of the uncertainty arises from the fact that the decomposition sequence of higher oxides, e.g. $Fe_2O_3 \rightarrow Fe_3O_4 \rightarrow FeO \rightarrow Fe$ is exactly that expected to be formed on a partly oxidised metal. Some oxides, e.g. Al_2O_3 or Cr_2O_3, are very resistant to such damage and Kim et al. [52] have suggested a relationship with free energy of formation. However the only safe policy at present is to examine the stability to the ion beam of each oxide likely to be present in the surface.

A further general problem is that of handling the large amount of data generated in ion etch sequences. Fig. 10 illustrates the oxygen and chromium peaks on a stainless steel sample after exposure to water at 160°C [72]. Each of the multiplets in such sequences must be deconvoluted and the resultant singlets treated separately and assigned a sensitivity factor. The data must then be assembled and tabulated in some form such as that in fig. 11. These curves give: the total normalised signal — used to identify short duration changes in the instrumental parameters such as X-ray flux; the carbon signal — used to monitor the condition of the spectrometer vacuum during the etch periods; the total signal from ionised species together with curves indicating the at% of individual ions and finally the total metal signal, with individual proportions. It is these curves in total which enable the locations of key interfaces to be identified even if rather subjectively, and thus their response to systematic changes in experimental variables can be discussed in the context of reaction kinetics [72]. Problems of interface definition and location will

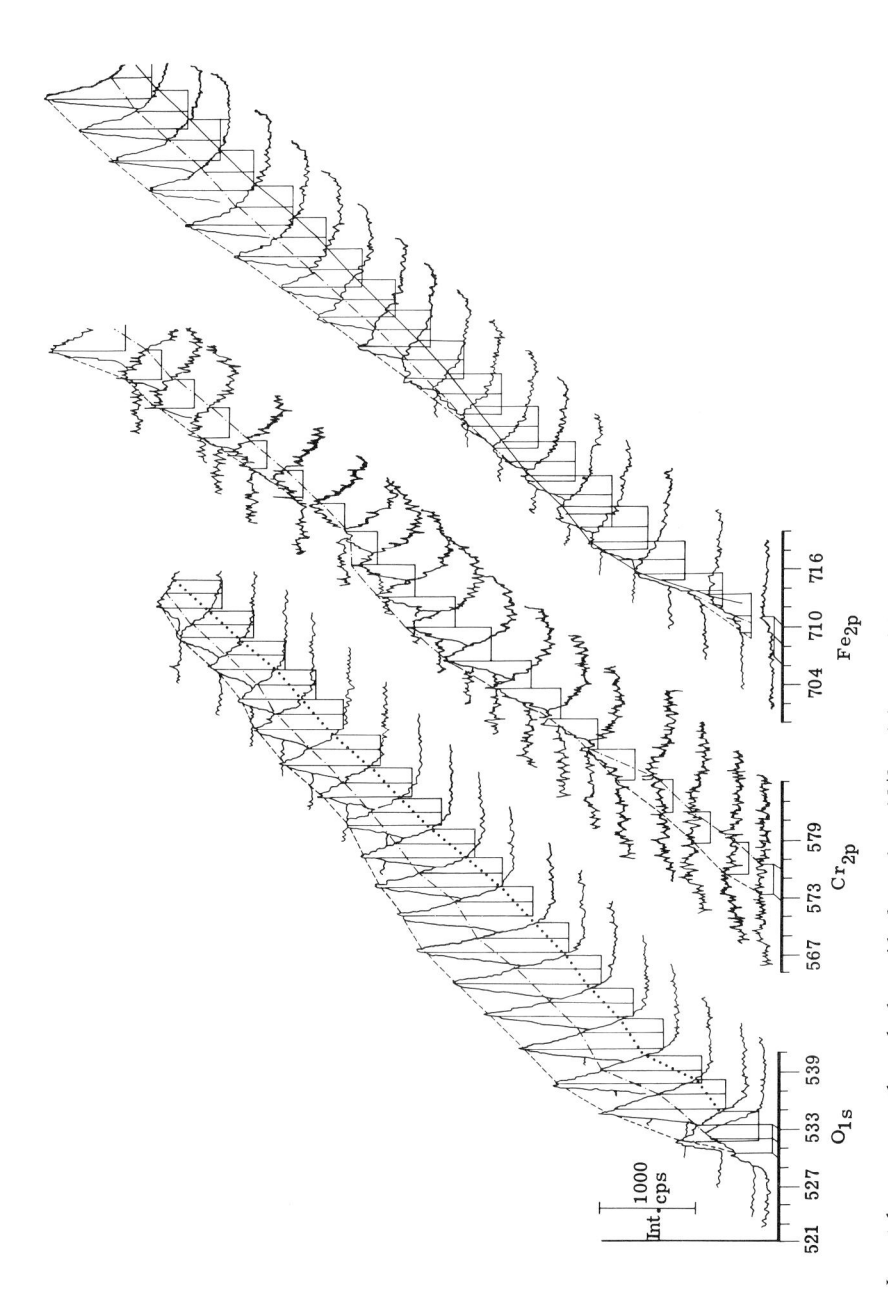

Fig. 10. Ion etch sequences through the oxide formed on 18/8 stainless steel in water at 433 K [72]. The oxide thickness is <5.0 nm. Each peak was deconvoluted into singlets at the positions shown by the vertical lines and the data used to produce fig. 11 (each step = 30 s).

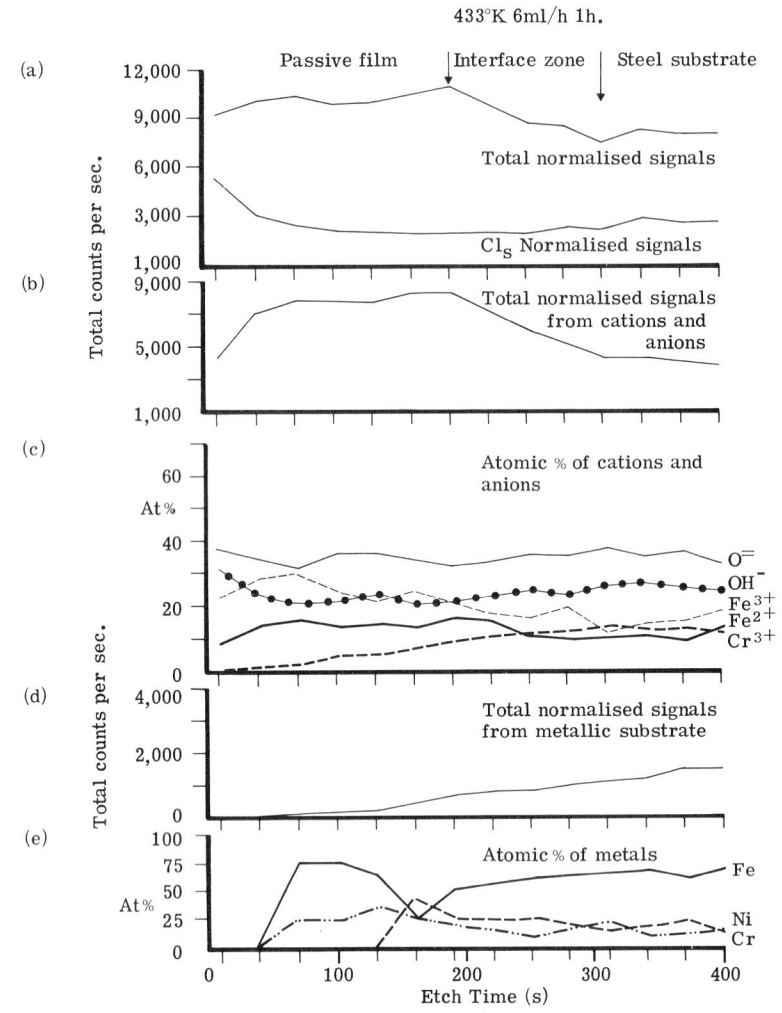

Fig. 11. Ion etch data derived from fig. 10. This form of presentation includes total normalised signals in addition to % values in order that contamination and instrumental performance can be monitored during progress of etch.

also be recognised and these are now the subject of study [73].

This review of necessity gives a foreshortened impression of the total field which now extends from dental amalgam [74] to nuclear reactors [71]. Much interesting work is not yet published because of commercial classification. However it will suffice to indicate to surface scientists and corrosion scientists the extent of their mutually overlapping interests and, hopefully, further extend the interdisciplinary research needed to understand and defeat corrosion.

Acknowledgement

It is a pleasure to acknowledge the help of the SRC who have been a major benefactor in our group activities, the Department and Faculty at the University of Surrey, Dr. L.B. Hazell, R. West for their capable handling of the spectroscopy and students M. Durbin, M. Nasserian-Riabi and C. Clayton whose work has been freely cited.

References

[1] E. Sokalowski, C. Nordling and K. Siegbahn, Phys. Rev. (1958) 776.
[2] K. Siegbahn et al., ESCA, Atomic Molecular and Solid State Structure Studies by Means of Electron Spectroscopy Uppsala (Almqvist and Wiksells, 1967).
[3] T.A. Carlson, Photoelectron and Auger Spectroscopy (Plenum, New York, 1975).
[4] P.M.A. Sherwood, in: Spectroscopy, Vol. 3, Eds. B.P. Straugham and S. Walker (Chapman and Hall, London, 1976) ch. 7.
[5] D. Hercules and S. Hercules, in: Characterization of Solid Surfaces, Eds. P.F. Kane and G.B. Larrabee (Plenum, New York, 1974) p. 307.
[6] See, e.g., J. C. Rivière, Contemp. Phys. 14 (1973) 513.
[7] W.H.J. Vernon, F. Wormwell and T.J. Nurse, J. Chem. Soc. (1939) 621.
[8] K. Asami, K. Hashimoto, S. Shimodaira and T. Masumoto, Corrosion Sci. 16 (1976) 909.
[9] T. Farrell, Met. Sci. 10 (1976) 87.
[10] C.R. Brundle, Surface Sci. 52 (1975) 426.
[11] G. Ertl and K. Wandell, Surface Sci. 50 (1975) 479.
[12] D. Chadwick and S. Bean, Initial oxidation of zirconium and its aqueous corrosion, private communication.
[13] F. Mansfeld and H.H. Uhlig, J. Electrochem. Soc. 117 (1970) 427.
[14] G. Okamoto, Corrosion Sci. 13 (1973) 471.
[15] J.B. Lumsden and R.W. Staehle, Scripta Met. 6 (1972) 1205.
[16] J. Kruger and J.R. Ambrose, Corrosion 28 (1959) 30.
[17] R.F. Roberts, J. Electron Spec. and Related Phenomena 4 (1974) 273.
[18] P. Fox and P. Boden, private communication.
[19] J.E. Castle and M.J. Durbin, Carbon 13 (1975) p. 23.
[20] J.E. Castle and M. Nasserian-Riabi, Corrosion Sci. 15 (1975) 537.
[21] K. Hirokawa and F. Honda, J. Electron Spectrosc. 6 (1975) 333.
[22] I. Olefjord, Corrosion Sci. 15 (1975) 687.
[23] I. Olefjord and H. Fischmeister, Corrosion Sci. 15 (1975) 697.
[24] R. Holm, Vakuum Tech. 23 (1974) 208.
[25] M.J. Durbin and J.E. Castle, Carbon 14 (1976) 27.
[26] J.S. Brinen, J. Electron Spectrosc. Related Phenomena 5 (1974) 377.
[27] T. Dickinson, A.F. Povey and P.M. Sherwood, J. Chem. Soc. Faraday I, 71 (1975) 298.
[28] C.E. Austing, A.M. Pritchard and N.J.M. Wilkins, Desalination 12 (1973) 251.
[29] J.E. Castle and D.C. Epler, Surface Sci. 53 (1975) 286.
[30] T. Dickinson and A.F. Povey, J. Chem. Soc. Faraday I, 72 (1976) 686.
[31] M. Gettings, J.P. Coad, F.S. Baker and A.S. Kinloch, J. Appl. Polymer Sci. (1977), to be published.
[32] J.E. Castle and R. Bailey, J. Mater. Sci. (1977) to be published.
[33] J.E. Castle and D.C. Epler, Proc. Roy. Soc. (London) A339 (1974) 49.

[34] K. Asami and K. Hashimoto, Corrosion Sci. (1977) to be published.
[35] G.C. Allen, M.T. Curtis, A.J. Hooper and P.M. Tucker, J. Chem. Soc. Dalton (1973) 1675.
[36] G.C. Allen, M.T. Curtis, A.J. Hooper and P.M. Tucker, J. Chem. Soc. Dalton (1974) 1525.
[37] O. Johnson, Chem. Scripta 8 (1975) 162.
[38] R. Holm and S. Storp, Appl. Phys. 9 (1976) 217.
[39] H. Nozoye, Y. Matsumoto, T. Onishi nd K. Tamaru, J. Chem. Soc. Faraday I, 72 (1976) 389.
[40] O. Masaoki and K. Hirokawa, J. Electron Spectrosc. Related Phenomena 7 (1975) 465.
[41] J.E. Castle, Nature (Phys. Sci.) 234 (1971) 93.
[42] J.C. Carver, in: Electron Spectroscopy, Ed. D.A. Shirley (North-Holland, Amsterdam, 1972) p. 781.
[43] K. Kishi and S. Ikeda, J. Phys. Chem. 78 (1974) 107.
[44] P.S. Belton and T.A. Clarke, Phil. Mag. 34 (1976) 157.
[45] N.S. McIntyre, T.E. Rummery, M.G. Cook and D. Owen, J. Electrochem. Soc. 123 (1976) 1164.
[46] K. Asami, K. Hashimoto and S. Shimodaira, Corrosion Sci. 16 (1976) 35.
[47] N.S. McIntyre and D.G. Zetaruk, Anal. Chem. (1977), to be published.
[48] L. Ramquist, J. Phys. Chem. Solids 30 (1969) 1835.
[49] C.D. Wagner, Anal. Chem. 47 (1975) 1201; Faraday Disc. Chem. Soc. 60 (1975) 291.
[50] C.D. Wagner and P. Biloen, Surface Sci. 35 (1973) 82.
[51] N.S. McIntyre and D.G. Zetaruk, J. Vacuum Sci. Technol. 14 (1977) 181. See also Proc. 7th Intern. Vacuum Congr., Vienna, 1977.
[52] K.J. Kim, W.E. Beitinger, J.W. Amy and N. Winograd, J. Electron Spectr. Related Phenomena 5 (1974) 351.
[53] R. Holm and S. Storp, Appl. Phys. 12 (1977) 101.
[54] B. Wallbank, C.E. Johnson and I.G. Main, J. Electon Spectr. 4 (1974) 263.
[55] C.S. Fadley, R.J. Baird, W. Siekhaus, T. Novakov and S.A.L. Bergstrom, J. Electron Spectr. 4 (74) 93.
[56] C.D. Wagner, Anal. Chem. 44 (1972).
[57] C.K. Jorgenson and H. Berthou, Disc. Faraday Soc. 54 (1974) 269.
[58] V.I. Nefedov, N.P. Sergushin, I.M. Brand and M.B. Trzhashovskaya, J. Electron Spectr. 2 (1973) 383.
[59] C.R. Brundle, J. Vacuum Sci. Technol. 11 (1974) 212.
[60] C.J. Powell, Surface Sci. 44 (1974) 29.
[61] J.E. Castle and C.R. Clayton, Corrosion Sci. 17 (1977) 7.
[62] S. Evans, J. Pielasek and J.M. Thomas, Surface Sci. 55 (1976) 644.
[63] D. Briggs, Faraday Disc. Chem. Soc. 60 (1975) 71.
[64] K.T. Ng and D. Hercules, J. Electron Spectr. 7 (1975) 257.
[65] R.S. Swingle II, Anal. Chem. 47 (1975) 21.
[66] H.W. Pickering and J.E. Holiday, J. Electrochem. Soc. 120 (1973) 470.
[67] K. Asami, K. Hashimoto and S. Shimodaira, Corrosion Sci. (1977) to be published. Several other papers by these authors are also in press.
[68] M.J. Durbin, Ph.D. Thesis, University of Surrey (1975).
[69] C.T.H. Stoddart, R.L. Moss and D. Pope, Surface Sci. 53 (1975) 241.
[70] J.E. Castle, D.C. Epler and D.B. Peplow, Corros. Sci. 16 (1976) 145. See also J.E. Castle, D.C. Epler 'XPS Study of Condenser Tube Surfaces', INCRA (New York) 1977.
[71] J.E. Castle, L.B. Hazell, R. Whitehead, J. Elec. Spec. and Related Phenomena 9 (1976) 247.
[72] C.R. Clayton and J.E. Castle, Paper for 1977. Passivity Meeting, Warrenton VA, U.S.A.
[73] J.E. Castle and L.B. Hazell, J. Elec. Spec. To be published 1977.
[74] G.W. Marshall, N.K. Sarkar, E.H. Greener, J. Dent. Res. 54 (1975) 904.
[75] J.S. Hammond and N. Winograd, J. Electrochem. Soc. 124 (1977) 826.

AUTHOR INDEX

Abon, M., G. Bergeret and B. Tardy, Field emission study of ammonia adsorption and catalytic decomposition on individual molybdenum planes   68 (1977) 305

Albers, H., W.J.J. van der Wal and G.A. Bootsma, Ellipsometric study of oxygen adsorption and the carbon monoxide–oxygen interaction on ordered and damaged Ag(111)   68 (1977) 47

Allen, G.C., P.M. Tucker and R.K. Wild, High resolution LMM Auger electron spectra of some first row transition elements   68 (1977) 469

Armand, G., see Garcia   68 (1977) 399

Backx, C., R.F. Willis, B. Feuerbacher and B. Fitton, Infrared vibration spectroscopy of molecular adsorbates on tungsten using reflection inelastic electron scattering   68 (1977) 516

Baker, B.G., see Price   68 (1977) 507

Barber, M., J.C. Vickerman and J. Wolstenholme, The application of SIMS to the study of CO adsorption on polycrystalline metal surfaces   68 (1977) 130

Battrell, C.F., C.F. Shoemaker and J.G. Dillard, A study of the interaction of sulfur-containing alkanes with clean nickel   68 (1977) 285

Benedek, G., see Garbassi   68 (1977) 286

Bergeret, G., see Abon   68 (1977) 305

Bertolini, J.C., G. Dalmai-Imelik and J. Rousseau, CO stretching vibration of carbon monoxide adsorbed on nickel(111) studied by high resolution electron loss spectroscopy   68 (1977) 539

Bertrand, P., F. Delannay, C. Bulens and J.-M. Streydio, Angular dependence of the scattered ion yields in $^4\text{He}^+ \to \text{Cu}$ scattering spectrometry   68 (1977) 108

Besocke, K., B. Krahl-Urban and H. Wagner, Dipole moments associated with edge atoms; a comparative study on stepped Pt, Au and W surfaces   68 (1977) 39

Binh, V.T., Y. Moulin, R. Uzan and M. Drechsler, Grain-boundary groove evolution in the presence of an evaporation   68 (1977) 409

Bonzel, H.P., The role of surface science experiments in understanding heterogeneous catalysis   68 (1977) 236

Bootsma, G.A., see Albers   68 (1977) 47

Bradshaw, A.M., P. Hofmann and W. Wyrobisch, The interaction of oxygen with aluminium(111)   68 (1977) 269

Browning, R., M.M. El Gomati and M. Prutton, A digital scanning Auger electron microscope   68 (1977) 328

Brundle, C.R., T.J. Chuang and K. Wandelt, Core and valence level photoemission studies of iron oxide surfaces and the oxidation of iron   68 (1977) 108

Bulens, C., see Bertrand   68 (1977) 108

Bullett, D.W., Localized orbital approach to chemisorption: H and O adsorption on Ni, Pt and W(001) surfaces   68 (1977) 149

Carrière, B., see Légaré   68 (1977) 348

Cassuto, A., see Housley   68 (1977) 277

Castle, J., The use of X-ray photoelectron spectroscopy in corrosion science   68 (1977) 583

Chuang, T.J., see Brundle   68 (1977) 459

Cinti, R.C., see Nguyen	68 (1977) 566
Coles, S.J.T. and J.P. Jones, Adsorption of gold on low index planes of rhenium	68 (1977) 312
Crljen, Ž., see Šunjić	68 (1977) 479
Crossley, A. and D.A. King, Infrared spectra for CO isotopes chemisorbed on Pt{111}: evidence for strong adsorbate coupling interactions	68 (1977) 528
Cyrot-Lackmann, F., see Gordon	68 (1977) 359
Dalmai-Imelik, G., see Bertolini	68 (1977) 539
Debe, M.K., D.A. King and F.S. Marsh, Further dynamical and experimental LEED results for a clean W{001}-(1 × 1) surface structure determination	68 (1977) 437
Delannay, F., see Bertrand	68 (1977) 108
Desjonquères, M.C., see Gordon	68 (1977) 359
Deville, J.P., see Légaré	68 (1977) 348
Dillard, J.G., see Battrell	68 (1977) 285
Dorn, R. and H. Lüth, The adsorption of oxygen and carbon monoxide on cleaved polar and nonpolar ZnO surfaces studied by electron energy loss	68 (1977) 385
Drechsler, M., see Binh	68 (1977) 409
Ducros, R., see Housley	68 (1977) 277
Echenique, P.M., see Soria	68 (1977) 448
El Gomati, M.M., see Browning	68 (1977) 328
Erlewein, J. and S. Hofmann, Segregation of tin on (111) and (100) surfaces of copper	68 (1977) 71
Feder, R., Spin-polarized LEED from low-index surfaces of platinum and gold	68 (1977) 229
Felsen, M.F. and P. Regnier, Influence of some additional elements on the surface tension of copper at intermediate and high temperatures	68 (1977) 410
Feuerbacher, B., see Backx	68 (1977) 516
Fitton, B., see Backx	68 (1977) 516
Foxon, C.T., see Laurence	68 (1977) 190
Fuggle, J.C., XPS, UPS and XAES studies of oxygen adsorption on polycrystalline Mg at ~100 and ~300 K	68 (1977) 458
Garbassi, F., G. Petrini, L. Pozzi, G. Benedek and G. Parravano, An AES study of the surface composition of cobalt ferrites	68 (1977) 286
Garcia, M., Threshold and Lennard-Jones resonances and elastic lifetimes in the scattering of atoms from a corrugated wall and an attractive well surface model	68 (1977) 408
Garcia, N., G. Armand and J. Lapujoulade, Diffraction intensities in helium scattering; topographic curves	68 (1977) 399
Gettings, M. and J.C. Rivière, Precipitation and re-solution of impurities at the surface of indium on traversing the melting-point	68 (1977) 64
Gordon, M.B., F. Cyrot-Lackmann and M.C. Desjonquères, On the influence of size and roughness on the electronic structure of transition metal surfaces	68 (1977) 359
Gunnarsson, O., see Hjelmberg	68 (1977) 158
Hamann, D.R., Theoretical studies of the electronic structure of semiconductor surfaces	68 (1977) 167
Heckingbottom, R., see Housley	68 (1977) 179
Heiland, W. and E. Taglauer, The backscattering of low energy ions and surface structure	68 (1977) 96

Heinz, K., see Wagner	68 (1977) 189
Hjelmberg, H., O. Gunnarsson and B.I. Lundqvist, Theoretical studies of atomic adsorption on nearly-free-electron-metal surfaces	68 (1977) 158
Hofmann, P., see Bradshaw	68 (1977) 269
Hoffmann, S., see Erlewein	68 (1977) 71
Holland, B.W., see White	68 (1977) 457
Holland, B.W., Theory of angle-resolved photoemission from localised orbitals at solid surfaces	68 (1977) 490
Holm, R., see Storp	68 (1977) 10
Housley, M., R. Heckingbottom and C.J. Todd, The interaction of Ag with Si(111)	68 (1977) 179
Housley, M., R. Ducros, G. Piquard and A. Cassuto, The adsorption of carbon monoxide on thenium: basal (0001) and stepped $\vert 14\ (0001) \times (10\bar{1}1)\vert$ planes	68 (1977) 277
Howard, J. and T.C. Waddington, An inelastic neutron scattering study of C_2H_2 adsorbed on type 13X zeolites	68 (1977) 86
Humblet, F., H. Van Hove and A. Neyens, Photoemission studies on ZnO (0001), (000$\bar{1}$) and (10$\bar{1}$0)	68 (1977) 178
Janssen, A.P., see Laurence	68 (1977) 190
Jona, F., Past and future surface crystallography by LEED	68 (1977) 204
Jones, J.P., see Coles	68 (1977) 312
Joyce, B.A., see Laurence	68 (1977) 190
Kern, R., see Le Lay	68 (1977) 346
King, D.A., see Debe	68 (1977) 437
King, D.A., see Crossley	68 (1977) 528
Kinniburgh, C.G., see Walker	68 (1977) 221
Krahl-Urban, B., see Besocke	68 (1977) 39
Kreutz, E.W., E. Rickus and N. Sotnik, Oxidation properties of InSb(110) surfaces	68 (1977) 392
Kuijers, F.J. and V. Ponec, The surface composition of the nickel–copper alloy system as determined by Auger electron spectroscopy	68 (1977) 294
Lang, N.D., see Williams	68 (1977) 138
Lapujoulade, J., see Garcia	68 (1977) 399
Laurence, G., B.A. Joyce, C.T. Foxon, A.P. Janssen, G.S. Samuel and J.A. Venables, Adsorption–desorption studies of Zn on GaAs	68 (1977) 190
Légaré, P., G. Maire, B. Carrière and J.P. Deville, Shapes and shifts in the oxygen Auger spectra	68 (1977) 348
Le Gressus, C., D. Massignon and R. Sopizet, Low beam current density Auger spectroscopy and surface analysis	68 (1977) 338
Le Lay, G., M. Manneville and R. Kern, Desorption kinetics of condensed two-dimension phases on a single crystal substrate	68 (1977) 346
Lewerenz, H.J., see Sass	68 (1977) 429
Lindell, G., Exchange corrections to the density–density correlation function at a surface	68 (1977) 368
Lloyd, D.R., C.M. Quinn and N.V. Richardson, The oxidation of a Cu(100) single crystal studied by angle-resolved photoemission using a range of photon energies	68 (1977) 419
Lloyd, D.R., C.M. Quinn and N.V. Richardson, Experimental aspects of angle-	

resolved photoemission from clean and adsorbate-covered metal surfaces 68 (1977) 547
Lundqvist, B.I., see Hjelmberg 68 (1977) 385
Lüth, H., see Dorn 68 (1977) 385

Maire, G., see Légaré 68 (1977) 348
Manneville, M., see Le Lay 68 (1977) 346
Marsh, F.S., see Debe 68 (1977) 437
Martin, G. and B. Perraillon, A model for morphological changes driven by
    step–step interaction on clean surfaces 68 (1977) 57
Massignon, D., see Le Gressus 68 (1977) 338
Matthew, J.A.D., see Walker 68 (1977) 221
Matthew, J.A.D., see Wille 68 (1977) 259
Moñoz, M.C. and J.L. Sacedón, AES analysis of oxygen adsorbated on Si(111)
    and its stimulated oxidation by electronic bombardment 68 (1977) 347
Moran-Lopez, J.L. and A. ten Bosch, Changes in work function due to charge
    transfer in chemisorbed layers 68 (1977) 377
Moulin, Y., see Binh 68 (1977) 409
Müller, K., see Wagner 68 (1977) 189

Netzer, F.P., see Wille 68 (1977) 259
Neyens, A., see Humblet 68 (1977) 178
Nguyen, T.T.A. and R.C. Cinti, Directional uv photoemission from clean and
    sulphur saturated (100), (110) and (111) nickel surfaces 68 (1977) 566

Parravano, G., see Garbassi 68 (1977) 286
Perraillon, B., see Martin 68 (1977) 57
Petrini, G., see Garbassi 68 (1977) 286
Piquard, G., see Housley 68 (1977) 277
Ponec, V., see Kuijers 68 (1977) 294
Pozzi, L., see Garbassi 68 (1977) 286
Price, G.L. and B.G. Baker, Angle-resolved UPS measurements in a modified
    LEED system 68 (1977) 507
Prutton, M., see Browning 68 (1977) 328
Prutton, M., see Welton-Cook 68 (1977) 436

Quinn, C.M., see Lloyd 68 (1977) 419
Quinn, C.M., see Lloyd 68 (1977) 547

Regnier, P., see Felsen 68 (1977) 410
Rhead, G.E., Surface defects 68 (1977) 20
Richardson, N.V., see Lloyd 68 (1977) 419
Richardson, N.V., see Lloyd 68 (1977) 547
Rickus, E., see Kreutz 68 (1977) 392
Rivière, J.C., see Gettings 68 (1977) 64
Rousseau, J., see Bertolini 68 (1977) 539

Sacedón, J.L., see Moñoz 68 (1977) 347
Sacedon, J.L., see Soria 68 (1977) 448
Samuel, G.S., see Laurence 68 (1977) 190
Sass, J.K., S. Stucki and H.J. Lewerenz, Plasma resonance absorption in inter-
    facial photoemission from very thin silver films on Cu(111) 68 (1977) 429
Shoemaker, C.F., see Battrell 68 (1977) 285

Šokčevic, D., see Šunjić	68 (1977) 479
Sopizet, R., see Le Gressus	68 (1977) 338
Soria, F., J.L. Sacedon, P.M. Echenique and D. Titterington, LEED study of the epitaxial growth of the thin film Au(111)/Ag(111) system	68 (1977) 448
Sotnik, N., see Kreutz	68 (1977) 392
Southon, M.J., see A.R. Waugh	68 (1977) 79
Storp, S. and R. Holm, ESCA investigation of the oxide layers on some Cr containing alloys	68 (1977) 10
Streydio, J.-M., see Bertrand	68 (1977) 108
Stucki, S., see Sass	68 (1977) 429
Šunjić, M., Ž. Crljen and D. Šokčevic, Photoelectron spectroscopy of localized levels near surfaces: scattering effects and relaxation shifts	68 (1977) 479
Taglauer, E., see Heiland	68 (1977) 96
Tardy, B., see Abon	68 (1977) 305
Ten Bosch, A., see Moran-Lopez	68 (1977) 377
Titterington, D., see Soria	68 (1977) 448
Todd, C.J., see Housley	68 (1977) 179
Tucker, P.M., see Allen	68 (1977) 469
Uzan, R., see Binh	68 (1977) 409
Van der Wal, W.J.J., see Albers	68 (1977) 47
Van Hove, H., see Humblet	68 (1977) 178
Van Ooij, W.J., The role of XPS in the study and understanding of rubber-to-metal bonding	68 (1977) 1
Venables, J.A., see Laurence	68 (1977) 190
Vickerman, J.C., see Barber	68 (1977) 130
Waddington, T.C., see Howard	68 (1977) 86
Wagner, H., see Besocke	68 (1977) 39
Wagner, P., K. Müller and K. Heinz, Adsorption studies of Cs on Si(111)	68 (1977) 189
Walker, J.A., C.G. Kinniburgh and J.A.D. Matthew, LEED calculations of exchange reflections from antiferromagnetic NiO(100)	68 (1977) 221
Wandelt, K., see Brundle	68 (1977) 459
Waugh, A.R. and M.J. Southon, Surface studies with an imaging atom-probe	68 (1977) 79
Webb, C. and P.M. Williams, Angle resolved photoemission studies of the band structures of semiconductors: PbI_2	68 (1977) 576
Welton-Cook, M.R. and M. Prutton, Calculations of rumpling in the (100) surfaces of divalent metal oxides	68 (1977) 436
White, S.J., D.P. Woodruff, B.W. Holland and R.S. Zimmer, A LEED study of the Si(100) (1 × 1)H surface structure	68 (1977) 457
Wild, R.K., see Allen	68 (1977) 469
Wille, R.A., F.P. Netzer and J.A.D. Matthew, Electron energy loss spectrum of cyanogen on Pt(100)	68 (1977) 259
Williams, A.R. and N.D. Lang, Atomic chemisorption on simple metals: chemical trends and core-hole relaxation effects	68 (1977) 138
Williams, P.M., see Webb	68 (1977) 576
Willis, R.F., see Backx	68 (1977) 516
Wittmaack, K., The use of secondary ion mass spectrometry for studies of oxygen adsorption and oxidation	68 (1977) 118

Wolstenholme, J., see Barber 68 (1977) 130
Woodruff, D.P., see White 68 (1977) 457
Wyrobisch, W., see Bradshaw 68 (1977) 269

Zimmer, R.S., see White 68 (1977) 457

SUBJECT INDEX

Acetylene

J. Howard and T.C. Waddington, An inelastic neutron scattering study of C_2H_2 adsorbed on type 13X zeolites   68 (1977) 86

C. Backx, R.F. Willis, B. Feuerbacher and B. Fitton, Infrared vibration spectroscopy of molecular adsorbates on tungsten using reflection inelastic electron scattering   68 (1977) 516

Adhesion

W.J. van Ooij, The role of XPS in the study and understanding of rubber-to-metal bonding   68 (1977) 1

Adsorption

P. Wagner, K. Müller and K. Heinz, Adsorption studies of Cs on Si(111)   68 (1977) 189

J.C. Bertolini, G. Dalmai-Imelik and J. Rousseau, CO stretching vibration of carbon monoxide adsorbed on nickel (111) studied by high resolution electron loss spectroscopy   68 (1977) 539

D.R. Lloyd, C.M. Quinn and N.V. Richardson, Experimental aspects of angle-resolved photoemission from clean and adsorbate-covered metal surfaces   68 (1977) 547

Alkali halides

N. Garcia, G. Armand and J. Lapujoulade, Diffraction intensities in helium scattering; topographic curves   68 (1977) 399

Alkali metals

J.L. Moran-Lopez and A. ten Bosch, Changes in work function due to charge transfer in chemisorbed layers   68 (1977) 377

Alloys

W.J. van Ooij, The role of XPS in the study and understanding of rubber-to-metal bonding   68 (1977) 1

S. Storp and R. Holm, ESCA investigation of the oxide layers on some Cr containing alloys   68 (1977) 10

F.J. Kuijers and V. Ponec, The surface composition of the nickel–copper alloy system as determined by Auger electron spectroscopy   68 (1977) 294

P. Légaré, G. Maire, B. Carrière and J.P. Deville, Shapes and shifts in the oxygen Auger spectra   68 (1977) 348

Aluminium

A.M. Bradshaw, P. Hofmann and W. Wyrobisch, The interaction of oxygen
with aluminium (111)   68 (1977) 269

Ammonia

M. Abon, G. Bergeret and B. Tardy, Field emission study of ammonia adsorption and catalytic decomposition on individual molybdenum planes   68 (1977) 305

Atom scattering

N. Garcia, G. Armand and J. Lapujoulade, Diffraction intensities in helium scattering; topographic curves   68 (1977) 399

M. Garcia, Threshold and Lennard-Jones resonance and elastic lifetimes in the scattering of atoms from a corrugated wall and an attractive well surface model   68 (1977) 408

Atom-surface interaction potential

N. Garcia, G. Armand and J. Lapujoulade, Diffraction intensities in helium scattering; topographic curves   68 (1977) 399

M. Garcia, Threshold and Lennard-Jones resonances and elastic lifetimes in the scattering of atoms from a corrugated wall and an attractive well surface model   68 (1977) 408

Auger electron spectroscopy (AES)

H. Albers, W.J.J. van der Wal and G.A. Bootsma, Ellipsometric study of oxygen adsorption and the carbon monoxide–Oxygen interaction on ordered and damaged Ag(111)   68 (1977) 47

M. Gettings and J.C. Rivière, Precipitation and re-solution of impurities at the surface of indium on traversing the melting-point   68 (1977) 64

J. Erlewein and S. Hofmann, Segregation of tin on (111) and (100) surfaces of copper   68 (1977) 71

M. Housley, R. Heckingbottom and C.J. Todd, The interaction of Ag with Si(111)   68 (1977) 179

G. Laurence, B.A. Joyce, C.T. Foxon, A.P. Janssen, G.S. Samuel and J.A. Venables, Adsorption–desorption studies of Zn on GaAs   68 (1977) 190

A.M. Bradshaw, P. Hofmann and W. Wyrobisch, The interaction of oxygen with aluminium (111)   68 (1977) 269

M. Housley, R. Ducros, G. Piquard and A. Cassuto, The adsorption of carbon monoxide on rhenium: basal (0001) and stepped $|14(0001) \times (10\bar{1}1)|$ planes   68 (1977) 277

F. Garbassi, G. Petrini, L. Pozzi, G. Benedek and G. Parravano, An AES study of the surface composition of cobalt ferrites   68 (1977) 286

F.J. Kuijers and V. Ponec, The surface composition of the nickel–copper alloy system as determined by Auger electron spectroscopy   68 (1977) 294

R. Browning, M.M. El Gomati and M. Prutton, A digital scanning Auger electron microscope   68 (1977) 328

G. Le Gressus, D. Massignon and R. Sopizet, Low beam current density Auger spectroscopy and surface analysis   68 (1977) 338

G. Le Lay, M. Manneville and R. Kern, Desorption kinetics of condensed two-dimensional phases on a single crystal substrate   68 (1977) 346

M.C. Muñoz and J.L. Sacedón, AES analysis of oxygen adsorbated on Si(111) and its stimulated oxidation by electronic bombardment   68 (1977) 347

P. Légaré, G. Maire, B. Carrière and J.P. Deville, Shapes and shifts in the oxygen Auger spectra   68 (1977) 348

J.C. Fuggle, XPS, UPS and XAES studies of oxygen adsorption on polycrystalline Mg at ~100 and ~300 K   68 (1977) 458

G.C. Allen, P.M. Tucker and R.K. Wild, High resolution LMM Auger electron spectra of some first row transition elements   68 (1977) 469

Band structure

B.W. Holland, Theory of angle-resolved photoemission from localised orbitals at solid surfaces   68 (1977) 490

C. Webb and P.M. Williams, Angle resolved photoemission studies of the band structures of semiconductors: PbI_2 68 (1977) 576

Carbon monoxide

H. Albers, W.J.J. van der Wal and G.A. Bootsma, Ellipsometric study of oxygen adsorption and the carbon monoxide–oxygen interaction on ordered and damaged Ag(111)   68 (1977) 47

M. Barber, J.C. Vickerman and J. Wolstenholme, The application of SIMS to the study of CO adsorption on polycrystalline metal surfaces   68 (1977) 130

M. Housley, R. Ducros, G. Piquard and A. Cassuto, The adsorption of carbon monoxide on rhenium: basal (0001) and stepped |4(0001) × (1011)| planes   68 (1977) 277

R. Dorn and H. Lüth, The adsorption of oxygen and carbon monoxide on cleaved polar and nonpolar ZnO surfaces studied by electron energy loss spectroscopy   68 (1977) 385

C. Backx, R.F. Willis, B. Feuerbacher and B. Fitton, Infrared vibration spectroscopy of molecular adsorbates on tungsten using reflection inelastic electron scattering   68 (1977) 516

A. Crossley and D.A. King, Infrared spectra for CO isotopes chemisorbed on Pt{111}: evidence for strong adsorbate coupling interactions   68 (1977) 528

J.C. Bertolini, G. Dalmai-Imelik and J. Rousseau, CO stretching vibration of carbon monoxide adsorbed on nickel(111) studied by high resolution electron loss spectroscopy   68 (1977) 539

Catalysis

H.P. Benzel, The role of surface science experiments in understanding heterogeneous catalysis   68 (1977) 236

M. Abon, G. Bergeret and B. Tardy, Field emission study of ammonia adsorption and catalytic decomposition on individual molybdenum planes   68 (1977) 305

Cesium

P. Wagner, K. Müller and K. Heinz, Adsorption studies of Cs on Si(111)   68 (1977) 189

Chemisorption

H. Albers, W.J.J. van der Wal and G.A. Bootsma, Ellipsometric study of oxygen adsorption and the carbon monoxide–oxygen interaction on ordered and damaged Ag(111) — 68 (1977) 47

J. Howard and T.C. Waddington, An inelastic neutron scattering study of C_2H_2 adsorbed on type 13X zeolites — 68 (1977) 86

W. Heiland and E. Taglauer, The backscattering of low energy ions and surface structure — 68 (1977) 96

M. Barber, J.C. Vickerman and J. Wolstenholme, The application of SIMS to the study of CO adsorption on polycrystalline metal surfaces — 68 (1977) 130

F. Jona, Past and future surface crystallography by LEED — 68 (1977) 204

H.P. Bonzel, The role of surface science experiments in understanding heterogeneous catalysis — 68 (1977) 236

A.M. Bradshaw, P. Hofmann and W. Wyrobisch, The interaction of oxygen with aluminium (111) — 68 (1977) 269

M. Housley, R. Ducros, G. Piquard and A. Cassuto, The adsorption of carbon monoxide on rhenium: basal (0001) and stepped $|14(0001) \times (10\bar{1}1)|$ planes — 68 (1977) 277

C.F. Battrell, C.F. Shoemaker and J.G. Dillard, A study of the interaction of sulfur-containing alkanes with clean nickel — 68 (1977) 285

R. Dorn and H. Lüth, The adsorption of oxygen and carbon monoxide on cleaved polar and nonpolar ZnO surfaces studied by electron energy loss spectroscopy — 68 (1977) 385

S.J. White, D.P. Woodruff, B.W. Holland and R.S. Zimmer, A LEED study of the Si(100) (1 × 1) surface structure — 68 (1977) 457

C. Backx, R.F. Willis, B. Feuerbacher and B. Fitton, Infrared vibration spectroscopy of molecular adsorbates on tungsten using reflection — 68 (1977) 516

A. Crossley and D.A. King, Infrared spectra for CO isotopes chemisorbed on Pt{111}: evidence for strong absorbate coupling interactions — 68 (1977) 528

T.T.A. Nguyen and R.C. Cinti, Directional uv photoemission from clean and sulphur saturated (100), (110) and (111) nickel surfaces — 68 (1977) 566

Chemisorption theory

A.R. Williams and N.D. Lang, Atomic chemisorption on simple metals: chemical trends and core-hole relaxation effects — 68 (1977) 138

D.W. Bullett, Localized orbital approach to chemisorption: H and O adsorption on Ni, Pt and W(001) surfaces — 68 (1977) 149

H. Hjelmberg, O. Gunnarsson and B.I. Lundqvist, Theoretical studies of atomic adsorption on nearly-free-electron-metal surfaces — 68 (1977) 158

M.B. Gordon, F. Cyrot-Lackmann and M.C. Desjonquères, On the influence of size and roughness on the electronic structure of transition metal surfaces — 68 (1977) 359

J.L. Moran-Lopez and A. ten Bosch, Changes in work function due to the charge transfer in chemisorbed layers — 68 (1977) 377

Chromium

S. Storp and R. Holm, ESCA investigation of the oxide layers on some Cr containing alloys — 68 (1977) 10

Clusters

M.B. Gordon, F. Cyrot-Lackmann and M.C. Desjonquères, On the influence of size and roughness on the electronic structure of transition metal surfaces 68 (1977) 359

Copper

W.J. van Ooij, The role of XPS in the study and understanding of rubber-to-metal bonding 68 (1977) 1
J. Erlewein and S. Hofmann, Segregation of tin on (111) and (100) surfaces of copper 68 (1977) 71
P. Bertrand, F. Delannay, C. Bulens and J.-M. Streydio, Angular dependence of the scattered ion yields in $^4He^+ \to Cu$ scattering spectrometry 68 (1977) 108
F.J. Kuijers and V. Ponec, The surface composition of the nickel–copper alloy system as determined by Auger electron spectroscopy 68 (1977) 294
M.F. Felsen and P. Regnier, Influence of some additional elements on the surface tension of copper at intermediate and high temperatures 68 (1977) 410
D.R. Lloyd, C.M. Quinn and N.V. Richardson, The oxidation of a Cu(100) single crystal studied by angle-resolved photoemission using a range of photon energies 68 (1977) 419
J.K. Sass, S. Stucki and H.J. Lewerenz, Plasma resonance absorption in interfacial photoemission from very thin silver films on Cu(111) 68 (1977) 429
D.R. Lloyd, C.M. Quinn and N.V. Richardson, Experimental aspects of angle-resolved photoemission from clean and adsorbate-covered metal surfaces 68 (1977) 547

Corrosion

J. Castle, The use of X-ray photoelectron spectroscopy in corrosion science 68 (1977) 583

Cyanogen

R.A. Wille, F.P. Netzer and J.A.D. Matthew, Electron energy loss spectrum of cyanogen on Pt(100) 68 (1977) 259

Diffusion

M. Gettings and J.C. Rivière, Precipitation and re-solution of impurities at the surface of indium on traversing the melting-point 68 (1977) 64

Electron diffraction (see also LEED)

G. Laurence, B.A. Joyce, C.T. Foxon, A.P. Janssen, G.S. Samuel and J.A. Venables, Adsorption–desorption studies of Zn on GaAs 68 (1977) 190

Electron loss spectroscopy

R.A. Wille, F.P. Netzer and J.A.D. Matthew, Electron energy loss spectrum of cyanogen on Pt(100) 68 (1977) 259
R. Dorn and H. Lüth, The adsorption of oxygen and carbon monoxide on cleaved polar and nonpolar ZnO surfaces studied by electron energy loss spectroscopy 68 (1977) 385

C. Backx, R.F. Willis, B. Feuerbacher and B. Fitton, Infrared vibration spectroscopy of molecular adsorbates on tungsten using reflection inelastic electron scattering 68 (1977) 516

J.C. Bertolini, G. Dalmai-Imelik and J. Rousseau, CO stretching vibration of carbon monoxide adsorbed on nickel(111) studied by high resolution electron loss spectroscopy 68 (1977) 539

Electron microscopy

G. Laurence, B.A. Joyce, C.T. Foxon, A.P. Janssen, G.S. Samuel and J.A. Venables, Adsorption–desorption studies of Zn on GaAs 68 (1977) 190

R. Browning, M.M. El Gomati and M. Prutton, A digital scanning Auger electron microscope 68 (1977) 328

Electron spectroscopy

C. Le Gressus, D. Massignon and R. Sopizet, Low beam current density Auger spectroscopy and surface analysis 68 (1977) 338

Electron stimulated adsorption

M.C. Muñoz and J.L. Sacedón, AES analysis of oxygen adsorbated on Si(111) and its stimulated oxidation by electronic bombardment 68 (1977) 347

Electron stimulated desorption

C. Le Gressus, D. Massignon and R. Sopizet, Low beam current density Auger spectroscopy and surface analysis 68 (1977) 338

Ellipsometry

H. Albers, W.J.J. van der Wal and G.A. Bootsma, Ellipsometric study of oxygen adsorption and the carbon monoxide–oxygen interaction on ordered and damaged Ag(111) 68 (1977) 47

A.M. Bradshaw, P. Hofmann and W. Wyrobisch, The interaction of oxygen with aluminium (111) 68 (1977) 269

Epitaxy

M. Housley, R. Heckingbottom and C.J. Todd, The interaction of Ag with Si(111) 68 (1977) 179

G. Laurence, B.A. Joyce, C.T. Foxon, A.P. Janssen, G.S. Samuel and J.A. Venables, Adsorption–desorption studies of Zn on GaAs 68 (1977) 190

S.J.T. Coles and J.P. Jones, Adsorption of gold on low index planes of rhenium 68 (1977) 312

F. Soria, J.L. Sacedón, P.M. Echenique and D. Titterington, LEED study of the epitaxial growth of the thin film Au(111)/Ag(111) system 68 (1977) 448

Field emission microscopy

M. Abon, G. Bergeret and B. Tardy, Field emission study of ammonia adsorption and catalytic decomposition on individual molybdenum planes — 68 (1977) 305

S.J.T. Coles and J.P. Jones, Adsorption of gold on low index planes of rhenium — 68 (1977) 312

Field evaporation

A.R. Waugh and M.J. Southon, Surface studies with an imaging atom-probe — 68 (1977) 79

Field ion microscopy

A.R. Waugh and M.J. Southon, Surface studies with an imaging atom-probe — 68 (1977) 79

Gallium arsenide

G. Laurence, B.A. Joyce, C.T. Foxon, A.P. Janssen, G.S. Samuel and J.A. Venables, Adsorption–desorption studies of Zn on GaAs — 68 (1977) 190

Gold

K. Besocke, B. Krahl-Urban and H. Wagner, Dipole moments associated with edge atoms; a comparative study on stepped Pt, Au and W surfaces — 68 (1977) 39

R. Feder, Spin-polarized LEED from low-index surfaces of platinum and gold — 68 (1977) 229

S.J.T. Coles and J.P. Jones, Adsorption of gold on low index planes of rhenium — 68 (1977) 312

G. Le Lay, M. Manneville and R. Kern, Desorption kinetics of condensed two-dimensional phases on a single crystal substrate — 68 (1977) 346

F. Soria, J.L. Sacedón, P.M. Echenique and D. Titterington, LEED study of the epitaxial growth of the thin film Au(111)/Ag(111) system — 68 (1977) 448

Grain boundary

V.T. Binh, Y. Moulin, R. Uzan and M. Drechsler, Grain-boundary groove evolution in the presence of an evaporation — 68 (1977) 409

Halides

C. Webb and P.M. Williams, Angle resolved photoemission studies of the band structures of semiconductors: PbI_2 — 68 (1977) 576

Hydrocarbons

C.F. Battrell, C.F. Shoemaker and J.G. Dillard, A study of the interaction of sulfur-containing alkanes with clean nickel — 68 (1977) 285

C. Le Gressus, D. Massignon and R. Sopizet, Low beam current density Auger spectroscopy and surface analysis — 68 (1977) 338

Hydrogen

D.W. Bullett, Localized orbital approach to chemisorption: H and O adsorption on Ni, Pt and W(001) surfaces — 68 (1977) 149

S.J. White, D.P. Woodruff, B.W. Holland and R.S. Zimmer, A LEED study of
the Si(100) (1 × 1)H surface structure   68 (1977) 457

Indium

M. Gettings and J.C. Rivière, Precipitation and re-solution of impurities at the
surface of indium on traversing the melting-point   68 (1977) 64

Indium antimonide

E.W. Kreutz, E. Rickus and N. Sotnik, Oxidation properties of InSb(110) surfaces   68 (1977) 392

Infrared spectroscopy

C. Backx, R.F. Willis, B. Feuerbacher and B. Fitton, Infrared vibration spectroscopy of molecular adsorbates on tungsten using reflection inelastic
electron scattering   68 (1977) 516
A. Crossley and D.A. King, Infrared spectra for CO isotopes chemisorbed on
Pt{111}: evidence for strong absorbate coupling interactions   68 (1977) 528

Ion scattering

W. Heiland and E. Taglauer, The backscattering of low energy ions and surface
structure   68 (1977) 96
P. Bertrand, F. Delannay, C. Bulens and J.-M. Streydio, Angular dependence
of the scattered ion yields in $^4He^+$ Cu scattering spectrometry   68 (1977) 108

Iron

C.R. Brundle, T.J. Chuang and K. Wandelt, Core and valence level photoemission studies of iron oxide surfaces and the oxidation of iron   68 (1977) 459

Iron oxide

C.R. Brundle, T.J. Chuang and K. Wandelt, Core and valence level photoemission studies of iron oxide surfaces and the oxidation of iron   68 (1977) 459

Low energy electron diffraction (LEED)

G.E. Rhead, Surface defects   68 (1977) 20
J. Erlewein and S. Hofmann, Segregation of tin on (111) and (100) surfaces of
copper   68 (1977) 71
P. Wagner, K. Müller and K. Heinz, Adsorption studies of Cs on Si(111)   68 (1977) 189
F. Jona, Past and future surface crystallography by LEED   68 (1977) 204
J.A. Walker, C.G. Kinniburgh and J.A.D. Matthew, LEED calculations of
exchange reflections from antiferromagnetic NiO(100)   68 (1977) 221
R. Feder, Spin-polarized LEED from low-index surfaces of platinum and gold   68 (1977) 229
M. Housley, R. Ducros, G. Piquard and A. Cassuto, The adsorption of carbon
monoxide on rhenium: basal (0001) and stepped |14(0001) ×(1011)|
planes   68 (1977) 277

P. Légaré, G. Maire, B. Carrière and J.P. Deville, Shapes and shifts in the oxygen Auger spectra 68 (1977) 348
E.W. Kreutz, E. Rickus and N. Sotnik, Oxidation properties if InSb(110) surfaces 68 (1977) 392
M.K. Debe, D.A. King and F.S. Marsh, Further dynamical and experimental LEED results for a clean W{001}-(1 × 1) surface structure determination 68 (1977) 437
F. Soria, J.L. Sacedon, P.M. Echenique and D. Titterington, LEED study of the epitaxial growth of the thin film Au(111)/Ag(111) system 68 (1977) 448
S.J. White, D.P. Woodruff, B.W. Holland and R.S. Zimmer, A LEED study of the Si(100) (1 × 1)H surface structure 68 (1977) 457

Magnesium

J.C. Fuggle, XPS, UPS and XAES studies of oxygen adsorption on polycrystalline Mg at ~100 and ~300 K 68 (1977) 458

Metals

M. Barber, J.C. Vickerman and J. Wolstenholme, The application of SIMS to the study of CO adsorption on polycrystalline metal surfaces 68 (1977) 130
A.R. Williams and N.D. Lang, Atomic chemisorption on simple metals: chemical trends and core-hole relaxation effects 68 (1977) 138
H. Hjelmberg, O. Gunnarsson and B.I. Lundqvist, Theoretical studies of atomic adsorption on nearly-free-electron-metal surfaces 68 (1977) 158
F. Jona, Past and future surface crystallography by LEED 68 (1977) 204
H.P. Bonzel, The role of surface science experiments in understanding heterogeneous catalysis 68 (1977) 236
M.B. Gordon, F. Cyrot-Lackmann and M.C. Desjonquères, On the influence of size and roughness on the electronic structure of transition metal surfaces 68 (1977) 359
J.L. Moran-Lopez and A. ten Bosch, Changes in work function due to charge transfer in chemisorbed layers 68 (1977) 377
N. Garcia, G. Armand and J. Lapujoulade, Diffraction intensities in helium scattering; topographic curves 68 (1977) 399
N. Garcia, Threshold and Lennard-Jones resonances and elastic lifetimes in the scattering of atoms from a corrugated wall and an attractive well surface model 68 (1977) 408
G.C. Allen, P.M. Tucker and R.K. Wild, High resolution LMM Auger electron spectra of some first row transition elements 68 (1977) 469
M. Šunjić, Ž. Crljen and D. Šokčević, Photoelectron spectroscopy of localized levels near surfaces: scattering effects and relaxation shifts 68 (1977) 479
J. Castle, The use of X-ray photoelectron spectroscopy in corrosion science 68 (1977) 583

Molecular beam studies

G. Laurence, B.A. Joyce, C.T. Foxon, A.P. Janssen, G.S. Samuel and J.A. Venables, Adsorption–desorption studies of Zn on GaAs 68 (1977) 190

Molybdenum

A.R. Waugh and M.J. Southon, Surface studies with an imaging atom-probe 68 (1977) 79
M. Abon, G. Bergeret and B. Tardy, Field emission study of ammonia adsorption and catalytic decomposition on individual molybdenum planes 68 (1977) 305

Neutron scattering

J. Howard and T.C. Waddington, An inelastic neutron scattering study of C_2H_2 adsorbed on type 13X zeolites — 68 (1977) 86

Nickel

W. Heiland and E. Taglauer, The backscattering of low energy ions and surface structure — 68 (1977) 96

D.W. Bullett, Localized orbital approach ro chemisorption: H and O adsorpadsorption on Ni, Pt and W(001) surfaces — 68 (1977) 149

C.F. Battrell, C.F. Shoemaker and J.G. Dillard, A study of the interaction of sulfur-containing alkanes with clean nickel — 68 (1977) 285

F.J. Kuijers and V. Ponec, The surface composition of the nickel–copper alloy system as determined by Auger electron spectroscopy — 68 (1977) 294

G.L. Price and B.G. Baker, Angle-resolved UPS measurements in a modified LEED system — 68 (1977) 507

J.C. Bertolini, G. Dalmai-Imelik and J. Rousseau, CO stretching vibration of carbon monoxide adsorbed on nickel(111) studied by high resolution electron loss spectroscopy — 68 (1977) 539

D.R. Lloyd, C.M. Quinn and N.V. Richardson, Experimental aspects of angleresolved photoemission from clean and adsorbate-covered metal surfaces — 68 (1977) 547

T.T.A. Nguyen and R.C. Cinti, Directional uv photoemission from clean and sulphur saturated (100), (110) and (111) nickel surfaces — 68 (1977) 566

Nickel oxide

J.A. Walker, C.G. Kinniburgh and J.A.D. Matthew, LEED calculations of exchange reflections from antiferromagnetic NiO(100) — 68 (1977) 221

Oxidation

S. Storp and R. Holm, ESCA investigation of the oxide layers on some Cr containing alloys — 68 (1977) 10

K. Wittmaack, The use of secondary ion mass spectrometry for study of oxygen adsorption and oxidation — 68 (1977) 118

M.C. Muñoz and J.L. Sacedón, AES analysis of oxygen adsorbated on Si(111) and its stimulated oxidation by electronic bombardment — 68 (1977) 347

P. Légaré, G. Maire, B. Carrière and J.P. Deville, Shapes and shifts in the oxygen Auger spectra — 68 (1977) 348

E.W. Kreutz, E. Rickus and N. Sotnik. Oxidation properties of InSb(110) surfaces — 68 (1977) 392

D.R. Lloyd, C.M. Quinn and N.V. Richardson, The oxidation of a Cu(100) single crystal studied by angle-resolved photoemission using a range of photon energies — 68 (1977) 419

J.C. Fuggle, XPS, UPS and XAES studies of oxygen adsorption on polycrystalline Mg at ~100 and ~300 K — 68 (1977) 458

C.R. Brundle, T.J. Chuang and K. Wandelt, Core and valence level photoemission studies of iron oxide surfaces and the oxidation of iron — 68 (1977) 459

J. Castle, The use of X-ray photoelectron spectroscopy in corrosion science — 68 (1977) 583

Oxides

K. Wittmaack, The use of secondary ion mass spectrometry for studies of oxygen adsorption and oxidation — 68 (1977) 118

F. Garbassi, G. Petrini, L. Pozzi, G. Benedek and G. Parravano, An AES study of the surface composition of cobalt ferrites — 68 (1977) 286

M.R. Welton-Cook and M. Prutton, Calculations of rumpling in the (100) surfaces or divalent metal oxides — 68 (1977) 436

Oxygen

H. Albers, W.J.J. van der Wal and G.A. Bootsma, Ellipsometric study of oxygen adsorption and the carbon monoxide–oxygen interaction on ordered and damaged Ag(111) — 68 (1977) 47

D.W. Bullett, Localized orbital approach to chemisorption: H and O adsorption on Ni, Pt and W(001) surfaces — 68 (1977) 149

A.M. Bradshaw, P. Hofmann and W. Wyrobisch, The interaction of oxygen with aluminium (111) — 68 (1977) 269

P. Légaré, H. Maire, B. Carrière and J.P. Deville, Shapes and shifts in the oxygen Auger spectra — 68 (1977) 348

R. Dorn and H. Lüth, The adsorption of oxygen and carbon monoxide cleaved polar and nonpolar ZnO surfaces studied by electron energy loss spectroscopy — 68 (1977) 385

J.C. Fuggle, XPS, UPS and XAES studies of oxygen adsorption on polycrystalline Mg at ~100 and ~300 K — 68 (1977) 458

Palladium

D.R. Lloyd, C.M. Quinn and N.V. Richardson, Experimental aspects of angle-resolved photoemission from clean and adsorbate-covered metal surfaces — 68 (1977) 547

Photoelectric emission

F. Humblet, H. Van Hove and A. Neyens, Photoemission studies on ZnO (0001), (000$\bar{1}$) and (10$\bar{1}$0) — 68 (1977) 178

J.K. Sass, S. Stucki and H.J. Lewerenz, Plasma resonance absorption in interfacial photoemission from very thin silver films on Cu(111) — 68 (1977) 429

B.W. Holland, Theory of angle-resolved photoemission from localised orbitals at solid surfaces — 68 (1977) 490

D.R. Lloyd, C.M. Quinn and N.V. Richardson, Experimental aspects of angle-resolved photoemission from clean and adsorbate-covered metal surfaces — 68 (1977) 547

T.T.A. Nguyen and R.C. Cinti, Directional uv photoemission from clean and sulphur saturated (100), (110) and (111) nickel surfaces — 68 (1977) 566

C. Webb and P.M. Williams, Angle resolved photoemission studies of the band structures of semiconductors: PdI_2 — 68 (1977) 576

Photoelectron spectroscopy (PES)

W.J. van Ooij, The role of XPS in the study and understanding of rubber-to-metal bonding — 68 (1977) 1

S. Storp and R. Holm, ESCA investigation of the oxide layers on some Cr containing alloys — 68 (1977) 10

M. Gettings and J.C. Rivière, Precipitation and re-solution of impurities at the surface of indium on traversing the melting-point — 68 (1977) 64

A.M. Bradshaw, P. Hofmann and W. Wyrobisch, The interaction of oxygen with aluminium (111) — 68 (1977) 269

C.F. Battrell, C.F. Shoemaker and J.G. Dillard, A study of the interaction of
sulfur-containing alkanes with clean nickel   68 (1977) 285
D.R. Lloyd, C.M. Quinn and N.V. Richardson, The oxidation of a Cu(100)
single crystal studied by angle-resolved photoemission using a range of
photon energies   68 (1977) 419
J.C. Fuggle, XPS, UPS and XAES studies of oxygen adsorption on polycrystalline Mg at ~100 and ~300   68 (1977) 458
C.R. Brundle, T.J. Chuang and K. Wandelt, Core and valence level photoemission studies of iron oxide surfaces and the oxidation of iron   68 (1977) 459
M. Šunjić, Ž. Crljen and D. Šokčević, Photoelectron spectroscopy of localized
levels near surfaces: scattering effects and relaxation shifts   68 (1977) 479
B.W. Holland, Theory of angle-resolved photoemission from localised orbitals
at solid surfaces   68 (1977) 490
G.L. Price and B.G. Baker, Angle-resolved UPS measurements in a modified
LEED system   68 (1977) 507
J. Castle, The use of X-ray photoelectron spectroscopy in corrosion science   68 (1977) 583

Physisorption

R. Dorn and H. Lüth, The adsorption of oxygen and carbon monoxide on
cleaved polar and nonpolar ZnO surfaces studied by electron energy loss
spectroscopy   68 (1977) 385

Platinum

K. Besocke, B. Krahl-Urban and H. Wagner, Dipole moments associated with
edge atoms; a comparative study on stepped Pt, Au and W surfaces   68 (1977) 39
D.W. Bullett, Localized orbital approach to chemisorption: H and O adsorption on Ni, Pt and W(001) surfaces   68 (1977) 149
R. Feder, Spin-polarized LEED from low-index surfaces of platinum and gold   68 (1977) 229
R.A. Wille, F.P. Netzer and J.A.D. Matthew, Electron energy loss spectrum of
cyanogen on Pt(100)   68 (1977) 259
P. Légaré, G. Maire, B. Carrière and J.P. Deville, Shapes and shifts in the
oxygen Auger spectra   68 (1977) 348
A. Crossley and D.A. King, Infrared spectra for CO isotopes chemisorbed on
Pt{111}: evidence for strong absorbate coupling interactions   68 (1977) 528

Rhenium

A.R. Waugh and M.J. Southon, Surface studies with an imaging atom-probe   68 (1977) 79
M. Housley, R. Ducros, G. Piquard and A. Cassuto, The adsorption of carbon
monoxide on rhenium: basal (0001) and stepped |4(0001) × (10$\bar{1}$1)|
planes   68 (1977) 277
S.J.T. Coles and J.P. Jones, Adsorption of gold on low index planes of
rhenium   68 (1977) 312

Rubber

W.J. van Ooij, The role of XPS in the study and understanding of rubber-to-metal bonding   68 (1977) 1

Secondary ion mass spectroscopy (SIMS)

K. Wittmaack, The use of secondary ion mass spectrometry for studies of oxygen adsorption and oxidation — 68 (1977) 118

M. Barker, J.C. Vickerman and J. Wolstenholme, The application of SIMS to the study of CO adsorption and polycrystalline metal surfaces — 68 (1977) 130

Semiconductors

D.R. Hamann, Theoretical studies of the electronic structure of semiconductor surfaces — 68 (1977) 167

Silicon

K. Wittmaack, The use of secondary ion mass spectrometry for studies of oxygen adsorption and oxidation — 68 (1977) 118

D.R. Hamann, Theoretical studies of the electronic structure of semiconductor surfaces — 68 (1977) 167

M. Housley, R. Heckingbottom and C.J. Todd, The interaction of Ag with Si(111) — 68 (1977) 179

P. Wagner, K. Müller and K. Heinz, Adsorption studies of Cs on Si(111) — 68 (1977) 189

G. Le Lay, M. Manneville and R. Kern, Desorption kinetics of condensed two-dimensional phases on a single crystal substrate — 68 (1977) 346

M.C. Munoz and J.L. Sacedón, AES analysis of oxygen adsorbated on Si(111) and its stimulated oxidation by electronic bombardment — 68 (1977) 347

P. Légaré, G. Maire, B. Carrière and J.P. Deville, Shapes and shifts in the Auger spectra — 68 (1977) 348

S.J. White, D.P. Woodruff, B.W. Holland and R.S. Zimmer, A LEED study of the Si(100) (1 × 1)H surface structure — 68 (1977) 457

Silver

H. Albers, W.J.J. van der Wal and G.A. Bootsma, Ellipsometric study of oxygen adsorption and the carbon monoxide–oxygen interaction on ordered and damaged Ag(111) — 68 (1977) 47

W. Heiland and E. Taglauer, The backscattering of low energy ions and surface structure — 68 (1977) 96

M. Housley, R. Heckingbottom and C.J. Todd, The interaction of Ag with Si(111) — 68 (1977) 179

G. Le Lay, M. Manneville and R. Kern, Desorption kinetics of condensed two-dimensional phases on a single crystal substrate — 68 (1977) 346

J.K. Sass, S. Stucki and H.J. Lewerenz, Plasma resonance absorption in interfacial photoemission from very thin silver films on Cu(111) — 68 (1977) 429

F. Soria, J.L. Sacedón, P.M. Echenique and D. Titterington, LEED study of the epitaxial growth of the thin film Au(111)/Ag(111) system — 68 (1977) 448

Sulphides

W.J. van Ooij, The role of XPS in the study and understanding of rubber-to-metal bonding — 68 (1977) 1

Sulphur

T.T.A. Nguyen and R.C. Cinti, Directional uv photoemission from clean and sulphur saturated (100), (110) and (111) nickel surfaces   68 (1977) 566

Surface composition

F. Garbassi, G. Petrini, L. Pozzi, G. Benedek and G. Parravano, An AES study of the surface composition of cobalt ferrites   68 (1977) 286

F.J. Kuijers and V. Ponec, The surface composition of the nickel–copper alloy system as determined by Auger electron spectroscopy   68 (1977) 294

Surface conductivity

E.W. Kreutz, E. Rickus and N. Sotnik, Oxidation properties of InSb(110) surfaces   68 (1977) 392

Surface defects

G.E. Rhead, Surface defects   68 (1977) 20

K. Besocke, B. Krahl-Urban and H. Wagner, Dipole moments associated with edge atoms; a comparative study on stepped Pt, Au and W surfaces   68 (1977) 39

H. Albers, W.J.J. van der Wal and G.A. Bootsma, Ellipsometric study of oxygen adsorption and the carbon monoxide–oxygen interaction on ordered and damaged Ag(111)   68 (1977) 47

G. Martin and B. Perraillon, A model for morphological changes driven by step–step interaction on clean surfaces   68 (1977) 57

M. Housley, R. Ducros, G. Piquard and A. Cassuto, The adsorption of carbon monoxide on rhenium: basal (0001) and stepped $|14(0001) \times (10\bar{1}1)|$ planes   68 (1977) 277

Surface diffusion

V.T. Binh, Y. Moulin, R. Uzan and M. Drechsler, Grain-boundary groove evolution in the presence of an evaporation   68 (1977) 409

Surface energy

G. Lindell, Exchange corrections to the density–density correlation function at a surface   68 (1977) 368

M.R. Welton-Cook and M. Prutton, Calculations of rumpling in the (100) surfaces of divalent metal oxides   68 (1977) 436

Surface plasmons

A.M. Bradshaw, P. Hofmann and W. Wyrobisch, The interaction of oxygen with aluminium (111)   68 (1977) 269

G. Lindell, Exchange corrections to the density–density correlation function at a surface   68 (1977) 368

M. Šunjić, Z. Crljen and D. Šokčević, Photoelectron spectroscopy of localized levels near surfaces: scattering effects and relaxation shifts   68 (1977) 479

Surface roughness

G. Martin and B. Perraillon, A model for morphological changes driven by step–step interaction on clean surfaces — 68 (1977) 57

N. Garcia, G. Armand and J. Lapujoulade, Diffraction intensities in helium scattering; topographic curves — 68 (1977) 399

Surface segregation

M. Gettings and J.C. Rivière, Precipitation and re-solution of impurities at the surface of indium on traversing the melting-point — 68 (1977) 64

J. Erlewein and S. Hofmann, Segregation of tin on (111) and (100) surfaces of copper — 68 (1977) 71

Surface states

D.R. Hamann, Theoretical studies of the electronic structure of semiconductor surfaces — 68 (1977) 167

M.B. Gordon, F. Cyrot-Lackmann and M.C. Desjonquères, On the influence of size and roughness on the electronic structure of transition metal surfaces — 68 (1977) 359

Surface structure

G.E. Rhead, Surface defects — 68 (1977) 20

W. Heiland and E. Taglauer, The backscattering of low energy ions and surface structure — 68 (1977) 96

D.R. Hamann, Theoretical studies of the electronic structure of semiconductor surfaces — 68 (1977) 167

F. Jona, Past and future surface crystallography by LEED — 68 (1977) 204

M.R. Welton-Cook and M. Prutton, Calculations of rumpling in the (100) surfaces of divalent metal oxides — 68 (1977) 436

M.K. Debe, D.A. King and F.S. Marsh, Further dynamical and experimental LEED results for a clean W{001}-(1 × 1) surface structure determination — 68 (1977) 437

Surface tension

M.F. Felsen and P. Regnier, Influence of some additional elements on the surface tension of copper at intermediate and high temperatures — 68 (1977) 410

Tantalum

K. Wittmaack, The use of secondary ion mass spectrometry for studies of oxygen adsorption and oxidation — 68 (1977) 118

Thermal desorption

P. Wagner, K. Müller and K. Heinz, Adsorption studies of Cs on Si(111) — 68 (1977) 189

G. Laurence, B.A. Joyce, C.T. Foxon, A.P. Janssen, G.S. Samuel and J.A. Venables, Adsorption–desorption studies of Zn on GaAs — 68 (1977) 190

M. Housley, R. Ducros, G. Piquard and A. Cassuto, The adsorption of carbon

monoxide on rhenium: basal (0001) and stepped |14(0001) × (10$\bar{1}$1)|
planes 68 (1977) 277
C.F. Battrell, C.F. Shoemaker and J.G. Dillard, A study of the interaction of
sulfur-containing alkanes with clean nickel 68 (1977) 285
G. Le Lay, M. Manneville and R. Kern, Desorption kinetics of condensed two-
dimensional phases on a single crystal substrate 68 (1977) 346

Thin films

J.K. Sass, S. Stucki and H.J. Lewerenz, Plasma resonance absorption in interfacial photoemission from very thin silver films on Cu(111) 68 (1977) 429
F. Soria, J.L. Sacedon, P.M. Echenique and D. Titterington, LEED study of
the epitaxial growth of the thin film Au(111)/Ag(111) system 68 (1977) 448

Tin

J. Erlewein and S. Hofmann, Segregation of tin on (111) and (100) surfaces
of copper 68 (1977) 71

Titanium

R. Browning, M.M. El Gomati and M. Prutton, A digital scanning Auger electron microscope 68 (1977) 328

Tungsten

K. Besocke, B. Krahl-Urban and H. Wagner, Dipole moments associated with
edge atoms; a comparative study on stepped Pt, Au and W surfaces 68 (1977) 39
W. Heiland and E. Taglauer, The backscattering of low energy ions and sur-
face structure 68 (1977) 96
D.W. Bullett, Localized orbital approach to chemisorption: H and O adsorption on Ni, Pt and W(001) surfaces 68 (1977) 149
M.K. Debe, D.A. King and F.S. March, Further dynamical and experimental
LEED results for a clean W{001}-(1 × 1) surface structure determination 68 (1977) 437
C. Backx, R.F. Willis, B. Feuerbacher and B. Fitton, Infrared vibration spec-
troscopy of molecular adsorbates on tungsten using reflection inelastic
electron scattering 68 (1977) 516
D.R. Lloyd, C.M. Quinn and N.V. Richardson, Experimental aspects of angle-
resolved photoemission from clean and adsorbate-covered metal surfaces 68 (1977) 547

Work function

F.E. Rhead, Surface defects 68 (1977) 20
K. Besocke, B. Krahl-Urban and H. Wagner, Dipole moments associated with
edge atoms; a comparative study on stepped Pt, Au and W surfaces 68 (1977) 39
P. Wagner, K. Müller and K. Heinz, Adsorption studies of Cs on Si(111) 68 (1977) 189
A.M. Bradshaw, P. Hofmann and W. Wyrobisch, The interaction of oxygen
with aluminium (111) 68 (1977) 269
J.L. Moran-Lopez and A. ten Bosch, Changes in work function due to charge
transfer in chemisorbed layers 68 (1977) 377

Zeolites

J. Howard and T.C. Waddington, An inelastic neutron scattering study of C_2H_2 adsorbed on type 13X zeolites 68 (1977) 86

Zinc

W.J. van Ooij, The role of XPS in the study and understanding of rubber-to-metal bonding 68 (1977) 1

G. Laurence, B.A. Joyce, C.T. Foxon, A.P. Janssen, G.S. Samuel and J.A. Venables, Adsorption–desorption studies of Zn on GaAs 68 (1977) 190

Zinc oxide

F. Humblet, H. Van Hove and A. Neyens, Photoemission studies on ZnO (0001), $(000\bar{1})$ and $(10\bar{1}0)$ 68 (1977) 178

R. Dorn and H. Lüth, The adsorption of oxygen and carbon monoxide on cleaved polar and nonpolar ZnO surfaces studied by electron energy loss spectroscopy 68 (1977) 385